THE SURVEILLANT SCIENCE

Remote Sensing of the Environment

THE SURVEILLANT SCIENCE
Remote Sensing of the Environment

Edited by

ROBERT K. HOLZ

The University of Texas at Austin

HOUGHTON MIFFLIN COMPANY • BOSTON

Atlanta • Dallas • Geneva, Illinois • Hopewell, New Jersey • Palo Alto

Printed in the United States of America

Library of Congress Catalog Card Number 72–7922
ISBN: 0–395–14041–2

Cover illustration from a color thermogram courtesy of Barnes Engineering Company, Eric Wormser, and Applied Optics

INTRODUCTION

MAN IS a big, strong, aggressive, alert animal. Even without his technology, man would be a dominant species on the earth. But of all the animals, man is apparently the only one who can think in the abstract and perceive phenomena beyond the range of his senses. The search for the unknown is strong—almost inherent—in man. There is a human urge to visualize the incredibly small and to discern the unimaginably large. The motivation to know what is behind the next hill, through the forest, across the sea, or beyond the clouds is a powerful goad to our restless quest for new horizons and fresh challenges. It is not surprising then that man should use his technical innovations to help him sense the world around him. I believe that through the ages man's greatest technologic breakthroughs have resulted from his curiosity about a phenomenon he could not experience directly with his senses. Therein lies the essence of *remote sensing of the environment.*

Remote sensing is generally defined as "sensing an object or phenomenon without having the sensor in direct contact with the object being sensed." Without direct contact, some means of transferring information through space must be utilized. In remote sensing, information transfer is accomplished by the use of electromagnetic radiation.

Perception of phenomena by the use of electromagnetic energy is not some startling new discovery by man. The first one-celled animals that crawled through the primordial ooze of prehistoric times had little capability for sensing their environment remotely. Witness the one-celled amoeba, which reacts to a strong light stimulus but which may have to come into direct contact with a strong acid before sensing it. That single contact may destroy the animal—almost certainly will damage it. The amoeba blunders along, finding its food by chance. It is not surprising then, that as single-celled animals evolved into more complex multicellular organisms, the most successful were those that developed specialized groups of cells that could experience stimuli from the environment remotely.

The animal eye is a perfect example of a remote sensor. Electromagnetic energy reflected from an object in the environment to the eye is turned into a nerve impulse. This impulse is recorded in the brain as an image. It is not certain exactly when the first animal evolved with a group of cells that could record light stimuli, but animals with true eyes have been on earth for several hundred million years, several hundred times longer than manlike creatures. Those animals which could best perceive their environment were undoubtedly the most successful. They survived, and their progeny pushed evolution onward.

As you read these words, a great deal of energy is flowing around you. The most obvious is the reflected light, either natural or artificial, by which your eye records this text. But a considerable amount of additional electromagnetic energy is also present. To prove this, you need only turn on a portable radio or television receiver. These receivers transform radiation into a sound and/or image we can experience. We cannot perceive the waves because their wavelength and frequency are beyond the reception capabilities of human sensors. The body itself is a source of radiation and sensors can be designed to display the pattern of infrared or thermal emission from the human body. Such patterns can be helpful in medical diagnosis.

We live in a world full of radiation; some of it we can perceive, much of it we cannot. The job of remote sensing is to utilize this radiation to present an image or stimulus which can be experienced by man's more limited sensing capability.

The electromagnetic spectrum extends over a tremendous range of wavelength and frequency. The spectrum is usually displayed as a continuum with short wavelengths at high frequency at one end and long wavelengths at lower frequency toward the other end. Usually no lower or upper limits are given for the spectrum because wavelengths become progressively smaller or larger extending beyond our capability to measure them. At these extreme limits very few wavelengths are generated naturally. Gamma rays are usually shown as the beginning of the spectrum of the shorter wavelengths. Gamma rays merge with x rays around 0.03 Angstroms. (The length of an Angstrom is so small it defies human comprehension—it is one ten-millionth of a millimeter.) At the other end of the spectrum are very long radio waves, over a kilometer in length. Visible light—that which we sense with our eyes—represents a tiny fraction of the total spectrum that is delineated at the extremes by gamma and radio waves. Invisible ultraviolet energy merges gradually with visible light on the shortwave side. The shortest waves visible to most human normal vision are about 0.3 microns (a micron is one millionth of a meter and is denoted by the Greek letter mu, μ), or the preferred usage today would be 300 nanometers. (A nanometer is a billionth of a meter.) Visible light melds into the invisible infrared around 0.7 microns, or 700 nanometers. That tiny portion of the spectrum from 300 to 700 nanometers is all the energy we can utilize with our prime sensors—our eyes. It is not surprising that the more successful animals on earth developed sensors which operate in this range of the spectrum. On the surface of the earth, the peak of the sun's radiant power occurs at about 500 nanometers, or almost exactly in the middle of the visible light range. Human and most animal eyes take advantage of the greatest amount of energy available on earth by sensing in this portion of the spectrum.

Remote sensing uses more of the electromagnetic spectrum or focuses on very narrow portions of it to give us a different and sometimes startling view of our environment. To utilize more of the spectrum, at shorter or longer wavelengths, various sensors have been developed which react to radiation at different carefully selected wavelengths. The stimulation of these sensors by energy is then used to produce an image which we can experience with our eyes. Because remote sensing is wavelength dependent, this parameter is the most logical approach to the organization of readings pertaining to the subject and is the method followed here.

This book has ten parts. The first two are concerned with introductory essays dealing with the electromagnetic spectrum and a general introduction to remote sensing. These sections provide the necessary background for the more technical articles which follow. The next six parts deal with specific regions of the electromagnetic spectrum and methods of their utilization for remote sensing. We begin at the shorter wavelengths—the ultraviolet—and progress through the spectrum to the longer wavelengths. Each part begins with a general article explaining the capabilities and function of the sensors in that region of the spectrum, and the remaining articles detail specific applications to environmental problems.

Part Nine deals with multispectral sensing, a technique that splits the spectrum into narrow bands or parts. The exact environmental scene viewed by the sensor is produced on two or more images. These are not true duplicate images because, being produced at slightly different wavelengths, they display image variations as the result of different spectral responses by the objects in the environment.

The final part deals with the controversial subject of social implications of remote sensing—a topic which has been of little concern to scientists conducting research in remote sensing. In our anxiety to bring these remarkable new tools to use in man's study of the environment, the broader social questions concerning repeated massive surveillance have been ignored. There is nothing inherently immoral in remote sensing but the potential for human misuse does exist. We have not yet addressed ourselves to this problem at the national or international level.

At the end of each part is a selective list of suggested readings.

The need for a book of this type became apparent to me when I attended a National Science Foundation Institute at the University of Michigan during the summer of 1968. Experts lectured on all aspects of remote sensing; they presented articles gleaned from many varied, and sometimes obscure, sources. The lack of an organized, systematic treatment of the literature in the field of remote sensing was pressingly felt.

Any scholar who sets out to select articles for a book of readings on remote sensing does so at great risk. The field changes rapidly—almost daily. Further, it encompasses so wide a range of knowledge that no one person can be competent in its entire breadth. Realizing this, in the early planning for this book I sought the counsel of experts in the various specialties of remote sensing. Over two hundred letters were sent to people who are recognized as important contributors to research; more than half of them responded. In some specialized areas there was remarkable agreement on the best or most useful articles. In other areas opinions varied so widely that each respondent cited different articles. For the final selection I take full responsibility. At the same time I extend deep and heartfelt thanks to all who took the time to respond to my inquiries.

Grateful appreciation is tendered to the authors whose articles are used in this publication; without exception they have been extremely cooperative. The task of compilation would have been impossible without their considerable help, collaboration, and patience. I am also indebted to the various publishing agencies who were most gracious in extending permission to use the works that appear here.

Remote sensing, by its very nature, makes use of many photographs, images, charts, graphs, and other visuals. These provided a major problem in producing this book. The individual authors had gathered their illustrations from many sources and over considerable periods of time and some of the original pieces have since become damaged, lost, or destroyed. When the original graphics were not available, I have where possible substituted others which portray the same intent; each substitution is noted in the text. A few illustrations were excluded because of space limitations.

The articles presented were chosen because they are good examples of research and writing in specific areas of remote sensing. The reader should note the original publication dates and keep in mind that the selections describe the state of the art at that time. In the very rapidly changing field of remote sensing, future or suggested research has often been accomplished by the time an article appears in print. Some very good articles had to be abandoned because they had become too dated.

Before each article an editor's introduction appears. These are not intended as summaries of the articles but rather as descriptive introductions to serve as bridges to the research presented.

ROBERT K. HOLZ

CONTENTS

CONTRIBUTORS
and Their Current Affiliations

The number in brackets indicates the page on which the contributor's selection begins.

AMSBURY, DAVID L. Earth Resources Division, NASA Manned Spacecraft Center, Houston, Texas [161]

BARNES, R. BOWLING Barnes Engineering Company, Stamford, Connecticut [236]

BARRINGER, A. R. Barringer Research, Rexdale, Toronto, Ontario, Canada [47]

BLIAMPTIS, E. E. Terrestrial Sciences Laboratory, Air Force Cambridge Research Laboratories, Hanscom Field, Bedford, Massachusetts [67]

CAIGER, J. H. Irene, Transvaal, Republic of South Africa [114]

COLWELL, ROBERT N. School of Forestry and Conservation, Agricultural Experiment Station, University of California, Berkeley, California [322]

COUTTS, JOHN W. Department of Chemistry, Lake Forest College, Lake Forest, Illinois [2]

CRONIN, J. F. Terrestrial Sciences Laboratory, Air Force Cambridge Research Laboratories, Hanscom Field, Bedford, Massachusetts [67]

DAVIES, SHANE Department of Geography, University of Texas at Austin, Austin, Texas [386]

DOYLE, FREDERICK J. Topographic Division, U.S. Geological Survey, McLean, Virginia [153]

DUDDEK, MANFRED Wild Heerbrugg Instruments, Inc., Farmingdale, Long Island, New York [130]

EDGERTON, A. T. Microwave Division, Aerojet-General Corporation, El Monte, California [269]

ESTEP, SAMUEL D. Law School, University of Michigan, Ann Arbor, Michigan [362]

ESTES, JOHN E. Geography Department, University of California, Santa Barbara, California [381]

FEINBERG, GERALD Department of Physics, Columbia University, New York, New York [10]

FISCHER, WILLIAM A. U.S. Geological Survey, Washington, D.C. [120]

FRITZ, NORMAN L. Eastman Kodak Company, Rochester, New York [185, 199]

GOLOMB, BERL Environmental Ecology Group, Santa Barbara, California [381]

GRAY, DWIGHT E. American Institute of Physics, Washington, D.C. [2]

GUMERMAN, GEORGE J. Center for Man and Environment, Prescott College, Prescott, Arizona [193]

HELLER, ROBERT C. Pacific Southwest Forest and Range Experiment Station, U.S. Forest Service, Berkeley, California [138]

HEMPHILL, WILLIAM R. U.S. Geological Survey, Washington, D.C. [83]

HIRSCH, S. N. Forest Fire Laboratory, U.S. Forest Service, Riverside, California [228]

HOLTER, MARVIN R. Willow Run Laboratories, Institute of Science and Technology, University of Michigan, Ann Arbor, Michigan [78, 247]

HOLZ, ROBERT K. Department of Geography, University of Texas at Austin, Austin, Texas [375, 386]

HUFF, DAVID L. Departments of Marketing Administration and Geography, University of Texas at Austin, Austin, Texas [375]

INNES, RICHARD B. Radar and Optics Division, Environmental Research Institute of Michigan, Ann Arbor, Michigan [282]

KOLIPINSKI, MILTON C. U.S. Geological Survey, Washington, D.C. [134]

KRUCKEBERG, R. F. Northern Forest Fire Laboratory, U.S. Forest Service, Missoula, Montana [228]

LEONARDO, EARL S. Goodyear Aerospace Corporation, Litchfield Park, Arizona [43]

LEWIS, ANTHONY J. Department of Geography and Anthropology, Louisiana State University, Baton Rouge, Louisiana [297]

LISTON, THOMAS C. Control Data Corporation, Minneapolis, Minnesota [291]

LOWMAN, PAUL D., JR. Goddard Space Flight Center, Greenbelt, Maryland [145, 170]

LYON, R. J. P. Department of Geology, Stanford University, Palo Alto, California [340]

MADDEN, F. H. Northern Forest Fire Laboratory, U.S. Forest Service, Missoula, Montana [228]

MALIDA, WILLIAM A. Willow Run Laboratories, Institute of Science and Technology, University of Michigan, Ann Arbor, Michigan [349]

MARKLE, DAVID A. Perkin-Elmer Corporation, Norwalk, Connecticut [83]

MAYFIELD, ROBERT C. Department of Geography, Boston University, Boston, Massachusetts [375]

MOLINEUX, C. E. Terrestrial Sciences Laboratory, Air Force Cambridge Research Laboratories, Hanscom Field, Bedford, Massachusetts [67]

NEELY, JAMES A. Department of Anthropology, University of Texas at Austin, Austin, Texas [193]

NORMAN, GERALD G. Division of Plant Industry, Florida Department of Agriculture, Winter Haven, Florida [199]

NUNNALLY, NELSON R. Department of Geography, University of Oklahoma, Norman, Oklahoma [18, 314]

OLSON, CHARLES E., JR. School of Natural Resources, University of Michigan, Ann Arbor, Michigan [95]

PAARMA, HEIKKI Rautaruukki Oy, Pakkahuoneenkatu, Oulu, Finland [202]

PARKER, DANA C. Deceased [29]

POLCYN, FABIAN C. Institute of Science and Technology, University of Michigan, Ann Arbor, Michigan [349]

RAY, RICHARD G. National Science Foundation, Washington, D.C. [120]

ROONEY, T. P. Terrestrial Sciences Laboratory, Air Force Cambridge Research Laboratories, Hanscom Field, Bedford, Massachusetts [67]

SAPP, CECIL D. Control Data Corporation, Minneapolis, Minnesota [291]

SCHNEIDER, WILLIAM J. Water Resources Division, U.S. Geological Survey, Washington, D.C. [134]

SPANSAIL, NORMA A. Willow Run Laboratories, Institute of Science and Technology, University of Michigan, Ann Arbor, Michigan [349]

SPURR, STEPHEN H. University of Texas at Austin, Austin, Texas [103]

STAELIN, DAVID H. Department of Electrical Engineering, Massachusetts Institute of Technology, Cambridge, Massachusetts [256]

STINGELIN, RONALD W. Environmental Sciences Branch, HRB-Singer, Inc., State College, Pennsylvania [220]

STOERTZ, GEORGE E. U.S. Geological Survey, Washington, D.C. [83]

TALVITIE, JOUKO Rautaruukki Oy, Pakkahuoneenkatu, Oulu, Finland [202]

TAYLOR, JAMES I. Department of Civil Engineering, Pennsylvania State University, University Park, Pennsylvania [220]

THACKREY, DONALD E. Office of Research Administration, University of Michigan, Ann Arbor, Michigan [209]

TREXLER, D. T. Geology–Geography Department, Mackay School of Mines, University of Nevada, Reno, Nevada [269]

TUYAHOV, ALEX Office of Information Services, The Governor's Office, Austin, Texas [386]

VIKSNE, ANDRIS Engineering and Research Center, U.S. Bureau of Reclamation, Denver, Colorado [291]

WENDEROTH, SONDRA Science Engineering Research Group, Long Island University, Greenvale, New York [334]

WILLIAMS, R. S., JR. U.S. Geological Survey, Washington, D.C. [67]

WOLFF, MICHAEL F. *Science and Technology*, Stamford, Connecticut [29]

WORMSER, ERIC M. Barnes Engineering Company, Stamford, Connecticut [38]

YOST, EDWARD F. Science Engineering Research Group, Long Island University, Greenvale, New York [334]

PART ONE

THE ELECTROMAGNETIC SPECTRUM

Energy for information transfer

MAN DID NOT easily learn about the existence of radiation at wavelengths shorter or longer than those of visible light. Almost by accident early experiments with light indicated the existence of this shorter- and longer-wave "invisible" radiation. The problem was that sensors did not exist that could accurately measure or record this energy and then produce an image from it. It was accepted that whatever light was, it extended beyond the limits of man's sensory capabilities.

An associated problem was the way this energy was transported. About half a century after the discovery of invisible radiation, the wave theory of electromagnetic energy was developed. A few years later the existence of waves like light, but with longer wavelengths, was proved theoretically. Almost 90 years after the discovery of invisible radiation, experiments verified the existence of these longer waves. Since these early discoveries, man has been probing constantly to measure, record, and utilize this great mass of available energy which is far greater in total range than the visible portion of the spectrum.

1-The Electromagnetic Spectrum

DWIGHT E. GRAY
JOHN W. COUTTS

THE TERM "spectrum" has been applied to the continuous series of colors obtained when white light is separated into its component colors of red, orange, yellow, green, blue, indigo, and violet—a series that taken together is called the *visible spectrum.* (As a memory device the initials of these colors can be combined to form the acronym ROY G. BIV.) On the basis of the wave theory of light, color varies with the radiation's wavelength or frequency. In the visible spectrum, red has the longest wavelength (the lowest frequency) and violet has the shortest wavelength (the highest frequency), light of other colors falling between in the sequences noted above.

In 1800 the English astronomer Sir William Herschel (1732–1822) discovered that if he moved a thermometer along the visible spectrum from violet to red, the temperature readings increased progressively. Then he found that if he moved the thermometer on into the dark region beyond the red the temperatures continued to rise, indicating the presence there of radiation similar to light. Two years later, other investigators who were studying the chemical effects of light demonstrated the existence of radiation in the region beyond the violet end of the visible spectrum. These experiments, plus many others performed since that time, have shown that the visible spectrum, the defined boundaries of which depend upon limitations of the human eye, is but a small, centrally located segment of a very much greater band of radiation, all of which travels with the speed of light and exhibits such light-related characteristics as reflection, refraction, interference, and polarization. This total range of radiation is called the *electromagnetic spectrum.*

From *Man and His Physical World,* 4th ed. Princeton, N.J.: D. Van Nostrand Company, Inc., 1966. Chap. 22, pp. 407–423. Reprinted with permission of the authors and the publisher.

Table 1, to which we shall refer frequently in this chapter, shows a breakdown into subgroups of the radiation in this overall spectrum. Dividing lines between the categories are arbitrarily defined and should not be regarded as sharp, there being at each boundary a small overlapping range of wavelengths that might legitimately be placed in either group. In the rest of this chapter these subdivisions of the electromagnetic spectrum are considered separately.

Waves Longer Than Visible Light

The Herschel experiment occurred some years before the wave theory of light was accepted. At the time it was performed, therefore, it simply proved that whatever the nature of light might be, the phenomenon extended beyond the range to which the human eye was sensitive. By 1850 the problem of the nature of light was believed finally solved in favor of the wave view. A few years later, the British physicist, James Clerk Maxwell (1831–1879), predicted, on purely theortical grounds, the existence of waves like light but with longer wavelengths. In 1888 this prediction was verified experimentally by the German physicist Heinrich Hertz (1857–1899). His experimental setup is shown diagrammatically in Figure 1. The small spheres of spark gap A were connected to a high-voltage source of alternating current. By means of the wire slide connecting the two

Table 1 The Electromagnetic Spectrum, Showing Approximate Wavelengths of Waves of Various Types

Kind of Waves	Approximate Wavelength Range—British Units	Approximate Wavelength Range—Metric Units	Number of Octaves[a]
Very long electromagnetic waves	Hundreds of miles to 10 miles	Hundreds of kilometers to 17 kilometers	5 or 6
Radio, TV, and microwave range	10 miles to 0.01 in.	17 kilometers to 0.025 cm	26
Infrared or heat rays	0.01 in. to 2.8×10^{-5} in.	0.025 cm to 7×10^{-5} cm	9
Visible light	2.8×10^{-5} in. to 1.4×10^{-5} in.	7×10^{-5} cm to 3.5×10^{-5} cm	1
Ultraviolet rays	1.4×10^{-5} in. to 1×10^{-7} in.	3.5×10^{-5} cm to 2.5×10^{-7} cm	7
X rays	1×10^{-7} in. to 4×10^{-10} in.	2.5×10^{-7} cm to 1×10^{-9} cm	8
Gamma rays	4×10^{-10} in. to 2.4×10^{-11} in.	1×10^{-9} cm to 6×10^{-11} cm	4

[a] An octave is the interval from a given frequency (or wavelength) to either half or twice that frequency (or wavelength).

sides of spark gap B, the length of the conducting path between the spheres at gap B could be varied. Hertz found that when sparks were jumping across gap A as a result of the alternating voltage applied to it, there was a particular position of the slide for which sparks also would jump across gap B, although that circuit contained no conventional source of electric current. Under these conditions the A gap radiated energy that traveled through space and set up an oscillating current in the B circuit. Hertz demonstrated that this radiation exhibited wave characteristics which, except for wavelength, were the same as those manifested by light waves. The term "Hertzian waves" is frequently applied to radiation generated in this manner by an oscillating current. Since Hertz's time, a variety of generating methods have been introduced and waves can now be produced that range in length from a fraction of a millimeter to hundreds of kilometers.

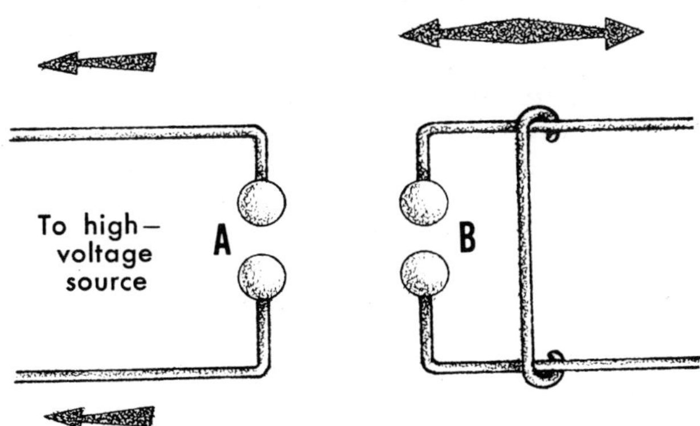

Figure 1 The essential parts of Hertz's experimental apparatus.

Very Long Electromagnetic Waves

The upper limit of the very-long-wave group is highly arbitrary. No wavelength can be established as representing an absolute maximum. For example, when a coil is rotated in a magnetic field, electromagnetic waves are generated that have the same frequency as that of the coil's rotation; theoretically, such a coil can be rotated as slowly as one wishes. The longer waves in this group have no particular applications at present. Those in the wavelength range of approximately 12 to 19 miles (about 20 to 30 kilometers) find some use in radio telegraphy.

Waves Used Principally for Communication

This broad band of some 26 octaves covers the range of electromagnetic wavelengths employed primarily for radio, television, and related media of communication. Applications listed in Table 2, together with frequency and wavelength ranges, are typical although by no means all-inclusive.

Hertz made no attempt to apply the radiation that bears his name in the field of communication; in effect, however, the apparatus shown in Figure 1 was a very simple form of radio transmitter and receiver. It remained for the Italian inventor Guglielmo Marconi (1874–1937) to realize that waves of this kind could be used to transmit information from one place to another. In the early days of radio communication, known then as wireless telegraphy, a spark gap actually was employed in the transmission. A spark discharge at the gap sent surges of electrons back and forth between the ground and an antenna, causing long-wavelength electromagnetic radiation to spread out through space. When these waves were cut by the aerial of the receiving set, electric currents were induced in its circuit and signals could be heard in the headphones. Obviously, this primitive kind

Table 2 Typical Applications of Radio Waves

Applications	Frequency Range	Wavelength Range
Ship's radio, aeronautical beams, coastal telegraph	17.6–550 kc[a]	17 kilometers–545 meters
Standard AM broadcast	550–1600 kc	545–187 meters
Amateurs, aeronautical, Loran	1.6–6.0 mc[a]	188–50 meters
International short-wave broadcast and amateurs	6–54 mc	50–5.6 meters
VHF television (Channels 1–6)	54–88 mc	5.6–3.4 meters
FM radio broadcast	88–108 mc	3.4–2.8 meters
Aircraft radio, airdrome control	108–132 mc	2.8–2.3 meters
Police calls, taxi radio	150–160 mc	2.0–1.9 meters
VHF television (Channels 7–13)	174–216 mc	1.7–1.4 meters
UHF television (Channels 14–83)	470–890 mc	64–34 cm
Meteorological aids (Radiosonde)	1660–1700 mc	18.1–17.7 cm
Radio telephone, remote TV pickup and TV relay	2,000–11,700 mc	15–0.26 cm
Radar	300–30,000 mc	1 meter–1 cm
Experimental	30,000–100,000 mc	1–0.3 cm

[a] 1 kc ≡ 1 kilocycle = 1000 cycles, or vibrations per second.
1 mc ≡ 1 megacycle = 1,000,000 cycles, or vibrations per second.

of equipment was suitable only for transmitting and receiving code.

Modern radio, television, and related communication systems were made possible by the invention of the vacuum tube (radio tube), one of the outstanding technological advances of this century. The basic forms of this device are the diode and the triode. The vacuum tube permits the generation of waves of widely varying frequencies, the production of very powerful radiations that can be transmitted over long distances, and the amplification of weak electromagnetic signals into useful currents. It plays an essential role in both the generation and the reception of radio and television waves.

The path of propagation of radio waves depends upon the frequency (wavelength) of the radiation. Rays having frequencies in the standard broadcast band may travel from the transmitter to the receiver by any of the paths shown in Figure 2. Those following a direct path to the receiver constitute the so-called *ground wave.* Another part of the radiation travels skyward and is partially reflected back to earth from the Kennelly-Heaviside layer of ionized gases in the upper atmosphere. These reflected skywaves may travel many times the distance traversed by ground waves before they become too weak for detection. Since long-wave reflection from the ionosphere is largely limited to the evening hours, the geographic area reached by a radio station is much greater at night than in the daytime.

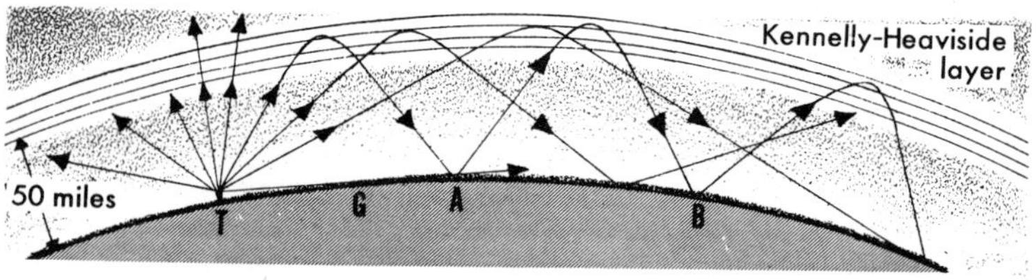

Figure 2 How radio waves are refracted (reflected) by the ionized gas layers in the earth's atmosphere.

The ionosphere is much less effective as a reflector of the higher frequencies (shorter wavelengths) used in FM and TV broadcasting. Accordingly these waves must reach receivers largely by direct paths, and their ranges are limited by the curvature of the earth. Microwaves, used in radiotelephony and in cross-country TV relay networks, can be beamed with high accuracy over straight-line paths (called line-of-sight paths).

When radio waves encounter objects in their paths, varying amounts of the radiation are absorbed, reflected, and allowed to pass through, the nature of the material and the wavelength of the radiation determining which process predominates. Brick and wood, for example, are relatively transparent to waves in the commercial broadcast range, as indicated by the fact that home radios do not require outside aerials. Metals, on the other hand, absorb a high proportion of such radiation, as evidenced by the extent to which, for example, steel-framed bridges blank out automobile radio reception. Reflection from solid objects is very much greater in the microwave (high-frequency) region than for long waves. This fact is utilized in *radar* detection by which an object can be located, and the distance to it determined, through its reflection of microwave radiation.

Though the approximately 26 octaves of electromagnetic radiation are employed *primarily* for communication purposes, other uses should not be entirely ignored. For example, radiation having a wavelength of several meters is effective in killing certain weevils, worms, and other pests when they are in the egg or larva stage. Artificial fevers that have proved valuable in treating various diseases are generated by subjecting the patient to very short radio waves. So-called diathermy machines are really short-wave transmitters. Microwave radiation is used in fundamental research on the structure of matter —for example, determining spatial configurations of various molecules.

Infrared or Heat Rays

Infrared radiation (Table 1) covers the wavelength range between radio waves and red light. Whereas radiation in the Hertzian band and beyond is due to oscillating currents in electrical conductors, infrared rays stem from thermal rotations and vibrations of molecules in the source. About half of the sun's radiation that reaches the earth's surface falls in the infrared range. It is also a familiar fact that any object at a higher temperature than its surroundings radiates heat—heat that can be felt even though the temperature of the source is far below that at which it becomes incandescent and emits light.

Although infrared rays are not visible to the human eye, photographic film sensitive to them exists and makes possible some of their most interesting modern applications. For example, objects can be photographed by the "light" of an electric iron or stove in a room that appears to be totally dark. Infrared-sensitive film is particularly useful for photographing distant objects. The longer the wavelength of light, the less it is scattered (that is, reflected at random) by particles of dust or mist. Consequently, infrared rays penetrate the atmosphere better than does visible light.

. . . .

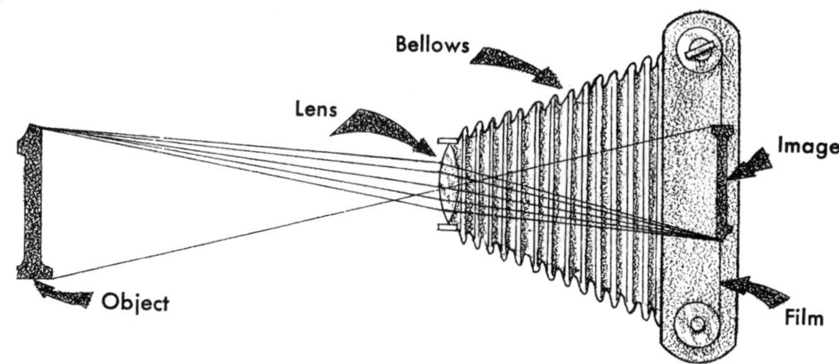

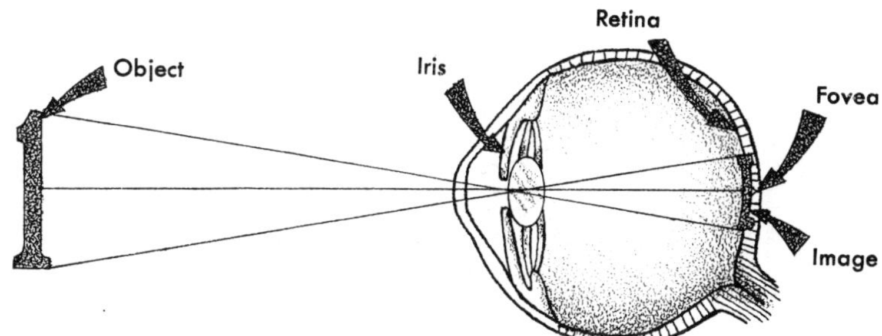

Figure 3 How an image is formed on the film in a bellows camera (upper) and on the retina of the human eye (lower).

Visible Light Waves

The characteristics of visible light and most of its applications are quite well known. Our discussion of it here is confined to brief consideration of the most widely used device for receiving light waves, the eye, and to a look at the physiological effects of variations in wavelength —that is, color.

Light waves result fundamentally from electrical and magnetic disturbances that occur within atoms and molecules when their electrons shift from one energy level to another. Such changes involve atomic structure and behavior.

The Physical Basis of Seeing

Of the approximately 60 octaves represented in the electromagnetic spectrum, the human eye is sensitive to only one—the wavelength range from about 0.000035 cm (violet) to 0.000070 cm (red). Our ears do somewhat better, letting us hear some 9 or 10 octaves of sound.

Comparison of the Human Eye with a Camera

The two parts of Figure 3 illustrate schematically how an image is formed by a bellows-type camera and by the eye. Both have convex (converging) lenses, the retina of the eye being analogous to the camera's film. In each case, an inverted image of the object is formed at the point where the rays from the object focus. The location of this image varies with the distance from the object to the lens. For a sharp picture or for sharp vision, this point of focus must be exactly on the film or on the retina. Necessary adjustments to achieve this goal are accomplished differently in the two cases.

In the camera the lens is moved toward or away from the film to adjust for different object distances. In the camera diagram the dotted line shows where parallel rays coming from a distant object would focus. To photograph such an object sharply, the lens would have to be moved back until the dotted-line position coincided with the film. For objects nearer the camera than the solid arrow, the lens would have to be moved out from its pictured position. This process is called *focusing*. In modern cameras of the candid type, the lens is screwed in or out of the mounting instead of being moved by a sliding bellows. The less the diameter of the lens opening, the greater the range of distance over which objects are in good focus. In so-called fixed-focus instruments, like box cameras, the lens opening is very small. A single converging lens, such as is shown in the diagram, is subject to a number of defects and is used only in very cheap cameras. In cameras of higher quality, shortcomings are eliminated by combinations of from three to five lenses.

In the human eye the adjustment for varying object distances is made by varying the curvature of the lens instead of by moving it in or out with respect to the retina. When one looks at nearby objects, certain muscles around the eye squeeze the lens and make it bulge out, bringing the point of focus nearer to it. For distant vision, other muscles pull at the edge of the lens making it thinner and moving the focal point back. This process, analogous to focusing in the camera, is called *accommodation*.

Correction of Defective Vision

When for any reason the lens of the eye hardens or the muscles grow weak, the accommodation process becomes difficult. If the lens grows hard in a shape that is

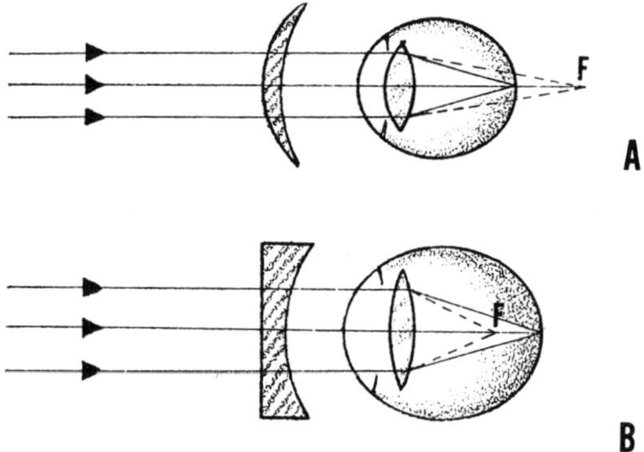

Figure 4 Correction of farsightedness (A) by a converging lens and of nearsightedness (B) by a diverging lens.

Figure 5 Test chart for astigmatism. The lines should all appear of equal blackness to the normal eye.

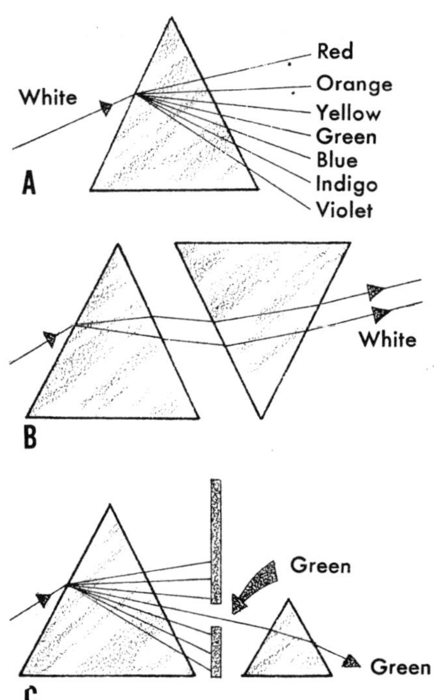

Figure 6 Newton's experiments which proved that white light consists of all colors. A, effect of passing white light through a prism. B, effect of passing white light through two prisms. C, effect of passing light of a single color through a prism.

too thin, *farsightedness* or *hyperopia* (also called *hypermetropia*) results. In this case, as shown in Figure 4, part A, the image forms behind the retina, and a converging eyeglass lens is necessary for correction. In the case of a lens that is too thick, as pictured in part B of Figure 4, where the image forms in front of the retina, the individual suffers from *nearsightedness* or *myopia,* and a diverging lens must be introduced to correct the condition. Bifocal eyeglasses, in which lenses of two different focal lengths are used, partially correct for an inability of the eye muscles to adjust for objects at different distances.

A third common eye defect is *astigmatism,* a condition that results when the lens of the eye is not of uniform curvature in all directions. Figure 5 is one kind of chart for identification of this defect; to the astigmatic eye, lines in different directions will appear to be of varying degrees of blackness. The remedy for astigmatism is eyeglasses whose lenses are unevenly curved in a manner that will just counteract the lack of uniformity of the natural eye's curvature.

Color Phenomena

Variations in color represent the interpretation man's eyes and brain make of differences in the wavelength or frequency of light. In other words, color always is associated with the way we see things.

Newton and the Nature of Color

Among the outstanding experiments in all the history of science were those by which Newton demonstrated that white light is composed of all colors of the visible spectrum. The first of the series, pictured in Figure 6A, had been performed before Newton's time, the results being explained on the assumption that the prism effected a fundamental change in the nature of the white light. Parts B and C of Figure 6 illustrate two experiments that Newton performed to test this theory. In B an inverted second prism was placed behind the first; when the continuous spectrum that emerged from the first prism passed through the second one, the resulting radiation proved to be white light again. It was argued, however, that the second prism, being inverted, simply reversed the action of the first prism and changed all of the colors back to white. Then Newton performed the experiment illustrated in C. All colors to the right of the first prism were masked out except one (green in the example) and this one was allowed to fall on the second prism; in this case, the beam was found to pass through unchanged, ruling out the possibility that the second prism actually converted other colors to white. These experimental facts proved to Newton's satisfaction that white light is actually a combination of the various colors, and that the first prism simply separated them. The rest of the scientific world did not readily accept his conclusion, however, and its announcement engendered so bitter a controversy that for a time Newton threatened to give up scientific work entirely.

The effect we have been describing is an example of refraction, the separation of colors occurring because different colors are refracted different amounts in going from air into glass and back into air again. In the case of a rainbow, the most familiar natural case of the splitting of white light, drops of water take the place of the prism. The so-called secondary rainbow, in which the sequence of colors is reversed, results when refracted rays emerge after two reflections within a raindrop.

Halos around the sun and moon, commonly called rings, have a similar cause, the refracting "prisms" in this case being tiny crystals of ice in the upper atmosphere.

Color of Materials

Newton's experiments dealt with the color of light. Next, we may ask just what it means to say an *object* is a certain color. A definition of the color of materials can be established by the following tests.

1. Various colors of light are allowed to fall in sequence on the kind of a screen we call white. The result is that each color is faithfully reflected, indicating that a white object or material can be defined as one that reflects all colors or wavelengths.

2. White light is permitted to shine on a material of the kind, for example, that we call red and, of course, we see the substance as red. At this point, two explanations are possible. Since white light is composed of all colors, either the red object changes all of these colors to red or it reflects only the red and absorbs the rest. To test these alternatives, we can use some other single color of light—say, blue or green—to illuminate the red object. If it changes all colors to red, the object should still appear red, but if it absorbs all colors but red, it should now appear black. The latter is found to be the case. Therefore, *the color of a material can be defined as that of the light which it reflects.*

Color Mixing

There is a difference between the effects of mixing light of different colors and the effects of mixing different-colored pigments. As for mixing light of different colors, in the second of Newton's experiments illustrated in Figure 6 the right-hand prism mixed together the seven colors of the rainbow and produced white light. Similarly, if a so-called color disk, containing segments of these seven colors of the appropriate relative sizes, is rotated, the sensation of almost pure white light can be produced. But it is not necessary to mix all seven colors to obtain white light; the result will be the same if blue, green, and red light are properly combined. These three are called *primary colors* or *additive primaries* because it is possible to obtain any hue by mixing them together in certain well-defined relative amounts. This fact is used in stage lighting where, by varying the relative intensities of groups like blue, green, and red bulbs, resultant light of any desired color can be obtained. There are also a number of pairs of hues, called *complementary colors,* which together can give white light. Examples include:

Yellow and blue	Red and bluish green
Orange and greenish blue	Violet and greenish yellow

The additive primaries and their combinations are shown at the upper left in Figure 7[1], their component spectral colors are pictured at the lower left.

Mixing paints or pigments gives results quite different from those obtained when different colors of light are combined, because the color of a substance is that of the light which it *reflects.* It is not surprising that there are mixtures of different-colored pigments that absorb all colors and therefore appear black. Analogous to the three additive primaries are three *subtractive primaries* which, properly mixed together, produce black. They are magenta, yellow, and cyan (a blue-green), shown at the upper right in Figure 7; at the lower right in the same figure are their spectral component colors. For both the additive and subtractive primaries, the result of mixing the colors in pairs can be seen from the areas in Figure 7 where only two circles overlap.

Color Vision

Many attempts have been made to explain color vision but so far none has been completely successful. The one that most nearly accounts for all experimentally observed facts is the Young-Helmholtz theory, advanced by Thomas Young (1773–1829) of wave-theory fame and amplified later by the German scientist Hermann von Helmholtz (1821–1894). This hypothesis postulates the existence in the retina of the eye of three kinds of tiny cones, each type being sensitive to one of the three additive primary colors—red, green, and blue. When we look at an object, this theory holds that one or more of the sets of cones are excited to varying degrees of activity. The brain then takes the three responses and mixes them to produce the net sensation that we call the object's color. This mixing process permits us to distinguish among thousands of hues and shades with but the three kinds of cones.

On the Young-Helmholtz theory, the absence of or a deficiency in one or more of the three sets of cones produces the condition known as *color blindness.* There are many kinds and degrees of this defect, the two most common being *protanopia* and *deuteranopia.* Both concern primarily the longer wavelengths of the visible spectrum. In protanopia, the red-sensitive cones are believed to be defective so that the individual with this defect sees only hues that can be produced by mixtures of blue and green. To the person having deuteranopia, on the other hand, while the visible spectrum is not shortened at the ends, there is confusion among the long-wavelength colors and he tends to see only yellow, blue, or white hues.

The principal weakness of the Young-Helmholtz theory of color vision is that so far no one has produced any *anatomic* evidence of the actual existence in the eye of the three different kinds of cones this hypothesis postulates.

Color from White Light

Two methods of producing color from white light have already been described. (1) White light passed through a prism separates into its constituents because different colors are refracted by different amounts. (2) When white light falls on an object, light the color of that object is reflected and all other colors are absorbed.

A third source of color is *interference.* Light waves from two sources, falling on a screen, can produce a series of alternating bright and dark bands—bright at places where the two light paths differ by an even number of half-wavelengths, and dark when the path differences are odd numbers of half-wavelengths. Monochromatic (that is, single-wavelength) light is assumed.

Now consider the following experiment. Two pieces

[1] Not reproduced here.

of glass are arranged so that there is a thin edge of air between them and white light is allowed to shine on one plate. Part of the light is reflected from the rear surface of the first plate; part is reflected from the front surface of the second plate; and the two reflected light-wave trains combine. But the reflection points are separated by the thickness of the air wedge, and this thickness increases from the top to the bottom of the combination. As a result, at some points there will be reinforcement, and at others interference, producing once again the series of alternating bright and dark bands characteristic of an interference pattern. But since different colors of light have different wavelengths, the bright bands for the several constituents of white light occur at slightly different places, and there is a separation of the white light into colors. Colors seen in a soap bubble result from interference of this kind. Because of gravitational attraction, the bubble is thinnest at the top and thickest at the bottom, thus forming the very thin reflecting wedge necessary for color separation by interference. Other examples include the colors in oil films on water and those seen in mother-of-pearl. The wedges must be very thin for this effect to occur; otherwise the separated colors overlap so extensively that they combine and one sees only white light.

Another phenomenon productive of color is *scattering* —the irregular random reflection of light in all directions. The color effect of this phenomenon results from an observed fact noted earlier in the discussion of infrared rays—namely, that the longer the wavelength of the light, the less it is scattered. Consider the sunlight that reaches earth early in the morning or late in the afternoon. Sunlight enters the atmosphere containing all visible wavelengths. But the shorter wavelengths—those in the blue-violet end of the spectrum—are scattered more than the longer ones by air molecules and dust particles. Consequently, in the morning and evening, when the atmospheric air path of sunlight is a maximum, the sun appears unusually red. In the middle of the day when the sun is overhead, the light travels a much shorter distance in the atmosphere and scattering is much less. The blue appearance of the sky results principally from the scattered short wavelengths in the sun's rays. Astronauts and stratosphere balloonists report that viewed from a point above the atmosphere, the sky appears a purplish black. The blue color of wood smoke is another example of color resulting from scattering of light; the carbon particles are very small and the scattering color effect overbalances the actual color of the pigment. In coal smoke, on the other hand, the carbon particles are larger, the pigment color is dominant, and the smoke appears black.

Waves Shorter Than Visible Light

The upper wavelength limit of radiation in this general category (Table 1) is 1.4×10^{-5} cm. Approximately 19 octaves are involved.

Ultraviolet Rays

This radiation, falling next below violet in wavelength, is believed to be generated in the same manner as visible light. In wave theory, this means shifts in electron energy levels of greater magnitude than those required for visible light. From the other viewpoint, one says that the quanta which make up ultraviolet radiation have a greater energy content than do those that constitute visible light. Ultraviolet rays, like infrared and visible radiation, can be detected by photography.

The beneficial physiological effects of ultraviolet rays have received a great deal of attention in recent years. For example, lack of vitamin D can be compensated for to a large extent by exposure to ultraviolet rays. Young rats fed on a diet that is adequate in all respects except vitamin D soon exhibit the symptoms of rickets. Exposure to as little as 20 minutes of sunshine a day, however, prevents this disease from developing. The reactions of human beings are similar, although by no means would one wish to claim that today's vast and devout cult of sun bathers is motivated primarily by a fear of rickets. Because these rays are both invisible and penetrating, one can acquire a severe case of sunburn on a day when the sun appears to be safely hidden behind clouds—a phenomenon about which most of us have obtained personal experimental evidence. Fortunately for our well-being, much of the sun's ultraviolet radiation is absorbed in the upper atmosphere, primarily by ozone in the layer known as the ozonosphere, and so does not reach the earth's surface.

Fluorescent lighting is a familiar modern application of ultraviolet radiation. When electric current flows through the mercury vapor in a fluorescent light, it "excites" the electrons in the mercury atoms, and resulting radiation is almost entirely in the ultraviolet range. The inside of the glass tube is coated with a substance that fluoresces in ultraviolet light—that is, has the ability to absorb rays of one wavelength and emit radiation of longer wavelength. In this case the absorbed ultraviolet radiation is re-radiated as visible light, the color of which depends upon the type of powder used to coat the tube's inner surface.

Numerous related applications depend upon the fact that the fluorescent effect can be obtained even though the powder and the ultraviolet source are some distance apart. Such applications include various personnel identification techniques, outdoor advertising, theatrical scenery and costumes, laundry marking, and many others.

X Rays

Of shorter wavelength than ultraviolet light, x rays are generated when electrons traveling at high speeds strike a target of some heavy metal—for example, platinum or tungsten. The principal applications of x rays result from their great penetrating power. Everyone is familiar with x ray photographs, an example of which is shown in Figure 8, and with their medical uses. The destructive effect that x rays have on living tissue makes them useful in the treatment of cancer. Within three months after their discovery in 1895 by the German physicist Wilhelm Roentgen (1845–1923), x rays were being used by surgeons to help in setting broken bones, a striking example of an unforeseen (and in fact unforeseeable) practical application of the results of experimentation in

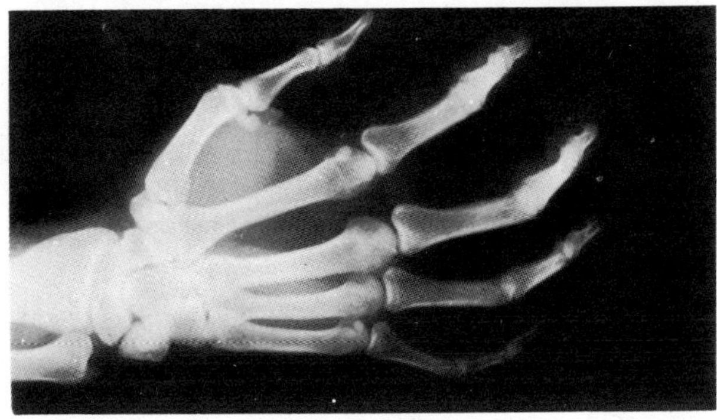

Figure 8 X ray photograph of the human hand.

very fundamental research. Had someone in Roentgen's time started a deliberate search for a tool to aid in bone-setting, a less likely place to look than Roentgen's electrical laboratory would have been difficult to imagine.

Modern x ray research has developed equipment with which exposures as short as a millionth of a second are possible, permitting x ray study of moving parts of machinery. In this way, in rotating shafts, flywheels, and the like, flaws can be detected that would not be apparent with the machine at rest.

Gamma Rays

These rays, the shortest in wavelength of any in the electromagnetic spectrum,[2] constitute one of the three kinds of radiation given off by radioactive substances. As such, gamma rays constitute a major hazard of atomic or nuclear explosions.

Being of shorter wavelength than x rays, gamma rays are even more penetrating. They find medical uses similar to those of x rays, being employed in cases where the trouble is too deep-seated to be reached by the longer rays. Gamma ray photographs are useful for studying the internal structure of large pieces of metal. In this technique a capsule containing a small amount of radioactive substance is placed on the inside of, say, a steel casting. The emitted gamma rays penetrate several inches of steel and register on the film strapped around the outside. In the gamma ray photograph of steel shown in Figure 9, the dark spots indicate air bubbles and impurities. From such pictures, engineers can detect flaws and prescribe corrective treatment.

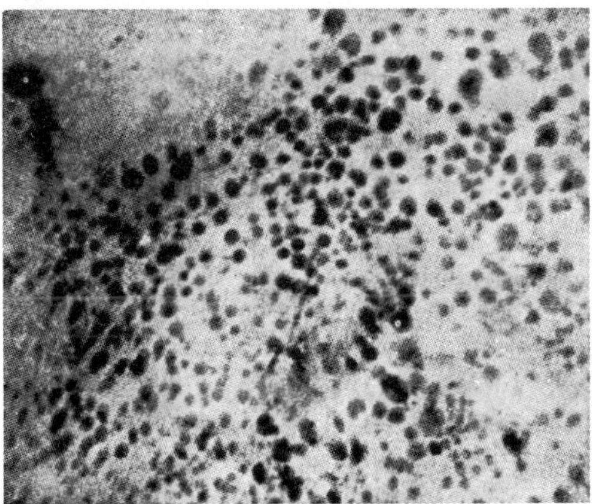

Figure 9 Gamma ray photograph of a piece of steel. The dark spots are internal air bubbles and impurities. (Westinghouse Electric and Mfg. Co.)

[2] Cosmic rays were once believed to be similar in nature to gamma rays but of shorter wavelength. Tests in recent years have shown, however, that they can be deflected by magnetic fields, indicating that they must consist of charged particles and therefore should not be included in the electromagnetic spectrum.

THE RESEARCHER *working in remote sensing must concern himself with the physical nature and characteristics of light. But the term "light" is a bit confusing or at least misunderstood. To a person not versed in physics or electrical engineering, light is the energy arriving from the sun, the reflection of which makes vision possible. Our eyes are capable of sensing only a very narrow band of the total radiation produced by the sun. Yet this electromagnetic energy has general characteristics that are the same throughout its continuum. The term "light" could be and is applied to all electromagnetic radiation.*

Another confusing characteristic is the apparent dual nature of light. It has both wave and particle properties. In certain aspects of remote sensing it is useful to deal with electromagnetic energy as waves. This is especially true in the study of optics and of image-recording over a broad band of the spectrum. The laser, on the other hand, is based in large measure on particle theory. In reality light acts in much more complex ways than can be explained by even this dual nature, but the two theories do give useful approximations for the study and practical applications of electromagnetic radiation.

2-Light

GERALD FEINBERG

THE PREVAILING view of the nature of light has changed several times in the past three centuries. Each time the answer to the question "What is light?" has assumed more fundamental importance in the physicist's picture of the universe.

Isaac Newton (in his *Opticks*, printed in 1704) described light as a stream of particles, partly because it "travels in a straight line." Out of his experiments with color phenomena in glass plates ("Newton's rings") he also recognized the necessity for associating certain wavelike properties with light beams. These properties he called "fits of easy reflection and easy transmission." Careful not to make hypotheses, he let the matter rest. His authority was so compelling, however, that the corpuscular theory of light held sway for a century, his successors being more persuaded to this view than Newton himself.

Early in the 19th century the notion that light consists of waves, a view already expressed by Christiaan Huygens in the 17th century, came into ascendance. A decisive experiment performed in 1803 by Thomas Young, a London physician, demonstrated that a monochromatic beam of light passed through two pinholes would set up an interference pattern resembling those observed "in the case of the waves of water, and the pulses of sound." At about this time Augustin Jean Fresnel and Dominique François Arago came forward with the correct interpretation of an experiment performed by Huygens. They showed that the light transmitted by Huygens' blocks of calcite crystal is polarized and that light waves therefore cannot be longitudinal compression waves as Huygens had thought but must be transverse waves oscillating at right angles to their direction of propagation (see Fig. 2).

This elucidation of the wave nature of light fit nicely into the electromagnetic theory of light propounded later in the century by James Clerk Maxwell. In Maxwell's equations light is described as a rapid variation in the electromagnetic field surrounding a charged particle, the variations in the field being generated by the oscillation of the particle. As such a varying field, light takes its place beside a number of other forms of radiant energy that were discovered in the 19th century. The different kinds of electromagnetic radiation—radio waves on one side of the spectrum of visible light and x rays on the other—correspond to different rates of variation of the field. Thus in Maxwell's theory light appears not as an independent element in nature but rather as an aspect of the fundamental phenomenon: electromagnetism.

The momentous developments in physics in this century have reopened and then resolved the old wave-particle controversy. Whereas the association of light with electromagnetism remains valid, the interpretation of this connection has changed. It has been shown that such wave properties as interference and polarization, so well demonstrated by light, are also exhibited under suitable circumstances by the subatomic constituents of matter, such as electrons. Conversely, it has been shown that light, in its interaction with matter, behaves as though it is composed of many individual bodies called

From *Scientific American* 219:50–59, 1968. Reprinted with permission of the author and *Scientific American*.

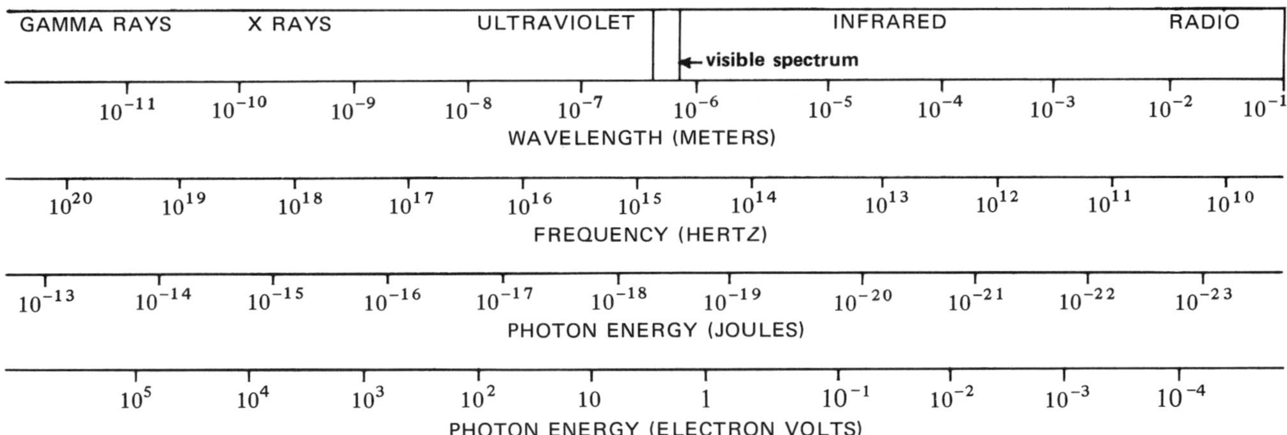

Figure 1 Electromagnetic spectrum, of which the visible spectrum . . . is only a small part, is a continuum of electromagnetic radiation, the energy of which is carried in the quanta called photons. This diagram extends from gamma rays to high-frequency radio waves. The upper scales give the radiation's frequency (in hertz, or cycles per second) and wavelength; their product at any wavelength is the speed of light. The lower scales give the energy of the corresponding photons first in joules, or watt-seconds, and then in terms of the electron volt, the energy imparted to an electron falling through a potential difference of one volt.

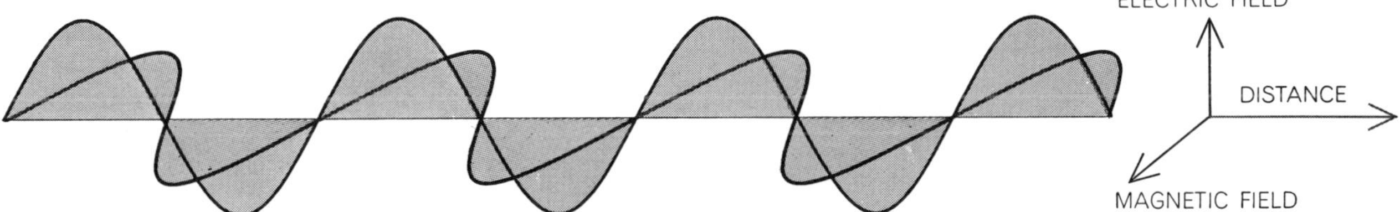

Figure 2 Electromagnetic waves, including light waves, are transverse; the electric and the magnetic fields are each at right angles to the direction of propagation. This illustration is a perspective view of a graph of the two fields at a given instant (electric field is vertical, magnetic field horizontal). The intensity of the radiation (light, for example) varies with the square of the peak amplitude of the electric field and is proportional to the number of photons in the field. The color of the light is governed by the wavelength.

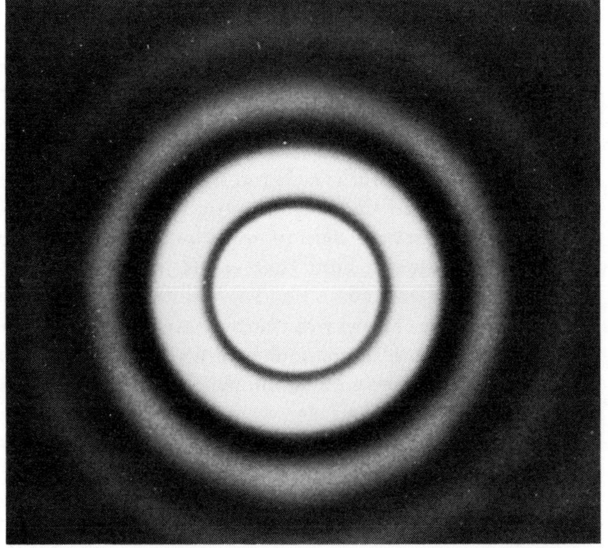

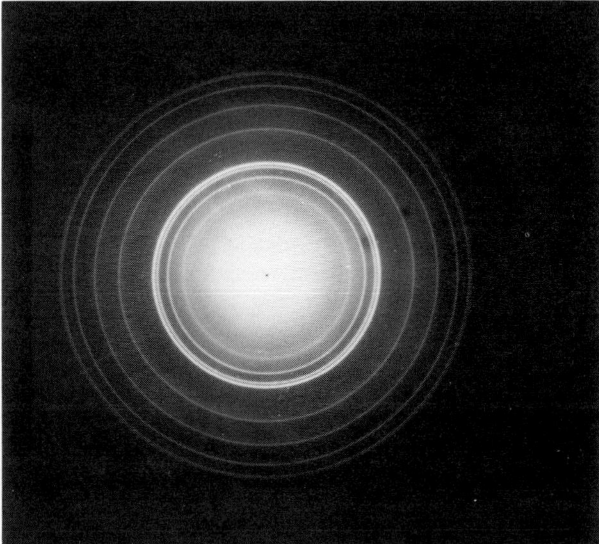

Figure 3 Wave nature of light and of electrons is demonstrated by diffraction effects. The diffraction image (left) of light from a point source was caused by interference among waves from different parts of a 0.2-mm aperture. The photograph was made with coherent light by Brian J. Thompson of Technical Operations, Inc. [now of The Institute of Optics, University of Rochester]. An electron diffraction pattern (right) is created when a beam of electrons passed through a crystal lattice, in this case beryllium, forms an image. The photograph is from the RCA Laboratories.

Figure 4 Newton described light as corpuscular but noted wavelike properties in color phenomena such as the rings formed when glass plates are placed in contact. Lines *AB* and *CD* are glass surfaces in this diagram from the *Opticks.* The light, he suggested, undergoes "fits" of reflection and transmission as it passes through varying distances of air between the two plates. (William Vandivert)

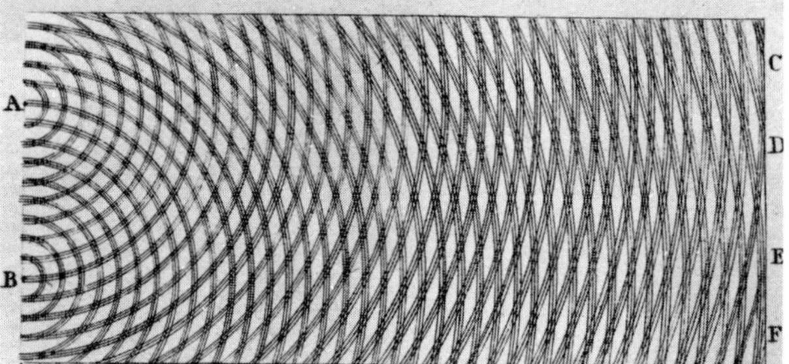

Figure 5 Young explained interference among light waves by analogy with water waves, as in this illustration from a book of his lectures published in 1807. Where the two sets of waves from two apertures are in phase (where the curves intersect), they reinforce each other. (William Vandivert)

photons, which carry such particle-like properties as energy and momentum.

As a result of these developments most physicists today would answer the question "What is light?" as Newton would have: "Light is a particular kind of matter." The differences between light and bulk matter are now thought to flow from relatively inessential differences between their constituent particles. Particles of both kinds—of all kinds—exhibit wave properties.

Much of this understanding has been acquired, of course, by means of light. In this issue of *Scientific American* Pierre Connes observes that the analysis of light "provides the best evidence for our belief in the homogeneity of the universe." In a wider context Victor F. Weisskopf shows that sight is our most important link to the world around us. Indeed, life itself is a manifestation of radiant energy in the visible spectrum; Sterling B. Hendricks describes how light starts up life, governs growth and stimulates behavior by the excitation of specialized light-absorbing molecules. The occasion for these observations and for the dedication of this issue to the topic of light is provided by an unexpected departure in the classic discipline of optics. That is the discovery of ways to synchronize the oscillation of electrons and thus produce coherent light, waves of the same length propagating in step. The laser has given physics a powerful instrument for study of the interaction of light and matter. . . . In technology laser light is finding uses in surveying and metrology, in cutting and welding, in communication and information storage. . . .

The operation of the laser exploits one of those inessential differences between photons and other particles. Let us now examine more closely the similarities and differences between the particles of light and ordinary matter and see how what is known about light can be understood in such terms.

One phenomenon that serves this purpose well is diffraction. It plainly demonstrates the wave properties of light and matter. If a beam of monochromatic light from a small source, or a stream of particles such as electrons, is directed at a screen with a small hole in it, the light or the particles that get through the hole will produce a characteristic pattern on a second screen placed beyond the first. The pattern is easy to understand in the case of light regarded as a wave, and it was cited in the 19th century as evidence in favor of the wave theory of light. Diffraction shows that light waves do not exactly travel in straight lines but diffract, or spread, as other waves do, and so find different pathways to the collecting screen. The interference of wavelets that in consequence arrive out of phase with one another at the collecting screen sets up the diffraction pattern.

In order to demonstrate the simple form of interference described above, a source of monochromatic light is required. Ordinary light sources are not monochromatic. Light from a luminous gas, for example, is emitted independently by many atoms in the gas. Furthermore, because collisions between the atoms excite and de-excite them, the light is emitted in pulses that have a finite length in time rather than an infinite length as is suggested by the textbook sine wave. Such a pulse can be resolved into a sum of pure sine waves of different wavelengths. The range of wavelengths in the pulse is inversely proportional to its duration; therefore the shorter the pulse in time, the greater the spread in wavelengths. Interference patterns are not usually detected in light from such a source. The reason is that the different wavelengths will interfere constructively in different places and the overall pattern will approximate constant illumination. Light of this kind is said to be incoherent. Conversely, light that yields interference patterns is said to be coherent. To obtain coherent light from natural sources one must narrow the range of wavelengths by means of a monochromatic filter and reduce the size of the source to a small area, as with a pinhole. With the advent of the laser we can now obtain highly coherent light without the loss of intensity involved in these procedures.

The exact shape of the diffraction pattern obtained from a given light source depends on the color of the light, objectively measured as wavelength. In the case of electrons the wavelength, and therefore the pattern, depends on the electrons' energy. In either case if diffraction is to be observed, the hole must be small with respect to the wavelength. For visible light, with wavelengths from 400 to 700 nanometers (4×10^{-7} to 7×10^{-7} meter), deviations from straight-line propagation are small unless the hole is very small. A barely visible pinhole will just produce a perceptible diffraction pattern. Electron and other subatomic particles usually have wavelengths of 10^{-9} meter or less, so that diffraction of these particles can be demonstrated only with crystals in which the "holes" are spaces between atoms about 10^{-10} meter apart.

It is this difference in the characteristic wavelength that explains why wave properties were so easy to demonstrate for light beams and so much harder to discover for matter beams. Diffraction was first shown in electrons in 1927. All the wave properties characteristic of light, however, have now been demonstrated in electron and neutron beams, and there is little doubt that they hold also for other particle beams.

An important step in establishing the underlying resemblance of the particles of matter to the particles of light was the recognition that for both kinds of particle wavelength is related to the momentum, and hence to the energy, of the particles constituting the beam. The same equations show for all cases that wavelength is inversely proportional to momentum (see Fig. 6). The equations also bring out the difference between photons and the particles of ordinary matter. In the case of the electron, for example, the energy must include the rest mass of the particle. Photons, on the other hand, have zero rest mass, and so the term for rest mass drops out of the equations.

It is the rest-mass energy ($E = mc^2$) of the electron and other matter particles that gives them wavelengths so much shorter than light beams. A photon of blue light has an energy of about 3×10^{-19} joule, which corresponds to a wavelength of 4×10^{-7} meter. If this photon energy is transferred as kinetic energy to an electron, which starts with a rest-mass energy of 8×10^{-14} joule, the total energy is changed very little (by less than 10 parts in a million) and the wavelength is about 10^{-9} meter. The wavelength is still shorter, of course, for particles that have larger rest-mass energies.

The inverse proportionality of wavelength to momentum is governed in these equations by the constant h, known as Planck's constant. It may be helpful to recall how this constant entered physics; the story is a significant chapter in the recognition of the particle nature of light. In 1900 Max Planck was concerned to explain the relation between wavelength and intensity in the radiant energy from a hot body. According to classical electromagnetic theory, the intensity was supposed to increase as the square of the frequency; by this reckoning an infinite amount of energy should be radiated at the higher frequencies or shorter wavelengths. Actual measurement had shown a quite different distribution of intensity with respect to wavelength for any given temperature (see Fig. 7). Planck found an empirical formula that described this distribution. It contained a constant, the value of which Planck chose in order to produce the best agreement with the observations. To explain why this formula should work he had to assert that light must exchange its energy with the matter of the hot body in quanta, or packets. His equation showed that the amount of energy in each quantum would be equal to the frequency multiplied by the constant $h = 6.63 \times 10^{-34}$ joule-second.

		GENERAL CASE		PHOTONS (m = 0)
	$p =$	$\frac{\sqrt{E^2 - m^2c^4}}{c}$	$=$	$\frac{E}{c}$
$\lambda =$	$\frac{h}{p} =$	$\frac{hc}{\sqrt{E^2 - m^2c^4}}$	$=$	$\frac{hc}{E}$
$v =$	$\frac{pc^2}{E} =$	$c\sqrt{1 - \frac{m^2c^4}{E^2}}$	$=$	c

Figure 6 Light differs from other forms of matter primarily in that photons have zero rest mass. The relationship between momentum *(p)* and energy *(E)* takes on a special form for photons (top row). Wavelength (λ) depends on momentum and Planck's constant *(h)*; when this universal relation is reexpressed as a dependence on energy, two forms result (middle). Similarly, particle velocity *(v)* depends on energy except in the case of photons (bottom).

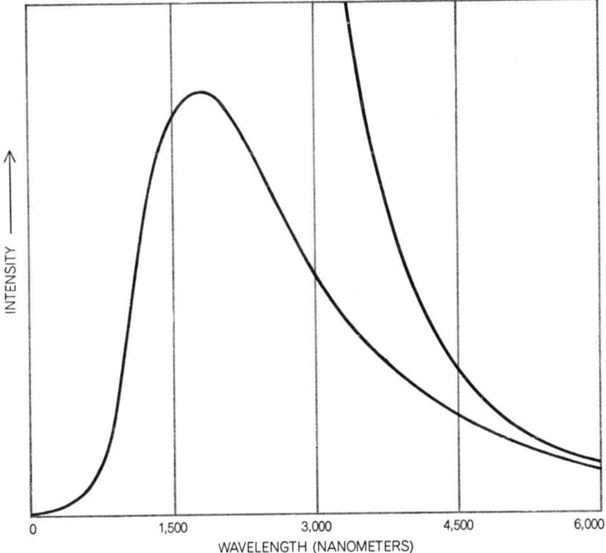

Figure 7 Quantum nature of light was recognized as a result of Max Planck's explanation of the spectral distribution of electromagnetic radiation from a hot, black (fully absorbing) body. The classical theory predicted infinite radiation at short wavelengths (right curve). The observed distribution (left curve) was explained by introduction of quantum constant *h*.

(The frequency is equal to the speed of light divided by the wavelength.) The constant h has since proved to be a fundamental constant of nature.

In 1905 Albert Einstein was prompted by another failure of classical electromagnetic theory to extend Planck's quantum concept further. Einstein asserted that not only is the energy of the light exchanged in quanta; the energy of the light beam is itself always divided into discrete quanta. His argument was based on his analysis of the photoelectric effect. It had been observed that negatively charged plates of certain metals lose their charge when they are exposed to ultraviolet radiation; in other metals the reaction could be triggered by visible light. It is now known that every metal has a

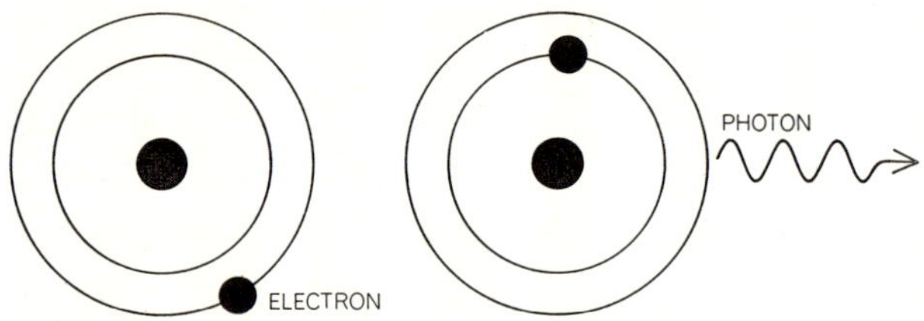

Figure 8 Light is emitted when an electron drops from a higher energy state to a lower state in an atom or a molecule. This process is reversed in most cases in which light is absorbed.

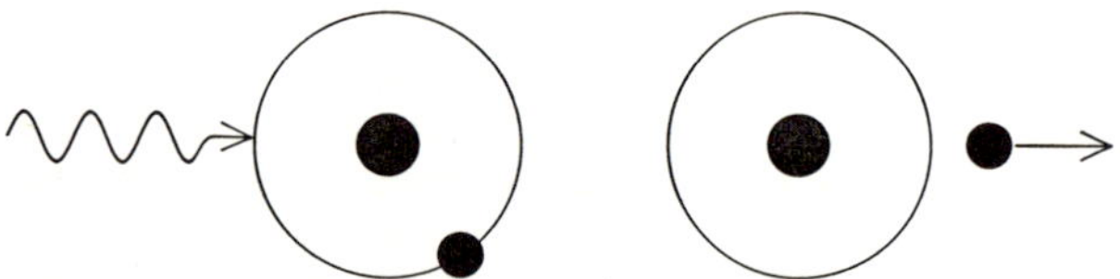

Figure 9 Photoelectric effect is an alternate means of light absorption, in which an electron is knocked out of an atom or a molecule by a photon. It was Einstein's explanation of the photoelectric effect as the absorption of a quantum of energy and the emission of an electron carrying the same amount of energy that established the quantum nature of light.

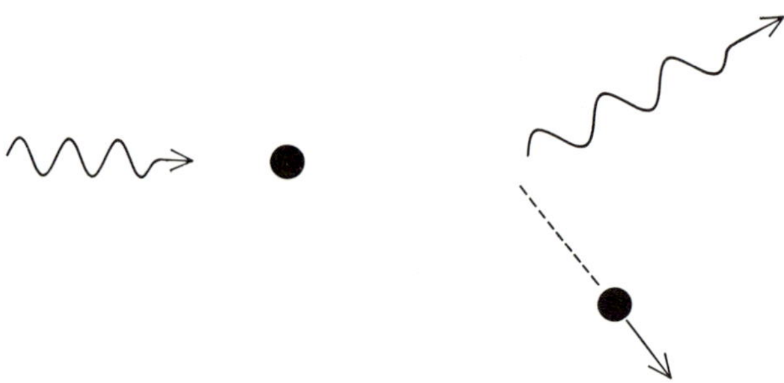

Figure 10 Compton effect explained how X rays passing through matter may increase in wavelength. An X ray photon that strikes an electron is deflected and loses energy; the wavelength shift and scattering angle are related by the dependence of wavelength on energy.

critical wavelength for the effect. The emission of electrons will occur only on exposure of the metal to light of this wavelength or a shorter one. The effect depends entirely on wavelength and is independent of the intensity of the light. Furthermore, even for very weak light sources, the ejected electrons may come out simultaneously with the incidence of the light, without any time being required for the accumulation of energy.

It was impossible to understand these aspects of the photoelectric effect on the basis of a model in which the energy of light is uniformly distributed over the whole of an incident wave. In light of low intensity there would be insufficient energy at any one place in the beam to eject an electron. On the other hand, the observed results follow directly from Einstein's description of the photoelectric effect, according to which each quantum of light, or photon, carries an energy inversely proportional to the wavelength of the light, the proportionality being governed by Planck's constant. In this model a light beam contains a large number of photons (about 10^{18} per second in the beam of a flashlight), and the photoelectric effect occurs when a given photon is absorbed by a particular electron, with the total energy of the photon being transferred to the electron. The relation between the energy transferred to the electron and the wavelength of the light has been precisely measured and found to be in accord with the prediction from Einstein's hypothesis.

The Compton effect provides further evidence that electrons interact with light through encounters with discrete bodies in the light beam that carry momentum and energy. Here it had been observed that x rays passing through matter often increase in wavelength. This was interpreted by Arthur Holly Compton as a loss of energy due to collisions between the highly energetic x-ray photons and the electrons. From the equation showing the relation of wavelength to energy Compton argued that the wavelength shift in the x rays would have a simple dependence on scattering angle, and this was in fact observed. It was shown soon afterward by the use of coincidence counting techniques that an electron recoils from each scattered photon and carries off the energy and momentum given up by the photon. These billiard-ball collisions plainly show that an x-ray beam behaves like a stream of particles. Similar behavior has been demonstrated for electromagnetic radiation of other wavelengths.

Although photons obey the same equations governing the relation of momentum and energy as other particles do, the relation is a special one in the case of the photon, owing to its vanishing rest mass (see Fig. 6). The relative prominence of the wave properties of photons is therefore a consequence of the vanishing of the rest mass, rather than a qualitative difference between photons and other particles. The same characteristic accounts for the fact that the speed of light is independent of its energy. The velocity of particles with nonzero rest mass increases as their energy increases. For the photon velocity does not change with energy at all.

The discovery that both light and matter have wave and particle properties has made it easier to understand how these properties can exist together in either light or matter. This understanding is set out in the new description of nature, perfected in the 1920's, known as quantum mechanics. The basic objects described by the quantum mechanics of either light or matter are particles that are at least somewhat localized in space. The wave aspects of light and matter express the fact that these particles do not obey deterministic laws of motion, as they would in classical mechanics. Instead the laws they obey govern only the relative probabilities of motion at different speeds in different directions, even for a single particle in a known field of force. The waves associated with light and matter are a way of describing these probabilities. Hence when a light beam passes through a hole, there is some probability, related to the wavelength of the light, that the photons in the beam will not go straight through, giving a geometrical image of the hole,

Figure 11 Interference fringes from an experiment like Young's were photographed by Brian Thompson. Two apertures 1.4 mm in diameter, 22 mm apart, were set in front of a lens (focal length 1.52 meters) and illuminated with mercury-arc light.

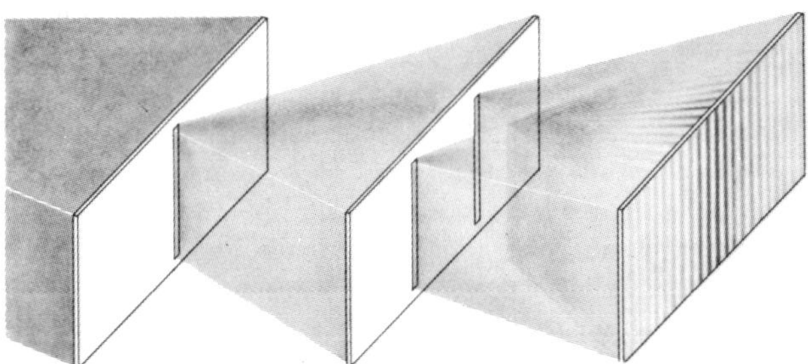

Figure 12 Formation of fringes is evident in this diagram, in which slits are used instead of holes. Light from a source is passed through a single slit to attain some coherence and then diffracted into fans by two slits. The fans interfere to form fringes (right).

Figure 13 Wave explanation of fringes is evident in a plan view of the two sets of wave fronts that emerge from the slits. As the wave fronts move outward, "beams" of high intensity (black diamond shapes) develop where wave crests from both slits travel together and are therefore in phase (as are the troughs also along the same directions). At right is a plot of the resulting wave intensity: the fringes.

but instead will be deflected, ending up in the geometrical shadow region. The intensity of the waves in the diffraction pattern is a measure of this probability.

Let us see how the probabilities that quantum mechanics associates with particles (of light or other matter) explain the Young experiment that historically "proved" the wave nature of light. The experiment is usually performed with parallel narrow slits, to yield an interference pattern of alternating bright and shadowed lines. If the photons followed classical trajectories, the total probability of a photon's hitting a given point on the collecting screen would be the sum of the probabilities of this happening for each path; in other words, the light pattern would be a simple sum of the independent intensities of the two parts of the divided beam. Instead the collecting screen displays an interference pattern. The pattern is perfectly intelligible according to laws governing wave motion, in which the intensity at any given point is the sum of the amplitudes of the waves (whose squares are the intensities) reaching that point along paths of different length and thus in different phase. In quantum theory this experiment is interpreted as indicating that the motion of the photons depends on the complete physical system. If we allow the photon to go through either slit without determining which, the experiment will yield the interference pattern that reflects the probabilities—or rather the interference of probabilities—that a photon will find this or that path to the collecting screen. If we could instead monitor the photons to find out which slit they passed through on their way to the screen, the interference pattern would disappear and we would see the sum of the independent intensities (see Figs. 14 and 15).

The development of the pattern does not depend at all on the intensity of the beam, that is, on the number of photons going through the slits. As long as 50 years ago G. I. Taylor of the University of Cambridge performed an experiment with a light source so attenuated that most of the time there was no more than one photon on its way to the collecting screen. Yet after an exposure time of several months the photographic plate showed the interference pattern!

A similar experiment, employing laser light, has recently been performed by Robert L. Pfleegor and Leonard Mandel of the University of Rochester. Monitoring the arrival of each photon as they came through one by one, the counters showed that each photon found a random location at the detector. Yet when a sufficient number of photons had come through, they formed the expected interference pattern. Thus the number of photons arriving at a given position on the screen is proportional to the intensity of the interference pattern at that point, calculated according to the wave theory for the wavelength of the light in question. This shows that

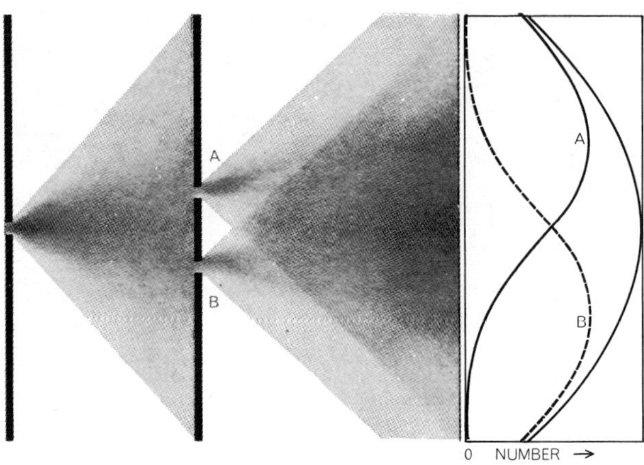

Figure 14 Classical particle description of the two-slit experiment would say that the distribution of all particles that arrive at the screen is the sum (right curve) of the distribution curves for particles from the upper slit (curve A) and lower slit (curve B).

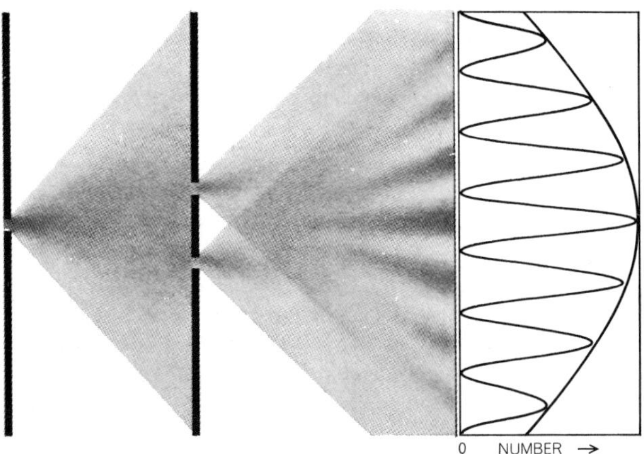

Figure 15 Quantum particle description says the distribution of particles will show a distribution pattern typical of wave phenomena, but only if *each* particle can go through *both* slits. Observation shows clearly that this description, and not the classical one, is correct.

the wave properties must be associated with each photon rather than with the entire beam.

The wave properties of light are examples of a universal behavior of objects, contained in the quantum-mechanical description of nature. From this standpoint contemporary physicists interpret experiments such as Young's interference experiment as showing not that light is a moving wave but rather that the probabilities of various photon motions are described by a wave equation. We might say that photons are the components of a light beam whereas the wave is a description of it. The waves are not vibrations of a new substance distinct from matter, as in the old ether theories. Rather they are a means of mathematically describing the probabilities that particles will do various things. No etherlike carrier is needed for them. Nor is there any paradox involved in the occurrence of both wave and particle phenomena in light. The particles composing light and matter do not follow classical laws. If anything is surprising, it is that the behavior of the particles can be described by a concept that is as familiar as a wave satisfying a simple equation.

It is natural to ask how this picture of light as a stream of photons can be made to conform with the relation between light and electromagnetism, the discovery of which was a high point of 19th-century physics. Perhaps the most instructive way to view the relation is to say that, rather than considering light as an aspect of electromagnetism, we now think of electromagnetic phenomena as one manifestation of photons. It is easy enough, in accordance with this policy, to think of the transmission of radiant energy from place to place as being due to photons traveling across the gap. With a little more difficulty we can visualize the state electric and magnetic forces that occur between charges and currents as being due to an exchange of photons between them. In this latter case the photons are called virtual photons because the relation between their energy and their momentum is different. Quantum physics invokes this notion of virtual-particle exchange to account for forces between particles in cases other than electromagnetic forces. Nuclear forces, for example, are described as being due to the exchange of virtual mesons. Here again we are dealing with a special case of a general phenomenon.

Carrying the generalization further, we can say that all electromagnetic fields are to be thought of as being composed of photons. One result is that electromagnetic fields cannot always be taken as having determined values in space and time. Instead their values generally have associated indeterminacies; it is impossible to measure exactly the value of both the electric and the magnetic fields at the same point in space and time. The exact knowledge of one generates a necessary uncertainty in the value of the other.

Although this reduction of electromagnetism to optics can be carried through generally, it must be recognized that in many circumstances there is justification for thinking of the electromagnetic field as a special entity, as Faraday and Maxwell did. Again this is a result of specific properties of photons. Because the photon has zero rest mass it is possible for a physical system to contain many low-energy photons without its total energy becoming very great. That is the situation, for example, with the electric field between macroscopic charges. If we analyze such a field into photons, there will be a high probability of finding many photons of low energy, and also varying probabilities that different large numbers of photons of other energies are present. In these cases it is more useful to use the older field description, since the addition or subtraction of a single photon makes little difference in a state containing many photons. To put the statement another way, the fluctuations in the field strength arising from the quantum properties of the photons that compose it are small compared with the average value at the field. In short, the particle aspects of electromagnetism and light are unimportant in many cases.

There is another property of photons that is important in situations involving many similar photons. It is termed Bose statistics. Most other stable subatomic particles, such as electrons, follow Fermi statistics. As a consequence electrons must satisfy the Pauli exclusion principle, which forbids the existence of more than one electron with any given value of momentum and angular momentum at any given time. For photons, on the other hand, not only is it allowed but also there is a tendency for photons to be produced in this way—in large numbers of the same momentum. Of all the stable elementary particles (except for the hypothetical graviton, or quantum of gravity), photons alone can occur in the combinations necessary to produce classical fields that have well-defined values over large regions. It is therefore not surprising that electromagnetism is the only case where the field aspects were recognized first.

This is the property, incidentally, that is exploited in the laser. What the laser does is to produce vast numbers of particles of exactly the same energy and wavelength. With no other stable particle but the photon is such a feat possible. The laser beam's remarkable macroscopic properties arise from the fact that its constituent photons are precisely identical. Whether the laser could have been invented without quantum mechanics is an interesting question!

In order to fully understand the properties of light it is important to know about the fundamental interaction process through which light and matter particles influence each other. Although photons do interact with most other subatomic particles, whether charged or neutral, it is believed (it is not completely proved) that the basic interaction is between photons and charges. According to this model, sometimes called Ampère's assumption, the fact that photons can be emitted and absorbed by electrically neutral objects such as neutrons is a consequence of the fact that these objects, although neutral as a whole, have a structure of opposite charges distributed throughout their volume. Supposedly it is these internal charges that do the emitting and absorbing of photons.

The simplest charged particle to consider is the electron, which does not have any known structure. An electron at rest can be taken as a point charge, and its fundamental interaction with light is the emission and absorption of photons, one at a time, from this point. If there is a photon at the point where the electron sits, there is some probability that the photon will be absorbed by the electron and so will disappear. Similarly, an electron can spontaneously emit a photon even if no photon was present. The probability of these events is proportional to the square of the charge of the electron.

It should be noted that in the fundamental interaction process the number of photons changes while the number of electrons remains the same. There is no conservation law for photons as there is for charged particles. As a consequence of this fact, and of the fact that photons can have arbitrarily low energy because of their vanishing rest mass, it is easy to produce them in very large numbers, as in a flashlight beam. All of the many photons in such a beam, however, are produced one at a time by individual electrons in the atoms of the flashlight's incandescent-lamp filament.

Starting with Ampère's assumption that photons interact only with charges, and applying the principles of special relativity and quantum mechanics, a mathematical theory known as quantum electrodynamics has been developed that provides detailed and accurate predictions for the entire range of phenomena involving light and electrons. Some of these predictions (for example, the response of hydrogen atoms to microwaves) have been verified to one part in a billion. Indeed, of all the theories physicists have constructed, the quantum electrodynamics of electrons is rivaled only by the gravitational theory of planetary motion in its agreement with observation. Since a large fraction of the phenomena involving ordinary matter occur through the interaction of light with matter, or through the play of the electromagnetic forces between charges that this interaction produces, we can agree with P. A. M. Dirac in his statement that this theory "explains all of chemistry and most of physics." As the present discussion has emphasized, the theory uses the same criteria to describe and distinguish photons as it does to specify the other particles. The success of the quantum electrodynamics of electrons can therefore be taken as testimony to the idea that all matter is similar, insofar as it is described by the same general principles.

The interaction of light with other charged particles, such as protons, is less well understood theoretically. These particles have a spatial structure arising from their rapid conversion into one another through the strong interaction that generates the force holding the atomic nucleus together. The details of the scattering of light by protons have not been adequately derived from any theory. Most physicists, however, consider this a result of our inability to deal with strong interactions rather than of a failure in our understanding of the properties of light. It is believed the interaction of light with charges is basically simple and universal, even when these charges have strong interactions as well. Techniques have been devised in the past few years to circumvent in part our inability to deal with the strong interactions, and some progress has been made in calculating the interaction of light with all the other charged particles. Whether or not our understanding of this interaction will approach our understanding of the interaction of light with electrons remains to be seen.

At several times in the past physicists have thought they understood the fundamental nature of light, and they were mistaken. Is it possible that our present view is also mistaken, that it will be radically modified in the future? Predicting the future of physics is a hazardous task; the most likely outcome of such predictions is that they will provide a source of amusement for future physicists. The question nevertheless must be faced. Since our present description holds light to be similar to other forms of matter, it seems likely that any fundamental modification in the description would include all kinds of matter, not just light. The need for such a new description is perhaps indicated by the bewildering number of types of matter and of laws governing their behavior that the study of "elementary" particles has revealed. It could be that particles are not the ultimate constituents of light and matter after all but rather manifestations of something deeper.

At present the photon theory gives an accurate description of all we know about light. The notion that light is fundamentally just another kind of matter is likely to persist in any future theory. That idea is the distinctive contribution of 20th-century physicists to the understanding of light, and it is one of which we can well be proud.

MUCH OF *the research done in science must be accomplished through remote reconnaissance. For example, the biologist cannot see microscopic forms of life without the magnification provided by a microscope. Time-lapse photography is extremely important to the botanist in the study of the mechanics of plant development from seed to mature plant. The forester utilizes air photographs to delineate major forest associations, estimate timber volume, and check for evidence of disease. Scientists frequently have need to detect and identify phenomena, conditions, or temporal changes that occur in the physical world but are impossible to detect with normal human senses. This detection must come from sensors that are usually not in direct contact with the phenomenon being investigated. This is* remote sensing, *and it is accomplished through the transfer of energy that is either emitted or reflected from the radiating object to the sensors. The sensors in turn reconstitute this energy into an image detectable by normal human sense organs, usually the eye.*

The major concern for us here is the fundamental characteristics of energy and matter as they relate to images created by remote sensors. Our knowledge of remote sensing must be grounded in a thorough understanding of the way electromagnetic energy is transferred and how it interacts when it encounters matter and other energy.

3-Introduction to Remote Sensing: The Physics of Electromagnetic Radiation

NELSON R. NUNNALLY

Introduction to Electromagnetic Theory

The Nature of Energy

Energy can be defined as the ability to do work. Energy may exist in a number of forms as chemical, electrical, heat, or mechanical energy. When work is done on a body and a system, it gains mechanical energy. This energy may be of either one or two kinds—kinetic or potential. *Kinetic energy* is the energy possessed by a body due to its motion with respect to some inertial frame of reference. *Potential energy,* on the other hand, is energy possessed by a body due to its position relative to another body in the system. Potential energy can be converted into kinetic energy or into electrical, chemical, or heat form.

Prepared for National Science Foundation 1969 Summer Short Course in the Geographic Applications of Remote Sensing at the University of Tennessee, Knoxville, Tenn. Reprinted with permission of the author and the Association of American Geographers.

The Transfer of Energy

In the process of doing work, energy must be transferred frequently from one body to another or from one place to another. The three basic ways in which energy can be transferred should be well known to geographers, since all are important in atmospheric heat transfer. The three are conduction, convection, and radiation.

Conduction occurs when one body (molecule or atom) transfers its kinetic energy to another by colliding with it. Collisions involving a transfer of kinetic energy are called inelastic and may take place between either like or unlike bodies. The speed and efficiency of the process of conduction is partly a function of the conductivity of a substance, its size, and the temperature gradient.

In *convection,* the kinetic energy of bodies is transferred from one place to another by physically moving the bodies. This is sometimes referred to as corpuscular transfer of energy. It is especially significant in fluid substances like water and air.

The transfer of energy by electromagnetic *radiation* is

of much more significance to us at the moment. Electromagnetic radiation is the only form of energy transfer that can take place through free space. The importance of this topic dictates that we give it special consideration.

The Nature of Electromagnetic Radiation

It is common practice to use models based on familiar phenomena in attempting to explain the unfamiliar. In the case of electromagnetic radiation two models are required to yield an understandable and workable knowledge of the topic.

The Wave Model

The first model is the wave model. According to this theory, energy is propagated through space at the velocity of light (a form of electromagnetic radiation) in a wave pattern analogous to sound waves in the air. An electromagnetic wave consists of two fluctuating fields, one electric and the other magnetic. Each field has a vector associated with it. The two vectors are mutually orthogonal, and both are perpendicular to the direction of propagation. Thus the wave is, in reality, made up of two parts, both transverse wave forms which spread from the source at the speed of light (Fig. 1).

Electromagnetic radiation is generated whenever an electrical charge is accelerated. In any mechanical system the period of its wave (and therefore, inversely its frequency) is dependent upon the length of time required for the oscillating body to complete one cycle of vibration. The wavelength of electromagnetic radiation depends upon the length of time that the charged particle is accelerated and the frequency on the number of accelerations per second.

The relationship between frequency and wavelength of electromagnetic radiation is shown by the following formula, where λ is wavelength, f is frequency, and c is the universal constant, the speed of light.

$$\lambda f = c \qquad \text{or} \qquad \lambda = \frac{c}{f} \tag{1}$$

The formula indicates that frequency is inversely proportional to wavelength and directly proportional to wave velocity. The latter is unimportant except when electromagnetic radiation passes from one substance to

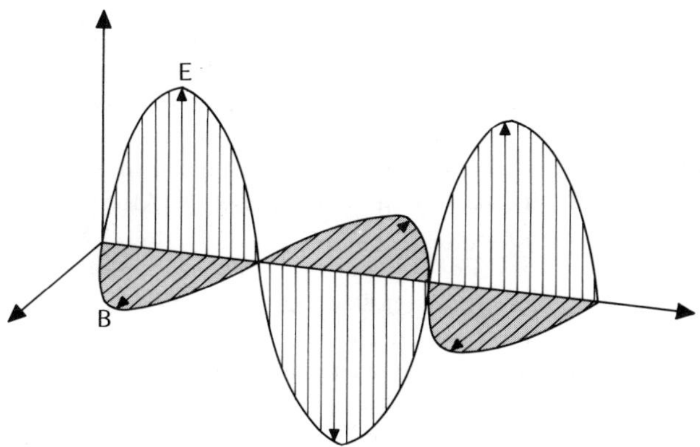

Figure 1 Electric (E) and magnetic (B) vectors of an electromagnetic wave.

another of different density. Once electromagnetic radiation has been generated, then when wave velocity (the speed of light) changes through the passage of the waves through media of differing densities, it is wavelength which is altered in response to the velocity change, and never frequency. More will be presented about wave theory and its ramifications later after a brief look at the particle idea.

The Particle Theory

Before 1800 it was generally thought that all electromagnetic radiation was transferred as particles or corpuscles of energy. The discovery of the wave theory did not entirely do away with the particle idea. It was realized that in some situations the particle theory still was a better approach to explanation. This is especially true when considering the spectral distribution of energy from blackbody radiators. Max Planck, in 1900, recognized the discrete nature of exchanges of radiant energy and proposed the quantum theory of electromagnetic radiation. This theory states, in effect, that energy is transferred in discrete units called *quanta* or *photons*. The relationship between the frequency of radiation expressed by wave theory and the theory content of a quantum is expressed as

$$E = hf \tag{2}$$

where E is the energy of a quantum, h is the Planck constant (6.624×10^{-27} erg sec), and f is the frequency of the radiation.

Referring back to formula (1), we can multiply the equation by h/h, or 1, without changing its value.

$$\lambda = \frac{hc}{hf} \tag{3}$$

Then by substituting E for hf (from formula 2), we can express the wavelength associated with a quantum of energy as

$$\lambda = \frac{hc}{E} \qquad \text{or} \qquad E = \frac{hc}{\lambda} \tag{4}$$

What these formulas tell us, simply, is that the energy of a photon varies directly with frequency and inversely with wavelength—the longer the wavelength, the lower the energy, and the higher the frequency, the greater the energy.

The particle theory assumes special importance when radiation and matter interact. Some examples of this in remote sensing are the exposure of film, the photoelectric effect, spectrum lines, and other absorption or emission effects. These are significant in understanding target-energy interactions, as well as detection and measurement of electromagnetic radiation.

We have said that at times electromagnetic radiation behaves as a particle and that other times it exhibits wave characteristics. It has also been demonstrated that some subatomic particles of matter, such as electrons, also exhibit wave properties. Thus we might assume that the photon is simply yet another kind of basic particle which differs from all others in that it has zero rest mass.

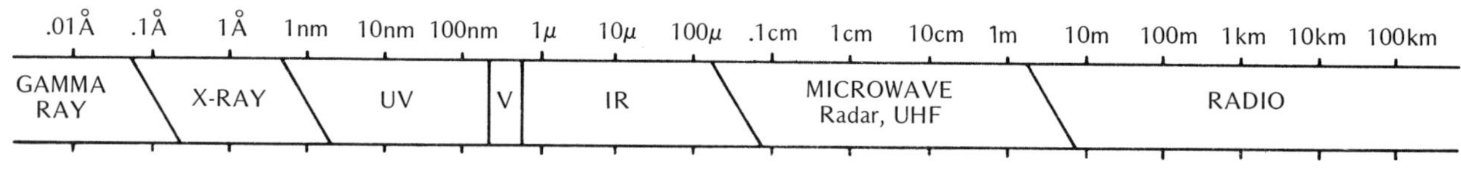

Figure 2 The regions of the electromagnetic spectrum. (Subcommittee I, Photo Interpretation Committee, American Society of Photogrammetry, *Photogrammetric Engineering*, 29:761–799, 1963)

The Spectrum and Its Regions

Electromagnetic radiation ranges from long, very low frequency radio waves to very short, extremely high frequency gamma and cosmic waves. Between these extremes there is a continuum of frequencies, wavelengths, and quanta. Figure 2 shows the spectrum and the relationships expressed in formulas (1) through (4).

Normally the spectrum is subdivided into the following regions: cosmic rays, gamma rays, x rays, ultraviolet, visible, infrared, microwave (radar), radio, and AC. Radar may be differentiated as a separate region occupying the shorter radio and longer microwave sections. The radio portion may be subdivided into UHF, VHF, Radio (HF), LF, and ULF frequencies. None of the regions has exact boundaries. There is considerable overlap in the regions and in the terminology applied to them.

Questions begin to arise at this point concerning the origins of electromagnetic radiation. These questions are sufficiently important that they deserve special attention, and the next section is devoted to the matter of sources of electromagnetic radiation.

The Energy Cycle in the Sensing Process

Virtually all energy detected by remote sensing systems undergoes certain fundamental processes. It is radiated by a source, is propagated through the atmosphere, interacts with a target, is reemitted, and passes back through the atmosphere before being sensed. The exceptions would include examples in which the target acts as its own source, such as radioactive decay, heat generation by friction or chemical processes, or heat conducted to a target from the earth's interior. Examining each process in the sequence in turn should contribute to a more complete understanding of the physics involved as well as helping to simplify some of the more difficult concepts.

Generation and Sources of Electromagnetic Radiation

Introduction

A short discussion of sources of electromagnetic radiation, especially as they relate to wavelength and frequency, seems appropriate. An electromagnetic source can emit either a continuum of wavelengths, a spectral band, or a single wavelength (actually a very narrow band), and the source may be natural or manmade. In each case the intensity of radiation (and in some cases the wavelength) can vary through time.

As stated earlier, electromagnetic radiation is emitted whenever an electric charge is accelerated, the wavelength depending on the length of time over which the acceleration occurs and the frequency on the number of accelerations per second. The electrically charged particles involved in generating electromagnetic radiation are electrons, ions, and molecules.

Atoms, electrons, and ions. Electrons are the tiny negatively charged particles which move around the positively charged nucleus of an atom. Atoms of different substances are made up of varying number of electrons arranged in different ways.

Electrons, like light, sometimes behave like particles and at other times like waves. The interaction between the positively charged nucleus and the negatively charged electron keep the electron in orbit. While its orbit is not explicitly fixed, each electron's motion is restricted to a definite range from the nucleus. The allowable orbitable paths of electrons about an atom might be thought of as energy classes or levels. In order for it to climb to a higher class, work must be done. However, just any amount of work cannot be done on an electron. Unless an amount of energy is available to move the electron up at least one energy level, it will accept no work. If a sufficient amount of energy is received, the electron will jump to a new level and the atom is said to be excited. Once an electron is in a higher orbit, it possesses potential energy. After about 10^{-8} seconds, the electron falls back to the atom's lowest empty energy level or orbit and gives off radiation. The wavelength of radiation given off is a function of the amount of work done on the atom or the quantum of energy it absorbed to cause the electron to be moved to a higher orbit (Fig. 3).

Occasionally the atom will absorb enough energy to cause the electron to jump all the way out of orbit, much as a rocket can escape the earth's gravitational field. When this happens, an atom remains with a positive

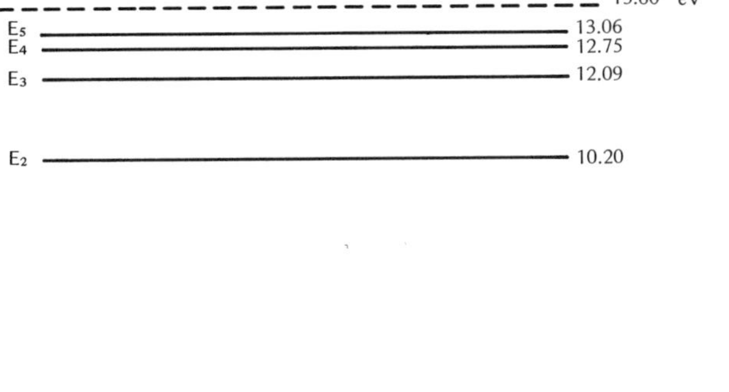

Figure 3 Energy levels of hydrogen. Energies required for lifting electrons to various levels given in electron volts. Dashed line represents binding energy. (© 1959, 1967, by Harcourt Brace Jovanovich, Inc., and reproduced with their permission from *College Physics* by Franklin Miller, Jr.)

charge equal to the negatively charged electron which escaped. The electron becomes a free electron and the atom is called an ion.

MOLECULES. Atoms are joined together to produce molecules of substances. The forces which hold the atoms together may be of several types: polar, covalent, or metallic. *Polar bonding,* sometimes called electrovalent or ionic bonding, occurs in molecules where the atoms carry opposing electrical charges and are separated by a small distance. H_2O molecules are good examples. The negative oxygen ion is bound to two positive hydrogen ions. *Covalent bonding* takes place between atoms sharing two electrons, generally, so that both are stabilized by having eight electrons in their outer shells. In *metallic bonding,* atoms pack close together so that each is equidistant from a number of others, and each atom contributes an electron or electrons to be shared by all its neighbors over a short time interval. These electrons, which are free to move about, account for the high electrical conductivity of metals.

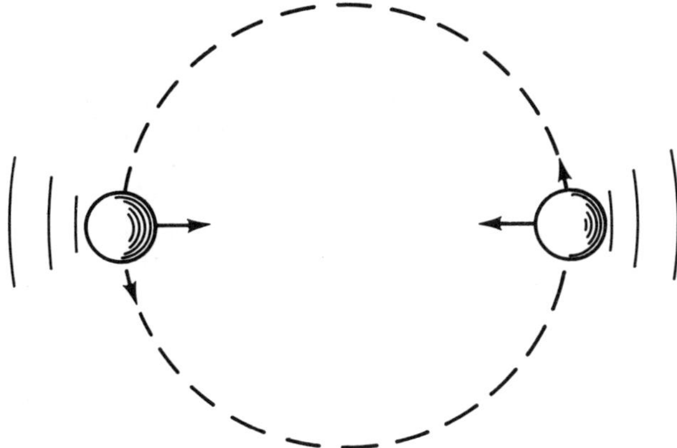

Figure 4 Vibration and rotation of a diatomic molecule.

The atoms in molecules vibrate about an equilibrium position, and the molecule itself may rotate. In addition, molecules of a fluid may have random translational motion. A translational motion is one in which all parts of a particle or body move in the same direction at the same velocity as opposed to the vibrational and rotational motions where the different parts may move in different directions or at different speeds (Fig. 4).

Vibration of molecules may involve stretching and bending. Stretching can best be seen in examples such as diatomic gases. Here the two atoms can be conceived of as balls connected by a spring which resists both compression and stretching and vibrates at some characteristic frequency. When exposed to the right frequency of electromagnetic radiation, the atoms can absorb quanta of energy and the vibrational motions are accelerated in discrete increments. The bending vibrations which many polyatomic molecules undergo are usually of a lower frequency but are also quantized. The same is true of rotational motion which dipolar molecules undergo. There is no continuum of rotational velocities for a particular molecule. Either the molecule does not rotate at all or else there are certain fixed angular velocities allowed, similar to the allowable orbitable levels of atoms. A molecule gains one quantum of energy at a time as it speeds up. As it loses speed, it loses a quantum at a time. However, only the nonrandom, periodic motions of a molecule are quantized. Other motions, translational motions and ordered motions which result from movement of the entire mass due to the application of exterior force, are nonquantized.

Sources of Energy

Since electromagnetic radiation is generated whenever charged particles are accelerated, it remains for us to investigate the manner in which these accelerations can take place. In the earlier discussion of electrons and atoms it was stated that an atom could absorb a quantum of energy, thereby raising an electron to a higher orbit, and that the excited atom could radiate a quantum of energy as the electron fell back to the lowest empty energy shell.

The question might be legitimately posed as to why all atoms don't continually emit electromagnetic radiation since the electrons are undergoing constant centripetal acceleration. Bohr explained this dilemma by postulating that electrons are restricted to only a few of the many possible orbits around an atom, and that while in these orbits they may not emit radiation. However, an electron may jump to a lower orbit, emitting a light quantum equivalent to the energy difference of the two levels.

X RAYS. If an atom absorbs enough energy to kick an inner electron completely out of orbit, allowing an outer valence electron to drop down into the empty level, it emits an x ray. The energy required to remove the electron (called the binding energy) may come from absorbing another x ray, but usually it comes from inelastic collisions of high-speed electrons with atoms in a metal plate. There are several ways to accelerate the electrons and many types of metal plates can be used as targets. The wavelengths produced depend on the type of atom of which the plate is made and the particular electron shell from which an electron is knocked out. Most high-voltage cathode tubes used in x radiation give off a spectrum which is a mixture of continuous and band spectra.

ULTRAVIOLET AND VISIBLE. In the ultraviolet and visible portions of the spectrum, radiation is produced by changes in the energy levels of the outer, valence electrons. The wavelengths of energy produced are a function of the particular orbital levels of the electrons involved in the excitation process.

If the atoms absorb enough energy to become ionized and if a free electron drops in to fill the vacant energy level, then the radiation given off is unquantized and a continuous spectrum is produced rather than a band or a series of bands. This is especially true of the ionized atoms of which metal is composed. The conduction electrons in metals are not bound to a particular atom and their energy changes are not quantized. Thus, metals and other solids radiate a continuous spectrum when heated.

Heating of solids speeds up the vibrational motions of the molecules about the equilibrium position and increases the energy of the free electrons. As inelastic collisions occur, atoms or ions are excited, and radiation is produced. Heating a solid may produce light which appears to be some specific color rather than white light. This occurs in spite of the fact that a continuum of wavelengths is produced because there is more radiation of that wavelength than any other. Filaments in light bulbs, glowing metals, and even chemically reacting systems such as the sun and flames produced by burning are all examples of phenomena which produce visible and ultraviolet radiation as well as heat (infrared radiation). Wavelengths shorter than ultraviolet cannot be produced in practice by thermal means because the high energy required causes the molecules of most substances to dissociate.

Both ultraviolet and visible radiation may be produced by causing a stream of electrons to pass through gases at low pressure. The electrons and the gas molecules undergo inelastic collisions causing the atoms to become excited and give off radiation. The type of gas determines the wavelengths produced. Some substances, such as mercury vapor, which radiate closely spaced spectral lines appear to radiate one dominant color.

Visible light as well as infrared can be produced by *lasers*. Lasers operate on the principle that electromagnetic radiation of the proper wavelength can cause atoms to absorb energy and become excited or, if they are already excited, can stimulate them to emit radiation.

Some substance is chosen which will emit radiation at a desired wavelength when its excited atoms drop to a lower energy level. The substance must be metastable (i.e., the excited state must last longer than usual so that spontaneous emission does not occur before the atoms can be stimulated to emit radiation).

Laser action is begun by applying intense energy of the proper wavelengths which, upon being absorbed, excites the atoms of the laser material (a crystal, gas, or semi-conductor normally). The process is called pumping. Normally a substance has far more atoms in lower energy states; thus the pumping creates a population inversion whereby more atoms are in the higher energy state than in the lower. When an atom emits a photon spontaneously, a chain reaction occurs whereby other atoms are stimulated to emit photons in phase with and traveling in the same direction as the stimulating photons. Bouncing the photons back and forth between two flat parallel mirrors magnifies the process, and by making one of the mirrors partially transparent (5 percent or thereabouts), a brief, intense pulse of light is emitted. Since the photons are all in phase and are all traveling in the same direction, the pulse emerges as a coherent, narrow beam of light with an extremely small angle of divergence. The rays are, in effect, parallel. The atoms of the substance eventually revert to a normal condition with more atoms in the lower energy state than the higher as energy is lost. In addition to radiation emitted as a pulse, all radiation produced which moves in a direction not perpendicular to the mirrors is lost, and some energy is lost as heat. The process builds up to a maximum and declines unless the population inversion is maintained or recreated. Continuous lasers which are also possible operate somewhat differently in detail (Fig. 5).

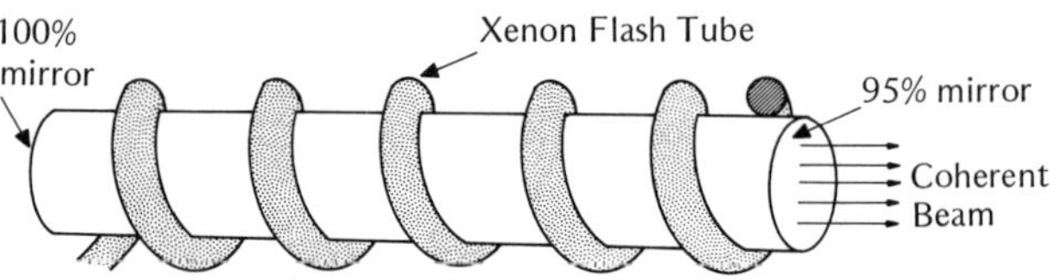

Figure 5 Simplified laser diagram.

One other phenomenon, *fluorescence*, ought to be mentioned at this time. Fluorescence occurs when certain substances absorb a frequency radiation which excites the atoms by raising electrons several energy levels at one time. As the electrons fall back to their former energy levels, they fall back one step at a time. Each step gives off radiation of a different wavelength, each of which is longer than the wavelength originally absorbed.

INFRARED. Some infrared radiation, especially the near infrared up to wavelengths of several microns, can be generated from excited atoms and ions. Thus, infrared can be generated by lasers in the manner described in the previous section. Other lasers operate on a molecular basis whereby excited molecules can be stimulated to emit radiation.

Molecules, especially of solids, are composed of atoms or ions and electrons which are bound together so as to exhibit characteristic frequencies of vibration and/or rotation. Because of these movements these molecules become, in effect, miniature oscillators capable of emitting photons whose frequency corresponds to the molecule's natural resonance. The molecules of the substance are distributed among energy states similar to atoms with the proportions in the higher states being a function of temperature. If the molecules of different energies can be separated or if a population inversion can be created. then the molecules can be stimulated to emit radiation of infrared or even microwave wavelengths. In general, vibrational motions produce near or middle infrared spectra while rotational motions produce far infrared or microwave radiation.

Infrared radiation is most commonly produced by thermal means. When energy is absorbed, it can be stored in potential form (such as latent heat) or in kinetic form as vibrational, rotational, or translational motion. Only the random translational motions of a liquid or gas and the vibrational motions in a solid contribute to temperature (temperature can be defined as the mean kinetic energy of these motions).

Any substance with a temperature above zero degrees Kelvin displays molecular motion and thus is capable of emitting heat radiation directly. In gases where there is little interaction among the atoms and molecules because because of the low density, the emitted radiation produces an arrangement of lines or bands which is characteristic of the gas. As mentioned earlier, the lines or bands which are in the near and middle infrared result primarily from molecular vibration while the longer wavelengths come from the rotational motions of the gas molecules. In a solid or liquid, or even a dense gas, the density of atoms and molecules produces an interaction among the molecules which in turn generates a continuous spectrum.

Most thermal generators of infrared are either cavity radiators or solid radiators. While the basic physical processes involved in generation are similar, there are some differences between the two types of generators.

Cavity Radiators. Cavity radiators produce the nearest thing to *blackbody* radiation that man is capable of producing. A blackbody is a theoretical concept. It is the perfect absorber and emitter of radiation, absorbing and emitting all wavelengths equally well. Because of this ability it would reflect no light, hence the black color.

No known substance behaves like a blackbody, although some substances such as lampblack or powdered graphite come close. They fail because they do not absorb or emit equally well over a wide spectral range. There is a way to generate blackbody radiation. An isothermal cavity constructed from an opaque substance and containing an aperture produces blackbody radiation by continually emitting, absorbing, reflecting, and reemitting radiation from the heated cavity walls. Through these processes the radiation in the cavity arrives at thermal equilibrium with the walls (Fig. 6).

The total emitted radiation (radiant emittance) from a blackbody is proportional to the fourth power of its absolute temperature. This is known as the *Stefen-Boltzmann law* and is expressed as:

$$W = kT^4 \qquad (5)$$

where W is emittance, k is the constant, and T is temperature in degrees Kelvin. Both the amount and the wavelength distribution of blackbody radiation are functions of temperature. Figure 7 shows that as a blackbody's temperature increases, the dominant wavelength shifts toward the short wavelength end of the spectrum. This is a graphical illustration of *Wien's displacement law:*

$$\lambda_{max} = C_3/T \qquad (6)$$

where λ_{max} is the dominant wavelength, C_3 is a constant, and T is temperature in degrees Kelvin.

Solid Radiators. Since no real body is a perfect emitter, its emittance is less than that of a blackbody. All solid radiators fall into this group. The ratio of a real body's emittance to that of a blackbody (W/Wb) is called its *emissivity*. Emissivity, then, is a measure of a body's radiating efficiency. Since a blackbody is a perfect absorber as well as emitter, absorptance and emittance are equal. The same is true for opaque nonblackbodies. These bodies conform to the same general radiation laws as the blackbodies—the hotter they are, the more they radiate and the shorter the dominant wavelength. However, since the emissivities are different, the radiation curves at different temperatures may differ considerably from the blackbody curves discussed in the previous section. When speaking of emittance or absorptance at a particular wavelength (actually over a narrow wavelength interval), *spectral emittance* is a useful concept. Spectral emittance is a radiant emittance per unit area over a designated wavelength interval.

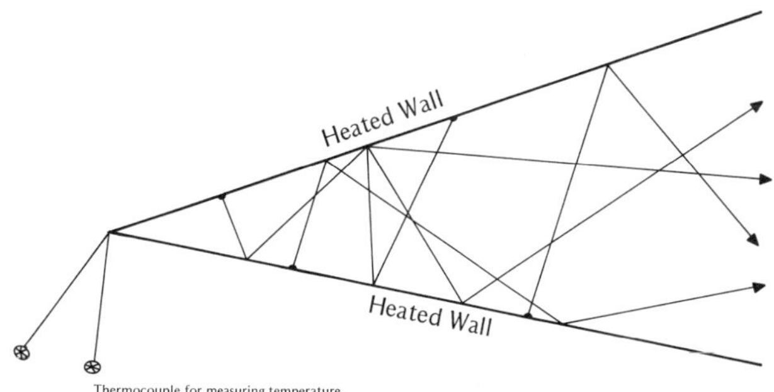

Figure 6 Diagram of a cavity radiator. (Modified from *Infrared Radiation,* by Ivan Simon. © 1966 by Litton Educational Publishing, Inc. Reprinted by permission of Van Nostrand Reinhold Company)

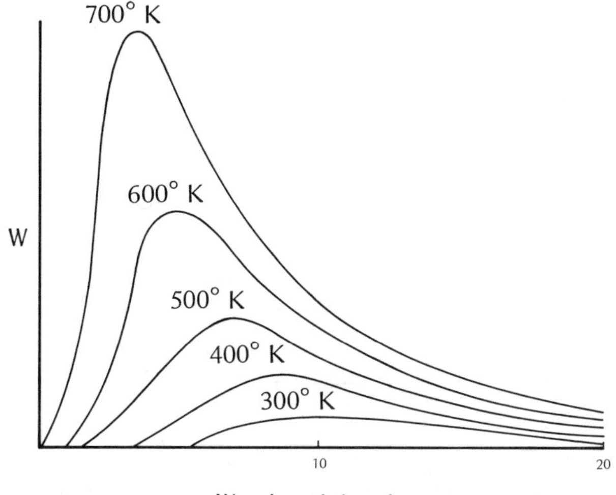

Figure 7 Blackbody radiation curves for various temperatures. W = emittance.

Gray bodies have different spectral emittance curves than blackbodies of the same temperature. Spectral emittance curves of gray bodies differ from those of blackbodies and from each other largely because of differences in atomic and molecular structure. Gray bodies may reflect some radiation or transmit some (Fig. 8). These energy interactions are discussed in more detail later.

MICROWAVE. Microwave radiation can be generated thermally, but it is generally not practical, especially for longer wavelengths. The reason is that as blackbody temperature increases, there is relatively little difference in the amount of microwave radiation emitted. Thus, for real bodies whose emissivities are less than blackbodies, generated radiation is easily lost against the background of naturally emitted microwave radiation. Likewise, naturally emitted radiation is difficult to detect with thermal detectors in the microwave range.

Microwaves, like visible light and infrared, can be generated by stimulating radiation. As a matter of fact, the first such devices operated in the microwave range and were caller *masers*. The theory and operation of masers are very similar to those of lasers, so will not be discussed again.

Microwave radiation can be amplified by a *cavity resonator*. The cavity resonator is a metal enclosure with reflecting surfaces at either end. By spacing the reflectors some integral wavelength apart, a standing electromagnetic wave can be produced, which is somewhat analogous to the standing wave produced by the string of a musical instrument. Like the wave produced by a string, the electromagnetic wave will gradually die out due to friction unless additional energy is added. By piercing one of the reflector walls, a small amount of energy in phase with the standing wave can be allowed to move into the cavity to amplify the standing wave. Once it is built up to the desired amplification, it can be released as an intense pulse of short duration or as a less intense pulse of a longer duration.

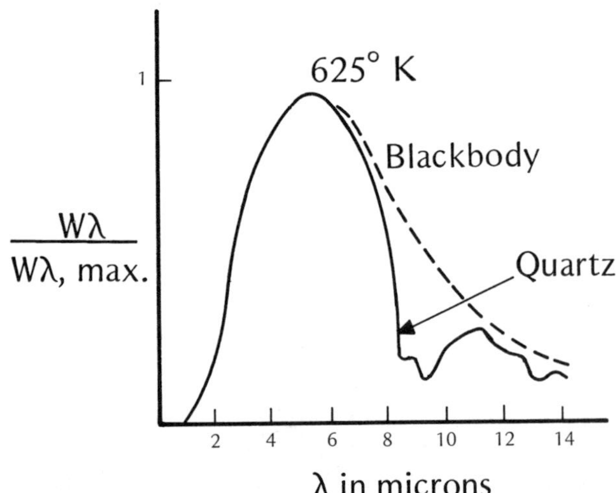

Figure 8 Spectral emittance of quartz compared to a blackbody. (From *Infrared Radiation,* by Ivan Simon. © 1966 by Litton Educational Publishing, Inc. Reprinted by permission of Van Nostrand Reinhold Company)

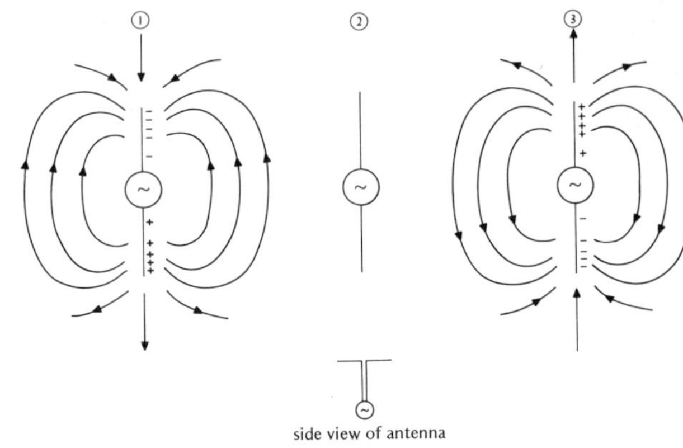

Figure 9 Radio waves generated by a simple antenna. (© 1959, 1967, by Harcourt Brace Jovanovich, Inc., and reproduced with their permission from *College Physics* by Franklin Miller, Jr.)

RADIO WAVES. Shorter wavelength high-frequency radio waves can be produced by resonators also (Fig. 9.) Most radio waves of longer wavelengths are generated by accelerating free electrons. Perhaps the most common means of doing this is to use an electric oscillator attached to an antenna. When voltage is applied to opposite ends of a metal wire, free electrons float toward the positively charged end. The oscillator periodically reverses the charges, causing the free electrons to migrate back and forth along the antenna wire. Since each time an electron stops and restarts it must deaccelerate and reaccelerate, electric-magnetic radiation is generated. The frequency of the alternating current determines the frequency of the radiation from the antenna.

NATURAL RADIATION. It has been stated several times that any object with a temperature above absolute zero emits some of all wavelengths of radiation. Yet at the beginning of this section we stressed the fact that all radiation results from the acceleration of electrical charges. Superficially, these two statements may be hard to reconcile. However, upon further analysis they are not.

It was explained in the section on infrared how the molecules of substances may vibrate or rotate. And since these molecules contain electrically charged particles (ions and electrons), they act as small oscillators which accelerate the electrical charges to produce infrared and microwave radiation. Actually, all of the radiation generated by warm or hot bodies is a result of molecular motions. Energy transferred in molecular, ionic, electronic, and atomic collisions is responsible for generating wavelengths longer and shorter than infrared. Recall that when molecules bump into each other, collisions can be either elastic or inelastic, i.e., they may bounce apart without either having lost kinetic energy to the other, or one may give up a quantum or quanta of energy to the other.

The higher the temperature, the more collisions occur and the more violent they are. This accounts not only for the increase in emitted energy but also for the shift in dominant wavelength toward the higher frequencies. Remember that the distribution of molecules among energy states is a function of temperature. All particles do not increase their motion equally. Rather, some particles increase their energy while other particles do not, and the net effect is that at higher temperatures there are proportionately more molecules in high energy states.

When atoms or ions absorb energy from these collisions, the result depends on how violent the collision was. The more violent collisions involving higher energy molecules might produce x radiation in the manner described in the section on x rays. Less violent collisions might produce ultraviolet or visible light.

Longer wavelengths of microwave or radio frequencies may result from one of several processes. Shorter microwaves are produced largely by molecular motions within the substance. However, the acceleration of the free electrons through inelastic collision and the fluctuations in the electric and magnetic fields of molecules contribute to both the microwave and radio spectra.

It is obvious, from the examination of the blackbody curves and Figure 9 that the amount of radio and microwave radiation emitted by any natural substance as a function of its temperature is small. Likewise, the amount of radiation emitted in a shorter ultraviolet and x ray portion of spectrum is also minimal regardless of temperature.

Atmospheric Effects on Radiation

Once electromagnetic radiation is generated, it is propagated through the earth's atmosphere almost at the speed of light in a vacuum. Unlike a vacuum in which nothing happens, the atmosphere may affect not only the speed of radiation but also its frequency, its intensity, its spectral distribution, and its direction.

Scatter

One very serious effect of the atmosphere is the scattering of radiation by atmospheric particles. Scatter differs from reflection in that the direction associated with scatter is unpredictable, whereas the direction of reflection is predictable. There are essentially four types of scatter although one, Raman scatter, is less important than the other three. The four are: Rayleigh scatter, Mie scatter, nonselective scatter, and Raman scatter.

RAYLEIGH SCATTER. Rayleigh scatter is largely due to molecules and other very small particles many times smaller than the wavelength of radiation under consideration. This type of scatter is inversely proportional to the fourth power of the wavelength. All scatter is accomplished through absorption and reemission of radiation by atoms or molecules in the manner described in the earlier sections. It is impossible to predict the direction in which a specific atom or molecule will emit a photon.

Rayleigh scatter involves reemission of radiation by atoms. And since the energy required to excite an atom is associated with short-wavelength, high-frequency radiation, this explains why Rayleigh scatter is proportional to the inverse of the fourth power of wavelength. Ultraviolet light, since it is about one-fourth of the wavelength of red light, is scattered 16 times as much and blue light four times as much. Rayleigh scatter is most obvious on clear days when there are few water vapor and dust particles.

RAMAN SCATTER. Raman scatter is much less common than the other types. It takes place when a photon has a partially elastic collision with a molecule, giving up energy to rotation or vibration of the molecule or gaining energy from the molecule. The wavelength of radiation is altered by an amount comparable to the amount of energy given up or received by the photon.

MIE SCATTER. Mie scatter takes place when there are essentially spherical particles present in the atmosphere with diameters approximately the wavelength of radiation being considered. For visible light, water vapor, dust, and other particles ranging from a few tenths of a micron to several microns in diameter are the main scattering agents. The amount of scatter is greater than Rayleigh scatter and the wavelengths scattered are longer.

NONSELECTIVE SCATTER. When there are particles in the atmosphere several times the diameter of the radiation being transmitted, the scattering process is nonselective with respect to wavelength. Thus, water droplets, which make up clouds, scatter all wavelengths of visible light equally well, causing the cloud to appear white (a mixture of all colors of light in approximately equal quantities produces white).

Scatter is an important consideration in remote sensing. An understanding of the types of scatter is necessary before one can choose sensor systems or filter systems to accomplish special effects or to prevent degradation of the image under certain conditions.

Absorption

Certain wavelengths of radiation are affected far more by absorption than by scatter. This is particularly true of infrared and wavelengths shorter than visible light. Figure 10 illustrates the absorption-transmittance effects of the atmosphere as well as the principal substances responsible for the absorption bands.

Absorption occurs when energy of the same frequency as the resonant frequency of an atom or a molecule is absorbed, producing an excited state. If, instead of reemitting a photon of the same wavelength (an unlikely event for bulk matter), the energy is transformed into heat motion and is subsequently reemitted at a longer wavelength, absorption occurs. When dealing with a medium like air, absorption and scatter are frequently combined into an *extinction coefficient*. This coefficient sometimes is called the attenuation coefficient. Transmission is inversely related to the extinction coefficient times the thickness of the layer. More will be said about absorption and transmission under Target Interactions.

Refraction

Refraction refers to the bending of light when it passes from one medium to another. Refraction occurs because the media are of differing densities and the speed of electromagnetic radiation is different in each. For the higher frequencies the speed is somewhat dependent on frequency also.

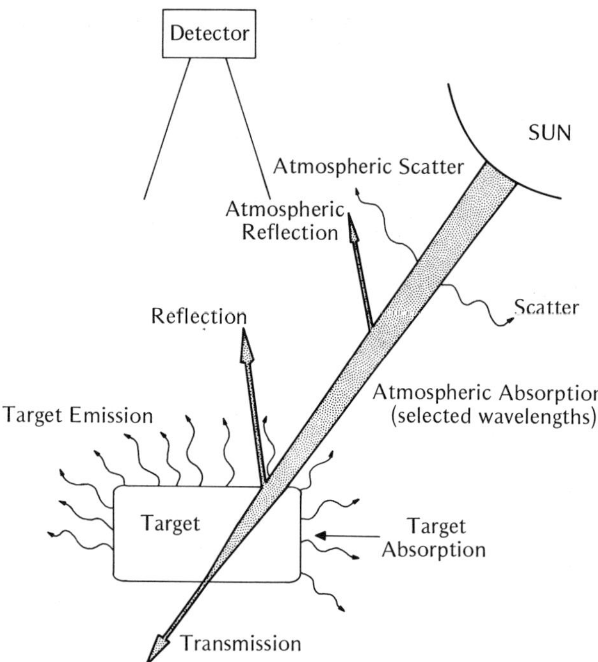

Figure 10 Generalized diagram of energy flow for passive sensor system.

The *index of refraction* is a measure of the optical density of a substance. This index is the ratio of the speed of light in a vacuum, c, to the speed of light in the substance, c_n. Thus, if n is the index of refraction,

$$n = c/c_n \qquad (7)$$

Refraction can be described by *Snell's law,* which states simply that for a given frequency of light (we must use frequency here since it does not change when the speed of light changes as does wavelength), the product of the index of refraction and the sign of the angle between the ray and a line normal to the interface is a constant.

$$n_1 \sin \Theta_1 = n_2 \sin \Theta_2 \qquad (8)$$

From Figure 11 the significance of refraction in the atmosphere should be obvious. A nonturbulent atmosphere can be thought of as a series of layers of gases, each with a different density. Anytime that energy is propagated through the atmosphere for any appreciable distance at any angle other than vertical, refraction occurs.

The amount of refraction will be a function of the angle made with the vertical, the distance involved (in the atmosphere, the greater the distance, the more changes in density), and the densities of the air involved (higher near sea level). Serious errors of location due to refraction can occur in images formed from energy detected at high altitudes or at acute angles. However, these locational errors are predictable by Snell's law and thus can be removed. Much more of a problem is refraction in a turbulent atmosphere. Since the turbulent motions are random, the bending of the wave front is unpredictable.

Target Interactions

Several things may happen when electromagnetic radiation strikes a potential target. Figure 11 illustrates what may take place. Some of the energy is reflected or scattered, and the rest enters the target to be propagated as a refracted wave front. If the target is opaque to the incident energy, that which enters the target is absorbed. If the substance is transparent to the energy, the refracted wave is transmitted through the target to emerge at the opposite side. A small amount would be reflected back into the substance upon striking the boundary, but most of the energy would emerge. If energy is absorbed, that part which is reflected in temperature change will be reemitted as heat radiation.

Absorption and Transmission in Targets

The proportions of incident radiation which are involved in reflection, absorption, or transmission depend partially on the radiant spectral distribution, the angle of incidence, the optical properties of the target (the refractive index and the absorption coefficient), and its thickness.

The ability of a substance to absorb radiation is known as *absorptance* and is expressed as the natural logarithm of the ratio of intensity of incident radiation at the surface to intensity of radiation at depth x in the medium.

$$A = \ln I_o/I \qquad (9)$$

where A is absorptance, I_o is incident radiant intensity, and I is radiant intensity at depth x. Absorptance can also be expressed in terms of the absorption coefficient α, which is an index of the degree of attenuation of radiation in an absorbing medium over a unit distance. When absorptance is expressed in terms of the absorption coefficient, the formula is:

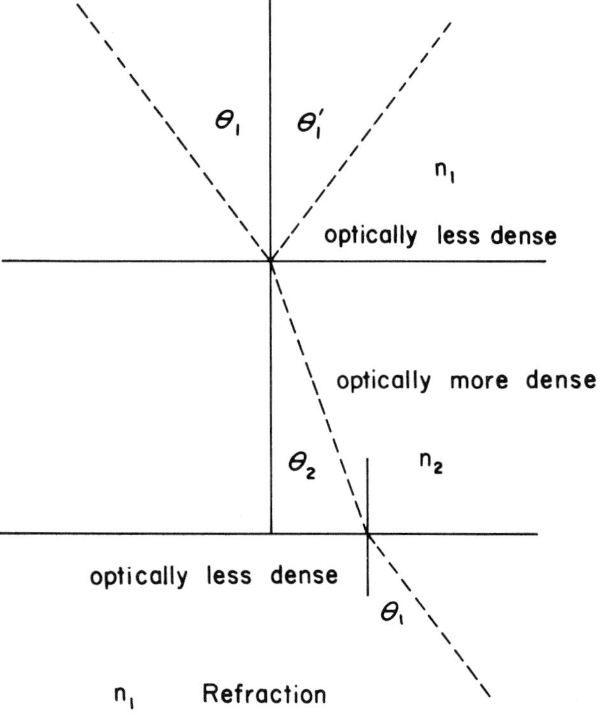

Figure 11 Reflection and refraction.

$$A = \alpha \lambda x \quad (10)$$

where $\alpha \lambda$ is the absorption coefficient for wavelength (measured outside the medium) and x is the thickness of the medium. When dealing with gases and liquids in which a substance may be suspended, the above term must be multiplied by a factor C, which is a measure of the concentration of the suspended substance.

Transmittance is defined as the ratio of radiation at distance x to incident radiation.

$$T = \frac{I}{I_o} \quad (11)$$

Transmittance can also be defined in terms of the absorption coefficient.

$$T = e^{-\alpha \lambda x} \quad (12)$$

where e is the base of natural logarithms, $\alpha \lambda$ is the absorption coefficient, and x is the distance.

Absorptance and transmittance both vary with wavelength. This is basically a function of the natural resonances of the atoms, molecules, lattice vibrations, and free electrons which make up a substance. Thus, a target may behave very differently if exposed to radiation of different wavelengths. It might, for instance, be highly absorptive in the visible range and transparent in the infrared, like some semi-conductors, or it might be like glass—transparent in the visible and opaque in the infrared. The net effect of absorption of radiation by most substances is that most of the energy is converted into heat, causing a subsequent rise in the substance's temperature.

Reflection

Reflection refers to the process whereby radiation "bounces off" an object. Actually the process is considerably more complicated, involving reradiation of photons in unison by atoms or molecules in a layer one-half wavelength deep. Reflection exhibits fundamental characteristics which are important in remote sensing. First, the incident radiation, the reflected radiation, and a vertical to the surface from which the angles of incidence and reflection are measured all lie in the same place. Second, the angle of incidence and the angle of reflection are always equal (Fig. 11).

The reflectivity of a surface is dependent upon three major factors: angle of incidence, the refractive index, and the extinction coefficient. The smaller the angle of incidence, the lower is the proportion of energy reflected. This can easily be observed by shining a flashlight on a smooth water surface at night; the more vertically the light strikes the water, the greater the penetration. Reflectivity at any given angle is directly proportional to the refractive index. When light strikes a transparent medium at vertical angles, the amount of reflection is almost solely a function of the refractive index. Since the extinction coefficient is proportional to the absorption coefficient, the greater the absorption the higher the reflectivity. This helps to explain the high reflectivity of some metals.

SPECULAR VERSUS DIFFUSE REFLECTION. The comments made in the preceding section apply equally well to all reflection. However, the effects described are more easily observed for spectral or "mirror" reflection. In specular reflection the surface from which the radiation is reflected is essentially smooth (the average surface profile or roughness is several times smaller than the wavelength of radiation striking the surface).

If the surface is rough, the reflected rays go in many directions, depending on the orientations of the smaller reflecting surfaces. This process, called diffuse reflection, does not yield a mirror image, but instead produces diffused radiation. White paper, sugar, salt, white powders, and other materials reflect visible light in this manner. If the surface is so rough that there are no individual reflecting surfaces, then scatter may occur.

Summary of Target Interactions

At the risk of oversimplification it is, perhaps desirable to generalize about the effects of reflection, absorption, and transmission. Obviously the sum of the three must be equal to the total incident radiation. The greater the ability of a substance to absorb, the higher is its reflectivity and the lower is its transmission, all other things equal. Reflectivity at a given wavelength varies in proportion to angle of incidence, Absorption decreases at wavelengths immediately above and below the maximum absorption bands of a substance—for the shorter wavelengths, as absorption decreases reflectivity decreases and transmission increases; for larger wavelengths, both reflectivity and transmission increase over a narrow band-width, then reflection begins to decrease. The amount of energy reflected back to a sensor, then, would depend on the wavelength and intensity of incident energy, the surface roughness, the orientation of the target, and the target's absorption characteristics with respect to wavelength.

SUGGESTED READINGS FOR PART ONE

Beckmann, P., and Spizzichino, A. *The Scattering of Electromagnetic Waves from Rough Surfaces.* Pergamon, 1963.

Clarke, G. L. The Significance of Spectral Changes in Light Scattered by the Sea. *Remote Sensing in Ecology.* Athens, Ga.: University of Georgia Press, 1969.

Colwell, R. N. et al. Basic matter and energy relationships involved in remote sensing. *Photogrammetric Engineering* 29:761–799, 1963.

Gates, D. M. The energy environment. in which we live. *American Scientist* 51:327–348, 1963.

Gates, D. M. et al. Spectral properties of plants. *Applied Optics* 4:11–20, 1965.

Kimball, H. H., and Hand, I. F. Reflectivity of different kinds of surfaces. *U.S. Monthly Weather Rev.* 58:280–282, 1930.

Ogle, K. N. *Optics,* 2nd ed. Springfield, Ill.: Thomas, 1968.

Olson, C. E., Jr. The Energy Flow Profile in Remote Sensing. *Proceedings of the Second Symposium on Remote Sensing of Environment.* Ann Arbor: Univ. of Michigan, Inst. of Sci. and Technol. Willow Run Laboratories, 1963.

Weisskopf, V. How light interacts with matter. *Scientific American* 219:60–71, 1968.

Wolfe, W. L., and Nicodemus, F. E. Radiation Theory. *Handbook of Military Infrared Technology.* Washington, D.C.: Office of Naval Research, 1965.

PART TWO

REMOTE SENSING

Acquiring information about a phenomenon while still at some distance from it

As you read this article, considerable quantities of invisible energy flow around you. To prove this, you need only to snap on a portable radio or television. This sensor *will pick up the unseen waves emitted by a remote transmitter. The radio or television receiver is a remote sensor because it is not in direct physical contact with the transmitter.*

All matter at temperatures above absolute zero (0°K) radiates electromagnetic waves because of the actions of atomic and molecular vibrations. As an object gets hotter, more radiation is emitted. This radiation will have a power peak at a particular wavelength which is dependent upon the absolute temperature of the radiating object. If a sensor can be designed that will receive only those electromagnetic wavelengths at the object's peak of power radiation, there is an excellent chance that we could sense it from some distance even if it were carefully hidden. This can be done as long as emitted radiation reaches the sensor. Water vapor, certain gases, ionized particles, even dust might attenuate this energy in the atmosphere and the object would then remain invisible to the sensor.

Certain wavelengths of electromagnetic energy pass through the atmosphere easily. These wavelengths are mainly distributed from just beyond the shorter wavelengths of visible light, in the ultraviolet, to the longer waves of infrared, radar, and radio, although in the infrared there are broad wavelength bands which are completely opaque in the earth's atmosphere. In the infrared there are narrow wavelength bands at which radiation moves through the atmosphere with little attenuation. Two of the more important openings are at about 3–5.5 microns (μ) and 8–14 μ. These radiation openings are called atmospheric windows, *and it is in these transmission windows that sensors must operate.*

4-Remote Sensing

DANA C. PARKER
MICHAEL F. WOLFF

Almost all the turmoil of the world around us is beyond the reach of the sensors we are born with. Our eyes respond to light in the minute portion of the spectrum between 4000 and 7000 angstroms (Å), our ears to sound between 16 and 20,000 cycles/sec, and our skin to relatively large changes in temperature. The rest goes by unheeded.

Over the years we have extended our limited capabilities for perceiving at a distance by building a variety of specialized instruments. Thus, we use cameras, infrared detectors, and radio-frequency receivers to detect electromagnetic radiation, seismometers to detect acoustical energy, and scintillation counters to detect radioactivity. We measure force fields with magnetometers and gravity meters, and we acquire still more information about unknown objects with "active" systems like radar and sonar. All these devices allow doing what is today labeled remote sensing—the acquisition of information about an object (or phenomenon) which is not in intimate contact with the information-gathering device.

From *International Science and Technology* 43:20–31, 1965. Reprinted with permission of *International Science and Technology*.

The impetus for developing much of this remote sensing equipment was provided by military requirements, but within the past few years there has been a growing awareness of how it might be used in a broader sense to increase our knowledge of the world in which we live, the atmosphere surrounding our planet, and the space and celestial bodies beyond, particularly with remote sensors installed in aircraft and spacecraft.

Such data are of considerable importance to geologists, geographers, oceanographers, meteorologists, agriculturists—in short, to anyone concerned with his natural environment. For example, theories pertaining to crevasse detection have been substantiated by infrared measurements over snow surfaces in Michigan and in the Arctic, and an airborne technique has been devised for finding snowbridged crevasses that often cannot be seen by visual means. Infrared data acquired in Yellowstone National Park provided the incentive for exploiting infrared techniques over volcanos in Hawaii and

Costa Rica, and this led to discovering new surface expressions of underground volcanic activity applicable to future geothermal mapping around the world.

Although potential applications are really just beginning to be investigated, and we must, therefore, be temperate in our enthusiasm, there is evidence that remote sensing should be important in many practical nonmilitary tasks. Some of these are: detection of forest fires, measurement of ice thickness and distribution in shipping lanes, and spotting of diseased crops and timber.

What You See

At temperatures above absolute zero all objects radiate electromagnetic energy by virtue of their atomic and molecular oscillations. The total amount of emitted radiation increases with the body's absolute temperature and peaks at progressively shorter wavelengths. With its average temperature of 300°K, the earth has a radiant power curve that falls off rapidly toward the visible from a peak near 10 μ (10^5 Å) and more gradually toward the microwave region in the other direction. This gives a relatively broad region at wavelengths longer than about 3.5 μ in which we can do remote sensing by detecting emitted radiation, while at wavelengths shorter than 3.5 μ our good fortune in having a sun with a sharp power peak around 0.5 μ allows us to capture reflected light with conventional (and some not so conventional) cameras and films.

The basic strategy for sensing electromagnetic radiation is clear. Everything in nature has its own unique distribution of reflected, emitted, and absorbed radiation. These spectral characteristics can—if ingeniously exploited—be used to distinguish one thing from another or to obtain information about shape, size, and other physical and chemical properties. Insofar as we know the spectral characteristics, we can pick the appropriately sensitive detector to make the desired measurement, remembering that for a given collector diameter we get our greatest spatial resolution where wavelengths are shortest and energies greatest, and that these energies decrease at longer wavelengths and distances.

Would that life were so simple! In actuality most natural materials are not the perfect or even near-perfect radiators to which blackbody curves apply. The exact shape of the curve of wavelength distribution of emitted power—from which we hope to deduce structure and chemical composition—depends not only on temperature but on the radiating efficiency factor, or emissivity. Unfortunately, there are great gaps in our knowledge of spectral emissivity and reflectivity for natural materials, especially at wavelengths outside the visible portion of the spectrum. Many of the existing data have been acquired in the laboratory and are inadequate for describing the complex spectral characteristics of soils, rocks, vegetation, etc. Data need to be acquired by in situ measurements under different weather and climate conditions, and although such studies are becoming more common, published data for wavelengths longer than 10,000 Å are by no means abundant.

In addition to the complexities of the radiating and reflecting surface we must also contend with the atmospheric path interposed between source and detector. This atmosphere is a variable absorber, emitter, reflector, and scattering medium at different wavelengths, and consequently imposes severe limitations on our ability to observe terrestrial objects. Of course, if the atmosphere and its constituents are the subject of our remote sensing study, then these limiting characteristics may actually be assets. But we are concerned here primarily with looking at the earth's surface, so we must deal with losses from scattering, absorption, and reflection by gases, smoke, haze, water vapor, and water droplets. These losses are a function of the path length and the number, size, and energy of the particles within this path length, as well as the wavelength of observation. Some of the more important absorption regions are shown in Figure 3. . . .

Looking for the Windows

From this brief introduction it should be obvious that we cannot just pick and choose our observational frequencies at random up and down the many decades of the electromagnetic spectrum. We are forced by nature to channel our attention to those spectral regions where the atmosphere will transmit sufficient radiation to our detector. The most familiar and widely used of these is, of course, the visible region between 4000 and 7000 Å. Here, as well as on the fringes down to 2900 Å in the ultraviolet and out to 10,000 Å in the infrared, we can use photochemical emulsions to capture reflected light from the sun, the moon, or some artificial source like a mercury flash lamp.

With the variety of films and filters available today, cameras—especially aerial cameras—provide one of our most powerful tools for remote sensing. Aerial photography in the visible region gives the most accurate information on the size, shape, and relative position of objects of any sensor. A nice example of capability was given by Yale Katz of the Rand Corp. at the first symposium on remote sensing of environment in 1962. He showed pictures taken 100,000 feet above Boston's Fenway Park baseball stadium in which the 6-inch-wide strips between the spaces in the parking lot could be seen.

Cameras also exist that permit taking pictures by moonlight, while some television cameras have been developed that are claimed to be useful on cloudy moonless nights. Interestingly enough, such systems have some nonmilitary applications too. For example, zoologists can get more useful information from an animal census conducted at night.

In short, there exists such a wide variety of camera systems and film-filter combinations in both the visible and the near infrared that you have considerable freedom of selection within these spectral regions. Traditionally, however, the resulting pictures have been used in essentially one way: objects are identified on the basis of being able to *recognize* them.

Today increasing attention is being paid to how you distinguish among objects that have the same photographic gray-scale values and therefore look the same—as do grass, concrete, asphalt, and soil runway surfaces on a high-altitude panchromatic photograph, for ex-

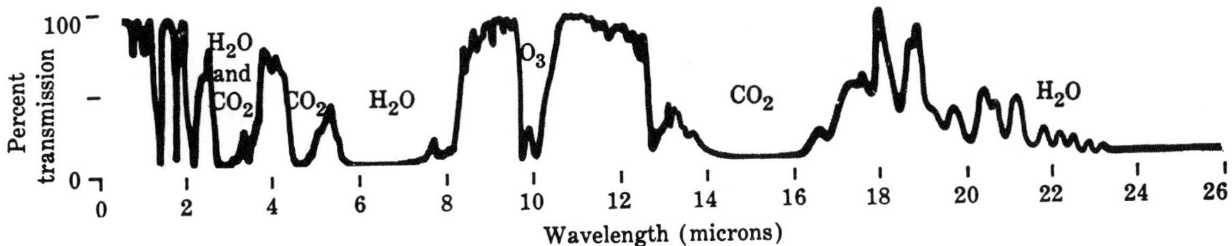

Figure 3A In the ultraviolet region of the electromagnetic spectrum, atmospheric absorption of radiation is nearly complete out to about 0.3 μ. Our small window in the adjacent visible region (0.4–0.7 μ) is relatively free from absorption, but one must still cope with considerable scattering by atmospheric particles. In the infrared out to 25 μ there are many absorption bands of water vapor, carbon dioxide, carbon monoxide, nitrous oxide, and ozone. So complete is the absorption in some of these regions that they are for all practical purposes opaque. Absorption by water vapor virtually closes the atmosphere from 25 μ to the more-or-less arbitrary start of the microwave region at 1000 μ.

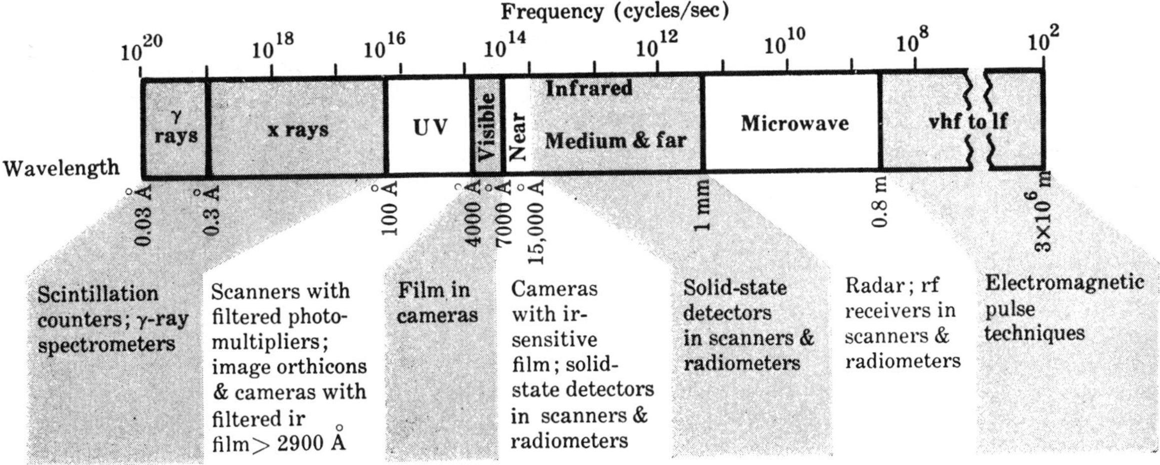

Figure 3B Remote sensors for the electromagnetic spectrum.

ample. Attempts to solve this problem by more analytic use of photographs have led to the concept of multiband spectral reconnaissance.

Where the Whole May Exceed the Sum

The idea behind multiband reconnaissance is that by comparing two or more photographs of the same object made in different regions of the spectrum we may learn something about the object we could not learn by studying the tonal values on just one photograph. Take the runway example, which has been discussed by Robert Colwell, of the University of California, who is one of the leading proponents of this concept: Light reflectance curves obtained with a spectrophotometer show there is no one spectral band between 4000 Å and 9000 Å wherein each of the four runway surfaces exhibits a reflectance that would cause it to be registered in a unique tone on a black and white photograph. Yet from these curves we can predict that if panchromatic film with a light red filter is exposed in the range 6000 Å to 7000 Å, then grass and asphalt will appear dark on positive prints while concrete and soil will be light. If we then expose infrared film with a deep red filter in the range 7000 Å to 9000 Å, asphalt and concrete will maintain their respective tones but grass will appear light and soil dark. Thus, while we cannot positively identify the four surfaces in *either* photograph, we can if we study *both* photographs.

Colwell has reported success with this technique in detecting diseased crops before they would be discovered by an observer on the ground or by using panchromatic or color films. The illustrations in Figure 1 show some typical examples. [See color insert.]

One of the newest cameras—the multilens camera—has been developed to permit fuller exploitation of the multiband concept. Figure 4[1] shows one that takes simultaneous exposures in nine different regions between 4000 Å and 9000 Å. This model was originally developed by Itek Corp. for the Vela Uniform program to detect physical changes in vegetation and terrain surface features due to underground nuclear explosions. It is now being investigated by the Air Force for such tasks as evaluating prospective aircraft landing sites.

A nine-lens camera is also being used along with other sensors in a University of Michigan program to collect multispectral imagery between 3000 and 14,000 Å. The program is sponsored by NASA, which wants to study differences between vegetation and its terrain background, and the Department of Agriculture, which

[1] Not available for reproduction.

is interested in establishing whether this technique is a way to determine land use, crop yield, crop identification, and the age and health of crops. The type of information obtained with the camera is illustrated in Figure 5.[2] You can see that it's possible to detect major changes in tonal contrast, but whether these differences can be used to identify some specific crop features is still unanswered.

Intriguing as multiband photography is, it should be obvious that, except for simple cases where only a few things are being compared, the addition of so much detail does not suddenly change the photo-interpreter's lot to a happy one. One of the big tasks at present is to eliminate the need for eye-weary observers to compare tonal values on a multitude of photos. A primary goal of the Michigan program is to see if the information content in the multiple channels can in some way be preprocessed and combined electronically to produce an image with enhanced target-to-background contrast.

Various approaches are also being investigated to generate composite color images that would give particular emphasis to areas on the original negatives whose reflectances differed between spectral bands. Figure 2[2] illustrates such a composite image, formed from selected negatives of the nine made with the Itek camera by a technique in which the subtle tonal differences of interest have been greatly enhanced and the similarities suppressed.

Looking in the Ultraviolet

Multiband photography has also been extended into the ultraviolet—an intriguing place to look because certain objects like unpainted metal roofs and calcareous terrain materials such as limestone outcrops show remarkably greater contrast than in other regions of the spectrum. Actually not much can be said about ultraviolet sensing, for its use has been inhibited by the strong atmospheric attenuation at wavelengths shorter than 2900 Å. This is an example of a place where persistence might pay off, however. Allen Feder says his group at Texas Instruments has used a detector sensitive down to 2400 Å and has built up a file of data in the natural solar-reflection region between 3100 Å and 4000 Å including situations where ground humidity conditions of 70–90 percent existed. This ultraviolet data collection was done flying a filtered photomultiplier scanning system at up to 5000 feet.

Another approach to ultraviolet sensing is atomic absorption spectrophotometry, which is being investigated at Barringer Research Ltd. in Canada. The majority of the atomic absorption lines lie in the ultraviolet, and the ability to use the technique to sense vapors of elements like iodine and mercury could be important in oceanography, as well as in petroleum and mineral exploration.

Naturally, ultraviolet techniques look promising for the vacuum around the moon, and studies are underway to develop such equipment for a lunar orbiter.

Similar spectrophotometric techniques are being developed at Texas Instruments for the higher-frequency gamma ray region. Emphasis is on airborne gamma ray spectrometers to determine composition and conditions like moisture content of terrain materials on the basis of measured ratios of radioactive daughter products. Operation of the gamma ray spectrometer, which measures the energy level and flux of gamma rays within preselected wavelength ranges, is roughly analogous to that of an optical spectrometer in that the number of gamma counts in each band corresponds to the intensity of the lines on an optical spectrograph. This technique is still in the early stages of development and there are problems, such as poor resolution, that remain to be overcome.

The Thermal World

As we have indicated, conventional photography is excellent for resolving recognizable objects, but may not tell all we want to know about their physical or chemical composition. Moreover, it is generally restricted to employment during daylight and clear weather. So it's not surprising that a great deal of effort (mostly military) has been expended on systems which, because they sample emitted radiation from terrestrial objects, are free of certain temporal and atmospheric restrictions and at the same time provide us with unique data about the nature of the object. Systems sampling infrared radiation are good examples. Second only to the radio spectrum in the extent of its wavelength span, the infrared region allows us to perceive a totally different world—a world of thermal radiation, which we can view day or night, through haze and smoke, provided we stay within our atmospheric windows.

A common method of obtaining aerial infrared imagery is with an optical-mechanical device in which a rotating mirror scans the terrain in continuous strips transverse to the line of flight. Radiation from each element of the scene is reflected off the face of the mirror and focused on a photoelectric detector such as indium antimonide or impurity-doped germanium. The detector output is amplified electronically and may be recorded on magnetic tape or used to intensity-modulate a cathode ray tube trace or a light source for recording directly on film. The resulting image is a strip map that superficially resemble an aerial photograph. The vital difference is that the tone in the conventional aerial photograph is a function of light reflectance whereas the tone in the infrared image represents variations in the intensity of the emitted radiation resulting from variations in the product of emissivity and surface temperature. (Below 3.5 μ during daylight the tone is more characteristic of reflected infrared radiation.) The interpreter of infrared images must, therefore, accustom himself to seeing things differently from his familiar visual world. For example, the warmer objects in a scene will appear lighter in tone than the cooler objects.

Existing scanners can discern subtle differences in surface thermal radiation—a unique capability that permits doing such things as detecting snow-bridged crevasses and tracing flow patterns in water that are undetectable by surface observations. Because much of the research and development on infrared equipment has been conducted for military applications, detailed specifications are classified, as is much of the equipment itself. However, through agreements between the Department

[2] Not available for reproduction.

of Defense and other government agencies, it has been possible to conduct certain nonmilitary research with this equipment. Within the past several months some data have been declassified regarding infrared surveys to detect the onset of volcanic eruptions in Hawaii, formation and decay of sea ice in the Canadian Arctic, geothermal activity in Yellowstone National Park, and forest fires in the western United States.

In these surveys infrared scanning has provided very useful information on such things as the area of thermal features and ice-water boundaries that are not available with more conventional techniques (see Figs. 6[3] and 7). As an example of how this can be economically important, the U.S. Forest Service needs to detect promptly incipient fires, which cannot be spotted by present visual methods, and also to delineate the perimeters of large fires which are obscured by smoke. Stanley Hirsch, of the Northern Forest Fire Laboratory, reported that during the 1963 and 1964 fire seasons more than 20 uncontrolled forest fires were mapped with an infrared scanner. The imagery was in some cases dropped immediately to the fire fighters and proved so valuable in helping them suppress the fires that infrared mapping of large forest fires is expected to become part of fire-fighting routine in the near future.

Absolute measurements of thermal radiation flux are made with a nonscanning device which is simply a radiometer. It looks at one element in a scene (the way a photometer does in the visible region) and, because it is referenced to some known temperature, gives quantitative data that permit calculating apparent and, in some cases, actual temperature. Radiometers can, therefore, be used to determine isotherms of water surfaces and contour underground coal fires, for example. Because an airborne radiometer can have a much slower data-acquisition rate than a scanner, it can be more sensitive; measurements of radiation fluctuations corresponding to temperature fluctuations of 0.01°C are not uncommon. Also resolution can be greater. Although the theoretical resolution angle for a scanner with a 1-foot-diameter aperture is given by the Rayleigh resolution criterion as less than 1 milliradian (not as high as in the visible), this resolution is degraded by the components in an actual system. This is a good place to remark that infrared scanners—like most of the new equipment we are describing—are much more complex and expensive than the ordinary general-purpose aerial camera. A small scanner can cost more than $50,000, and its care and feeding require special skills in electronic troubleshooting and cryogenics.

Beyond 25 μ

At wavelengths longer than 25 μ, atmospheric absorption severely limits what can be done here on earth with present infrared techniques. But if we go to wavelengths longer than 1 mm—which is our arbitrary boundary between infrared and microwaves—then we can detect emitted thermal radiation with the highly sensitive radiometric techniques which the radio astronomers have perfected. Since this equipment requires sophisticated electronic receivers and antennas, the question "Why bother?" deserves to be answered. First, we're better off using microwaves instead of infrared in cloudy weather because lots of clouds are relatively transparent at microwave frequencies. Secondly, there is no major atmospheric attenuation at wavelengths longer than 18 mm and there are several windows between 1 mm and 18 mm. Furthermore, we are for the first time dealing with natural energy that is emitted, in part, from below the surface of the ground, a fact that makes many people think they can learn more about conditions, such as moisture, at depths of several centimeters in solid materials and possibly a few meters in loose materials like sand. Finally, certain materials exhibit noticeable variations in emissivity with frequency that may provide a means of differentiation. For example, emissivity is a function of many factors including dielectric constant, frequency, polarization, surface roughness, and the angle at which you observe the surface. The fact that ice has a small dielectric constant at microwave frequencies, while water has a large dielectric constant, results in a difference in emissivity that has been utilized in experiments to distinguish between icebergs and open water.

[3] Not available for reproduction.

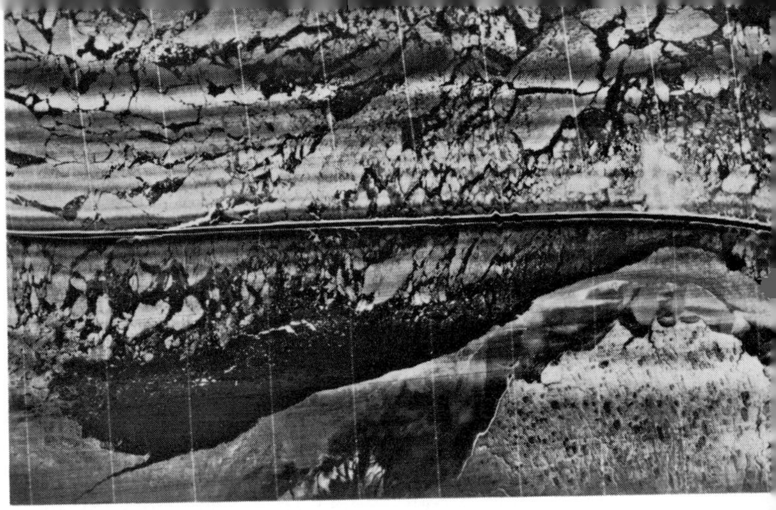

Figure 7 Infrared scanner imagery of Harrison Bay, Alaska. (Substitution for original imagery which was not available for reproduction.)

In contrast with nonscanning microwave radiometers, equipment for mapping at these wavelengths is in a rather primitive state. This is primarily because it has not had the military support infrared has had. Present systems have low resolution and low sensitivity resulting from the requirements for rapid scanning in high-speed, low-altitude work. Nevertheless, the potential is attractive enough to warrant further development.

Probing with Radar

The area of technology within the microwave region upon which considerable research effort has been expended is for aerial imaging in the active mode—i.e., with radar, in which we look at the reflected component of a radio-frequency pulse we generate ourselves. With its ability to operate effectively day or night, at certain frequencies, through all but extremely heavy clouds and rain, radar lets us do continuous strip imaging of large areas more economically than present aerial surveying

Figure 8 Side-looking radar image of South Cascade Glacier region. (Substitution for original imagery which was not available for reproduction.)

methods. Such images provide information on geologic structures and landscape forms not easily obtainable on the ground or readily identifiable in aerial photographs (see Fig. 8). Side-looking radars (particularly those which use data-processing techniques to synthesize the effect of an extremely long antenna), give remarkable resolution. In a report on the recently declassified AN/APQ-56 system, which was an early side-looking radar, George Loelkes of Raytheon said it had been possible to separate returns from objects less than 50 feet apart and to obtain images suitable for constructing charts at a scale of 1:250,000.

Radarscope photos can be displayed free from the foreshortening you get in taking an oblique photograph of a segment of terrain; yet, because the illumination angle is oblique, small textural differences are enhanced. This makes it easier to detect fractures and faults. In addition, it appears possible to penetrate certain vegetation with radar and get a return that is indicative of the underlying surface configuration.

As with other sensors, we should be able to identify objects by analyzing the characteristics of the reflected signal. It's possible to see certain things on the basis of long-wavelength reflectivity alone that you might not see on a small-scale aerial photograph—like transmission lines and fences. Moreover, small reflectivity variations can be detected, a feature that makes radar a potential sensor for studying sea state and mapping crop types. Also, it might be possible to obtain characteristic "signatures" for various natural materials on the basis of the amplitude and state of polarization of the reflected wave. Signatures of water, forest, desert, and farmland can be differentiated that have been obtained with nonimaging radar-type systems which measure terrain-scattering coefficient as a function of the angle of incidence. Finer differentiation may be possible but this awaits further experiments.

Right now, not enough is known about natural materials to be certain of the role radar can play in identification, but numerous studies are underway. For example, the U.S. Army Engineer Waterways Experiment Station in Mississippi has been experimenting for a few years now with a radar at four different frequencies (as well as with other sensors like gamma ray spectrometers) to analyze soil conditions as far down as 18 inches in order to "fingerprint" properties like soil type, moisture content, and density. The U.S. Army Engineer Geodesy, Intelligence and Mapping Research and Development Agency has a radar-mapping test facility at Willcox Dry Lake, Arizona, which it is supplementing with various surface and buried rock patches for follow-on investigations in this area.

The ability to penetrate the surface is, of course, an important reason for being interested in radar for remote sensing. Penetration increases as frequencies get lower—at several meters wavelength you might get down on the order of feet, depending on the terrain conditions.

A wide variety of geologic, agricultural, geomorphic, oceanographic, hydrologic, and other uses of radar are being investigated by a NASA-sponsored team coordinated by the Center of Research in Engineering Science at the University of Kansas. Multiple-frequency, multiple-polarization images, and scattering coefficients are being obtained from several military and NASA aircraft for targets selected and studied by earth scientists.

Perhaps the potential of radar can best be appreciated if you consider the many underdeveloped areas of the world for which there are poor maps, few, if any, aerial photographs, and little understanding of the distribution of natural resources. Data derived from radar imagery could provide an efficient tool for initial planning in such areas.

Electromagnetic Sensing

As we continue toward longer and longer wavelengths, we gradually slip out of the spectral region in which conventional microwave radar sets are used and—at around 400 mc—enter an area that is sometimes loosely termed electromagnetic sensing by its practitioners. The even greater subsurface penetration possible in this frequency range makes it look attractive for studying moisture distribution in the top few feet of soil, contouring the land surface beneath continental ice caps, and measuring the thickness of fresh water ice and sea ice.

A key problem is, of course, the degradation of resolution as we go to longer wavelengths. Azimuth resolution depends to a large extent on using a narrow beam, and this is hard to get at low frequencies since it requires a high ratio of antenna aperture to wavelength. Also, distance resolution depends on using very short pulses; so we must be clever with the electronics in order to effect a suitable compromise between short pulse length and long wavelength.

The longer wavelengths in this region are being used by several groups for radio soundings of ice caps in the Arctic and Antarctic. One group—under the direction of A. H. Waite at the U.S. Army Electronics Laboratories—has reported sounding through 1500 feet of cold ice from low-flying helicopters with a 7-W 400-mc pulsed radar system and as much as 9300 feet with a 400-W 30-mc system. Accuracy compares very favorably with seismic measurements wherever the two have been made coincidentally. Such equipment is believed capable of penetrating the thickest ice caps in the world, thereby providing information on internal structure and subsurface contour that is not available economically from present techniques, which are based on gravity measurements and seismic spot soundings.

At lower frequencies, around a few kilocycles say, electromagnetic pulses will induce eddy currents in conductive bodies underneath the ground. This phenomenon has been exploited in the search for mineral deposits for several years now. The basic technique is to generate the pulses in a loop surrounding an airplane. These pulses will generate eddy currents in any conductive bodies within range, even though they may lie at some depth. The eddy currents produce a secondary field whose rate of decay is a function of the conductivity of the material. This effect has been utilized extensively in looking for conductive bodies of sulphides, and there is some evidence that it might be useful for detecting the important class of copper deposit known as porphyry coppers.

Force-Field Sensors

With the substantial economic stakes involved in mineral and petroleum exploration, it should not be surprising that people in this business have been among the pioneers in remote sensing. While airborne radio techniques are relatively new, magnetometers and gravity meters (which fall into our category of force-field sensors) are proven tools in geophysical prospecting, and the various forms these instruments take are well documented in the literature.

Airborne magnetometers have been used routinely for years and the newest of these devices, the very sensitive alkali vapor magnetometer, has proven valuable in several rocket and satellite experiments. The airborne gravity meter, a newcomer to the instrumentation field, employs conventional gravity meters on special platforms that isolate them from accelerations caused by the carrying vehicle. A system developed by the Air Force Cambridge Research Laboratories has already demonstrated a capability for measuring gross variations in the strength of the earth's gravitational field. The limitation on the technology at present is caused by the need to dampen those acceleration components not caused by gravity.

The use of these force-field sensors in geophysical exploration demonstrates something that should be said for remote sensing in general; namely, that only in certain specialized applications will one sensor provide all you want to know—usually complementary techniques will be required. For example, magnetometers have been used with great success in locating iron ore deposits directly, yet searching for oil is a different situation since oil-bearing rocks are generally not strongly magnetic. Here, magnetometers can be used in conjunction with gravity meters to find anomalies indicative of structures favorable to oil accumulation, while further exploration—to more accurately locate and profile the structure—may involve seismic techniques. Profiling can be done to depths of thousands of feet with a seismic wave generated by an explosive charge or other means. This allows us to zero in on a suspected oil-bearing region but is as far as remote sensing can take us before it's necessary to drill.

Seismic techniques, which have long been used to study properties of continental land masses and shelves, can also be integrated with aeromagnetic and gravity measurements for detailed mapping of the crustal formation of continents and oceans. For instance, the U.S. Geological Survey has made transcontinental profiles in which airborne magnetometer and surface gravity measurements confirmed the results of seismic surveys showing differences in crustal structure on either side of the Rocky Mountains.

Unlike these traditional tools of geophysical exploration, the complementary use of the newer electromagnetic sensors is only just beginning to be investigated. To see how this might work out in what seems to us to be an application that is within the state of the art, let's take a look at that sailor's nemesis—sea ice.

Sensing Sea Ice

If the distribution, thickness, and perhaps type of ice along major shipping routes like the St. Lawrence Seaway could be charted rapidly, it might be possible to extend the shipping season by days or even weeks. How might this be done? The basic requirements—the need to measure over large areas and furnish data in the form of daily synoptic reports or maps—force us to abandon surface techniques in favor of long-range reconnaissance aircraft or satellites. Furthermore, since we want to know distribution, we should choose a system with a mapping type of readout. This means aerial photography, infrared, passive microwave imaging, or radar.

Between 3000 Å and 10,000 Å there are regions where the difference in reflectivity between snow-covered or clear ice and water is sufficient to produce a photographic record that contrasts these two substances nicely and permits easy determination of the percentage of ice in a given water background. Also, the very high resolution attainable in this spectral region allows recording a tremendous amount of detail that would be useful in determining the type of ice-rafting conditions and the amount of pressure ridging and, consequently, allows estimating thickness. Such a permanent record is lacking in the current practice of sea ice reconnaissance which is to fly routine patrols with an observer stationed at a good vantage point in the aircraft marking the estimated distribution of ice on a chart.

The drawback to photographic systems is, of course, the limited time available for aerial photography in high latitudes during winter and the fact that imaging must be done below clouds. For all practical purposes this will eliminate aerial photography as the sole tool for ice mapping, forcing us to think next about the infrared region of the spectrum. Here the difference of temperature that usually exists between the various snow, ice, and water surfaces will produce sharp contrasts on imagery generated at these wavelengths. Also, since we are not restricted to daylight, an infrared system could provide a 24-hour, year-round capability for imaging at high latitudes during good weather. Present infrared systems also have sufficient resolution to give enough detail that various types of ice could still be identified.

Furthermore, since heat transfer through ice and overlying snow is a function of thickness, it's possible from the infrared record to delineate areas of different thickness. In many cases it has been noted that areas of very thin ice can be detected readily, whereas in a photograph they are masked by an overlaying layer of snow. However, the infrared imaging system is weather-limited and the data readout is not in planimetric form. Also, the necessity for flying at low altitudes below cloud layers can severely restrict the lateral coverage of the scanner and thus limit the area that can be observed.

Microwave devices are, as we have seen, not limited either by time of day or weather. At certain frequencies within the microwave region, propagation is not appreciably attenuated even by thick layers of clouds. Consequently, an all-weather, 24-hour-a-day, year-round capability can be realized. Both the passive microwave systems and radar produce a maplike readout so that we can see distribution. Furthermore, the emissive and reflective properties of snow, ice, and water at these frequencies are such that they produce sufficient contrast on the imagery to permit distinguishing ice from water. Also, there is evidence that the radiometric temperature of ice layers on water in the 2- to 3-cm region can be related to thickness.

Thus, with microwave systems we have a large-area imaging capability and the possibility of doing some qualitative assessment of relative ice thickness from the passive microwave imagery in the same manner as with infrared imagery. The most limiting feature is the resolution of detail with present systems when the imaging is done from an altitude high enough to give the large area coverage desired. Consequently, classification of ice types would be more limited than by infrared or aerial photography.

Profiling Techniques

So far, all the imaging devices discussed have been able to give us ice distribution data and, in some cases, qualitative information concerning ice thickness. However, none (with the exception of photographic stereo) can give surface or subsurface profile information or absolute ice thickness to a fair degree of accuracy. So other than being able to say the ice is thinner here than there or that this ice should be approximately so many feet thick by reasoning based on a knowledge of ice type, we are unable to make any further statement as to true thickness.

It is at this point that we reach the limit of present remote sensing technology and must look at equipment that is just evolving. There are in existence radar systems that will provide information on the profile of the snow-ice surface over which you are flying, but for high accuracy they must be flown at relatively low altitudes. Aero Services and Spectra Physics are incorporating a gas-laser ranging device and a barometric pressure transducer into a system that has demonstrated a capability for accurately profiling the surface over which the aircraft is being flown (accuracy from recent tests is claimed to be on the order of a few feet at 10,000 feet altitude). Since height of pressure ridging and pressure ridge density are indicative of the depth of ice below the surface at these points, such information would be quite meaningful in assessing the ice conditions.

What is really desired, however, is a device that would give absolute ice thickness at many points along a line below the aircraft. Such data could be used in assigning thickness values based on the gray-scale renditions of the ice in the thermal or microwave region if used in conjunction with such an imaging device. Vhf radar appears to be just such a system. While still experimental, it shows promise of being able to provide absolute ice thickness with a fair degree of accuracy from measurements of the echo-delay time between the two reflecting interfaces. Barringer has been able to measure the thickness of 21 inches of lake ice to an accuracy of about 5 percent using vhf radar techniques at the surface. Similar accuracies using a vhf radar system are reported by the U.S. Army Cold Regions Research and Engineering Laboratory. If such accuracy could be maintained from an aircraft, it would undoubtedly be sufficient for most applications.

It seems, therefore, that the remote sensing technology is close to the point where a problem such as sea ice reconnaissance could be handled by a system of complementary sensors. In fact this is typical of how the field of remote sensing as a whole is evolving. The type of remote sensing studies that whet the appetite of the environmental scientist—heat and water budget at the earth's surface, crop conditions, lunar geology, etc.—will more often than not require simultaneous imaging of both surface and subsurface radiation in different regions of the spectrum. Furthermore, we may find, as some investigators have suggested, that identification of a particular rock type or plant species will depend more on comparing the changes in its radiant emittance with time at different wavelengths than on any one-shot measure of emittance. Sometimes correlation with magnetometer and gravity measurements will be necessary; invariably we'll need to add one or more of the innumerable specialized sensors that we have not even mentioned here, like scintillation counters, for example.

Sensing from Spacecraft

The practical application of the so-called multisensor systems is perhaps ultimately realized in high-altitude aircraft or spacecraft, where vast amounts of data could be acquired for many different users. Experimental systems have been used by several commercial and military organizations in aircraft, and systems for spacecraft are presently being planned.

Observation from spacecraft represents the glamor side of remote sensing. There is considerable optimism that the comprehensive synoptic coverage of vast land and ocean masses obtainable at orbital altitudes will provide important new information on many aspects of the terrestrial environment. Global weather, heat and water budget, physical oceanography, crustal and subcrustal geology as well as the interaction of man with his surroundings are only a few of the areas in which a good deal of the information we need just cannot be obtained from the surface of the earth. For instance, our earthbound micrometeorology stations only give isolated point data on temperature and precipitation, whereas a knowledge of the heat energy budget at the surface which is so important for pollution studies, agriculture, etc., requires measurements averaged over thousands of square miles. Similarly, oceanographers are essentially operating with data taken from widely separated points at different times. Spacecraft observations are expected to give them their first integrated view of many oceanographic phenomena, like temperature distribution, ocean currents, and wave height.

NASA has studied the requirements for experiments for the 30–60-day manned earth and lunar orbital flights scheduled to follow the initial Apollo lunar landings. Some of the sensors being investigated are listed in

Table 1 Possible Applications for Orbiting Sensors

Experimental Technique	Agriculture, Forestry	Geology	Hydrology	Oceanography	Geography
Visual photography	Crop and soil identification, identification of plant vigor and disease	Identification of surface structure	Identification of drainage patterns	Identification of sea state, beach erosion, off-shore depth and turbidity	Urban and rural land use, transportation routes and facilities; terrain and vegetation characteristics
Multispectral photography		Identification of surface features	Soil-moisture content	Sea color as indicative of living organisms	
IR imagery and spectroscopy	Terrain composition, plant vigor and disease condition	Mapping thermal anomalies, mineral identification	Detection of areas cooled by evaporation	Mapping of ocean currents, sea-ice investigations	Surface energy budgets, near-shore currents and land use
Radar imagery	Soil characteristics	Surface roughness, tectonic mapping	Measurement of soil-moisture content, identification of run-off slopes	Sea state, ice flow and ice penetration, Tsunami warning	Land and ice mapping, cartographic and geodetic mapping
RF reflectivity		Subsurface layering, mineral identification	Moisture content of soils	Sea ice thickness and mapping, sea state	Land ice mapping and thickness, penetration of vegetation cover
Passive microwave radiometry and imagery	Brightness-temperature map of terrain	Dielectric constant measurement indicative of subsurface layering	Snow and ice surveys	Sea ice and ocean current mapping	Snow and ice measurements
Absorption spectroscopy (remote geochemical sensing)	—	Detection of mineral deposits, trace metals and oil fields	—	Detection of concentrations of surface marine flora	—

Table 1; in most of these instrument areas teams of scientists and engineers are currently cooperating with groups of potential users to explore applications and help define additional instrument requirements. The different sensing techniques and applications are being evaluated by means of aircraft test flights over ground sites selected to permit attacking one problem at a time; for example, the first flight—over Pisgah Crater, Calif., in February—was to determine the capability of infrared sensors to solve geologic problems. NASA has also held initial discussions with several countries, including Mexico and Australia, concerning the establishment of test sites. Known ground sites are a crucial part of any remote sensing program because they permit gathering the characteristic spectral "signatures" and images against which data gathered on any flight over new territory must be compared in order to know what you're looking at.

Needless to say, the cost of instrumenting and executing these global-scale projects is such that they can only be conducted under government sponsorship or through the cooperative efforts of many institutions and agencies having a vested interest in the data to be acquired. However, the more local airborne and ground-based applications for remote sensing can be handled by individual researchers or small private concerns or academic institutions. Local geology, ecology, and weather can, for example, be studied with sensors from helicopters, balloons, hilltops, water towers, and even an extension ladder in your backyard.

Handling the Data

Regardless of the program, however, there is the critical problem of what to do with these volumes of wonderful data we gather. Our ability to acquire data is so far ahead of our ability to interpret and manage it that there is some question as to just how far we can go toward realizing the promise of much of this remote sensing. Probably 90 percent of the data gathered to date have not been utilized, and, with large multisensor programs in the offing, we face the danger of ending up prostrate beneath a mountain of utterly useless films, tapes, and charts.

What is needed is a means of programming our sensors to provide the interpreter only with the data he needs for his particular purpose. Even in the relatively old field of photographic interpretation, automatic discrimination techniques—usually based on recognizing spatial relationships—have not been particularly fruitful. As we go into other regions of the spectrum, we need, first, more basic information on physical properties and spectral characteristics of natural and manmade objects so that we can properly select our sensors and interpret the resultant images. Next we need an automatic means of feeding the interpreter only the information relevant to his particular problem. This last involves a whole program of data management—storage, processing, retrieval, display, and decisions on such questions as centralized storage and handling versus regional data centers, computer time-sharing, and the like. Some studies are underway on these problems but they are just a beginning. We have a long way to go before being able to obtain maximum benefit from our ability to "see" at a distance.

REMOTE SENSING *in its broad sense is not new. The first air photographs were made more than a century ago, and in the 1930s the U.S. Government began an extensive program of photographing forest and crop lands. The advent of World War II and the development of high-speed aircraft, rockets, and missiles led to new surveillant devices which made use of a greater portion of the electromagnetic spectrum. Postwar reconstruction, the Cold War, increasing population, rising expectations, and a diminishing resource base have placed enormous pressures on natural resources and the quality of our environment.*

New sources of raw materials must be found. More accurate inventories of existing supplies have to be determined. Monitoring urban and industrial growth has to be carried out more effectively. To accomplish these tasks, new techniques of surveillance are needed. New devices which supplement and enhance the aerial camera offer significant advantages in detecting, inventorying, monitoring, and measuring the rate of change of natural resources and quality of the environment.

Thermal scanners have proved effective in spotting wildfires in remote forest areas and in tracing sources of thermal pollution in streams. Color infrared photography has been used to locate disease and insect infestations in agricultural crops and forests. With space photographs it is possible to measure urban and industrial growth more rapidly than by conventional methods. This new research suggests that we are on the threshold of a significant breakthrough in methods of understanding the environment in which we live.

5-Sensing the Invisible World

ERIC M. WORMSER

THE HUMAN eye is in many ways an optical instrument of remarkable performance (Hardy and Perrin, 1932). It is equipped with an automatically adjustable pupil, the iris, which varies its entrance aperture from 2 mm to 8 mm, depending on the level of illumination. In a healthy person, it provides for automatic focusing from about 25 cm to infinity. In the central portion, its angular resolution is limited about equally by diffraction, aberrations, and the structure of the retina to approximately 1 min of arc. The eye's ability to accommodate and distinguish varying levels of brightness is truly remarkable. It can adapt to levels from 10,000 candles/meter^2 to 0.01 candles/meter^2 and distinguish brightness levels in excess of that range. In all of these respects, the capability of the eye exceeds that of most optical instruments. Yet, in its ability to sense optical radiation of varying wavelengths, the eye is quite limited, its spectral response extending only from 450 mμ to 650 mμ. While this spectral response is well matched to the peak region of luminance of the sun, the range of the eye's sensitivity is very small and covers less than one octave of wavelength out of the complete optical spectrum, which

From *Applied Optics* 7:1667–1672, 1968. Reprinted with permission of the author, Barnes Engineering Company, and the journal editor.

extends over eight orders of magnitude from 10 mμ in the short uv to wavelengths as long as 1000 μ in the infrared (ir).

If we compare the spectral response of the eye to the energy emitted by objects at various temperatures, which is shown in a logarithmic plot in Figure 1, we see how limited indeed is the spectral range of visible light. This plot shows that the eye responds to only 20 percent of the energy radiated by the sun having a color temperture of 6000°K, and to only 8 percent of artificial tungsten illumination, having a color temperature of 3000°K. The eye is not at all responsive to the energy emission of objects cooler than 800°K, which is the temperature of objects that are barely red hot.

We live on the planet Earth, where in the temperate zones the ambient temperature around us varies only over a narrow range up and down from an average level close to 300°K (27°C). Almost all of the objects surrounding us also have temperatures within a narrow range around 300°K. As shown in Figure 1, objects at or near 300°K emit ir radiation extending from 2 μ to 100 μ and peaked at 10 μ. Thus, all of the objects surrounding us are *incandescent* or *glowing due to heat* but unfortunately, owing to the limited wavelength response of our eyes, this *ir incandescence* is invisible to us. In order to make the invisible ir glow visible and measurable, we have to employ techniques and instruments that detect and measure ir radiation or provide image translation of the invisible ir glow into visible images.

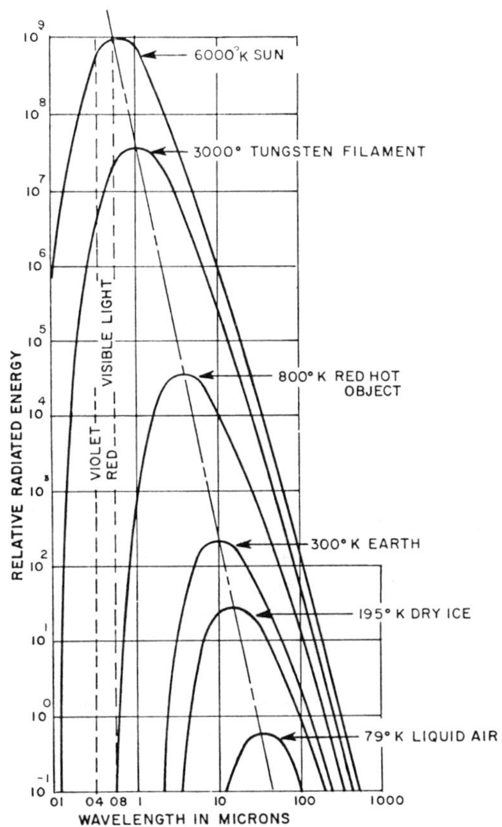

Figure 1 Emission of optical radiation for blackbodies at various temperatures. $W = \epsilon \sigma T^4$, $\lambda_{max} \sim 3000/T$.

Conventional Methods of Optical Image Translation

The simplest and best known technique of image translation is the photographic process. On the short wave end of the invisible optical spectrum it can be made responsive to x rays (1–10^{-1} mμ), as well as gamma rays (10^{-3} mμ) and uv rays (10–400 mμ). On the long wavelength side, photography employing conventional ir film is capable of extending human vision only very little, to about 1.2 μ.

Photoemissive devices, such as the ir image converter tubes (Wiseman and Klein, 1959) developed for the military during World War 2, for such equipment as the snooperscope and sniperscope, which are now commercially available, also have wavelength response to approximately 1.2 μ.

Visual examination by the eye aided by magnifying glasses and microscopes is the most common and most extensively used technique of nondestructive inspection and examination.

Photography of all types, including photography using x ray or gamma ray illumination, is also a very common and widely used technique in nondestructive testing.

As mentioned previously, only as objects exceed temperatures of 800°K (527°C) does their glow become visible. Measuring temperature and examining temperature distribution by measuring or observing this visible glow of very hot objects above 800°K is again a well-established method of nondestructive measurement and examination.

All of the above methods of examination and image translation have been in use for many years and are adequately covered in the existing literature.

Examination of surface temperature distribution, which is related to heat generation and heat flow in both inanimate and living objects, is one of the most fundamental of nondestructive examination or testing techniques.

While glow or emission due to heat is at all temperatures related to the surface temperature of the object, unfortunately, as previously stated, the spectral response of the human eye, the photographic plate, and photoemissive image conversion devices are all limited in long wavelength response to about 1.2 μ and, hence, this ability to see, measure, and examine is limited to very hot objects about 800°K (or 527°C). This left invisible or undetectable, prior to the introduction of modern ir measuring and imaging techniques, the glow or incandescence of the wide variety of objects having temperatures at or near the earth's ambient temperature.

Methods of Infrared to Visible Image Translation

In order to make visible an object's long wavelength glow or ir emission beyond 2 μ, we have to rely on various types of ir-to-visible image translation devices. Most of these instruments employ ir detectors in optical mechanical scanning devices and have associated display or imaging systems. However, before surveying these instruments, we briefly examine two types of heat-sensitive layers.

Image Translation Using Heat-Sensitive Layers

The phenomenon of differential evaporation of thin liquid films was first utilized in 1840 by Sir John F. W. Herschel (the son of Sir William Herschel who discovered ir) in further exploring the ir portion of the spectrum and thus obtaining what he called a *thermograph.* The use of thin oil films that change color with differential evaporation due to incident ir radiation has been described by Czerny (1929).

A nonscanning image translation device known as the evapograph (Ovrebo et al., 1959) utilizes this phenomenon. It was developed under military sponsorship and has been commercially available for over ten years.[1] The evapograph has been used as a nondestructive testing device (Bobo, 1965) for qualitative observation of hot spots. It is difficult to make the color changes obtained in the evapograph layers repeatable, and, hence, it is difficult to obtain quantitative observation of two-dimensional temperature phenomena with this instrument.

Recently, certain types of liquid crystal films having a greater temperature sensitivity have become available. Cholesteric liquid crystals can be utilized for the visualization of two-dimensional temperature patterns by applying these layers in direct contact with the surface of the object to be examined.

The cholesteric heat-sensitive layers could also be utilized in noncontact thermal imaging systems in a manner analogous to the use of photographic film in conventional cameras, and their use in this manner has been suggested by Fergason and others; however, certain difficulties associated with this application should be pointed out.

The conventional photographic camera employs in its simplest form a pinhole aperture and a black box, i.e., a chamber which is black or not glowing, together with conventional photosensitive film. For ordinary light-sensitive film, this is very simply accomplished by making a light-tight box or enclosure. In order to make an ir camera, we have to utilize a photosensitive layer (such as the evapograph or the cholesteric layers) sensitive at long ir wavelengths, around 10 μ. At these wavelengths, an ordinary black box enclosure glows.

A nonglowing enclosure would have to be one that is uniformly cooled to near zero °K, which is very difficult and cumbersome to accomplish. Alternately, techniques have to be devised to prevent the heat-sensitive layer in a long-wavelength ir camera from seeing the camera enclosure. As a result of these and other difficulties, sensitive long-wavelength ir cameras employing cholesteric heat-sensitive layers have not yet become commercially available.

Optical–Mechanical Scanning

Most work in extending human vision by means of image translation from longer wavelengths in the ir has employed optical-mechanical scanning devices utilizing various types of ir detectors having electrical outputs. One of the image translation scanners, known as thermograph T–2, was developed by Astheimer and Wormser about 1954.

[1] From Baird-Atomic, Inc., Cambridge, Mass.

Figure 2 Visual photograph and thermogram of woodland scene at Fort Belvoir, Virginia. The camouflaged soldier cannot be seen in the visual photograph, but stands out clearly due to his body temperature in the thermogram.

This T–2 thermograph was designed for use with themistor bolometers (Wormser, 1953), and most of the over 100 units, which are in varied uses today, employ uncooled immersed thermistor bolometers (De Waard and Wormser, 1959) having nearly uniform sensitivity from 2 μ to 15 μ in the ir. Some units have employed other types of ir detectors.

The T–2 thermograph image translation scanner was originally developed for use by the military for ir target and background studies, and Figure 2 illustrates this application, as well as showing the capability of the thermograph in general.

In Figure 2 are shown a visible photograph, as well as a thermogram, of woodland terrain at Fort Belvoir, Virginia. In the visible photograph, the camouflaged soldier positioned in the underbrush cannot be seen, while in the thermogram the soldier can be clearly distinguished, since his body surface temperature of 25°C to 35°C glows distinctly brighter at ir wavelengths than the tree and foliage background, which are below 20°C.

The thermograph infrared camera was designed in 1954 utilizing themistor bolometers having time constants of about 1 millisecond, which were the fastest long-wavelength ir detectors then available. It transforms emitted long-wavelength ir radiation, which over a narrow temperature range is proportional to temperature differentials, quantitatively into tones of gray recorded on photographic film. A high-resolution thermogram containing 60,000 data points with a minimum resolvable temperature differential of 0.1°C can now be produced in 4 minutes using immersed thermistor bolometers.

A comparable thermogram utilizing thermistor bolometers without germanium immersion optics required up to 20 minutes when the T–2 thermograph was first developed around 1954.

The full range of temperature contrasts recordable in a single thermogram can be varied from a minimum of 1°C, which makes a minimum differential of 0.1°C readable, to a full range of several hundred degrees in many steps.

Through use of an adjustable internal blackbody reference source and electronic temperature offset or brightness controls, the temperature contrast range in the T–2 thermograph can be set to any desired level independent of ambient temperature variations.

Modifications of the above model, designated the T–4 thermograph for general ir nondestructive testing applications, and the M–1A thermograph[2] for medical applications, are in extensive use today.

The T–4 thermograph is equipped with PbS detectors. The instrument provides for 0.5-mrad optical resolution, giving around 240,000 image points per picture frame, and it is used to study underpaintings in medieval works of art. The underpainting becomes visible owing to the relative transparency of paints at ir wavelengths near 2 μ.

As pointed out by Barnes [see Diagnostic Thermography article], medical diagnosis is the area of application of ir thermography that holds out the greatest promise for mankind. Since 1957, when the author first assisted Ray Lawson of the Royal Victoria Hospital of Montreal, Canada, in applying one of the early military models of the thermograph to the investigation of malignant breast lesions, tremendous strides have been made in this field. Nearly 200 publications on the subject of medical applications of thermography have appeared in many parts of the world. Many thousands of patients have been examined and thermography has been found useful in examining many parts of the body for a wide variety of diseases.

As related by Barnes, the application of ir thermography has been promising in the early detection of breast cancers. It was particularly pleasing to the author to learn on a recent trip to Scandinavia that the first mobile van is now being equipped there with a combination of high-speed ir thermography and x ray mammography equipment. The van's purpose is to provide for regular screening and, hopefully, for early detection and treatment of breast tumors in a manner similar to the screening for tuberculosis by chest x rays that has become commonplace in this country.

While in the final phases of preparing this article the author had an opportunity to check on the effectiveness of medical thermography on himself. He contracted a rather painful case of bursitis of the upper right arm. Before proceeding to get medical attention, a color thermogram, Figure 3, was taken of the author's upper right arm and shoulder, which showed an increase of 5°C of the biceps muscle of the right arm as compared with the same area of the left arm. This thermogram helped the attending physician to identify the location of the bursitis. [See color insert for Figs. 3 and 4.]

The thermography equipment mentioned so far was developed starting in 1954 utilizing thermistor type ir bolometers that had become available as a result of military development work carried on during and after World War 2 (Brattain and Becker, 1946).

Since that time, a variety of other uncooled and cooled ir detectors of higher sensitivity and greater speed have become available.

A new type of uncooled long-wavelength ir detector, the pyroelectric detector, has recently been developed (Astheimer and Beerman, 1966). This detector has both a higher-frequency response and less low-frequency noise. This has led to the development of a simpler and faster ir scanning camera, designated the T–6 thermograph. This scanner provides themograms in 30 seconds covering a field of view of 10° × 10° with 10,000 resolution elements subtending an angle of 0.1° each, and having a minimum resolvable temperature difference of 0.2°C.

As a result of military research and development in the United States and abroad, a number of cryogenically cooled photoconductive detectors have recently become available. The development and characteristics of these detectors are covered in a number of publications (Wolfe, 1966).

Amongst these cooled photoconductive detectors, InSb at liquid nitrogen temperature provides submicrosecond response approaching the theoretical detection limit to 6 μ, while helium-cooled mercury-doped germanium extends the response out to 15 μ.

Two types of high-speed ir cameras[3] utilizing liquid nitrogen-cooled InSb detectors have become commercially available from Sweden. It is interesting to note that the early development work on both of these high-speed ir cameras was undertaken at the Swedish Defense Research Institute (FOA).

One ir camera provides a scan rate of 16 frames/sec utilizing an eight-sided silicon optical scanning tube in the converging optical beam. This type of device, while highly efficient in its optical scanning performance, limits the deflection of the optical beam to a relatively small scan angle of ±2.5°. Thus, an ir picture covering only 5° × 5°, with angular resolution of 1 mrad and containing 10,000 picture elements, is presented on a cathode ray screen at a repetition rate of 16 frames/sec. The minimum detectable temperature difference is 0.2°C.

The other ir camera utilizes a six-sided mirror drum for horizontal scanning in the nearly parallel beam of radiation from the focusing mirror to the object being scanned.

This provides for a considerably wider optical scan angle. In this high-speed camera, an angle of 25° in the horizontal direction by 12.5° in the vertical direction is scanned with an angular resolution of 0.1° and a repetition rate of 4 frames/sec. Temperature resolution of 0.2°C is obtained. Presentation is on a medium-persistence cathode ray screen.

An ir camera in which three wavelength regions, one in the visible and two in the ir, are translated into three visible colors is now available. This camera also utilizes a six-sided mirror drum in the near parallel beam for high-speed scanning, this time in the vertical direction, with the other scan motion provided by tilting this mirror drum.

[2] The thermographs T–4, M–1A, and T–6 are products of Barnes Engineering Co., Stamford, Conn.

[3] Products of AGA AB, Roslags-Nasby and AB Bofors, Bofors, Sweden.

On a recent visit to England, the author saw a prototype of a high-speed ir camera developed at the Royal Radar Establishment utilizing a multi-element cooled InSb detector. This camera employed an eight-sided mirror drum for horizontal high-speed scanning in the near parallel optical beam and had a picture repetition rate of 30 frames/sec.

A somewhat different method of high speed ir scanning has been developed by Kutzscher and Zimmerman for nondestructive testing examination of aerospace structures. This scanner resolves 3600 picture elements in ⅓ second, and has a minimum detectable temperature of 0.2°C.

A wider-angle ir line scanner for airborne mapping of natural resources has been developed. This ir line scanner, as well as similar instruments recently declassified, are finding important uses in airborne surveying of natural resources. To back up airborne surveys on plants and soils, as well as manmade materials, ground truth measurements are being made, using thermography (Colwell, 1968).

All of the ir image translation devices (except the multicolor device described in Nichols and Lamar) were originally developed utilizing black and white image presentations, resulting in black-white photographs of ir scenes.

The resolution capability of the photographic process is excellent, and photographic film generally exceeds the capability of the eye in this respect. The dynamic range of energy response of the photographic process, however, is rather limited and the number of gray tones distinguishable by the unaided human eye in a black-to-white photograph is even more limited to about ten shades of gray. A number of the developers of ir cameras have realized during the last year that advantages could be gained in presenting the wide range of temperatures recorded in each thermogram in color, where different hues represent different levels of temperature. This is advantageous, since the unaided human observer can more easily distinguish hues of color as compared with shades of gray, and can thus more easily distinguish small temperature differentials recorded in color thermograms.

An example of the capability of thermography in color is shown in Figure 4. R. Bowling Barnes and the author are seated on two of three adjacent chairs. The warm areas of the exposed skin are clearly shown in red in the upper half picture. The range of temperature from blue to red in this thermogram is 10°C. The thermogram on the lower half of Figure 4 , which clearly shows the thermal imprints left on the two chairs, was taken 15 minutes after both persons left the chairs. In the lower thermogram, the range of temperature from blue to red is 2.5°C. Additional thermograms of the empty chairs were taken at intervals for a period of 1 hour, and distinct thermal imprints could be detected throughout this period.

Conclusions

I have confined my remarks in this introductory article to a discussion of the various ir image translation methods and image translation scanners that are presented in the September, 1968, issue of *Applied Optics* on infrared techniques for nondestructive testing. This is based on my close personal involvement over the past 20 years in much of the research, development, and application work involved in developing thermistor-type ir detectors, as well as the thermograph-type image translation scanners, and in helping to apply these to a truly amazing variety of problems.

While the work of the past has led to a very wide variety of most useful applications, I am convinced that future applications of the techniques and instruments are only limited by the imaginations of future investigators.

This paper is not meant to constitute a complete review of all ir image translation devices. In addition to those discussed here, several other scanners have recently been developed in the U.S., in England, and in France. My introductory remarks are confined to papers on ir image translation devices that scan or make visible two-dimensional surface temperature patterns.

A great deal of important nondestructive testing can, however, be accomplished with ir radiometric instruments that measure the temperature or ir radiation of a single point or a limited area.

REFERENCES

Astheimer, R. W., and Beerman, H. P. *Proc. IRIS* 10:13, 1966.

Astheimer, R. W., and Wormser, E. M. *J. Opt. Soc. Amer.* 49:179, 1959.

Bobo, S. N. *Proc. Soc. NDT,* Feb., 1965.

Brattain, W. H., and Becker, J. A. *J. Opt. Soc. Amer.* 36:343, 1946.

Colwell, R. N., *Sci. Amer.* 218:54, 1968.

Czerny, M. *Z. Phys.* 53:1, 1929.

De Waard, R., and Wormser, E. M. *Proc. IEEE* 47:1508, 1959.

Hardy, A. C., and Perrin, F. H. *The Principles of Optics.* New York: McGraw-Hill, 1932. Chap. 10.

Herschel, J. F. *Phil. Trans. Roy. Soc. London* 131:52, 1840.

Lawson, R. N. *Canad. Service Med. J.* 13:517, 1957.

Ovrebo, P. J., Sawyer, R. R., Ostergren, R. H., Powell, R. W., and Woodcock, E. L. *Proc. IEEE* 47:1643, 1959.

Wiseman, R. S., and Klein, M. W. *Proc. IEEE* 47:1604, 1959.

Wolfe, W. L. *Handbook of Military Infrared Technology.* Washington, D.C.: Government Printing Office, 1966.

Wormser, E. M. *J. Opt. Soc. Amer.* 43:15, 1953.

When technological breakthroughs were made on new sensing devices in the 1950s and 1960s, some enthusiasts proclaimed the older techniques associated with conventional air photography were obsolete. Indeed, some suggested that the new sensors might replace the camera. As the field developed, careful studies into the capabilities and limitations of remote sensors indicated that for certain purposes the aerial camera system is still preeminent. The camera is the best sensing device we have for looking in the visible portion of the electromagnetic spectrum. The new sensors have added new identification signatures for both natural and cultural features in spectral areas beyond the cameras' capabilities. It is obvious now that these new sensors will supplement and enhance, but not replace aerial cameras. We have not yet developed a sensor that has all the requirements for an optimal device in all parts of the electromagnetic spectrum. In the final analysis it seems that remote sensing is accomplished best by a system of devices, each contributing its special information for interpretation and analysis.

6-Capabilities and Limitations of Remote Sensors

EARL S. LEONARDO

Since World War II the term "remote reconnaissance" has involved more than better cameras, sharper lenses, or faster films. No longer are image analysts restricted to aerial photography alone. Instead, their analyses span the electromagnetic spectrum by means of infrared, radar, and other even more sophisticated systems. They extract data from sections of the electromagnetic spectrum several million times wider than that available to conventional camera systems.

Of what value is this additional coverage? Can it really provide information not obtainable with good aerial cameras? How do images generated by infrared and side-looking radar systems compare with conventional photography? Will they, as some claim, make aerial photography obsolete?

The added coverage has revealed new identification signatures for both natural and cultural features in spectral areas beyond cameras' capabilities. Under certain conditions, infrared and radar systems can produce imagery of specific subjects of nearly photographic quality. But they will only supplement, not replace aerial cameras—at least, not for a few years. A comprehensive system still needs several sensors to satisfy the needs of military reconnaissance or commercial exploration.

To realize the capabilities and limitations of these sensors, one must first understand the electromagnetic spectrum.

. . . .

From *Photogrammetric Engineering* 30:1005–1010, 1964. Reprinted and edited with permission of the author and the American Society of Photogrammetry.

Each sensor reacts only to energy bands of specific frequency and wavelength; radar receivers cannot detect visible light; transmitted microwaves are invisible to infrared scanners.

Figure 1 shows how airborne remote sensors are either passive or active. Passive systems (some infrared devices, cameras) detect radiations that would be present whether or not the sensor was operating.

Active sensors, like radar, record echoes of reflected electromagnetic energy which they themselves transmit.

Aerial cameras produce their best imagery on cloudless, hazefree days, but with new techniques and equipment they do obtain reasonably good imagery on clear nights. Figure 2 shows that radar and infrared systems can overcome these limitations.

Infrared systems also produce good daytime imagery. However, since they respond to energy radiated from beyond the visible spectrum, night infrared missions with middle and far infrared sensitivity yield excellent results. For most purposes, far infrared data flights obtain their best imagery after dark when there is no interference from solar insolation. Military needs for nighttime operations are obvious.

Infrared radiation may penetrate dust and haze, depending on the size of the aerosol particles, but clouds, high surface winds, and rain greatly reduce image quality.

Radar, an active sensor, provides its own source of energy. Therefore, it too is independent of time-of-day. Its longer wavelengths penetrate fog, haze, and clouds with minimum signal loss. Rains attenuate the signal, but the extent depends on system wavelength and rain-

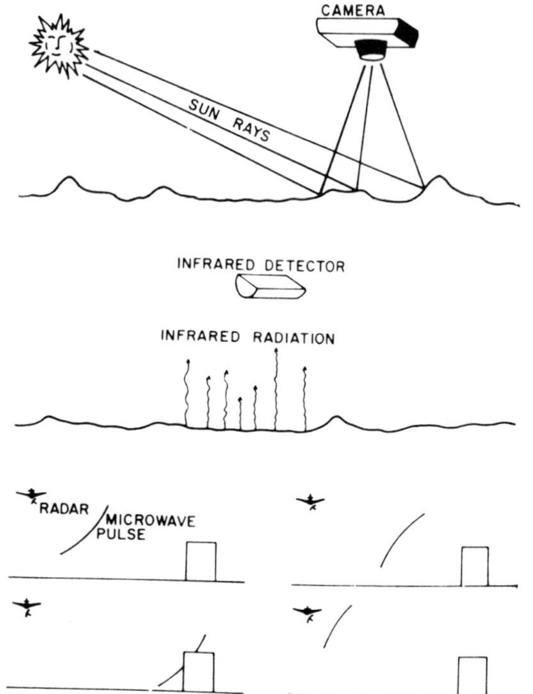

Figure 1 Difference between passive and active sensors.

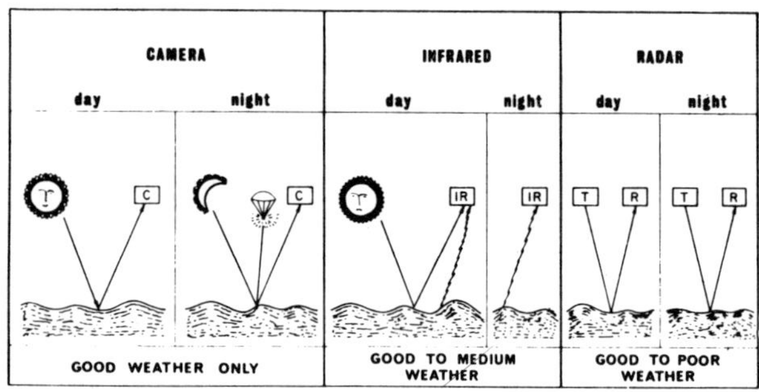

Figure 2 Time-weather capabilities of remote sensors.

fall rate. Thick, moisture-laden clouds, however, can effectively block transmitted waves. To what extent these factors affect radar imagery depends on several system parameters.

All three systems can be "tuned" to be more selective to specific frequencies within their operational bands. Narrow-band film-filter combinations enable cameras to record spectral responses of one color. Filters are often added to infrared systems to eliminate effect of solar reflection below the middle or far infrared range, depending on the system.

Figure 3 shows infrared energy transmission through a standard atmosphere and spectral sensitivity of a mercury-doped germanium detector. The atmosphere's composition allows relatively undisturbed transmission of infrared radiations in the 2–5 μ and 7–15 μ ranges. However, the detector, a standard in the industry, transmits from the edge of the visible spectrum to about 15 μ without the interruption between 5 and 7 μ. During daylight hours, therefore, it transmits some solar energy.

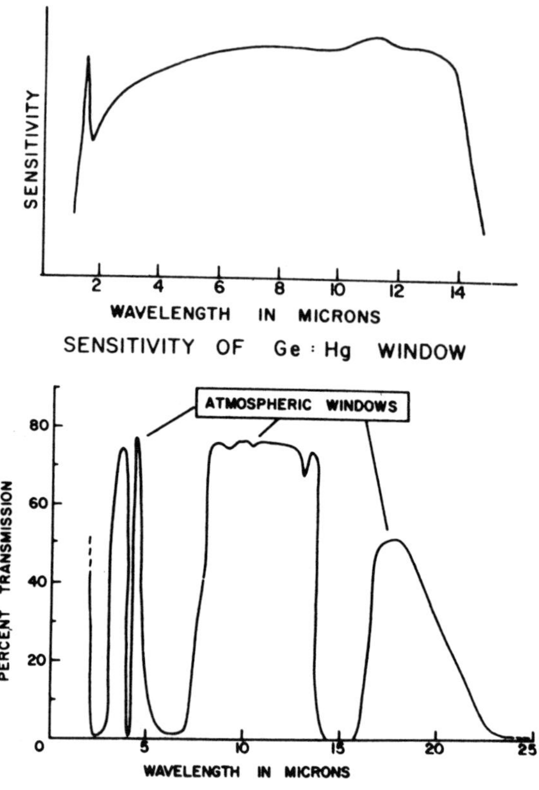

Figure 3 Spectral detectivity of mercury-doped germanium infrared detector compared to infrared energy transmission.

Despite excellence of nighttime infrared imagery, some mineral surveys are best conducted during the day because of the rate of absorption and transmission of infrared radiation. However, the infrared system should include filters sensitive to infrared radiations only, to effectively eliminate reflections from the rest of the spectrum.

Radar system resolution and sensitivity are functions of pulse repetition rate, polarization, power, wavelength, and gain setting. For example, weather radars (which require only moderate resolution) use long-wavelength systems; terrain reconnaissance and mapping radars use much shorter wavelengths (comparatively) to record natural and cultural features with almost photographic clarity.

System components (such as cathode ray tube and film) cannot record with equal discrimination all signal levels received at the antenna. Whether the film records maximum differences between high or low intensity signals depends on the settings according to mission's main purpose. In Figure 4, if most of the available cathode ray tube and film density range is used to record differences between low-level signals, strong intensity returns are compressed into a small density range. Much of the signal is, in reality, "clipped" and recorded at one density level. Conversely, setting the gain to discriminate between strong signals loses low-level separation. Farmland, orchards, and open country are typical low-level returns. Urban and industrial areas give strong returns.

An object's surface smoothness and orientation affects its radar image. Surfaces smoother, i.e., with irregularities smaller than the wavelength of the impinging electromagnetic energy, will reflect most energy specularly, or mirror-like, while rough surfaces tend to scatter reflections. The smooth surface of Figure 5 (upper) acts like a mirror: the angle of reflection equals the angle of incidence. In this case, impinging energy is reflected away from the transmitter-receiver. The multifaceted rough surfaces of Figure 5 (lower) scatter energy unequally (diffusely) in all directions. Some energy eventually returns to and is recorded by the receiver.

Because visible spectrum wavelengths are so short, most surfaces reflect light diffusely regardless of orientation. Longer microwaves create more specular reflections off the same surfaces. For example, aim a flashlight at a wall, first perpendicularly, then at an angle. One sees the wall equally as well regardless of the illuminating angle. A radar system illuminating a similar wall shows much stronger returns for head-on orientations than oblique.

On aerial photographs, images of bodies of water frequently vary in tone. Wide density ranges often occur on the same negative. Many times, these density changes vary with respect to water depth; other times, to the sun angle.

With radar, water's smooth surface reflects most transmitted microwave energy specularly; it gives "no-return" image, unless the transmitter is perpendicular to the water's surface. Then most energy will be reflected back to the antenna.

Radar can record some of water's phenomena, however. Surfaces broken by breakers, waves, or submerged rocks are often detectable. Contrast between normal no-return images and slight returns off broken water usually are sufficient to assure surface detection of submerged features.

Infrared systems can extract considerable information from bodies of water. Hot effluents discharged into streams or lakes are readily detectable because temperature differences between them yield varying density ranges on the film. Densitometric analyses show how far and in what direction effluents travel before the stream or lake absorbs them. Chemical wastes or other polluted discharges should be as readily detectable at all hours because of the temperature anomalies—and therefore density changes—they create.

No single sensor possesses all requirements for an optimum device. But neither does any sensor have so many disadvantages that it is valueless. Table 1 summarizes the advantages and disadvantages of remote sensors. Cameras possess the best resolution and give geometrically accurate reproductions; infrared systems record minute temperature differences; side-looking airborne radars operate independently of the clock in almost all weather and maintain constant image quality (up to 3rd order planimetric accuracy) over extremely long ranges.

But some remote sensor problems require the interpreter to be very cautious. One of infrared's most confusing situations occurs when an object's temperature is the same as its background. Then the two cannot be separated. Properly positioned metal corner reflectors

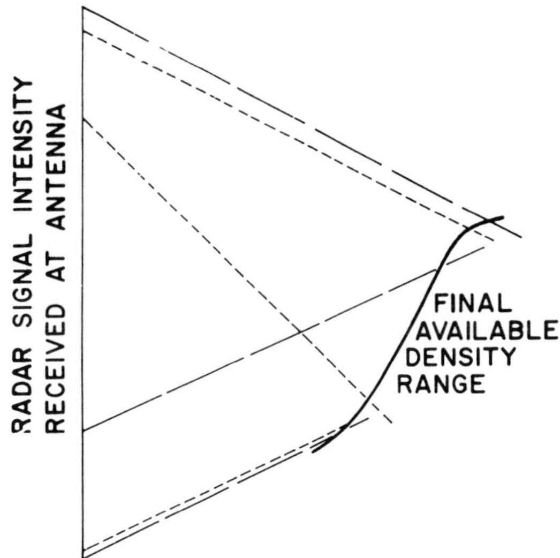

Figure 4 Effects of gain setting on negative density.

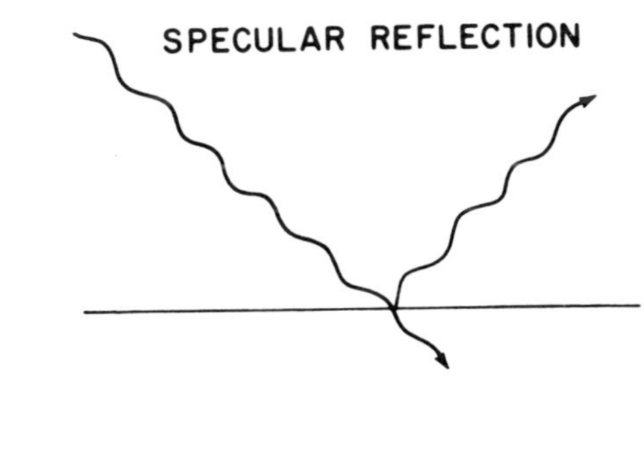

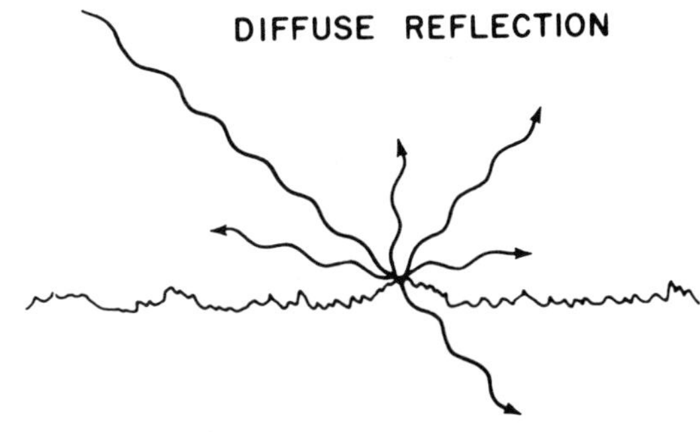

Figure 5 Specular vs. diffuse (or scattered) reflections.

Table 1 Remote Sensor Comparison

Parameter	Camera	Infrared	Radar
Day/night	5	10	10
Haze-fog penetration	3	7	10
Cloud penetration	1	2	9
Temperature discrimination	2	10	1
Subsurface detection	4	6	3
Stereo capability	10	6	6
Geometric fidelity	9	7	8
Long-range capability	7	4	9
Ground resolution	9	7	5
Interpretability of imagery	9	6	6
Availability of equipment	10	6	6

Poor = 0; good = 10.

Figure 6 Photo index, Arbuckle Mountains, Oklahoma.

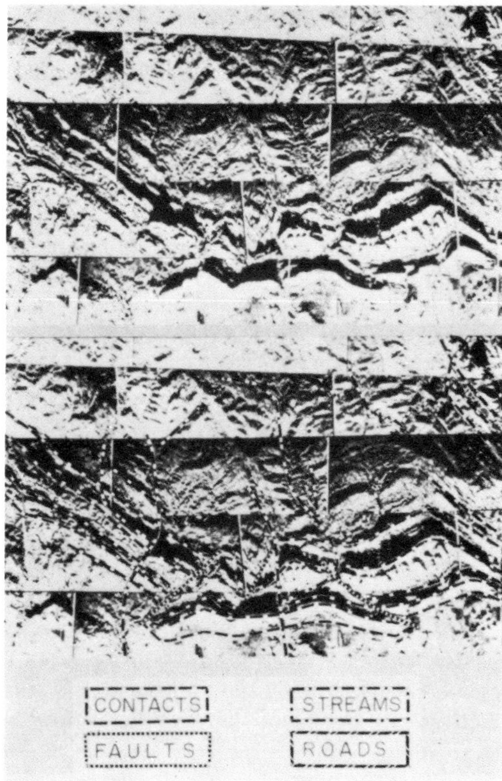

Figure 7 Radar mosaics, Arbuckle Mountains, annotated and unannotated.

Figure 8 Infrared image of Dallas, Texas.

produce radar reflections as strong as large industrial complexes. However, to hide a large object—like a factory—from radar detection presents more serious problems.

Figure 6 is a photo index mosaic of the Arbuckle Mountains, Oklahoma. Detail is sharp, but at this scale it shows only relatively gross objects. (Of course, stereo analysis of the contact prints will reveal much data.) Figure 7 shows unclassified duplicate radar images of the Arbuckles collected by what is now a primitive radar system. Many features of geologic interest (annotated on the lower photo) are readily apparent. Most important, the radar image could have been taken day or night in virtually any weather that allowed the aircraft to fly. This entire geologic mapping program took less than 6 hours.

Figure 8 is an unclassified high-altitude nighttime infrared image of Dallas, Texas. One can see a complete road pattern, stream, open areas, parks, and many individual buildings across the entire format. Many of the streams and roads are below tree canopies that would preclude their being seen in the visual spectrum.

Optimum systems, then, consist of more than one sensor, each contributing its special information.

To exploit the advantages of multisensor reconnaissance, Texas Instruments converted a B-25 twin-engine aircraft into a multisensor platform. It was equipped with a side-looking radar, a far-infrared scanner, two aerial cameras, and two hypersensitive radiometers. With this aircraft they have acquired, under government and commercial contracts, multisensor imagery and performed interpretations of various cultural and natural features in ten states and one foreign country and hope to outfit a much larger aircraft with an even more sophisticated, integrated multisensor system. This system will have four radar systems (side-looking, 360° scan, vertical incidence, and terrain avoidance), a far-infrared detector, several aerial cameras, three radiometers, and a magnetometer array. The vertical incidence radar system is based upon Texas Instruments terrain analysis research performed for the Army Engineers at Vicksburg. With these sensors they will be able to study many decades of the electromagnetic spectrum and broaden its application to remote reconnaissance or exploration problems.

In 1972 the total world population was estimated at 3.63 billion people, and the world rate of annual population increase at 2.0 percent. That means about 73 million new people are added to the world's population each year—truly a frightening statistic. A few other facts make it even more frightening. Just 3 to 4 years from now the world's population will pass the 4 billion mark. It will double (7.2 billion) in 35 to 40 years. The total area of the earth's surface is about 197 million square miles. However, over 70 percent is water; approximately 57.3 million square miles (29 percent) is land. Equating the present total world population to total land surface indicates a population density of about 64 per square mile. Because land resources are static and because population is rapidly increasing, we can expect the world population density to double in 35 to 40 years.

These new millions will need food, clothing, housing. Generally, around the world, there is a rising expectation in the demands for a higher standard of living. Not only will there be more people, but these people will demand more material goods. The developing nations will push toward industrialization and the developed nations will be forced to increase production to help meet the rising demands.

Where will we find the minerals and other resources needed to meet this increased production? Older industrial countries such as the United States and those in Western Europe already have exploited the richest, most accessible resource deposits. In many cases these countries are becoming more dependent upon imported resources. In the future, developing nations will be less willing to export resources that could be used in their own industrial growth.

The sea offers one area for possible exploration of new mineral deposits, but at the present level of technology we cannot venture far beyond the waters of the continental margins for mineral extraction.

At the present time a more fruitful area of exploration is to reexamine the land surface of the earth using new remote sensing techniques that will allow rapid surveys of large areas and detection of deeply buried or hidden mineral deposits. It is in the dynamic field of mineral prospecting that remote sensing may make one of its most important contributions.

7-Remote Sensing Techniques for Mineral Discovery

A. R. BARRINGER

The term "remote sensing" is used here to describe detection and mapping techniques which can be flown from aircraft or spacecraft. Thus, this definition includes measurements made from helicopters flying 100 feet above the ground and orbiting spacecraft operating at a height of 100 miles. The term is usually taken to include aerial photography, and therefore remote sensing can be said to date back over 100 years to the time of the American Civil War when General George B. McClellan used balloon photography for intelligence purposes. Since that time, military interest in the development of remote-sensing techniques has continued to be a major one, and most of the systems today, with a few notable exceptions, have been greatly influenced by military research.

From *Mining and Petroleum Geology*. Jones, M. J. (ed.), *Proceedings of the 9th Commonwealth Mining and Metallurgical Congress, 1969*, Vol. 2. London: The Institution of Mining and Metallurgy, 1970. Pp. 649–690. Reprinted with permission of the author and The Institution of Mining and Metallurgy.

The impact of new technology in general, and remote sensing methods in particular, on mineral exploration during the past 15 years has resulted in an almost total swing from the old methods of prospecting on foot to new approaches which draw heavily on the fields of applied physics and chemistry as well as basic geologic

Table 1 Remote Sensing Techniques Applicable to Mineral Exploration

MAGNETIC

(1) Standard total field aeromagnetic survey (±1 gamma, fluxgate, proton)
(2) High-resolution total field aeromagnetic survey (±1/10 gamma effective, optical pumping)
(3) Vertical field (±25 gammas, gyro, stabilized fluxgate)
(4) Vertical gradient (optical pumping)

ELECTROMAGNETIC (NONOPTICAL)

Imaging	**Nonimaging**
Side-look airborne radar (SLAR) (1) X-band (2) K-band	*Inductive field continuous wave, audio frequency* Fixed-wing, rigid Tx-Rx systems Fixed-wing, non-rigid Tx-Rx (bird) systems Helicopter Rigid Tx-Rx towed-bird systems Rigid Tx-Rx aircraft-mounted system Servo-alignment systems *Audio-frequency pulse system* Fixed-wing induced pulse transient system *Radiated field* VLF radio-frequency systems Audio-frequency natural field systems

OPTICAL

Imaging	**Nonimaging**
Photographic (3000–9000Å) Ortho, pan, color, multiband Black and white infrared, color infrared	*(a)* Gas and vapor detection by spectral correlation *(b)* Fluorescence by Fraunhofer line discrimination *(c)* Rock identification by far infrared spectral correlation

Video

High-resolution TV (return beam vidicon)
Multiband TV (WISP)

Mechanical line scanning

Principally infrared, but also ultraviolet and VIS

GAMMA RAY

(1) Total gamma ray count rate
(2) Simple gamma ray spectroscopy
(3) Advanced gamma ray spectroscopy

AIR SAMPLING METHODS

Mercury vapor, iodine, SO_2, aerosols

knowledge. Mineral prospecting remains a dynamic field, and the development of new sensors and improved techniques continues on a steep upward curve. Present-day usage is mainly confined to the combination of aerial photography, aeromagnetic mapping, and electromagnetometer surveying, gamma ray detection methods being included in the case of prospecting for radioactive ores. In the future we can look forward to the integration of orbiting satellite imaging methods to supplement existing airborne surveys and the airborne surveys themselves will be carried out with a much broader range of mapping and detection systems, which are currently emerging from the laboratory.

The remote sensing systems which are applicable to mineral exploration and which are either in use today or are well advanced in development are summarized in Table 1. They have been grouped under the headings Magnetic, Electromagnetic (nonoptical), Optical, Gamma ray, and Air sampling methods; where applicable, these headings have been subdivided into imaging and nonimaging techniques. It is proposed to discuss salient features of these systems, to give some idea of their areas of applicability, and to forecast development trends. The subject matter will therefore range through a review of published work to a brief description of new unpublished research still in progress. A summary of specific remote sensor applications in mineral exploration is given in Table 2.

Airborne Magnetic Systems

Airborne magnetometers have already been well covered in the literature and the list of references given in this paper represents ample coverage of the subject. Most aeromagnetic surveys are carried out with systems which measure the total magnetic field with an accuracy of ±1 gamma. In the past the main equipment used has been the fluxgate airborne magnetometer, perfected during the second world war for submarine detection purposes and first tested for geologic applications by H. E. Hawkes of the U.S. Geological Survey in 1943. Gulf Research and Development made the equipment available for commercial aeromagnetic surveys, and large-scale use commenced shortly after the war (Wyckoff, 1948). Packard and Varian announced in 1954 the first prototype nuclear precession magnetometer, and attempts were made to introduce this for airborne surveys. For a long time, however, fluxgate equipment dominated the field, due to the preference of oil exploration geophysicists for the smooth profile curve obtained on the fluxgate charts compared to the stepped output of the proton magnetometer. Early acceptance of the proton precession magnetometer was also hindered by instrumental problems associated with relay switching in the "polarize head." The attitude towards the proton precession magnetometer has recently changed, owing to the greatly improved performance associated with solid state circuitry and the development of electronic conversion methods which provide a direct reading of the absolute field in gammas, as well as the adoption by the oil exploration industry of digital processing techniques.

Table 2 Specific Remote Sensor Applications in Mineral Exploration

Target	Applicable Sensor or Techniques
Identification of major structures	Aerial and satellite black and white and color photography Infrared line-scan imagery Side-look radar Radiated field systems (VLF and audio frequency) Natural fields Aeromagnetic mapping
Classification of rock types and identification of hydrothermal alteration	Aerial and satellite color photography Vapor detection by remote sensing (I_2) Air sampling (Hg) Aeromagnetic mapping VLF radio-frequency mapping Far infrared spectral correlation Gamma ray spectroscopy
Identification of geobotanical criteria associated with specific rock types, hydrothermal alteration and geochemical anomalies	Aerial and satellite color photography Color infrared photography (camouflage detection film)
Direct detection of ore or weathered mantle over ore	Audio-frequency inductive field EM systems Radiated field audio- and radio-frequency EM systems Aerial and satellite color photography Airborne magnetometer (magnetite, pyrrhotite) Gamma ray detection (radioactive ores) Air sampling (e.g., Hg vapor over precious and base metal deposits)

The drift-free and inherently digital output of the proton magnetometer has become very attractive, especially when it is used in conjunction with computerized data handling. Furthermore, the use of integrated circuits in proton magnetometers has allowed their size to be reduced to the point at which they can be operated very economically in the smallest of aircraft. A typical modern lightweight airborne magnetometer is shown in Figure 1.[1]

The current trend to digital processing is perhaps the most significant development in aeromagnetic mapping. The introduction of digital recording in flight allows not only total field contoured maps to be prepared at great speed but also other types of map of considerable potential value. Thus two-dimensional digital filtering can be used to suppress near-surface effects if these are troublesome, or, conversely, they can be accentuated to assist in surface mapping. Two-dimensional filtering can also be used to accentuate linear features such as faulting, and techniques of digital Fourier analysis can be applied to determine depth to basement. Once airborne digital recording is adopted for automatic preparation of contoured maps, little extra cost is involved in producing stereo pairs of these maps so that they can be viewed stereoscopically by the interpreter. This method shows some promise as an aid to geologic interpretation of aeromagnetics (Gay, 1968).

Turning to high-resolution areomagnetic mapping methods, the oil industry is starting to use such methods for the determination of subtle structural features in sedimentary basins, where rock susceptibilities are low. At the present time, such surveys have nearly always been carried out with optically pumped magnetometers, which have noise levels in the vicinity of 1/100 of a gamma. Effective precision, however, is of the order of $\pm 1/10$ of a gamma due to practical problems of navigation, maintenance of altitude, etc. Whether this type of precision will warrant the additional acquisition cost is highly doubtful from the mineral exploration standpoint, but the application of high-resolution sensors to vertical gradient measurements may prove significant in mining work. For some applications of detailed geologic mapping a vertical gradient aeromagnetic map has the advantage of defining contacts more precisely—and providing a map which is easier for the geologist to relate to the ground surface. Even in this area, however, airborne digital recording and the computerized production of filtered maps may prove to be a better solution.

Airborne magnetic surveys have been carried out with vertical fluxgate sensors mounted on gyroscopes (Lundberg, 1947), but accuracy is limited to the order of ± 25 gammas due to the sensitivity of the fluxgate to misalignment when it is not operated in line with the earth's field. The use of vertical field measurements has some interpretative advantages in equatorial regions where the dip is flat, but the scope of this magnetometer is severely limited by its noise level.

Aeromagnetic surveys remain an absolute prime tool in mineral exploration. Originally thought of purely in terms of prospecting for magnetic ores, today their geological significance has become fully appreciated. In Precambrian Shield areas, for example, they help to delineate greenstone belts, the pattern of distribution of intrusive bodies, major faults, and many other geologic features which often cannot be adequately covered by the use of conventional mapping methods. The value of this type of geologic information to exploration geologists is immense, and experience in Canada has shown that the publication of regional aeromagnetic maps has time and again proved to be the precursor of a massive exploration effort by the mining industry in the area in question.

The full value of aeromagnetic maps can only be achieved when large areas are covered and localized

1 Not reproduced here.

geology can be brought into perspective with the broad regional patterns which appear on large-scale aeromagnetic compilations. The picture that has been unfolding of the crustal structures in the Canadian Precambrian Shield as the Canadian aeromagnetic surveys have been published has been dramatic (Kornick and McLaren, 1966), and it is clear that there is a great requirement for similar large-scale surveys to be carried out in all countries interested in developing their mineral and oil resources. It is noteworthy that the impact of magnetic surveying has also been very considerable in offshore areas. The publication by Raff and Mason (1961) of the remarkable striped pattern of the ocean-floor magnetics off the Pacific coast of America has been followed by worldwide magnetic studies and an integration of magnetic, bathymetric, and seismic data into a theory of ocean-floor spreading and continental drift which is starting to convince many former opponents of continental drift. The possibility of further integration of these data (in terms of broad crustal tectonics) into comprehensive theories of ore genesis is very intriguing and could be far from academic in its potential importance in providing leads to the exploration geologist.

Electromagnetic Survey Systems

Under this heading are included both imaging and nonimaging techniques, ranging from microwave to low audio frequencies. Optical and gamma ray methods, both of which are electromagnetic (EM), are given under separate headings, because they belong to branches of physics which are normally dealt with individually.

Nonimaging Systems

The low-frequency inductive field airborne EM systems are the techniques which, so far, have achieved the most spectacular success in the application of remote sensing methods to mineral exploration (Pemberton, 1962; Ward, 1967). These methods are unusual among remote sensing techniques in that they have received almost no attention from the military; the initial reduction to practice of an operational airborne system was, in fact, carried out by International Nickel Company of Canada in 1952 (Cartier et al., 1952). Following the early lead by INCO, the subsequent years saw the introduction of a whole range of airborne EM systems, all of which depended on creating a low-frequency inductive field in the vicinity of the aircraft, inducing eddy currents in conductive orebodies, and detecting the secondary fields reradiated by the conductive bodies.

An excellent review of most of the airborne electromagnetic systems in use up until 1963 was given by Pemberton (1962), and since that time there have been only a limited number of new developments with regard to airborne systems which employ a locally generated field. Perhaps the most notable success with airborne EM was the discovery of the Texas Gulf Sulphur deposit at Timmins, Ontario, in which a major copper–silver–zinc orebody of reputed value in excess of two thousand million dollars was located 40 feet beneath the surface. The equipment, manufactured by Varian Associates, featured a rigid boom, 60 feet long, attached to a helicopter. The transmitting coil was in the front and the receiving coil in the rear, and compensation techniques were built into the boom in order to achieve constant separation as the boom bent under varying *g* loads. The system worked well, achieving noise levels in the vicinity of 12 ppm, but was difficult to fly. In the year following the discovery, the pilot and the company geologist, Mr. Peter White, were killed by an accident while on survey. Further systems of this type have not been constructed.

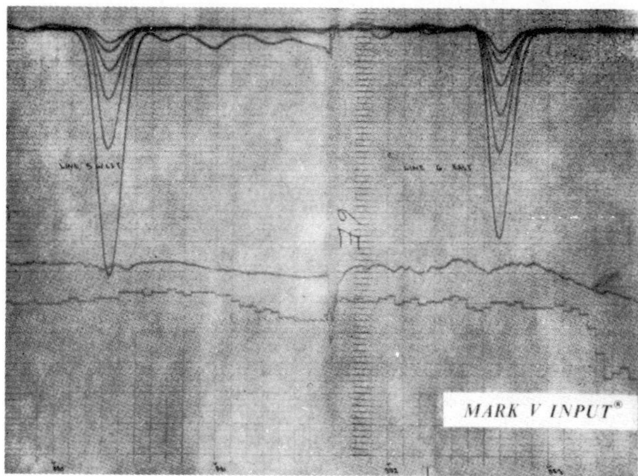

Figure 2 INPUT profile over West MacDonald orebody.

Another recent discovery of topical interest is that of a copper–silver–zinc deposit in the Uchi Lake area of northwestern Ontario. This survey was carried out with the induced pulse transient (INPUT) system by Questor Surveys, operating on behalf of Selco Exploration, the Canadian subsidiary of Selection Trust, London. The initial drill-holes at the time of writing have indicated high-grade copper-, zinc-, and silver-rich ore of relatively short strike length, which makes the body a relatively poor airborne target. As in the case of the Texas Gulf Sulphur discovery in the Timmins area, there was no airborne direct magnetic anomaly associated with the ore, which once again stresses the dangers of the use of coincident aeromagnetic and electromagnetic anomalies as key criteria in selecting drilling targets. Further targets in the immediate vicinity of the initial conductor remain to be tested at the time of writing. A typical profile with the INPUT system over the West MacDonald orebody of northern Quebec is shown in Figure 2.

The main areas of endeavor which appear important at the present time in airborne electromagnetic inductive field systems are related to increasing sensitivity and improving interpretation methods. The key problem in the use of airborne electromagnetic systems in many areas is the separation of signals generated by eddy currents in conductive overburden from fields generated by conductive orebodies. This becomes particularly difficult in regions where the surface conductivity is exceptionally high, owing to climatic conditions. In semi-arid regions, where rainfall is less than about 14 inches per year and where the distribution of rainfall is erratic, development of salt pans and saline washes is common. Furthermore, the soil profile in these regions frequently develops a hard pan, or calcrete horizon, which is rich in calcium and other salts and is highly conductive in the presence of the small amount of moisture normally trapped in the soil even in desert conditions.

One approach to enable the discrimination between conductive overburden and ore deposits is to use a multiple component system, which measures at least two components of the field (and preferably three), and to compute the dip angle of the circulating eddy current induced by the transmitter. In this way it is possible to identify anomalies caused by edge effects over flat sheets of conductivity and to separate them from anomalies associated with steeply dipping bodies. It is apparent that a multiple-component EM system used in conjunction with suitable data processing and computing methods and computer facilities should be capable of making discriminations against overburden on the basis of the three-dimensional behavior of the secondary fields (intensity and orientation).

Geophysical Engineering and Surveys, Ltd., and Barringer Research have been developing jointly a digital airborne electromagnetic system for use in helicopters, with the express purpose of measuring three components and using computer data handling procedures to achieve improved interpretation as compared to conventional single-component electromagnetic systems. The transmitting and receiving coils are mounted on a bird 30 feet long which is towed beneath a helicopter. A transmitting coil operating at 900 cycles has a horizontal axis in line with the direction of flight, and three orthogonal receiving coils are mounted at the back end of the bird. Two of the coils are in null coupling and are mounted on gymbals, which are connected to the front end of the bird through crossed wires made of invar. The technique of mounting a null-coupled coil in order to maintain precise null coupling at all times is shown in Figure 3.[2] When the ratio of the spacing of the crossed wires between the connection points on the gymbals and the connection points at the transmitting coil is correctly adjusted, the receiving coils track in null coupling with the transmitted field as the bird bends under varying g loads. With this method effective null coupling with the transmitted field is maintained within less than 2 seconds of arc while flying at 70 knots with the bird bending through many minutes of arc, Overall noise levels on all three coils are maintained below 5 ppm, and they frequently fall well below 4 ppm. Research is in progress to further reduce these noise levels and thereby increase penetration. The system is designed for use with inflight magnetic tape recording and subsequent digital processing. Back-up to the tape recording is provided by a multiple pen chart recorder (Figure 4).

Parallel developments have been proceeding with the Induced pulse transient (INPUT) system and experimental surveys have been carried out in cooperation with the Canadian Geological Survey. Part of this work, under the direction of L. Collet and A. Becker of the Survey, has been aimed at identifying the presence of aquifers in gravels in the clay environment of the Canadian Prairies, and is also relevant to some placer exploration problems. Additional studies on groundwater detection have been sponsored by the Government of Saskatchewan. A special version of the INPUT system has been used for these surveys and is described briefly. A standard INPUT survey aircraft and Mark V transmitting system is used as shown in Figure 5.[2] Half sine shaped current pulses are circulated in alternate polarity around a three-turn transmitting loop at a peak current of just under 400 Å. The repetition rate is 288 pulses/sec, and the pulses are detected by a vertical axis coil carried in a bird towed at the end of 500 feet of cable. A vertical-axis receiving coil is substituted for the usual horizontal axis for this type of work, due to the improved coupling obtained with horizontally layered deposits. The current pulses in the transmitting loop generate large pulses of electromagnetic fields, which, in turn, cause eddy currents to circulate in the conductive clays and silts of the underlying terrain. The eddy current decays after the termination of each pulse and the associated decaying secondary field is detected in the towed bird. This transient signal is sampled at 12 different delay points by the receiver electronics (Fig. 6). The integrated and smoothed outputs of each sampling channel are recorded on FM magnetic tape together with positional information and other required data. The FM magnetic tapes are subsequently converted to computer-compatible digital tapes in a data handling facility and geophysical maps are prepared based on distortion parameters in the decay transient. Over a flat earth of homogeneous conductivity, the decay transient assumes a specific shape which is approximately hyperbolic. When the ground becomes layered, the decay curve departs from this shape in a manner which is definitive according to the type of layering. Near-surface inhomogeneities tend to affect the early part of the decay curve, whereas deep layering

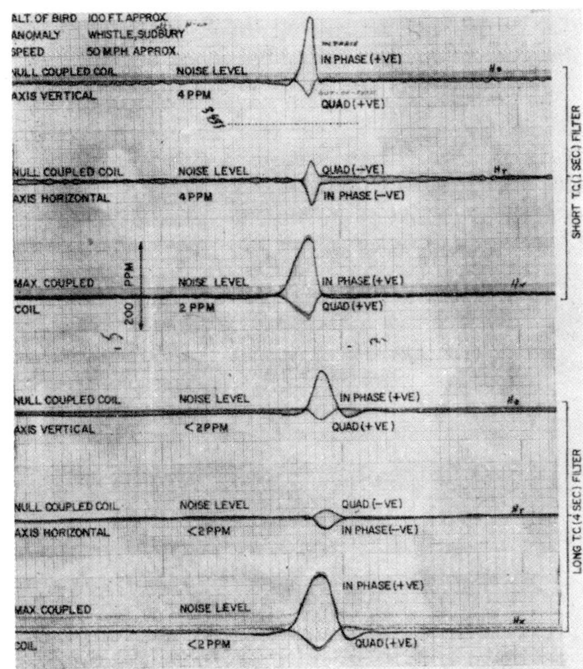

Figure 4 Typical flight record with multicomponent helicopter EM system.

[2] Not reproduced here.

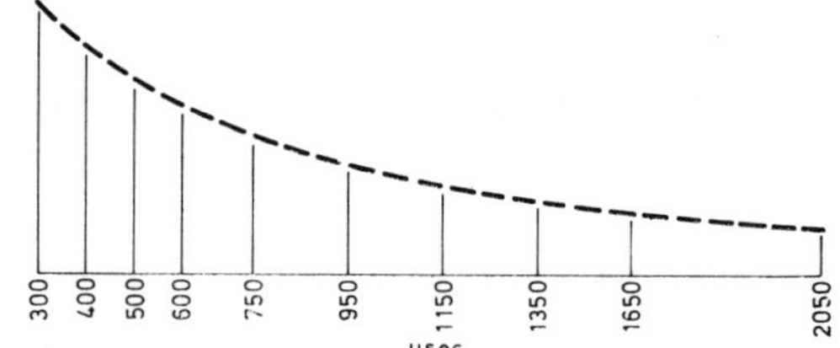

Figure 6 INPUT sampling delays on 10-channel version.

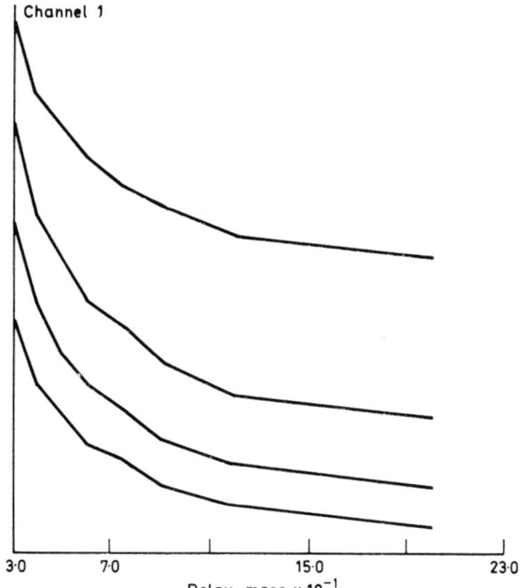

Figure 7 INPUT transient decay curves—Saskatchewan Pleistocene geology.

influences the shape of the delayed or tail portion of the decay curve (Fig. 7).

The electromagnetic systems discussed so far belong to a class in which low-frequency inductive fields are generated locally. Another class of systems relies on the use of radiated fields originating from distant sources. In a radiated field the energy is distributed equally between the electric and magnetic components, and in the far fields propagate as a ground wave and penetrate into the the transmitting source, the electric and magnetic components are in phase. At the lower frequencies these fields propagate as a ground wave and pentrate into the ground to a depth which depends on their wavelength and on the ground conductivity. During the last decade a chain of transmitting stations has been constructed by the United States Navy for communication with submarines, which operate at an exceptionally low radio frequency (in the vicinity of 20 kHz). At this wavelength signals can be reliably detected at ranges of thousands of miles, and skin depths of penetration into the earth vary between 50 and 1000 feet, according to ground conductivity (Table 3). The United States Station NAA, operating from Maine, has been detected as deep as 240 ohmmeters below ground at Kiruna in northern Sweden, more than 3000 miles from the transmitter site (Paál, 1965). In addition to the U.S. Navy other government agencies in the U.S.A., Great Britain, and the U.S.S.R. also make regular transmissions in these VLF bands. Much of the surface of the earth is therefore now covered by these signals.

Table 3 Penetration Depth (ohm-meters)

	$\epsilon = 5$	$\epsilon = 80$	$\epsilon = 15$
180 Hz (ELF)			
$\rho(\Omega\text{—m}) = 10^4$	3760	3760	3760
10^3	1187	1187	1187
10^2	376	376	376
10	119	119	119
1	37.6	37.6	37.6
18 kHz (VLF)			
$\rho(\Omega\text{—m}) = 10^4$	383	400	542
10^3	119	120	124
10^2	37.6	37.6	37.6
10	11.9	11.9	11.9
1	3.75	3.75	3.75

$\epsilon =$ dielectric constant; $\rho =$ resistivity, Ω— m.

The depth of penetration of VLF radio fields is adequate to make them of considerable interest for use in conductivity mapping and in direct prospecting for conductive ores. One of the first commercial systems for use on the ground was developed by V. Ronka in Toronto, and this equipment has demonstrated quite successfully that, used with care, it is possible to define conductive zones and trace certain types of geologic structure.

Experimental airborne equipment was placed into operation by the writer and his colleagues in September, 1967, and a continuous program of development has been going on since that time to perfect the equipment and achieve operating and interpretative experience (Barringer and McNeill, 1968). The system has been called Radiophase, since it depends on the use of the vertical component of the electric field as a phase reference against which to measure changes in the magnetic field. It can be shown that the electric component of a radiated field is far less affected by underlying terrain than the magnetic component. This is due to the fact that substantial eddy currents are induced in conductive bodies in the ground by radiated fields, but the secondary fields generated by these conductive bodies are almost entirely of the inductive-field type in which the magnetic component predominates. Thus, as terrain is traversed, fluctuations of phase and amplitude will arise in the total VLF magnetic field due to contributions from the secondary field adding to the primary field. The secondary electric fields, on the other hand, are smaller in amplitude and vary much less in phase than the magnetic field, and the resultant electric field is therefore relatively phase-stable. This fact was initially confirmed in modeling tests and has been substantially verified in field tests.

The airborne equipment carries three orthogonal receiving coils, which act as magnetic field detectors, and a vertical whip antenna to provide the electric field phase reference (Fig. 8).[3] Seven channels are recorded, which include the in-phase and out-of-phase components of the magnetic field with respect to the electric field, together with the amplitude of the vertical electric field.

[3] Not reproduced here.

These outputs are placed both on paper recording charts and magnetic tape—the first to provide a quick scan and monitoring capability and the second to facilitate automatic data processing. Data handling can be carried out manually, but the computer has numerous advantages. For example, digital filtering is employed to remove the dc component from the in-phase recordings, and the computer determines the resultant horizontal magnetic field strength derived from the outputs of the two orthogonal horizontal coils. Measurement of this quantity makes the system insensitive to changes in heading of the aircraft, and since the receiving coils are in maximum coupling, negligible effects are caused by small amounts of pitch and roll. The system has a great advantage, therefore, in that it can be flown without noticeable degradation in almost any V.F.R. weather conditions.

Computer programs have been prepared which enable not only the total intensity of horizontal fields to be mapped but also the vectors of the in-phase and out-of-phase components of the magnetic field. The full potential of these measurements has not yet been studied, but intriguing possibilities result from the ability to look at the differential flow patterns between in-phase and out-of-phase currents. Two types of current flow are associated with the VLF fields—one connected with the direct induction of eddy currents in conductive sheets and bodies, and the other associated with the horizontal current sheet which flows to and from the transmitting antenna in a conductive earth, even if this conductivity is homogeneous. The latter current flow pattern is disturbed by variations in conductivity of the earth, the current flow being channelled and bunched into more conductive zones. The secondary field patterns arising from this cause are somewhat different from the secondary field distribution associated with circulating eddy currents in conductive bodies. The two effects are inevitably superimposed upon each other so that the interpretative schemes used for conventional airborne electromagnetic systems employing locally generated fields cannot be applied to far field systems.

Data processing methods are at present being worked out to separate eddy current effects from those effects caused by inhomogeneities in the ground current sheet. If this can be achieved, it could represent a major advance in the application of the system to the direct detection of ore.

Figure 9 shows an example of survey with the Radiophase system over an area of known geology north of Noranda in Quebec. Careful study of the map in relation to geologic, aeromagnetic, and photo-interpretative maps indicates that some of the conductive trends are contiguous with the geology and the photo-lineaments, whereas in other areas conductive trends cut across geologic and aeromagnetic features and appear to be related to shear zones seen on the aerial photographs. It is this direction of shearing which is associated with mineralization in the region and it would seem that in this case the plumbing zone of the mineralization may have become conductive due to wallrock alteration and possible circulation of groundwaters in the shear zones. Thus the conductivity pattern, when used in conjunction with aeromagnetics, photo-interpretation, and known geology, can provide structural information on an area which may prove of considerable value in assessing drilling priorities on conventional electromagnetic anomalies.

Additional Radiophase surveys have been flown in the Highland Valley region of British Columbia in an area of much deeper overburden than was present in Noranda. The most striking difference in the Radiophase results was the appearance of regional variations in response which extended over tens of miles and which were, as might be expected, quite insensitive to changes in aircraft flying altitude from 150 to 1500 feet. Such large variations were completely absent in surveys carried out in the Precambrian Shield.

A different approach to making airborne VLF measurements has been employed by McPhar Geophysics, who have recently introduced a dip-angle system. The principal difference in this system is that tilt angles of the resultant field are measured, and discrete conductive bodies are represented by crossovers in which the tilt in the direction of the flight line swings from one side of the vertical to the other. The form of map produced by such a technique is clearly different, and little is known at the present time about the interpretative schemes which are proposed.

Systems which employ low-frequency natural electromagnetic fields are obviously closely related to the VLF systems. They carry, however, a number of unique problems. The signal sources are electrical storms in which lightning discharges carry currents of the order of 20,000 Å down to the ground. These discharges behave as vertical antennas and radiate fields which cover a spectrum from a few hertzes up to million-hertz frequencies. Energy at frequencies below about 500 Hz and between 5 and 20 kHz can be detected at ranges of up to several thousand miles, and can therefore be used as a source for geophysical measurements.

The original system which utilizes these fields is known as AFMAG and was developed by Ward, Cartier, Harvey, McLaughlin, and Robinson in 1958. The airborne version measures the tilt angles of the resultant magnetic fields in the direction of the survey flight line (Ward, 1967). The system has not seen particularly widespread use, partly because it was initially looked upon as a competitor to conventional airborne EM systems with the primary function of directly locating orebodies. Due to the nature of the radiated fields, AFMAG gives strong responses over large-scale structures, which tend to dominate the records as compared with the anomalies obtained over discrete conductive bodies. Nevertheless, these types of measurements are of considerable significance when used in conjunction with other airborne methods, and the value of the AFMAG approach used in this fashion has been, with certain exceptions, largely overlooked by the mining community. The significant feature about the natural electromagnetic fields in the low-frequency region below 500 cycles is their great penetration. Skin depths are frequently in excess of 1000 feet and responses to major geologic structures can be expected down to at least that depth. These fields therefore potentially form an ideal complement to aeromagnetic surveying in assisting in structural and geologic interpretations. Such structural interpretations are considered to be of

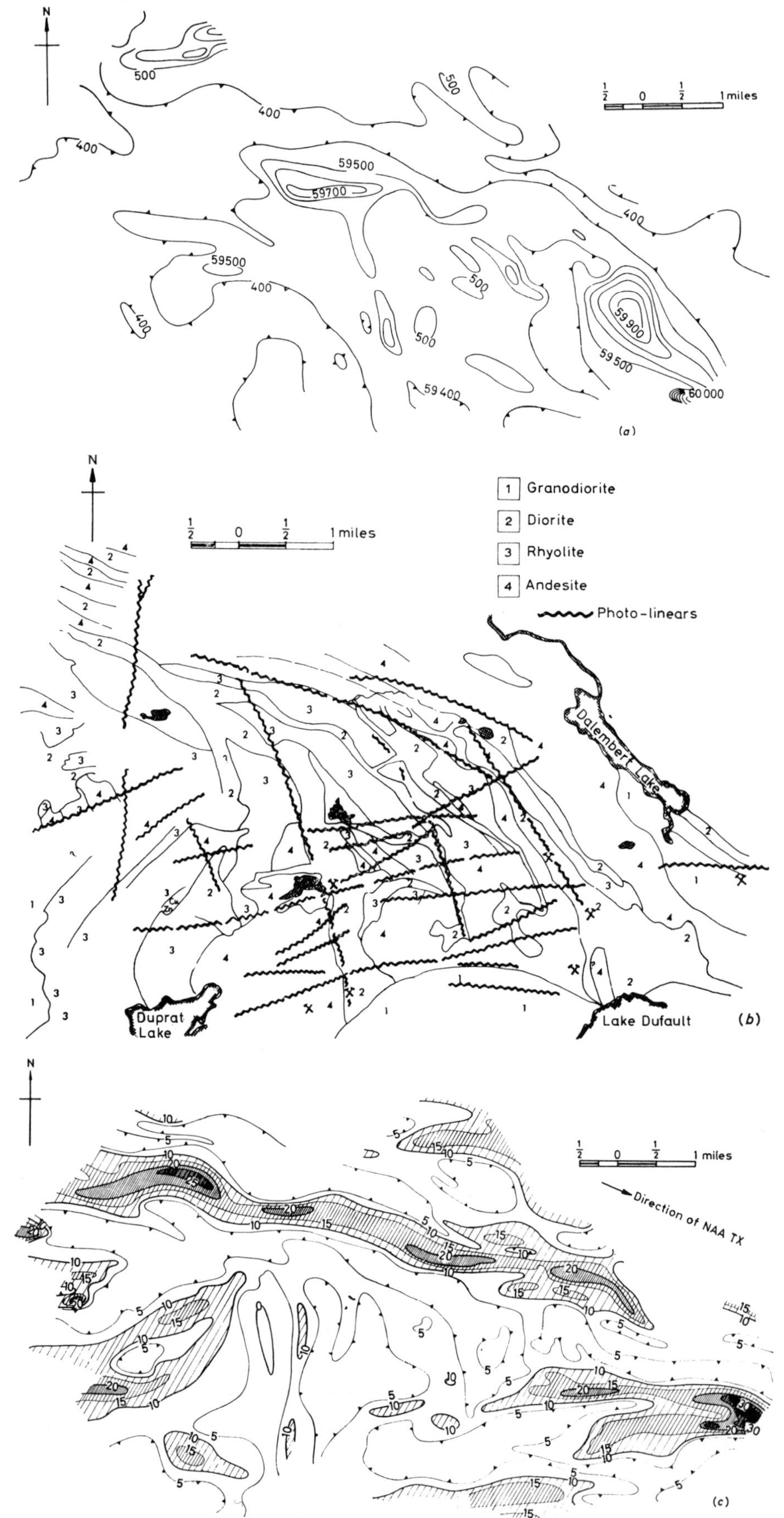

Figure 9 Lac Dufault area, Quebec: (A) Aeromagnetic survey. (B) Mines and geology. (C) Airborne Radiophase survey. (Contours of conductivity response in arbitrary units)

major importance due to the well-known association of orebodies with structures. Major faulting frequently shows up as a conductive anomaly, particularly if faulting has acted as a channelway to hydrothermal activity or has been the locus of extensive fracturing and repeated movement. Such faults may not always show up on aeromagnetic maps if they do not displace rocks of high susceptibility, and, in any event, the presence of conductivity in a fault zone is a significant item of geologic information not available from aeromagnetic surveying.

One of the principal problems of the AFMAG system is the fact that signals come in from a variety of directions and the dip angles of the resultant fields over conductive zones are measured without regard to the direction of the primary field. The dip angle can therefore vary drastically from signal to signal according to the degree of coupling of the primary field with the conductive zone, with the result that the records tend to be erratic and noisy. Certain interpretation approaches are also made virtually impossible by lack of knowledge of the primary field direction. The writer and his colleagues have therefore been developing a new natural field system known as Teltran (an abbreviation for Telluric Transients) which is licensed under the original AFMAG patents, but which utilizes three-component measurements of the natural fields and a base station, which records on magnetic tape. It has been proved to be feasible to synchronize airborne records with base station records (Fig. 10) and to use the computer for normalizing the airborne records against the base station. Residual fields in three components can be determined, which, in turn, allow the computation of relative strengths of the secondary field and the dip and strike of the fields. All of these readings can be related to selected directions of primary field, so the system is capable of providing large amounts of data which can be related to geometric factors in the computer. Its principal drawback remains the operational confinement to the summer periods when there is sufficient electrical storm activity within range of the survey area.

A comparison of natural field systems with VLF techniques shows that the former have the advantage of producing much simpler patterns due to the fact that a far smaller number of structures respond as conductors to the much lower frequencies used in these systems.

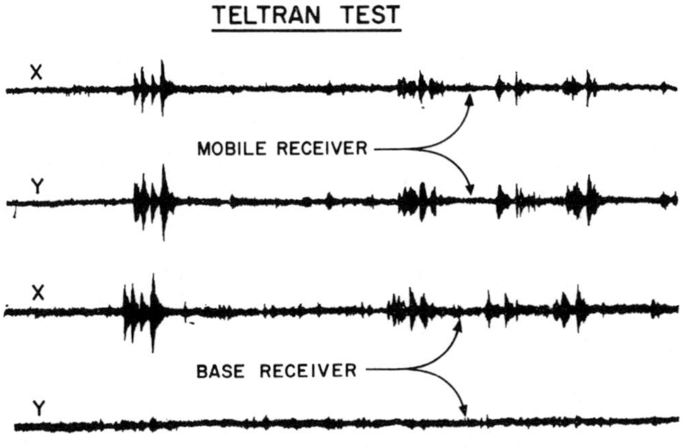

Figure 10 Mobile receiver at mine site.

The signals, however, are much more difficult to use due to the fluctuations in amplitude and time and direction of arrival. Whereas the VLF systems are completely compatible with the airborne magnetometer and can be flown on a year-round basis in any weather in which the magnetometer can be used, the natural field systems are very limited in time of usage, due to requirements for relatively low air turbulence and adequate natural field signal strengths.

Imaging Systems

The electromagnetic systems referred to above measure field strengths at low altitudes and are used to produce maps by flying traverses at intervals which may vary between 1 mile and ⅛ mile, according to the detail required. The presentation usually takes the form of anomalies spotted on the map with suitably coded symbols, or of a contoured map representing variations in intensity or phase. These systems, therefore, are of rather coarse resolution and cannot be said to produce an actual picture of the terrain. One electromagnetic technique, however, which is of considerable importance, uses a radar approach to acquire high-quality photographic-like images of the ground. These imaging radars have certain special features of interest to the geologist and they have the additional advantage that surveys can be made through cloud cover that makes normal aerial photography impossible. The technique, known as side-looking radar (SLAR), is an entirely military development which for a long time remained classified. In recent years, however, excellent imagery has been released which has demonstrated the considerable potential of side-looking radar as a remote sensing tool.

Fundamentally, the method relies upon the propagation of short radar pulses in a fanlike beam beneath the aircraft with the plane of the fan perpendicular to the flight direction of the aircraft. The radar returns from beneath the aircraft constitute the first signal arrivals, whereas the radar returns from the ground some distance off to the side of the aircraft track are delayed. By coherently adding the return signals as the aircraft moves along the flight path, it is possible to effectively synthesize a radar of very large aperture, and thus of high resolving power, while actually using antennas of quite limited size. The use of a scanning print-out system which is synchronized with the pulse transmissions and which is responsive in intensity to the amplitude of the radar returns makes it possible to produce an image whose density varies with changes in radar reflectivity of the terrain. Sophisticated methods of signal processing have been developed to the point where remarkable resolutions have been achieved, and to the layman some radar imagery is quite hard to distinguish from aerial photography. An example of such imagery taken with the Westinghouse AN/APQ 97 is shown in Figure 11.

The principal advantages of SLAR may be listed as follows. (1) Good-quality imagery can be obtained through cloud cover, allowing pictures to be made of regions of the earth where aerial photography is hampered by persistent cloud conditions. (2) Long, broad strips of imagery can be flown, providing continuous unbroken coverage over very much larger areas than can

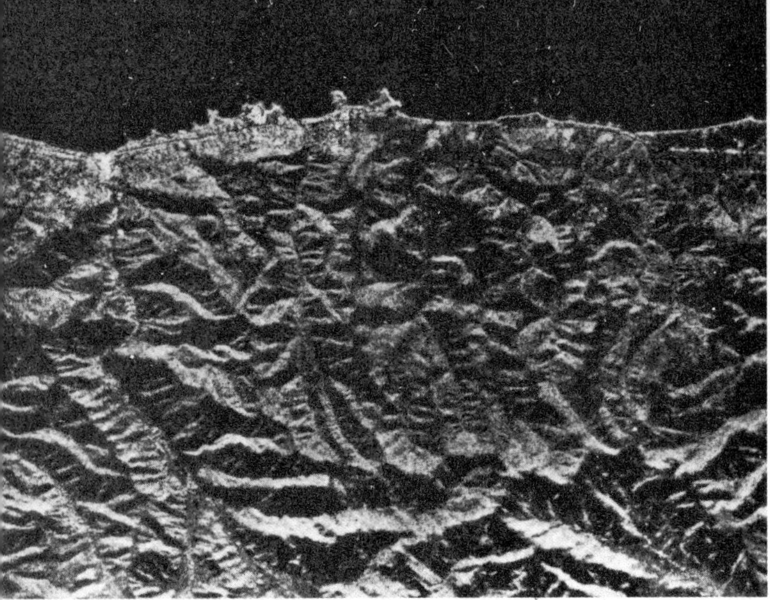

Figure 11 Westinghouse AN/APQ 97 side-looking radar imagery.

be covered in aerial photographs. The lack of tonal discontinuities normally associated with air photo mosaics is an attractive feature, although sometimes overstressed since the latest electronic dodging techniques can produce aerial photographic prints of extremely good tonal match for mosaic purposes. (3) SLAR illumination often accentuates linear features in the topography, and defines faulting patterns. The large regional aerial coverage obtainable from high-altitude aircraft often enables major structures to be traced for great distances with relative ease. Delineation of linear features is particularly marked when they lie parallel to the flight track and, conversely, they are not so apparent when they are perpendicular. (4) Subtle changes in lithology are sometimes depicted quite clearly in SLAR imagery, and this can be further accentuated when return signals are separated into two different images—one representing the component which has the electric vector polarized in the same direction as the transmitted signal and the other component which is perpendicularly polarized with respect to the outgoing signal. Some types of terrain texture have the ability to depolarize the return signal more than others, thus providing another factor useful for differentiation between soil and rock units. (5) Radar signals can penetrate a short distance into vegetation and also into dry soil of low dielectric constant. However, this fact has been greatly overstressed in the past by some writers since at the X and K band frequencies normally used for SLAR these penetrations tend to be in the millimeter or fractional millimeter range. In most cases the apparent penetration of vegetation noted from time to time is probably related to geobotanic expressions of the underlying soils and lithology.

Some of the disadvantages associated with SLAR are as follows. (1) The equipment is expensive and the support facilities required behind the operation of this equipment are substantial and currently beyond the means of most commercial aerial survey organizations. Costs for surveys are likely to be regarded as too high in relation to aerial photography to be attractive to mining companies, although they may be acceptable to governments carrying out national surveys. (2) High-altitude color aerial photography or color infrared aerial photography in the hands of a skilled photo-interpreter carries a higher information content, which heavily overlaps radar imagery. The imagery is more familiar to the eye of the geologist and well-known features associated with mineralization such as gossan and iron staining are immediately discernible on color air photographs, whereas they may be either invisible on radar imagery or only identifiable after careful cross-checking against ground truth sites.

Summarizing the potential role of SLAR applied to mineral exploration, it is felt that it may have certain limited, but nevertheless important, areas of applicability. In some areas or countries, such as Panama, where cloud cover is a major problem, the U.S. Government has provided complete SLAR coverage; such pictures could prove of great value to geologists in defining major structural features of possible significance in mineralization. This type of coverage is at present limited, but may well be extended to cover many regions where the same problems occur. A similar situation exists with regard to earth resource satellite programs proposed for the future. The immediate plans call for photographic and video imagery in the visible and near infrared spectrum, but ultimately SLAR will be developed for space which will be able to provide pictures of many areas almost continuously under cloud. However, it should be stressed that although clouds obviously constitute a major interference when one examines pictures of large portions of the earth taken from very high-altitude spacecraft, this interference is greatly reduced when a photographic satellite is in orbit for long periods and can gradually accumulate coverage of the earth by photographing areas at opportune times when clouds are absent. For some types of information, real-time data are required, but fortunately such data are not a necessity for the exploration geologist. Thus the value of SLAR from space compared with color photography of the type we have already seen from the Gemini spacecraft and Apollo VI may prove to be doubtful, if worldwide coverage of the latter type can gradually be accumulated over a period of time.

Optical Remote Sensing Systems

Nonimaging Systems

A variety of nonimaging optical techniques are under development which have potential application in mineral exploration. These include methods for detecting trace gases liberated from regions such as oxidizing ore deposits, hydrothermally altered zones, oil-field source rocks, etc., spectral correlation techniques being used to identify absorption spectra present in reflected solar radiation. Also included are Fraunhofer line detection techniques for measuring fluorescence associated with mineralogic types in the presence of daylight and the identification of characteristic infrared spectra emitted by various types of rock.

The work with which the writer and his colleagues have been closely associated is that of remote vapor detection. Interest in this field was stimulated by the known fact that emission of mercury vapor occurs over some types of mineral deposit. Mercury vapor is detected optically at 2537 Å, and at this wavelength solar radiation is prevented from reaching the earth's surface because of absorption by ozone in the upper atmosphere. Similarly, it is not possible to look through the upper atmosphere at this wavelength from a spacecraft. Therefore, in order to develop passive remote sensing systems which do not require built-in powerful sources of light, and which can be operated from either aircraft or spacecraft, it is necessary to look for those vapors of interest which have absorption bands in that part of the solar spectrum reaching the surface of the earth. The most obvious substance that came to light on initial examination was iodine vapor, since iodine is considerably more abundant than mercury in the crust of the earth and has 50 times the vapor pressure.

Iodine has a strong absorption spectrum in the visible region between 5000 and 6000 Å (Fig. 12), so remote sensing should be entirely feasible; in addition, it has geologic associations of considerable interest. A study of fluid inclusions in ores and associated wallrocks indicates that Na-Ca-Cl have been dominant components of the ore-forming fluid. Such evidence for the presence of chlorine is further supported by widespread studies of the behavior of natural brine solutions in the geologic environment and by considerations of the physical chemistry of ore deposition (White, 1968). It is becoming apparent that chlorine has played a major role in the ore-forming fluid, and it is therefore inevitable that the other halogens will be present in significant amounts. Very few analyses exist for halogens in the wallrocks of orebodies, but circumstantial evidence points strongly to their presence, and actual occurrences of iodine minerals have been reported in a number of mineral deposits (Chilean Iodine Educational Bureau, 1956). The possible association of the halogens with such deposits has therefore focused attention on iodine vapor detection. In addition, other important associations of iodine include oil-field brines and oil source rocks, and also marine life of the type which concentrates in primary fish food areas in the oceans (Goldschmidt, 1954; Rankama and Sahama, 1960). Thus the monitoring of iodine from orbiting spacecraft, with its multiple economic and scientific facets, has proved to be an attractive possibility from the outset and has provided the initial spur for the development of new equipment capable of remotely sensing spectra of this type.

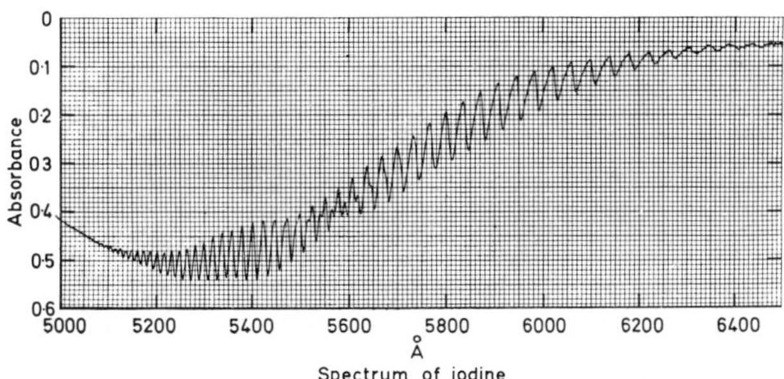

Figure 12 Absorption spectrum of iodine.

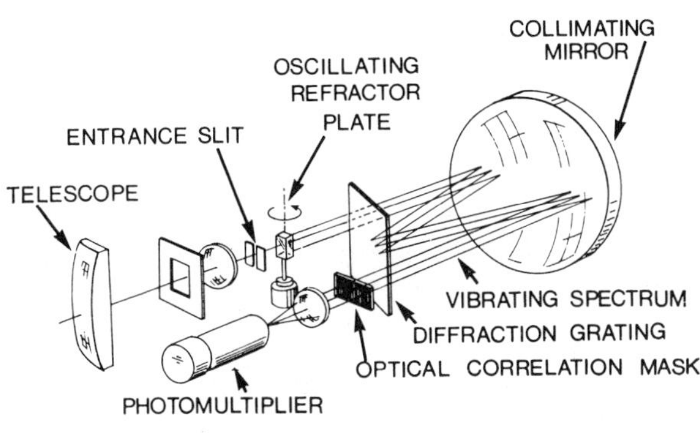

Figure 13 Correlation spectrometer.

The basic instrument developed, called the Correlation Spectrometer, is best understood by reference to Figure 13. Solar radiation reflected from the surface of the earth is picked up by a telescopic objective looking vertically down at the earth from an aircraft or spacecraft and is passed through an entrance slit of a spectrometer of the Ebert type. The light is dispersed by the spectrometer and is made to exit via a multi-slit mask, which replaces the usual single exit slit of the spectrometer. This mask is, in reality, a photographic replica of the spectrum of iodine or whatever gas is being detected. The dispersed light output of the spectrometer is made to vibrate across the mask by vibrating the grating or a refractor plate placed just behind the entrance slit. If iodine vapor is present in the lightpath of the incoming radiation, then the absorption spectrum will correlate with the photographic replica of the spectrum on the exit mask. As this spectrum is vibrated across the mask, a beat will be obtained which can be detected on a photomultiplier placed behind the mask. The presence of a beat signal can be monitored at high sensitivity with phase detection techniques, and, in fact, iodine contents of 1 ppm spread over a 1-meter path length can be readily measured by this technique. This means that concentrations of a part per billion can be measured over a 1000-meter path length; thus if a deep layer of iodine vapor exists over the surface of the earth, it can be detected at very high sensitivity. The method is applicable to a wide range of gases and, in particular, to two key air pollutants, SO_2 and NO_2. Because of the intense interest in air pollutants, much of the government finance obtained in support of this development has been directed towards the air pollution problem (Barringer, Newbury, and Moffat, 1968). A major effort has been expended in developing an airborne system, which is mounted in an Aero Commander aircraft (see Fig. 14). The spectrometer views the surface of the earth via a double-mirror system which folds the optical path of the system through a port in the side of the aircraft. The side-mounted mirror is rotatable so that the instrument can be calibrated by looking vertically to the sky prior to making vertical measurements looking at the earth's surface. The ground track is followed with a TV camera mounted in the nose and, in addition, the operator is provided with instruments to monitor meteorologic conditions such as barometric pressure and temperature. Considerable success has been scored with the system in flying profiles across urban

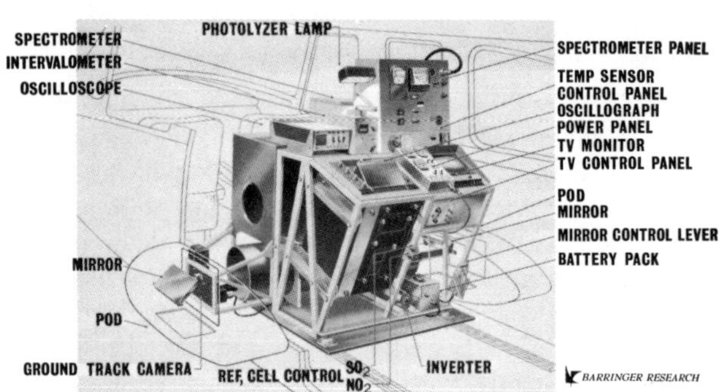

Figure 14 Remote sensing airborne correlation spectrometer-aircraft installation.

areas such as Washington, D.C., Los Angeles, San Francisco, Chattanooga, and Toronto. Measurements over cities have been confined to SO_2 and NO_2; profiles for NO_2 over Chattanooga are shown in Figure 15. Preliminary airborne surveys of iodine vapor have already been carried out over kelp beds off the coast of Maine, and results of the iodine profiles are shown in Figure 16. The preliminary offshore surveys for iodine have indicated a requirement for greater sensitivity if detailed facets of the iodine profiles are to be recorded. Nevertheless, clear-cut responses were obtained over seaweed concentrations and the general coastline of Maine showed a build-up of free iodine vapor in the atmosphere as compared with inland areas. This completely corroborated the predictions for this experiment and searches are now being conducted in oil-field regions for iodine vapor. Much of this work has been sponsored by the National Aeronautical and Space Administration and the next phase of the development program will include a high-altitude balloon flight operating at a height of 120,000 feet to verify the feasibility of making measurements of the distribution of gases at the earth's surface through the greater part of the atmospheric column and the ozonosphere. The aims of NASA are to ultimately develop an orbiting spacecraft system which can be used to monitor air pollution and the distribution of trace vapors associated with various types of earth resources. This approach, however, has equally significant application for airborne surveys, particularly since high resolution can be obtained at lower altitudes.

Looking to the future, there are many gases of potential interest as well as iodine and sulphur dioxide. Little is known of the oxidation by-products of sulphide orebodies, and the well-known sulphur smell associated with underground workings and with dumps of sulphide ore has not been precisely defined in terms of chemical composition. The Finnish Geological Survey has for some time been making use of dogs in exploring for buried boulders of sulphide ore as an aid to the float boulder prospecting techniques applied so successfully in Finland. It appears that these dogs smell a substance which the Finns claim is not sulphur dioxide. The possibility exists that there is an organic vapor which is the metabolic by-product of the bacteria invariably present in oxidizing sulphides. The entire vapor spectrum present in the vicinity of mineral deposits requires a careful investigation to determine the composition and distribution of derived gases.

Apart from sulphide mineral deposits, vapor methods could have an important place in looking for micro gas-seeps associated with natural gas fields. The remote detection of methane appears feasible, and possibilities exist for detecting ethane and some of the higher hydrocarbons, although the quantities present are very much smaller than for methane. Scanning techniques can even be applied to the correlation spectrometer to provide a form of low-resolution imagery, which could be of use in examining a fairly broad path beneath the aircraft for micro-seep emanations.

Development work on remote sensing techniques for spectral correlation is continuing at an energetic level, and a correlation interferometer is now being tested which has major potential for remote sensing in the infrared region. This interferometer carries a magnetic recording of the Fourier transform of the gas spectra, and correlation is carried out in real time against the transform output of the scanning interferometer. The advantage of the use of an interferometer rather than the dispersive spectrometer in the infrared lies in the much larger light throughput. This becomes important in view of the very much lower sensitivity of detectors in the infrared compared with the extremely high sensitivity of the photomultiplier in the visible and ultraviolet regions. Tests of the remote detection of carbon monoxide generated by automobile exhausts have already shown the considerable promise of this instrument.

Moving from the subject of the remote detection of gases, a technique for the remote detection of fluorescence in daylight is being developed for NASA by the U.S. Geological Survey and has some potential application in mineral exploration (Hemphill, 1968). This technique depends upon the fact that the normal solar spectrum contains numerous Fraunhofer lines, which represent deep absorption bands caused by the presence of elemental gases in the chromosphere of the sun. The depths of these absorption lines remain reasonably constant over short periods and the presence of these lines can be measured not only by directly viewing the sun but also by looking downward at the surface of the earth from an aircraft or satellite and detecting them in the solar radiation reflected from the surface of the earth. Provided there are no fluorescent substances present at the surface, the percentage depth of an individual Fraunhofer line in relation to the illumination at immediately adjacent wavelengths is not modified by reflection. If, however, there is a contribution from fluorescence radiation in the reflecting surface, this changes the ratio of the reflected light intensity at the center of the Fraunhofer line compared with the light intensity at the line wings. Hemphill has used this technique to study the fluorescence of rhodamine B solutions in water, and has demonstrated the feasibility of detecting as little as 3.5 ppb of rhodamine B in a depth of 0.5 meter of water in a tank with natural sunlight as a source. A number of ore metals, such as scheelite and willemite, as well as some asphalts and bituminous compounds, are fluorescent, so this represents a possible aid to geologic exploration by a remote sensing technique. Furthermore, chlorophyll in plants is fluorescent and there is an interesting possibility that this fluorescence will be modified by

Figure 15 Chattanooga area, airborne NO_2 survey.

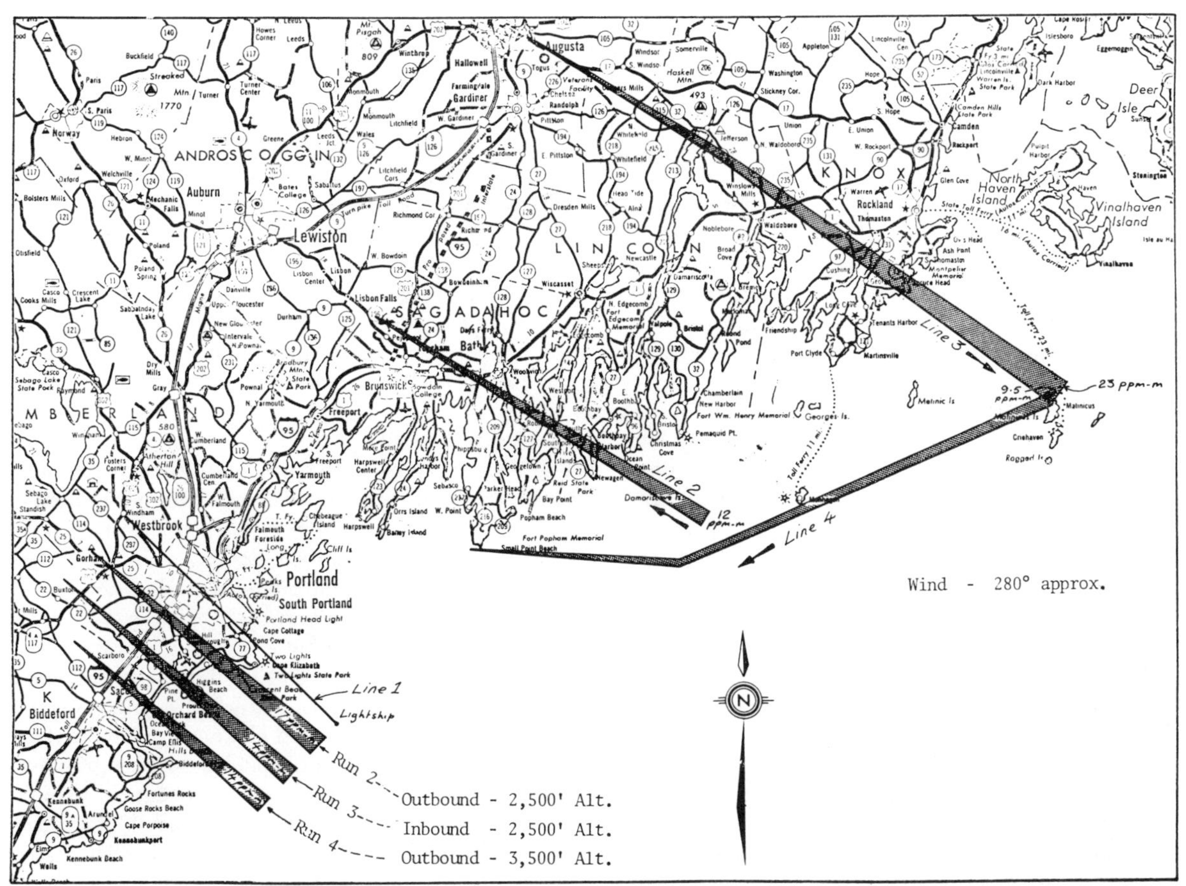

Figure 16 I_2 search, coast of Maine, U.S.A., 12 August, 1968, flight 5. Scale, 1 inch = 10 miles (approx.).

anomalous quantities of trace elements in the plants. The most important application of the Fraunhofer line method could prove to be a measurement of geobotanic effects rather than the direct detection of fluorescent minerals.

The present equipment used for Fraunhofer line fluorescence investigations has been a very narrow band special type of the Fabry-Perot filter. However, it is also possible to use the identical spectral correlation techniques described above for detecting the presence of trace gases. This approach is advantageous in that it enables a number of Fraunhofer lines to be used simultaneously, and would consist of measuring the correlation of the solar spectrum against a mask of the Fraunhofer lines, and measuring departures of the correlation from figures obtained when looking directly at the sun. The system, therefore, is very similar to that used for making gas measurements, except that a somewhat higher-resolution spectrometer is required, since some of the Fraunhofer lines are less than 1 Å wide.

A further interesting nonimaging remote sensing, optical technique which shows considerable promise has been developed by R. J. P. Lyon of Stanford University (Lyon and Burns, 1963; Lyon, 1964, 1965, 1966; Lyon and Paterson, 1966; Vickers and Lyon, 1967). The method operates on the measurement of the infrared emission spectra of rocks and soils in the 8–13 μm wavelength region (where the atmosphere is essentially transparent). It depends upon the shift in wavelength of the silicon-oxygen fundamental vibration transition with changing composition of silicate minerals in rocks. This transition, occurring near 9 μm in the middle infrared region, shifts approximately 2 μm (8.8–10.8 μm) between rock compositions of granite and dunite, respectively. Field studies have shown that natural terrain free of vegetation will also produce identifiable spectral signatures due to the Si–O vibration absorption of the constituent minerals in the surface rocks. The effects of surface roughness are considerable, however, and the method loses much of its precision in dealing with finely powered materials. The surface geometry determines the detailed photometric character of the radiation emitted. Targets rough on a scale of a wavelength appear gray spectrally. A variety of instrumentation has been constructed by Lyon for field testing and, most recently, he has evolved a rotating filter wheel system able to make rapid spectral scans six times per second of underlying terrain during aircraft traversing. The system traces a line profile beneath the aircraft covering a track 15 feet wide when flying at 2000 feet with a resolution of 40 feet along the track. Thus an outcrop about the size of a railway boxcar would fill the effective aperture of the instrument at 2000 feet (R. J. P. Lyon, personal communication). Magnetic tape recording and digital data processing are used to handle the data. Typical spectra of a wide range of rock types are shown in Figure 17. At the present state of the art, airborne classification of rock type into one of several possible groupings can be achieved with about an 80 percent reliability.

The system shows considerable promise as a supplementary aid to the interpretation of colored air photographs and aeromagnetic maps. Spot identifications on

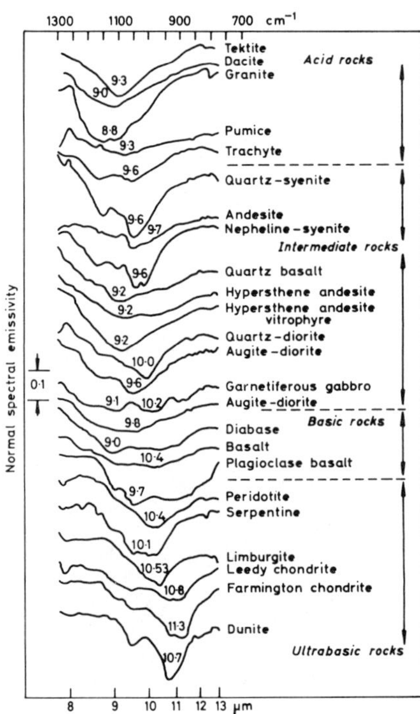

Figure 17 Infrared thermal emission spectra of a variety of rock types. (After Lyon, 1966)

exposed areas can provide calibration points on aerial photographs, which can then be extrapolated to assist in the rapid production of geologic maps with minimum entry on the ground. In particular, the system could prove valuable in identifying key geologic targets, such as differentiated suites of volcanic rocks, where marked contrasts in composition are present and rock differentiation is exhibited as a substantial shift in wavelengths of the silicon-oxygen fundamental vibration. Some of the other types of rock, such as ultrabasics, which have significant associations with certain types of mineralization, can also be identified by this method. This optical technique can therefore considerably broaden the scope of remote sensing applied to mineral exploration if it is used judiciously with complementary methods.

Imaging Systems

Photography remains the principal optical imaging system and the original remote sensing technique. Volumes have been written on this subject and it is therefore proposed not to cover photogeologic methods in any detail but rather to discuss certain points regarding modern emulsions and satellite photography.

The majority of air photo interpretation that has been carried out in the past has utilized black and white prints on a 9 × 9 format. The advent of high-speed color emulsions and color infrared film has broadened the scope of color photography and opened up a requirement for photohandling equipment and procedures which can take full advantage of these materials. Color prints on paper lose much of the tonal range and brilliance of transparencies, and are costly. Operating in stereo on the original transparencies produces imagery of outstanding quality, which enables the photo-interpreter to look for many subtle details of fracturing, discoloration caused by hydrothermal alteration, geobotanic variations, etc. The value of photogeologic interpretation for mineral exploration purposes is also greatly enhanced by the simultaneous study of aeromagnetic maps. As an example of the value of this procedure, faulting may be very clear on some parts of an aeromagnetic map and totally obscured on another portion. A similar situation exists with aerial photographs; when the two are brought together they may assist each other in tracing fully the pattern of faulting and shearing. The presence of intrusive bodies is often quite clear on aeromagnetic maps, but its precise boundaries may be difficult to define. The aerial photograph can assist in mapping these boundaries by tonal changes and it may also be possible to make fairly accurate interpretations of rock type by a joint consideration of photographic color and magnetic susceptibility. Consideration of the joint value of aerial magnetics and air photo color transparencies points toward the requirement for specialized handling and projection equipment that allows for the production of interpretative maps with minimum operator fatigue.

Turning to the applications of color infrared films, there appears great potential for these emulsions in looking for geobotanic effects associated with mineralization and hydrothermal alteration. The most common film used in the West is Kodak Ektachrome Infrared Aero film, which is a threelayer film with layers sensitized to green, red, and infrared wavelength (Fig. 18). The film is normally exposed through a Wratten No. 12 filter, which eliminates blue light from the image. The dyes are chosen so that green objects, with the exception of vegetation which is highly infrared-reflective, appear blue, red objects appear green, and infrared-reflective objects (such as healthy vegetation) appear red. One effect is that the infrared energy of highest intensity produces the brightest reds in the photograph. If we examine the

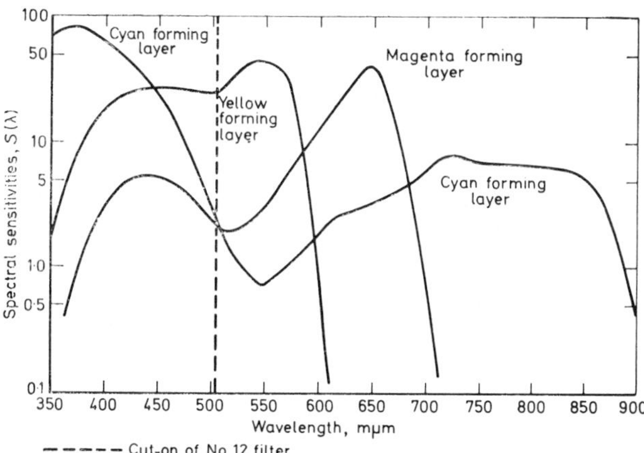

Figure 18 Sensitometric data: spectral sensitivity curves (Kodak Ektachrome Infrared Aero film, type 8443). (Spectral sensitivity, $S(\lambda) = 1/E(\lambda)$, where $E(\lambda)$ is the energy, ergs/cm^2, of monochromatic radiation at wavelength λ required to reduce the dye image density in the individual layer to an equivalent neutral density of 1.0 above minimum density. Data are adjusted to correspond to an effective exposure time of 1/100 sec.)

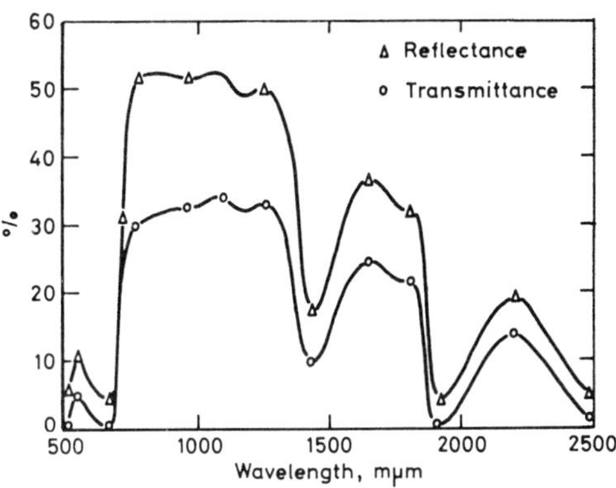

Figure 19 Reflectance spectrum of a cotton leaf. (After Myers et al., 1966)

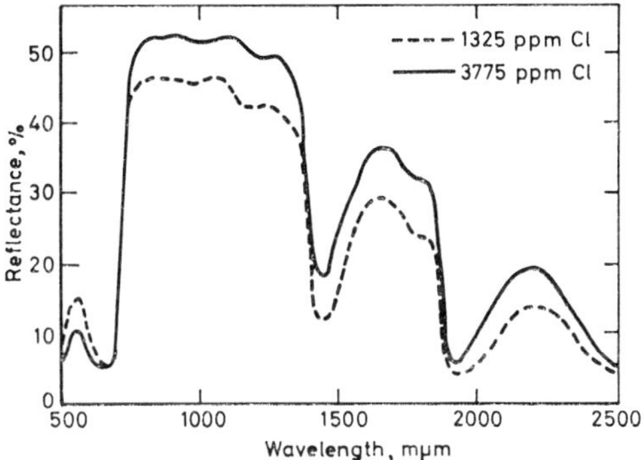

Figure 20 Effect on cotton leaves of variations in chloride content in soil. (After Thomas et al., 1966)

typical reflectance spectrum of a plant, such as in Figure 19, it will be noted that the reflectance of the plant increases very markedly between 0.7 and 0.8 μm (Myers et al., 1966). The high reflectance in the 0.8 μm region records as red on the color film, giving the characteristic red color to vegetation. In unhealthy plants, which suffer from chlorosis, for example, reflection in the infrared is substantially reduced, whereas reflectivity in the yellow and red regions may be increased. This changes the balance of colors between the emulsions and causes marked changes in hue. Whereas the eye is able to discriminate only a limited range of shades of gray, it is able to separately distinguish between thousands of different hues, and is therefore able to see very subtle changes in the spectral reflectance characteristics of plants on color infrared film. Such reflectivity changes can be caused by a large range of factors, including infestation by pests, disease, and excess or deficiency of certain nutrients. Figure 20 shows the effect on cotton leaves of variations in the chloride content in the soil (Thomas et al., 1966). Excess copper in soils has been reported as causing chlorosis in plants (Cannon, 1960), and in Zambia and Katanga copper concentration in soils has been so intense as to inhibit the growth entirely of trees and create the so-called copper clearings (Duvigneaud, 1958). Obviously, there is important potential for the use of color infrared film as a means of indicating subtle changes in growth associated with unusual trace-element concentrations in the soil. In addition, the general variations in types of vegetation tend to show up more clearly on color infrared and can be used for mapping of geologic boundaries in areas of vegetation.

Russian geologists have made extensive use of spectrazonal films, which are color infrared films employing two layer emulsions—one sensitive to the infrared and the other to the red end of the visible spectrum. These films are very effective in the identification of stress symptoms in vegetation and the differentiation between deciduous trees and conifers. In prospecting for diamond deposits in the Daldyn region of the U.S.S.R., it was noted that the soil that was derived from the weathering of kimberlite was rich and supported a dense forest of alder and larch. This was in dramatic contrast to the sparse taiga growth which covered the generally poor soils of the region. During the diamond exploration program the entire area was photographed with panchromatic, infrared, and spectrazonal films at a scale of 1:15,000. Aeromagnetic surveys were made and photointerpretation was carried out in conjunction with magnetic interpretation. As a result three new pipes were discovered (Kobets and Komarov, 1959).

Infrared color film, however, may offer no advantage—and possibly some disadvantage—when it is used in areas of minimal vegetation. Rock types and soils all assume strange unfamiliar colors, which means that the photo-interpreter has to adjust himself to a whole new set of colors to carry out geologic mapping.

One important advantage of infrared color photography is its excellent haze penetration. Removal of blue light by the Wratten No. 12 filter used with this film, plus the addition of the near infrared portion of the spectrum, considerably lengthens the average wavelength used and greatly reduces the effects of Rayleigh scattering and aerosol scattering in the atmosphere. Aerial photographs taken with color infrared film are brighter and show improved contrast as compared with color photography taken on standard emulsions. This renders the economics of infrared color photography attractive, since satisfactory pictures can be taken in a much broader range of atmospheric conditions. It is particularly true in some forested areas where aerosol hazes caused by the turpenes in the trees create a blue scattering which seriously degrades standard color photography.

Much has been written about the use of multi-band photography in which up to eight separate lenses are used to take simultaneous separate photographs through different relatively narrow band filters. Although the spectral information obtained with this approach can be more detailed than that acquired with color or color infrared photography, it has the disadvantage that it calls upon the human eye to make large numbers of comparisons between photographs of a given area. Automated and semi-automated data handling procedures may be developed, but at the present time color photography remains more attractive as a practical tool. The ability of the eye to discriminate between slight changes in the balance between two different colors makes it particularly effective in discriminating small spectral changes. In comparing pairs of filtered photographs, this sensitivity of discrimination cannot be achieved unless special methods are employed to combine or differentially scan the photographs.

Satellite photography shows major promise for the future from the point of view of the exploration geologist. The photographic experiments on the Gemini series of satellites orbited by NASA indicated that even handheld photographs taken on a 2¼-inch square format with a Hasselblad camera were capable of producing remarkably good results (Lowman, 1966).

Studies of the Gemini series of photographs have shown that many major structural features can be recognized which have not been recorded on any existing geologic maps. It is made clear by a detailed comparison of these photographs with group mapping that excellent photogeologic interpretation can be obtained from satellite photographs. More recently, vertical photography has been carried out from the Apollo VI spacecraft and once again the results demonstrate how effective this photography is and the contribution that will be made to the geologic sciences when worldwide coverage with this type of photography is available.

A prime sensor apparently planned by NASA for the initial earth resources satellite is the high-resolution TV camera made by RCA and known as the Return Beam Vidicon. Line resolutions of the order of 5000 are obtained with this camera, and telemetered imagery of the earth will be substantially better than any other types of video imagery derived from satellites to date. This imagery will be of considerable value for interpreting major structures and is therefore pertinent in our discussion of remote sensing applied to mineral exploration. However, it will probably not be too effective in detecting subtle changes in tone associated with hydrothermal alteration, gossanization, etc., and will hopefully be replaced ultimately by color photography.

Another video system has been proposed by TRW Systems and is referred to by the code name WISP. It is an imaging system which can scan simultaneously in one or more spectral lines. As in color photography, therefore, a considerable amount of spectral, as well as spatial, information is available. Resolution, however, is considerably poorer than the Return Beam Vidicon.

Yet another class of imaging instrument which has resulted from military research, and is now available commercially, is the mechanical infrared line scanner. These systems operate either in the 3–5 μm atmospheric infrared window or the 8–12 μm window. A rotating mirror assembly optically scans the ground and reflects the point image of the ground to an infrared sensor. The output of this sensor is amplified and is printed as a variable-intensity image on a synchronously scanning print-out system. The imagery produced by modern equipment is of quite adequate resolution for geologic interpretation purposes and represents the thermal emission from the terrain. As such, the imagery is subject to many variables, such as the time of day or night that it is acquired, the thermal capacity of the surface materials, the topographic relief, which affects the sun exposure of different portions of the ground, the thermal emissivity of the surface materials, the moisture content of the soils, etc. Thus a particular problem of interpreting thermal infrared imagery is that the picture obtained over a given area can vary considerably. In the early hours of the morning a body of water may appear bright and, hence, warmer than its surroundings, due to the heat-retaining capacity of the water. On the other hand, after the sun has risen the water may appear cooler due to the fact that it warms up more slowly than the surrounding land. In some instances fault zones show up more clearly on thermal infrared imagery than they do on conventional aerial photography, due to the thermal effect of the presence of increased soil moisture over the fault zone. However, in the writer's opinion, the fairly high cost of acquiring thermal infrared imagery may not, in general, be worthwhile for mineral exploration purposes, except in specialized instances. These may be, for example, a requirement for locating geothermal areas and fumaroles as part of a broader program to locate mineral deposits associated with geothermal areas. It has been suggested that the heat developed by oxidizing ore deposits may be detectable by these methods, but, unfortunately, infrared scanners are responsive to the top surface layer of the ground only, whereas geothermal effects from oxidation tend to be detectable only at some distance below the surface where there is isolation from ambient meteorologic conditions.

Mechanical line scanners have been applied experimentally by the University of Michigan (Lowe, 1968) scanning at 17 discrete wavelengths from the ultraviolet right through to the far infrared. This highly sophisticated instrument, when used in conjunction with appropriate and relatively complex data handling methods, has many applications in remote sensing. Nevertheless, it has not been established that the high cost of data acquisition relative to the low cost of photographic techniques is compensated for by the additional information obtained which can be used for mineral exploration purposes.

Gamma Ray Detection Techniques

Gamma ray methods certainly rate as remote sensing systems and have considerable potential outside the obvious application of the direct detection of radioactive ores. Total count systems have become increasingly replaced by gamma ray spectrometers, which discriminate among the energy levels of incoming gamma rays and allocate them to specific channels. When proper computational procedures are used, it is possible to approximately separate the uranium, thorium, and potassium response. Broadly speaking, two classes of gamma ray spectrometers have been used. One is a relatively simple class of instruments in which data processing is carried out in real time in the aircraft, and rate meters provide analog outputs for each channel. A much more advanced system has been built by Texas Instruments (Foote, 1968), in which multiple channels are recorded on magnetic tape for subsequent processing by computer. Somewhat similar systems have been installed by the U.S. Geological Survey and, more recently, by the Canadian Geological Survey (Darnley and Fleet, 1968). These installations are capable of providing reasonably accurate estimates of the surface concentrations of uranium, thorium, and potassium, and are therefore potential aids to geologic mapping. Results from these systems can be contoured in much the same way as an areomagnetic map, and they provide complementary information to this type of map. In some areas, such as granitic regions, where rock susceptibility may be low, potassium counts can be high and structural detail appears on the radiometric maps of these areas which may be missing on aeromagnetic maps.

The gamma ray spectrometer can to some extent be regarded as an airborne geochemical tool, and can be used for tracing the dispersion of radioactive elements from a radioactive orebody as well as for direct detection of ore. A careful analysis of radiometric results in terms of the drainage of a region and groundwater movement could prove important in locating blind radioactive orebodies.

Similarly, it may be feasible to trace the potassic phases of wallrock alteration around base metal or gold deposits, and to use this as an exploration guide (Moxham, Foote, and Bunker, 1965).

Air Sampling Techniques

The possibility of detecting ores by vapors being given off to the atmosphere appears first to have been raised by Williston for the case of mercury vapor (Williston, 1964, 1968). Williston published a profile in 1964 of atmospheric mercury over a mercury prospect, but does not appear to have published any further work on air sampling over mineralization since that time. Field tests carried out by Barringer Research with air sampling have also indicated a substantial mercury anomaly in the atmosphere over a large soil mercury anomaly. At the present time, the U.S. Geological Survey is carrying out an extensive program concerned with the measurement of mercury vapor over gold deposits and porphyry coppers. They appear to have achieved very encouraging results, but only limited information has been released at the time of writing (Canney, 1968). It is understood in personal communications with J. H. McCarthy of the U.S. Geological Survey that values up to ten times background have been obtained over some porphyry coppers such as Silver Bell and Ajo in Arizona. Furthermore, strong anomalies have been obtained over gold deposits buried beneath as much as 50 feet of gravel in Nevada.

All of this work has been done with mercury absorption techniques with precious metal filters. Although the approach is a sound one for experimental surveys, it is not very suitable for production survey operations. A new version of a truck-mounted air-sampling mercury spectrometer (Barringer, 1965) is now under construction for use in a small fixed-wing aircraft, which will allow continuous profiling of background values of mercury vapor with a response time of only a few seconds.

The outgassing of mercury vapor, which evidently takes place under appropriate meteorologic conditions from some ore deposits, opens up the possibility of detecting a number of other items. Weiss has already worked upon the detection of aerosol particles of high metal content as direct indicators of ore (Weiss, 1967), and there are other possible direct and indirect detection methods based upon both gases and aerosols. The problems at the present time arise because of the necessity for achieving extreme sensitivity. These problems do not seem insurmountable and it is predicted that air sampling methods may become an important new airborne geochemical tool which will provide a valuable supplement to other remote sensing techniques.

Conclusions

The techniques of aeromagnetic and electromagnetic surveying have already proved to be extremely valuable in the discovery of new ore deposits. However, major new methods are required to assist in area selection in order to focus the more detailed and expensive ground methods of geophysics and geochemistry into areas where the probability of finding ore is high. The greatest opportunity for achieving this aim lies in integrating new remote sensing techniques with the methods that are already well established.

The complementary nature of the remotely sensed information that can be obtained from satellites and aircraft is such that, taken together, these data are capable of having a major impact on mineral exploration. The effectiveness of this approach will be even greater as our knowledge of the ore environment improves and we are better able to recognize the geologic factors favorable to mineralization.

REFERENCES

Barringer, A. R. Developments Towards the Remote Sensing of Vapors as an Airborne and Space Exploration Tool. *Proceedings of the Third Symposium on Remote Sensing of Environment,* 2nd ed. Ann Arbor: University of Michigan, 1965. Pp. 279–292.

—, and McNeill, J. D. Radiophase—a New System of Conductivity Mapping. *Proceedings of the Fifth Symposium on Remote Sensing of Environment.* Ann Arbor: University of Michigan, 1968. Pp. 157–167.

—, Newbury, B. C., and Moffat, A. J. Surveillance of Air Pollution from Airborne and Space Platforms. *Proceedings of the Fifth Symposium on Remote Sensing of Environment.* Ann Arbor: University of Michigan, 1968. Pp. 123–155.

Canney, F. C. Geochemical exploration—state of the art (abstract). *Society of Exploration Geophysicists 38th Annual International Meeting Booklet.* Denver: The Society, 1968. P. 122.

Cannon, H. L. Botanical prospecting for ore deposits. *Science* 132:591, 1960.

Cartier, W. O., et al. System of airborne conductor measurements. U.S. Patent 2,623,924; 1952.

Chilean Iodine Educational Bureau. *Geochemistry of Iodine.* London: The Bureau, 1956.

Darnley, A. J., and Fleet, M. Evaluation of Airborne Gamma Ray Spectrometry in the Bancroft and Elliot Lake Areas of Ontario, Canada. *Proceedings of the Fifth Symposium on Remote Sensing of Environment.* Ann Arbor: University of Michigan, 1968. Pp. 833–853.

Duvigneaud, P. La vegétation du Katanga et de ses sols metallifères. *Bull. Soc. R. Bot. Belg.* 90:127, 1958.

Foote, R. S. Application of Airborne Gamma Radiation Measurements to Pedologic Mapping. *Proceedings of the Fifth Symposium on Remote Sensing of Environment.* Ann Arbor: University of Michigan, 1968. Pp. 855–875.

Gay, S. P., Jr. Morphological study of geophysical data by viewing in three dimensions (abstract). *Society of Exploration Geophysicists 38th Annual International Meeting Booklet.* Denver: The Society, 1968. P. 121.

Goldschmidt, V. M. *Geochemistry.* Oxford: Clarendon, 1954.

Hemphill, W. R. Remote detection of solar-stimulated luminescence. Paper presented at 19th Congress of the International Astronautical Federation, New York, No. AS-156, 1968.

Kobets, N. V., and Komarov, V. B. Use of aerial methods in exploration for kimberlite pipes. *Trudy Lab. Aerometodov.* 8:120, 1959.

Kornick, L. J., and McLaren, A. S. Aeromagnetic study of the Churchill-Superior boundary in northern Manitoba. *Can. J. Earth Sci.* 3:547, 1966.

Lowe, D. S. Line Scan Devices and Why Use Them. *Proceedings of the Fifth Symposium on Remote Sensing of Environment.* Ann Arbor: University of Michigan, 1968. Pp. 77–101.

Lowman, P. D., Jr. The earth from orbit. *Natl. Geogr. Mag.* 130:644, 1966.

Lundberg, H. Magnetic surveys with helicopters. *Trans. Inst. Min. Metall.* 57:69, July 1947.

Lyon, R. J. P. *Evaluation of Infrared Spectrophotometry for Compositional Analysis of Lunar and Planetary Soils.* NASA Rept. CR–100, 1964.

— Analysis of rocks by spectral infrared emission (8–25 microns). *Econ. Geol.* 60:175, 1965.

— *Field Infrared Analysis of Terrain.* Rept. NGR-05-020-115, 1966.

—, and Burns, E. A. Infrared Spectral Analysis of the Lunar Surface from an Orbiting Spacecraft. *Proceedings of the Second Symposium on Remote Sensing of Environment.* Ann Arbor: University of Michigan, 1963. Pp. 309–326.

—, and Paterson, J. W. Infrared Spectral Signatures—Field Geological Tool. *Proceedings of the Fourth Symposium on Remote Sensing of Environment.* Ann Arbor: University of Michigan, 1966. Pp. 215–230.

Moxham, R. N., Foote, R. S., and Bunker, C. M. Gamma ray spectrometer studies of hydrothermally altered rocks. *Econ. Geol.* 60:653, 1965.

Myers, V. I., et al. Remote Sensing in Soil and Water Conservation Research. *Proceedings of the Fourth Symposium on Remote Sensing of Environment.* Ann Arbor: University of Michigan, 1966. Pp. 801–813.

Paal, G. Ore prospecting based on VLF radio signals. *Geoexploration* 3:139, 1965.

Packard, M., and Varian, R. Free nuclear induction in the earth's magnetic field (abstract). *Phys. Rev.* 93:941, 1954.

Pemberton, R. H. Airborne electromagnetics in review. *Geophysics* 27:695, 1962.

Raff, A. D., and Mason, R. G. Magnetic survey off the west coast of North America, 40°N. latitude to 52°N. latitude. *Bull. Geol. Soc. Am.* 72:1267, 1961.

Rankama, K., and Sahama, T. G. *Geochemistry.* Chicago: University of Chicago Press, 1960.

Thomas, J. R., et al. Factors Affecting Light Reflectance of Cotton. *Proceedings of the Fourth Symposium on Remote Sensing of Environment.* Ann Arbor: University of Michigan, 1966. Pp. 305–312.

Vickers, R. S., and Lyon, R. J. P. Infrared Sensing from Spacecraft—a Geological Interpretation. Pap. Am. Inst. Aeron. Astronaut., No. 67-284, 1967.

Ward, S. H. The Electromagnetic Method. *Mining Geophysics,* Vol. II. Tulsa, Okla.: Society of Exploration Geophysicists, 1967. Pp. 224–372.

— Electromagnetic Theory for Geophysical Applications. *Mining Geophysics,* Vol. II. Tulsa, Okla.: Society of Exploration Geophysicists, 1967. Pp. 10–196.

—, et al. Prospecting by use of natural alternating magnetic fields of audio and sub-audio frequencies. *Trans. Can. Inst. Min. Metall.* 61:261, 1958.

Weiss, O. Method of aerial prospecting which includes a step of analyzing each sample for element content, number, and size of particles. U.S. Patent 3,309,518; 1967.

White, D. E. Environments of generation of some base metal ore deposits. *Econ. Geol.* 63:301, 1968.

Williston, S. H. The mercury halo method of exploration. *Engng. Min. J.* 165:98, May 1964.

— Mercury in the atmosphere. *J. Geophys. Res.* 73:7051, 1968.

Wyckoff, R. D. The Gulf airborne magnetometer. *Geophysics* 13:182, 1948.

SUGGESTED READINGS FOR PART TWO

Alexander, R. H., Bowden, L. W., Marble, D. F., and Moore, E. G. Remote Sensing of Urban Environments. *Proceedings of the Fifth Symposium on Remote Sensing of Environment.* Ann Arbor: University of Michigan, Institute of Science and Technology, Willow Run Laboratories, 1968.

American Astronautical Society. *Scientific Experiments for Manned Orbital Flight,* IV. Washington, D.C., 1965.

American Society of Photogrammetry. *Manual of Color Aerial Photography.* Falls Church, Va.: American Society of Photogrammetry, 1968.

Avery, T. E. Remote Sensing Techniques. *Interpretation of Aerial Photographs.* Minneapolis: Burgess, 1962.

Avery, T. E. Introduction to Photography and Photogrammetry. *Interpretation of Aerial Photographs.* Minneapolis: Burgess, 1962.

Barath, F. T. Microwave Radiometry and Its Applications to Oceanography. *Oceanography from Space.* Woods Hole, Mass.: Woods Hole Oceanographic Institution, Ref. 65–10, 1965.

Barton, D. K. *Radar System Analysis.* Englewood Cliffs, N.J.: Prentice-Hall, 1964.

Bauer, G. *Measurement of Optical Radiations.* London: Focal Press, 1965.

Belcher, D. J. Five facets of aerial photography. *Photogrammetric Engineering* 19:746–752, 1953.

Borchert, J. R. Remote Sensors and Geographical Science. *Proceedings of the Fifth Symposium on Remote Sensing of Environment.* Ann Arbor: University of Michigan, Institute of Science and Technology, Willow Run, Laboratories, 1968.

Brock, G. C. *Physical Aspects of Air Photography.* London: Longmans, Green, 1952.

Brock, G. C. *Physical Aspects of Aerial Photography.* New York: Dover, 1967.

Colwell, R. N. To measure is to know—or is it? *Photogrammetric Engineering* 29:71–83, 1963.

Combs, A. C., Weickmann, H. K., Mader, C., and Tebo, A. Applications of infrared radiometers to meteorology. *Journal of Applied Meteorology* 4:253–262, 1965.

Conti, M. A. *Evaluation of Nimbus I High-Resolution Infrared Radiometer (HRIR) Imagery.* U.S. Geological Survey Technical Letter NASA-35, 1966.

Donner, W. Gas Chromatography as a Remote Sensing Device. *Proceedings of the Second Symposium on Remote Sensing of Environment.* Ann Arbor: University of Michigan, Institute of Science and Technology, Willow Run Laboratories, 1963.

Ewen, H. I. State of the Art of Microwave and Millimeter Wave Radiometric Sensors. *International Symposium on Electromagnetic Sensing of the Earth from Satellites.* Brooklyn, N. Y.: Polytechnic Press, 1967.

Gates, D. M. Remote sensing for the biologist. *Bioscience* 17: 303–307, 1967.

Gawarecki, S. J., Lyon, R. J. P., and Nordberg, W. Infrared Spectral Returns and Imagery of the Earth from Space and Their Application to Geologic Problems. Badgley, P. C. (ed.), *Scientific Experiments for Manned Orbital Flight,* Vol. 4. Washington, D.C.: American Astronautical Society, 1965. Pp. 13–33.

Haack, P. M. Evaluating color, infrared and panchromatic aerial photos for the forest survey of interior Alaska. *Photogrammetric Engineering* 28:592–598, 1962.

Heller, R. C. Imaging with Photographic Sensors. *Remote Sensing with Special Reference to Agriculture and Forestry.* Washington, D.C.: National Academy of Sciences, 1970.

Heller, R. C., Aldrich, R.C., and Bailey, W. F. Evaluation of several camera systems for sampling forest insect damage at low altitude. *Photogrammetric Engineering* 25:137–144, 1959.

Hodgin, D. M. Characteristics of Microwave Radiometry in Remote Sensing of Environment. *Proceedings of the Second Symposium on Remote Sensing of Environment.* Ann Arbor: University of Michigan, Institute of Science and Technology, Willow Run Laboratories, 1963.

Holter, M. R. Imaging with Nonphotographic Sensors. *Remote Sensing with Special Reference to Agriculture and Forestry.* Washington, D.C.: National Academy of Sciences, 1970.

Lancaster, J. Geographers and remote sensing. *Journal of Geography* 67:301–310, 1968.

Larmore, L. *Introduction to Photographic Principles.* New York: Dover, 1965.

Latham, J. P. Remote sensing of environment. *Geographical Review* 56:288–291, 1966.

Lowman, P. D., Jr., and Chang, T. L. Hyperaltitude Photography and Its Applications. *Proceedings of the Third Symposium on Remote Sensing of Environment.* Ann Arbor: University of Michigan, Institute of Science and Technology, Willow Run Laboratories, 1966.

McLerran, J. H. Infrared thermal sensing. *Photogrammetric Engineering* 33: 507–512, 1967.

Mees, C. E. K., and James, T. H. *The Theory of the Photographic Process.* New York: Macmillan, 1967.

Myers, V. I., Wiegand, C. L., Heilman, M. D., and Thomas, J. R. Remote Sensing in Soil and Water Conservation Research. *Proceedings of the Fourth Symposium on Remote Sensing of Environment.* Ann Arbor: University of Michigan, Institute of Science and Technology, Willow Run Laboratories, 1966.

National Academy of Sciences. *Remote Sensing With Special Reference to Agriculture and Forestry.* Washington, D.C., 1970.

Noble, V. E., Ketchum, R. D., and Ross, D. B. Some aspects of remote sensing as applied to oceanography. *Proceedings of the Institute of Electrical and Electronic Engineers* 57:594–604, 1969.

Olson, C. E. Elements of photographic interpretation common to several sensors. *Photogrammetric Engineering* 26:651–656, 1960.

Olson, C. E. The Energy Flow Profile in Remote Sensing. *Proceedings of the Second Symposium on Remote Sensing of Environment.* Ann Arbor: University of Michigan, Institute of Science and Technology, Willow Run Laboratories, 1963.

Page, R. M. *The Origin of Radar.* New York: Doubleday, 1962.

Park, A. B. Remote Sensing of Time-Dependent Phenomena. *Proceedings of the Sixth International Symposium on Remote Sensing of Environment.* Ann Arbor: University of Michigan, Institute of Science and Technology, Willow Run Laboratories, 1969.

Parker, D. Some Basic Considerations Related to the Problem of Remote Sensing. *Proceedings of the First Symposium on Remote Sensing of Environment.* Ann Arbor: University of Michigan, Institute of Science and Technology, Willow Run Laboratories, 1962.

Rouse, J. W., Jr., Waite, W. P., and Walters, R. L. Use of Orbital Radars for Geoscience Investigations. Center for Research in Engineering Science, University of Kansas Report, No. 61–68, 1966.

Simon, I. *Infrared Radiation.* Princeton, N.J.: Van Nostrand, 1966.

Skolnik, M. I. *Introduction to Radar Systems.* New York: McGraw-Hill, 1962.

Smith, J. T., Jr. Color—a new dimension in photogrammetry. *Photogrammetric Engineering* 29:999–1013, 1962.

Smith, N. Radar Technology and Remote Sensing. *Proceedings of the First Symposium on Remote Sensing of Environment.* Ann Arbor: University of Michigan, Institute of Science and Technology, Willow Run Laboratories, 1962.

Sorem, A. L. Principles of aerial color photography. *Photogrammetric Engineering* 33:1008–1019, 1967.

Suits, G. H. The nature of infrared radiation and ways to photograph it. *Photogrammetric Engineering* 26:763–772, 1960.

Weaver, K. F. Remote Sensing, New Eyes to See the World. *National Geographic Magazine* 135:46–73, 1969.

Wheeler, G. J. *Radar Fundamentals.* Englewood Cliffs, N.J.: Prentice-Hall, 1967.

PART THREE

REMOTE SENSING IN THE ULTRAVIOLET

Sensing at wavelengths shorter than those we can see

THE LEAST *explored and utilized major division of the electromagnetic spectrum is the ultraviolet. This is somewhat surprising because the ultraviolet region is contiguous to the visual limits of man's eyesight. In fact, some evidence suggests that a few unusual individuals can see into the longest wavelengths of the ultraviolet. This visual capability is apparently exploited more fully by the common honeybee. Another characteristic of the ultraviolet is that it is the region of the electromagnetic spectrum in which the silver halides of photographic emulsions are most sensitive to light. The ultraviolet comprises an extensive part of the electromagnetic spectrum. Normally it is defined as energy from 4000 Å to 100 Å. (One angstrom is one ten-billionth of a meter; it is precisely determined by the wavelength of the red spectrum line of cadmium divided by 6438.4966.) Toward the shorter wavelengths the ultraviolet merges gradually with x rays. Toward the longer wavelengths the ultraviolet changes gradually into the violet of visible light. The ultraviolet spectrum is sometimes divided into three parts based on wavelength: the* near *(4000 Å to 3000 Å), the* far *(3000 Å to 2000 Å), and the* extreme *(2000 Å to 100 Å) ultraviolet.*

8-Ultraviolet Radiation and the Terrestrial Surface

J. F. CRONIN
T. P. ROONEY
R. S. WILLIAMS, JR.
C. E. MOLINEUX
E. E. BLIAMPTIS

IN 1801, the year after Sir William Herschel observed that the solar spectrum extended beyond human vision (>7000 Å), J. W. Ritter discovered the ultraviolet region when he observed that silver chloride blackened more rapidly beyond the violet end of the spectrum (<4000 Å) than it did in the visible region. In subsequent years, it was found that energy in the ultraviolet, visible, and infrared regions differed only in frequency, that all three regions were in the electromagnetic spectrum, and that there were no sharp boundaries between contiguous spectral regions. By the early part of the twentieth century Lyman had extended the spectrum to 500 Å after it had become known that quartz, fluorite, and other substances allowed ultraviolet radiation to be observed at shorter and shorter wavelengths.

The ultraviolet frequencies have since been used for research in such diverse fields as astronomy, medicine, biology, atmospheric physics, and aerial reconnaissance. To the terrestrial scientist the more significant investigations have been those in the areas of (1) the luminescence of organic and inorganic materials, (2) the atmospheric ozone, and (3) the spectral and photographic studies of Mars and the Moon by astronomers and physicists during this century. It was not until after World War II that any thought was given to looking back at Earth at bandwidths other than those of the so-called photographic spectrum. Since that time, however, much interest has been generated in the development of remote sensing techniques at all wavelengths of the electromagnetic spectrum.

Office of Aerospace Research, U.S. Air Force, AFCRL 68–0572. 1968. Reprinted with permission.

The impetus toward a coordinated program of *nonmilitary* sensor research of terrestrial surfaces originated in 1964 in the offices of P. C. Badgley at NASA Headquarters in Washington, D.C. when Badgley, with W. A. Fischer of the U.S. Geological Survey, R. J. P. Lyon of Stanford Research Institute, J. F. Cronin of Air Force Cambridge Research Laboratories, and a few other terrestrial scientists set out to establish a set of techniques suitable for the exploration of lunar and planetary bodies (including Earth) from orbiting spacecraft. The entire electromagnetic spectrum was to be scanned for regions proved or potentially useful to scientists seeking to extend their knowledge of the solar system. This program has to date elicited only one or two sustained efforts of nonmilitary research in the ultraviolet spectrum.

Although the ephemeral variations in the composition, physical properties, and distribution of the earth's atmosphere have been intensely studied by atmospheric physicists for several decades, it is only since manned orbitors were launched by NASA that the spectral character of the earth's surface has provoked much curiosity. Indeed, it is now becoming apparent how very little is actually known of the reflection, absorption, polarization, and emission properties of the earth's vegetative, rock, and sedimentary cover. Environmental spectral data during the diurnal and annual cycles are particularly sparse.

Among the conventional subdivisions of the electromagnetic spectrum—gamma ray, x ray, ultraviolet, visible, infrared, microwave, radio, and audio—perhaps the least explored and least utilized is the ultraviolet. This is so in spite of the fact that the ultraviolet region is contiguous to man's visual limits—and the region where silver halides of photographic emulsions are most light-sensitive.

Ultraviolet Radiation

Ultraviolet radiation is characteristically defined as that part of the electromagnetic spectrum from 4000 Å to 100 Å, or between the visible region and the x ray region. The ultraviolet spectrum may be divided into the near (4000 Å to 3000 Å), far (3000 Å to 2000 Å), and extreme (2000 Å to 100 Å) ultraviolet. The primary source of energy at these wavelengths is the sun.

The total solar irradiance of Earth is essentially constant, although there is a seasonal variation of –3.27 percent at aphelion and +3.42 percent at perihelion. In the far ultraviolet, solar flares cause relatively large fluctuations in intensity, but these are of little consequence at the earth's surface since the atmosphere absorbs solar energy at wavelengths shorter than 2900 Å.

Solar irradiance within the earth's atmosphere is, to some extent, a function of the integrated density of atmosphere in the path of the solar beam. Solar ultraviolet intensity depends on such factors as elevation above sea level, latitude, time of day, and time of year.

The intensity of solar ultraviolet radiation within the earth's atmosphere or at sea level is not simply a function of the air mass through which the solar energy has traversed, but also of the turbidity, optical properties, chemical composition, component distribution, and, especially, the amount of total ozone. Measurements of the ultraviolet flux, after the incoming solar radiation has passed through the atmosphere and been reflected from a surface, depend on the optical thickness of the atmosphere, the zenith angle of the sun, the zenith angle and azimuth of the direction of observation, and the spectral reflection, absorption, emission, and polarization of the terrestrial surface.

The physical processes that effect changes in ultraviolet radiation are primarily scattering and absorption by the atmosphere. Scattering is the angular redistribution of radiation incident upon the atmosphere. It is of two types: Rayleigh, or molecular, scattering; and Mie, or aerosol, scattering. The Rayleigh molecular law of scattering may be written as:

$$\beta_{\phi\lambda} = \frac{2\pi^2}{N\lambda^4}\,(\mu_\lambda - 1)^2\,(1 + \cos^2\phi)$$

where β_{ϕ_λ} is the Rayleigh scattering coefficient at angle ϕ and wavelength λ; N is the number of scatterers per unit volume, and μ_λ is the wavelength-dependent index of refraction of the scatterers. A popular description of Rayleigh scattering is that it is inversely proportional to the fourth power of the wavelength.

For the larger Mie particles, the intensity of the scattered light is a complicated function of λ, μ, ϕ, and the radius of the scatterers. Field measurements seldom agree with classical theory largely because of the variable concentrations of aerosols in the atmosphere.

All of the major atmospheric gases—nitrogen, oxygen, water vapor, carbon dioxide, and ozone—and some of the minor constituents—such as nitrous oxide, methane, and carbon monoxide—demonstrate various degrees of absorption. The most important of these is ozone since it is the ozone in the earth's atmosphere that effectively terminates solar irradiation of the earth's surface below 2900 Å.

Ozone

The first to note the important relationship between the ultraviolet region and the atmosphere was Cornu (1879), who suggested that the abrupt termination of the solar spectrum in the ultraviolet at ground level was due to absorption in the atmosphere. Hartley (1880), delineating what are now termed the Hartley absorption bands, suggested that ozone was responsible for this absorption in the ultraviolet. Several decades later, Fabry and Buisson (1913, 1921) proved the existence of ozone in the atmosphere when they discovered that the absorption pattern of sunlight in the ultraviolet matched their earlier laboratory measurements of the absorption spectrum of ozone. With G. M. B. Dobson's (1930) development of the ozone spectrophotometer, a worldwide network of stations was eventually established to provide continuous daily measurements of ozone.

Absorption by Ozone

There are three principal absorption bands in the ozone spectrum: the Hartley bands in the far ultraviolet, the Huggins bands in the near ultraviolet, and the Chappuis bands in the visible region (Figs. 1, 2, 3). The Hartley bands, which are centered at 2553 Å and dominate the ozone spectrum with a very strong continuum, generally block out all radiation shorter than 3000 Å. The Huggins bands overlap but are weaker than the Hartley region, and extend to 3400 Å. Between 3400 Å and 4500 Å there exists a relatively transparent window. In the region from 4500 Å to 7500 Å there are the Chappuis bands, the weakest bands of the ozone spectrum, with a peak absorption of about 7 percent for solar radiation traversing two air masses. Although the Chappuis absorption is small, its optical effect at twilight is significant.

Since absorption of ultraviolet radiation by ozone is an exponential function of the quantity of ozone along the path length of irradiation (Bouger's law), relatively small variations of total ozone in wavelength regions of strong absorption may cause large changes in ultraviolet intensity. At the earth's surface, variations in solar ultraviolet radiation, which are related to variations in total ozone, are small. Seasonal changes in total ozone increase poleward and are about ±35 percent in the middle and high latitudes. The standard deviation of total ozone about the seasonal mean is approximately 10 to 15 percent at all latitudes; the standard deviation of total ozone about the monthly mean is less than 15 percent for all latitudes (Hering and Borden, 1967).

From 3000 Å to 3700 Å the standard deviation of solar UV at the earth's surface, owing to fluctuations in ozone, is approximately 5 percent day to day, and 10 percent seasonally (W. S. Hering, private communication). Relative changes in intensity due to changes in ozone are somewhat larger at noon and somewhat smaller near sunset, owing to the increase in optical depth with increasing solar zenith angle.

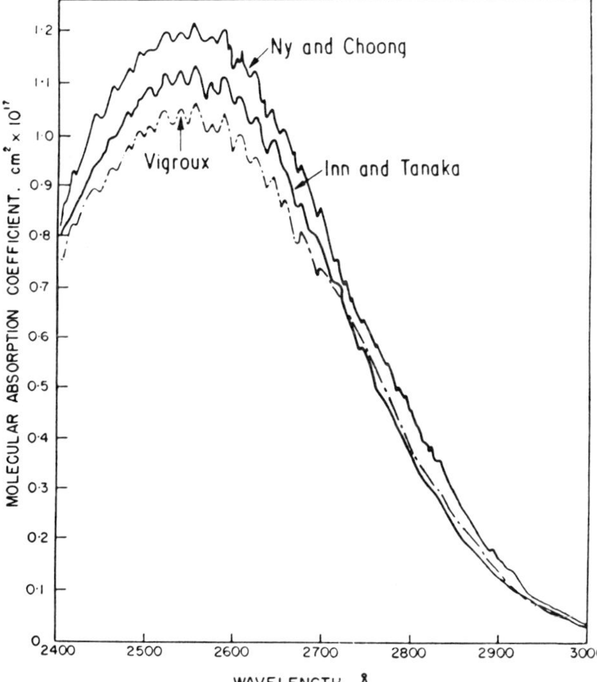

Figure 1 Molecular absorption coefficients in the Hartley bands of ozone, at 18°C. (Goody, 1964)

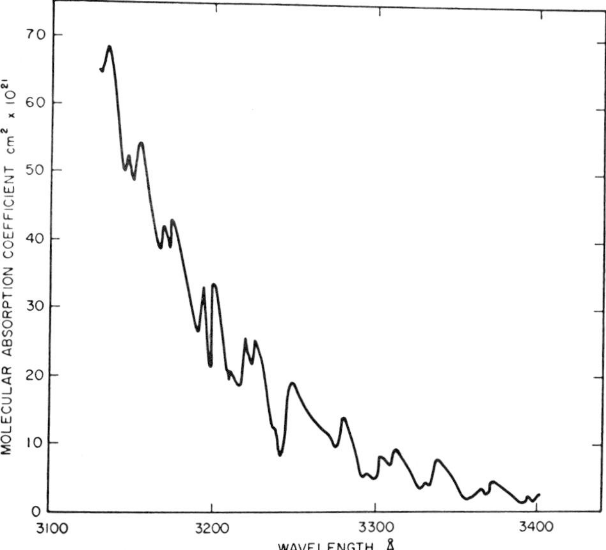

Figure 2 Molecular absorption coefficients in the Huggins bands of ozone, at 18°C. (Goody, 1964)

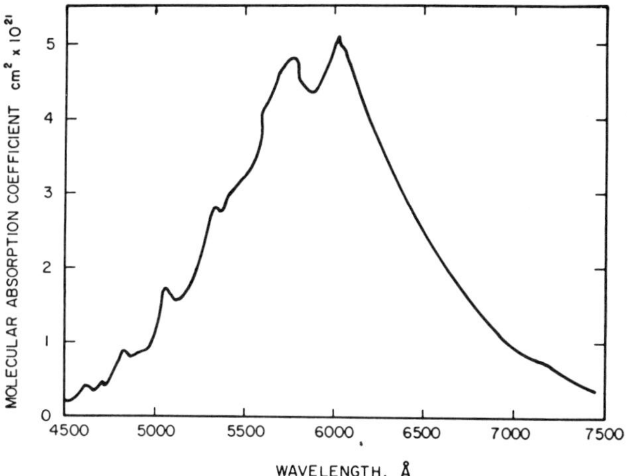

Figure 3 Molecular absorption coefficients in the Chappuis bands of ozone, at 18°C. (Goody, 1964)

Total Ozone

Ozone is formed primarily by photodissociation of molecular oxygen and subsequent recombination of atomic and molecular oxygen. The primary source region is near 30 km. Total ozone and changes in total ozone are least at the equator, but both increase with increasing latitude. In the northern hemisphere the amount of total ozone varies with latitude and season; it is greatest in the spring and least in the fall. In addition, there are daily or short-period variations in total ozone that are the result of vertical or horizontal air motions.

Numerous studies (Johnson, 1954; Craig, 1965; and others) have shown that there is a definite relationship between total ozone and the cyclonic-anticyclonic circulation of the lower stratosphere and the troposphere. In the trailing area of a cyclonic trough, ozone both subsides and increases because of horizontal advection from the polar regions and the effect of the vertical motions that have been described by Reed (1950) and Normand (1951). Conversely, in the leading area of the trough, or the region of the anticyclonic circulation, ozone is observed to ascend and decrease.

Ozone Distribution

F. W. Paul Götz (1931, 1934) discovered that with the decrease in the zenith angle of the sun shortly after culmination, the decrease in intensity at the ultraviolet wavelengths is more rapid than at the longer wavelengths; when the sun is low the reverse is true, the decrease in intensity at the shorter wavelengths being slower than at the longer ones. G. M. B. Dobson confirmed this in 1930 when he found that the ratio of ultraviolet to longer wavelengths does increase at first as the sun rises, then becomes constant, and eventually shows the normal decrease expected with the increasing height of the sun.

Götz developed the so-called umkehr method, which depends on a simple measurement of the ratio of the intensity of two ultraviolet wavelengths in the Hartley ozone band. (See Fig. 4.) It gives an accurate measure of the average height of the ozone, but only a general

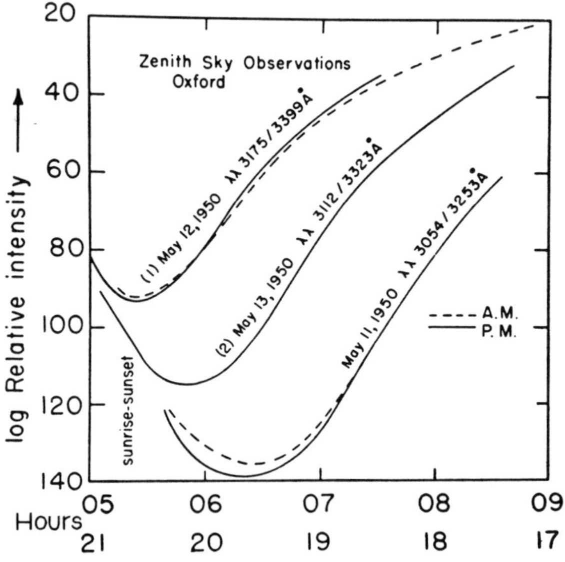

Figure 4 Measurements of zenith skylight near sunrise and sunset. (Johnson, 1954)

estimate of its vertical distribution. The umkehr method is used only on clear days, and normally when the zenith angle of the sun is between 90° and 60°.

The development in 1960 of the bubbler type of electrochemical (wet chemical) ozonesonde by Brewer, and the chemiluminescent (dry chemical) ozonesonde by Regener (1960, 1964) has considerably increased our ability to measure vertical ozone distribution. The chemical ozonesondes can be used at any time of the day or night since they do not need solar radiation to be activated. They also determine the variation of ozone with height in greater detail than was possible with earlier methods.

The seasonal and latitudinal variations of ozone in a vertical direction through the atmosphere have been quantitatively described by Paetzold (1963), who defined the ozone profile for the polar and equatorial latitudes as follows:

Equatorial ozone profile

1. Only 10 percent of the ozone is between ground level and 18 to 20 km.
2. The increase of ozone density does not begin at the tropopause, but at 1 to 3 km above it.
3. The profile of ozone density in the atmosphere has a single sharp maximum at about 26 to 28 km.
4. There are little, if any, seasonal variations.

Polar ozone profile

1. 60 percent of the ozone is below 20 km in the spring; only 20 percent is below 20 km in the fall.
2. The increase of ozone begins immediately above the tropopause.
3. There is a marked seasonal variation in the total amount of ozone below 20 km. It is greatest in the spring and least in the fall.

Terrestrial Surfaces

Although visual and photographic contrast definition or contrast attenuation is most significant, determinations of the photometric properties or optical properties of the features or objects whose contrast is significant are often neglected. A few casual measurements of some of the properties of paint, concrete, grass, or glass have been reported. The background is usually described as some kind of formation, or as a clump of vegetation, with no indication of the optical properties. Many studies of polarization and reflectance are greatly diminished in value when the material is described simply as a desert sand or a white soil. Measurements, for instance, of such important parameters as the composition, size, shape, and orientation of the grains composing the sediment or soil, or of the microrelief of the surface, are utterly lacking. Most investigators seem to be unaware of the complexity of an apparently uniform terrestrial surface.

Organic matter, a major constituent of soils, is intimately mixed with the water, air, and mineralogic materials of soils. Rock surfaces and the regolith also support a variety of life forms, primarily bacteria and actinomycetales, and various algae, fungi, and lichens. Again, many investigators appear to be unaware of the ubiquity of such microorganisms. For instance, no research has been directed toward the study of the optical properties of the natural, undisturbed, weathered surfaces of rocks and sediments and their organic surface materials. Further, although 75 percent of the earth's surface is composed of soil, unconsolidated sediments, or sedimentary rock, studies in situ of the reflection, absorption, polarization, and emission properties of such lithologic materials are almost unknown.

Rocks and Minerals

In the visible and infrared regions the laboratory spectrophotometric characteristics of rocks and minerals are well known, but in the ultraviolet such information is scanty. Two recent reports (Greenman et al., 1967; Thorpe et al., 1966) are the first to provide data in the far-ultraviolet region (2000 Å to 3000 Å). As for the near-ultraviolet region (3000 Å to 4000 Å), little data is available. To fill the need, two of the writers of this report are obtaining spectra in the near-ultraviolet region of a considerable number of rocks, minerals, and sediments.

In the near-ultraviolet and the shorter-wavelength regions of the visible spectrum, the carbonates, phosphates, and evaporites are usually more reflective than other lithologic materials. Acidic rocks such as granite and rhyolite show little reflectance in the ultraviolet but considerable reflection in the visible; basic rocks such as basalt show little reflection in either the ultraviolet or visible (Fig. 5).

Hemphill (1968) has shown that at 2500 Å in the ultraviolet, limestone is 2½ times more reflective than granite and 10 times more reflective than rhyolite.

Greenman (1967) has measured the reflectance in the ultraviolet from 2000Å to 3000 Å of the silicate rocks, granite, gabbro, and serpentine, both solid and granular (Fig. 6). Reflectance declined gradually from the longer to the shorter wavelengths. No differences attributable to composition were discovered. Spectroscopy of a higher resolution than normally used is suggested by the authors if further investigation is warranted.

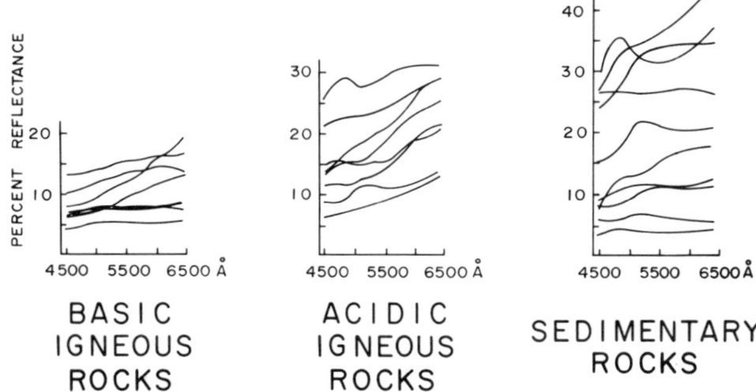

Figure 5 Spectral reflectance of the principal rock types. (After V. V. Sharonov, 1956)

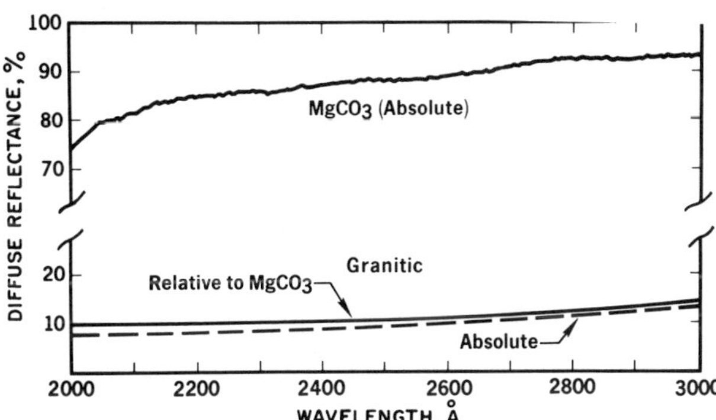

Figure 6 Diffuse reflectance of solid granitic sample and of $MgCO_3$ standard. (Greenman, Burkig, and Young, 1967)

It should be noted that neither Hemphill's nor Greenman's studies were of weathered or naturally occurring lithologic surfaces.

AFCRL Studies

Experiments at AFCRL (Air Force Cambridge Research Laboratories) have been conducted to provide a basis for more extended interpretations of ultraviolet reflectance from geologic materials. Only freshly exposed surfaces have been measured to date but studies of weathered surfaces are scheduled. The influence of composition and grain size on spectral reflectance has been assessed for a few common rocks and minerals. Spectra were obtained from 2400 Å in the far-ultraviolet to 1.8μ in the infrared region. Measurements were made on a Cary model No. 14 spectrophotometer equipped with a model No. 1411 diffuse reflectance accessory.

The Cary No. 14 is a ratio-recording double-beam instrument using two monochromators, a fused-silica foreprism and a 600 line/mm echelette grating. The spectral bandwidth is not constant since the width of the monochromator slits is varied automatically in order to maintain constant energy in the reference beam.

A ring collector is used to measure the diffuse reflectance. The sample is illuminated at normal incidence by monochromatic light. A phototube, mounted at the top of the ring collector, alternately receives reflected radiation from the annular mirror viewing the sample at 45° ± 7° and from the reference beam. A screen allows the reference beam to be standardized to a reference sample. The reference for all measurements is a freshly prepared $MgCO_3$ surface.

The ring collector makes it possible to measure diffuse reflectance of horizontally mounted powdered samples, a considerable advantage over methods that require the samples to be mounted vertically. Vertically mounted powders, unless they are very finely divided, have to be covered. Reflection losses of samples covered by glass are 10 to 15 percent between 4000 Å and 7000 Å. In addition, below 3500 Å the glass begins to absorb strongly, although this might be avoided by the use of an optical fused-silica cover. (See Table 1.)

The reflectance of coarse powders may be affected by the packing arrangement of the grains. Reflectance curves for compacted and uncompacted samples of quartz grains with diameters of 420μ to 500μ and <53μ are given in Figure 7. The compacted samples of larger grain size show 2 to 5 percent higher reflectance throughout the 3000 Å to 7000 Å range, with more pronounced differences at the longer wavelengths. The upper curves of Figure 7 show that packing has little influence on the reflectance of <53μ powders.

Based on these results, a standard loading technique for powders was developed. The sample holders were filled to above the top of the sample holder, a straight edge was drawn across the holder to level the powder to the top of the holder, a small additional amount of powder was added, and a 500-gm weight was then placed on top of the powder and rotated until the powder surface was smooth and level. This procedure gave highly reproducible results.

Table 1 Wavelength Range, Light Sources, and Detectors

Wavelength Range (A)	Source	Detector
2400–3500 Å	Nester H_2 lamp	Dumont 7664[a]
3200–7000 Å	Quartz halogen tungsten lamp[b]	Dumont 7664
7000 Å–1.8μ	Quartz halogen tungsten lamp[b]	Dumont Pbs

[a] s-13 response, fused-silica window.
[b] Operated at 70 V.

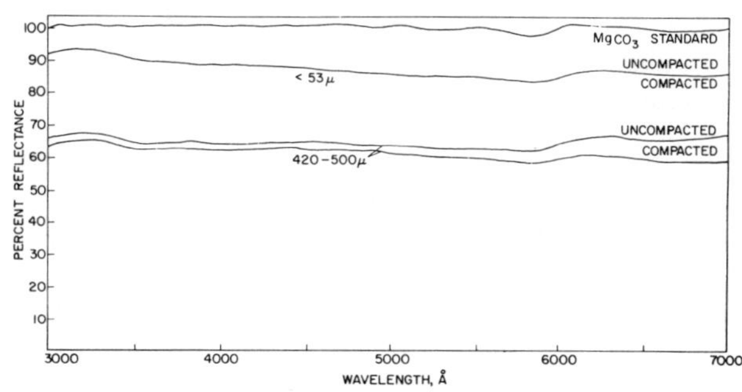

Figure 7 Compaction effects on the spectral reflectance of powdered quartz.

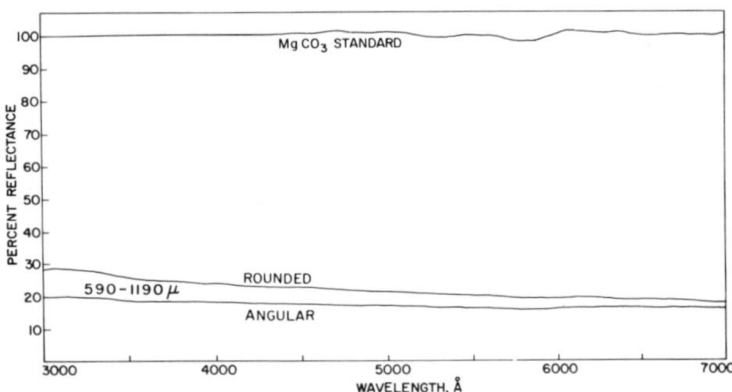

Figure 8 Influence of particle shape on reflectance. The upper curve represents the average of three tests of rounded particles whose grain diameters are from 590 μ to 1190 μ. The lower curve represents the average of three tests of angular particles whose grain diameters are from 590 μ to 1190 μ.

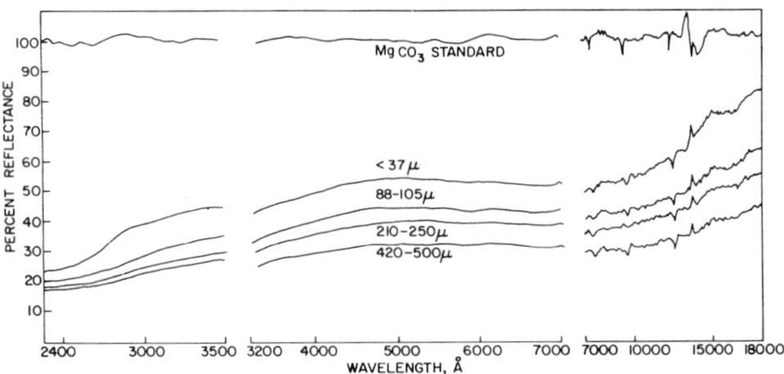

Figure 9 Spectral reflectance of alkali granite.

Particle shape or preferred orientation may also bear importantly on the reflectance response of natural materials. Some common minerals such as clays have a flaky habit and thus in water they tend to settle out parallel to their prominent cleavage.

One experiment was conducted to measure differences in reflectance as a function of grain shape (Fig. 8). The National Crushed Rock Institute prepared several samples of crushed rock, carefully graded with respect to size and shape.[1] Three reflectance runs were made of rounded aggregate and three of angular grains. Both samples had grain diameters of 590μ to 1190μ. The rounded samples were more reflective by 2 to 9 percent, the greatest difference being in the ultraviolet region.

Figures 9 through 18 show the reflectance spectra of granite, rhyolite, obsidian, basalt, and peridotite, of the three iron oxides limonite, goethite, and hematite, and of quartz and calcite; the grain sizes are <37μ to 500μ. Several features are evident from examination of these spectra. The most salient features are:

1. No distinctive peaks are present in the wavelength range 2400 A to 1.8μ.
2. The reflectance intensity falls off from the near infrared to the ultraviolet. In the ultraviolet the reflectance level is generally from 15 to 30 percent, and the curves for various grain sizes tend to converge in this region. There is little difference in reflectance of light or dark rocks.
3. The reflectance level, especially at longer wavelengths, depends distinctly on grain size; the finer the grain size, the greater the reflectance.
4. The spectra of the three iron oxides (limonite, geothite, and hematite, Figs. 14 to 16) follow the same trends as noted for the rocks.
5. The spectral features of two transparent minerals (quartz and calcite, Figs. 17 and 18) depart somewhat from those of the rocks and iron oxides. The major difference is that the reflectance level is high and remains high throughout the range 2400 A to 1.8μ. The dropoff in the ultraviolet is slight, except perhaps for the coarsest grain size.

[1] Furnished by Prof. H. F. Winterkorn, Princeton University.

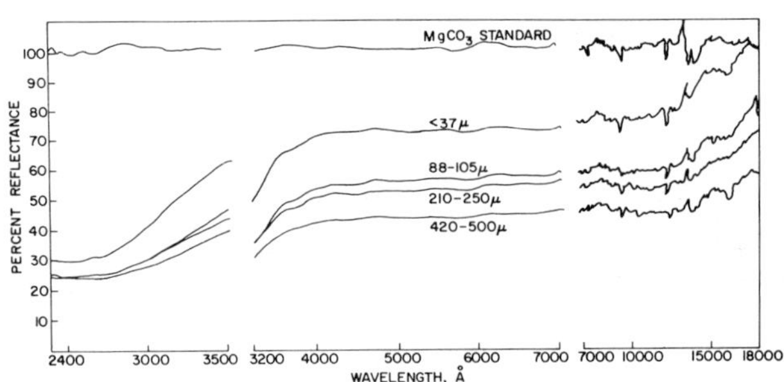

Figure 10 Spectral reflectance of rhyolite.

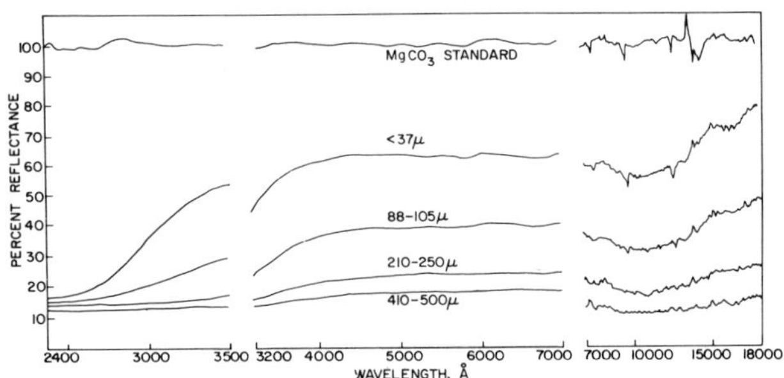

Figure 11 Spectral reflectance of obsidian.

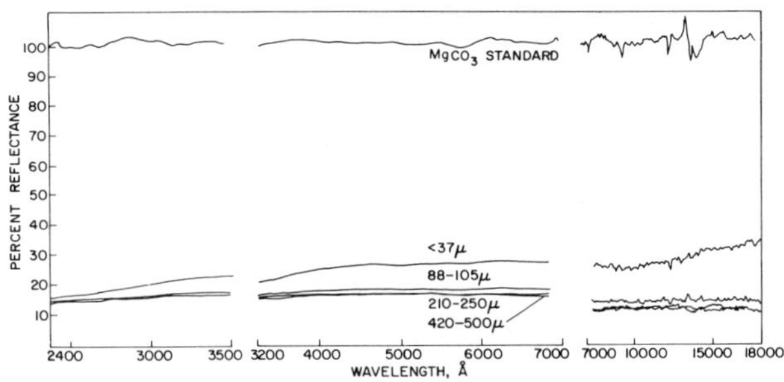

Figure 12 Spectral reflectance of basalt.

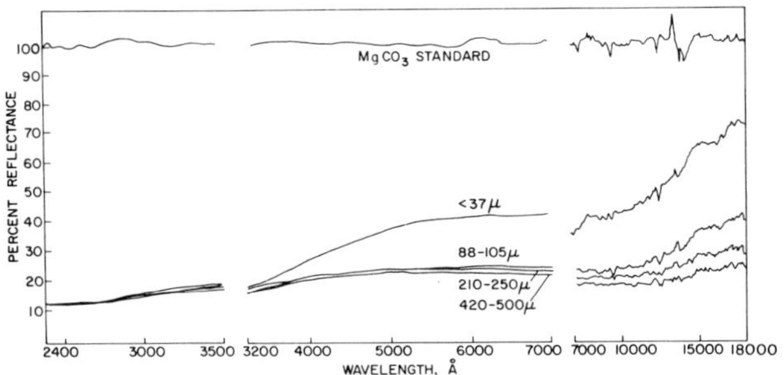

Figure 13 Spectral reflectance of peridotite.

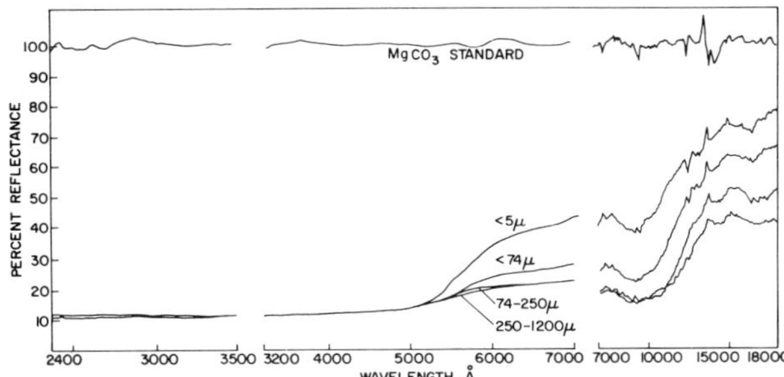

Figure 14 Spectral reflectance of limonite.

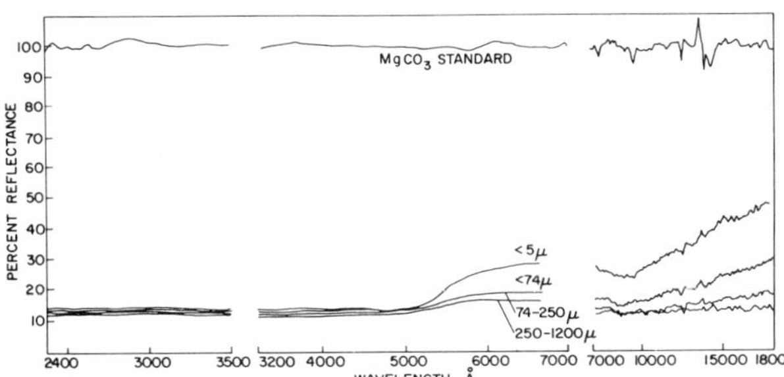

Figure 15 Spectral reflectance of goethite.

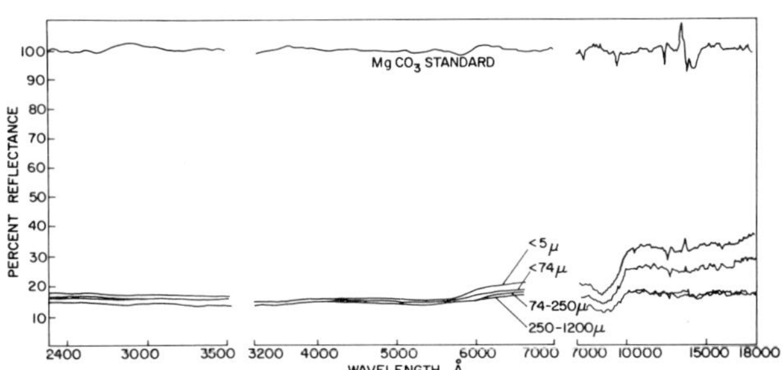

Figure 16 Spectral reflectance of hermatite.

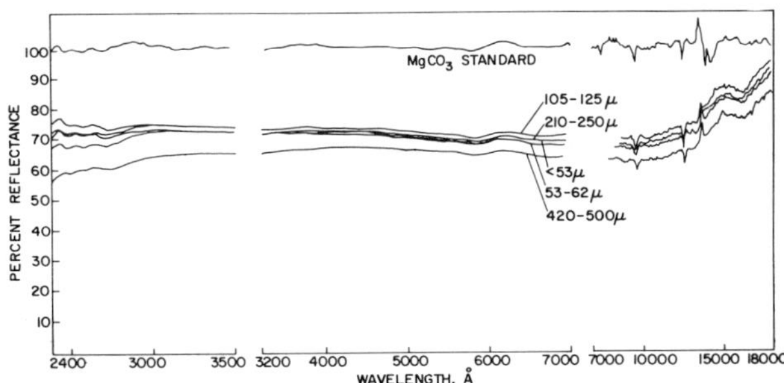

Figure 17 Spectral reflectance of quartz.

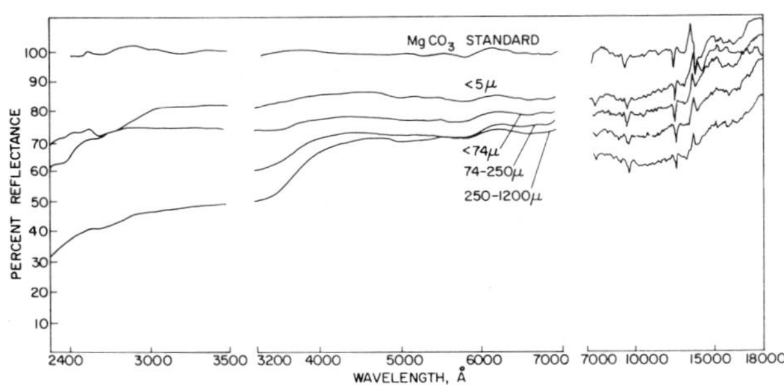

Figure 18 Spectral reflectance of calcite.

Vegetation, Soils, and Sediments

The current resurgence of interest in multispectral studies in the fields of forestry and agriculture is undoubtedly due to R. N. Colwell of the Department of Forestry of the University of California. Other investigation centers following his lead include the U.S. Department of Agriculture, the National Bureau of Standards, Purdue University, and the Willow Run Laboratories of the Institute of Science and Technology of the University of Michigan.

Support to Purdue and Michigan was provided in 1963 by the U.S. Army Electronics Command and in 1963 by the National Aeronautics and Space Administration. As Polcyn (1967) reported, the early work at Michigan and Purdue confirmed the differences between crops. Many of these differences had been pointed out earlier by Colwell (1956, 1961, 1963) in his many papers. Polcyn and his coworkers also observed that tonal differences in selected bands for certain crops depended on sun and view-angle geometry, state of maturity (which is in turn a factor of soil and moisture), herbicide treatments, irrigation, row direction, and wind damage.

One of their more interesting observations was the effect of scan angle and scan direction at the bandwidth 0.502 to 524μ (Fig. 19). It was found that the scan-angle dependence at this bandwidth was quite different from that at other bandwidths. For instance, it was as much as +60 percent of the nadir (0°) value of the bandwidth from 0.58 to 0.62μ but only +5 to +10 percent of the infrared channels 0.72 to 0.8μ and 0.8 to 1.0μ.

Michigan and Purdue concluded that calibrated, simultaneous, multispectral sensing does indeed provide a basis for automatic recognition, and recommended a broadband multichannel instrument, from the ultraviolet to the infrared. Their work, which had been confined to the visible and infrared regions, had not included the ultraviolet and therefore has not contributed to the little information that was heretofore available on the ultraviolet spectrum.

Spectral properties of leafy materials are strongly dependent on the characteristics of the cellulose in the cell walls, of water and solutes within the cells, of intercellular air spaces, and of pigments within the chloroplasts. The usual pigments, in order of abundance, are chlorophylls, xanophylls, and carotenes. Absorption spectra of plant pigments and water (Fig. 20), measured by Gates et al. (1965), indicate that all the predominant pigments absorb strongly in the vicinity of 4450 Å. Chlorophyll a and chlorophyll b, which are most frequent in the higher plants, also absorb significantly at 6450 Å. In the visable range, plants lacking cholorophyll show markedly less absorption than other plants.

Spectral absorption by chlorophyll converts absorbed energy into heat or fluorescence. The principal representatives of plant life known to be luminous are certain bacteria and fungi; their emission of light is continuous and independent of any stimulus. The Japanese, who have been particularly interested in the study of luminous plants and animals, grew luminous bacteria for use during blackouts in World War II.

In the tropics there are many species of luminescent fungi that appear to multiply exuberantly during the

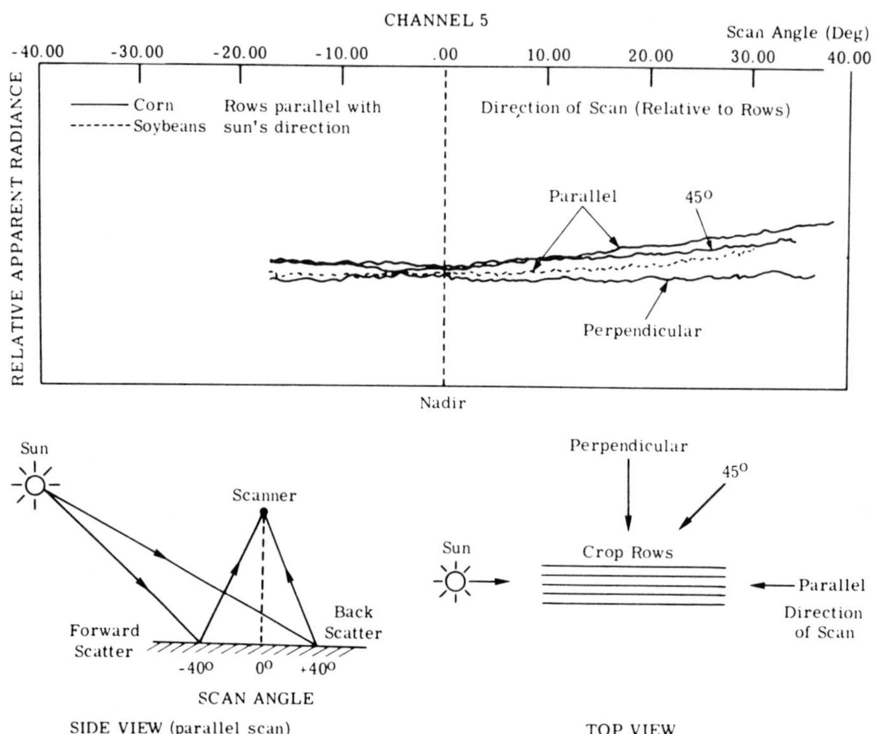

Figure 19 Effects of scan angle and scan direction on crop rows. (Polcyn, 1967)

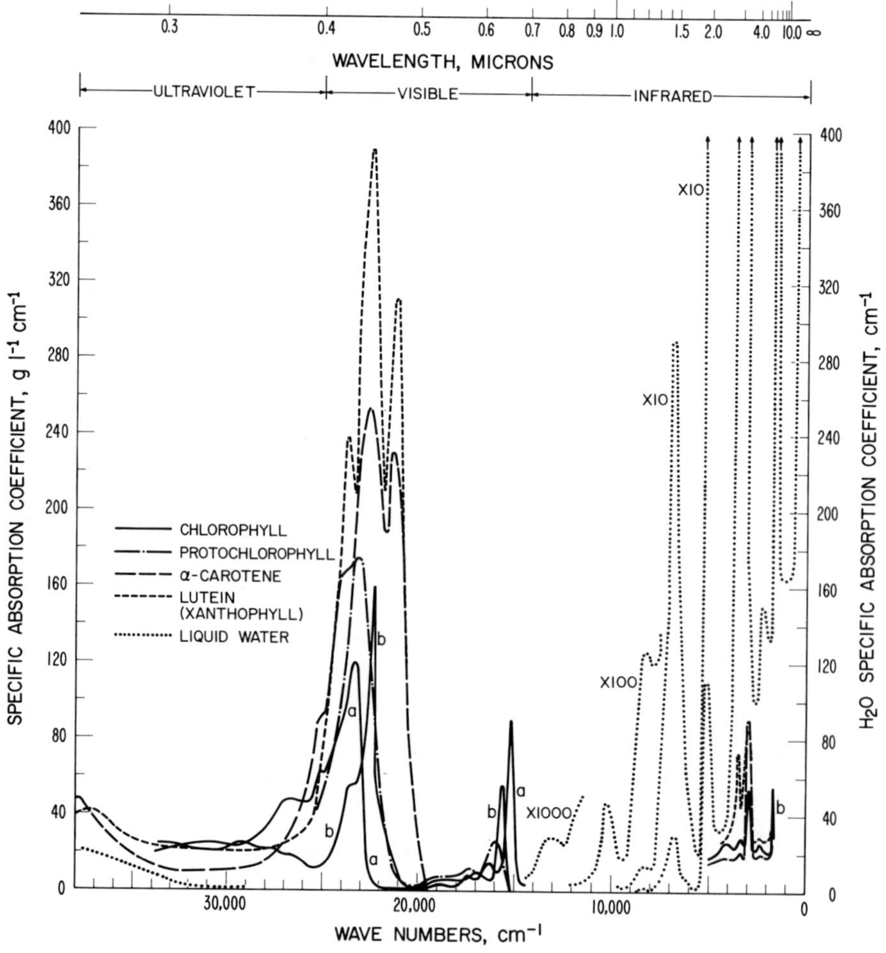

Figure 20 Absorption spectra of plant pigments and water. (Gates et al., 1965)

rainy season. Spectral data of luminescent fungi and bacteria, together with pertinent environmental information, are not known to the writers; information on bacterial luminescence versus pH, temperature, and salinity of environment, is available.

Plants possessing protochlorophyll and chlorophyll a and b may demonstrate a shift in spectral properties during growth. The red protochlorophyll reflectance band may shift to shorter wavelengths, and the development of chlorophyll at a later stage may be accompanied by a recurrent shift to shorter wavelengths. With further growth, however, the reflectance peak in the green diminishes and that at infrared wavelengths increases.

As is true of vegetation, soils also readily absorb energy in the ultraviolet and reflect and scatter it at longer wavelengths. Baumgartner (1953) found that the larger the grain size of soils, the deeper the radiation penetrates the soil. In the coarsest soils, significant amounts of radiation reach depths of 5 mm or more. Earlier, Sauberer (1951) had observed that at a depth of 1.8 mm in wet sand, solar energy at 6600 Å is 100 times that at 4250 Å.

Water, Snow, and Ice

The transmission of incident solar energy by both salt waters and fresh waters is strongly wavelength-dependent. Further, the transmission properties of coastal waters and inland waters vary greatly with turbidity and the plankton content of the water. For example, the wavelength of maximum penetration shifts from the blue in clear water to the red in muddy water. In a study of the penetration of solar and sky ultraviolet radiation in the Mediterranean Sea, Lenoble (1956) observed that at 4000 Å, the amount of energy at a depth of 30 meters is only 10 percent that at the sea surface. At 3200 Å, the 10 percent depth is 10 meters.

For snow and ice, penetration is greatest between 0.4 and 0.5μ, decreasing toward the shorter and longer wavelengths. The transmissivity of snow decreases with increasing water content. At a depth of 10 cm it may vary from 2 percent to 20 percent. Glacier ice of the same thickness may transmit as much solar radiation as pure water.

Spectrophotography

Spectral photography originated with astronomers' photographic studies of the Moon and Mars during the early part of this century. In photographs taken in the UV at 3230 Å nearly 60 years ago, R. W. Wood (1910) noted an anomalous area on the Moon, the so-called Wood's Spot, near the crater Aristarchus. In 1929 W. H. Wright of Lick Observatory published his lunar photography at six bandwidths, one of which was centered at 3600 Å.

Wood (1911) had also shown some curiosity about the use of filter techniques in terrestrial photography and was probably the first to note the greater reflectance of trees in the IR than at conventional wavelengths. Angerer (1930) also published some early examples of UV and IR photography of landscapes.

The first aerial spectrophotography of Earth may have been made in the middle 1930s by the Russian, E. L. Krinov (1947), and his associates. The first to have obtained spectrophotometry of the natural surfaces of vegetation and rocks, these workers are known to have also photographed features in the same bandwidths as those used in their photometry. So far as we know, however, they did not publish their photography.

In 1960 the American Society of Photogrammetry, under the editorship of R. N. Colwell, published a *Manual of Photographic Interpretation* that was at that time an excellent summary of developments in nonmilitary photo interpretation. This volume contained some of the first spectrophotographs taken from aircraft. By 1964, with the support of many representatives of the scientific community, the National Aeronautics and Space Administration had launched an intensive exploration of potential photographic and sensor techniques.

Discrete bandwidths in spectrophotography can be achieved by proper selection of film emulsion and sensitivity and appropriate filters. All terrestrial materials such as bedrock, sediments, soils, vegetation, and other organic matter have rather specific spectral reflection, absorption, emission, and polarization characteristics. Spectrophotography, with other sensor techniques, may therefore bring about more rapid and positive identification of terrestrial features from airborne sensor platforms.

Many factors are significant in the measurement in situ of the spectral characteristics of terrestrial material. Among these are the amount and distribution of water vapor, carbon dioxide, ozone, and aerosols in the atmosphere, the angle of solar illumination, the kind and amount of cloud cover, the microrelief of the surface, the micrometeorology at the site, and the amount and kinds of microorganisms that are supported on the weathered surfaces of rocks and sediments. Such information is not always obtainable in spectrophotographic and spectrophotometric studies. It is particularly important to make such measurements at the same time as the aircraft and satellites, with their sensors, are overhead.

A related subject of interest is the application of spectrophotographic techniques to submerged coastal areas and inland bodies of water.

. . . .

Instrumentation

Photographic or scanning imagery in the ultraviolet region of the spectrum is generally collected on ultraviolet-sensitive film used in (1) conventional cameras with ultraviolet-transmitting narrow-band filters, or in (2) specially designed high-resolution ultraviolet cameras; and with (3) opticomechanical scanners using photomultiplier tube detectors.

The most efficient narrow-band UV photography to date (Mangold, 1966), covering the 3000 Å to 4000 Å spectral range, was accomplished by means of a set of Fabry-Perot thin-film evaporated filters. The camera used was a standard Graflex fitted with an achromat lens of lithium fluoride and quartz. Several varieties of commercial UV film (Tri-X, Kodak 103-0, 1-N, SO-243) can be used.

Ultraviolet cameras have been commercially developed and used for some time. One such camera, which has been successfully used in a wide variety of UV photographic applications under both natural sunlight and artificial illumination, has an optical system whose f/1.5 lens of 6-inch focal length is capable of resolving 100 lines/mm at the focal plane, with maximum transmission and minimum internal scatter. It accommodates a variety of recording media, in both sheet and roll-film magazines, and uses the same filters as described above to provide a narrow-band input to the film. Further details of the development and test program are given by Perkin-Elmer Corp. (1965), who have announced advances in development of an airborne UV camera system incorporating rapid framing and wider shutter capability.

The proven ability of airborne imaging infrared radiometers (generally called *infrared scanners*) to generate high-quality imagery of terrain surfaces has led to their use in the ultraviolet region. Modification is readily made by incorporating an ultraviolet-sensitive photomultiplier detector that has an S-11 photocathode and using filters that limit the spectral response to wavelengths shorter than 4000 Å. Such a system has been successfully used by the U.S. Geological Survey (Hemphill, 1966) to generate high-quality UV imagery of geologic features from altitudes as high as 15,000 feet. For imaging features in terms of their own individual ultraviolet reflectance, rather than achieving integrated total reflectance from adjacent scan lines, the system must be operated without automatic gain control.

A similar conversion kit is being incorporated into the AFCRL airborne infrared scanner for future use. This scanner has a 70° field of view and resolution of 2 mrad, typical of present-day scanners. Further description of the scanning system is contained in Fisher (1965). For airborne ultraviolet photography, it should be remembered that the normal glass windows covering aircraft camera wells do *not* transmit ultraviolet. Fused silica or quartz windows must therefore be installed. A UV-transmitting fused lens is being incorporated by AFCRL in its multiband nine-lens camera.

Radiometers or spectrophotometers can readily be adapted to measure reflectance in the ultraviolet as well as in the visible or infrared regions of the spectrum. A triradiometer recently developed for AFCRL (Cronin and Noble, 1968) has a spectral range from 2600 Å to 11300 Å in 30 bands, consecutively sampling sun, sky, and target radiance. At each spectral setting, a detector views the radiation from a collector mirror directed into a reflectance sphere. Two detectors are used interchangeably to cover the entire spectral range. The radiometer fields of view are controlled from 1.6° to 0.1°. The amplified current outputs from the detectors are recorded on a paper chart and the entire system is self-powered for field operation.

Hemphill and Carnahan (1965) of the U.S. Geological Survey have demonstrated that in the field, such materials as talc and dolomite, when stimulated by an ultraviolet transmitter, may be imaged at a distance of several hundred feet by an image dissector sensitive to visible light. Their transmitter consisted of a cathode ray tube, UV phosphor, and a sequentially illuminating raster scan.

A technique demonstrated by the U.S. Geological Survey to be of potential value at night is to use a pulsed nitrogen gas laser, emitting at 3317 Å, to stimulate luminescence of various rocks and minerals. On the basis of the luminescence decay time, it is possible to discriminate between granites of differing mineralogic composition and also between various feldspars. The sodic varieties of feldspars appear to have the longest decay times, longer than those of calcic feldspars.

The most promising USGS technique, termed by Hemphill (1968) a Fraunhofer line-depth method, uses the sun as the ultraviolet source. The technique is one of observing the ratio between the central intensity of a selected Fraunhofer line and the continuum as reflected from a terrestrial material, and comparing it directly with the same ratio in the solar spectrum. When the terrestrial ratio exceeds that of the solar conjugate, the material is deemed luminescent. The prototype instrument designed for aircraft use will have sufficient sensitivity to detect rhodamine dye in concentrations of 20 ppb or less. It will be operational at the calcium Fraunhofer line at 3868 Å in the UV, where a variety of substances such as oil, detergents, phosphates, and other minerals are luminescent.

The Effects of Ozone and Ultraviolet Radiation on Vision

Lagerwerff, Kane, and Thornberg (1961) conducted research to determine what effects prolonged exposure to ozone had on several human visual parameters. Twenty-eight subjects were exposed to a variety of ozone concentrations for two different periods of time. Each exposure was preceded and followed by vision tests.

Photopic visual acuity, stereopsis, vertical phoria, and color vision were not affected at all or, in the case of a few subjects, very slightly. There were significant changes in lateral phoria determinations, however. The range of changes in the prism convergence tests was extremely wide; of 143 pairs of tests, only 9 pairs were identical.

One surprising result was that although there was no apparent change in the normal photopic visual acuity, 25 of the 28 subjects noticed an increase in peripheral vision. Two had a decrease. No explanation was provided.

Scotopic vision was also determined and all subjects indicated deterioration ranging from ¼ through 4½ brightness units.

To test whether deleterious effects from exposure to ultraviolet radiation could be demonstrated in human vision, Ludvigh and Kinsey (1964) exposed a group of subjects to radiation from a 1000-watt mercury vapor arc from which most of the visible rays and all of the ultraviolet radiation shorter than 3200 Å had been filtered out. Fixation was for 5 minutes at a distance of 30 cm. The difference in foveal sensitivity to light and the critical fusion frequency of both eyes of each subject had previously been determined. Results showed no difference between measurements of both eyes of the normal subjects or between measurements of any one eye before and after irradiation. It can therefore be assumed that ultraviolet radiation in the near ultraviolet at wavelengths longer than 3200 Å are not harmful to these two important functions of the human eye.

. . . .

REFERENCES

Angerer, E. V. Landschaftsphotographien in ultraroten und ultraviolettem Licht. *Naturwissenschaften* 18 (no. 17):361, 1930.

Baumgartner, A. Das Eindringen des Lichtes in den Boden. *Forstwiss. Zentr.* 72:172, 1953.

Colwell, R. N. Determining the prevalence of certain cereal crop diseases by means of aerial photography. *Hilgardia* 26 (No. 5):223, 1956.

Colwell, R. N. Some practical applications for multiband spectral reconnaissance. *Am. Sci.* 49 (No. 1):9, 1961.

Colwell, R. N. Aerial photo interpretation for the evaluation of vegetation and soil resources. In papers prepared for the United Nations Conference on the Application of Science and Technology for the Benefit of the Less Developed Areas. *Natural Resources* 2:314, 1963.

Cornu, A. *Comp. Rend.* 88:1101, 1285, 1879.

Craig, R. A. *The Upper Atmosphere: Meteorology and Physics.* New York: Academic Press, 1965.

Cronin, J. F., and Noble, R. H. Design and Application of Triradiometer for Terrestrial Studies. In preparation, 1968.

Dobson, G. M. B. Observations of the amount of ozone in the earth's atmosphere, and its relation to other geophysical conditions, Part IV. *Proc. Roy. Soc. of London,* Series A, 129:411, 1930.

Fabry, C., and Buisson, M. L'absorption de l'ultraviolet par l'ozone et la limite du spectre solaire. *J. Phys. Rad.* (Series 5)3:196, 1913.

Fabry, C., and Buisson, M. Etude de l'extrémité ultraviolet du spectre solaire. *J. Phys. Rad.* (Series 6)2:197, 1921.

Fisher, D. F., et al. *Airborne Infrared Scanning Systems M1A1.* Final Rpt., Contract AF19(628)-4038, U. Mich. (Conf. Rpt.), 1965.

Gates, D. M., Keegan, H. J., Schleter, J. C., and Weidner, V. R. Spectral properties of plants. *Appl. Opt.* 4:11, 1965.

Goody, R. M. *Atmospheric Radiation* (Vol. 1, Theoretical Basis). Oxford: Clarendon Press, 1964. (Figures appeared originally in Vigroux, *Annales de Phys.* 8)

Götz, F. W. Paul. Zum Strahlungsklima des Spitzbergensommers. *Gerlands Beiträge zur Geophysik* 31:119, 1931.

Götz, F. W. Paul, Meetham, A. R., and Dobson, G. M. B. The vertical distribution of ozone in the atmosphere. *Proc. Roy. Soc. of London* (Series A) 145:416, 1934.

Greenman, N. N., Burkig, V. W., and Young, J. F. Ultraviolet reflectance measurements of possible lunar silicates, *J. Geophys. Res.* 72:1355, 1967.

Hartley, W. N. *Chem. News* 42:268, 1880.

Hemphill, W. R. *Ultraviolet Absorption and Luminescence Studies.* U.S. Geological Survey, Progress Rpt. April-December 1967, Interagency Rpt., NASA-100, 1968.

Hemphill, W. R., and Carnahan, S. U. Ultraviolet absorption and luminescence investigations. U.S. Geological Survey, *Tech. Letter* (*NASA*) 6, 1965.

Hemphill, W. R., and Vickers, R. Geological studies of the earth and planetary surfaces of ultraviolet absorption and stimulated luminescence. U.S. Geological Survey, *Tech. Letter, NASA Supplement-33A,* 1966.

Hering, W. S., and Borden, T. R., Jr. *Ozonesonde Observations Over North America, Vol. 4.* AFCRL-64-30(IV), Environmental Research Paper No. 279, 1967.

Johnson, J. C. *Physical Meteorology.* Cambridge, Mass.: M.I.T. Press, 1954.

Krinov, E. L. *Speltral'naia Otrazhatel'naia Sposobnost' Prirodnykh Obrazovanii (Spectral Reflectance Properties of Natural Formations).* Moscow: Laboratoriia Aerometodov, Akad. Nauk USSR, 1947.

Lagerwerff, J. M., Kane, G. L., and Thornberg, G. H. Res. Rpt. No. 180, *The Effects of Repeated and Prolonged Exposure to High Concentration of Ozone on the Vision of Airline Pilots.* U. Mich., Rosemount Aero. Lab., 1961.

Lenoble, J. Etude de la pénétration de l'ultraviolet dans la mer; nouvelles mesures. *Ann. Geophys.* 12:16, 1956.

Ludvigh, E., and Kinsey, V. Effect of long ultraviolet radiation on the human eye. *Science* 104:246, 1946.

Mangold, V. L. *Narrowband Ultraviolet Filter Photography.* AFFDL-TR-66-140, Wright-Patterson Air Force Base, Ohio, 1966.

Normand, C. Some recent work on ozone. *Quart. J. Roy. Meteorol. Soc.* 77:474, 1951.

Paetzold, H. K. *Research on the Synoptical Measurements of the Vertical Ozone Distribution.* Final Rpt., Contract AF61 (052)-330, Institute of Geophysics and Meteorology, U. Cologne, 1963.

Perkin-Elmer Corp. *Ultraviolet Photographic Research.* AL-TDR-64-231 AFAL, Wright-Patterson Air Force Base, Ohio, 1965.

Polcyn, F. C. *Investigations of Spectrum-matching Sensing in Agriculture.* Final Rpt., Vol. 1 (NASA), Inst. Sci. and Tech., U. Mich., 1967.

Reed, R. J. The role of vertical motions in ozone-weather relationships. *J. Meteorol.* 7:263, 1950.

Regener, V. H. On a sensitive method for the recording of atmospheric ozone. *J. Geophys. Res.* 65:3975, 1960.

Regener, V. H. Measurement of atmospheric ozone with the chemiluminescent method. *J. Geophys. Res.* 69:3795, 1964.

Sauberer, F. Das Licht im Boden. *Wetter und Leben* 3:40, 1951.

Thorpe, A. N., Alexander, C. M., and Senftle, F. E. Preliminary ultraviolet reflectance of some rocks and minerals from 2000 Å to 3000 Å. U.S. Geological Survey, *Tech. Letter NASA-37,* 1966.

Wood, R. W. *Recent Experiments with Invisible Light.* Annual *Monthly Notices, Roy Astron.* Soc. 70:226, 1910.

Wood, R. W. *Recent Experiments with Invisible Light.* Annual Report to the Board of Regents, Smithsonian Institution, 1911. Pp. 151–166.

Wright, W. H. The moon as photographed by light of different colors. *Publ. Astron. Soc. Pac.* 41:125, 1929.

THE ATMOSPHERE attenuates absorbs, scatters, or reflects most of the ultraviolet energy which reaches the earth from the sun. The sun is our most important source of ultraviolet radiation; ultraviolet rays at the shorter wavelength can cause blindness or they can even be fatal. If it were not for the protective cover provided by atmosphere, life as we know it could not have developed on planet Earth. Two-thirds of the energy that impacts on the upper atmosphere reaches the earth's surface; only 10 percent of this is ultraviolet, 50 percent is concentrated in the visible portion of the spectrum, and 40 percent is in the infrared. Thus, for remote sensing in the lower atmosphere, far less ultraviolet energy is available than visible or infrared radiation. Most atmospheric absorption in the atmosphere is the result of ozone (O_3) and molecular oxygen (O_2), although dust and water vapor are important attenuating factors at certain wavelengths. A factor that further complicates sensing in the ultraviolet is that most window material is opaque to these shorter wavelengths. An ordinary camera lens will not pass ultraviolet wavelengths. It is for this same reason that a suntan cannot be obtained through a normal window glass. Sensors operating in the ultraviolet must have a special lens or window that will allow passage of these shorter waves.

9-Ultraviolet Imaging

MARVIN R. HOLTER

THE ULTRAVIOLET portion of the electromagnetic spectrum is, by definition, the region between 0.004 and 0.380 μ. It lies between the long-wavelength x rays and the violet portion of the visible spectrum. The ultraviolet region was discovered in 1801 by J. W. Ritter, who noted that in a prismatic spectrum the region beyond the violet caused a chemical action. He found that the rate of silver chloride blackening, which is caused by decomposition, increased when the silver chloride was exposed to this invisible energy. Experiments by Thomas Young in 1804, and later by others, showed that this invisible energy had, in fact, electromagnetic wave properties like the energy in the visible and infrared regions. A recent textbook by L. R. Koller (1965) presents a basic survey of ultraviolet technology.

Radiation Phenomena

The sun is the most important source of ultraviolet radiation. On the Kelvin scale, its surface temperature is about 6000°, and its calculated internal temperature is around 20 million degrees. The average value of solar radiation arriving outside the earth's atmosphere at the earth's mean solar distance is about 0.1396 W cm^{-2}. This quantity is called the solar constant, and to give the reader an idea of its magnitude, he can think of it as equivalent to 1170 watts or 1.56 horsepower per square yard. About two-thirds of this radiant energy actually reaches the earth's surface; the other third is lost by reflection, scattering, or atmospheric absorption. About 50 percent of this incident solar energy is concentrated in the visible spectrum, about 40 percent is in the infrared, and about 10 percent is in the ultraviolet. The spectral distribution of the ultraviolet portion of solar energy arriving outside the earth's atmosphere is shown in Figure 1. This plot illustrates how rapidly the radiation intensity diminishes toward the shorter wavelengths. The value of irradiance is actually down to 2×10^{-6} W cm^{-2} at 0.14 μ. The earth's atmosphere attenuates strongly in the ultraviolet at wavelengths shorter than 0.28μ, so much that the existence of solar radiation in this region was not experimentally verified until after World War II, when high-resolution solar ultraviolet spectroscopy was made possible by use of captured V-2 rockets. Since that time the region between 168 and 3000 Å has been studied with photographic instruments (Koller, 1965) and the region from 120 to 1300 Å has been studied by use of sensors with photoelectric detectors. These measurements have indicated that the solar radiation in the region from 1400 to 2800 Å is continuous and approximately follows the blackbody law given by the Planck equation. For wavelengths shorter than 1400 Å, the major portion of the solar radiation appears as discrete spikes of spectral irradiance. These spikes are caused by atomic emission lines, which have as their source the chromosphere of the sun, where a great deal of atomic activity takes place.

From *Remote Sensing with Special Reference to Agriculture and Forestry.* Washington, D.C.: National Academy of Sciences, 1970. Pp. 151–163. Reproduced with permission of the author and the National Academy of Sciences.

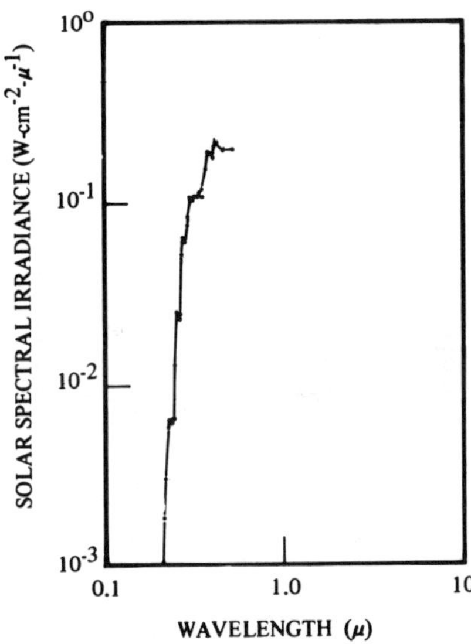

Figure 1 Plot of solar spectral irradiance at given wavelengths.

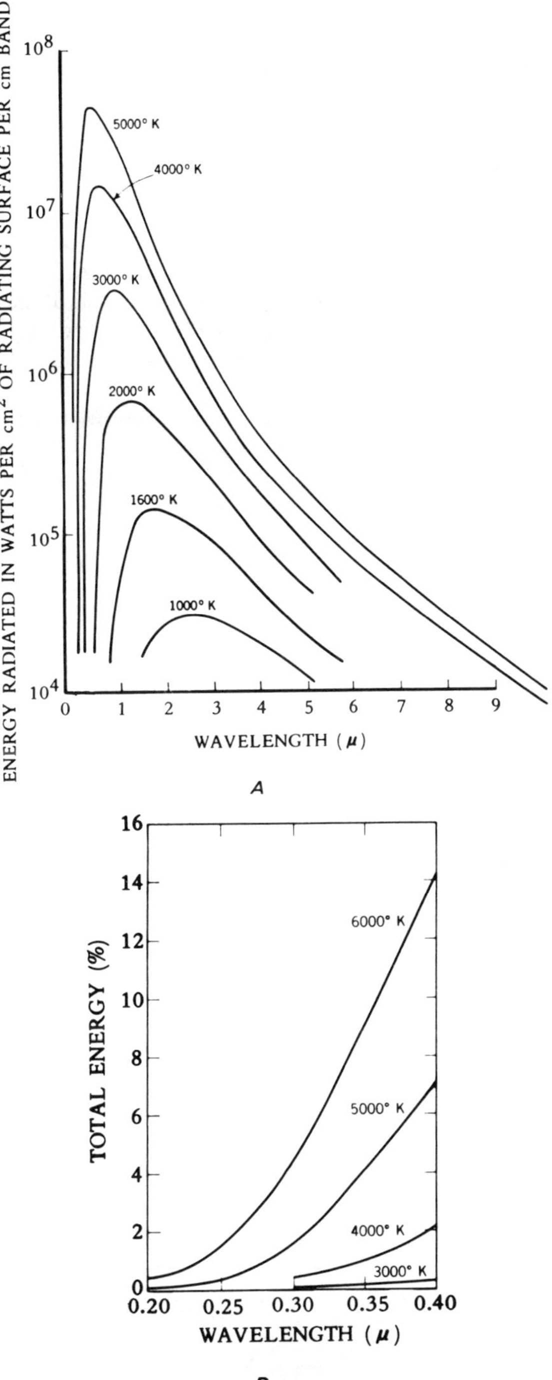

Figure 2 Blackbody energy. *A*, energy distribution for a blackbody at various temperatures. *B*, percentage of total energy radiation by a blackbody. (Koller, 1965)

Unlike the infrared region (for which remote sensors can be fabricated to observe the reflected radiation and the self-emitted radiation from an object, or both), remote sensors for the ultraviolet make use mainly of reflected radiation. This reflected radiation can have as its source the sun and sky during daytime, or some artificial lamp located at the remote sensor platform at night. For sufficient self-emission in the ultraviolet (see Fig. 2), the object of interest must have a temperature of about 1000°K or more. In order to detect the characteristics of agricultural materials such as crops and soils, which have temperatures around 300°K, it is reasonable to ignore the ultraviolet self-emission from the object because it contributes relatively little radiation in this part of the spectrum.

Four important properties must be considered in the design of ultraviolet remote sensors: the spectral characteristics of the radiation source, the effects of the intervening atmosphere, the spectral reflectance of the object of interest, and the spectral reflectance of surrounding objects that may be considered background materials. These properties are related to the spectral irradiance, H_λ, arriving at the remote sensor.

$$H_\lambda \sim \rho_\lambda \tau_\lambda S_\lambda$$

where ρ_λ is the spectral reflectance,
τ_λ is the spectral transmittance, and
S_λ is the source spectral irradiance.

Attenuation effects by the intervening atmosphere include primarily scattering and absorption. The fraction of radiation intensity transmitted through x kilometers of homogeneous air is given by $e^{-\sigma x}$, where σ is the attenuation coefficient, which is a function of wavelength λ. Values of σ at a given wavelength for two different samples of air can easily differ by a factor of 100. To illustrate further the variability, two samples of air having the same σ at one wavelength often have quite different values at another. The value of σ for air (Johnson, 1954) depends upon three additive factors: (1) σ_A, Rayleigh scattering by air molecules; (2) σ_B, scattering and absorption by airborne particles and droplets; and (3) σ_C, absorption by gases. Thus $\sigma = \sigma_A + \sigma_B + \sigma_C$. Rayleigh scattering is the dominant mode in the short-wavelength end of the visible region and in the ultraviolet regions. For this reason, the sky appears blue; i.e., solar radiation in the blue end of the visible spectrum is scattered about in the atmosphere and appears to arrive at the observer from all parts of the hemisphere. Scattering is quite strong in this portion of the electromagnetic spectrum, as evidenced by the fact that on a clear day the ultraviolet radiation due to scattered light from the sky, if observed as it falls on a horizontal plate on the earth's surface, may be greater in intensity than the amount directly radiating from the sun.

The primary causes of atmospheric absorption in the ultraviolet are ozone, O_3, and molecular oxygen, O_2. Ozone exists in layers a few kilometers thick about 30 km above the earth. This layer of gas strongly absorbs ultraviolet radiation between 0.2 and 0.28 μ. It has a weaker absorption band at 0.32–0.35 μ. Oxygen absorbs strongly in the region 0.13 μ to about 0.2 μ. In the region from 0.176 to 0.2 μ, the absorption spectrum of oxygen is made up of a number of strong absorption bands, whereas between the bands, oxygen is relatively transparent. Figure 3 presents some experimental results of measurements of σ versus wavelength on various days in Pasadena, California, in the year 1949 (Tousey, 1962). To provide the reader with some quantitative understanding of the data, consider the day September 20, 1949. Figure 3 gives a value of 1 km^{-1} for σ at 3200 Å. Since the percent transmittance is given as $e^{-\sigma x} \cdot 100$, we obtain a value of $e^{-1} \cdot 100$ for a 1-km atmospheric path (i.e., $x = 1$ km). The transmission is then about 37 percent. The theoretically derived Rayleigh scattering curve is also presented in Figure 3 to indicate that the observed attenuation is due to more than just molecular scattering.

For the most part, the ultraviolet-reflectance properties of agricultural materials and other materials that appear in natural scenes are not well known. Although laboratory spectrophotometers with attachments for measuring reflectance generally operate to wavelengths as short as 0.3 μ, reliable measurements at shorter wavelengths are virtually nonexistent. Figure 4, prepared at the University of Michigan, presents a sample of data in the 0.3–0.4-μ region. The data were obtained with a Beckman DK-2 spectrometer with reflectance attachment. Included in the figure are data on the reflectance of leaves and bark from a healthy apple tree and from a moisture-stressed apple tree. Scientists at the University of Michigan have produced ultraviolet imagery of an apple orchard containing both healthy and diseased trees. This imagery (not shown here) shows noticeable contrast differences between healthy and diseased trees in consonance with the differences present in the spectrometer measurements. That is, the diseased tree had fewer leaves in the crown, so more of the bark was exposed, and the diseased tree crown appears lighter in tone than the healthy crown.

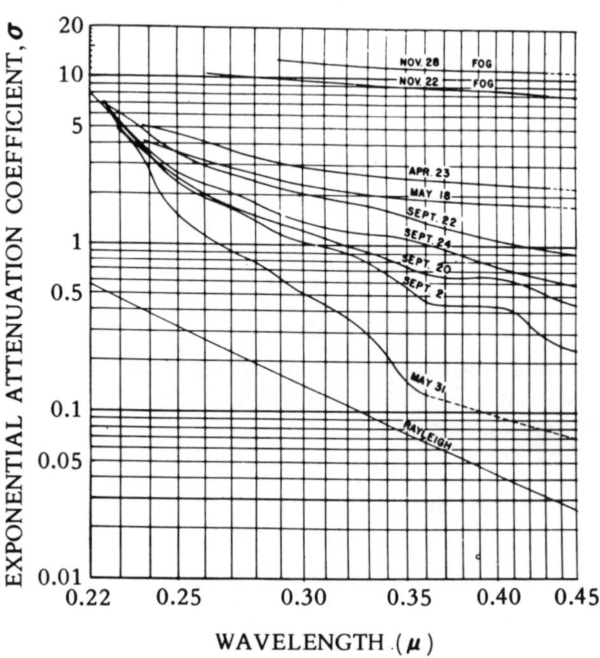

Figure 3 Typical attenuation curves illustrating the full range of atmospheric conditions, Pasadena, California, 1949. (Tousey, 1962)

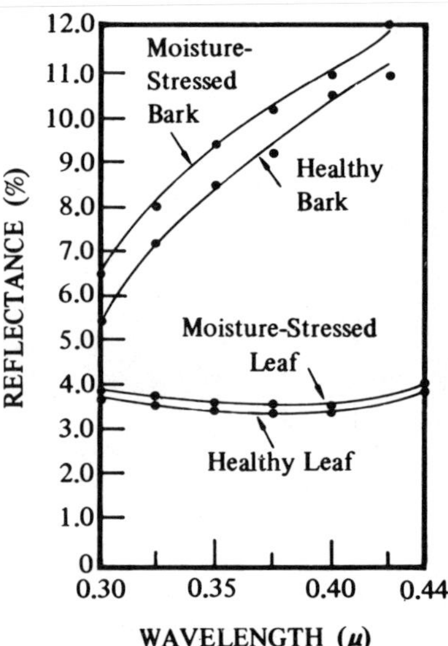

Figure 4 Spectral reflectance in the long-wavelength ultraviolet of bark and leaves of an apple tree. Data for both moisture-stressed and healthy materials are shown. (University of Michigan)

Ultraviolet Sensors

As compared to developments of radar and infrared nonphotographic sensors, relatively little has been done in the ultraviolet. Like infrared systems, ultraviolet systems employ such things as windows, mirrors and/or lenses, detectors, electronic amplifiers, and output-display devices. At wavelengths shorter than 0.3 μ, care must be taken in selecting a window material since many of them become opaque in this region. Some quartz glasses, quartz crystal, and several ionic salts such as lithium fluoride and barium fluoride are transparent at wavelengths as short as 0.12 μ. For wavelengths shorter than this, one is hard pressed to find a good window material.

For wavelengths longer than 0.28 μ, conventional lenses, as well as mirrors made of aluminum, or silver metal films are quite satisfactory for focusing the radiation onto a detector. At shorter wavelengths, however, the metal films become better absorbers than reflectors, and so refractive optics, i.e., lenses, may be more practical for focusing.

Ultraviolet radiation detectors are mainly photoemissive devices. This means that incident ultraviolet photons are absorbed by electrons in the sensitive film, imparting enough kinetic energy to the electron to allow it to escape from the surface. An electrically positive plate is placed sufficiently close to the sensitive surface to collect the emitted electrons. The electric current flow between the

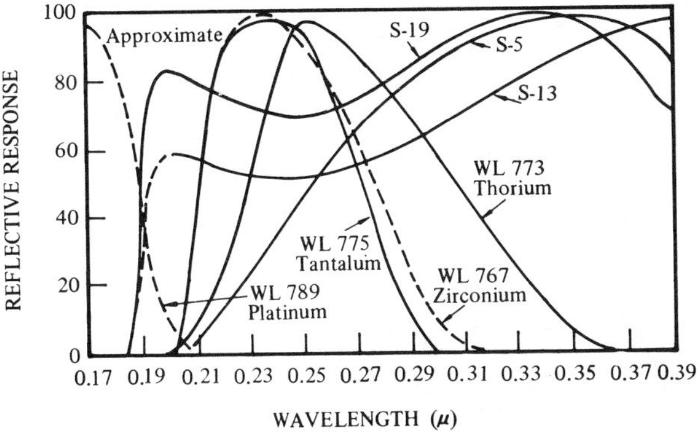

Figure 5 Spectral response curves of some ultraviolet-sensitive photoemissive devices.

collector and the sensitive surface is monitored to provide a measure of the incident radiation flux. Commercial ultraviolet-sensitive detectors have been made by Westinghouse with platinum, tantalum, zirconium, or thorium as the sensitive element. Each has its own characteristic spectral response. In addition, sensitive surfaces composed of metal mixtures have been made by RCA and are commercially available. Relative spectral response curves of these detectors are presented in Figure 5. Detectors operating at wavelengths shorter than 0.15 μ are research devices built for special applications. Apparently, the demand for detectors in this region has not been sufficient for industry to develop a commercial product.

In the part of the ultraviolet region of interest for remote sensing, i.e., where the atmosphere is transparent, scanning systems are of interest. It is worth noting that these scanning systems have an advantage over photographic methods in "seeing" through the high level of scattered radiation. In the scanning systems, part of the deleterious signal due to this scattered radiation can be subtracted electronically, permitting observation of more surface detail than can be recorded photographically.

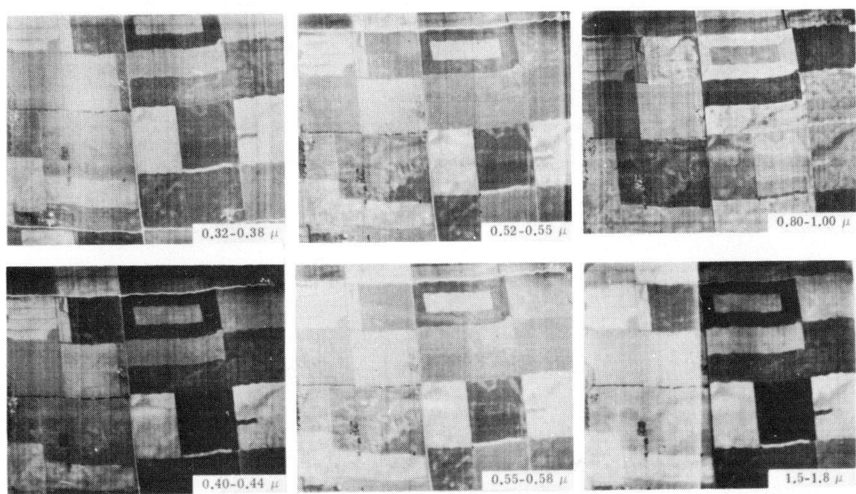

Figure 6 Imagery and photography compared, area 1. (University of Michigan)

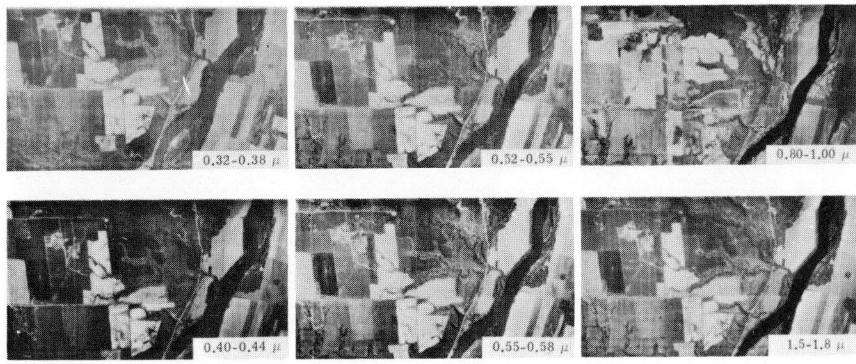

Figure 7 Imagery and photography compared, area 2. (University of Michigan)

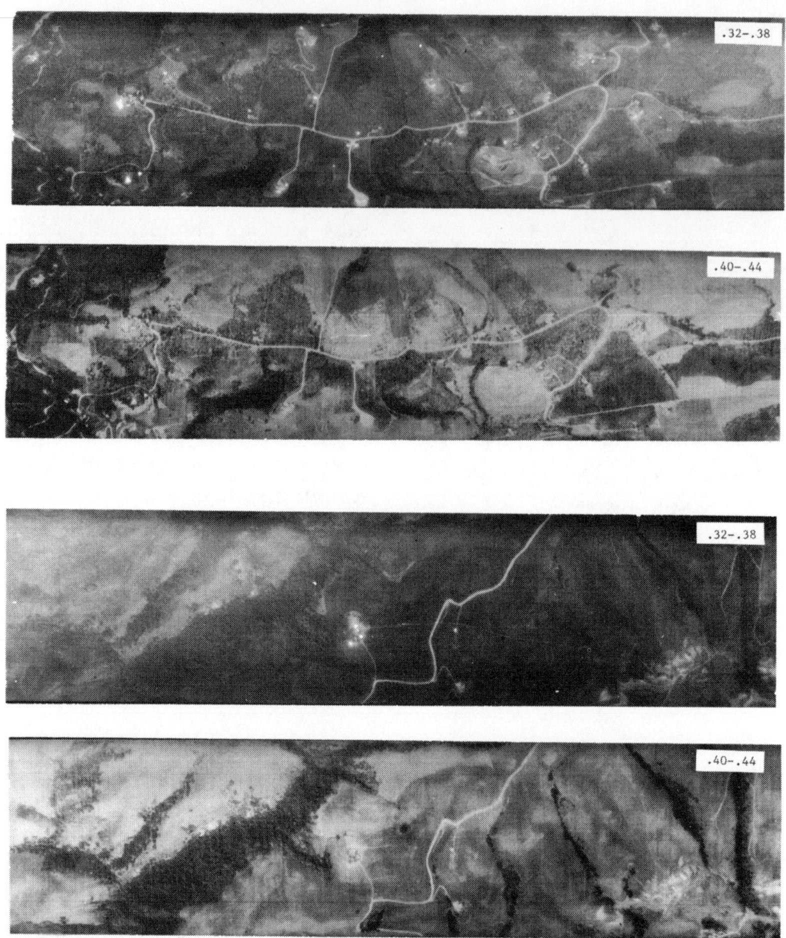

Figure 8 Imagery comparisons between ultraviolet (0.32–0.38 μ) and nearest adjacent visible band (0.40–0.44 μ), Hayward Fault, California, May 26, 1966, 2000-foot altitude, 2:40 P.M. (University of Michigan)

Ultraviolet Scanner Imagery

In Figures 6 and 7, scanner imagery in the ultraviolet region is shown in the band 0.32–0.38 μ. This demonstrates that imagery of near-photographic quality can be obtained with scanners. For areas 1 and 2, there are no strong differences in contrast as compared with the adjacent visible band 0.40–0.44 μ. This lack of difference between the ultraviolet and the nearest visible bands, evidenced by the imagery of areas 1 and 2, is not true for all areas, as shown in Figure 8. That figure shows the Hayward fault area in California in two bands, the ultraviolet and the nearest adjacent visible band. A number of differences are immediately apparent. The road network contrasts more strongly with the surrounding terrain in the ultraviolet. The same is true for the human structures. Close examination will also reveal other differences between the two bands in the appearance of the terrain.

REFERENCES

Johnson, F. S. The solar constant. *J. Meteorol.* 2:431, 1954.
Koller, L. R. *Ultraviolet Radiation.* New York: Wiley, 1965.
Tousey, R. The extreme ultraviolet—past and future. *Appl. Opt.* 1:679, 1962.

Color Illustrations

5—WORMSER

Figure 3 Color thermogram (temperature range blue-green to red is 5°C) of both shoulders and arms of E. M. Wormser. Hot area (red) in area of biceps muscle on right arm was diagnosed as bursitis. (Courtesy Barnes Engineering Company)

Figure 4 Thermogram of R. B. Barnes and E. M. Wormser seated (temperature range from blue to red is 10°C), and of the chairs 15 minutes after they got up (temperature range from blue to red is 2.5°C). (Courtesy Barnes Engineering Company)

Figure 6 Contact of rhyolitic sill and Bliss sandstone. Closely spaced joints in both the rhyolite and sandstone break the materials into cobbles and boulders that mantle the surface. The vegetation consists of light brown to gray-green grass and shrubs that completely cover portions of the slope.

Figure 18 Faulted, gullied cinder cone in central part of West Potrillo volcanic field. (Mapping camera; original scale, 1:20,000.)

16—SCHNEIDER AND KOLIPINSKI

Figure 1 Shoreline area east of Cutler, Florida, showing areas of fresh-water discharge; from regular color transparency at original scale of 1:12,000. (U.S. Geological Survey)

Figure 2 Part of Prairie Pothole area in North Dakota showing differences in salinity of potholes; from regular color transparency at original scale of 1:12,000. (U.S. Geological Survey)

Figure 3 Part of Great Salt Lake, Utah, showing underwater features; from regular color transparency at original scale of 1:12,000. (U.S. Geological Survey).

Figure 4 Lake Erie at Cleveland, Ohio, showing discharge of Cuyahoga River; from regular color transparency at original scale of 1:8,000. (U.S. Geological Survey)

Figure 5 Area of Florida Everglades showing rock reef; from infrared color transparency at original scale of 1:6,000. (U.S. Geological Survey)

Figure 6 Coastal area along Florida Bay showing mangroves and brackish-water marshes, from infrared color transparency at original scale of 1:10,000. (U.S. Geological Survey)

20—AMSBURY

Figure 1 Apollo 6 photograph of southern New Mexico, western Texas, and northern Chihuahua. (Original scale, 1:2,800,000.) North is at the top of the page on all spacecraft and aerial photographs in this article.

Figure 2 High altitude stereophotographs of southernmost Franklin Mountains. High-contrast beds 30 to 50 feet thick can be identified and mapped, and contacts between igneous rock and terrace gravels can be followed. (SO-136 film, 25A filter; original scale, 1:250,000.)

Figure 5 Outcrop of 50-foot-wide sill of fine rhyolite between dark reddish brown Bliss sandstone. Enough light-colored, highly reflective rhyolitic rubble shows through the vegetation that the sill is easily discerned on low-resolution photographs such as Figure 2.

21—LOWMAN

Figure 2A Apollo 7 hand-held photograph taken in 1968 with color film, with main physiographic and structural features shown on index map (p. 172).

Figure 6A Color infrared photograph of San Diego County, California; one of four views photographed simultaneously with different film-filter combinations as part of the SO65 experiment. (See index map, p. 175.) (Lowman, 1969)

24—NORMAN AND FRITZ

Figure 2 Aerial photograph made on KODAK EKTACHROME Infrared AERO Film showing areas infected with (A) Psorosis virus, (B) Tristeza virus, (C) early stages of decline from burrowing nematodes, (D) late stages of decline, and (E) footrot caused by fungi.

25—PAARMA AND TALVITIE

Figure 3 Original scale 1:60,000. Broad and ragged W-E lineaments seen as bogs, sharp and narrow NW lineaments as sets of lakes (black), small blocks (gray), and green vegetation (red). Fracture traces (near NNE) are reflected as orientation of some individual small lakes and veins of green vegetation (pink).

Figure 4 The sharpness and clarity of NW and W-E (northern part of the figure) trends of lineament seen as orientation of bogs. Reddish colors of surroundings cause a great contrast. Cf. interpretation map and Figure 5.

Figure 6 W-E series of lakes (black) and parallel boggy lineaments (gray). Orientation of individual small lakes is that of the fracture traces. Lengthy block between lineaments is probably broken along the fracture traces.

Figure 7 Enlargement (original scale 1:60,000). WNW trends of lineament are seen as combination of red and blue (foliage, other green vegetation, and bogs). Nearly vertically trending fracture traces are mostly short and sharp reddish lines. Cf. Figure 8.

29—BARNES

Figure 9 Ten-digit color thermogram of lower extremities. Each color change repreesnts a ΔT of 1°C, since thermograph was set at a sensitivity of 10°C. On left (anterior view), note skin temperature abnormality on right leg. On right (posterior view), note abrupt drop-off in temperature on left leg. (Infrared color thermogram courtesy of Barnes Engineering Company)

Figure 10 Peripheral vascular changes apparently caused by the inhalation of cigarette smoke. In each thermogram, view on right is of the palmar side of the arm, while left shows the dorsal side. The thermograms show the ten-digit color process with each color change representing a $\Delta T = 0.5$°C between 36°C (red) and 31°C (blue). (Left) Before smoking. (Right) Thermogram started 40 minutes after beginning of inhalation of half a cigarette. (Infrared color thermogram courtesy of Barnes Engineering Company)

32—EDGERTON AND TREXLER

Figure 6 Microwave images of the Salton Sea. Images were produced from brightness temperatures recorded from a 1.55 cm electronically scanned microwave radiometer from aboard the NASA Convair 990 Earth Resources aircraft during an overflight of the Salton Sea on June 7, 1969. Flight direction was from north to south; therefore, the southern end of the sea is at the top of the sequence. Flight altitude was 11 km. The second frame shows a marked difference in radiometric temperature of the sea surface corresponding to higher sea state conditions in the southern half of the sea.

37—COLWELL

Figure 2 Central Australia's characteristic topography appears in the photograph . . . , made from the Gemini V spacecraft on Aug. 27, 1965. The spacecraft was at an altitude of 165 miles; the area shown is west of Alice Springs. An experienced interpreter can use such a photograph to obtain information about a variety of natural resources and land uses. Some ways the photograph can be interpreted are illustrated in Figures 3 and A. (NASA)

Figure 16 Wild-land resources are studied in a test area in the Bucks Lake region of the Sierra Nevada. This photograph was made from a camera station on a rock 2000 feet above the lake. The station is used to make simulated aerial photographs of a known area. Such terrestrial photographs can be used as guides in interpreting similar photographs, obtained from aircraft or spacecraft, of regions where the wild-land resources need to be identified. A normal color photograph is only one of several types of photograph used for evaluating resources. Sensings can also be made in other wavelength bands of the spectrum to detect different physical features. (William Draeger)

Figure 17 Same area of the Sierra Nevada is photographed in infrared. The film used to make this photograph . . . contains a red dye that is sensitive to near infrared wavelengths of 0.7 to 0.9 μ. The false colors often provide more scope for interpretation than is possible with other films. Largest amounts of infrared energy produce the reddest color. (William Draeger)

43—ESTES AND GOLOMB

Figure 2 Union Oil Company's Platforms B and A, respectively, are in the foreground while four other platforms in the background trail off toward the Rincon Coast east of Santa Barbara (view looking northeast). Just east of Platform A is the active blowout, with oil pollutants bubbling up and spreading outward. (Photo by Dick Smith, *Santa Barbara New Press*)

Figure 3 If it were not for the boat wake across this picture, it would be difficult to tell that we are looking at an oil-polluted surface. The boat has cut through the thin surface layer of pollutants to reveal the underlying tone signature of the water. (Photo by Norman K. Sanders)

Figure 4 Center of the Santa Barbara Oil Spill, Feb., 1969. Active blowout sites are bubbling at the east of Platform A. To the west (background) is Platform B. Streamers of oil and slicks of varying thickness radiate out from the source area.

Figure 6 Areas in dark blue represent essentially open water; areas in color ranging from yellow through red and dark green to black depict varying thickness of oil.

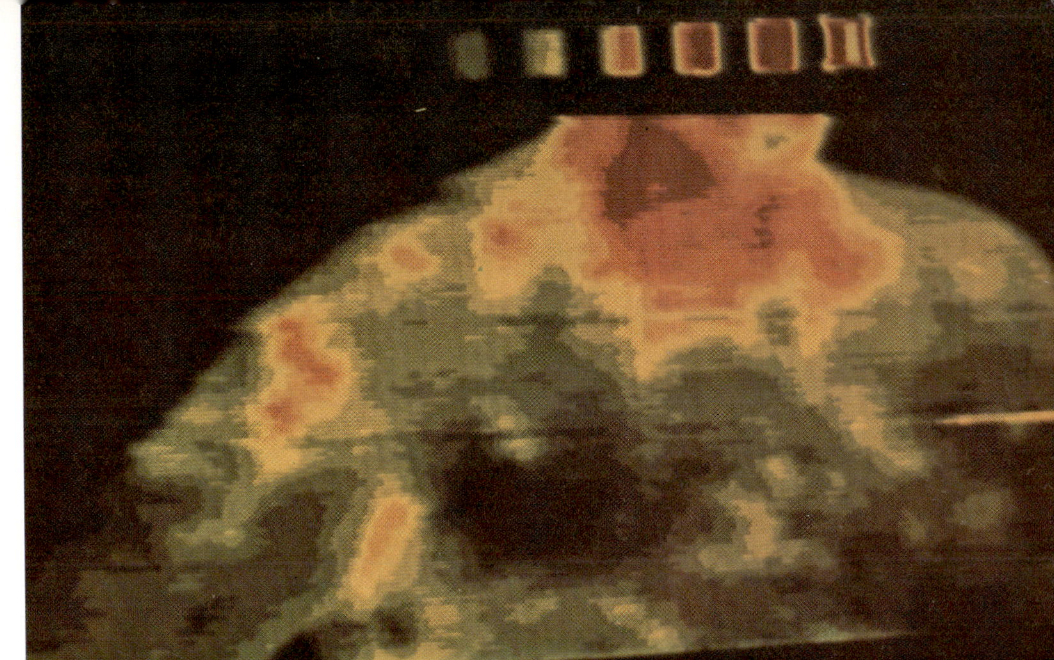

Figure 3

Figure 4

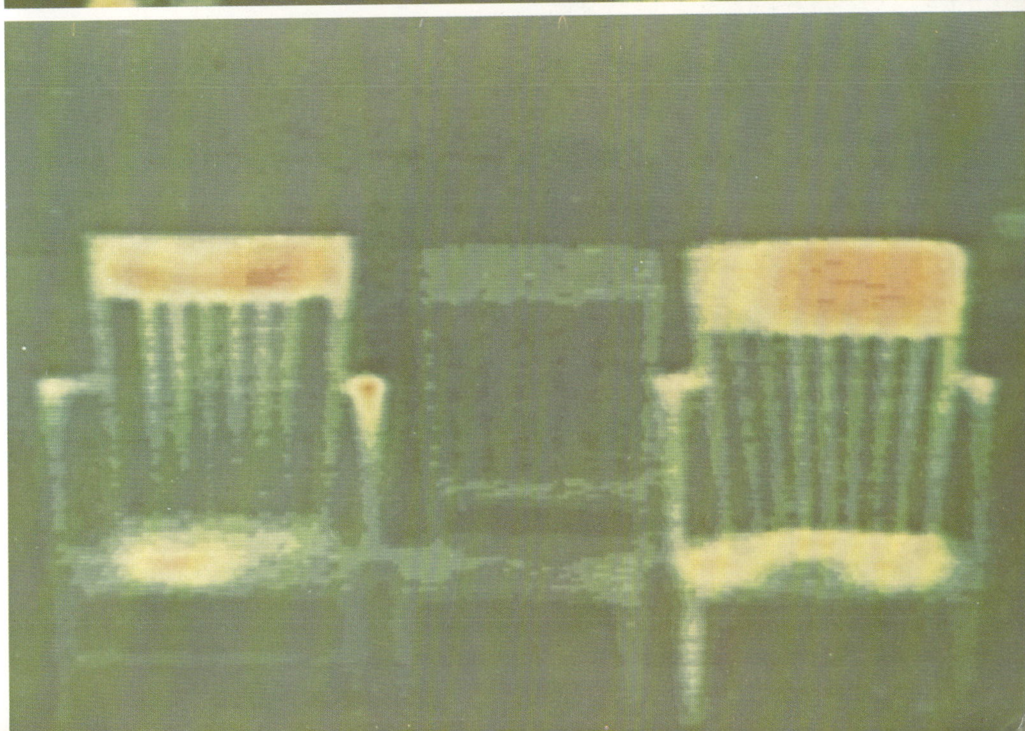

Figure 1

Figure 2

Figure 3

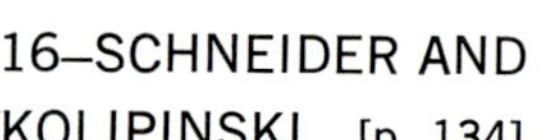

16–SCHNEIDER AND KOLIPINSKI [p. 134]

Figure 4

Figure 5

Figure 6

20–AMSBURY [p. 161]

Figure 1

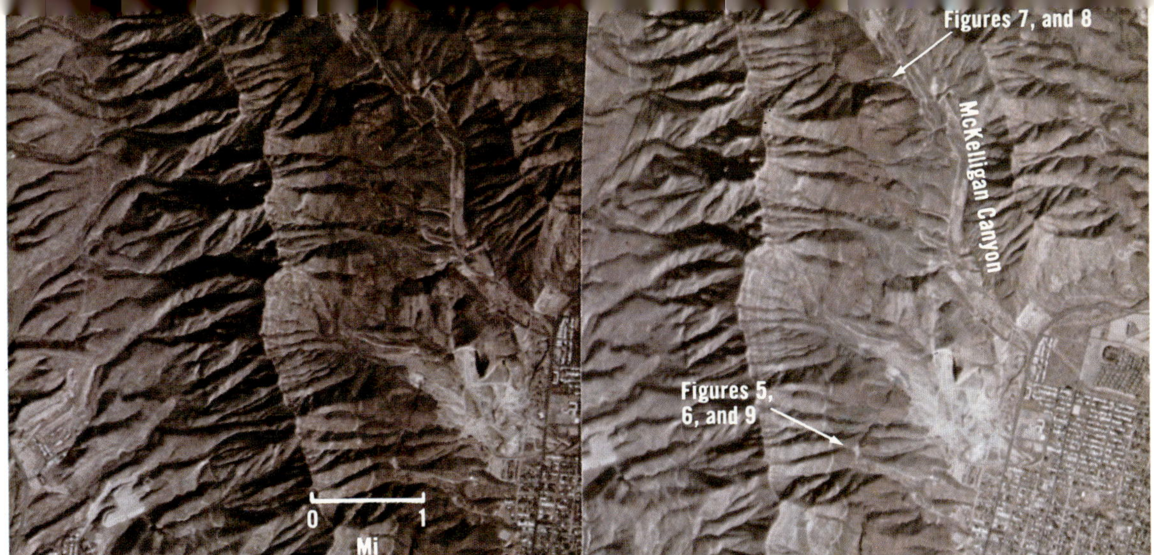

Figure 2

Figure 5

Figure 6

Figure 18

21–LOWMAN [p. 170]

Figure 2A

Figure 6A

24–NORMAN AND FRITZ [p. 199]

Figure 2

Figure 3

Figure 6

Figure 4

Figure 7

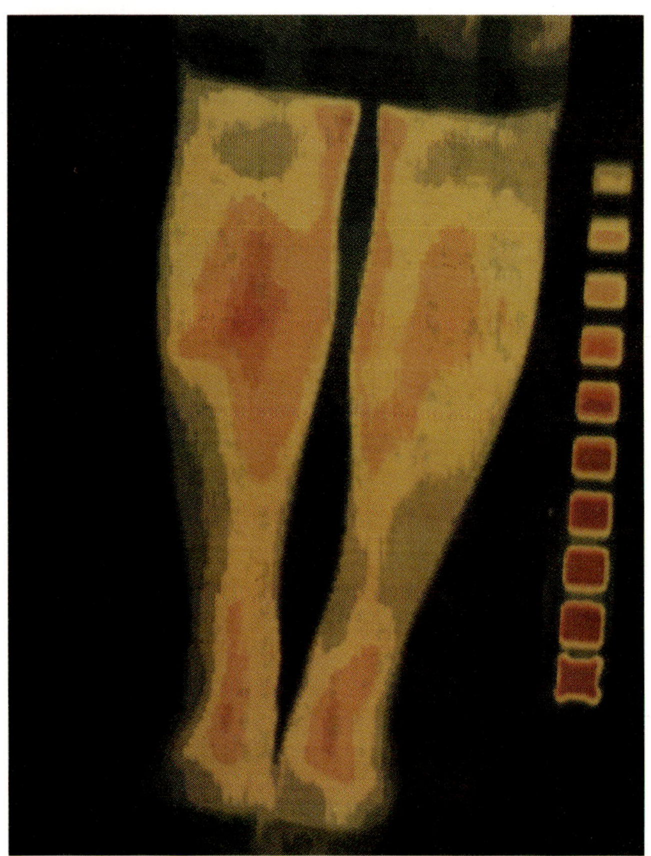

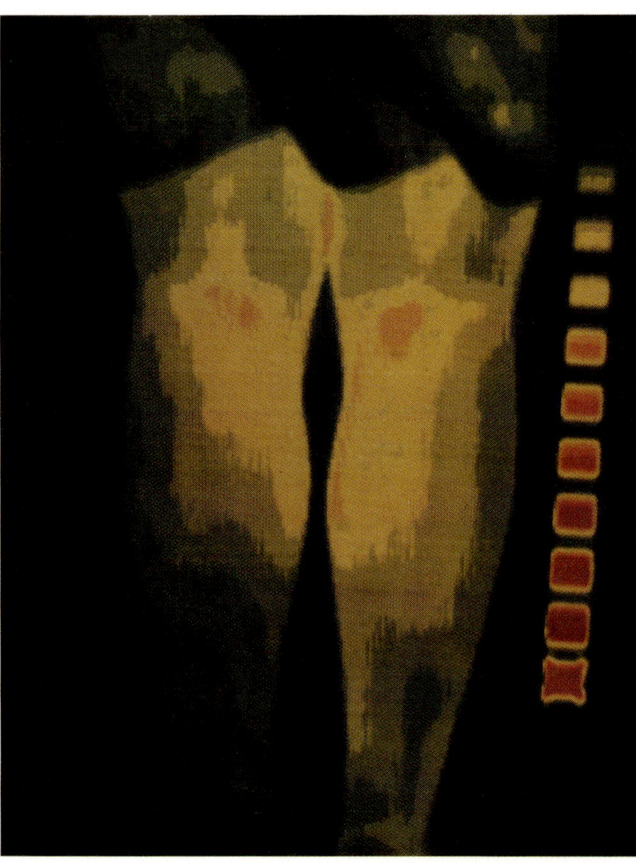

Figure 9

Figure 10

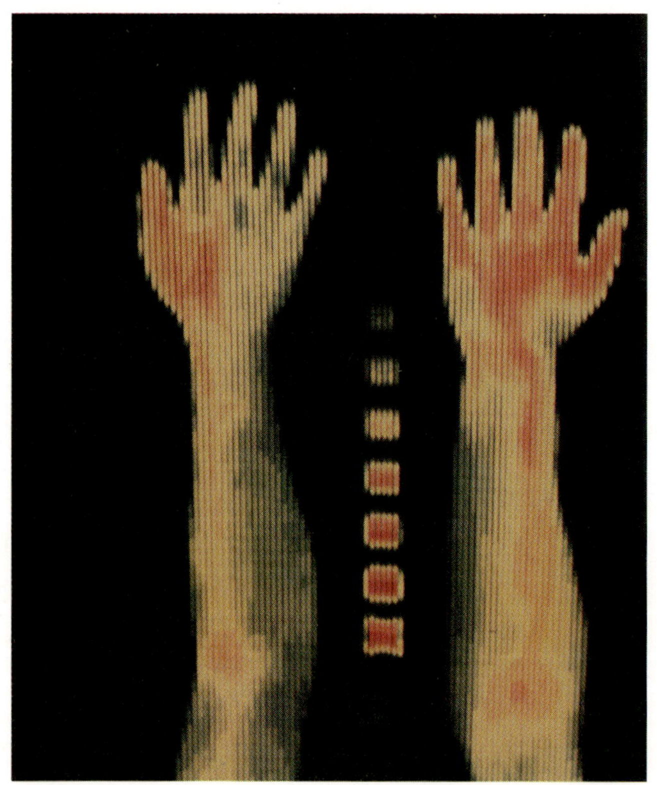

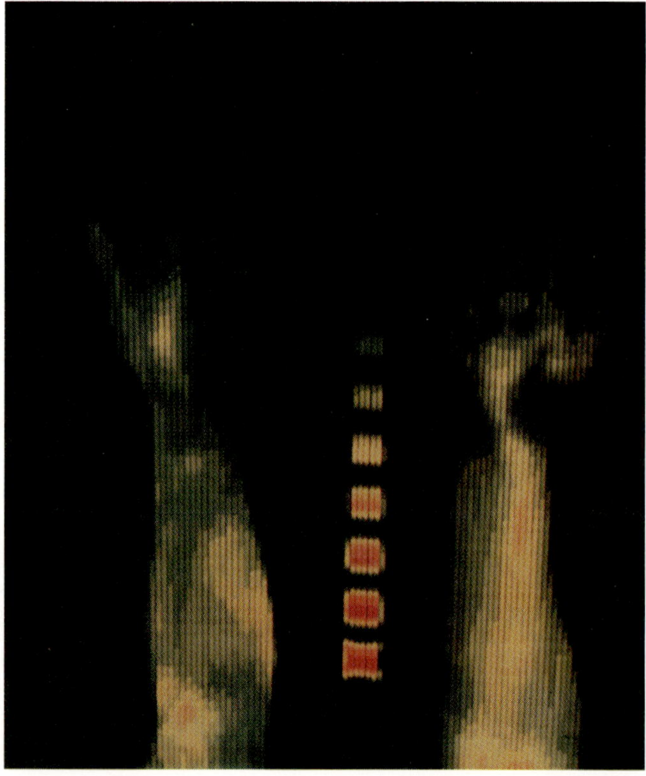

32–EDGERTON AND TREXLER [p. 269]

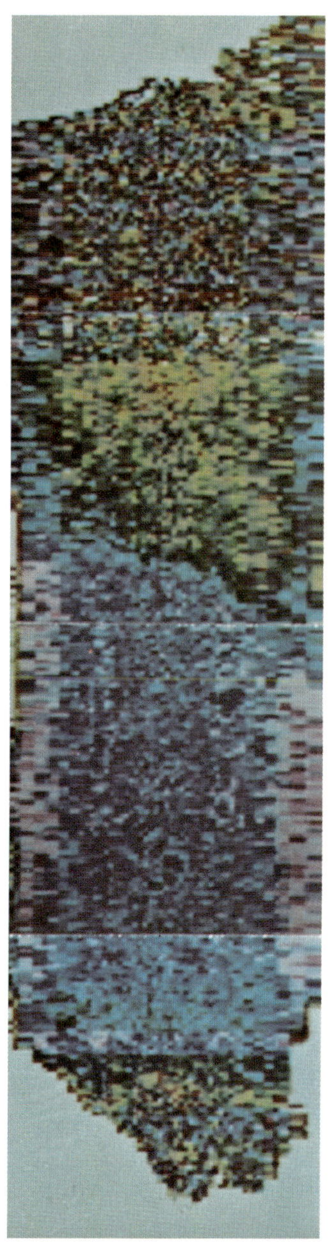

Figure 6

37–COLWELL [p. 322]

Figure 2

Figure 16

Figure 17

Figure 2

Figure 3

Figure 4

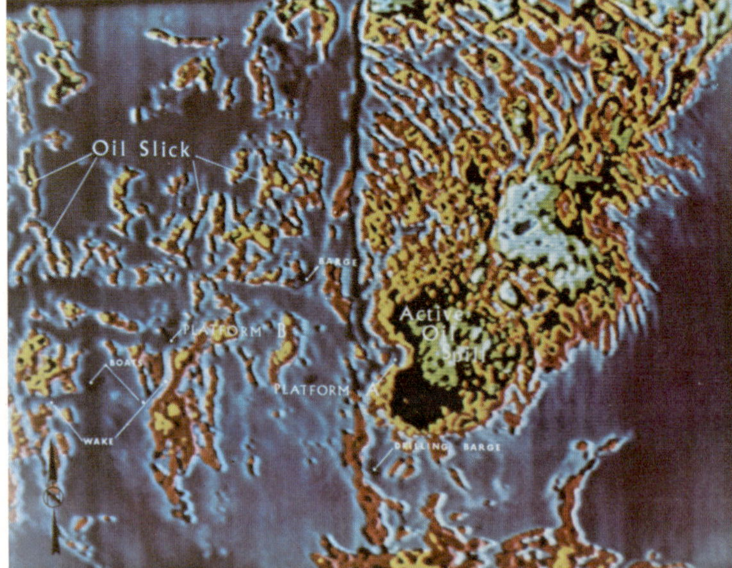

Figure 6

MANY PHENOMENA in the earth's environment are luminous to ultraviolet light; that is, when exposed to radiation at these shorter wavelengths, they luminesce or reflect light. Because of the change in wavelength that occurs in this reflected energy—to longer wavelengths in the visible—this modified ultraviolet light is reflected from luminous materials and can be experienced with the human eye. Most of us are familiar with "black light" exhibits of rocks and minerals on display at many museums. The rocks and minerals are exposed to ultraviolet light, usually in a darkened area, and the resulting colored glow is the luminescence of reflected ultraviolet light at visible wavelengths.

Ultraviolet-stimulated luminescence can be used to detect and to discriminate earth materials. For many years small hand-carried ultraviolet lamps have been used in prospecting for strongly luminescent rocks and minerals such as scheelite, talc, dolomite, and fluorite. This kind of mineral exploration is limited by the size of the portable lamp, which has low power output and very limited range. Work with these lamps must be accomplished at night in order to avoid the overwhelming background of reflected sunlight in the visible portion of the spectrum, which would mask the luminescence of materials illuminated. A recently developed technique suggests that the Fraunhofer line-depth method will prove effective in distinguishing luminescing materials at a distance, using the sun as an excitation source.

10-Remote Sensing of Luminescent Materials

WILLIAM R. HEMPHILL
GEORGE E. STOERTZ
DAVID A. MARKLE

PORTABLE ULTRAVIOLET lamps have been used in the field to detect luminescence of minerals such as fluorite (CaF_2), calcite ($CaCO_3$), and secondary uranium minerals. A truck-mounted cathode ray tube transmitter emitting in the ultraviolet, and a television receiver have been used experimentally to image luminescing materials (Hemphill et al., 1966). A pulsed ultraviolet nitrogen gas laser has been used in the laboratory to stimulate and measure phosphorescence decay times in the microsecond range of plagioclase feldspars and other materials (Hemphill, 1968).

Potential application of luminescence detection in mineral prospecting and other fields is limited, however, because these artificial sources are relatively low powered, and effective range is limited from a few feet with hand-carried lamps to a few hundred feet with a cathode ray tube system. The work must be conducted at night in order to avoid obscuring the low-intensity luminescence by bright sunlight. Eclipsing methods are adequate only for those materials that exhibit phosphorescence of sufficient duration to exceed the period of eclipse.

The Fraunhofer line-depth method uses the sun as an excitation source, and permits detection of luminescing materials during daylight, thus avoiding the power and distance limitations of artificial sources and the awkwardness of nighttime operations.

Successful use of the Fraunhofer line-depth method was first reported by Kozyrev (1956) who detected luminescence in the ray systems of the lunar craters Aristarchus and Herodotus. Other astronomers also reported positive results using this method, and many, like Kozyrev, used the H and K lines of calcium in the violet because the relatively broad width of these lines permitted better spectral definition than narrower Fraunhofer lines at longer wavelengths.

Dubois (1959) observed that luminescence intensities range from 3 percent to 25 percent of the normal lunar reflected sunlight in 11 lunar regions. Grainger and Ring (1962a, 1962b) detected luminescence of typically

From *Proceedings of the Sixth International Symposium on Remote Sensing of Environment.* Ann Arbor: University of Michigan, Institute of Science and Technology, Willow Run Laboratories, 1969. Pp. 565–575. Reprinted with permission of the authors and the publisher.

5 percent of the continuum level in four areas of the lunar surface, including Aristarchus. Spinrad (1964) compared the spectra of the Sun, Mars, and Jupiter with the Moon, and detected lunar luminescence that averaged 13 percent of the lunar continuum at 3950 A. Myronova (1965) detected luminescence intensities of about 12 percent in the interior of Aristarchus crater and attributed the phenomenon to excitation of luminescent minerals by corpuscular and ultraviolet radiation from the sun. McCord (1967) observed 24 lunar areas and found evidence of luminescence intensities of between 0.5 percent and 2 percent which varied both in lunar location and in time; he speculates that a large part of the lunar surface may continuously emit at intensities of less than 1 percent of the reflected solar continuum. No luminescence intensities observed by McCord exceeded 5 percent.

The feasibility of using the Fraunhofer line-depth method to detect luminescing earth materials was indicated in a series of outdoor experiments[1] conducted at the IIT Research Institute, Chicago, Ill., using the following Fraunhofer lines: D_2, a sodium line at 5890 A; G′, a hydrogen line at 4340 A; and H and K, two calcium lines at 3968 A and 3934 A, respectively. A laboratory grating spectrometer was used to spectrally scan sunlight reflected alternately from luminescing samples—calcite ($CaCO_3$), colemanite ($Ca_2B_6O_{11} \cdot 5H_2O$), and phosphate rock—and from a nonluminescing reference standard, magnesium oxide. Evidence of luminescence was obtained only in the H and K lines, but the results of the experiment were sufficiently definitive to (1) indicate the need for an instrument specifically designed for detection of luminescence by means of the Fraunhofer line-depth method, (2) indicate some of the design features that such an instrument would require, and (3) stimulate thought as to how and for what practical purpose a field-test prototype could be used.

This paper discusses the development and preliminary testing of a Fraunhofer line discriminator (FLD) suitable for detecting luminescence of selected earth materials from low-flying aircraft. The work was sponsored jointly by the U.S. Geological Survey and the National Aeronautics and Space Administration.

Principle of the Fraunhofer Line-depth Method

Fraunhofer lines are dark lines in the solar spectrum caused by selective absorption of light by gases in the relatively cool upper part of the solar atmosphere. Widths range from less than one-tenth angstrom to several angstroms and the central intensity of some lines is less than 10 percent of the adjacent continuum. The lines are sharpest, deepest, and most numerous in the near-ultraviolet, visible, and near-infrared regions of the electromagnetic spectrum.

The Fraunhofer line-depth method involves observing a selected Fraunhofer line in the solar spectrum, and measuring the ratio of the central intensity of the line to a convenient point on the continuum a few angstroms distant. This ratio is compared with a conjugate spectrum reflected from a material that is suspected to luminesce. Both ratios normally are identical, but luminescence is indicated where the reflected ratio exceeds the solar ratio. Reflectivity differences between the central intensity of the Fraunhofer line and the adjacent continuum can generally be ignored because variation of reflectivity with wavelength is negligible for most materials over spectral ranges of only a few angstroms.

In Figure 1, diagram 1 is an idealized Fraunhofer line observed in the solar spectrum directly, wherein the central intensity, B, and the continuum, A, may be expressed as the ratio, R_s, that is,

$$R_s = \frac{B}{A} \tag{1}$$

Diagram 2 shows the same Fraunhofer line in a spectrum reflected from a nonluminescing material. If the reflectively is γ, the ratio of the line depth in the spectrum reflected by the material is:

$$R_m = \frac{\gamma B}{\gamma A} = R_s \tag{2}$$

Diagram 3 shows the luminescence component, L. Because many materials exhibit broad-band luminescence, commonly half-widths of several hundred angstroms, the luminescence contribution may be considered to be constant across the width of one Fraunhofer line.

Diagram 4 shows the line profile of a material containing both reflected and luminescing components. The luminescence component is assumed to contribute equally to both the central intensity and the adjacent continuum. Thus,

$$D = \gamma A + L \tag{3}$$

$$C = \gamma B + L \tag{4}$$

$$R_m = \frac{\gamma B + L}{\gamma A + L} = \frac{C}{D}\ ;\ R_m > R_s \tag{5}$$

The difference between R_m and R_s is a measure of the amount of luminescence.

The luminescence component, L, may be measured in terms of the reflected component level, A, and expressed as a luminescence coefficient by the term ρ. Thus,

$$\rho = \frac{L}{A} \tag{6}$$

$$L = \rho A \tag{7}$$

Substituting (7) into (3) and (4),

$$D = \gamma A + \rho A \tag{8}$$

$$C = \gamma B + \rho A \tag{9}$$

These two equations can be solved to determine the two unknowns γ and ρ. Multiplying (8) by B and (9) by A yields:

$$BD = \gamma AB + \rho AB \tag{10}$$

$$AC = \gamma AB + \rho A^2 \tag{11}$$

Subtracting (10) from (11):

$$AC - BD = \rho A\,(A - B) \tag{12}$$

[1] Sponsored by the U.S. Geological Survey and the National Aeronautics and Space Administration.

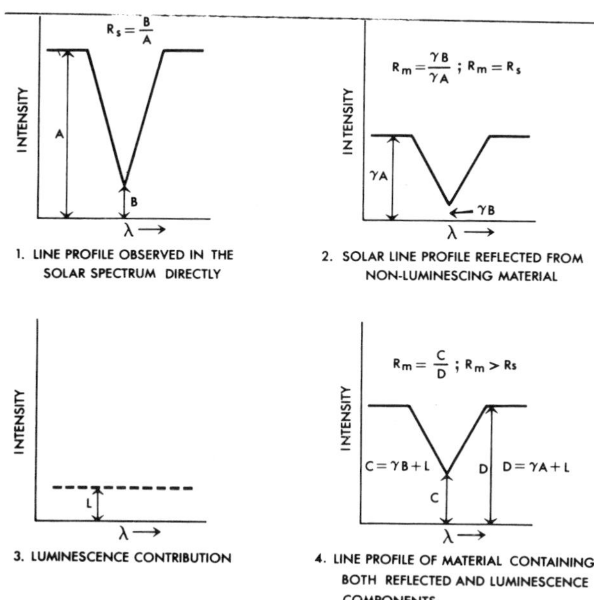

Figure 1 Diagrams illustrating the Fraunhofer line-depth method.

Solving for ρ:

$$\rho = \frac{AC - BD}{A(A - B)} \tag{13}$$

$$\rho = \frac{1}{A - B}\left[C - \frac{BD}{A}\right] \tag{14}$$

Equation (14) is useful because γ and L are eliminated and the expression of ρ is independent of incident solar flux; moreover, the expression is amenable to solution using an analog computer.

Perkin-Elmer Instrument

The first airborne instrument employing the Fraunhofer line-depth method was designed and constructed by the Perkin-Elmer Corporation under contract from the National Aeronautics and Space Administration. This instrument includes an optical unit which may be mounted in the tail section or on the side of an aircraft, and a control console for operation of the instrument from inside the aircraft cabin (Fig. 2)[2].

Optical Unit

The optical unit is comparable in size and shape to a small suitcase, and consists of two telescopes each directed toward a window or aperture on opposite sides (Fig. 3). The unit is positioned so that one telescope is directed upward toward skylight and sunlight and the other, with a 1° field of view, is directed downward toward the ground.

The sky-looking telescope views a convex mirror through a hole in the center of a horizontal circular disk. The disk is painted matte white in order to approximate a Lambertian surface, and it scatters into the telescope field an amount of light proportional to the total sunlight and skylight incident on a horizontal surface. This arrangement eliminates the need for a sun tracker, places no restriction on aircraft orientation in level flight, and permits the output signal to be normalized so that, within certain limits, it is independent of the sky brightness. In practice, some variation in signal strength is observed because the matte white disk and convex mirror are rigidly fixed to the aircraft and, therefore, do not remain perfectly horizontal during flight.

A rotating chopper near the focal plane of each telescope encodes each beam with a different frequency before both beams are combined on a partially transmitting surface. The combined beam is collimated and split-into halves, both of which are directed toward a narrow-band filter package and photomultiplier. The sky and ground signal components passing through each filter are synchronously detected to yield the four signals, A, B, C, and D referred to above.

The use of dielectric optical coatings results in very little light loss, but nearly all beamsplitters, including the dielectric variety, treat one polarization differently from the other. This could result in different proportions of light being sent to the two photomultipliers, depending on the polarization of light received at the aperture of the instrument. To prevent an erroneous variation in the beamsplitter ratio, a polarizer is included between the collimator lens and the beamsplitter.

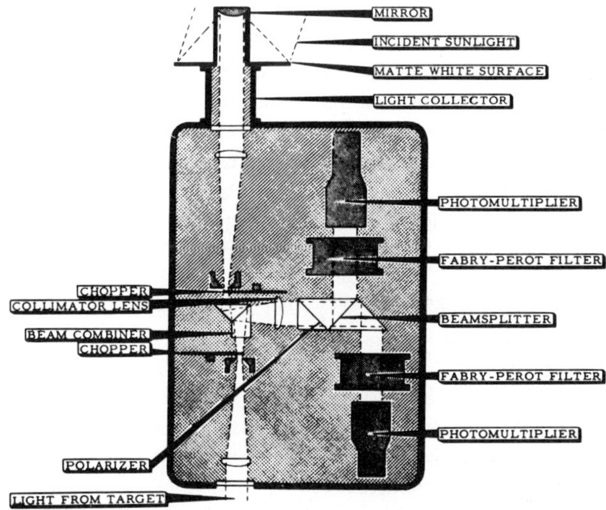

Figure 3 Diagram showing design of the optical unit of the prototype Fraunhofer line-discriminator.

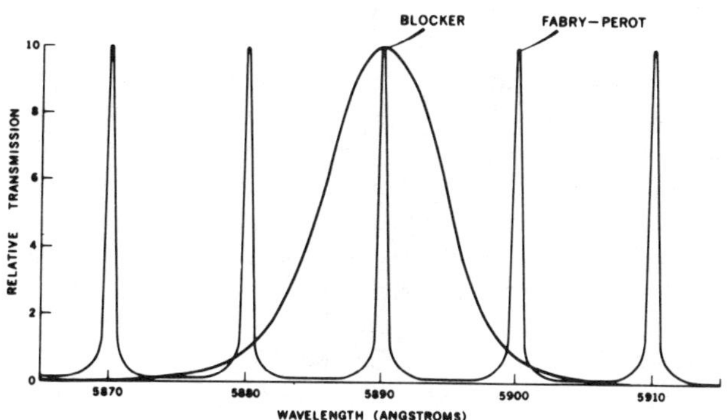

Figure 4 Spectral transmission of a Fabry-Perot spike filter and a narrow-band blocking filter.

[2] Not reproduced here.

Key components of the optical unit are two glass-spaced Fabry-Perot filters. These filters consist of a glass etalon spacer about 0.1 mm thick which produces a comb-shaped spectral response as shown in Figure 4. A blocking filter about 10 A wide eliminates adjacent spectral side bands but retains a single transmission spike with a half-width of about 0.5 A. The center wavelength may be tuned within a limited spectral range by means of a temperature-controlled housing around the filter package. At the red end of the spectrum, a 25°C change in temperature alters the index of the etalon sufficiently to shift the pass band about 1 A.

Control Console

In the control console the four demodulated signals corresponding to *A*, *B*, *C*, and *D* from the optical unit are filtered in separate low-pass amplifiers and processed in an analog computer to yield a luminescence component (ρ) with a 20 Hz bandwidth. The analog computer contains operational amplifiers for adding and subtracting, and analog log and antilog amplifiers to multiply and divide in real time. The output signal bandwidth can be reduced from the 20 Hz limit of the low-pass amplifiers to as low as 2 Hz by means of a variable low-pass output amplifier. The luminescence coefficient (ρ) and the B/A ratio can be read from meters, recorded from a low impedance output jack, or monitored on an oscilloscope built into the control console. Temperature control is provided for the computer elements to minimize temperature-induced bias errors and also for the optical filter packages to stabilize the center wavelength. The temperature of the continuum filter is not critical and the optimum temperature of the Fraunhofer filter is readily determined by adjusting a multiturn potentiometer on the temperature controller until a minimum B/A ratio is found.

Rhodamine Dye

Rhodamine dye is a water-soluble luminescent dye used by marine geologists and hydrologists to monitor current dynamics in rivers, estuaries, and coastal waters. As a minimum performance requirement, the prototype Fraunhofer line discriminator is designed to detect Rhodamine WT dye in 1 meter of water in concentrations of 20 ppb or less.

Detection of rhodamine dye was selected as a design objective mainly because the dye is available commercially, it lends itself to quantitative performance tests, and discrete concentrations of the dye provide predictable and easily repeatable luminescence levels. These advantages would be difficult to achieve with other materials such as outcrops of luminescent minerals, mineral oils, and other materials for which luminescence intensity and spectral distribution are less well known and may vary widely for the same material in different localities.

A practical benefit was also anticipated. Conventional methods of monitoring rhodamine dye are laborious and awkward. It was hoped that the technique could be of some immediate practical benefit in dye studies of large rivers and estuaries.

Figure 5 shows the excitation and emission spectra of rhodamine dye. The emission peak is near 5800 A, and, accordingly, the prototype line discriminator is filtered to look at the central intensity of the nearby D_2 Fraunhofer line of sodium at 5890 A (Fig. 6). The continuum filter, *C*, is centered at 5892 A.

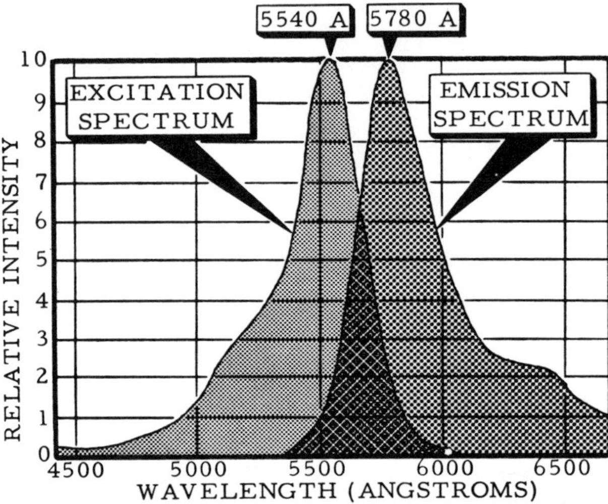

Figure 5 Excitation and emission spectra of rhodamine dye. (After spectrofluorometric analysis furnished by G. K. Turner Associates, Palo Alto, Calif.)

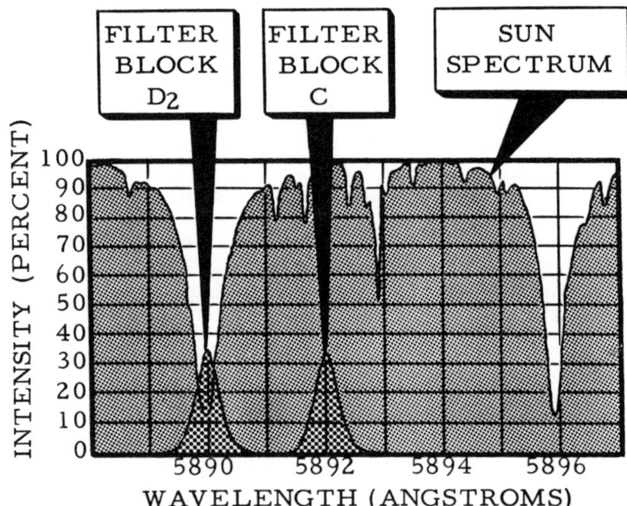

Figure 6 D_2 and C filter transmission superimposed on the solar spectrum.

Outdoor Tests

Tank Tests

Initial tank tests of the Fraunhofer line discriminator were conducted out-of-doors in Phoenix, Ariz., and Menlo Park, Calif. These tests established that all components were functioning as an integrated unit and that sensitivity of the instrument was adequate to detect concentrations of dye smaller than 20 ppb in 1 meter of water.

The tank test arrangement is shown in Figure 7.[3] The

[3] Not reproduced here.

optical unit is mounted on a construction scaffold, and the ground-looking aperture is directed into a stock tank filled with water to a depth of ½ meter. Sides of the tank are painted flat black. The tank is mounted on casters so that it may be moved to permit maximum solar illumination of the water throughout the day. A dual-channel strip chart recorder permits the luminescence intensity, ρ, and, optionally, the line-depth ratio in the solar spectrum, R_s (that is, B/A) to be plotted simultaneously as a function of time. Rhodamine dye was added to the tank in concentrations of less than 5 ppb. Luminescence of water samples collected from the tank was verified with a G. K. Turner Model 110 laboratory fluorometer and standard solutions for which the rhodamine dye concentration was known.

The tank tests show that rhodamine dye can be detected in ½ meter of water in concentrations well under 5 ppb. Figure 8 shows the increase in luminescence intensity that resulted when each of 14 increments of dye were added to the tank. Dye concentration of each increment was about 3 ppb as verified with the laboratory fluorometer. In another tank test, 25 successive increments averaging 1.3 ppb were clearly differentiated.

Analysis of all strip chart records show that the relation between luminescence intensity and dye concentration appears to be nearly linear. Monitoring R_s, that is, B/A, on more than 20 days throughout the summer and fall in Phoenix, and spring and early summer in Menlo Park show only a very gradual increase in B/A from early morning to late afternoon. The increase observed could be due to instrument drift.

Clouds and even thin overcast obscuring the sun adversely affected the sensitivity of the Fraunhofer line discriminator. On clear, sunny days the optimum time for operation is during the 3-hour period before and after midday, but detection of dye has been successful as late as 5:45 p.m. during August.

Absorption of incident and emitted light by suspended material in the water is more significant than absorption

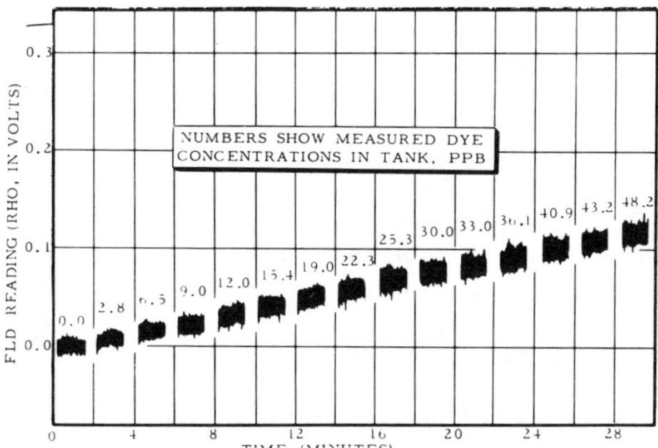

Figure 8 Chart showing results of controlled experiment over a tank of luminescent rhodamine dye, October 29, 1968. Numbers show measured dye concentrations in parts per billion.

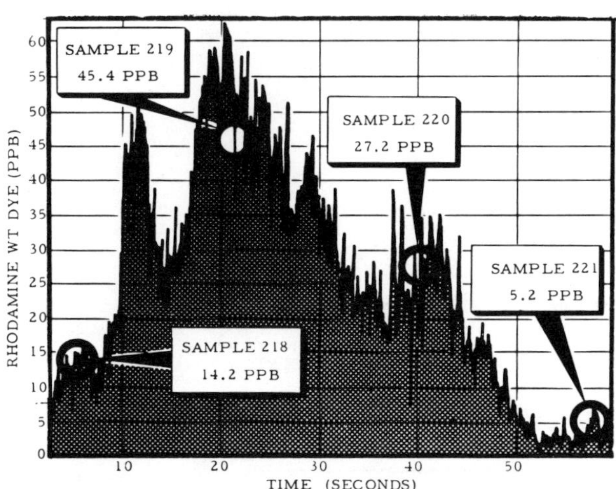

Figure 10 Part of a strip chart record of luminescence intensity in San Francisco Bay near Redwood City, California, May 20, 1969. Dye concentration in sample was verified with a laboratory fluorometer.

of excitation and emission light by the dye itself. Formulas have been derived by Stoertz (1969) relating attenuation of light angle of the sun, water temperature, and vertical distribution of the dye.

Shipboard Tests

The Fraunhofer line discriminator was operated as a shipboard fluorometer aboard the U.S. Geological Survey's ship *Polaris* in San Francisco Bay near Redwood City, Calif. A shipboard test was an essential part in the early evaluation of the instrument because it permitted water samples to be taken and precisely correlated on the chart.

The Fraunhofer line discriminator was suspended from a davit over the side of the ship (Fig. 9)[4] and luminescence intensity recorded as the ship moved slowly through a cloud of rhodamine dye. Part of a typical strip chart record is shown in Figure 10. Dye concentrations of the water samples were verified with a laboratory fluorometer; the positions of those samples shown in Figure 10 correlate well with the trace of luminescence intensity plotted on the chart record.

Helicopter Tests

Airborne tests of the Fraunhofer line discriminator were conducted in the San Francisco area during May, July, and August 1969 (Fig. 11). The optical unit was mounted on the side of a Sikorsky H-19 helicopter (Fig. 12)[4] about 10 feet behind the door, and sufficiently high on the fuselage so that no light scattered from the yellow side of the helicopter could enter the sky-looking aperture. An aluminum mounting bracket with four shock mounts supports the optical unit. An insulating jacket was used to cover the optical unit before takeoff to prevent overheating from exposure to direct sunlight. The control console and strip chart recorder were located inside the cabin.

[4] Not reproduced here.

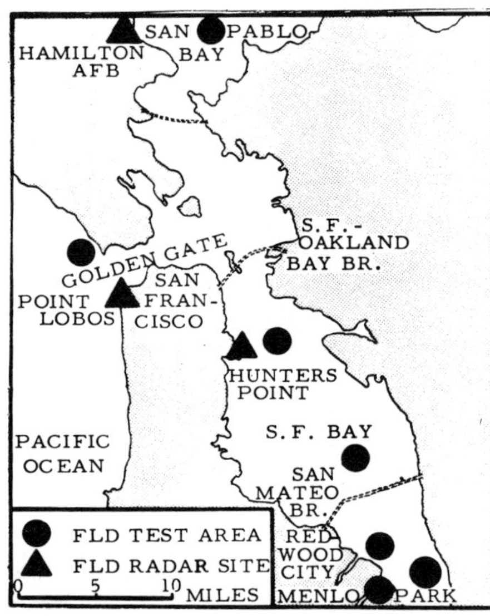

Figure 11 Areas where helicopter tests of the Fraunhofer line-discriminator were conducted during May, July, and August, 1969.

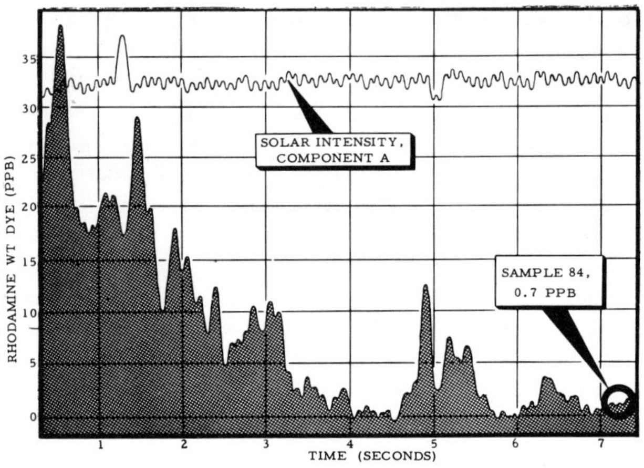

To monitor changes in instrument sensitivity due to changes in incident light levels, or drift in electronics, standard luminescent targets (Fig. 13)[5] were moved in and out of the field of view of the ground-looking aperture during flight by means of a cable operated from the cabin. The most satisfactory target was an acrylic resin cylinder, ½-inch deep, filled with rhodamine dye solution of known concentration.

A shore-based tracking radar (Fig. 14)[5] was used to plot the precise flight path of the helicopter and location of dye clouds, and was frequently useful in directing the helicopter to the dye when it was no longer clearly visible to the pilot. Locations of radar sites are shown in Figure 11. All phases of the radar operation are described by Howell (1969).

Luminescent dye was commonly dropped from the helicopter in a string of uniformly spaced patches while the aircraft moved at a uniform speed into the wind. The most suitable method for air dropping was found to be double plastic bags of dye which were released from a stiff cardboard carton. Breakage of the bag at the water surface was assured by excluding all air and by dropping from an altitude of at least 100 feet.

Water samples were taken during the tests, but the problems of collecting samples from a slow-moving or hovering helicopter precluded extensive sampling.

A segment of a strip chart record made during tests over San Francisco Bay, north of the San Mateo Bridge, is shown in Figure 15. The bay is extremely turbid and the column of water integrated by the Fraunhofer line discriminator probably does not exceed 20 to 30 cm. Figure 16 is a strip chart record made during tests over the Pacific Ocean, west of the Golden Gate, where the water is relatively clear. No subsurface samples were obtained, so the depth of the dye in the water column is not known. Surface samples analyzed with the laboratory fluorometer, however, showed concentrations of

[5] Not reproduced here.

Figure 15 Part of a strip chart made May 8, 1969, during helicopter tests over San Francisco Bay north of the San Mateo Bridge. Lower trace is luminescence intensity of the dye plotted as a function of time. Upper trace is a record of solar intensity (A).

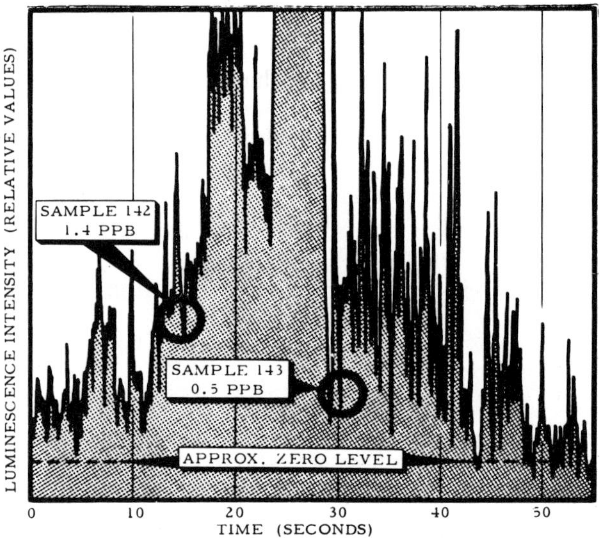

Figure 16 Part of a strip chart made May 14, 1969, over the Pacific Ocean west of the Golden Gate.

0.5 and 1.4 ppb. The locations of these samples in Figure 16 suggests that a sensitivity of 0.1 or 0.2 ppb was achieved. Comparison of Figures 15 and 16 shows that changes in sensitivity are appreciable from place to place, and illustrates the need of calibration by means of standard targets and/or water samples.

In other airborne tests, rhodamine dye was detected from altitudes of up to 5000 feet. As altitude increased, the highest peaks recorded were flattened due to increasing size of the field of view, but the dye was clearly detectable at the highest altitude reached.

Application to Operational Rhodamine Dye Studies

Two possible operational uses of the Fraunhofer line discriminator are suggested by the experimental work completed to date. One use would be in measuring the absolute concentration of dye from a low-flying helicopter or fixed-wing aircraft. This measurement would be based on the assumption that the dye was uniformly distributed in the column of water integrated by the instrument. It would be desirable to support this measurement with at least a minimal number of subsurface water samples from which to measure attenuation of light by suspended material in the water, and to verify the depth to which the dye is dispersed in the water.

The other operational use suggested by preliminary results acquired to date would be to measure the rate of dispersion of dye, or by analogy, a soluble pollutant in the water body being studied, by repeatedly traversing a single dye cloud over a period of several minutes or hours. The slopes of the curves acquired in this manner could be related to the rate of dispersion of the dye. Absolute concentrations of dye would not be needed for this measurement.

A limitation of the use of Fraunhofer line discriminator is that the method will not work under cloudy conditions, at least with the relatively narrow D_2 Fraunhofer line. It is conceivable that nighttime operations would be possible by using the line discriminator as a filter photometer detector together with an artificial excitation source, such as a pulsed gas laser operating near the excitation peak of rhodamine dye.

Other Applications

The prototype Fraunhofer line discriminator was designed to detect rhodamine dye largely as a matter of convenience. The dye is readily available, its luminescence properties are well known, and it lends itself to a controlled test. A modified or improved instrument could conceivably be designed to detect other materials.

The sodium D_2 line at 5890 Å (Fig. 17) is one of the narrower Fraunhofer lines, about 0.7 Å at half intensity. Successful use of this line in detecting rhodamine dye with the prototype instrument is a kind of "worst-case" test. At the other extreme are the two calcium lines, H and K, at 3968 Å and 3934 Å in the violet. The half-widths of these lines exceed 10 Å. Solar brightness in the blue and violet regions, moreover, decreases markedly with wavelength, and the H and K lines would provide a lower solar background against which to detect luminescence than Fraunhofer lines at longer wavelengths. Also of interest are the hydrogen lines, F at 4861 Å and C at 6563 Å. The widths at half intensity of these lines is more than twice that of the sodium D_2 line.

One of the original considerations prompting selection of the Fabry-Perot filter design is that these filters retain the two-dimensional structure of the scene being looked at. Conceivably, a videcon or image orthicon could replace the phototubes in the prototype, and luminescence could be imaged frame by frame in a future instrument.

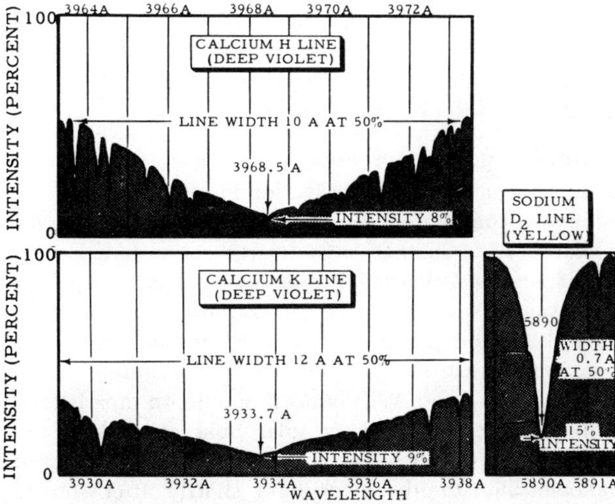

Figure 17 Profiles across the sodium D_2 and the calcium H and K Fraunhofer lines.

Lignin Sulfonate

Spent sulfite liquor is produced as a waste material in the process of making paper from wood pulp. Disposal of the material into rivers and estuaries constitutes a severe water pollution problem in areas where numerous paper mills are located, such as in the Pacific northwest. Similar materials are said to be injurious to commercial fish and oysters (Waldichuk, 1964). Lignin sulfonate is a constituent of the liquor that exhibits an inherent luminescence.

Christman and Minear (1967), in a preliminary study, used a laboratory spectrofluorometer and found that both pure lignin sulfonate and whole spent sulfite liquor, collected from three localities, exhibited identical excitation and emission peaks at 3400 Å and 4000 Å, respectively. The relation between luminescence intensity and concentration of the whole waste liquor was linear for those samples studied.

H. F. Smith and P. S. Flandreau (Perkin-Elmer Corp., written communication, 1969) observed a slight shift of the emission peak of sulfite liquor towards shorter wavelengths as concentration of the liquor in the water was reduced. They also noted that changing the pH of the solution from acidic to basic was accompanied by both a reduction in emission intensity and a 200–400 Å shift in the emission peak toward a longer wavelength.

In an attempt to relate the luminescence of lignin sulfonate to the sensitivity of the prototype Fraunhofer line discriminator, the authors ran excitation spectra on a sample of spent sulfite liquor collected at the outflow of an oxidation pond of a paper mill in Oregon (Fig. 18).

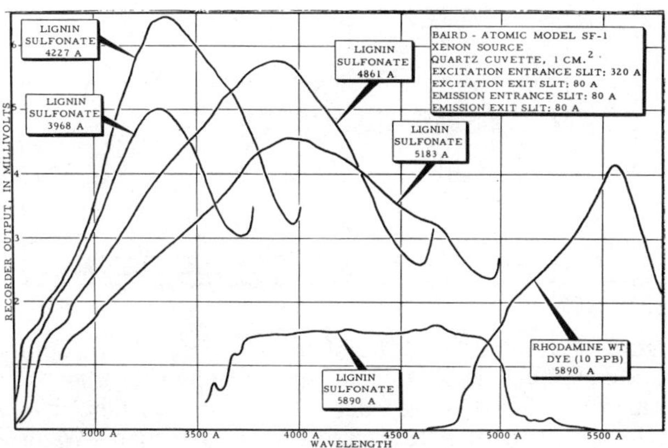

Figure 18 Excitation spectra of lignin sulfonate and Rhodamine WT dye (10 ppb).

The spectra were acquired with the excitation monochromator operating in the scanning mode, while the band pass of the emission monochromator was centered at the following Fraunhofer lines: 3968 Å, 4227 Å, 4861 Å, 5183 Å, and 5890 Å. The sample was contained in a 1-cm quartz cuvette. Using the same slits, a 1-cm cuvette of 10 ppb Rhodamine WT dye was run with emission monochromator centered at the D_2 Fraunhofer line at 5890 Å, which is near the peak emission of the dye.

Although the spectra are not corrected for wavelength variation in source intensity and detector sensitivity, the amplitude of the excitation spectra for the spent sulfite liquor compare favorably with the curve for rhodamine dye at 10 ppb, a concentration well within the sensitivity limits of the prototype Fraunhofer line discriminator.

Oil Spills and Seeps

Riecker (1962) notes that all crude oils luminesce, and he analyzes emission peaks of 115 samples from a wide variety of geologic settings in Colorado, Wyoming, and Alberta. Emission peaks of 12 percent of the samples were in the red at 6300 Å, 53 percent peaked at 5780 Å in the yellow, and 35 percent peaked in the blue-green and blue at 4820 Å, 4720 Å, 4520 Å, and 4420 Å. Riecker correlates the emission peaks with heavy hydrocarbon fractions which are selectively filtered during subsurface migration from the source shifting the emission peaks to shorter wavelengths.

The first attempt to use the prototype Fraunhofer line discriminator to detect crude oil was conducted in a tank test using a sample collected from the oil spill near Santa Barbara, Calif. The test was not successful partly because the emission peak of the Santa Barbara crude is at a shorter wavelength than the D_2 Fraunhofer line (5890 Å) where the prototype instrument operates, and partly because of lack of adequate sensitivity of the prototype instrument.

The excitation spectra shown in Figure 19 were run

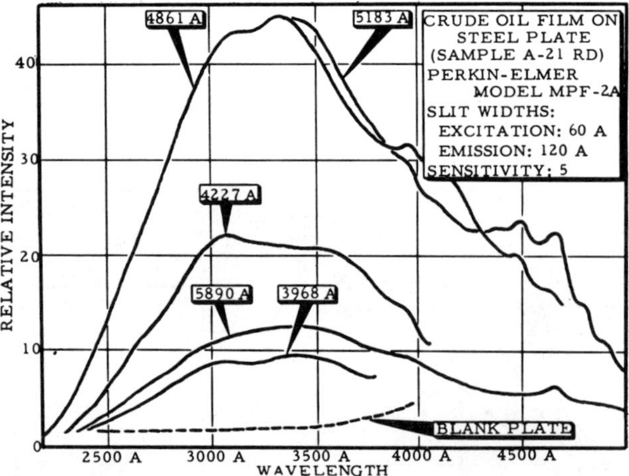

Figure 19 Excitation spectra of crude oil from the oil spill at Santa Barbara, California, and monitored by specific Fraunhofer lines. Oil was deposited as a thin film on a stainless steel plate. (Spectra courtesy of Federal Water Pollution Control Administration, Athens, Ga.)

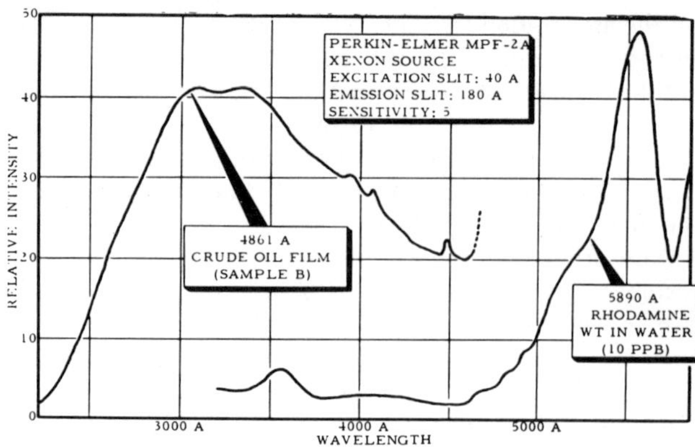

Figure 20 Excitation spectrum for film of crude oil monitored at 4861 A and Rhodamine WT (10 ppb) monitored at 5890 Å. (Spectra courtesy of Federal Water Pollution Control Administration, Athens, Ga.)

with a laboratory spectrofluorometer on crude oil from the Santa Barbara oil spill, and serve to identify luminescence in terms of specific Fraunhofer lines. The oil was deposited as a film on a stainless steel plate. The emission monochromator was centered at the following Fraunhofer lines: 3968 Å, 4227 Å, 4861 Å, 5183 Å, and 5890 Å. Peak excitation was near 3500 Å for all the lines; maximum amplitude was shown in the excitation spectrum monitored by the 4861 Å and 5183 Å lines.

In an attempt to relate luminescence of the Santa Barbara spill to the sensitivity of the prototype Fraunhofer line discriminator, Figure 20 plots the excitation spectrum of the oil film monitored at the 4861 Å line, together with the excitation spectrum of Rhodamine WT dye at 10 ppb in a 1-cm cuvette.

Although the amplitudes of both spectra are about the same, the luminescence detectivity of the two materials is not directly comparable. Crude oil occurs as a film on the water surface, but the dye is water soluble and is distributed to some depth. Detection of a 10 ppb concentration of dye with the prototype Fraunhofer line discriminator requires integrating a column of water of at least several centimeters, whereas this laboratory experiment was based on only a 1-cm cuvette of dye solution. Detection of oil spills and seeps might be possible with an improved Fraunhofer line discriminator featuring larger optics and an improvement in sensitivity of perhaps one order of magnitude.

Atmospheric Luminescence

Monitoring aerosols and other luminescent components in the atmosphere from the ground, high-altitude aircraft, or spacecraft is another possible application of a Fraunhofer line discriminator. Several workers have reported some success in detecting atmospheric luminescence using the line-depth method and techniques similar to those used by astronomers observing lunar luminescence. Shefov (1959) observed reduced line depth in scattered light compared with direct solar light at dusk in the blue-violet part of the spectrum. Wildey (1964) observed shallower line depths in skylight shortly after sunrise than for moonlight during the preceding night, and this indicated to Wildey "that the sky was definitely more luminescent than the moon. "Grainger and Ring (1962a, 1962b) compared night observations of the sunlit lunar surface with daytime observations of cloud and blue sky, and observed that the lunar surface exhibited the deepest line depths, and blue sky the shallowest; they attribute this phenomenon to the presence of a daylight airglow. Noxon and Goody (1965) corroborated Grainger and Ring's observations, suggested that the luminescence intensity is higher in the blue part of the spectrum than in the red, and attributed the phenomenon to the luminescence of aerosols below an altitude of about 20 kilometers.

Air Pollution

Carcinogenic hydrocarbons are found in urban air and, according to Sawicki, Elbert, et al. (1960), some of them are comparable in composition from city to city. Sawicki, Hauser, and Stanley (1960) found benzopyrene to be the only hydrocarbon to exhibit an emission maximum at 5480 Å, and this specificity permitted traces of impurities to be detected in what was thought to be pure benzopyrene. They used laboratory procedures to isolate the hydocarbons and to obtain their spectral emission, but it is suggested that relative concentrations of hydrocarbon pollution might be monitored with a Fraunhofer line discriminator capable of operating at several Fraunhofer lines selected in accordance with the emission spectra characteristics of the selected hydrocarbons to be detected.

Luminescent Tracers of Agricultural Chemicals

Yates and Akesson (1963) cite drift of sprayed agricultural pesticides and herbicides outside the area of intended treatment to be one of the major problems during aerial or ground application. They considered luminescent soluble dye added to the spray, and successfully used the dye Brilliant Sulpho Flavine as a luminescent tracer in two pesticides sprayed at a concentration of 7 to 10 gallons per acre. Coverage was verified with a laboratory fluorometer. Himel et al. (1965) have conducted similar experiments using insoluble luminescent particles of zinc cadmium sulfide suspended in insecticide sprayed from a helicopter.

Monitoring coverage of the spray in the field is conceivable provided that the tracers are applied with sufficient concentration to be detected with a Fraunhofer line discriminator.

Chlorophyll Luminescence

Absorption and emission of light by chlorophyll are intimately related to photosynthesis, the process of converting light energy into chemical energy, and a fundamental function of all living plants. Luminescence has been used as a tool by many workers for laboratory study of photosynthesis. Figure 21 shows excitation and emission spectra for four cultures of marine phytoplankton (Yentsch and Menzel, 1963). Maximum excitation occurs between 4300 Å and 4500 Å; maximum emission occurs between 6500 Å and 6750 Å. Green leaves exhibit similar excitation and emission maxima.

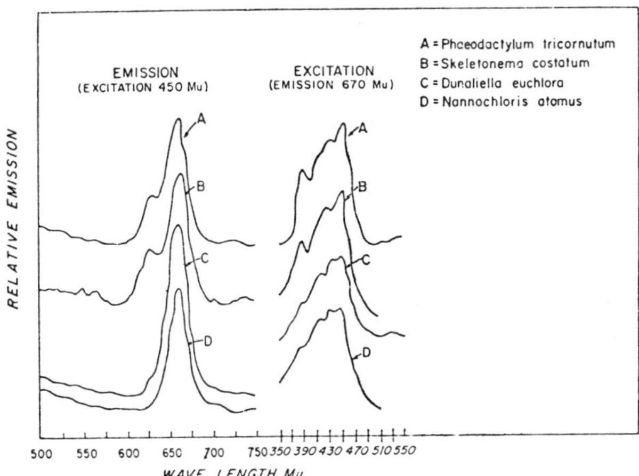

Figure 21 Excitation and emission spectra for several cultures of marine phytoplankton. (After Yentsch and Menzel, 1963)

T J. Porro et al. (Perkin-Elmer Corp., written communication, 1963) observed that green leaves of dogwood, sugar maple, African violet, and philodendron exhibit distinctive ratios between emission bands centered at 5550 Å and 6950 Å.

Pringsheim (1949) and others have noted that luminescence may be increased by introducing environmental conditions that impede photosynthesis, such as reduction in CO_2 concentration in the surrounding atmosphere, or application of material that is toxic to the plant. Udenfriend (1962) describes chlorophyll luminescence as evidence of inefficient use of absorbed energy by the photosynthesis process, and cites wide use of the luminescence spectra of intact plant cells as an index of the vigor of photosynthesis and plant metabolism in the laboratory.

A Fraunhofer line discriminator could be modified to operate in the region beyond 6500 Å, although it is not known whether sensitivity of the modified instrument would be adequate to monitor in vivo chlorophyll emission. Udenfriend (1962) notes that although luminescence efficiencies as high as 33 percent have been observed for chlorophyll a in solution, luminescence efficiency in live cells is no higher than 2 to 2.8 percent, and values as low as 0.15 percent have been observed. Latimer et al. (1956) found the quantum yields in the living plant to range from 0.017 to 0.027, about a factor of 10 less than the same material in solution. Thomas and Flight (1964) also report luminescence efficiency of chlorophyll in vivo to be about one-tenth as much per unit weight as the same material in solution. G. Weber and F. W. J. Teale (in Udenfriend, 1962) show that the quantum efficiency of chlorophyll a in ethanol is about 24 percent the quantum efficiency of rhodamine dye in the same solvent.

If measurable differences do occur, however, between the emission of healthy and stressed plants, monitoring emission from low-flying aircraft might help to delineate agricultural crop areas where plant disease and pest damage are in early stages. Another application might be in identifying areas of plant stress related to geochemical soil anomalies. Determining the presence and relative concentration of phytoplankton chlorophyll in the open ocean is another possible application, and, if feasible, would offer the advantage of speed and broad aerial coverage over the more quantitative laboratory and shipboard alternatives discussed by Yentsch and Menzel (1963) and by Lorenzen (1966).

Mineral Luminescence

Small hand-carried ultraviolet lamps have been used for many years in prospecting for luminescing minerals, notably scheelite, an ore of tungsten, and some minerals containing uranium. Detecting luminescing minerals with a hand-carried Fraunhofer line discriminator is conceivable, and would avoid the low power and awkward nighttime operations which limit the use of the hand-carried lamps.

Emission spectra of some luminescing minerals may vary widely for the same mineral, however, even for specimens collected from the same locality. Small outcrop areas, products of surface weathering, and thin veneer of covering material obscure luminescence that is prominent on a freshly fractured surface. These factors would greatly impede controlled testing of a prototype Fraunhofer line discriminator, and therefore, detection of mineral luminescence is not being emphasized in work now underway.

Bioluminescence

Bioluminescence of marine plants and animals has received increased laboratory and field study in recent years as photometers have been developed with adequate sensitivity to detect the low light levels that are necessary. Nicol (1960) reports that bioluminescence is generally restricted to wavelengths between about 4200 Å and 5400 Å, and that bioluminescence intensity ranges from 1×10^{-6} to 1×10^{-1} microwatt/cm^2 of surface receiving the light at a distance of 1 cm.

Work by Taylor et al. (1966) shows that at least some dinoflagellates (single-celled plantlike organisms) migrate to the surface only when certain temperature and salinity conditions are achieved. An image orthicon or other low-level video detector might be more appropriate than a Fraunhofer line discriminator, however, because surface population and luminescence capacity increase substantially after twilight. Nicol (1960) notes that bioluminescence of several forms is inhibited by daylight, and at least some near-surface forms show a diurnal rhythmicity and disappear during the day. Backus et al. (1961) also believe that diurnal fluctuation is due to the suppression of the bioluminescent process during daylight.

Conclusions

The Fraunhofer line-depth method uses the sun as a source to excite luminescence, and permits detection of luminescing materials during daylight against a high solar background from substantially greater range than can be achieved with artificial excitation sources. Tests of a prototype Fraunhofer line discriminator over a tank of water containing a known quantity of luminescent dye show that the instrument responds to dye concentrations as small at 1 ppb, exceeding original design expectations by about one order of magnitude. Results of helicopter tests over San Francisco Bay suggest at least limited operational use of the instrument in studies where luminescent dye, such as Rhodamine WT, are used to study current dynamics and dispersion rates in large rivers and estuaries. A modified version may be useful in distinguishing materials that exhibit natural luminescence such as hydrocarbons, fish oil, some water pollutants, and luminescing components in the atmosphere.

REFERENCES

Backus, R. H., Yentsch, C. S., and Wing, A. Bioluminescence in the surface waters of the sea. *Nature* 192:518, 1961.

Christman, R. F., and Minear, R. A. Fluorometric detection of lignin sulfonates. *Trend in Engineering* 19:3, 1967.

Dubois, J. Contribution à l'étude de la luminescence lunaire. *Ceskoslovenske Akad. Ved, Rozpravy* 69, Pt. 6, 1959.

Grainger, J. F., and Ring J. Anomalous Fraunhofer line profiles. *Nature* 193:762, 1962a.

— Lunar luminescence and solar radiation. *Space Res.* 3:989, 1962b.

Hemphill, W. R., Fischer, W. A., and Dornbach, J. E. Ultraviolet investigations for lunar missions. *Adv. Astronaut. Sci.* 20, Pt. 1:397, 1966.

Hemphill, W. R. Application of ultraviolet reflectance and stimulated luminescence to the remote detection of natural materials. U.S. Geological Survey, Open-File Report, 1968.

Himel, C. M., Vaughn, L., Miskus, R. P., and Moore, A. D. A new method for spray deposit assessment. U.S. Forest Service, Research Note PSW 87, Pac. Southwest Forest and Range Exper. Stat., Berkeley, Calif., 1965.

Howell, R. L. Equipment and techniques for low-altitude aerial sensing of water-vapor concentration and movement. *Remote Sensing of Environment* 1:13, 1969.

Kozyrev, N. A. The luminescence of the lunar surface and intensity of the solar corpuscular radiation. *Izvestia Krymskoi Astroizitcheskoy Observatorye* 16:148, 1956.

Latimer, P., Bannister, T. T., and Rabinowitch, E. Quantum yields of fluorescence of plant pigments. *Science* 124:586, 1956.

Lorenzen, C. J. A method for the continuous measurement of *in vivo* chlorophyll concentration. *Deep-Sea Res.* 13:223, 1966.

McCord, T. B. Observational study of lunar visible emission. *J. Geophys. Res.* 72:2087, 1967.

Myronova, M. M. Luminescence in the crater Aristarchus. *Akad. Nauk Ukrainskoi SSR, Main Astronomical Observatory* No. 4:455, 1965.

Nicol, J. A. C. Luminescence in marine organisms. *Smithsonian Inst. Ann. Report 1960,* 1960. Pp. 447–456.

Noxon, J., and Goody, R. M. Noncoherent scattering of skylight. *Akad. Nauk SSSR, Izvestiya, Atmospheric and Oceanic Physics Series* 1, No. 3:275, 1965.

Pringsheim, P. *Fluorescence and Phosphorescence.* New York: Interscience, 1949.

Riecker, R. E. Hydrocarbon fluorescence and migration of petroleum. *Am. Assoc. Petroleum Geologists Bull.* 46:60, 1962.

Sawicki, E., Elbert, W., Stanley, T. W., Hauser, T. R., and Fox, F. T. The detection and determination of polynuclear hydrocarbons in urban airborne particulates. *Internatl. J. Air Pollution* 2:273, 1960.

Sawicki, E., Hauser, T. R., and Stanley, T. W. Ultraviolet, visible and fluorescence spectral analysis of polynuclear hydrocarbons. *Internatl. J. Air Pollution* 2:253, 1960.

Shefov, N. N., *in* Noxon, J., and Goody, R. M. (1965) p. 275, 1959.

Spinrad, H., Lunar luminescence in the near ultraviolet. *Icarus* 3:500, 1964.

Stoertz, G. E., Fraunhofer line-depth sensing applied to water. U.S. Geological Survey, Open-File Rept., 1969.

Taylor, W. R., Seliger, H. H., Fastie, W. G., and McElroy, W. D. Biological and physical observations on a phosphorescent bay in Falmouth Harbor, Jamaica, W.I. *J. Marine Res.* 24:28, 1966.

Thomas, J. B., and Flight, W. F. G. Fluorescence responses of chlorophyll *in vivo* to treatment with acetone. *Biochim. Biophys. Acta* 79:500, 1964.
Udenfriend, S. *Fluorescence Assay in Biology and Medicine.* New York: Academic, 1962.
Waldichuk, M. Dispersion of kraft-mill effluent from a submarine diffuser in Stuart Channel, British Columbia. *Can. Fisheries Res. Board J.* 21:1289, 1964.
Wildey, R. L. Lunar luminescence. *Astron. Soc. of the Pacific Pubs.* 76:112, 1964.
Yates, W. E., and Akesson, N. B. Fluorescent tracers for quantitative microresidue analysis. *Am. Soc. Agric. Engineers Trans.* 6:104, 114, 1963.
Yentsch, C. C., and Menzel, D. W. A method for the determination of phytoplankton chlorophyll and phaeophytin by fluorescence. *Deep-Sea Res.* 10:221, 1963.

SUGGESTED READINGS FOR PART THREE

Fischer, W. A., and Daniels, D. L. Interim Report of Ultraviolet Absorption and Stimulated Luminescence. Investigations Being Undertaken in Cooperation with NASA. Part II. *Spectral Distribution of Ultraviolet-Stimulated Luminescence.* U.S. Geological Survey, Technical Letter, NASA-3, 1964.
Friedel, R. A., and Gibson, H. L. Infrared and Ultraviolet Visible Spectometry. Internal Report. U.S. Dept. of the Interior, Bureau of Mines, 30.JI, 1964.
Hemphill, W. R., and Carnahan, S. U. Ultraviolet Absorption and Luminescence Investigations Progress Report. U.S. Geological Survey, Technical Letter, NASA-6, 1965.
Hemphill, W. R., and Gawarecki, S. J. Interim Report of Ultraviolet Absorption and Stimulated Luminescence. Investigations Being Undertaken in Cooperation with NASA. Part I. *Ultraviolet Video Imaging System.* U.S. Geological Survey, Technical Letter, NASA-3, 1964.
Koller, L. R. *Ultraviolet Radiation.* New York: Wiley, 1965.
Olson, D. L., and Cantrell, J. L. Comparison of Airborne Conventional Photography and Scanned Ultraviolet Imagery. *Photogrammetric Engineering* 31:507, 1965.
Tousey, R. The Extreme Ultraviolet—Past and Future. *Applied Optics* 1:679, 1962.
Watts, H. V. Reflectance of Rocks and Minerals to Visible and Ultraviolet Radiation. U.S. Geological Survey, Technical Letter, NASA-32, 1966.

PART FOUR
VISIBLE LIGHT

Conventional photography, the oldest and most highly developed realm of remote sensing

SECTION A

Suborbital air photography, black and white

By what processes does man interpret the environment that surrounds him through the images formed by reflected light? The answers to this question are only partially known. Man has binocular vision and because of the parallax displacement of his eyes, he is able to view this world stereoscopically with three-dimensional vision. This stereoscopic vision and associated depth perception are important methods in the interpretation of the environment. Also, man gets an oblique or angular view of most objects he observes. The experiences, training, and education that an observer has developed in a lifetime of viewing his surroundings diminish in value—and are perhaps even a hindrance—when he observes and tries to interpret a vertical air photograph. The photograph is flat, or two-dimensional, unless viewed stereoscopically. In the black and white mode, objects have no revealing colors. Objects are imaged at greatly reduced scale. Finally, if the observer is looking at a vertical air photograph, he is presented with a nonoblique view of a world he can experience only from an airplane or spacecraft. As a result, some elements on the photograph assume a greater importance in image interpretation.

The interpreter must pay special attention to characteristics and clues which can be identified on the image. In the final analysis, although anyone can be trained to read a photograph, a really good interpreter has an almost intuitive understanding of the peculiar view presented to him on the imagery! We can define photo-interpretation as the identification of objects on photographs and the determination of their meaning or significance. In broader terms this definition applies to the entire discipline of remote sensing.

11-What is Photographic Interpretation?

CHARLES E. OLSON, JR.

Photographic interpretation is defined by the American Society of Photogrammetry (1960) as: "The act of examining photographic images for the purpose of identifying objects and judging their significance."

Almost every individual engages in photographic interpretation. Books, newspapers, magazines, billboards, and television all offer photographs to look at, and with each look the observer gains ideas or impressions about something. These ideas or impressions are actually photographic interpretations. Such interpretations may be conscious or unconscious, accurate or inaccurate, complete or partial, but interpretation is an essential part of the process through which information is obtained from photographic images.

A photographic interpreter, or photointerpreter, is an individual specially trained or skilled in photographic interpretation. Because of his training or skill, the photointerpreter is better able to identify objects from their photographic images and can more accurately and completely judge the significance of these objects than can an unskilled or untrained observer.

Prepared for 1969 Summer Short Course in Fundamentals of Remote Sensing at the University of Michigan, Ann Arbor, Mich. Reprinted with permission of the author.

Photographs as Graphic Records

Civilization and man's ability to communicate developed together. People may argue about the order in which communication media developed, but all agree that drawings came early. Man's first drawings may have been finger markings in the dirt which approximated the shape of things that he had seen. When some indication of size, or scale, was added, his drawings became more meaningful. Color and indications of fur, hide, or texture have been found in early drawings of the cave man, indicating that the importance of these qualities was recognized at an early date. As man and man's civilization advanced, his ability to communicate improved. Languages became more expressive and drawings more detailed. Such things as shading and shadows, repetition of patterns, and combinations of related features were added. These additions improved his graphics, but man was not satisfied. He wanted a complete picture of what he had seen, and he kept searching for some way to record what his eye actually saw.

The search for better graphic records led to the photographic process, but the road was a long one and progress was neither direct nor steady. Aristotle's concern with the nature of light (approximately 384–322 B.C.) was an early prelude to photography. Actually, the word "photography" was derived from Greek and means "to draw with light."

A photograph is nothing more nor less than a graphic record of energy intensities.

Today, and increasingly in the years to come, photographic interpreters should not limit their consideration to images or records produced with visible light. A photograph records the intensity of energy received at the focal plane of the taking camera. By varying the characteristics of the recording system, we can obtain records representing wavelengths of energy in many parts of the electromagnetic spectrum. Each part of the electromagnetic spectrum can provide useful information, and most photographic interpreters will have to work with graphic records from several parts of the spectrum. Their field of interest includes the entire electromagnetic spectrum and the images that can be produced with many sensors. Unfortunately, the mechanics involved in obtaining the photographic, or just graphic, images from several sensors cause us to lose sight of the basic situation that brought these sensors into existence. Man's search for better graphic records led to photography, and *any photograph, regardless of its origin or the type of images displayed, is nothing more nor less than a graphic record.*

Elements of Photographic Interpretation

Photographic records come in many shapes and sizes and represent energy in many parts of the spectrum. Photographic interpretation is essential to the effective use of these records, and interpreters should be prepared to handle them. Interpretation of these varied records is not as difficult as it may appear, for the elements of photographic interpretation apply to all graphic records regardless of the nature of the sensing system that produced them or the portion of the electromagnetic spectrum represented. While a thorough knowledge of the characteristics of the sensing system is of immense value, interpretations of graphic records are always based on one or more specific properties of the images making up the record at hand. These properties are sometimes called elements of photographic interpretation, for interpretation of photographic records requires either conscious or unconscious consideration of one or more of them, and preferably of all of them (Olson, 1960).

Nine elements of photographic interpretation are described in the following paragraphs. The discussion is not intended to be exhaustive, for a separate book could be written about each of them. Appreciation of the importance of these elements grows with experience and practice.

1. *Shape.* The shape or form of some objects is so distinctive that their images may be identified solely from this criterion. The Pentagon Building near Washington, D.C., is a classic example.

2. *Size.* In many cases, length, width, height, area, and/or volume are essential to accurate and complete interpretation. The volume of wood which could be cut from the stand in Figure 1 is dependent upon tree size, stand density, and size (or area) of the stand.

3. *Tone.* Different objects reflect and emit different amounts and wavelengths of energy. These differences are recorded as tonal, color, or density variations in the record. The stand of mixed hardwoods shown in Figure 1 was photographed in late October at the peak of the fall color change. Species differences show clearly in different tones or shades of gray.

4. *Shadow.* Shadows can help or hinder the interpreter, for they reveal invisible silhouettes but hide some detail. Shadows in Figure 2 provide information on the size and shape of this building which is not apparent from the image of the building alone. These same shadows obscure detail in the lawn and sidewalk areas in front of the building.

5. *Pattern.* Pattern, or repetition, is characteristic of many manmade objects and of some natural features. The land use pattern shown in Figure 3 is typical of areas of deep, windblown soils. Orchards and strip cropping are particularly conspicuous because of pattern.

6. *Texture.* The visual impressions of roughness or smoothness created by some images are often valuable clues in interpretation. Tree size is often interpreted on the basis of apparent texture. Smooth, velvety textures are commonly associated with young saplings, while rougher, cobbled textures usually indicate older trees of sawtimber size.

7. *Site.* The location of objects with respect to terrain features or other objects is often helpful. The open pond and the flatness of the area of dark-toned vegetation in Figure 4 both indicate a coniferous swamp. In this area of Michigan, northern white cedar, balsam fir, and black spruce are the predominant swamp conifers. The stand shown here is actually a mixture of these species.

8. *Association.* Some objects are commonly associated with other objects that one tends to indicate or confirm the other. In Figure 5, the two tall smokestacks, large building, coal piles, conveyors, and cooling towers are obviously related. This combination and arrangement of features identify the installation as a thermal power plant.

9. *Resolution.* Resolution depends on many things, but it always places a practical limit on interpretation. Some objects are too small, or otherwise lacking, to form a distinct image on the photograph. The lake in Figure 6 shows some interesting tonal patterns, but the exact location of the shoreline is not visible along much of the boundary of the lake. The shoreline failed to resolve because of insufficient tonal contrast between adjacent land and water surfaces.

Note that seven items on this list are qualities that man developed in his early drawings. The last three items are important to interpretation because they tend to integrate or set limits on the others. In 1954, Colwell presented a concise statement showing how integration of interpretations of several photographic images could lead to accurate interpretation of conditions not directly visible in the photograph. He called this approach the

Figure 1 Panchromatic-minus blue photograph of mixed hardwood stand at the peak of the fall color change. Lightest-toned tree crowns are sugar maple; dark-toned crowns are generally oak. (University of Illinois)

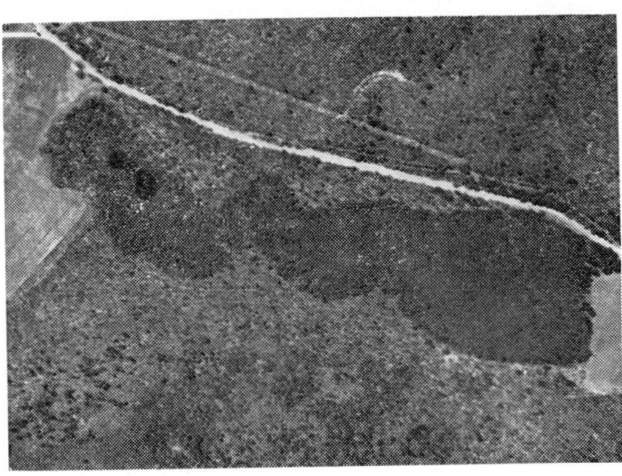

Figure 4 Black spruce, northern white cedar, and balsam fir in swampy area in Leelanau County, Michigan. (University of Illinois)

Figure 2 The shadow of this building shows the steeple more clearly than does the image of the steeple itself. (University of Illinois)

Figure 5 Thermal power plant at the University of Illinois. (University of Illinois)

Figure 3 Land-use pattern on loess in Calhoun County, Illinois. (U.S.D.A.)

Figure 6 Shallow water revealed by tonal contrast, although exact shoreline is not resolved; Leelanau County, Michigan. (University of Illinois)

convergence of evidence. Integration of several of the elements of photographic interpretation is essential to any interpretation and contributes to the convergence-of-evidence frame of mind. Association, in particular, is closely related to this convergence-of-evidence concept.

The Philosophy of Photographic Interpretation

In recent years, photographic interpretation has been called a new natural science (Lueder, 1959, 1961). An increasing volume of literature has appeared, and is appearing, that attempts to establish rules or procedures for photointerpretation (Kedar, 1958; Lueder, 1959; Stone, 1956). However, no single, best method of procedure has evolved which is acceptable to even a majority of photointerpreters. It is generally agreed that the interpreter should work methodically, should proceed from general considerations to specific details, and should proceed from known to unknown features (Stone, 1956). Methodical work and interpretation of known features before evaluating the new and unknown features are almost axiomatic in scientific endeavors. Proceeding from general to specific considerations is also desirable as long as the considerations of general (sometimes called regional) features do not bias the interpretation of the specifics. Specific, local considerations often provide the evidence needed to complete or confirm the broader regional pattern. In most cases, general and specific features must be considered together. To say that one must come before the other can be misleading. To go further than this and propose that interpretation should proceed from one specific group of features to another is an unwarranted channelization of the infinite variation that the photointerpreter encounters.

Drainage, landform, vegetation, and man's activities are thoroughly scrambled in most photographs, and each feature may be indicative of the nature of the others. For this reason, the photointerpreter should be terrain-conscious in the broad sense of terrain. He should understand that the images he interprets make up a terrain and that the terrain he deals with is a complex series of interrelated features academically catalogued as agriculture, botany, geology, engineering, etc. The importance of these interrelationships was clearly expressed by C. H. Summerson (1954) when he said:

> The subject matter of the photointerpreter is as broad as the earth itself;... upon the earth's surface is a tremendously varied pattern of natural features, upon which man has superimposed an equally varied cultural pattern.... Where there are so many varied objects to be considered, our first inclination is to classify. Classification is indeed necessary, but it must always be remembered that groupings are manmade and not natural.... Conclusions based on empirical identifications of isolated elements in a terrain may often be very inaccurate.... The knowledge that no region is a static thing greatly aids the interpreter in his understanding of the terrain.

Because aerial photographs record images of all types of terrain features, photointerpretation has become a valuable tool in each of the earth sciences, and in several other fields as well. However, photointerpretation cannot be confined to geology, forestry, soils, or any single field and remain effective. The photointerpreter must consider all of the factors making up any terrain, even when his special interest may be directly solely to geology, forestry, agronomy, or engineering. *Failure to give adequate consideration to all aspects of a terrain is a major cause of misinterpretations.* Belcher (1957) put it this way during a discussion of correlations between man's activity and his environment:

> ... the average interpreter does not know many things and he doesn't realize his lack of knowledge. If he places two elements of the environment together and pulls a boner, it isn't because the environment is wrong; instead it's because there were three other elements that he ignored or didn't know were there.

The Photographic Interpretation Process

Photographic interpretation is a deductive process and features that can be recognized and identified directly lead the photointerpreter to the identification and location of other features in the terrain under consideration. Even though all aspects of a terrain are irreversibly intertwined, the photointerpreter must start some place. He can't consider drainage, landform, vegetation, and manmade features simultaneously. He must start with one feature or group of features and then go on to the others, integrating each of the facets of the terrain as he goes. In all probability, no one starting point will satisfy, or be best for, all interpreters or for any one interpreter all of the time. For each terrain, the interpreter must find his own point of beginning and then consider each of the various aspects of the terrain in logical fashion.

The deductive process which is photographic interpretation requires conscious or unconscious consideration of the elements of photographic interpretation listed earlier. The completeness and accuracy of photointerpretation are proportional to the interpreter's understanding of how and why photographic images show shape, size, tone, shadow, pattern, and texture, while an understanding of site, association, and resolution strengthens the interpreter's ability to integrate the different features making up a terrain. For the beginning interpreter, systematic consideration of the elements of photographic interpretation should precede integrated terrain interpretation. Mastery of the elements of photographic interpretation is seldom possible, however, before the interpreter gains considerable experience in terrain interpretation.

Photographic Interpretation and Photogrammetry

Interest in photointerpretation has been stimulated during the last 15 years, or more, by outstanding successes achieved through photointerpretation of aerial photography for military and civilian purposes. Without question, early and continuing development of instruments permitting precise measurement and mapping from aerial photographs contributed to these photointerpretation successes. Since these instruments are usually considered tools of photogrammetry, it is sometimes difficult to distinguish between photogrammetry and photographic interpretation.

Photogrammetry is defined by the American Society of Photogrammetry (1960) as: "The science or art of obtaining reliable measurements by means of photographs."

Photographic interpretation was defined by the same source at the beginning of these notes as: "The act of examining photographic images for the purpose of identifying objects and judging their significance."

From these definitions, it is clear that the photogrammetrist (as the specialist in photogrammetry is called) is primarily concerned with measuring and the photointerpreter with identifying and judging significance. It should also be clear that the photogrammetrist cannot make a measurement unless he first identifies what he is going to measure, and that the photointerpreter must often measure to arrive at a final identification and his judgment of significance.

Interpretation of Aerial Photographs

Vertical aerial photographs—those taken with the optical axis of the camera pointed vertically downwards toward the earth—are the most common graphic records encountered by photointerpreters. This does not mean that vertical photographs are always best for interpretation purposes, for oblique photographs can provide tremendous amounts of information. It just happens that the advantages of vertical photographs for photogrammetric mapping purposes are so great, and the amount of mapping photography so plentiful, that vertical photographs are more readily available than all other types of aerial imagery.

Interpretation of vertical aerial photographs is sometimes difficult for the beginner. Everyone is used to the oblique perspective that we see every day with our own eyeballs, but the plan view observed when looking down on the tops of objects is much less familiar. Being less familiar, in plan view, many objects are more difficult to identify from vertical than from oblique air photos. Despite this, correct integration of clues gained from an assessment of the nine elements of interpretation mentioned earlier can lead to consistently accurate results. Brief looks at how each of the elements of interpretation can contribute to correct interpretations follow. The discussion is not intended to be exhaustive, and some elements are treated at greater length than others. Taken together, however, they provide a starting point from which interpretation of the complex interrelationships portrayed in aerial photographs can begin.

Shape

One of the most outstanding examples of shape as a factor in interpretation is found in the almost instantaneous recognition of the Pentagon Building in Washington, D.C., because of its shape. All shapes are not this diagnostic, but every shape is of significance to the photographic interpreter. Stated another way: The shape of any image is a factor that the interpreter should consider when interpreting that image.

Consideration of shape should include consideration of total form. Relative proportions of length to width, height to length or width, etc., are just as important as generalized outlines seen in plan or profile view.

When geometric shapes are present, man's activity is almost always indicated. Symmetric, straight, or perpendicular forms are rare in nature, but do occur in geologic features such as domes, faults, and bedding planes. Also, old fire scars and storm damage can give rise to straight and/or angular boundaries between vegetative types. In any case, the presence or absence of definite shapes should be observed and noted by the interpreter.

Linear Features

Linear features—those that are exceptionally long compared to their width—are some of the most obvious images in photographs. Roads, railroads, power lines, and streams are some of the most important of these, but the interpreter needs to be alert to even the less obvious linear elements in the terrain he is interpreting. Even distinguishing between the more obvious linear elements presents some difficulties, but the following observations may be helpful.

Railroads are characterized by broad sweeping curves and the complete absence of sharp corners, except at crossovers which are relatively rare. Grades are relatively flat, when compared with highways, and cuts and fills are common. The roadbed is continuous and bridges will be present where the railroad crosses streams and significant ravines.

Highways and roads resemble railroads, but have sharp corners and right-angle intersections. Grades are steeper than with railroads, and highways tend to follow the contour of the ground with fewer cuts and fills. Bridges are common (where needed), but many small streams have earth-fill over culverts instead of bridges. When side by side, highways often swing away from the railroad going through small towns, but the railroad usually goes straight through. While it is true that the newer superhighways have grade and alignment characteristics resembling railroads, the dual lanes with wide median strips and the presence of cloverleaf interchanges permit ready interpretation in almost all cases.

Electric transmission lines are straight, or composed of straight segments, and tend to cross the terrain with no reference to topography. Trees are cleared from rights-of-way and this leaves a narrow, but distinctive, scar through forested areas. Sometimes poles or pylons are visible, but not at the smaller scales often encountered by the photointerpreter. Pole locations can sometimes be identified from the mound of subsoil left after digging the postholes. In some cases, the wires themselves can be seen, even at scales as small as 1:20,000. Electric transmission lines cross streams and ravines without bridges, but maintenance roads may be visible along the otherwise clear pole line.

Gas pipelines resemble electric transmission lines, but seldom have maintenance roads along them and are less oblivious to topography. A continuous line of subsoil, left after digging the ditch for the pipe, persists as a distinctive feature for years. In some areas, pipelines can be seen in aerial photographs as light-toned streaks crossing corn fields even though the photographs were taken more than ten years after the pipeline was constructed. Even where cropping is continuous, the soil disturbance associated with the pipeline produces different light reflectances than the undisturbed areas adjacent to it, and these may well persist for more than ten years. To simplify tunneling under existing roads, pipelines often bend sharply towards the road just before the point where the tunneling begins. This jog in the path of the linear feature is rare with power lines.

Fence lines are difficult to identify except when land use is different on the two sides of the fence. Once land use has been established for a long period of time, the fence lines may be quite obvious in aerial photographs even decades after the land use pattern has been abandoned. This is particularly true of old farms that have been planted to trees or allowed to reseed naturally.

Streams are seldom confused with other linear features because of their irregular shape, varying width, and complete dependence on topography.

Drainage ditches are common in flat lands and are usually easy to separate from natural streams because of their distinct tendency to be straight and have uniform width. A trail of dredged material may be visible on one or both banks of the ditch. Some confusion may arise where existing stream channels are improved to facilitate drainage, with or without a realignment of the channel, but careful consideration of the total drainage pattern usually indicates the true situation. In some areas, ditches are used to transport irrigation water from mountain reservoirs to dryer valleys and plains. In these valleys, irrigation ditches may resemble the drainage ditches of other areas. Once again, complete consideration of the drainage system (or irrigation system) will clear the air.

Geologic faults are linear features that can be quite obvious when prominent rock structures are truncated or roads displaced. In other cases, and especially after extensive weathering, faults may be very difficult to recognize.

Dams are shorter than the features discussed above, but may well be mentioned at this point. Straight or smoothly curving sections of shoreline should be studied carefully to be sure that that piece of beach isn't actually the dam that indicates an artificial lake as opposed to a natural one. The presence of the dam may indicate the work of man and suggest other manmade features in the same area. However, all dams are not manmade and beaver dams more than a mile in length are known.

Tree Species

Shape is often an aid in interpreting tree species, for branching habits and total form vary from genus to genus, and in some cases between species within a genus. Shadows often provide a profile view which helps in tree identification.

Soil Texture

Soil texture can often be determined from aerial photographs if careful attention is paid to the cross-sectional shape of the small gullies that are present. Several factors have to be considered and it is easy to make errors, especially when hardpans or abrupt changes in soil texture with depth produce compound gully shapes. . . .

Size

The fact that some objects are bigger than others should need no special mention, but beginning interpreters often forget that relative sizes of different objects in a single photograph can provide valuable clues. Some objects are readily identifiable and can be used as guides to the approximate sizes of other objects that are not so easily recognized. In many cases, however, relative size is not enough and the interpreter must use some of the techniques of the photogrammetrist. These will not be discussed in detail, but some indication of the possibilities seems in order.

Distance

The distance between two points, or the length of an object, can be determined from an aerial photograph. The accuracy of such measurements is a function of photo scale and the accuracy of the photo measurement. On standard USDA photography at a scale of 1:20,000, a careful interpreter can usually determine distances to an accuracy of approximately $\pm$ 5 feet. When the ends of the distance(s) measured are at significantly different ground elevations, accuracy usually decreases.

Height

The heights of objects, or differences in elevation between ground points, can be determined from air photographs. Several methods are available, but the best is that based upon differences in parallax between two vertical photographs of an object taken from different points in space. Stereoscopic photography is required. Careful work can provide spot height data (i.e., height of a tree or building) accurate to within $\pm$ 1/2000 of the flying height. Topographic mapping or contouring can be performed to somewhat lower accuracy. Contouring requires continuous determination of spot elevations and is more difficult than determining the height of a single small object.

Area

Determination of the area of irregular parcels is more easily accomplished from vertical photographs than with ground methods. Accuracy is limited by scale and slope, but has been found acceptable for most natural resource purposes. In rough terrain, it is often easiest to transfer boundaries from the photographs to a planimetric map, and then determine the area of the parcels of interest from the map boundaries.

Volume

Many organizations with large quantities of raw materials routinely use aerial photographs to determine the volume of material in open piles. Photogrammetrically determined volumes have proven accurate enough to permit useful estimations of the quantity of the finished product produced by the plant.

Tone

Tone, whether the shade of gray in a black and white photograph or the combination of hue, chroma, and saturation in a color photograph, conveys more information to an alert, knowledgeable interpreter than any other single element of interpretation. In almost all cases, however, it is the difference in tone between objects or between an object and its background which is important. In fact, without a difference in tone between the background and the edge of an object, there can be no detectable image.

Relative tone can be quantified by measuring the amount of light transmitted through a photographic negative or other transparency. As the amount of silver present increases, the amount of light transmitted decreases and we say that we have a denser, or darker-toned, image. Microdensitometers exist that can scan photographs at reasonably high rates of speed. If the output of such an instrument were plotted for single scans across a photograph, we might get plots such as those shown in Figure 7. The total range of density is greater in trace A than in B; but, due to the uniform change in density across the photograph, no images would be detectable in the strip of the photograph represented by trace A. The second trace (B) shows two distinct images. Notice, however, that the slope of the trace indicates that there is a much sharper change in density at the edges of image one than at the edges of image two. The slope of the microdensitometer trace, or the *edge gradient* (Fox, 1957), indicates that image one will be sharper, and easier to resolve, than image two. The second image may blend with its background and appear as a small blob rather than an object of distinct shape.

Colwell (1959) described the significance of tone and edge gradient in much the same fashion when he listed three primary characteristics governing the quality of photographic images for photointerpretation. These were:

1. *Tone contrast,* or "the difference in brightness between an image and its background."
2. *Image sharpness,* or "the abruptness with which the tone or color contrast appears to take place on the photograph."

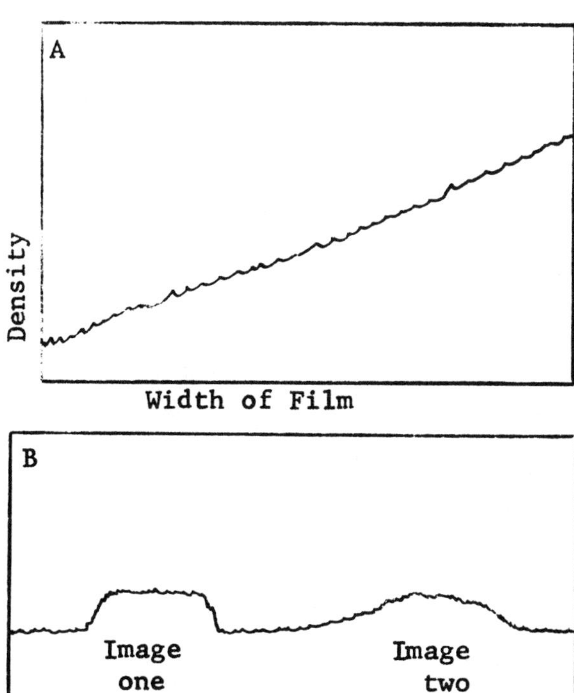

Figure 7 Potential microdensitometer traces at two different locations.

3. *Stereoscopic parallax,* or "the displacement of the apparent position of a body with respect to a reference point or system caused by a shift in the point of observation."

Colwell's first two characteristics have just been discussed, and the third was mentioned in the remarks under Size-Height.

. . . .[1]

Shadow

In addition to providing a silhouette, the presence or absence of shadows—and shadow length—provides an immediate index of the relative height of different objects in a photograph. Shadows are often the best indication of relative heights of objects when looking at a single photograph where the advantages of the stereoscopic view provided by overlapping photos are not obtainable. Compare the shadow lengths of the smokestacks and various buildings in Figure 5, as an example.

Pattern

Land use patterns are often distinctive. Agricultural practices, in particular, leave clues as to what has taken place that can provide positive indication of the crop present. Row patterns in corn and soybean fields separate them from wheat fields and pastures and lack such patterns. Wave patterns along coastlines often reveal the presence of submerged obstacles, such as sandbars and reefs, even though the obstacle cannot be imaged directly

Texture

Texture, like pattern, is often related to land use. Planted vegetation usually appears in smoother textures than natural vegetation. As abandoned agricultural areas revert to forest, encroachment by woody species results in a rougher texture of fields that may or may not be accompanied by obvious tonal differences. Eventually, this encroachment produces a mottling which is a distinctive disruption of the previous uniform texture of the active fields.

Site

Just as some tree species grow in swamps and others on dry upland ridges, so some manmade objects are found along rivers and others on hilltops. Thermal power plants need an abundant supply of coolant water and are often (but not always) found near major streams. Hydroelectric plants must have water under pressure and are always associated with a natural or manmade water supply. Early warning radar stations, on the other hand, are usually placed on high promontories to minimize terrain interference with the line of sight of the radars.

Association

Association is often one of the most helpful clues to the identity of manmade installations. The manufacture of aluminum requires large quantities of electric power.

[1] A section on factors controlling tone in aerial photographs has been deleted with the author's permission. This included a discussion of the energy flow profile for a camera system, light reflectance properties of trees, photographic tone of tree foliage, and photographic tone in water areas, and also Figures 8–14.

Absence of the power supply rules out this industry. Schools usually have associated playgrounds and athletic fields while churches do not. Large farm silos are an indication that livestock are (or were) present. Unexpectedly large culverts or bridge spans across small streams indicate that heavy runoff occurs frequently enough to require the engineering works necessary to cope with the water when it comes.

To the military interpreter, the presence of antiaircraft guns is almost positive proof of the presence of a target worth defending. If he hasn't found it, he had best look again.

Resolution

The ability of a photographic system (including lens, filter, emulsion, exposure, and processing, as well as other factors) to record fine detail in a distinguishable manner is referred to as the resolution, or resolving power, of the system. Unfortunately, resolution is one of the most misunderstood and misused qualities of the photographic system—at least by most photographic interpreters.

Resolving power depends on many factors. The first and often overlooked requirement is that a difference exists in the energy reaching the focal plane from the object and its background. If there is no such difference, then nothing that the recording system can do will create a discernible image.

The importance of tone contrast and edge gradient were included in the discussion of tone. It should be clear that resolution is related to edge gradient. If the difference in density across an edge is great enough to be detectable, then the steeper the edge gradient, the greater the probability that the density change will be detectable as a visible change in tone of the photograph.

Edge gradients are influenced by the total photographic exposure, or the total energy involved in producing a particular image. Steep gradients are usually associated with intermediate exposures, and lower gradients occur at both extremes. Thus, resolution falls off at high or low exposure levels, compared to the maximum resolution achieved at some intermediate exposure. This loss in resolution is one of the major reasons why significant over or underexposures are undesirable, and is a major problem encountered by the interpreter who must work with imagery containing both very light and very dark images in the some photograph. The washed-out appearance of broad-leaved trees imaged in infrared photographs is partially due to the loss of resolution that accompanies the overexposure of the trees when the overall photographic exposure is set to record less reflective objects occurring in the same terrain. Also, the loss of resolution observed in shadowed areas is due to the fact that objects in the shaded areas reflect so little light to the camera that they are underexposed.

. . . .

Photographic interpreters are often confused by errors in assessing resolution. When telephone wires resolve, it does not necessarily follow that any object the thickness of a telephone wire will resolve. The wire (or a paint line on a parking lot) is detectable because its length results in exposure to several silver halide grains, rather than because the camera system can resolve very small images. Resolution of the familiar USDA photography at a scale of 1:20,000 is such that round or square objects less than 2.5 feet across are seldom discernible. When measuring tree heights, interpreters usually do quite well with round-topped trees like large oak and maple, but poorly with trees having sharp-pointed crowns like balsam-fir or spruce. The crown of a balsam-fir may extend 10 or 15 feet higher than the point where its diameter narrows to 2.5 feet.

The tone of the background affects resolution of small objects. Photographic emulsions can act as a scattering media and some of the energy from highly reflective (bright) objects may cause exposure of silver halide grains adjacent to those at the geometric location of the object. Bright objects against a dark background will appear larger than they actually are, but dark objects against a bright background will appear smaller than they actually are. Small bushes on a white sand beach often fail to resolve because lateral irradiation of adjacent grains results in exposure of the grains that should have remained unexposed to mark the location of the bush.

REFERENCES

American Society of Photogrammetry. *Manual of Photographic Interpretation.* Washington, D.C., 1960.

Belcher, D. J. Remarks made during discussion of a paper by G. R. Heath (below) and published in *Photogram. Engng.* 23:114, 1957.

Colwell, R. N. A systematic analysis of some factors affecting photographic interpretation. *Photogram. Engng.* 20:433, 1954.

Colwell, R. N. The future of photogrammetry and photo interpretation. *Photogram. Engng.* 25:712, 1959.

Fox, R. F. Visual discrimination as a function of stimulus size, shape, and edge gradient. Boston University Technical Note No. 132, 1957. 52 pp.

Heath, G. R. Correlations between man's activity and his environment which may be analyzed by photo interpretation. *Photogram. Engng.* 23:108, 1957.

Kedar, Y. A geographic approach to the study of photographic interpretation. *Photogram. Engng.* 24:821, 1958.

Lueder, D. R. *Aerial Photographic Interpretation—Principles and Application.* New York: McGraw-Hill, 1959.

Lueder, D. R. Photo interpretation as a new natural science. Paper presented at the Annual Meeting of the American Society of Photogrammetry, Shoreham Hotel, Washington, D.C., March 21, 1961.

Olson, C. E., Jr. Elements of photographic interpretation common to several sensors. *Photogram. Engng.* 26:651, 1960.

Seymour, T. D. The interpretation of unidentified information—a basic concept. *Photogram. Engng.* 23:115, 1957.

Stone, K. H. Air photo interpretation procedures. *Photogram. Engng.* 22:123, 1956.

Summerson, C. H. A philosophy for photo interpreters. *Photogram. Engng.* 20:396, 1954.

Most air photographs taken of the earth's surface are exposed through a single lens on black and white film and in a vertical or near-vertical orientation. It is with this type of aerial photograph that the reader is most likely to have had experience. Vertical photographs are the most useful and simple kind of imagery from which to take measurements, but they present the interpreter an unfamiliar view of his world. Special training is frequently needed to interpret vertical photographs. Air photos can also be taken with the camera axis held at various angles between the horizon and the ground. These angular photographs give us a view of the earth which is closer to the normal way we see objects with our eyes. Angular views are called oblique photographs. A single oblique photo covers more of the earth's surface than a vertical photo, but scale distortions are introduced which make it more difficult to take measurements from it. Other variations can be introduced to these two basic photo orientations (vertical and oblique). The length of the camera lens can be varied, film emulsion and filters can be changed, an additional lens or lenses can be added, additional cameras can be used, or the type of camera can be varied. Each of the orientations and variations can produce a different type of photograph. The usefulness and validity of the photographs will depend upon the purpose for which they are intended.

12-Types of Aerial Photographs

STEPHEN H. SPURR

For the purpose of a cursory survey, aerial photographs may be classified according to:

1. Angle of photography
2. Specifications
3. Multiple-camera photography
4. Special-purpose photography
5. Forms in which photographs are used

Angle of Photography

Aerial photographs are of two general types: vertical photographs and oblique photographs. *Vertical photographs* are those made with the camera pointed down vertically, or as nearly vertically as is practicable in an aircraft. *Oblique photographs* are taken with the camera axis directed intentionally between the horizontal and the vertical; i.e., with the camera pointed obliquely (Fig. 1).

Oblique photographs are of two types. In *high obliques*, the apparent horizon is shown; in *low obliques*, it is not shown, although its position may be recorded on the film by means of a mirror attachment. The terms "high" and "low" in the above sense refer not to the elevation of the aircraft but solely to the angle of tilt of the photograph.

From *Photogrammetry and Photointerpretation, with a Section on Applications to Forestry,* 2nd ed. Copyright © 1960, The Ronald Press Company, New York. Chap. 2, pp. 13–36. Reprinted with permission of the author and the publisher.

Because vertical aerial photographs are used much more commonly in modern aerial photogrammetry and photointerpretation, the term "photograph" throughout this chapter, unless otherwise qualified, is used to refer to a vertical aerial photograph.

Vertical Photographs

Pictures taken with the camera axis vertical, or nearly so, closely resemble a map and may be used with a minimum of correction. They may readily be studied stereoscopically to produce a three-dimensional image. Systematic vertical coverage of an area is used for planimetric mapping, topographic mapping, and photointerpretation Although the costs of vertical coverage may be higher than those of comparable oblique coverage, the savings in analytical costs render vertical photography preferable under most conditions.

The relatively undistorted view and great amount of detailed information contained in vertical photographs is illustrated by Figure 2, a vertical photograph taken in central Massachusetts at an elevation of about 4200 feet above the ground and covering an area of somewhat less than 1 mile square.

High Obliques

High obliques were extensively taken, following World War I, in Canada and in other parts of the world. During a period when the capabilities of airplanes and aerial cameras were strictly limited, high obliques had the advantage of covering large areas with relatively few exposures taken at relatively low altitudes. Their use,

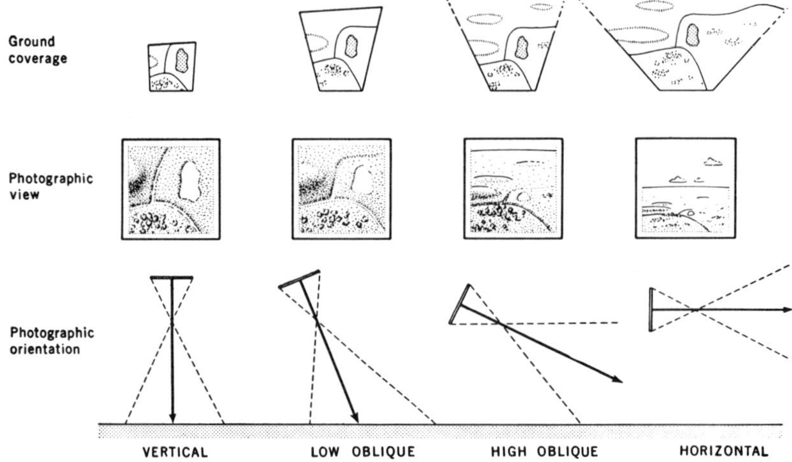

Figure 1 Characteristics of vertical and oblique photographs.

Figure 2 Vertical photograph. Petersham, Massachusetts, July 5, 1944. Panchromatic film. Negative size: 9 × 9 inches. Scale as reproduced [in original article]: 1:8800 [Reduced here by 29%] (Fairchild Aerial Surveys, Inc.)

however, is best adapted to relatively flat terrain, since hills, trees, and other elevations obscure detail on the far side. High obliques may be used over rough terrain, provided the flight strips are placed so that both sides of elevations and depressions are photographed.

Low Obliques

Low obliques present a less distorted view than do high obliques but are more difficult to use in mapping, inasmuch as the angle of tilt cannot readily be calculated. They may be useful where it is necessary to photograph a fairly large area with each exposure under conditions where high obliques cannot advantageously be employed. They are also valuable for illustrative purposes. Figure 3 covers the same areas as Figure 2. Although taken at a much lower altitude (the buildings are at a larger scale than in Figure 2), it covers about four times the area. In only the lower part of the picture, however, does the detail compare wth that of the vertical.

Oblique photographs are more readily comprehensible

Figure 3 Low oblique photograph. Petersham, Massachusetts, July 5, 1944. Panchromatic film. Negative size: 9 × 9 inches. [Reduced here by 29%] (Fairchild Aerial Surveys, Inc.)

than verticals, as they present a more familiar picture. They also register a distorted view of the landscape. It is possible, through ground control, tilt analysis, and restitution, to prepare an acceptable map from information appearing on an oblique photograph. Nevertheless, because of the variation in the amount of detail discernble within the oblique photograph, and because of the difficulty of map preparation from oblique photography, it is little used at present in comparison with vertical photography.

Specifications

Since the quality of the work done from aerial photographs depends upon the quality of the photographs themselves, the specifications of aerial photography are of the greatest importance. A brief review of the principal considerations is presented here. These deal with camera, lens, focal length of lens, scale, season of photography, and type of emulsion.

Camera

Aerial photographs taken for photogrammetric purposes (i.e., to be measured) must be taken with precision-built cameras costing many thousands of dollars. The camera must be mounted in an appropriate aircraft in such a way that vibration is reduced and that the camera can be kept oriented so as to take vertical photographs. Specialized equipment is necessary to permit accurate navigation of the aircraft and to space exposures at regular intervals so as to provide the desired amount of overlap.

Several makes of aerial cameras are in general use, but such items as focal length of lens, size of negative exposed, and specifications of accuracy have become fairly well standardized. In North America, most cameras take photographs 9 inches (23 cm) square, but other sizes of negatives are also taken, the next most common size being 7 by 9 inches. In Europe, negatives 18 cm (7 inches) square are commonly employed. Negative sizes of 14 cm and 12 cm square, however, are also widely used.

Special types of aerial cameras and special-purpose photography are considered later in this chapter.

Lens and Focal Length

The modern photogrammetric lens is an exceedingly expensive and exceedingly precise piece of optical equipment costing in the thousands of dollars and capable of producing photographs with an optical distortion of less than 4 μ.

Assuming a good-quality lens, the characteristic of principal interest to the user of the photographs is its focal length. Approximately, this is the distance from the lens to the film; precisely, it is the distance in the camera along the lens axis from the rear nodal point of the lens to the plane of best average definition over the entire field.

Various focal-length lenses are used, the most common for 9-inch (23 cm) cameras being the 6-inch (152 mm), 8¼-inch (210 mm), and 12-inch (300 mm). These are somewhat loosely termed wide-angle, standard, and telephoto lenses, respectively.

Short focal-length lenses permit wide coverage from low altitude. On photographs taken with such lenses, the three-dimensional stereoscopic effect is exaggerated, but image displacement is also increased. Wide-angle-lens photography is especially suitable for topographic mapping and other uses where vertical measurements of height and elevation are required. It is normally required when small-scale photography is to be taken. Because of the displacement of images on wide-range photographs, however, these photographs do not approximate maps if the terrain photographed is hilly or mountainous, and they should not be used under such conditions without photogrammetric correction.

In contrast, long focal-length lenses are preferable for very mountainous country and for large-scale photography in general. It follows that medium focal-length lenses are often the best compromise for average conditions.

Scale

The scale of an aerial photograph is determined by the focal length of the lens and the height above ground from which the picture is taken. Scale is not constant on any given photograph but varies according to the ground elevation and the tilt of the camera. It is usually expressed in terms of *scale ratio*, giving the number of units of distance on the ground equivalent to one unit on the photograph or map. Thus a scale ratio of 1:12,000 indicates that the ground distance is 12,000 times longer than the map distance. Dividing by 12 gives the scale of 1000 feet per inch.

Aerial photographs may be taken at a wide range of scales. Figure 4 was taken with a 6-inch lens 20,000 feet above the Sierra Nevada of California. The original 9-inch-square negative thus had a scale of 1:40,000 and pictured about 20,000 acres. In contrast, Figure 5 is a low-altitude, continuous-strip photograph with a negative scale of about 1:300 and covers little more than an acre. Even smaller and larger scales are entirely feasible.

Intermediate scales, however, are more often adopted. For small-scale mapping and for geologic reconnaissance, scales ranging from 1:20,000 to 1:40,000 are usual; for forestry and other natural resource photointerpretation and mapping, scales from 1:10,000 to 1:20,000 are commonly employed; while scales from 1:5,000 to 1:10,000 are frequently used for detailed photointerpretation and very large-scale maps.

Season of Photography

The season in which aerial photographs are taken greatly affects their value. The proper time for photography depends in part upon the local climatic cycle and in part upon the uses to which the photographs are to be put. The percentage of clear days is a very important item in aerial photography costs.

For photointerpretation, season is of principal importance in regions where the trees are deciduous—as in the winter in temperate zones or in the dry season in many tropical zones. During the period when the trees are leafless, mapping and photointerpretation of the ground surface is facilitated, provided the surface is not covered by snow or ice and provided the seasonal water levels in the streams and lakes are satisfactory for the purposes of the interpretation. At the same time, however, interpretation of the tree canopy is decidedly limited during the deciduous season because no part of the average leafless tree is large enough to be resolved in aerial photographs of average scales. Aerial photographs taken when the foliage is in full leaf are therefore usually required for forestry and botanic use.

Most photography taken during the full foliage season is taken during the period of normal green coloration.

Figure 4 Small-scale photograph. Sierra Nevada, California. About 20,000 acres are pictured. [This photo, scale 1:48,000, and reduced here by 41%, was substituted for the one that appeared in the original article, which was not available for reproduction.] (U.S. Forest Service)

Figure 5 Large-scale photograph. Low-altitude (150 ft.), continuous-strip photograph with a negative scale of about 1:400. Area covered is a little less than an acre. Photo reduced here by 41%. [This is a substitution for photo in original article.] (Chicago Aerial Survey)

Possibilities exist, however, for distinguishing between different types of vegetation by photography during the period the foliage is coming out, changing color, or dropping, as in spring and fall photography in temperate zones. Although excellent results can often be obtained from carefully timed photography at such seasons, the photography is apt to be exceedingly variable, and the chances of failure are often as great as the chances of success.

Type of Emulsion

The choice of the type of film—and the filter through which the film is exposed—is of great importance in photointerpretation and must be considered in all phases of the use of aerial photographs. Depending upon the place of photography, the weather, and the area to be photographed, film and filter combinations should be chosen that will provide the most detailed and sharpest images and will emphasize those aspects of the area in which interest is centered.

Panchromatic emulsions, capable of registering the entire visible spectrum in black and white, are used for most aerial photography. As normally used with light yellow filters to cut the haze, panchromatic film produces familiar black and white images with fine detail capable of being used for a wide variety of purposes. Unless otherwise indicated, all photography illustrated in this chapter was taken with panchromatic film.

Infrared emulsions are sensitive to longer wavelengths than are visible to the human eye. They are also usually sensitized to a portion of the visible spectrum. *True infrared* photography is taken with a deep red filter that cuts out the visible spectrum and allows only the infrared radiation to reach the emulsion. *Modified infrared* photography, in contrast, is taken with a light filter that allows a portion of the visible spectrum as well to reach the film. A yellow "minus blue" filter is frequently employed for this purpose.

Infrared film is used chiefly for the purposes of distinguishing between vegetation types, soil types, and moisture conditions which register in contrasting tones on it. Figure 6 contrasts the appearance of the same area photographed in panchromatic and infrared. On the former, all vegetation and soil types are registered in similar tones; on the latter, the conifers and moister soils register in dark tones in contrast to the broad-leaved trees and the lighter dried soils which register in light tones.

Color photography is more expensive and more difficult to take satisfactorily than black and white photography, but does of course contain information not otherwise attainable. At the present time [1960], its use is largely confined to relatively small areas and to specialized photointerpretation problems.

Multiple-Lens Cameras

Cameras with two or more lenses have been built from time to time in order to photograph large areas with a single exposure. The first such camera was built in Austria by Scheimspflug in 1904, and the first in the United States was built by Bagley in 1917 (Sanders, 1944). Among the most remarkable is a huge nine-lens camera, built by the U.S. Coast and Geodetic Survey, which photographs a field of 145 degrees and produces a 36-inch square transformed print. With the appearance of the first precision wide-angle lens in 1934, however, interest in multiple-lens cameras has been curtailed.

Multiple-Camera Photography

Although most aerial photographs are taken with single-lens cameras, used singly in the aircraft, increasing use is being made for special purposes of two or more cameras coupled in a single, rigid mounting. Combinations of two and three aerial cameras are most common. The use of two or more lenses attached to the same camera body achieves much the same results.

Figure 6 Panchromatic (left) and modified infrared (right) comparison of agricultural and swamp area near Plymouth, North Carolina. Note how infrared brings out moisture detail in farmland (A), shows scattered pines in dark tones among predominant hardwood forest (B), and brings out wetter swampland in darker tones (C). (Fairchild Aerial Surveys, Inc.)

Tri-Camera Photography

Tri-camera photography is not taken by any special type of camera but rather by an assembly of three cameras equipped with wide-angle lenses (usually 6-inch focal length) and mounted in a rigid frame. The central camera points vertically downward. The side cameras point obliquely at right angles to the aircraft so that at normal flying heights they will photograph both the horizon and a portion of the area covered by the vertical photograph. Film is exposed simultaneously in the three cameras at intervals along parallel lines of flight. Each set of three photographs provides a picture of the terrain from horizon to horizon (Fig. 7).

The tri-camera system, then, combines oblique and vertical photography. Efficient coverage is obtained by systematic exposures of cameras pointed at right angles to the lines of flight. The preparation of accurate maps from the photographs is generally simplified by the rigid structure of the assembly, by the appearance of the horizon in both the right-hand and left-hand photographs, and by the control afforded by the vertical photographs.

In the Tri-metrogon assembly as developed and used in North America, the 9-inch-square negatives are exposed through a 6-inch lens (originally the Metrogon lens) with the oblique cameras inclined at an angle of 60 degrees from the vertical. The system has been widely used for reconnaissance of undeveloped areas. Flight strips may be spaced approximately 25 miles apart, or six times as far apart as in conventional vertical photography, with the same type of camera at the same altitude of 20,000 feet. Precise navigation is not necessary, the need for ground control is greatly reduced compared to vertical photography, and the time and cost of compilation is held at a minimum. The Tri-metrogon system was developed by the U.S. Army Air Forces and the U.S. Geological Survey in 1941.

An adaptation of tri-camera photography that has proved of value in forestry work in Canada involves the use of three 12-inch lens cameras with the oblique cameras inclined at an angle of approximately 18 degrees from the vertical. The side pictures obtained are low obliques rather than high obliques, as in the Tri-metrogon system, but the horizon position may be registered along the edge of the oblique by means of a mirror attachment which reflects the horizon through the regular camera lens. This forestry tri-camera coverage is approximately three times greater than vertical camera coverage at the same scale. In other words, the same number of prints are required as for vertical coverage, but only one-third the number of flight strips are needed. The method has been used chiefly for forest photointerpretation from winter photographs under conditions where the obliquely viewed tree images are silhouetted against a light snow cover and can be measured readily (Robinson, 1948; Seely, 1949).

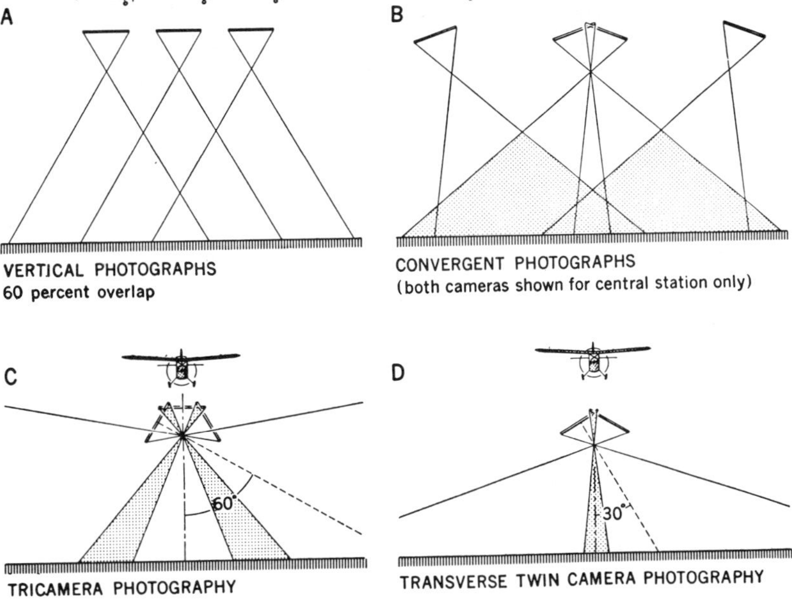

Figure 7 Multiple-camera photography. The conventional spacing of single-camera vertical photography to provide 60 percent overlap along the flight strip (A) is contrasted with convergent photography (B), tri-camera photography (C), and transverse twin-camera photography (D).

Figure 8 A convergent camera assembly. Two Zeiss cameras in a common suspension mount for 27-degree convergent photography. Equipped with 8¼-inch low-distortion lenses for taking 7 × 7 inch negatives. (Zeiss Aerotopograph München)

Transverse Twin-Camera Photography

Two cameras can be paired in tandem in the same manner as three. If the cameras are pointed obliquely transversely to the line of flight, the resulting coverage is similar to that obtained by tri-camera assemblies except that only two exposures are taken at each camera station and that a narrow-width strip is photographed. With recently developed plotting devices, such as the U.S. Geological Survey's Twinplex Plotter, however, maps can be prepared from such wing obliques without having vertical photographs tied in (Davey, 1954).

In one application of transverse twin obliques, both oblique and vertical coverage was obtained at the same time in an aerial survey of southeast Alaska. The verticals, taken with a 6-inch lens to a scale of 1:40,000, were used for mapping, while the twin obliques taken with 12-inch lenses to a scale of 1:20,000 were used for forest photointerpretation and forest-type mapping (Andersen, 1956).

Convergent Twin-Camera Photography

If, in contrast to transverse photography, the two cameras are pointed forward and backward along the line of flight, it is possible to separate the exposures at distance such that the area covered by the forward-pointing camera at one station is rephotographed by the backward-pointing camera at the next. By using convergent cameras (Fig. 8), successive pairs of photographs with 100 percent overlap can be obtained. A number of modern photogrammetric instruments can be utilized to produce maps of high accuracy from this photography.

The cameras are available from different manufacturers in different convergences, ranging from 18 to 40 degrees. For a convergent angle of 27 degrees, the ratio of taking base (distance between exposures) to flying height is approximately 1:1.7, resulting in extreme exaggeration of the three-dimensional image under the stereoscope and consequent accurate height determination by photogrammetry.

Special-Purpose Photography

The types of photographs thus far discussed are types taken with conventional, precision aerial cameras. A number of special-purpose cameras exist which are useful for large-scale aerial photography, for recording aerial views, and for casual aerial photography.

Continuous-Strip Photography

A recent development of considerable interest is the Sonne strip camera, which takes a continuous strip of photography at a low altitude (Kistler, 1946). The camera has no shutter but rather a slit in the focal plane, past which the film travels at a speed determined by the altitude and speed of the aircraft. The operation of the camera eliminates the necessity of stopping motion by high shutter speeds, thus permitting sharp, clear photography at very low altitudes. Stereoscopic photographs can be taken at very large scales, such as 1:300 and 1:600 (Fig. 9).[1] At the latter scale, a strip more than 20 miles long can be photographed with a single roll of film. Minute detail can be seen and measured on leafless trees photographed stereoscopically with this camera.

Continuous-strip photographs are distorted by any change in the orientation of the aircraft, by any variation in its ground speed from that upon which the speed of the film is based, and by any variation in its altitude above the ground. Height is the only photogrammetric measurement that can be made with the strip camera, and equations have been developed for its computation with some precision (Katz, 1952). As it is designed primarily for large-scale photography, a large amount of flying and film are needed to cover a given area, and the cost of extensive coverage is prohibitive. The potential value of continuous-strip photography lies in the possibility of its use in sampling areas photographically in minute detail, thus supplementing conventional photographic coverage. Strip photography may also be taken under weather conditions that do not permit the taking of normal vertical photography.

The application of continuous-strip photography to forest photointerpretation has been tested a number of times, including forest conditions in southern Alabama (Mignery, 1951), the northeastern United States, and southern Michigan. The principal difficulties experienced were with the variable scale resulting from variations in aircraft speed, film speed, flying height and camera orientation, the substantial variation in scale between even the top and the bottom of a single tree, and difficulty of obtaining sharp definition of both the tree crowns and the ground surface at the same time. In more recent tests by the University of Michigan, however, for which the specifications were drawn in the light of previous experience, good stereoscopic images and satisfactory measurement of tree images were obtained.

Image-Motion-Compensation Photography

In addition to the continuous-strip camera, other cameras have been developed that will compensate for the speed of the aircraft and produce photographs with a minimum of distortion due to movement of the aircraft during exposure of the film (Heller, Aldrich, and Bailey, 1959). With an image-motion-compensation magazine, the film is moved in a direction opposite to the line of flight during the instant that the shutter is opened. If the speed of movement is correctly determined to balance with air speed, sharply defined 9-inch square photographs can be taken at 500 feet above the ground at a speed of 150 miles per hour. Much faster aircraft speeds can be used with modern military equipment. On the other hand, complex and bulky equipment is necessary, no fiducial marks are available, and stereoscopic overlap cannot be obtained under the conditions outlined.

Flying speed can also be compensated for by panning the camera—moving the entire camera backward during exposure, either manually or mechanically. In a test of the panning camera for forestry purposes in Canada, sharply defined images were obtained on 1:1200 scale photographs taken with a 24-inch lens from a height of 2400 feet (Losee, 1953).

The continuous-strip, image-motion-compensation magazine, and panning cameras all have the capacity of halting image motion parallel to the line of flight but have little effect on blurring due to vibration of the aircraft, sway, and other factors. The best way to compensate for all types of motion is through very high shutter speeds. By using a camera (such as pictured in Figure 8) with a shutter speed of 1/1000 second and a fast *f*4.0 lens, it is possible to produce 1:1000 photographs with an 8¼-inch lens that are extremely sharp and distortion free.

Small Cameras

Although practically all mapping photography is done with large precision-built aerial cameras mounted in the floor of the aircraft, smaller cameras of many types can be used either in the hand or in simple mounts. Pinpoints or snapshots of points of special interest can be made with any good camera but can be used for mapping purposes only when the points have been located previously by more precise photography. A *pinpoint* is an isolated pair of photographs taken so as to give stereoscopic coverage of a specific place on the ground.

Among the most promising of the smaller cameras are 70-mm cameras developed for use in fighter aircraft or for tracking missile firings. Several of these cameras have been adapted successfully for reconnaissance or pinpoint aerial photography.

The 70-mm camera is available in a lightweight panoramic model. This camera, weighing 53 pounds fully loaded, employs scanning principles to obtain successive 180-degree panoramic photographs sweeping the ground from horizon to horizon. Figure 10 is a panoramic photograph taken 12,800 feet above Long Island, New York, at a speed of about 460 miles per hour. The scale at the center of the photograph is 1:51,200, and at 30 degrees below the horizon, 1:102,400. Bridges can be seen out to 12 miles and shorelines and rivers out to 40 to 50 miles.

It should be remembered that cameras equipped with focal-plane shutters, as well as all cameras in which the film is moved during exposure, cannot be used for precise mapping purposes because the scale is not constant on the negative, nor can the variation in scale be determined except by detailed ground checks. Most image-

[1] Not reproduced here. A large-scale, low-altitude, continuous-strip photograph is shown in Figure 5.

motion-compensation cameras and small aerial cameras, therefore, are primarily adapted for reconnaissance and photointerpretation rather than for mapping.

In one application, two identical hand cameras were placed in an improvised stereoscopic mount and used to take pinpoints, large-scale vertical stereoscopic photographs of forest stands from altitudes of 200 to 600 feet utilizing a helicopter (Avery, 1958). Although the information obtained was not sufficient to justify the general use of such low-altitude helicopter photography in forest survey work, the trials were of sufficient promise to justify further experimentation.

Figure 10 Panoramic section across Long Island, New York, taken with a 70 mm aerial reconnaissance panoramic camera at 12,800 feet. (The Perkin-Elmer Corporation)

Forms in Which Photographs Are Used

Aerial photographs are most commonly used as contact prints, either on paper or on transparent bases. The prints may also be *ratioed* to equate insofar as possible the scale of successive exposures, *rectified* to compensate for displacement due to tilt, or simply enlarged. A series of photographs may also be assembled into a single composite picture or *mosaic*. Finally, two overlapping photographs may be combined various ways into *three-dimensional prints* that can be viewed stereoscopically.

Paper Prints

Most aerial photographs are furnished to the customer in the form of contact paper prints, prints made directly from the negatives without the use of a projector and which are at the same scale as the negative except for paper shrinkage. Provided the photography is adequate, such prints are well suited for most uses and are, of course, the least expensive type of print available. They are of convenient size to handle in the field and to study under the pocket stereoscope.

The weight and other characteristics of the paper used for the base influences the quality of the resulting print. Other base media, such as aluminum sheets, may also be employed.

Positive Transparencies

Prints made on film, glass, or other transparent materials may be viewed with transmitted rather than reflected light and normally show finer detail and sharper definition than do paper prints. Positive transparencies should be used whenever the maximum amount of information is desired from the photographs. They may be viewed conveniently over homemade light tables fitted with a translucent surface. Commercially made light tables are also readily available. Color photography, in particular, is usually viewed in the form of positive transparencies. Light tables permitting the degree of illumination to be varied and various colored lights to be used have particular application in the interpretation of color photography (O'Neill and Nagel, 1952).

Positive transparencies are also known as *diapositives*, although the latter term is usually restricted to prints very carefully registered on glass for use in the higher-order plotting instruments. Glass is, of course, notable for its lack of grain, smooth surface, and dimensional stability.

Ratioed Prints

Because of variations in the flying height of the plane and in the elevation of the ground, the average scale of photographs may vary considerably in even a single flight strip. When the prints are photographically enlarged or reduced to a common average scale, they are said to be ratioed.

Rectified Prints

Photographs are generally taken with the camera axis tilted slightly, since a perfectly vertical position can seldom be obtained in a moving and vibrating airplane. Neither level bubbles, gyroscope, nor any other existing device has been developed to the point where truly vertical photographs can be consistently taken. If the amount and direction of tilt is known, rectified prints may be produced by recreating an appropriate tilt between the negative and the printing paper. Rectified prints, therefore, are photographs corrected to a horizontal reference plane. Since the amount of tilt must be determined by ground measurements and mathematical analysis, the preparation of rectified prints is relatively expensive.

Enlargements

Modern aerial photographs can be enlarged at least two to four times and yet produce images of high pictorial quality. Enlargements are sometimes used as office records, inasmuch as detail can be identified with ease, and data such as boundaries, burned-over areas, cut-over areas, roads, and telephone lines can be recorded directly on the picture. Enlargements may also be used as a base for extremely detailed mapping (Littleton, 1944). For these and similar purposes, enlargements are useful, but they have their disadvantages. They are too bulky for convenient handling in the field and for study under a simple stereoscope. Furthermore, they yield no information not obtainable under magnification of contact prints. The present trend in photointerpretation is not to enlarge photographically but rather to study the image on contact prints, enlarged optically by the use of a magnifying lens stereoscope. Future practice, however, may well make greater use of enlargements in photointerpretation, since photographic enlargements effectively increase the apparent depth of the third dimension in the stereoscopic image.

Mosaics

Mosaics are pictures made by assembling a number of photographs. The assemblage is usually rephotographed and may be reprinted at any scale. A very rough mosaic, prepared for the primary purpose of providing an index to the individual photographs is termed an index mosaic. The image of each photograph in an index mosaic is clearly labeled so that the observer can quickly determine which photograph covers a particular piece of ground. Index mosaics are commonly furnished, along with contact prints, under most aerial photography contracts. Figure 11 is an index mosaic for a small piece of photography covering an island in the Penobscot River in Maine. The photographs are of the modified infrared type at a scale of 1:12,000 so that the native spruce-fir forest shows clearly in dark tones.

Figure 11 Index mosaic. Orono and Old Town, Maine. July 16, 1946. Modified infrared.

In the strictest sense, however, a mosaic is an assemblage of photographs whose edges have been torn or cut and carefully matched to form a single, continuous photographic representation. If rectified and ratioed photographs are laid down over a control network, the assemblage is called a controlled mosaic; if the ground control is limited, the mosaic is termed semicontrolled. When carefully prepared, mosaics frequently appear as a single photograph. Figure 12 is a controlled mosaic covering one-ninth of a standard U.S. Geological Survey, 15-minute quadrangle in northern New Hampshire. In other words, it covers 5 minutes of latitude and 5 minutes of longitude. At its unreduced size of 17½ by 24 inches, it was at a scale of 1:15,840 compared with the scale of 1:20,000 of the contact prints.

Mosaics are photographic map substitutes which approach but seldom equal a carefully prepared map in accuracy, due to distortions inherent in single photographs. Their principal value is the fact that they may be used to control work where an adequate base map is not available. The mosaic in Figure 12, for instance, was

Figure 12 Controlled mosaic. Northern New Hampshire. 1944. Panchromatic film. (Fairchild Aerial Surveys, Inc.)

used in forestry to locate sample plots in a line-plot timber cruise. Mosaics are also of value in presenting a compact and continuous pictorial representation, convenient to store and to consult. Their disadvantages are their cost and the fact that they usually cannot be studied stereoscopically, although stereomosaics are possible. The cost of preparing controlled mosaics alone is frequently as great as the total cost of flying the area in the first place and of preparing both contact prints and an index mosaic.

REFERENCES

Andersen, H. E. Use of twin low-oblique aerial photographs for forest inventories in southeast Alaska. *Photogram. Engng.* 22:930, 1956.

Avery, G. Helicopter stereo-photography of forest plots. *Photogram. Engng.* 24:617, 1958.

Davey, C. H. Advances in Geological Survey photogrammetric techniques. *Photogram. Engng.* 20:701, 1954.

Heller, R. C., Aldrich, R. C., and Bailey, W. F. Evaluation of several camera systems for sampling forest insect damage at low altitude. *Photogram. Engng.* 25:137, 1959.

Katz, A. H. Height measurements with the stereoscopic continuous strip camera. *Photogram. Engng.* 18:53, 1952.

Kistler, P. S. Continuous strip aerial photography. *Photogram. Engng.* 12:219, 1946.

Littleton, J. W. Detail mapping by the use of enlarged aerial photographs. *Photogram. Engng.* 10:214, 1944.

Losee, S. T. B. Timber estimates from large scale photographs. *Photogram. Engng.* 19:752, 1953.

Mignery, A. L. Use of low-altitude continuous-strip aerial photography in forestry. *U.S. For. Serv., South. For. Exp. Sta. Occ. Pap. 118.* 19 pp. 1951.

O'Neill, H., and Nagel, W. The O'Neill-Nagel light-table (a multipurpose light-table). Its uses in photo-interpretation of color and other photography. *Photogram. Engng.* 18:134, 1952.

Robinson, J. M. Tri-camera winter photography cuts forest inventory costs. *Forestry* 46:643, 1948.

Sanders, R. G. Aerial cameras and photogrammetric equipment: a quarter-century of progress. *Photogram. Engng.* 10:136, 1944.

Seely, H. E. The forestry tri-camera method of air photography. *Photogram. Engng.* 15:461, 1949. Also *Canada Dominion For. Serv., For. Surv. Leaflet* 3. 12 pp.

A KEY ELEMENT in the *development of a region is the density and extent of the transportation net. This net is an expression of how man has organized an area functionally. The transportation system is so important in both cultural and economic spatial organization that Mark Jefferson, in his benchmark article,* The Civilizing Rails (Economic Geography 4:217–231, 1928), *suggests it is one of the best indicators of economic development. Building a road or railroad into an undeveloped area is not an easy or inexpensive task, even with modern technology. One of the most difficult and time-consuming tasks in route location is survey and terrain evaluation. The building project can be expedited if sources of construction material, such as gravel and sand, can be located near the right-of-way. Aerial photographic interpretation has direct engineering significance as a tool in route location, terrain evaluation, and prospecting for sources of construction materials. Air photo reconnaissance and other remote sensing techniques will prove increasingly valuable in aiding the planning and development of transport systems in developed and under-developed countries.*

13-Aerial Photographic Interpretation of Road Construction Materials in Southern Africa

J. H. CAIGER

FOR JUST over a decade now aerial photographic interpretation has been used to a greater or lesser extent in most countries of southern Africa for materials investigations of rural road projects. Despite this, many road engineers still appear hazy as to its true value and place in road planning. This could be due to a lack of appreciation of certain basic principles and, hence, to the misuse of aerial photographic interpretation on some projects. It is hoped that this paper will help to correct any misconceptions and, in particular, will draw attention to a little appreciated facet of aerial photographic interpretation of road construction materials, namely, its potential to influence actual route location in undeveloped territories. . . . .[1]

Route Location

The normal criteria taken into consideration as a matter of routine in the location of a route between two fixed points are: road length, topography, drainage, and local development. However, in undeveloped territories the distribution of low-cost subbase and base coarse materials can actually be a deciding economic factor. This possibility is seldom, if ever, taken into account when planning a route. Formerly this could be ascribed to the lack of a suitable method for both quickly and dependably obtaining the necessary information; but today this reasoning is no longer valid. By the correct application of aerial photographic interpretation it is now possible to both speedily and reliably assess the road building potential of large tracts of country.

The technique adopted in carrying out such an appraisal is to note all variations in the air photos and to field-check at least two or three different localities for each similar type of variation. Any likely sources of base or subbase material should be noted and, if time permits, sampled for laboratory testing. By interpolation and with adequate practice the potential of large areas of country can be evaluated quickly and fairly reliably, thus enabling a decision on the most economic route to be made with a reasonable degree of confidence.

To illustrate the foregoing technique, it is proposed to outline the salient features of two materials investigations undertaken in South West Africa, insofar as these investigations had a bearing on the problem of route location. The first of these two (Sukses–Otjiwarongo) shows how the distribution of high-quality road building materials did finally determine the location of the route. The steps taken in arriving at the final solution were, however, not

[1] A review of the use of aerial photographic interpretation in materials investigations of road projects in southern Africa, including Figure 1, has been deleted with the author's permission.

From *Photogrammetria* 25:151–176, 1970. Reprinted with permission of the author and Elsevier Publishing Company.

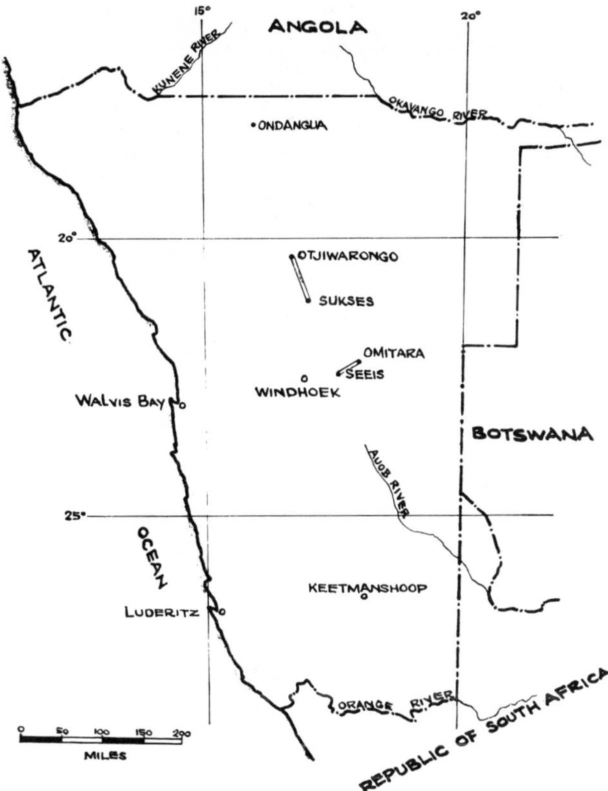

Figure 2 Locality plan of South West Africa.

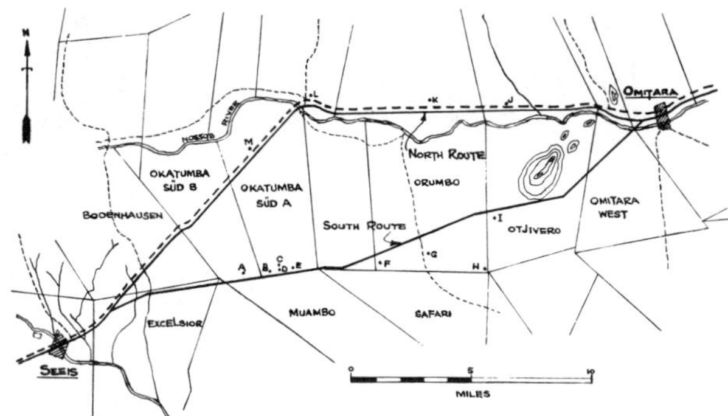

Figure 12 Locality plan, Seeis Omitara.

typical of materials evaluations of this nature. Hence, a second investigation (Seeis–Omitara), which comprised a classical assessment of the significant road building materials within an area of approximately 200 square miles in extent, is included to illustrate both the steps required in, and the relative speed of execution of, such a typical appraisal (see Fig. 2). It is hoped that in detailing these two projects, attention will be focused on the importance, under certain environmental conditions, of this particular aspect of route location.

. . . .[2]

Route Location between Seeis and Omitara in South West Africa

Terms of Reference

The location of the trunk road over a distance of approximately 27 miles between Seeis and Omitara resolved around the selection of one of two possible routes, the horizontal alignments of which were as much as 6.75 miles apart at one point. These alternatives were designated the northern and southern routes respectively. The northern possibility more or less followed the existing road, while the southern route ran along the boundary fence between the farms Bodenhausen and Excelsior, Okatumba Süd B and A and Muambo, to a point on the boundary fence between Silversands and Safari. From this point the route ran approximately northeast, traversing the farms Silversands and Orumbo before running immediately south of the mountains of Otjivero to finally cut across Omitara West and join the existing road and northern route on the farm De Hoop (see Fig. 12).

The southern route is about 1.5 miles shorter than its northern counterpart. This fact, in conjunction with drainage and other considerations, made the selection of the latter route an uneconomic proposition, unless the cost of road building materials was to be appreciably lower. A materials evaluation of the strip of country effectively covering the significant area along both routes was therefore undertaken. The area of country involved was of the order of 200 square miles in extent.

Environmental Conditions

The region investigated, which enjoys an average annual rainfall of about 370 mm, comprises mainly mica-schists and quartzites from the Khomas Series of the Damara System. Within these formations, veins and lenses of quartz occur. It is generally accepted that this quartz was developed during the metamorphic processes which resulted in the formation of the mica-schists and quartzites, and was not an igneous intrusion. The quartz, being more durable than either the mica-schists or the quartzites, has weathered at a considerably slower rate, thus forming residual and transported quartz gravel sources of varying thickness. The soil fines fraction of these gravels fluctuates considerably in plasticity. In this particular region the plastic fines have tended to gravitate downward by a process of leaching. Thus, the gravel occurrences in low-lying areas tend to be more plastic than those straddling the summit of hills. Likewise, the plasticity within a gravel deposit tends to increase with increase in depth. Both surface gravels and basic geologic formations are often covered by sand deposited during former flood periods. This sand cover is especially widespread over the area adjacent to the southern route.

The topography of the landscape is mainly flat, but occasional hills and undulations do occur (see Figs. 13–15). In the eastern section, the southern route skirts a quartzitic mountain range which then runs in a general northeasterly direction out of the area investigated (see Fig. 12). The whole region is covered by stunted acacia trees and, in certain localities, by scrub brush as well. In places where thick sand occurs, yellow wood *(Terminalia sericea)* and large acacia trees may be found. After rain the grass forms a dense carpet on the otherwise bare ground between the trees and bushes.

[2] The section on the route location between Sukses and Otjiwarongo in South West Africa, including Figures 3–11 and Table 1, has been deleted with the author's permission.

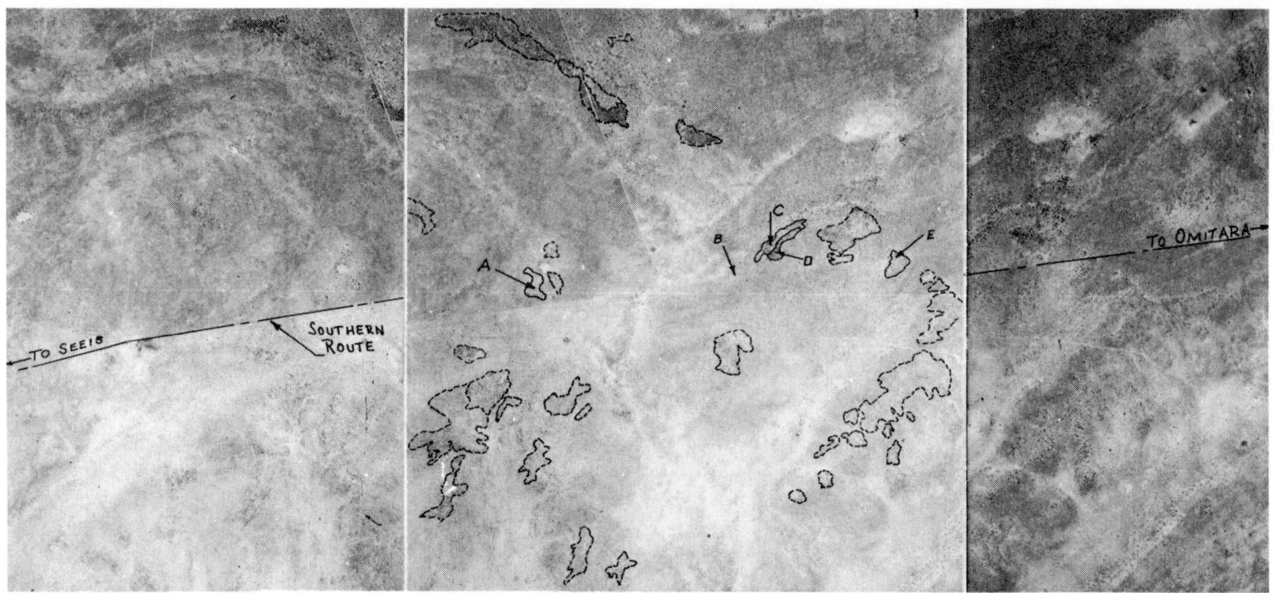

Figure 13 Stereotriplet showing a section of the southern route between Seeis and Omitara.

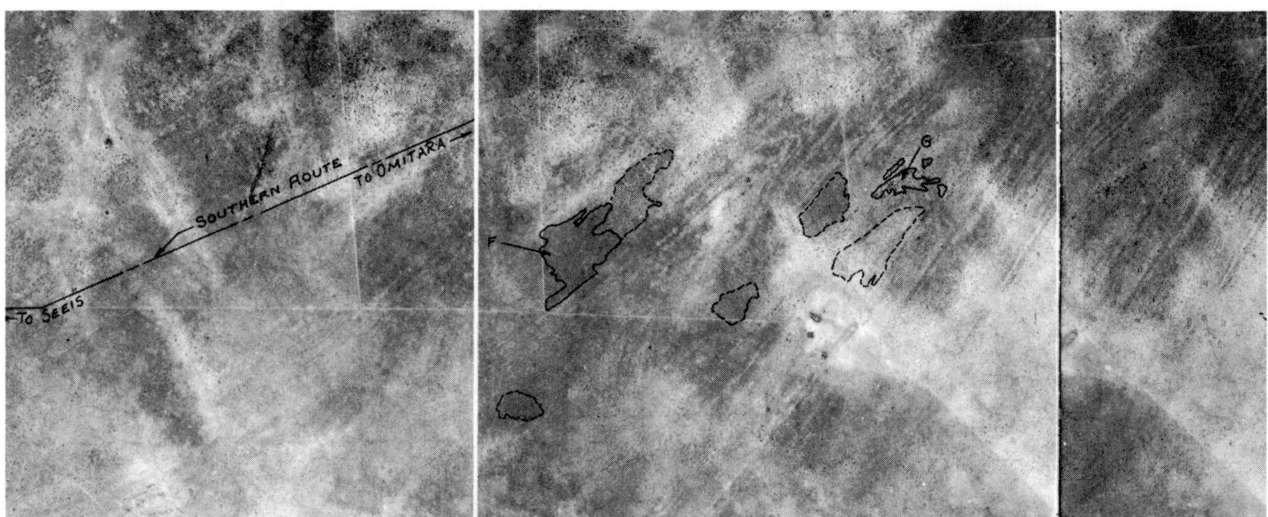

Figure 14 Stereotriplet showing a section of the southern route between Seeis and Omitara.

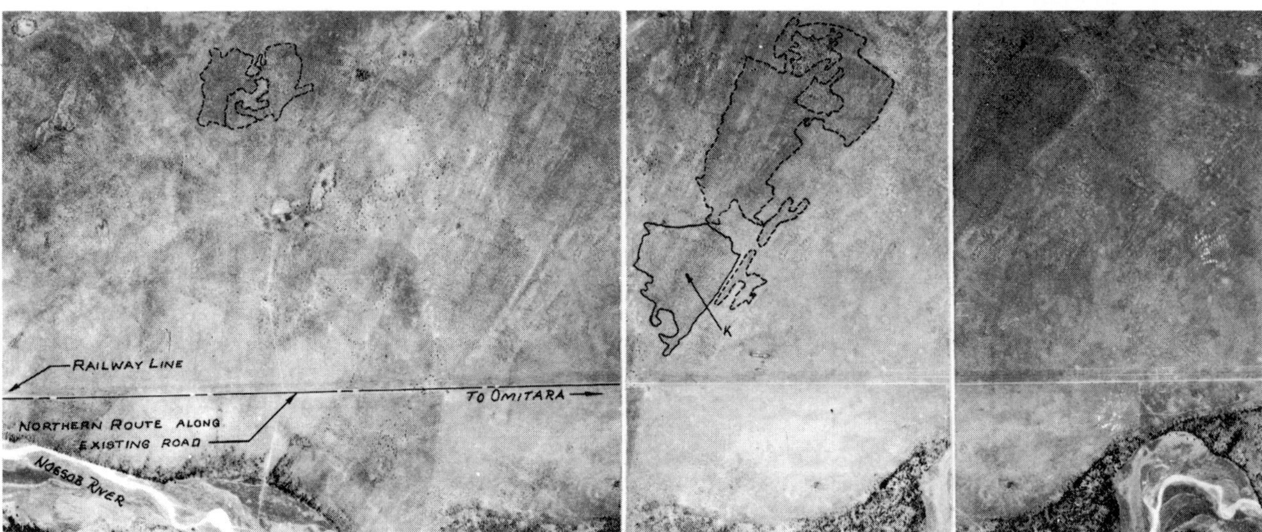

Figure 15 Stereotriplet showing a section of the northern route between Seeis and Omitara.

Table 2 Test Results of Typical Quartz Gravels between Seeis and Omitara

Description:	Southern Route Quartz Pebbles in Brown Soil										Northern Route Quartz Pebbles in Brown Sandy Soil			
Sample No:	1968	1967	1964	1966	1962	1969	1971	1973	1974	1977	1975	1978	1980	1981
Hole Notation:	A	B	C	D	E	F	G	G	H	I	J	K	L	M
Depth below ground level (inches)	13–26	16–27	0–24	32–48	2–31	12–30	9–35	35–47	10–24	4–20	0–36	11–31	0–25	8–36
H.R.B. classification	A1-a	A2-4	A1-a	A1-a	A1-a	A1-a	A1-a	A2-4	A1-a	A1-a	A1-a	A1-a	A1-a	A1-a
Screen analysis: % by wt. passing (inches) 1.5	64	68	99	86	86	81	97	86	100	83	96	94	78	87
0.75	42	32	86	64	70	60	80	59	75	55	75	80	58	64
0.185	19	17	43	29	29	30	33	25	33	26	32	39	29	33
0.078	17	16	34	22	18	23	25	21	28	24	22	27	20	26
0.0164	15	14	30	20	16	20	22	18	26	22	19	23	17	24
0.0029	4	4	11	4	4	5	6	5	5	5	4	5	5	7
0.0021	3	3	8	3	3	4	4	4	3	3	3	3	4	5
Lower liquid limit		22	20				18	21						
Plasticity index	S.P.	8	5	S.P.	S.P.	S.P.	5	7	S.P.	N.P.	S.P.	N.P.	S.P.	S.P.
Linear shrinkage	0.5	2.5	3.0	<0.5	<0.5	0.5	1.5	3.0	<0.5	0.0	<0.5	0.0	0.5	1.0
Optimum moisture content (%)			4.8		4.8	6.3	5.0				4.3			
Maximum density (lb. ft.^{3})			140.9		140.9	142.0	142.9				140.6			
C.B.R. values at: (Modified A.A.S.H.O. compactive effort) 100%			150		170	200	250				130			
95%			44		44	34	78				55			
90%			11		10	6	23				42			

Materials Evaluation of Both Routes

Using the technique referred to previously in Route Location, it was found that the most significant deposits in this area usually occur on high ground and straddle many, but not all, of the hills. These deposits consist almost entirely of quartz gravels, which occur in large quantities in layers of from 18 inches to 3 feet in thickness, and, which are often exposed or found under a thin layer of sand only. Similar occurrences lower down the slopes and on the flats generally were found to underlie a thicker sand cover, to occur in relatively thin layers, and to contain fines of markedly higher plasticity due to the process of downward leaching. It was further established that an abundance of good-quality road-building gravels are distributed along both routes and, in view of this, adoption of the southern route became the obvious choice (see Table 2 and Fig. 12).

Photointerpretation

Figure 13 shows where gravel samples were taken from holes *A, B, C, D,* and *E* along the southern route (see Table 2 and Fig. 12). The boundaries of the sources sampled were demarcated from their topographic position, texture of vegetation, and the lighter tone of materials generally, but not always, exhibited by the less plastic gravels. This change in tone is illustrated by the respective positions and the corresponding test results of holes *B, C,* and *D*. Also illustrated is the tendency of the plasticity to increase on the lower reaches. The dotted areas were marked by analogy with those areas actually sampled, and a few of the former which were field-checked were found to contain visually similar quartz gravel. Due to the widespread nature of the gravel occurrences, all likely sources are not necessarily delineated but only those expected to be of most significance.

Demarcated areas *F* and *G* in Figure 14 are two further sources sampled along the southern route. The deposit represented by *F* is primarily identifiable in the air photos from its relief and texture of vegetation, while that represented by *G* is similar to some of the sources demarcated in Figure 13, although it is not quite so well defined. The test results of the two samples from hole *G* illustrate again the tendency to downward leaching of plastic fines, resulting in an increase in the plasticity index with depth (see Table 2). A few areas where similar gravel deposits can be expected have been demarcated with dotted lines.

Figure 15 shows a section of the proposed northern route running along the existing road and adjacent to the Nossob River. Test results of samples from holes *J, K, L,* and *M,* together with the air photo indicators of the demarcated source represented by hole *K,* illustrate the similarity of the gravel occurrences along both routes (see Figs. 12 and 15 and Table 2). Dotted areas in Figure 15 again indicate some of the other possible gravel sources.

Time Factor

With the experience gained previously under similar environmental conditions in the Windhoek area, a period of two days was all that was required for the field identification of the significant air photo variations and for the sampling of typical potential road building gravels. After the test results became available, it was possible to annotate a mosaic of the whole area with detailed materials information similar to that shown in Figures 13–15.

This information was available, within a relatively short time, to those responsible for the final decision on the location of the route.

Conclusions

From the brief review of the use being made of aerial photographic interpretation for materials investigations of rural road projects in the territories of southern Africa, it is clear that the present major usage is in countries such as Rhodesia and South West Africa, where the need is greatest and where interpretation with a strong engineering bias has been, and still is being, applied. A further implication of the foregoing is that the greatest need occurs in relatively undeveloped countries which are confronted with a combination of difficult environmental conditions and in which, in addition, road development programs are in the process of implementation. It is also clear that, in order to solve the specific materials problems of different territories and to obtain optimum benefits from aerial photographic interpretation, the modus operandi must remain flexible and be adjusted, where required, to meet particular local circumstances. Thus, the method developed in Rhodesia, although apparently successful in its home Rhodesian environs, may not necessarily be the best, or even a suitable method for all territories and organizations connected with road planning. For example, its sophistication and hence personnel limitations, among others, would neither be desirable for, nor essential to, the solution of Malawi's materials problems.

Apart from flexibility in technique just mentioned, certain other basic factors must be recognized before optimum or even satisfactory results can be expected from materials investigations based on aerial photographic interpretation. Among the most important of these is that the photointerpreter be an expert in the particular field for which he is purporting to interpret. Coupled with this in importance is that interpretation, combined with adequate field-checking, should be in direct terms of engineering significance, rather than in terms of geologic and pedologic boundary lines which can be, and are on occasions, misleading from a road engineering standpoint. The necessity of this latter proviso is amply illustrated by the differing plasticities of adjacent areas of a granitic outcrop between Sukses and Otjiwarongo, and likewise, but to a lesser extent, by the differing plasticities of quartz gravel deposits in close proximity to each other between Seeis and Omitara. In southern Africa the importance of both these factors is not always fully appreciated, and this is tending to retard full acceptance of the validity of materials investigation methods based on aerial photographic interpretation and, hence, to retard expansion of the use of air photo interpretation in this sphere.

That materials investigations can achieve a remarkably high degree of reliability when aerial photographic interpretation is correctly applied is reflected by the actual completion data from the Sukses–Otjiwarongo project in South West Africa. When taking the difficult environmental conditions pertaining in this area into account, the standard of materials planning on this project bears closer inspection and comparison with the corresponding data from similar construction projects in which one or other of the more usual methods of materials investigation was applied.

The location of the road between Sukses and Otjiwarongo provides, further, a case in point in which the particular distribution of low-cost subbase and base course gravels over a critical distance of approximately 14 miles, strongly influenced the final choice of route. This example stresses the need, under certain environmental conditions, for a preliminary materials survey effectively covering the full area which could possibly have a bearing on the location of the route. Ideally, a full materials appraisal of just over 200 square miles, comprising approximately a 25-mile length with a variable width of 6–10 miles, would have been sufficient to cover the critical area in this particular case. In view of the vegetational cover coupled with the topographic and geologic nature of the country, it is also open to question whether the true areal extent and disposition of the subbase and bases coarse gravel deposits would have been revealed at all, had aerial photography and air photo interpretation not been used, in the manner described, for the detailed prospecting along the existing route.

The terrain evaluation between Seeis and Omitara provides an indication of the speed with which a typical assessment can be undertaken. In a more complex area of comparable size, the period required for field observations and sampling could, possibly, increase to a maximum of up to four days. With the use of conventional methods only, either a much longer period of field work, or substantially increased numbers of technical personnel with their accompanying equipment, would be required to accumulate the same information with an equivalent degree of dependability. Primarily because of this, such an investigation would then not normally be considered feasible.

In view of the possible advantages, some of which have been illustrated in the foregoing, engineers working in undeveloped territories with difficult environmental conditions should avoid the mistake of lightly underestimating the ultimate financial benefits which can be derived from the correct and timely application of air photo interpretation to both materials appraisals for route location purposes and to detailed prospecting along a chosen route.

REFERENCES

Atkinson, J. R., and Brown, N. B. Airphoto interpretation study for the Division of Roads and Road Traffic of the Southern Rhodesia Government. *Proc. Intern. Symp. Photo Interpretation, Delft, The Netherlands, 1962.*

Atkinson, J. R., and Haine, J. L. W. Identification of common landforms of Southern Rhodesia by airphoto interpretation. *Proc. Reg. Conf. Africa Soil Mech. Found. Eng., 3rd, Salisbury, Southern Rhodesia, 1963,* 1:119; 2:66.

Brink, A. B. A. Airphoto interpretation applied to soil engineering mapping in South Africa. *Proc. Intern. Symp. Photo Interpretation, Delft, The Netherlands, 1962.* Pp. 498–506.

Brink, A. B. A. *Report on the Proceedings of Working Group 9 (Engineering) at the International Symposium on Photo Interpretation held in Delft, The Netherlands, 1962.* To: The Director of the National Institute for Road Research, C.S.I.R., Pretoria, South Africa. (Internal report.)

Brink, A. B. A., and Partridge, T. C. Kyalami land system: an example of physiographic classification for the storage of terrain data. *Proc. Reg. Conf. Africa Soil Mech. Found. Eng., 4th Cape Town, South Africa, 1967,* 1:9.

Caiger, J. H. *Report on Geological Mapping by Airphoto Interpretation for Major Road Construction Projects, 1960.* In: Files of the Chief Roads Engineer of the South West Africa Administration, Windhoek, South West Africa.

Caiger, J. H. *The Use of Airphoto Interpretation as an Aid to Prospecting for Road Building Materials in South West Africa.* Thesis, Univ. of Cape Town, South Africa, 1964, unpublished.

Caiger, J. H. The use of airphoto interpretation in materials investigations for rural road projects. *Proc. Reg. Conf. Africa Soil Mech. Found. Eng., 4th, Cape Town, South Africa, 1967,* 1:15.

Holden, A. Engineering soil mapping from airphotos. *Proc. Reg. Conf. Africa Soil Mech. Found. Eng., 4th, Cape Town, South Africa, 1967,* 1:23.

Holden, A. Discussion on Caiger's paper: Aerial photography and aerial reconnaissance applied to gravel road construction in the Okavango territory of South West Africa. *Civil Engineer South Africa, 1969,* 2:29.

Kantey, B. A., and Williams, A. A. B. The use of soil engineering maps for road projects. *Civil Engineer South Africa, 1962.* 8:149.

Keeble, D. H. L. Discussion on Caiger's paper: The use of airphoto interpretation in materials investigations for rural road projects. *Proc. Reg. Conf. Africa Soil Mech. Found. Eng. 4th, Cape Town, South Africa, 1967,* 2:326.

Koch, J. H. Discussion on Caiger's paper: The use of airphoto interpretation in materials investigations for rural road projects. *Proc. Reg. Conf. Africa Soil Mech. Found. Eng., 4th, Cape Town, South Africa, 1967,* 2:326.

Kraft, B. W. Discussion on Caiger's paper: The use of airphoto interpretation in materials investigations for rural road projects. *Proc. Reg. Conf. Africa Soil Mech. Found. Eng., 4th, Cape Town, South Africa, 1967,* 2:323.

Mollard, J. D., and Pryor, W. T. Photointerpretation in engineering. In: R. N. Colwell (Ed.), *Manual of Photographic Interpretation.* Washington, D.C.: Am. Soc. Photogrammetry, 1960. Pp. 403–456.

Schofield, A. N. The use of aerial photographs in road construction in Nyasaland. *Dept. Sci. Ind. Res., Road Res., Overseas Bull.,* 4, 1957.

Schofield, A. N. Nyasaland laterites and their indications on aerial photographs. *Dept. Sci. Ind. Res., Road Res., Overseas Bull.,* 5, 1957.

MAN LIVES *at the bottom of the atmospheric ocean on the surface of the earth. Except in specialized instances he does not penetrate into the upper atmosphere or deeply into the earth's surface. Man's home then is the surface of planet Earth, and it is from this surface that he must support his life by extracting food, fiber, minerals, and all the other resources he needs. Because man is dependent upon the earth's surface for resources, it is little wonder that early in his existence he turned to the use of rock, the main constituent of the lithosphere.*

One of the earliest periods into which the study of man is divided is the Paleolithic (Old Stone Age), which is marked by the exclusive use of chipped stone implements. Man's early concern for the use of rock gradually evolved into the discipline of geology. But until the advent of the balloon and the airplane, geologists were restricted to the surface of the earth for their observations, explorations, and mapping. The field geologist, insignificantly small in comparison to the earth, was faced with not seeing or misinterpreting surface geology because of this scale difference. Further, no map provides the field worker with all the information and clues needed to assess the geology of a region properly.

With the advent of the aerial camera, the geologist was presented with a new tool for observing, measuring, and mapping the earth's surface. The scale of the photography can be controlled. Large-scale, detailed images of remote or complex areas can be generated, or smaller-scale, more generalized views provide the opportunity for regional overviews, unmarred by too much detail. The air photo has become an important tool in the hands of the field geologist.

Many keys trigger recognition of geologic phenomena on photographs. Some are almost intuitive and are based on subtle associations that are almost subconscious and difficult to articulate. Photointerpreters have found that the most important recognition elements are: size, shape, tone, color, texture, pattern, and relation or association to other features. Careful attention to these elements on photographs allows the geologists to make complex interpretations of rock type and underlying lithologic structure.

14-Geology from the Air

RICHARD G. RAY
WILLIAM A. FISCHER

AERIAL PHOTOGRAPHS are one of the most important tools of the modern geologist. However, it is only in recent years that aerial photographs have received extensive use in geologic mapping and exploration and that their significance as a source of geologic information has gained wide recognition.

The economic importance of aerial photographs in geologic mapping and exploration has been demonstrated many times in recent years. Noteworthy in this regard was the extensive use of photographs for geologic interpretation of the central Iranian basin, the site of one of the most spectacular oil discoveries of 1956 (Mostofi and Gansser, 1957). Some years earlier, aerial photographs played a significant part in the discovery of Cerro Bolivar, the iron-ore bonanza of Venezuela (Lake, 1950). Today, a number of private Canadian companies are collaborating in mapping part of the Canadian Shield in one of the most extensive geologic mapping programs yet undertaken with aerial photographs (Guild, 1957).

Aerial photographs, when properly used in geologic work, add speed, economy, and accuracy to geologic mapping as well as certain geologic information impossible, difficult, or economically infeasible to obtain by routine field mapping methods. These assists to geologic mapping are a result of many factors. Aerial photographs permit views of large areas and hence reveal to the geologist over-all geologic relationships that could not readily be seen otherwise. In addition, the geologist obtains a plan view of the terrain, similar to the planimetric presentation of geologic maps. Photographs may also show the geologic terrain in a way that it cannot be seen by the naked eye, that is, as in infrared, camouflage-detection, or other special photography; hence, geologic information is revealed that might otherwise be obscure.

Furthermore, all features, both natural and cultural, that are clearly expressed on aerial photographs can be easily measured. The accuracy of measurement depends primarily on the scale of the photography. The use of photogrammetric instruments not only permits measurements, which are important in interpretation, but also increases mapping accuracy and to a lesser extent increases the speed and economy of mapping. The degree to which aerial photographs are used varies widely, but to whatever extent photo techniques are employed, they must remain principally an aid, or tool, in geologic mapping and not a substitute for field mapping.

From *Science*, 126:725–735, Oct. 18, 1957. Reprinted with permission of the authors and *Science*.

Viewing the Photographs

Vertical photographs are most commonly used for geologic study; these are photographs taken with the camera lens pointing vertically down from the airplane. Normally, aerial photographs are taken from positions so spaced that each image within the field of the camera appears on at least two photographs. When two photographic images of the same object, taken from different positions, are combined optically by means of some sort of stereoscopic viewing device, the familiar 3-D effect is seen. The viewing device may be a simple lens stereoscope (Fig. 1)[1], and mirror stereoscope (Fig. 2)[1], or a precision mapping instrument such as the Kelsh plotter (Fig. 3)[1]. Aerial photographs, of course, can be studied in two-dimensional plan view by using single prints of aerial photographs or groups of prints mosaicked together, but the value of three-dimensional stereoscopic examination of the aerial photographs as compared with the value of examination of the two-dimensional plan view cannot be overemphasized. Whereas conspicuous geologic features are commonly visible on single aerial photographs or mosaics of aerial photographs, the wealth of information shown in a stereoscopic view is many times greater. Details, such as fine lines or textural differences not readily seen on single photographs, or even on the ground, are commonly shown clearly in the stereoscopic model. Such clarity is in many places a direct result of the common association of fine lines and textures with relief changes, which are exaggerated in most stereoscopic models.

The value of the 3-D effect in geologic interpretation is increased by the vertical exaggeration, or relief exaggeration, that commonly occurs in stereoscopic viewing of aerial photographs. This exaggeration results from the wide spacing of camera positions at the time of exposure, as contrasted to the spacing of the human eyes; it is of particular value in interpreting the angle of dip of sloping surfaces, such as sedimentary beds, and thus low dips of 1 to 2 degrees, which may be especially significant in petroleum exploration, may be readily interpreted from the aerial photographs. In addition, minor topographic differences, which may reflect underlying geologic structure, are exaggerated and in turn may be easily recognized. The exaggeration of relief in a stereoscopic model of 1:20,000-scale photographs taken with a 6-inch-length lens is such that a geologist is enabled to differentiate differences in elevation as small as 1 foot.

[1] Not available for reproduction.

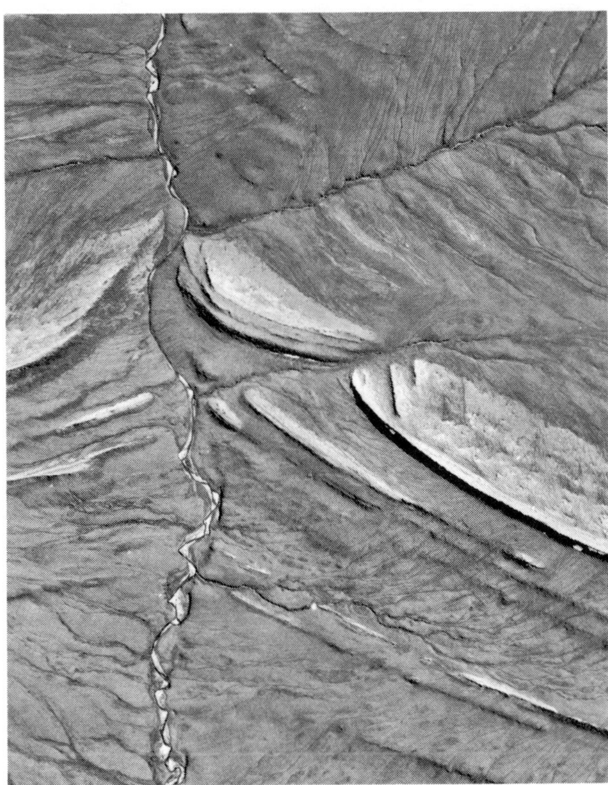

Figure 4 Vertical aerial photograph showing striking delineation of light-colored sandstone in a shale-sandstone sequence.

Kinds of Information

There is hardly any terrain which will not yield some geologic information from a study of aerial photographs. The amount of information naturally varies with the kind of terrain and the climatic environment.

Figure 4, showing a sequence of shales and sandstones in western United States[2], demonstrates convincingly the usefulness of photogeologic procedures in mapping the distribution of rock types. The clearcut geologic contacts can be accurately mapped from photographs, thus eliminating much time-consuming effort in the field.

In a different climatic but geologically similar area, Arctic Alaska, gently folded and faulted rocks in a potentially oil-bearing region are very poorly exposed (see Fig. 5)[3]. Differential resistance to erosion of sedimentary rock units combines with structural attitude to reveal bedding by slight topographic breaks and minor changes in vegetation type. The topographic breaks are exaggerated in stereoscopic view and permit the interpreter to detect information not readily apparent in ground study. Although the rocks do not actually crop out, the structural setting, so important to the petroleum industry, can be easily interpreted.

An extreme example of the value of aerial photographs in interpreting oil structures was a photogeologic study of the Square Lake area in northern Alaska. Fischer (1953) reports that field parties have spent entire seasons within and immediately adjacent to the Square Lake area without detecting the Square Lake anticline. No outcrops are present, but small hills in the area are elongate in the same direction as the regional structural trend. Normal to this trend, other hills slope gently in a direction of postulated dip. Also, streams on the postulated dip slope have a slightly less well-developed dendritic pattern than those on the opposite slope. There also seems to be a possible correlation between the slopes of the tops of cutbanks and the direction of dip. Plotting of all such data gave a consistent apparent structural pattern—a plunging anticline—of unknown reliability. A subsequent seismic study corroborated the photo study, and test drilling in turn confirmed the photo and seismic work. In another similar terrain, photointerpretation based on minor topographic variations and stream patterns indicated a structural axis, later substantiated by seismic work (Fischer, 1953). Drilling indicated a huge gas reservoir.

A study of heavily forested terrain of southern Alaska further indicates how aerial photographs and photo procedures may be used in geologic study and mapping. In the Prince William Sound area a detailed study of fracture systems was made, and locations of fractures were plotted with respect to fractures mapped by ground methods and known to be associated with ore deposits, primarily copper. Figure 6 shows how fractures are visible on aerial photographs even within the heavily forested areas. Such fractures are observed only with great difficulty and expenditure of effort by ground traverse. The association of certain fractures with ore deposits in turn suggests areas that might be prime target areas for ore search. Similar studies have been made in

Figure 6 Vertical aerial photograph showing fractures in heavily forested terrain.

[2] A similar photo of nothern Alaska is used here.
[3] Not available for reproduction.

British Columbia by a leading mining company, and many square miles have been eliminated as primary target areas prior to any field study. Another leading mining company in eastern United States follows this photointerpretation technique with geophysical surveys of areas of favorable structural setting for ore deposits.

Of more subtle nature, but of considerable economic significance, is the study of patterns, particularly stream patterns, resulting from the adjustment of streams to underlying geologic structure. Even in jungle areas, invaluable data relating to structure may be obtained from aerial photographs by a technique colloquially termed "creekology," the analysis of stream pattern. Figure 7[4] shows how stream patterns can define a geologic structure, such as an anticline. Note how streams flow away from the suspected elongate domal structure. In other areas soil or vegetation patterns may suggest underlying structure.

It is clear, then, that aerial photographs can be an extremely useful tool in geologic mapping and exploration, but one must realize that the technique has its limitations as well as its advantages. The interpreter can only guess at the composition of rock types; he cannot identify mineral type or absolute ages of rocks, nor can he obtain such information as paleontologic data.

The importance of aerial photographs in geologic study may differ with respect to the geologic terrain being mapped and the objectives to be obtained. In many studies, it would seem logical to make preliminary photogeologic studies just as it is normally a preliminary step to investigate available literature before going into the field. In reconnaissance mapping, especially of remote or inaccessible areas, photogeologic procedures can be used as the principal mapping technique, and preliminary maps can be compiled solely on the basis of photo study. In detailed mapping, photogeologic procedures provide a supplemental mapping technique and assist the detailed field investigation. But, in any event, a look at photographs early in a mapping program is desirable and in certain investigations is necessary to effective follow-up study in the field. Commercial companies are relying more and more on preliminary photo study to delineate potential oil-bearing or mineralized areas in which to concentrate field investigations.

Interpretation—Recognition Elements

Interpretation of aerial photographs is based on recognition elements—characteristics of the photograph that result from the scale selected, the color of the rocks and other elements of the terrain photographed, the kind of film and filters used, the processing of the film, and similar related factors. The most significant recognition elements for geologic interpretation are relative photographic tone, color, texture, pattern, and relation to associated features. Size and shape may also be diagnostic recognition elements in certain geologic problems.

Photographic Tone

Because of the ability of the human eye to differentiate subtle tone changes, relative photographic tone is a significant asset in geologic interpretation of aerial photographs. In areas of good exposures, bedding is characteristically recorded on the aerial photograph by differences of photographic tone. These differences may be abrupt, as illustrated by white-weathering marlstone beds in a sandstone sequence, or they may be subtle, as shown by two successive similarly colored sandstone units. Faults may be indicated by a change in photographic tone as shown on opposite sides of a straight or gently curving line; and alluvial fans, pediments, lava flows, intrusive rocks, and many other geologic features are usually identified, at least in part, by relative differences of tone.

Color

When color aerial photography is available, recognition is greatly facilitated as a result of the ability of the human eye to differentiate about 1000 times as many tints and shades of color as it can tints and shades of gray, characteristics of black and white photography. Subtle differences on black and white photographs may thus become obvious on color transparencies, as in rock units of only slightly different lithologic composition. The advantages of color aerial photography are generally present even though some color fidelity may be lost in the transparencies. Special color film, in which colors are distorted, as in camouflage-detection film, may be singularly useful in the study of certain areas.

Texture

Texture is the composite appearance of a combination of features too small to be clearly seen individually; thus the scale of the photographs has an important bearing on what is called texture. For example, a network of fine lines referred to as a texture on high-altitude photographs may well be recognizable as a network of joints on low-altitude photographs. Likewise, a mottled texture on high-altitude photographs, resulting from small amoeba-like outlines, may be clearly the result of kettle holes that are distinctly discernible on low-altitude photographs. Intrusive igneous rocks commonly have a distinctive texture owing to a crisscrossing of the many joints almost universally present in such rocks.

Pattern

Pattern, or arrangement of geologic or other features, is especially significant in geologic interpretation from aerial photographs. A common use of this recognition element is the analysis of stream pattern, which may be a significant aid in interpreting the underlying geologic terrain. Patterns of joints may suggest certain rock types, or a knowledge of fault patterns of an area may be helpful in locating faults in a similar nearby terrain. Patterns resulting from particular distributions of lines are common, but a single line, or lineation, may be a special illustration of pattern. For example, a lineation may result from an orderly arrangement of stream segments, trees, depressions, or other features. This arrangement may be a continuous alignment of geologic, topographic, or vegetation features, but more commonly it is a discontinuous alignment. Lines are especially representative as expressions of faults, but they may also represent a variety of other geologic phenomena.

[4] Not available for reproduction.

Relation to Associated Features

The relation to associated features is commonly important because a single feature may not be distinctive enough to permit its identification. Thus, for example, identification of depressions in surficial deposits such as kettle holes may be possible because of the presence of associated glacial ice nearby.

Size

The term "size," used as a general recognition element covering all interpretation fields, is more appropriately considered in geologic interpretation in relation to thicknesses of strata, amounts of offset along faults, or other finite measurements. These measurements may be directly related to topographic expression. If the thickness of a formation is known, this knowledge may aid in identification, and determining the range in thicknesses may be essential to understanding the regional geology.

Shape

Shape as a recognition element in geologic interpretation is of significance primarily only in its broadest definition, involving relief or topographic expression. In this regard, it may be important for recognizing geologic features in certain areas. For example, the bold cliff face of one formation in contrast to a lesser angle of slope across an underlying formation locally may be of considerable importance in differentiating the rock units. Rectilinear depressions are expressions of faults in many areas. However, in its strictest definition, as a spatial form with respect to a relatively constant contour or periphery, shape is of little importance as a recognition element because nature may reveal the same geologic feature in an infinite number of different shapes.

Interpretive Process

The interpretation of geology from aerial photographs involves many of the same mental and physical processes as the interpretation of geology from field observations. If the full value of the photographs is to be utilized, preliminary field reconnaissance combined with cursory study of the aerial photographs must be undertaken prior to detailed interpretation. Detailed interpretation, including measurement of features considered to be geologically significant, is then undertaken; rigorous application of the interpretive process is made in this phase of applying aerial photography to geologic study. Photographs should be used in the field not only for recording locations of observations but for contrasting geologic features with their images on the photographs to provide a basis for further interpretation. Whenever possible, geologic data derived from study of the aerial photographs should be checked and evaluated in the field.

The initial phase of the interpretive process is an observational phase wherein recognition of geologic features or characteristics of the terrain is involved. Recognition of terrain features is commonly based on combined use of fundamental recognition elements such as photographic tone and pattern. Data thus observed are then interpreted with regard to geologic significance, and the geologic history of an area is deduced, insofar as possible, from the distribution and relationships of the features recognized. Many features are expressed on aerial photographs. Sorting those features that are of significance to a particular problem and properly relating these features one to another provides a measure of the ability of the interpreter. In geologic interpretation, this ability depends primarily on the geologist's background training in geology, such as his understanding of structure and natural processes operative on the rocks, and secondarily on his experience in viewing aerial photographs. The geologist may also make use of photogrammetric instruments for making measurements, which in turn become the basis for interpretation of the geologic significance or history of an area.

Interpretation is a multistep operation, and hence a final comprehensive interpretation of the regional geology may be a synthesis of many lesser but specific interpretations, such as the direction of dip of beds in a sedimentary sequence. On the other hand, the immediate recognition of regional or large-scale geologic features as a result of the over-all aerial view, permitted particularly by small-scale photography, is commonly the basis for interpretation of smaller specific features. For example, a geologist may immediately recognize from the general land form that an area has been glaciated. This basic information would facilitate recognition of specific glacial features, such as kames and morains.

Example of Interpretive Process

The interpretation and mapping of the distribution of younger lava flows in an older igneous-metamorphic terrain of southeastern Alaska by W. H. Condon (1957) provides an excellent example of the interpretive process in geologic study. The area has a maximum relief of slightly more than 2000 feet. Outcrops are masked almost completely by a heavy forest growth, largely coniferous trees. In poorly drained sections, a grassy swamp vegetation or muskeg has developed.

The area is underlain by highly folded and faulted phyllites, schists, and gneisses that have been intruded by granitic and dioritic igneous rocks. Younger basaltic lava flows have been extruded onto the older igneous-metamorphic complex.

Criteria used in photogeologic analysis were derived from photo study of locations where lava flows had been reported in the field. The most important criteria were based for the most part not on actual observation of the bedrock, but rather on the effects of rock types, structure, and geologic processes on terrain expression. These effects involve vegetation and drainage, which commonly are expressed in significant patterns. Specifically, these patterns involve (1) linear features, (2) minor streams and abundance of small shallow ponds, (3) density and type of vegetation, and (4) vegetation expressed as lobate outlines by denser, darker growth. Further analysis was made for (1) the possible relation of the volcanic rocks to the pattern of probable faults, (2) terrain forms expected in volcanic rocks, and (3) the controlling influence of existing topography on the distribution of lava flows.

Linear features of terrain appear as long narrow gullies or troughs of varying depth and are generally well

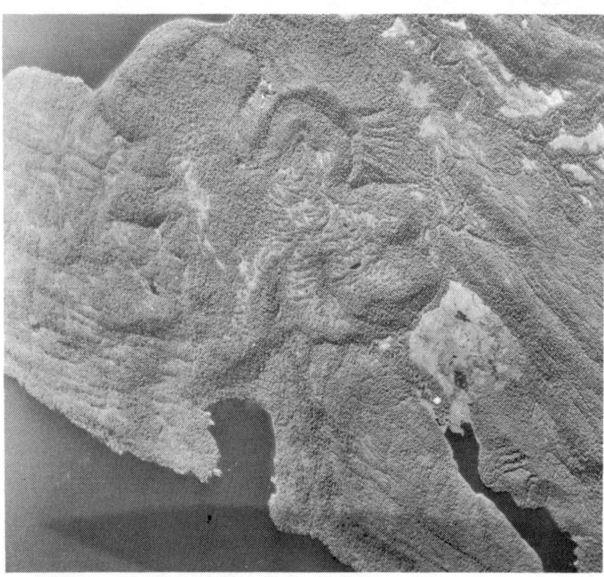

Figure 8 Part of vertical aerial photograph showing linear features across ridges of older rocks but not in the area occupied by lava flows (upper left).

expressed in areas of the older rock complex. In many places they may be seen on aerial photographs to terminate abruptly at the edges of the younger volcanic rocks across which they cannot be traced. The pattern of linear features is most probably the expression of faults, although at least locally the possibility of joints, bedding, schistosity, or, perhaps, glacial gouging, the gross effect of which cannot be overlooked completely. Any of these possible alternate interpretations, however, would serve similarly to date the volcanic rocks with respect to the older metamorphic-igneous complex. Figure 8 shows linear features clearly expressed across ridges of older rocks but not traceable across the intervening valley, interpreted as occupied by lavas.

Drainage characteristics and stream patterns have developed in response to the rock type, structure, and slope of terrain surfaces, and to the degree of fracturing of the bedrock. In adjusting to terrain, streams on the complexly fractured older rocks tend to be channelized to the system of linear features and to form a pattern of parallel drainage. In contrast, streams that developed on the slightly sloping surfaces of the volcanic flows, or on the steeper sides of volcanic cones, show an irregular, somewhat dendritic or radial dendritic pattern of drainage not controlled by a fracture system. Furthermore, upon the nearly flat, poorly drained, muskeg-covered surfaces of the volcanic flows there is an abundance of small, shallow, swampy ponds not characteristic of the generally better-drained surfaces of the older rock complex. Figure 9[5] shows the contrast of parallel drainage pattern of the older rock complex to the somewhat radial-dendritic stream pattern on the slopes of a probable volcanic cone. . . .

The density and type of vegetation, and thus the pattern, seem to be greatly influenced by the drainage conditions of the surface. Where drainage is good, as on the steeper slopes of volcanic cones and the generally well-drained igneous-metamorphic terrain, the surface is heavily timbered with coniferous trees. Where drainage is poor, as on the nearly flat surfaces of the volcanic flows, a swampy muskeg with a sparse and patchy tree cover has developed. On these nearly flat lava surfaces a patchy vegetation pattern results from the contrast between the lighter-toned, low, grassy swamp vegetation and the taller brush and trees, which tend to be concentrated only along stream courses.

Lobate patterns of vegetation formed by dense growths of coniferous trees contrast sharply in height and in photographic tone with the low, light-toned swamp vegetation. These patterns are interpreted as marking the raised edges and fronts of the most recent lava flows, and they are particularly useful in indicating direction of flow. The development of such patterns is believed to be caused by better drainage along the raised edges of the flows in contrast to poorer drainage within the muskeg-covered central areas.

A comparison of the distribution of rocks interpreted as volcanic with the pattern of linear features in the older rocks is believed to be significant. Volcanic rocks were probably extruded along or very close to continuous and prominent linear features, interpreted as faults, trending north-northeastward to northward. Along the extension of one major fault a second area of volcanic rocks was interpreted from aerial photographs. The presence of these rocks was verified in the field.

A further analysis of terrain was made for topographic features to be expected in volcanic rocks, such as remnants of cones and volcanic necks or plugs. Two sizable crescentic features were observed, one definitely a remnant breached cone, and the second probably also a breached cone.

A final consideration was made for the controlling influence of the existing topography on the distribution of the volcanic flows during the time they were being poured out on the earth's surface. Field workers in southeastern Alaska have postulated that the topography has not been greatly changed in either form or relief since the last general glaciation. It has been stated also, with some reservations, that the young volcanic rocks within this area are of late- or post-glacial Quarternary age. As seen on aerial photographs, lava flows not disturbed by glacial erosion exist on ridge slopes and occupy the bottoms of U-shaped valleys, presumably glaciated valleys. Their undisturbed appearance is anomalous within glaciated valleys, unless they were not subjected to the last general glaciation. Thus a post-glacial age is strongly suggested.

Photogrammetry

Photogrammetric Measurements

The important task of interpretation cannot always be accomplished by mere observation of photographs. It may be necessary to plot geologic data and to measure geologic features by photogrammetric methods in order to arrive at a sound interpretive conclusion. Photogrammetry is the science of obtaining reliable measurements by means of photography. Photogrammetric instruments used in conjunction with the three-dimensional stereo-

[5] Not available for reproduction.

scopic model formed by overlapping aerial photographs thus provide a tool of significant use to the geologist. It is commonly possible to make necessary measurements to compile isopach and structure-contour maps of well-exposed areas without going into the area of study, although to assure best results a thorough field check of all work should be made. *Isopach maps* are those that show lines of equal thickness of parts of a rock formation—information that is vital in many commercial studies such as the search for petroleum and for some ore deposits. *Structure-contour maps* are maps that show lines of equal elevation on the top of a formation and help delineate geologic structures that may be important to the petroleum industry.

Example of the Use of Photogrammetry

An isopach-mapping project in Monument Valley, Arizona, illustrates the photogrammetric value of aerial photography in geologic study (Hemphill, 1957). In Monument Valley favorable sites for uranium minerals occur in paleostream scours or channels in the top of the Moenkopi formation. In these channels the overlying Shinarump member of the Chinle formation has thickened and provides loci of deposition of uranium minerals. The Shinarump member is particularly resistant to erosion and normally forms broad benches or the capping unit of buttes and mesas, a relation that aids in its identification on aerial photographs. By measuring a stratigraphic interval below the Shinarump member with a suitable photogrammetric instrument, it has been possible to show local thinning of the underlying rock units and inferred thickening, or channel formation, in the overlying Shinarump member. The isopach intervals—local thicknesses of the Moenkopi formation and underlying Hoskinnini tongue of the Cutler formation—were measured with a Kelsh plotter by using photography of approximately 1:20,000 scale. The Moenkopi formation and Hoskinnini tongue of the Cutler formation were measured as a unit because the base of the Hoskinnini tongue provided a more reliable structural datum, and was more readily identifiable on aerial photographs. Numerous altitude measurements were made at the base of the Hoskinnini tongue and at the top of the Moenkopi formation. Corrections were made for tilting of the rock units, and true local thicknesses were computed, which then served as a basis for isopach compilation—that is, connecting all points of equal unit thickness.

The resulting maps showed several lines, or contours, that closed in an elongate pattern, indicating the presence of a channel, possibly uranium-bearing, in the overlying Shinarump member of the Chinle formation. It is significant that the map was compiled entirely from aerial photographs; no ground control was used. The channels delineated by photogrammetric methods agreed closely with those located by field study. Although details of the aforementioned study have necessarily been omitted, the results demonstrate the significance of photogrammetric procedures in geologic interpretation.

Types of Instruments

The instruments used in making measurements for geologic purposes are varied; the choice of instrument depends not only on the geologic objectives sought but also on a complex set of factors, including photography available, accuracy desired, time available, reliability of ground control, character of topography, and character of vegetation and cover. Measurement of differences in elevation can be made with the simple stereometer or parallax bar from paper prints (Fig. 10)[6]. These measurements are thus made economically, but some reliability of results must be sacrificed, for no correction can normally be made for tilt that may be inherent in the photography. Corrections for tilt can be made if precision instruments (see Fig. 3) are used, and thus correct thicknesses of rock units and correct altitudes can be obtained. Local accuracy of 2 feet in elevation readings may be possible with the common 1:20,000-scale photography, but accuracy throughout the entire stereoscopic model will be less than this.

As the scale of photography decreases, so does the reliability of measurements. Hence scale of photography is important in considering objectives to be attained, particularly if only one type of instrument, such as a parallax bar, is available for use in measuring. Greater over-all accuracy may be obtained from a given scale of photography with precision photogrammetric equipment (see Fig. 3) than with simpler instruments.

The procedures described may be considered as routine in using aerial photographs in geologic study; that is, interpretation is accompanied by collection of metric data which in turn may be of further use in final interpretation.

New Techniques

In addition to these routine procedures, many new avenues of study are being tested. The new procedures involve experimentation with photographic systems, high-altitude photography, color aerial photography, new photogrammetric applications, and orthophotography (Fischer, 1957; Ray, 1957).

Photographic Systems

Based on the premise that aerial photographic tone or color of an object should be predictable for any particular film and filter combination if certain factors are known, a photographic system may be devised to accomplish a specific objective, such as differentiating certain rock units. The wavelength and intensity of light reflected from a surface are the main requirements in devising a photographic system. Spectrophotometer analyses are made of the reflective spectra of rocks in question, and resulting spectral curves are used as a basis for devising a film emulsion that will differentiate the reflected light of each sample, thus recording a different photographic tone for each rock photographed.

High-Altitude Photography

High-altitude photographs taken from approximately 30,000 feet above mean terrain have been used recently with much success in geologic mapping and measurement. The three-dimensional view provided by this photography may provide the geologist with a single stereoscopic view of as much as 50 square miles. This is a much larger ground area than was heretofore covered

[6] Not available for reproduction.

by a single stereoscopic model. This view of a larger area enables the geologist to begin his study with a broader understanding of the relations of the general geologic features.

The advantages of extensive areal coverage in a single stereoscopic model are realized even further in the technique of using twin low-oblique photographs, in which the overlap area of photographs taken at 30,000 feet flying height can cover more than 100 square miles (Radlinski, 1952). Experience indicates that in many areas, regardless of the complexity of the geology, a fundamental orderly arrangement of geologic structures can be discerned on photographs, provided that the scale is small enough. Because of the large area covered per stereoscopic model with high-altitude photographs, precision photogrammetric equipment may be used economically in some photogeologic mapping, depending on the geologic terrain and metric requirements of the job. If camera lenses of the same focal length are used, reliability of measurement varies inversely with the altitude of the aircraft taking the photograph.

Color Aerial Photography

Recent experimental color aerial photography has demonstrated significant uses in geologic interpretation not possible with available black and white photography. In Death Valley, California, it has been possible to differentiate lava flows of similar color but of different ages and to differentiate certain lake sediments from lava flows, all of which appear similar on black and white photographs. Many of the stratigraphic units in this area have characteristic colors. When the units are in normal stratigraphic sequence, the characteristic colors are likewise arranged in a normal sequence. Interruptions in the normal sequence of colors have suggested thrust faults, later verified by field check, that were not interpreted from black and white photographs.

Perhaps of greater significance is the ability to differentiate some zones of alteration on color aerial photographs; these zones of alteration may be significant in mineral exploration. In a study of the Tonopah and Goldfield areas in Nevada, it was found that early stages of rock alteration could be distinguished by color and were characteristic of certain rock types. Intense alteration, however, although readily identified on color photographs, tended to produce the same color regardless of the original composition of the rock. Such intensely altered zones may well be significant with regard to ore deposition, however. With regard to color photography, it is interesting to note that the Canadians plan to "fly" the Sudbury, Blind River, and Bancroft districts with color film to determine whether clues to mineralized areas can be picked up which may guide in the search for new districts (Guild, 1957).

Color aerial photography in geologic study has received only limited use, presumably because of its relatively high cost compared with the cost of black and white photography. The high cost is ascribed by many to technical limitations of color film, such as the limited latitude in photographic conditions, the need for lenses of long focal length, and the requirement of low altitudes of flight. However, many of these limitations have been overcome in recent developments, and color aerial photography may be expected to become more competitive with black and white photography for purposes of interpretation. And, in any event, an evaluation of cost should logically be made in terms of results of use, as in an exploration program, and not solely in terms of the cost of black and white photography.

New Photogrammetric Applications

The introduction of projection-type stereoscopic plotters, such as the Kelsh, multiplex, and ER-55 plotters, has facilitated geologic mapping and study, for the geologist uses these instruments to combine interpretation, measurement, and plotting in a mutually supporting operation. These instruments increase the accuracy and soundness of interpretation (1) by presenting the terrain in proper orientation so that features are in correct relation one to another, (2) by allowing features to be plotted orthographically during the process of interpretation so that their relationship to features previously interpreted from adjacent stereoscopic models can be continuously studied, and (3) by allowing measurements to be made quickly and easily so that measurement becomes a closely integrated tool of interpretation. With stereoscopic plotting instruments, isopach or structure-contour maps of some areas may be made primarily or entirely by photogrammetric means; the isopach map of part of Monument Valley, which has been described, is an example of the usefulness of photogrammetry in geologic mapping. The accuracy of stereoscopic instruments is considered particularly significant with respect to plotting features that are visible on aerial photographs but that are difficult to locate on the ground. Correct plotting of positions of features aids in their subsequent location and study in the field.

Several new instruments have recently been devised to aid in geologic interpretation. One of these is a tilting platen, or viewing surface, used with projection-type stereoplotting instruments. The surface of the tilting platen can be made to coincide with a sloping surface in the stereoscopic model, and the angle of tilt can be directly measured with a clinometer or other measuring device. A profile plotter also has been constructed; it not only permits an accurate profile of the terrain to be drawn in any orientation of the stereoscopic model but also permits exaggeration of this profile, as desired, at the time of plotting. Another instrument is being made to measure directly the thickness of inclined rock formations shown in stereoscopic view.

It is expected that projection-type stereoplotting instruments will be useful in geophysical studies in determining the altitudes of gravity stations and measuring the mass of the topography surrounding gravity stations for terrain correction. Statistical methods used in conjunction with the plotting instruments may eliminate many of the laborious computations of terrain corrections from topographic maps.

Some stereoplotting instruments can be equipped with coordinate-measuring devices so that any point or object on the photograph may be located quickly in a three-dimensional grid system. Thus the photograph becomes an ideal starting point for translating positions into a form usable in electronic computers. Furthermore, these instruments allow a model to be deliberately inclined; in this way regional dips may be introduced or removed, and the model may be studied and measured in any desired hypothetical orientation. Stereoplotting instruments provide quantitative information easily and thus allow closer integration of photogeologic interpretation with an over-all exploration program. In some areas the photographs may be the prime source of metric data.

Orthophotography

Orthophotography is photography that has the position and scale qualities of a map plus the abundant imagery of photographs. Conventional vertical aerial photographs are perspective views, and, in this form, all images are displaced radially from the center of the photograph. This displacement of relative position of features makes it difficult to transfer data accurately from a photograph to a map.

Figure 11 (left) is a direct copy of a part of a perspective aerial photograph; it shows the straight path of a power line as it is distorted by normal relief displacement. Figure 11 (right) is an orthophotograph made with a device, the orthophotoscope (Bean and Thompson, 1957), that removes relief displacement; note that the power line is straight. An orthophotograph is in itself an excellent planimetric map; data may be transferred directly from the orthophotograph to a topographic or planimetric map. Orthophotographs also provide a means for rapidly determining altitudes in the field by reading vertical angles with an alidade, or similar instrument, and scaling the horizontal distances directly from the orthophotograph.

Figure 11 (Left) Part of perspective aerial photograph showing distortions in power line caused by relief displacement of image points. (Right) Orthophotograph of the same area. Relief displacement has been eliminated in printing. Note that the power line is straight.

Historical Summary

Flying advances in World War I were influential in stimulating aerial photography for commercial use during the 1920s. Since that time aerial photographs have been used increasingly in geologic study. One of the first important uses of aerial photographs for geologic study was in compiling mosaics, which were used for general interpretation purposes, as planning maps, and as general base maps for plotting geology. Interpretations were generally made by viewing the over-all mosaic rather than by stereoscopic inspection of photo pairs.

Subsequently, stereoscopic study of photographs was undertaken in a rather extensive way, by use of simple viewing devices. Among the first important geologic interpretation studies from aerial photographs was a reconnaissance study of 35,000 square miles in New Guinea, begun in 1935 by the Dutch. (Helbling, 1949). Although the final maps were compiled by simple methods, and although positioning errors were present, the results met the requirements of reconnaissance study and demonstrated convincingly for the first time the potential of aerial photographs in a petroleum exploration program, from the standpoint both of information obtained from the photographs and of the great saving in time and money in completing the job.

Yet no extensive use of photographs was made by the petroleum industry as a whole prior to World War II. It was only after World War II, and as a result of techniques and interests developed during the war, that the use of aerial photographs began to rise spectacularly in commercial studies. Primarily the petroleum companies began extensive use of photogeologic maps, but the mining industry also indicated an increasing interest in photogeologic procedures.

Noteworthy since 1945 has been the increased use of photogrammetric instruments in compiling geologic data interpreted from aerial photographs. Although the advantages and limitations of many photogrammetric instruments have only recently received wide attention, the desirability of reliable compilation of photogeologic information has long been recognized, and the Dutch study of New Guinea in 1935 was followed by a test of the A-6 precision stereoplotting instrument for photogeologic purposes (Helbling, 1949); plotting with stereoplotting instruments was found to have many advantages. But use of instruments lagged until after World War II. Within the past few years, however, stereoplotting instruments, such as the Kelsh and multiplex, have come into increasing use, both for interpreting and for plotting geologic data.

In a general way it may be said that use of aerial photographs in geologic study in the United States has evolved through the following stages: (1) emphasis on uncontrolled mosaics and use of single views for interpretation and plotting; (2) use of stereoscopic pairs of prints for interpretation together with simple procedures for plotting these data; (3) use of stereoscopic pairs of prints for interpretation together with rectification of positioning these data with simple instruments such as the radial planimetric plotter; and (4) use of stereoscopic pairs of diapositive glass plates in precision photogrammetric instruments for interpretation and plotting. In addition, with increased use of aerial photographs in recent years, new instruments have been devised especially for geologic study, and new avenues of research are actively being pursued in interpretation studies.

The recent use of precision photogrammetric instruments in photogeologic study presages a closer integration of photogeologic and field studies. Because of the reliability of geologic measurements and positioning with precision stereoplotting equipment, a greater amount of photogeologic data may be expected to be incorporated in the geologic maps of the future.

REFERENCES

Bean, R. K., and Thompson, M. M. *Photogram. Engng.* 23:170, 1957.

Condon, W. H. Interpretation of aerial photographs in mapping Cenozoic volcanic rocks of a part of Revillagigedo Island, near Ketchikan, Alaska. Paper presented at 23d annual meeting of the American Society of Photogrammetry, Washington, D.C., March, 1957. Discussion of interpretive process is based on this paper.

Fischer, W. A. Photogeologic Studies of Arctic Alaska and Other Areas. *Selected Papers on Photogeology and Photointerpretation.* Washington, D.C.: Research and Development Board, 1953. Pp. 207–209.

Fischer, W. A. Some new photogeologic techniques, their advantages and limitations. Oral presentation at the meeting of the Northwest Mining Assn., Spokane, Wash., March, 1957.

Guild, P. W. *Mining Engng.* 9:203, 1957.

Helbling, R. *Studies in Photogeology.* N. E. Odell, trans. Zurich: Federal Institute of Technology, 1949. Chap. 10.

Hemphill, W. R. Application of photogrammetry to isopach mapping and structure contouring. Paper presented at the annual meeting of the American Association of Petroleum Geologists, St. Louis, Mo., April, 1957. Discussion of the use of photogrammetry in isopach mapping in Monument Valley, Ariz., is based on this paper.

Lake, M. C. *Engng. Mining J.* 151:79, 1950.

Mostofi, B., and Gansser, A. *Oil Gas J.* 55:78, 1957.

Radlinski, W. A. *Photogram. Engng.* 18:591, 1952.

Ray, R. G. A look at color aerial photography in geologic study. Oral presentation at the meeting of the Northwest Mining Association, Spokane, Wash., March, 1957.

SUGGESTED READINGS FOR PART FOUR, SECTION A

Allison, G. W., and Breadon, R. E. *Timber Volume Estimates from Aerial Photographs.* British Columbia Forest Service. Forest Survey Note 5, 1960.

Austin, R. M. Aerial war against cereal crop disease. *Industrial Photography* 7:38–40, 1958.

Avery, T. E. Engineering Applications. *Interpretation of Aerial Photographs.* Minneapolis: Burgess, 1962.

Avery, T. E. Measuring land use changes on USDA photographs. *Photogrammetric Engineering* 31:620–624, 1965.

Axelson, H. Effect of photo scale on the use of aerial photographs in Swedish forestry. *Norrlauds Skogsvardsforbunds Tidskriff* 252–292, 1956.

Belecher, J. Determination of soil conditions from aerial photographs. *Photogrammetric Engineering* 14:482–488, 1948.

Brunnschweiler, D. H. Seasonal changes of the agricultural pattern: A study in comparative airphoto interpretation. *Photogrammetric Engineering* 23:131–139, 1957.

Cissna, V. J. Photogrammetry and comprehensive city planning for the small community. *Photogrammetric Engineering* 29:681–684, 1963.

Colwell, R. N. Aerial photography—a valuable sensor for the scientist. *American Scientist* 52:17–49, 1964.

Cross, B. Aerial photos: New weapon against pollution. *Chemical Engineering* 69 (7):42–43, April 2, 1962.

Frost, R. E. *Evaluation of Soils and Permafrost Conditions in the Territory of Alaska by Means of Aerial Photography.* Lafayette, Ind.: Purdue Engineering Experimental Station, Vols. 1 and 2, 1950.

Herrington, R. B., and Tocher, S. R. *Aerial Photo Techniques for a Recreation Inventory of Mountain Lakes and Streams.* U.S. Forest Service Res. Paper INT–37, 1967.

Kelly, M. G., and Conrod, A. Aerial Photographic Studies of Shallow Water Benthic Ecology. *Remote Sensing in Ecology.* Athens, Ga.: University of Georgia Press, 1969.

Kohn, C. F. The use of aerial photographs in the geographic analysis of rural settlements. *Photogrammetric Engineering* 17:759–771, 1951.

Mumbower, L., and Donoghue, J. Urban poverty study. *Photogrammetric Engineering* 33:610–618, 1967.

Richter, D. M. Sequential urban change. *Photogrammetric Engineering* 35:764–770, 1969.

Rosenfeld, A. Automatic recognition of basic terrain types from aerial photographs. *Photogrammetric Engineering* 28:115–132, 1962.

Strandberg, C. H. Water quality analysis. *Photogrammetric Engineering* 32:234–248, 1966.

Wray, J. R. Photo Interpretation in Urban Area Analysis. *Manual of Photographic Interpretation.* Washington, D.C.: American Society of Photogrammetry, 1960.

Section B
Suborbital air photography, color

Even before *World War II, experiments were being conducted using color aerial photographs. In general, these met with disappointing results. It was not until the late 1950s that improved technology renewed interest in air photographs generated in color. This reawakening was due to three technological achievements:*

1. New, faster emulsions could produce an improved color response; these emulsions were then coated on plastic base materials which are dimensionally stable. This resulted in polyemulsion-layered color films, as stable chemically as the single panchromatic emulsion used in older black and white films.

2. Use of the color film required production of special processing equipment and chemicals for development of the polyemulsion layers. Careful processing and stability of the emulsions give these new films metric qualities as good as, or better than, the older black and white films.

3. Neither the new film nor new processing techniques would have been of any value without corresponding innovations in camera systems which made it possible to expose the film properly. New wide-angle-lens cameras were developed; experiments with lenses and their component materials produced the color-corrected camera lenses which are used with the color films. Finally, breakthroughs in the type and use of in-front-of-the-lens filters further improved the response of the films.

In 1964 the American Society of Photogrammetry recognized the increasing use and interest in color aerial photography by designating a former working group as the technical committee on color.

15-Lenses and Techniques for Aerial Color

MANFRED DUDDEK

Color aerial photography forces us to pay greater attention to the factors that should be familiar to us from black and white photography, but which we have come to disregard—to some extent with good justification. A more profound knowledge of the various influences that may have an adverse effect on the results achieved in color photography constitutes the real difference between the techniques of black and white and color photography.

Such familiarity with these influences and their interrelationship are of twofold importance in color photography: first, in the successful planning and execution of an aerial photographic mission; second, in the assessment of the results obtained and, where necessary, in the taking of the proper corrective steps for subsequent assignments.

It must be said at this early stage that the possibilities of influencing the results are limited; these limitations are based primarily on atmospheric effects. Moreover, it must be considered that the atmospheric conditions are judged visually, thus introducing great difficulties into the entire system of color photography, mainly with regard to the reproducibility of a given result.

From *Photogrammetric Engineering* 36:58–62, 1970. Reprinted and edited with permission of Wild Heerbrugg Instruments, Inc., and the American Society of Photogrammetry.

What should be required as the results of a properly executed photographic mission? The requirement should be defined as color photographs whose color composition harmonizes substantially with the natural color perception of our visual apparatus, taking into consideration the physiologic condition of the observer. According to this definition, a picture with a dominant color bothering our color perception—in other words, with a color cast—cannot be called a good result. Of course, this does not apply to false-color films or to films where special color effects or color differentiations were attempted through the use of color filters. This definition of results must be qualified with the statement that a picture with a color cast is often unavoidable because of atmospheric influences. The information content of such a picture can nevertheless still be better than of one which is black and white.

In the following, a number of interesting points will be treated that have an influence on the results and must therefore be noted.

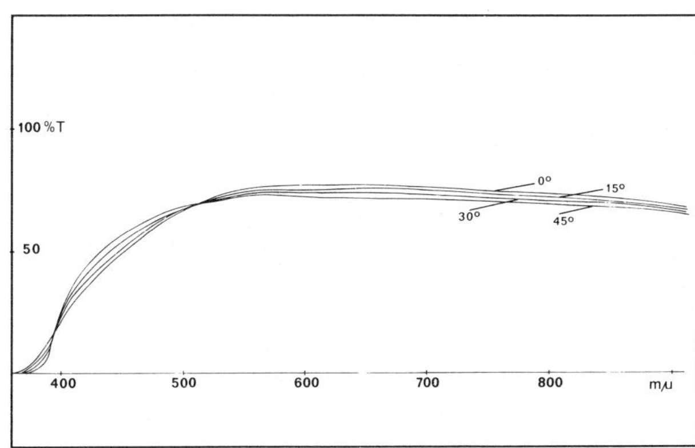

Figure 1 Spectral transmission of the 6-inch Aviogon and Universal Aviogon lenses.

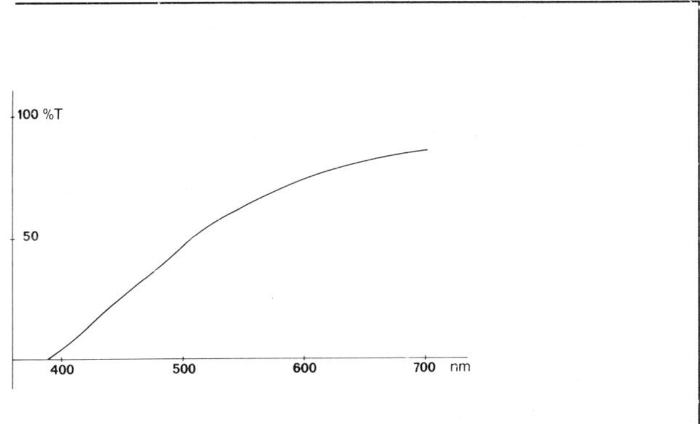

Figure 2 Spectral transmission of the 88-mm Super Aviogon lens.

Aerial Cameras

Of special interest for color photography are the optical properties of the taking lenses with regard to: (1) chromatic correction; (2) spectral transmission; and (3) light distribution in the plane of the picture.

Chromatic Correction

Aviotar, Aviogon, and Super-Aviogon lenses are chromatically corrected for the spectral range from 450 to 650 nm.[1] The 6-inch Universal Aviogon and the new 3.5-inch Super-Aviogon II of the RC-10 aerial camera are chromatically corrected for the spectral range from 450 to 850 nm. All these lenses are therefore suited for color photography. If films are to be used whose sensitivity range extends into the long-wave region, such as Kodak Ektachrome Infrared Aero Film 8443, the Universal Aviogon and Super-Aviogon II lenses are to be preferred.

Spectral Transmission

The spectral transmission of the Aviogon or Universal Aviogon lenses and of the Super-Aviogon lens can be seen in Figures 1 and 2. These curves indicate that these lens types—like all high-performance lenses of photogrammetric cameras—have little transmission in the short-wave spectral range. This slight filter effect is of advantage also in color photography because the undesirable short-wave portion of the radiation is thereby eliminated.

Light Fall-off

The unavoidable light fall-off of the lens types under discussion is problematic as the physical initial light fall-off, the field angle of the lens, and the atmospheric conditions are interconnected. Consideration should be given to the fact that the atmospheric light scatter counteracts the physical initial light fall-off dependent upon the field angle. In the Aviogon and Super-Aviogon lenses, for example, the light fall-off at $f/8$ corresponds approximately to the cosine of the third power (Figure 3). The often-quoted law that the light fall-off follows the fourth power of the cosine is valid only for lenses with a front aperture diaphragm and does not take into account the design characteristics of modern lenses. To compensate the light fall-off, filters with a vacuum-deposited layer of diminishing density from the center toward the edge are placed in front of the lens.

Because color films have little exposure latitude, and partial density differences are easily noticeable, the lightening effect must be taken into account in the selection of the graded-density filter. The atmospheric light scattering directed towards the lens (aerosol) causes the uneven illumination in the plane of the picture depending on the solar altitude and the field angle of the lens. As the amount of light scatter upward depends on the volume of "aerosol" between the terrain and the lens, some means must be found to evaluate this atmospheric influence. It is understandable that under equal atmospheric conditions the lightening effect at, say, 1000 meters is less than at, say, 4000 meters, and that a graded-density filter which is correct for 1000 meters flying height is too dense for a flying height of 4000 meters. The result of this example would be that the photographs taken from a height of 4000 meters would suffer from noticeable underexposure in the center of the picture.

A possibility to determine the maximum density of the graded density layer in practice has been shown by Duntley (1948). An optical standard atmospheric is a prerequisite there, and the maximum density of the graded density layer is obtained as a function of horizontal visibility and flying height.

In connection with the graded density layer, the spectral composition of the scattered light and its registering on the film are of consequence. The "aerosol" particles are selective in their spectral dispersion, depending on their size. It can be said that the short-wave portion of the scattered light grows in direct proportion with the visual range. In this simplified explanation, the spectral composition of the scattered light that depends on the solar altitude is not taken into consideration. Assuming that in practice color photographs are taken under meteorologic conditions where horizontal visibility exceeds 50 kilometers and the solar altitude ranges from 40° to 50°, then the blue portion in the scattered light will be approximately three times greater than the red portion. As mentioned earlier, it is the upward-directed portion of this scattered light that is responsible for the additional illumination (lightening)

[1] nm = nanometer = 10^{-9} meter.

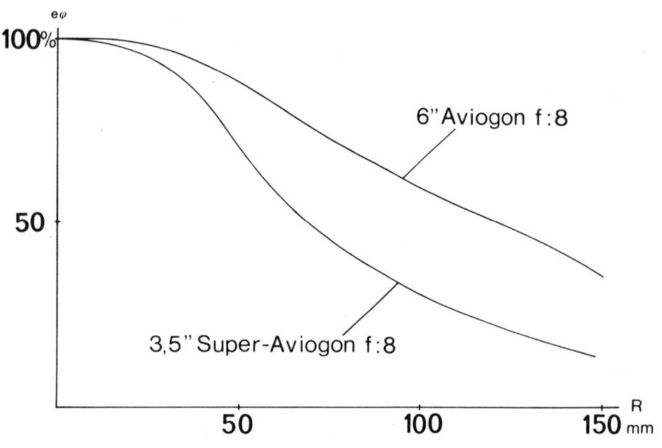

Figure 3 Relative illumination in the negative plane.

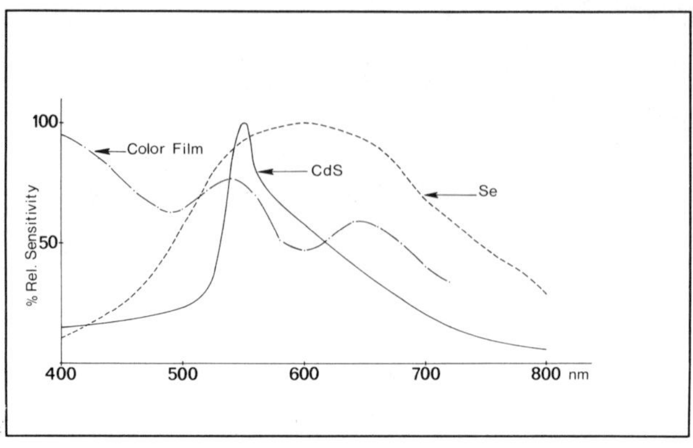

Figure 4 Relative sensitivity of the cadmium sulfide and selenium cells and color film.

in the plane of the picture. Because conventional color films also register the blue range of the spectrum, whereas in false-color films the blue is cut off by means of a Wratten No. 12 filter (500 nm), it is understandable that a graded-density filter with a 2.2× anti-vignetting coating yields a more satisfactory compensation over a greater range of flying heights with false-color film than with conventional color film.

Determination of Exposure

The Kodak Aerial Exposure Computer usually gives good results for color photography, provided that the atmospheric conditions are properly evaluated. The figures given for the speed of Kodak color films are merely index figures for use in combination with the Kodak Aerial Exposure Computer. Should an exposure meter be used, the Aerial Exposure Computer can be useful to determine the USASI rating of a certain type of film. In using an exposure meter one should be aware that there must be compatibility between the spectral sensitivity of the respective film emulsion and the meter measuring cell.

Figure 4 shows that the difference in spectral sensitivity between cadmium sulfide cells and selenium cells is considerable. A comparison of the spectral sensitivity characteristics of the meter cells and of the color film indicates that selenium cells—as used, for instance, in the Weston Master exposure meter—are better suited than CdS cells. The CdS cells, however, have a much greater total sensitivity: it is therefore possible to place a filter combination before the meter cell to reduce the greater sensitivity peak between 525 and 580 nm. This results in a more even spectral sensitivity over a wider range. The USASI value must of course be adjusted in accordance with the density of the filters used.

It is interesting to note that these cells (CdS) have a relatively low sensitivity between the range of 400 and 500 nm, which is the range where normally scattered light is most intensive under normal conditions.

As mentioned previously, the composition of the atmosphere, the sun angle, and the flying height all contribute extensively to the amount of scattered light. For this reason, it is suggested that the USASI rating of normal color film be increased with the flying height. Experience has shown that it is best to base the USASI rating on a 1000 meter flying height and to increase it by 30 percent for each additional 1000 meters.[2]

Atmospheric Conditions and Filter

The question of choice of proper filter in front of the taking lens arises if a normal color film such as Kodak 2448 is to be developed as a positive transparency. In this instance the best possible color balance is required to utilize the inherent color distribution of the film.

Color films are sensitized for a mean daylight of 5400°K. In aerial photography, however, the composition of daylight changes depending on atmospheric conditions and the sun angle. Three factors must be taken into consideration to compensate for this:

What is the spectral composition of the rays falling on the subject?

What is the amount of spectral reflectance of the subject?

What influences affect the rays between the objective and the subject?

It is known that the subject or terrain is illuminated by a mixture of sunlight and skylight. The skylight is scattered by particles in the atmosphere (aerosol). On a clear day at a horizontal visibility of 100 kilometers and 40° solar altitude, the ratio of sunlight is four to one, this rapidly changes to three to one at 40 kilometers visibility. At the same time, the spectral composition of the entire illumination changes. A decreasing solar altitude will also change the spectral composition of the entire illumination. In short, the proportion of the predominantly red (direct sunlight) and blue (skylight) changes continuously. For the sake of simplicity, let us ignore the influence of the spectral reflectance effect (the second factor) on the subject. There remains the influence of the haze layer between the subject and the lens which, depending on its thickness (flying height) and density (particle size) superimposes principally short-wave light on the image-forming

[2] The [*Photogrammetric Engineering*] Reviewer understands this to mean if the USASI is 100 at 1000 meters, it becomes USASI 130 at 2000 meters; a USASI of 160 at 3000 meters; a USASI of 190 at 4000 meters, and at 5000 meters a USASI of 220. The Reviewer also believes that this formula is not intended to be carried through to infinity and that the probable cutoff point is around 5000 meters of flight altitude. *Reviewer's comment.*

rays. In addition, the spectral composition of the image-forming rays changes with the atmospheric influence (haze) and depends on the angular field of the objective.

This short analysis of the complex conditions of illumination encountered in aerial photography shows that we must be rather tolerant with respect to the color balance we can expect in the picture and that we cannot expect wonders from the use of filters during the picture-taking, especially at high altitudes. In practice, the only remaining possibility of reducing the increasing amount of short-wave encountered with greater flying heights is through the use of light yellow filters (CC 05, CC 10, CC 15) (see Determination of Exposure above).

Choice of Film

From the problems listed previously, one can understand that the reversal process is beset with numerous problems during the picture-taking stage. It is recommended extensively where film diapositives are to be used directly for photogrammetric compilation.

Wider possibilities are offered, however, through the use of reversal film developed and processed to the negative, such as Kodak 8442 or 2448. Compared with Ektacolor negative film, it has the advantage that it is higher in contrast and therefore more suitable for flying at higher altitudes. Experiments with Ektacolor film have shown that the maximum flying height with respect to color contrast is around 1500 to 2000 meters. The development to the negative has many advantages, such as skipping color balance filtering of the picture. The color balance is adjusted in the darkroom following the copying process. Color copies and enlargements are easy to produce. The Wild VG-1 Enlarger and E-4 Rectifier can now be delivered with a color head for the additive filter process. Existing instruments can be retrofitted. This accessory is easily exchangeable with the standard lamp housing.

Black and white prints and enlargements can either be produced on standard bromide or panalure paper. Prints made on panalure paper have a somewhat extended tonal contrast over comparable prints made on normal bromide paper.

With reference to Kodak color films developed to the negative, one has the choice between Ektachrome Aero Film 8442 and Ektachrome MS Aerographic Film 2448. Both types of film can be developed in the normal or modified C-22 process. (Kodak recommends only type 2448 and the modified C-22 process.) The advantages of 8442 films are: four times higher speed (Aerial Exposure Index 25, approximately ASA 125); and somewhat harder gradation of the emulsion, therefore more suitable for higher altitudes. The advantages of the 2448 compared to the 8442 are: Estar base, 35 percent higher resolution measured at 1000:1 and 43 percent at 1.6:1. The somewhat lower speed of the 2448 (Aerial Exposure Index 6, approximately USASI 30) is still sufficient for most practical flying conditions. In addition, one can lengthen the first development time by approximately 30 percent and gain about one stop without noticeably losing quality. The gradation does not increase with longer developing time; that is to say, with color one cannot partially compensate for loss in contrast due to high altitude through longer developing as is the case with black and white film.

For flying from low to medium altitudes, the 2448 has more advantages than the 8442; the latter should be given preference for higher altitudes.

Ektachrome Infrared Aero Film 8443 (False Color Film) holds a unique position among the color films. The *original* colors of the terrain are reproduced falsely on this reversal color film. If the original radiation (with the 500 nm filter) was, for instance, blue, it will be registered as black, green as blue, red as green, green and infrared as magenta, and infrared as red.

From the above remarks it becomes evident that this film is especially suited for forestry, agriculture, resources, and irrigation purposes. Technically, the film is easy to handle because the 500 nm filter eliminates the atmospheric influences to a large extent. A certain disadvantage is that processing to a negative in the C-22 process produces unacceptable results. The contract of the cyan layer (the layer which is infrared sensitive) becomes very flat due to reduced differentiation. The film is suitable for use with other filters or filter combinations, in addition to the 500 nm filter for studies of specific terrain features.

REFERENCES

Duntley, S. Q. The reduction of apparent contrast by the atmosphere. *J. Opt. Soc. Am.* 38, No. 2, Feb. 1948.

Duntley, S. Q. The visibility of distant objects. *J. Opt. Soc. Am.* 38, No. 3, March 1948.

COLOR IS difficult to define in qualitative terms. For example, what words could be used to explain the color red to a blind person? Color is a psychophysiologic response that each person must experience for himself. Yet color is one of the important recognition characteristics by which we identify phenomena in our surrounding world. It is obviously an advantage in remote sensing if images can be displayed in hues that approach their true color or color as it is humanly perceived in the natural environment. Color photography presents the observer with a reasonably faithful rendition of the color world. But color has other advantages over a black and white format. The human eye can distinguish thousands of variations in color and purity but at most only a few score variations in gray tones. Interpretation becomes more meaningful, more precise, and easier when working with color aerial photographs. For hydrologic studies, color film has two distinct advantages: its ability to record reflected light which has penetrated through deep layers of water, and its ability to discriminate the land/water interface and other indicators such as vegetation.

16-Applications of Color Aerial Photography to Water Resources Studies

WILLIAM J. SCHNEIDER
MILTON C. KOLIPINSKI

AERIAL PHOTOGRAPHY is a valuable and recognized tool in the investigations of the resources of the United States. Panchromatic photography has been used for many years in the studies for both photogrammetric and photointerpretive measurements of such characteristics as stream patterns, drainage areas, and other physical features. During the past decade, however, the horizon has expanded tremendously with the advent of color photography. The development of color film of reliable quality, both in stability and color fidelity, have made possible advances in the use of aerial photography not heretofore possible for water resources studies.

Color aerial photography has proven its worth in numerous applications. Although the experiences are recent and perhaps limited as compared to other applications, it is obvious that color aerial photography is indeed a valuable tool to the expert in this field.

Color aerial photography has several distinct advantages in applications to water resources studies. Perhaps the most fundamental advantage is that color photography presents a picture of the terrain and the water features approaching their true color. Water is blue, trees are green, and so on. This is, of course, a distinct advantage in interpretation, especially since most water resource investigators have little or no experience in the use of aerial photography as an interpretive medium for water resources studies. Color photography also offers the water resource expert a synoptic view of his problem. While this is also true of all aerial photography, color photography is especially useful in that interpretation capabilities are enhanced and intensified.

Both color and panchromatic photographic films cover the visible spectrum and both record in approximately the same region from the blue through red. With color photography, however, the user does not have to interpret between often barely discernible gray tones, but has for his interpretation criteria a range of color shades and tones that are recorded on the photography. Because of the affinity to the natural color, interpretation thus becomes easier and more meaningful.

For water resources studies, color film also has a distinct advantage in its ability to penetrate through water. This capability is particularly important in studies of lakes and open water areas where knowledge of subsurface conditions such as types and distributions of vegetation and sediment are important factors in the understanding of the hydrologic regimen.

Perhaps, though, the most significant identification key in the use of color photography for water resources studies is the discrimination of subtle features possible

From *New Horizons in Color Aerial Photography.* Falls Church, Va.: American Society of Photogrammetry, 1969. Pp. 257–262. Reprinted and edited with permission of the authors and the American Society of Photogrammetry.

through the use of various types of color films and through the use of selective filters. Color infrared film is an especially valuable tool where vegetation is closely allied to the water regimen. The sharp discrimination of vegetation on this film has proven useful in numerous studies. In many studies, this close relationship between the plant ecologic communities, which can be recorded very distinctly on color infrared film, and the water regimen has enabled water resource experts to develop sound hypotheses and arrive at significant conclusions in water resources studies. In fact, in numerous studies in the United States, the capability has provided the key by which data collected at a point has been extrapolated and interpreted on a regional scale to provide knowledge of the water resources not obtainable through standard techniques.

Meaningful interpretation, however, depends upon adequate ground control data. Usually such data must be obtained in the field at the time of the photography. However, where vegetative indicators are used to infer hydrologic conditions, the collection of the ground control data may be spanned over a period of several weeks or longer, depending upon the stability of both the plant communities and the hydrologic regimen. Aerial photography cannot be used as a substitute for adequate ground data; it is, however, an extremely valuable tool for the extension of hydrologic data. In water resources studies, the most successful applications of aerial photography have been in projects where the photography was used to expand and supplement the normal data collection processes rather than substitute for them. In these instances, interpretation of the photography has been made by the scientists and engineers who are thoroughly familiar with the water resources problems of the area and have available to them sufficient hydrologic data to make interpretation meaningful. Without such data, the usefulness of color photography, or any photography, loses its objectivity and is greatly diminished.

The degree of interpretation, of course, is related to the scale of the photography. To date, only gross inferences are possible from high-altitude photography from aircraft or spacecraft. For water resources studies, photography at scales of 1:8000 to 1:12,000 are generally large enough to permit detailed interpretations of water features for most water resources studies. However, where extremely detailed studies of small areas are involved, scales as large as 1:1000 may be necessary. In general, though, the type of data necessary for quantitative water resource appraisals can be obtained from photography at scales of 1:15,000 or larger.

Until recently, color aerial photography was used mainly for extension of ground-control data by photointerpretation, for overall appraisal of the hydrologic regimen, and for general location of specific hydrologic phenomena. Few, if any, photogrammetric measurements were made. However, during the present decade, manufacturers of photogrammetric products and photogrammetric instruments have introduced color film of stable metric quality and moderately priced direct-viewing stereoscopic compilation instruments. Although scientists have taken but little advantage of applying photogrammetry to water resources studies, the potential for broad application has been demonstrated in several studies. For example, changes in the ecology of the Florida Everglades related to the water regimen were determined photogrammetrically within 2 percent, and plant communities were delineated from color aerial photography for use in the preparation of experimental orthophoto maps.

The successful use of color aerial photography depends upon the proper understanding of the problem and the knowledge of the capabilities and limitations of the aerial photography toward the solution of the problem. Rarely, if ever, has an indiscriminate use of aerial photography yielded benefits to justify the expenditure. Rather, carefully planned applications have provided knowledge and understanding of water resources situations that by conventional methods could be obtained only at greatly increased costs, and in some cases would have been unobtainable at any cost.

Color aerial photography has been used in such diversified studies as the mapping of offshore springs, inventories of coastal estuaries, determination of pollution, effects of urbanization, and determination of water availability and formulation of water management programs. Some of the better-known areas where these studies have been conducted are Great Salt Lake, the prairie potholes of North Dakota, Lake Erie, the Delmarva Peninsula along the Atlantic Coast, the Potomac River, and the Florida Everglades. In these studies, as in all water resources studies where color aerial photography has been used successfully, photoidentification keys were correlated with known hydrologic data to provide the means of extending the data on regional bases.

In the studies of offshore springs and general offshore discharge of fresh water, several photoidentification keys are clearly registered on color aerial photography. Where springs discharge water with sufficient velocity, turbidity is generally present and easily observed on the photography. At low tide, the continued seaward flow of water through rivulets along a sloping beach also gives photographic evidence of ground-water seepage. In areas of offshore shallow waters, color aerial photography can be used to distinguish areas of fresh-water outflow through the differentiation of attached algae and sea grasses. *Diplanthera*, a grass that prefers only moderate salinity and, therefore, grows only in coastal waters where the salt water is diluted by fresh-water discharge, can be differentiated on large-scale color aerial photography from *Thalassia*, a grass which thrives in the high salinity such as that of normal sea water.[1]

Grasses also serve as indicators of salinity in studies of the prairie potholes of North Dakota. These potholes are depressions of various sizes and depths which were formed when retreating glaciers left huge blocks of ice in their wakes. Today these depressions contain water which ranges in salinity from near zero (fresh water) to more than three times that of sea water. Some preliminary interpretations of color aerial photography of these potholes indicate that shallow fresh-water potholes can

[1] The figures accompanying this article appear in the color insert.

be identified in the photography by the presence of several distinctive grasses. Also, highly saline potholes are easily identified by a complete lack of aquatic vegetation and the presence of easily distinguished white salt deposits along the shorelines of the ponds.

Ground-water inflow to some of the potholes can be identified in the photography by the presence of trees along the shorelines of the potholes; springs which feed some potholes are readily distinguished by clumps of trees in an otherwise grassland community. In some potholes, a salinity gradient across the potholes can be inferred from the presence of fresh-water indicators on one side of the potholes and the presence of salt deposits on the opposite shoreline.

In the water resources studies of the Great Salt Lake, color aerial photography has been used to identify circulation patterns in the lake movement and deposition of sediments in the lake, and areas of ground-water discharge of the lake. Of particular interest in the photographs of the lake are the bioherms growing on the lake bottom. These bioherms are coral-like reefs several feet high formed by minute living organisms such as corals and algae. Their occurrence in formations similar in size and shape to desiccation cracks formed in soils suggests that at one time these areas of lake bottom were exposed, dried out, and cracked, and that following subsequent inundation, these organisms occupied the cracks and started growing.

Color aerial photography is also useful in identifying some pollutants and the hydrologic environment into which they are being introduced. Streams polluted by acid mine drainage can be identified on color photographs by characteristic deposits of bright orange ferric oxide commonly called "yellow-boy." The acidity of the stream can also be inferred from the photography. Water in the reaches of the stream above the deposits will be strongly acid and hold iron in suspension. However, in the vicinity of the deposits, the water will be less acid, with a pH of about 6.0, generally as a result of neutralization from the introduction of alkaline waters either from tributary streams or from solution of limestone or other alkaline bedrock beneath the local reach of the stream. Many other stream pollutants can be readily identified on color photographs: paper mill wastes, industrial discharges, sewage effluent, and many others.

Perhaps more important from the water resources aspect is the fate of these pollutants rather than the mere identification of them. Color aerial photography is excellent for determining dispersion and mixing characteristics of the body of water into which the pollutants are being introduced. In some cases, water temperatures can be inferred from the layering of the pollutants, as the warmer water overrides the cooler water. Such hydrologic observations are especially important in selection of sites for water quality monitoring of the water resource, so that representative sites may be selected rather than sites biased by local conditions.

In some areas the use of color aerial photography is a continuing part of the water resources study. Perhaps the largest continuing use of color aerial photography for water resources studies is in south Florida and adjacent coastal regions. Continued use of color aerial photography over the area since 1962 has resulted in advances in knowledge of the water resources of this region that would not have been possible through the standard hydrologic data-collection techniques. Prior to 1962, only limited synoptic observations of hydrologic data were available for appraisal of the water resources of the Everglades. At that time, despite intensive efforts, little was known of the overall water regimen of the area. A lack of topographic maps, a flat terrain, and a hostile prairie-marsh environment had severely inhibited ground travel and the subsequent collection of hydrologic data. Only through the use of color aerial photography have the regional characteristics and the complex interrelations between the hydrology and the ecology in the Everglades been understood.

The first color aerial photography in the Everglades was obtained in 1962 over a 40-mile transect as a pilot endeavor. Immediately the value of color photography became apparent. Limited ground data showed clearly that the readily identifiable hydrologic and ecologic features in the photographs were intimately related to the water regimen of the entire Everglades. As a result, additional color aerial photographs were obtained over the entire area of 2400 square miles. Since then, supplemental color aerial photography has been obtained.

In these photographs, observations on the shapes and locations of the tree islands (heads and hammocks) in the glades show clearly both the patterns and relative velocities of flow. In the so-called Shark River Slough—the main drainage course through the lower Everglades—the tree islands show an elongation with the major axis parallel to the direction of flow. The adjacent sawgrass communities surrounding the tree islands also show orientation indicative of velocity movement of the water which seasonally inundates the Everglades, in its movement toward the sea. The wet prairie or open-water areas amid the sawgrass marshes show interconnections typical of braided stream channels, leading one to conclude that the Everglades is indeed a river flowing through south Florida on its way to the sea.

In the Everglades, relative ground elevations can also be determined from the photography. The lowest ground elevations occur in the wet prairies where the open water is occupied by sparsely scattered aquatic grasses, sedges, and rushes. At elevations only a few inches higher than the wet prairies are found the sawgrass communities which develop on the peat and marl soils where the accumulation of peat and organic detritus is augmented by the decayed vegetation. From the aerial photographs and from ground observations, one can conclude that distribution of plants and the hydrologic regimen are so closely related that the difference of merely inches in elevation governs the succession of ecologic communities whose extent can be measured in acres or even in square miles. Recently, changes in the ecologic communities reflecting hydrologic changes have been measured photogrammetrically within an accuracy of 2 percent from color aerial photography.

In the Everglades, ground control data indicate that soil depths can be interpreted from the color photography. In the southern part of the Everglades, soils range from 3 to 5 feet below the sawgrass and open-water areas. However, beneath the tree islands, soil depths may vary as deep as 14 feet in the limestone bedrock as a result of solution of the underlying bedrock by acids formed by the decaying plant material.

In the coastal regions adjacent to the Everglades, the inland penetration of salt water is clearly evident on color photography. The sharp line of demarcation between the fresh-water area of the Everglades and the ecotone is marked by the presence of dwarf red mangrove (*Rhizophora mangle* L.) in the ecotone. Also present in this transitional brackish area are halophitic grasses such as salt-grass (*Distichlis spicata* L.) and blackrush (*Juncus Roemerianus* Scheele). Although not apparent in the photographs, hydrologic data indicate that this is an extremely dynamic zone in which the zonation of the grasses is severely dependent upon the delicate interrelated effects of inundation by tides and fresh-water drainage from the interior of the Everglades.

In many places the zonation or distribution of plants in the vicinity of an intertidal zone serve to indicate the mean high-water level as well as other levels that may be of interest. A specific zonation of tropical trees occurs along the underdeveloped shorelines of Biscayne Bay and along the other south Florida coasts. The typical zonation pattern consists of red mangrove (*Rhizophora mangle* L.) and black mangrove (*Avicennia nitida* Jacq.) through the intertidal zone and in the upper tidal reaches, and buttonwood (*Conocarous erectus* L.) in the supratidal region. The delineation and mapping of these species can be valuable to those concerned in planning and in settling problems relating to the legal definitions of shorelines or mean high-water lines that concern ownership or the seaward limit allowable for landfill and bulkhead lines.

As indicated by the examples cited here, color aerial photography has proved its value in water resources studies. The benefits derived in numerous applications have shown clearly that if properly used, it is a valuable tool for the water resource expert. Its full potential in water resources studies has not been explored. Properly utilized as a technique to augment rather than supplant the conventional methods of water resources investigations, color aerial photography both by itself and in combination with other types of photography and imagery will permit broad interpretations based on limited synoptic hydrologic data. Indications are that we have only begun to scratch the surface of this potential in water resources studies.

ONLY IN *the last few years have broad segments of the population concerned themselves with the vexing hazards associated with air and water pollution. Individual man is on earth an instant in time and his main concern is with the pollution problems that directly affect him or his immediate family. Most of us miss the subtle build-up of pollutants, both atmospheric and hydrologic. It is estimated that atmospheric sulfur dioxide is increasing in the United States at the rate of 6–7 percent annually. Yet our main sensors, our eyes, cannot detect sulfur dioxide in the air. It takes special sensing devices such as a correlation spectrometer to record this corrosive gas. We know very little about the full extent of pollution, the rate at which it is increasing, dissemination vectors, relative concentrations, and local, regional, and international distribution. Perhaps most important we do not yet understand the insidious effects of the long-term increase of environmental pollution.*

One method of measuring the temporal results of pollution is to record the changes that occur in the health, vigor, reproduction, and distribution of long-lived members of the earth's ecosystem. Trees make an excellent experimental subject for this type of study. They have a long life span and we know a good deal about their ecologic requirements. In addition, their image signature can be recorded with a number of sensors. Further, as health or vigor of the plant changes, observable variations in this signature also change. If these changes are the result of pollution, which must be measured by "ground truth" examination, we have a measure of its temporal and spatial effects on vegetation.

17-Large-Scale Photo Assessment of Smog-Damaged Pines

ROBERT C. HELLER

THE GROWING concern over pollutants in our environment has spawned a growing list of scientific studies. The economic effects of pollutants is now being studied by the U.S. Public Health Service. Biological organisms affected include man, fish, wildlife, and vegetation. To date, few studies have been made of the effects of air pollution on longer-lived vegetation, such as trees. Most investigations have reported effects on sensitive agricultural crops, such as citrus groves, and the resulting lower yields. And they have been concerned with the effects on people, and the resulting discomfort and inconvenience.

The Remote Sensing Project of the Pacific Southwest Forest and Range Experiment Station, at Berkeley, California, was recently asked to participate in a study to evaluate injury to trees in the mountains near Los Angeles. The Station's experience in assessing tree injury from insects and diseases by airborne sensing techniques appeared particularly applicable to the problem.

From *New Horizons in Color Aerial Photography.* Falls Church, Va.: American Society of Photogrammetry, 1969. Pp. 85–98. Reprinted and edited with permission of the author and the American Society of Photogrammetry.

Studies by Miller, Parmeter, Taylor, and Cardiff (1963), Parmeter, Bega, and Neff (1962), and Stark et al. (1968) have suggested that airborne oxidants can cause decline and mortality in ponderosa pine (***Pinus** ponderosa* Laws). In controlled studies of this species, Miller (1968) introduced oxidants identical to smog (nitrous oxide and ozone principally) in Los Angeles to seedlings growing in greenhouses. He found that chlorosis and shortened needles that resulted were similar to damage in natural growing pine stands. Middleton (1968) considers the automobile to be the major contributor to smog in Los Angeles. He found that nearly 95 percent is caused by motor vehicles, and only 5 percent is of industrial or miscellaneous origin.

Ponderosa pine is both the major tree species growing in the San Bernardino and San Gabriel mountains in southern California, and also the most susceptible species to smog injury. Smog injury to foliage is a fairly subtle process to discriminate—even on the ground. For example, small yellowish lesions or mottle appear first on the needles. Needles become shorter, change color from green to yellow, and fall off sooner. Consequently, more bare branches than healthy foliage are exposed (Fig. 1). A healthy pine will have up to six years of foliage remaining on each branch; an advanced smog-affected tree may have only foliage from the current growing season.

Some ponderosa pines are more susceptible to oxidant injury than their neighbors. Almost completely foliated green trees can be found standing next to dying and nearly defoliated yellow-red trees. Variation in resistance to oxidants by trees is a factor now being studied by geneticists. Bark beetles also play a part in killing the smog-weakened trees, but only ground examination can determine the cause, insects or disease, of mortality.

Before an economic assessment could be made, we considered a feasibility study necessary to determine (1) the best photographic sensor, and (2) the photographic scales needed to detect smog injury and to assess tree damage accurately. Detection of smog injury, by our definition, means to identify that ponderosa pine is present and is being affected; whereas assessment (or evaluation) means to compare accurately ground conditions of all levels of smog injury by photographic interpretations. Comparison is the more difficult task because subtle changes are lost with small scale and poorer resolution. Only photographic systems were tried because they give better resolution than other imaging systems.

Four study areas were established on the Angeles and San Bernardino National Forests (Fig. 2). Elevations ranged from 5000 to 6000 feet, and advanced

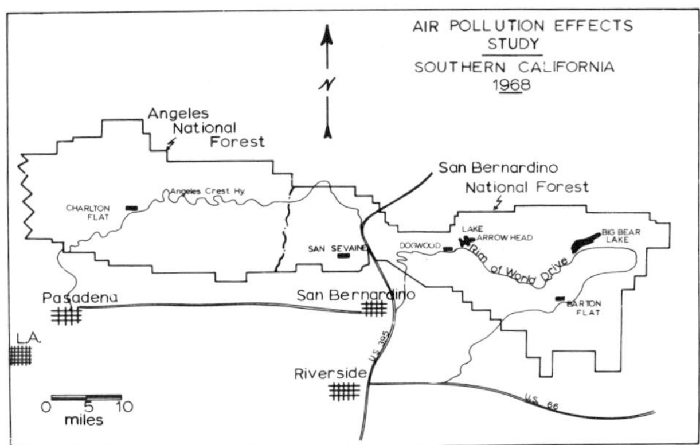

Figure 2 Four study sites located on the Angeles and San Bernardino National Forests, near metropolitan Los Angeles, California.

oxidant injury varied from 15 to 50 percent of the study trees. The remaining study trees at each location were either not damaged or had an intermediate level of damage.

Figure 1 (Left) Advanced smog injury shows only shortened current foliage on the ends of branches. This tree probably will not survive another year. (Right) Intermediate smog injury indicates sparse foliage on two center ponderosa pines. Note bare branches in midcrown of left-center tree, and apparently healthy pine in right foreground.

Methods

Ground Procedures

At each study site, rectangular plots—each about 100 feet wide by 660 feet long—were set up. Large white panel markers were placed at each corner of the rectangle to help in the orientation during photographic flights and to establish photographic scale accurately on the photos. At least 50 ponderosa pine trees were examined at each study site, in June, August, and December, 1968. The location of each tree was plotted on a plane table.

A forest pathologist categorized each tree, by vigor class: (1) advanced damage, (2) intermediate damage, or (3) healthy or no damage.

To quantify smog damage to each tree, number ratings were given to each of the damage characteristics which make up the human subjective judgment: (a) the number of years (0 to 6) that needles were retained on branches; (b) needle condition: ranging from green (4) through chlorotic mottle (2) to yellow necrosis (0); (c) needle length: average (1) to shorter than average (0); (d) branch mortality: average (1) to greater than average (0); (e) foliage color: assigned by viewing sun-illuminated crown with Munsell cards; and (f) negative (1) or positive (0) evidence of bark beetle attack.

In using the scoring system, we gave higher numbers to pines that showed healthy characteristics. For example, a healthy tree having 5 years of needle retention scored *5*, green needles *4*, average needle length *1*, average branch mortality *1*, 5 green-yellow hue (5GY) Munsell notation *10*, and no evidence of bark beetle *1*—for a total of *22*. A smog-injured tree might be rated: 2 years of needle retention for *2*, chlorotic needle mottle *2*, shorter than average needles *0*, above average branch mortality *0*, 10 yellow (10Y) Munsell notation *5*, and

Table 1 Films, Filters, and Scales Used in Feasibility Study to Assess Oxidant Injury to Pondersosa Pine in Southern California

Film and Filter	Scales				
	1:1584	1:3960	1:7920	1:16,000	1:32,000
		June, 1968			
Anscochrome D/200 HF3-HF4 haze filters Didymium filter[a]	×	×	×	×	×
Ektachrome Infrared (Type 8443) #12, minus blue filter	×	×	×	×	×
Plus X Aerographic #25 (A) filter	×	×	—	—	—
Aerographic Infrared #89 b filter	×	×	—	—	—
		August, 1968			
Anscochrome D/200 HF3-HF4 haze filters Didymium filter[a]	×	—	×	—	—
		December, 1968			
Anscochrome D/200 HF3-HF4 haze filters Didymium filter[a]	×	—	×	—	—

[a] Only at 1:1584 scale.

no evidence of bark beetle *1*—for a total of *10*. This kind of scoring permits discriminant analysis procedures to place the tree into appropriate damage classes. Advanced decline might be 0–8, intermediate 8–14, and healthy 15+.

Foliage samples were also clipped from three trees in each damage class for a total of nine trees. Within each tree, samples were obtained from the upper, mid, and lower crowns. These samples were packed in refrigerated containers and shipped by air to the University of Michigan, where the reflectance spectrum of each damage class from the ultraviolet (0.38 μ) to the near infrared (2.7 μ) was obtained. The reflectance curves indicate likely portions of the spectrum to exploit by selective filtration.

Aerial Procedures

From earlier photographic studies on insect and disease damage (Roth, Heller, and Stegall, 1963; Croxton, 1967; Heller, Aldrich, and Bailey, 1959; Heller, 1968), we knew that certain films and filters showed more promise than others. When the information is coupled with that obtained from the spectrophotometric curves, we arrived at a combination of film, filter, and scales (Table 1).

In addition to photographing sites in June, we repeated the best film, filter, and scale combinations in August and December to determine seasonal effects—both from the standpoint of smog injury and the phenologic development.

All photographs were exposed through identical 70 mm cameras which were triggered simultaneously (Fig. 3)[1]. In December, a K-17 12-inch focal length camera (9- × 9-inch format) was used to obtain larger area coverage at the 1:8000 scale.

Photointerpretation

Two photointerpreters examined all photographs. The color films were viewed as transparencies over a balanced light source by using a 2.25× lens stereoscope. Prints made from the panchromatic and infrared films were examined stereoscopically under reflected light.

Both interpreters were given training on affected and healthy trees which were outside the study areas. They were also shown how to discriminate ponderosa and Jeffrey pine from other tree species, including sugar pine (*Pinus lambertiana* Dougl.), big-cone Douglas fir (*Pseudotsuga macrocarpa (Mayr)*, white fir (*Abies concolor* Lind. and Gord.), incense cedar (*Libocedrus decurrens* Torr.), and various oak species.

Each interpreter was given clear acetate templates with only the site boundaries shown. He was told to circle and number all ponderosa pine trees on each site on all films at the two largest scales (1:1584 and 1:3960). After identifying the items examined on the ground, such as needle retention, needle condition, needle length, branch mortality, and Munsell hue, value and chroma, he gave them appropriate scores. On scales 1:7920 and smaller, he was told to identify suspected smog-injured pines by a dot on each template. Other characteristics could not be distinguished at these scales because of loss of resolution.

[1] Not reproduced here.

Results

Each interpreter had a possible 3000 pine trees to evaluate when all combinations of film, scale, and filter were examined for the photography taken in June. Only 400 decisions were needed for the August and December interpretations because many of the films and scales were rejected.

Detection of smog-affected trees on small-scale transparencies depends entirely upon identifying color changes in the foliage. All pines in the advanced category and about 10 percent of the intermediate category were faded (or discolored). The distribution of the numbers of trees found on the ground in each vigor class from a total of 200 is shown in Table 2.

Detection

To detect advanced smog injury, we found that the most efficient combination was normal color film exposed at a scale of 1:8000. Accuracies were equally good at the two larger scales (1:7920 and 1:3960) (Table 3). On an operational survey, the smaller scale (1:7920) is more economical because each photo covers four times the area of larger scale (1:3960).

The color transparencies proved slightly more accurate than the false-color film (Ektachrome Infrared Aero). An interpreter related normal color film with the colors of ground vegetation quite readily and did not have to learn what the colors on false-color film represented. Consequently, we chose Anscochrome D/200 as the most useful film.

No attempt was made to discriminate between levels of smog injury on the small-scale transparencies. And a large number of trees (88) were smog-affected but not discolored; they also were not detected at small scales. For this type of evaluation we would need better resolution, requiring larger images and scales.

Table 2 Distribution of Trees by Vigor Class

Foliage Color	Vigor Class: Advanced Damage	Intermediate Damage	Healthy	Total
Faded	54	11	0	65
Not faded	0	88	47	135
Total	54	99	47	200

Table 3 Detection of Discolored Ponderosa Pine Affected by Air Pollution at Small Photographic Scales

Scale	Faders Correctly Detected[a]: EKT IR	ANS D/200	Actual Faders on Ground
1:30,000	17	16	65
1:16,000	29	34	65
1:7920	47	52	65
1:3960	49	53	65

[a] Numbers are averages of the two photointerpreters examining June transparencies.

Table 4 Percentage Agreement of Photointerpretation with Ground Assessment of Tree Vigor[a] (Summary of four films at 1:1584 scale for all sites, June, 1968)

Vigor Class on Ground	ANS D/200 Didymium Filter	EKT IR	Plus X	AERO IR
Advanced injury	70	65	24	20
Intermediate injury	60	62	50	35
Healthy — no injury	55	49	35	33

[a] Average of two interpreters.

Evaluation

We determined that normal color film exposed through a didymium filter at the largest scale was best for classifying all vigor classes (Table 4). The percentages of correct classification were based on the arbitrary rating system described earlier. The rating system provides a basis for comparing films but does not indicate the true potential for using large-scale aerial color photography for smog-injury evaluation. For example, we found that the weights used did not always reflect the items that the photointerpreters could see on the photographs. In fact, some items—for instance, color—should have been given higher weights or scores. Such a weighting system put some trees on the borderline in one vigor class by photo inspection when they fell in the next higher or lower vigor class by ground inspection. This question is resolved elsewhere in the paper. We could, however, judge the best films and filters.

Surprisingly, the Corning didymium filter provided more contrast between green healthy foliage and the diseased yellow-red foliage than conventional haze filters. The didymium filter has very sharp-cutting transmission bands (Fig. 4). It is commonly used in calibrating spectrophotometers because of its stability through all visible and near-ir wavelengths. The filter transmits strongly the UV, blues, greens, and reds which coincide with the peak sensitivities of the three emulsion layers of color film (Fig. 4). However, its use should be confined to low-altitude photography where only a small slice of the atmosphere needs to be penetrated. Ultraviolet, violet, and blue light are readily transmitted and permit strong atmospheric scattering of light in these wavelengths at high altitude.

The Ektachrome Infrared Aero film at the largest scale, 1:1584, again showed no advantage over the normal color film. On the false-color transparencies, smog-injured pines, which appeared in various hues of green-yellow, yellow, and yellow-red (Munsell notation)

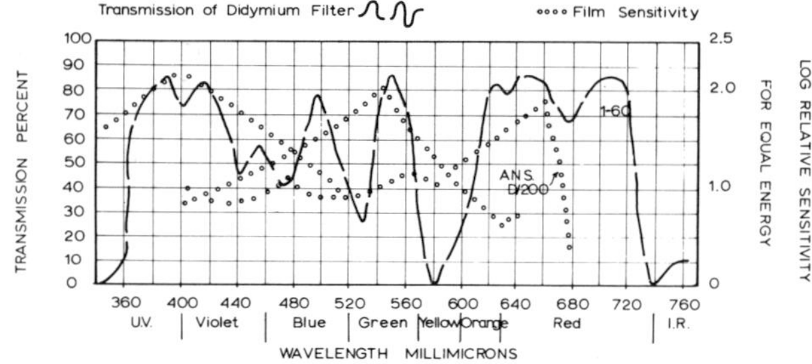

Figure 4 Transmittance characteristics of Corning didymium filter (1–60) is shown on solid line. Film sensitivity of three-layer Anscochrome D/200 film is shown by small circles. The didymium filter transmits strongly at the peak of the color film sensitivity.

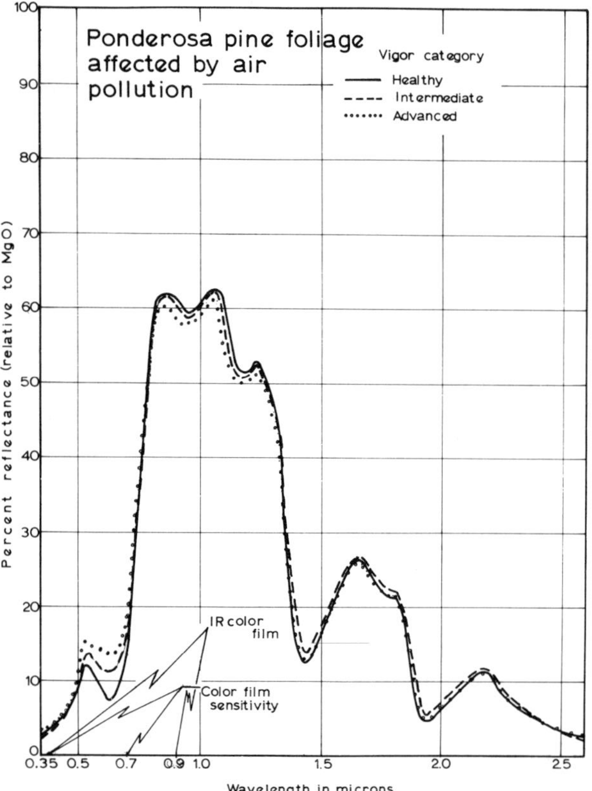

Figure 5 Reflectance spectra from Beckman DK-2 spectrophotometer. Major reflectance differences among the three vigor classes are most apparent within the visible part of the electromagnetic spectrum.

on normal color film, appeared in various lighter values or tones of magenta or red-purple hues. Thus, fewer hue discriminations were available on the false-color film than on the color film. Also, the smog-injured pines in the near infrared wavelengths, 0.7–0.9 μ, where Ektachrome Infrared film is sensitized, showed little difference in reflectance from healthy trees on the spectrophotometer traces (Fig. 5). The principal differences show up within the visible wavelengths (0.38–0.70 μ) where color film is sensitized.

The Plus X Aerographic and Aerographic Infrared films fared poorly (Table 4). On these films, the interpreters could see such fine details as branch mortality and needle retention as well as they could on the color transparencies, but needle and crown color were missing. Evidently color is a vital attribute for successful identification of smog injury. Consequently, these black and white films were not considered for further testing.

Season of Photography

On the basis of the June photointerpretation results, only color film (Anscochrome D/200) was used at two scales, 1:1584 and 1:7920, with the didymium and appropriate haze filters, respectively, in August and December.

On the August transparencies, we were surprised to find that the photointerpreters located fewer smog-affected trees than they did in June. This was particularly startling because air pollution reports indicate that air pollution levels are higher in summer and early fall than in spring. Even ground observers have reported that affected trees appear healthier in August than in June. We found that new foliages of the current year (1968) had not elongated from their buds by early June; thus, our June interpretations were made on 1967 foliage which had been subjected to a whole year of air oxidant fumigation. By August, the new 1968 foliage had elongated and covered over older foliage or bare branches, making the tree appear healthier both on the ground and on the color transparencies than it really was. Furthermore, the new foliage had only about two months of oxidant fumigation. From this normal phenologic development of the tree, we have ruled out the use of airborne sensors in July, August, or September in southern California.

Photointerpretation results from the December color transparencies showed the best correlation ($r = 0.97$) with the ground observations of any time period tested. December is normally a poor month to take aerial photographs because of the low sun angle. However, the very large-scale color transparencies viewed over a light table permitted good shadow penetration not possible at small scale or on prints.

Our December data were subjected to discriminant analysis techniques to learn which recognizable features on the ground and air photos contributed most to correct identification of vigor classes. Using data from 200 trees in the study, we could predict the vigor classes from the photointerpretations (Table 5).

Table 5 Prediction of Vigor Class from Photointerpretation

Predicted Vigor Class	Actual Ground Vigor Class[a]			
	Healthy	Intermediate Damage	Advanced Damage	Dead
Healthy	92	—	—	—
Intermediate	—	95	—	—
Advanced	—	—	80	—
Dead	—	—	—	100

[a] Percentage of actual cases in December, 1968.

Using this kind of analysis for these kinds of data has several advantages. The predictability of vigor classes improves as more data from paired ground and photointerpretation are introduced into the basic formula. The technique permits the photointerpreter to merely identify the image he is viewing and to check off the condition of each significant feature he can see onto a tally form or mark-sense card. The raw data can then be put through the discriminant analysis program, and the trees will be properly classified and summed into vigor classes. The interpretor need not decide the vigor class at the time of viewing. And this procedure should speed up the photointerpretation job.

We are encouraged over the use of the large-scale color transparencies to evaluate smog injury (Anscochrome D/200 scale 1:1584). In May, 1969, we made a sampling survey over 100,000 acres of the San Bernardino National Forest by using the techniques described in this paper. We feel we could not have executed this job efficiently without using aerial color photography as the main sensing tool.

REFERENCES

Croxton, R. J. *Detection and Classification of Ash Dieback on Large-Scale Color Aerial Photographs.* U.S. Forest Service Research Paper PSW-35. Berkeley, Calif.: Pacific SW. Forest and Range Exp. Sta., 1966.

Heller, R. C., Aldrich, R. C., and Bailey, W. F. Evaluation of several camera systems for sampling forest insect damage at low altitude. *Photogram. Engng.* 25:137, 1959.

Heller, R. C. Previsual detection of ponderosa pine trees dying from bark beetle attack. Proc., Fifth Symposium on Remote Sensing of Environment, Institute of Science and Technology, University of Michigan, 1968. Pp. 387–433.

Middleton, J. T. Air pollution: Enemy of man and agriculture. *Agric. Sci. Rev.* 6(2):1, 1968.

Miller, P. R., Parmeter, J. R., Jr., Taylor, O. C., and Cardiff, E. A. Ozone injury to the foliage of *Pinus ponderosa. Phytopathology* 53:1072, 1963.

Miller, P. R. Personal communications on inducement of smog injury on seedlings from his controlled greenhouse studies, 1968.

Parmeter, J. R., Jr., Bega, R. V., and Neff, T. A chlorotic decline of ponderosa pine in southern California. *Plant Dis. Reporter* 46:269, 1962.

Roth, E. R., Heller, R. C., and Stegall, W. A. Color photography for oak wilt detection. *J. Forestry* 61:774, 1963.

Stark, R. W., et al. Photochemical oxidant injury and bark beetle (Coleoptera: Scolydidae) infestation of ponderosa pine. Incidence of bark beetle infestation in injured trees. *Hilgardia* 39:121, 1968.

SUGGESTED READINGS FOR PART FOUR, SECTION B

Becking, R. W. Forestry applications of aerial color photography. *Photogrammetric Engineering* 25:559–565, 1959.

Carneggie, D. M. Interpretation of Color Photography for Range Management. *Manual of Color Aerial Photography.* Falls Church, Va.: American Society of Photogrammetry, 1968. Pp. 408–409.

Colwell, R. N. Uses of Aerial Color Photography in Agriculture. *Manual of Color Aerial Photography.* Falls Church, Va.: American Society of Photogrammetry, 1968. Pp. 382–383.

Cooper, C. F., and Smith, F. M. Color aerial photography: Toy or tool. *Journal of Forestry* 64:373–378, 1966.

Duddek, M. Practical experiences with aerial color photography. *Photogrammetric Engineering* 33:1117–1125, 1967.

Fischer, W. A. Color aerial photography in geological investigations. *Photogrammetric Engineering* 28:133–139, 1962.

Harp, E., Jr. Anthropological Interpretation from Color. *Manual of Color Aerial Photography.* Falls Church, Va.: American Society of Photogrammetry, 1968. Pp. 384–385.

Heller, R. C., Lowe, J. H., Jr., Aldrich, R. C., and Weber, F. P. A test with large-scale aerial photographs to sample balsam woolly aphid damage in the Northeast. *Journal of Forestry* 65:10–18, 1967.

Leedy, D. L. The Inventory of Wildlife. *Manual of Color Aerial Photography.* Falls Church, Va.: American Society of Photogrammetry, 1968. Pp. 422–423.

Mintzer, O. W. Soils. *Manual of Color Aerial Photography.* Falls Church, Va.: American Society of Photogrammetry, 1968. Pp. 425–430.

Mollard, J. D. Landform Analysis. *Manual of Color Aerial Photography.* Falls Church, Va.: American Society of Photogrammetry, 1968. Pp. 406–407.

Parry, J. T., Cowan, W. R., and Heginbottom, J. A. Soil studies using color photos. *Photogrammetric Engineering* 35:44–56, 1969.

Pryor, W. T. Photographic Interpretation for Highway Engineering. *Manual of Color Aerial Photography.* Falls Church, Va.: American Society of Photogrammetry, 1968. Pp. 402–403.

Swanson, L. W. Shoreline Mapping. *Manual of Color Aerial Photography.* Falls Church, Va.: American Society of Photogrammetry, 1968. Pp. 410–411.

Wear, J. F. Interpretation methods and field use of aerial color photos. *Photogrammetric Engineering* 26:805–808, 1960.

Welch, R. Color Aerial Photography Applied to a Study of a Glacier Area. *Manual of Color Aerial Photography.* Falls Church, Va.: American Society of Photogrammetry, 1968. Pp. 400–401.

Section C

Earth orbital photography

SEVENTY-FIVE *percent of the earth's atmosphere is in the troposphere below about 10 miles in altitude. This layer also contains almost all of the atmosphere's water vapor, dust, and organic debris. Suborbital photography is generated in this turbulent layer and aircraft crews frequently have to wait long periods for weather to clear over a target area. It is expensive and often difficult to get aircraft and crew to remote areas. In contrast, an orbiting satellite has several advantages over ground-based photographic platforms. It offers a considerably larger perspective, allowing the observer to see broad regional associations which would be possible from suborbital photography only through time-consuming and costly mosaics. Because the spacecraft may be in orbit for years and travel at high speed, the earth's surface could be imaged very rapidly. If the satellite is in a high-inclination or polar orbit, it will repeatedly pass over every point on the earth's surface. The Earth Resources Technology Satellite (ERTS-A) launched in 1972 permits repetitive observations of each "scene" (an area 100 × 100 nautical miles) every 18 days. Weather permitting, each spot on the earth can be imaged every eighteenth day. This is an incredibly rapid way to image the entire earth. Speed of coverage is one advantage but its repetitiveness may prove to be of even greater value, for the planned life of this satellite is one year. Research of a temporal nature never before possible can be accomplished from space. It will be possible to monitor monthly, seasonal, and yearly changes of phenomena such as land use, smog levels, urban growth, ocean currents, and snow cover.*

18-Space Photography—A Review

PAUL D. LOWMAN, JR.

SINCE THE development of large rockets in World War II, thousands of photographs of the earth have been taken from altitudes of 50 miles or higher, i.e., from space. The purpose of this paper is to review briefly the history, present status, and unique capabilities of what may be called space photography, or hyperaltitude photography. In addition, its potential applications will be reviewed.

Stress will be on pictures of the earth's surface rather than on those of cloud patterns, since such photography is of interest primarily to meteorologists. Military space photography is excluded from the scope of this paper.

History of Space Photography

Although photographs were reportedly taken from rockets before World War I (Katz, 1963), space photography began in earnest with the use of small cameras carried by V-2 rockets fired from White Sands Proving Ground after World War II. Since that time, numerous flights which obtained photographs have been made by a variety of sounding rockets, ballistic missiles, satellites, and manned spacecraft; these are listed in Table 1. Representative pictures taken during these flights are presented in Figures 1, 2, 3, 4, 5, 6, and 7. It should be kept in mind, in judging the quality of these pictures, that nearly all of these flights were made for purposes other than photography, which was usually an auxiliary experiment.

Useful references on certain aspects of space photography include papers by Katz (1963), Merifield (1964), Bird and Morrison (1964), Rochlin (1962), and Lowman (1964).

. . . .[1]

Unique Capabilities of Space Photography

It is apparent, from the examples presented here, that the scale numbers and coverage per picture of available space photographs are orders of magnitude greater than those of conventional air photos. We now ask what space photography can offer which aerial photography cannot. Before discussing this question, however, it must be

[1] A section on Current Projects has been deleted with the permission of the author and the American Society of Photogrammetry.

From *Photogrammetric Engineering* 31:76–86, 1965. Reprinted and edited with permission of the author and the American Society of Photogrammetry.

Table 1 Summary of Successful Space Photography Flights

Vehicle	Date	Cameras	Film (Filter)[a]	Area	Altitude (mi.)	Reference
V-2	1946	35 mm motion picture	Super XX (25A)	SW USA	76	Holliday, 1954
V-2	1947	K-25 aircraft	Infrared Reconnaissance Base (25A)	SW USA	100	Bergstrahl, 1947
Aerobee and V-2	1946–50	K-25 aircraft	Aerographic Super XX (25A)	SW USA	60–80	Holliday, 1954 Newell, 1953 Newell, 1959
		35 mm motion picture	Eastman IN Spectroscopic (29A)	SW USA		
		16 mm gunsight	Kodachrome	SW USA		
Vikings 11 and 12	1954–55	K-25 aircraft	Eastman Hi-Speed Infrared	SW USA	Up to 158	Baumann and Winkler, 1955 Baumann and Winkler, 1959
Atlas	1959	16 mm time-lapse	Recordak Fine-gr. Panchromatic	Atlantic Ocean SE of Atlantic Missile Range	Up to 230	Lathrop and Rush, 1959
Aerobee	1960	Maurer 220 70 mm aerial	Kodak IR Aerographic (88A) Kodak Experimental Ektachrome (8778) Kadak High-Definition Negative (3)	North-Central Canada, Hudson Bay	47–140	Evans, Baumann, and Andryshak, 1962
Mercury Flight MR-1[b]	December, 1960	Maurer 220G 70 mm	Super Anscochrome	Florida, Bahama Islands	Maximum over 130	

[a] Wratten filter numbers.
[b] MR refers to suborbital flights with a Redstone launch vehicle; MA to flights with an Atlas launch vehicle. The MR-2 flight carried the chimpanzee Ham; MR-3, A. Shepard; MR-4, V. Grissom; MA-4, a simulated man; MA-5, the chimpanzee Enos; MA-6, J. Glenn; MA-7, M. Carpenter; MA-8, W. Schirra; MA-9, L. Cooper.

pointed out that space photography can duplicate, to an unknown but probably high degree, the functions of aerial photography in producing large-scale, high-resolution pictures of the ground. Katz (1963), for example, points out that a ground resolution of 2.4 feet should be obtainable from 150 miles altitude on a photograph with resolution of 100 lines/mm, using a 120-inch focal length camera. Neglecting this possible duplication of aerial photography, however, we see that space photography from orbiting vehicles offers the following unique advantages.

1. Greater perspective
2. Wider coverage
3. Greater speed
4. Rapid repetition of coverage

The *perspective* afforded by the great altitude of orbiting spacecraft is, of course, the most striking characteristic of space photographs, permitting one to see entire orogenic belts, drainage basins, and wrench fault systems at a glance. This perspective affords a continuity of observation which might permit, for example, the detection of large geologic structures unnoticed on large-scale air photos. In Figures 1 and 4, many lineaments scores of miles long are easily seen. The value of small-scale photographs has, of course, been recognized before (Hemphill, 1958; Cameron, 1961), and is demonstrated by the wide use of mosaics. Space photographs, however, have an advantage over mosaics in showing the terrain as it is, without the necessity for dodging, which must destroy many of the tonal clues to structure (Miller, 1961). This advantage may be especially useful in delineating structure in heavily vegetated areas, where interpretation must depend largely on subtle tone or color differences.

Worldwide coverage can be provided for space photography by high-inclination or polar orbits. The importance of this advantage for photography of the polar regions is obvious, but it may be pointed out that camera-carrying satellites will also cover large areas, such as the central Pacific, which would be very difficult and expensive to photograph from aircraft. A related characteristic of space photography from orbiting vehicles is the *speed* of areal coverage which is possible. Rochlin (1962) points out that one satellite at a 300-mile altitude in polar

Table 1 *(cont'd)*

Vehicle	Date	Cameras	Film (Filter)[a]	Area	Altitude (mi.)	Reference
Mercury Flight MA-3	April, 1961	Maurer 220G 70 mm	Super Ansco-chrome	Not known	Low-altitude abort flight	
Mercury Flight MR-3	May, 1961	Maurer 220G 70 mm	Super Ansco-chrome	Florida, Bahama Islands, mainly cloud covered		
Mercury Flight MA-4 (1961αα1)	September, 1961	Maurer 220G 70 mm	Super Ansco-chrome	First orbit flight path: Atlantic Ocean, North and Central Africa	86–128	
Mercury Flight MA-5 (1961α∼1)	November, 1961	Maurer 220G 70 mm Milliken DBM7, 16 mm (periscope observer camera)	Super Ansco-chrome Kodachrome EK Type II	SE USA, West Coast of Mexico North Africa	86–128	
Mercury Flight MA-6 (1962γ1)	February, 1962	Ansco autoset, 35 mm	Eastman Color Negative	Florida, North Africa	87–141	Glenn, 1962
Mercury Flight MA-7 (1962 τ 1)	May, 1962	Robot recorder, 35 mm	Eastman Color Negative	West Africa, Atlantic Ocean, and other areas	87–145	Carpenter, 1962
Mercury Flight MA-8 (1962 βδ 1)	October, 1962	Hasselblad 500C, Modified 70 mm	Anscochrome 200	Western USA, Mexican Gulf Coast, South Atlantic Ocean	87–145	Schirra, 1962
Mercury Flight MA-9 (1963 15A)	May, 1963	Hasselblad 500C, Modified 70 mm	Anscochrome 200	South-Central Asia, Philippine Islands, Pacific Ocean, Middle East, North Africa	87–144	Cooper, 1963

orbit could photograph the entire surface of the earth in about 4½ days. This extremely rapid coverage will also permit, if the satellite stays up for a few weeks or months, *rapid repetition of coverage* which would be very difficult to achieve with aircraft. This would permit the repeated photography of cloud-covered areas and detection of seasonal changes in features such as vegetation, snow fields, and ocean currents.

Potential Applications of Space Photography

The unique benefits offered by space photography, coupled with the possibility of duplicating conventional aerial photography with long focal length cameras, suggests applications in many areas, such as the following.

Geologic Reconnaissance

The most obvious application of space photography to geology is the photomapping of remote areas not previously mapped, such as parts of the interior of Antarctica. An even more interesting possibility is that of photographing previously mapped areas to show large geologic structures unnoticed on conventional air photos (Figure 4 is of interest in this connection). For example, the existence of transcontinental fracture zones, such as the Clipperton-Vema lineament crossing Venezuela (Fuller, 1964), might be investigated by space photography. Other geologic applications are suggested by the fact that color film adds a negligible amount to the cost of space photography; the spectacular pictures taken by Cooper during the MA-9 flight demonstrated the practicability of color space photography (Lowman, 1964).

Topographic Mapping

The Army Map Service study previously referred to (Spooner, 1959) concluded tentatively that 1:1,000,000 scale topographic mapping might be done with satellite photography with good accuracy, and that somewhat larger scales might be possible with lower accuracy. The requirements for attitude and altitude control of a cartographic satellite are considerably more stringent than those for one used only for reconnaissance mapping. However, even the present Mercury spacecraft would meet most of these requirements.

Figure 1 Viking 12 photo of southwest U.S.A. from about 140 miles altitude. View to southwest.

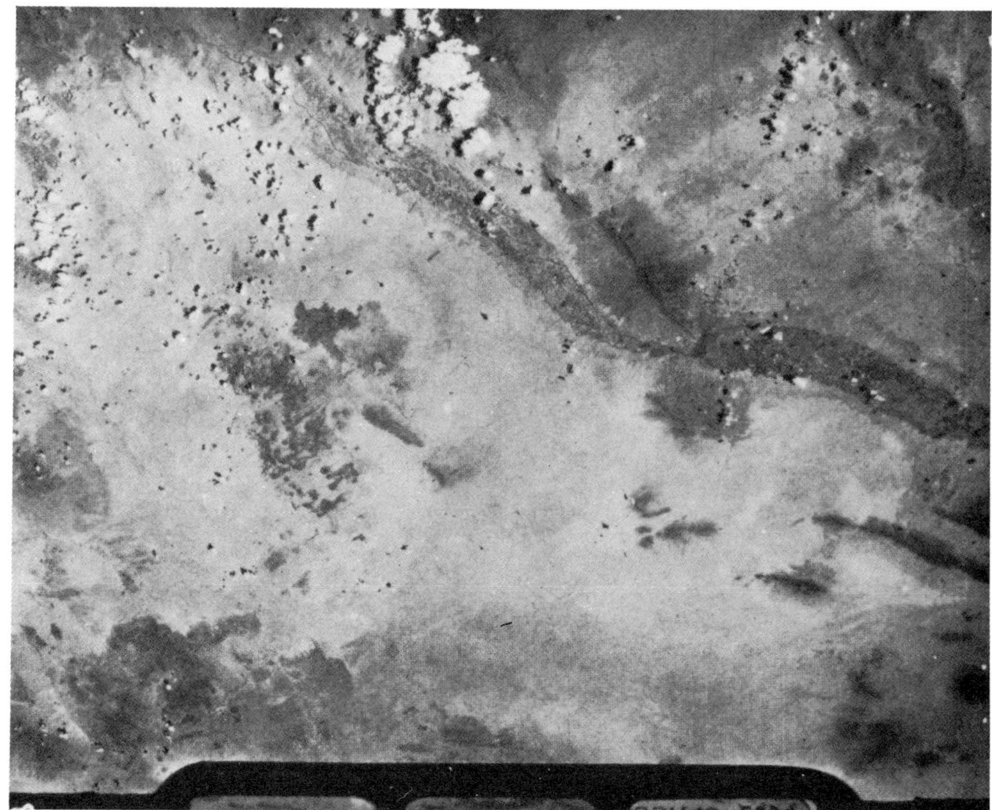

Figure 2 Viking 11 photo of El Paso and Rio Grande Riva from about 158 miles altitude. North at top left.

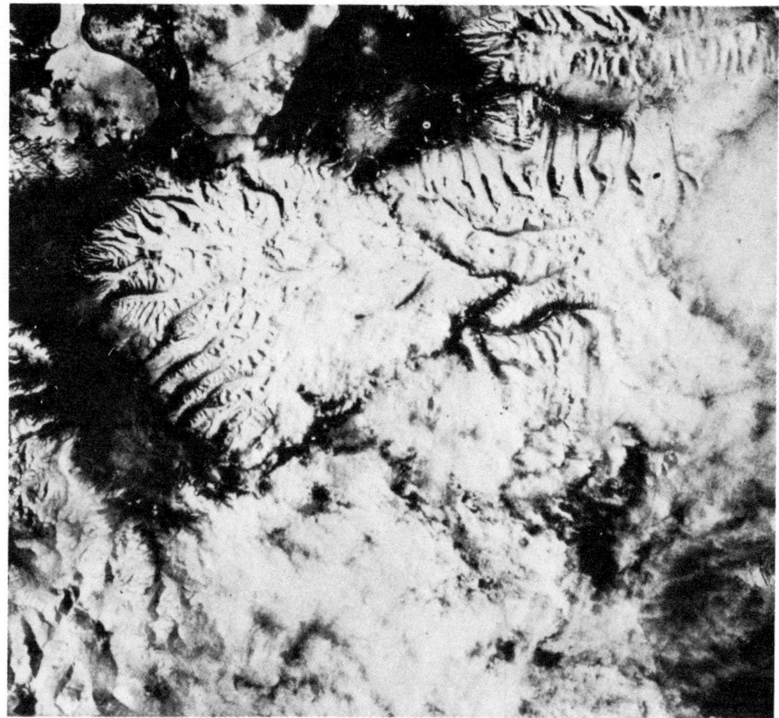

Figure 3 Black and white print of a 70 mm color transparency taken by L. G. Cooper during MA-9 flight, showing southwest Tibet. Lakes in upper left of photo are Rakas Tal (left; about 10 miles in east-west width) and Manasarowar (right); snow-covered mountain in upper left center of photo is Gurla Mandhata (25,335 feet). North at top.

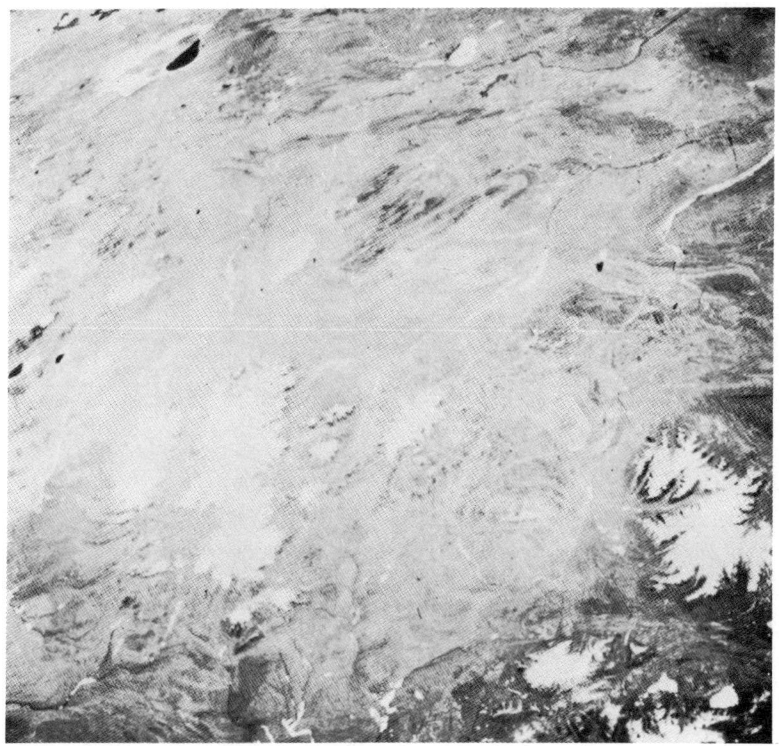

Figure 4 Black and white print of a 70 mm color transparency taken by L. G. Cooper during MA-9 flight, showing central Tibet. North at top.

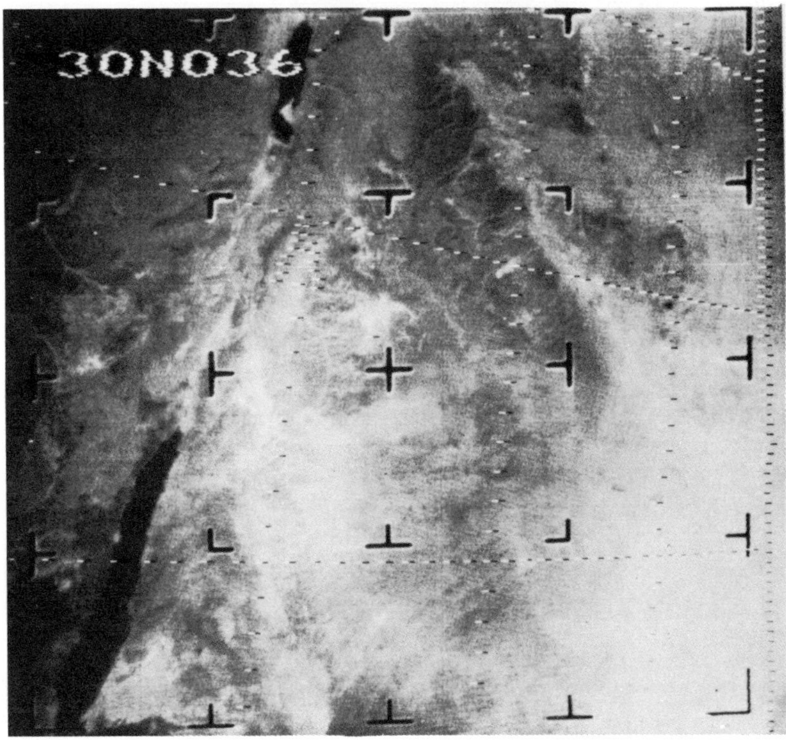

Figure 5 Nimbus I AVCS picture of Dead Sea (top left) and Red Sea showing rift valley. Altitude 371 miles.

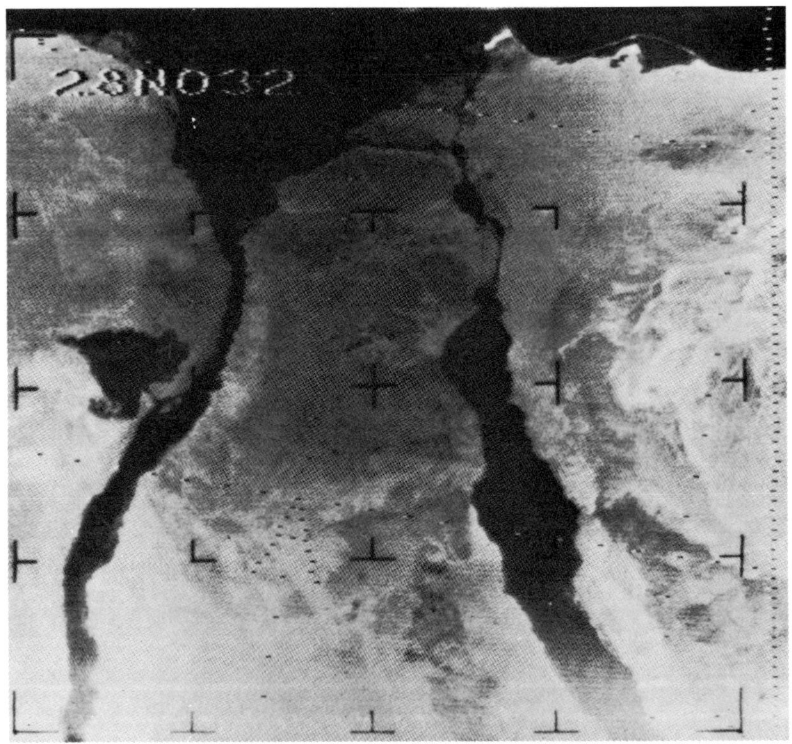

Figure 6 Nimbus I AVCS picture showing Gulf of Suez, Suez Canal, and Nile River. Altitude 371 miles.

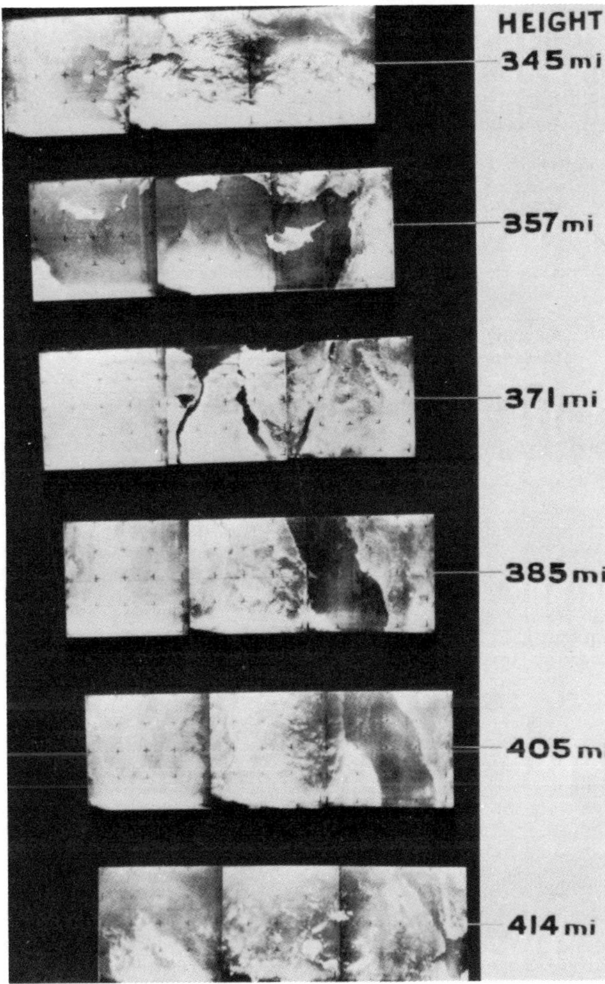

Figure 7 Mosaic of Nimbus I pictures including areas shown in Figures 5 and 6.

Forestry

The fact that aerial photography has become a nearly indispensable tool of the modern forester makes it seem likely that space photography will have application in this field. The scale and resolution possible with small cameras would prevent the use of space photographs for detailed studies such as crown counts, but reconnaissance forest mapping might be possible. The use of color film for space photography would increase its value in forestry. The great potential value of multispectral photography (Colwell, 1961) in detecting changes in vegetation suggests that it would be useful to carry out such photography from space using instruments such as the recently-developed 9-lens Multiband camera (Yaffee, 1963).

Ice Pack Reconnaissance

It was discovered shortly after Tiros I was put into orbit that sea and river ice could be seen on the telemetered images despite their relatively low resolution. This led rapidly to the joint Canadian-American Project Tirec to investigate the application of weather satellites to ice reconnaissance. Using Tiros photographs in conjunction with aircraft photography and ground observations, the project demonstrated that satellite ice reconnaissance in areas such as the Gulf of St. Lawrence was clearly feasible and of great potential value (Balites and Neiss, 1962); Singer and Popham (1963) report that as much as $1,700,000 might have been saved in 1961 by the United States and Canada through ice observations from a Nimbus satellite had one been in orbit. It may be pointed out that a figure of this sort is misleadingly conservative; an operational Nimbus satellite could provide similar ice reconnaissance over the approaches to western Europe, Russia, and Antarctica at very little extra cost.

Film photography would have the advantage of greater resolution than television. But in day-to-day ice pack monitoring, methods of rapid image retrieval would have to be utilized, such as facsimile transmission of films developed in flight. It seems safe to say that both television and photography from orbiting vehicles promise to be immensely useful in ice studies.

Hydrology

Although relatively little application has been made of space photography to hydrology, it might be useful in several ways. One of these is the measurement, over large areas, of snow cover; the National Weather Satellite Center and other agencies in the United States and Canada are currently investigating the use of Tiros pictures for this purpose. As shown by the MA-9 photographs, deep snow, light snow, and valley glaciers can be distinguished easily, indicating that methods involving film recovery should be valuable because of the high resolution of the film.

The fact that entire drainage basins of major rivers can be photographed quickly from satellites suggests that many other hydrologic applications can be found for space photography.

Supplemental Weather Photography

The greater resolution obtainable with film-recovery methods of space photography makes space photographs valuable for synoptic studies of the fine structure of cloud systems (S. Soules and K. Nagler, personal communication), similar to those already conducted from aircraft (Malkus, 1963). The Arctic Meteorology Photo Probe (Evans, Baumann, and Andryshak, 1962) further demonstrated the usefulness of rocket photography in supplementing and supporting meteorologic satellites.

These applications would not be considered photogrammetry in the usual sense, but are worth mentioning because a surprisingly large part of the earth's surface is covered with clouds at any one time. This will obviously hamper terrain photography, but the pictures of the cloud cover itself will be of value.

Oceanography

The ability to take individual photographs covering scores of thousands of square miles should prove invaluable in oceanographic studies. In addition to the obvious benefits of weather observations over remote oceanic areas, the following applications may be possible:

1. Multispectral photography covering the near infrared, visible, and ultraviolet can show the distribution of currents and possibly of areas with differing salinity.

That such photography is possible even from space vehicles was suggested by Glenn's ability to see the Gulf Stream during the MA-6 flight (Glenn, 1962). A knowledge of the structure of such major near-surface currents would obviously be of value to the fishing and shipping industries and from a broader viewpoint to nations whose climate is strongly influenced by these currents, such as Iceland, England, and Chile.

2. The discovery by Cameron (1952, 1962) that water currents in oceans, bays, and rivers could be mapped by pseudostereoscopic time-lapse air photography opens another possible application of space photography. Small areas can, of course, be mapped with low-altitude photography, but to map large currents, such as those in the Bay of Fundy, Cameron found it necessary to use photographs with scales of 1:85,000. He suggests extension of the method to major currents such as the Gulf Stream (and to large physiographic features) by the use of 1:270,000 photographs taken from altitudes of 80,000 feet or higher, or by the use of satellite photography.

3. L. G. Cooper, during the MA-9 flight, noticed striking color differences in the water around islands in the Bahamas, which he attributed, presumably correctly, to water depth. Conventional air photos have been used to study this, and it is interesting to note the possibility raised by Cooper's observation that space photographs can also be used to map bottom topography.

Extraterrestrial Photointerpretation

The most immediate extraterrestrial use for space photography of the earth is its application to the study of pictures of other planetary surfaces, that of Mars in particular. It is interesting to note that nearly all the efforts to interpret the thousands of available pictures have been made in essentially complete ignorance of what the earth would look like under similar conditions. The problem of deducing the nature of the Martian surface is complicated by the fact that experience gained by the study of lunar features cannot be applied reliably to Mars, because the existence of a Martian atmosphere and an intermittent hydrosphere may have produced physiography more nearly terrestrial than lunar. An interesting example of the use of space photography in the study of Mars is presented by Gifford (1964).

Summary

It is, of course, obvious that photography from orbital distances cannot replace aerial photography, especially for applications requiring extremely large scales. Nevertheless, it seems clear that the uniquely great coverage and perspective possible with space photography will make it an invaluable tool for many purposes, and may uncover broad features of the earth's structure whose existence has been only conjectured.

REFERENCES

Badgley, P. C., and Lyon, R. J. P. Lunar Exploration from Orbital Altitudes, presented at the Conference on Geological Problems in Lunar Research sponsored by the New York Academy of Sciences, May 16–19, New York, to be published by the Academy.

Balites, M.D., and Neiss, H. Conference on Satellite and Ice Studies. Meteorological Satellite Lab. Rept. No. 20, U.S. Weather Bureau, National Weather Satellite Center, 1962.

Baumann, R. C., and Winkler, L. Rocket Research Report No. XXI. Photography from the Viking 12 Rocket at Altitudes Ranging up to 143.5 Miles. U.S. Naval Res. Lab. Report R-4489, Washington, D.C., February, 1955.

Bergstrahl, T. A. Photography from the V-2 Rocket at Altitudes Ranging up to 160 Kilometers, Naval Res. Lab. Rept. No. R-3083, April, 1947.

Bird, J. B., and Morrison, A. Space photography and its geographical applications. *Geographical Review* (in press).

Cameron, H. L. The measurement of water current velocities by parallax methods. *Photogrammetric Engineering* 18:99, 1952.

—. Interpretation of high-altitude small-scale photography. *Can. Surveyor* 15:567, 1961.

—. Water current and movement measurement by time-lapse air photography. *Photogrammetric Engineering* 28:158, 1962.

Carpenter, M. S. Pilot's Flight Report. In: Results of the Second United States Manned Orbital Space Flight, May 24, 1962, NASA Special Report SP-6, pp. 69–75, 1962.

Colwell, R. N. Some practical applications of multiband spectral reconnaissance. *Am. Scientist* 49:9, 1961.

Cooper, L. G. Astronaut's Summary Flight Report. In: Mercury Project Summary Including the Results of the Fourth Manned Orbital Flight, May 15 and 16, 1963, NASA Special Report SP-45, pp. 349–358, 1963.

Cronin, J. F. Terrestrial Features of the United States as Viewed by Tiros. Resp. AFCRL-63-664, U.S. Air Force Cambridge Research Labs; Rep. ARA-T-9219-4, Aracon Geophysics Co., July, 1963.

Evans, H. E., Baumann, R. C., and Andryshak, R. J. The Arctic Meteorology Photo Probe. NASA Technical Note D-706, February, 1962.

Fuller, M. D. Expression of E-W fractures in magnetic surveys in parts of the U.S.A. *Geophysics,* 29:602, 1964.

Gifford, F. A., Jr., The Martian canals according to a purely Aeolian hypothesis. *Icarus,* 3:130, 1964.

Gill, J. R., and Gerathewohl, S. J. Gemini Science Program. *Aeronautics and Astronautics,* in press.

Glenn, J. H., Jr. Pilot's Flight Report. In: Results of the First United States Manned Orbital Space Flight, February 20, 1962. NASA, pp. 119–136, 1962.

Hemphill, W. R. Small-scale photographs in photogeologic interpretation. *Photogrammetric Engineering* 24:562, 1958.

Holliday, C. T. The Earth As Seen from Outside the Atmosphere. In: *The Earth As a Planet.* G. P. Kuiper, ed. Chicago: Univ. of Chicago Press, 1954.

Katz, A. Observation Satellites. In: *Space Handbook* (R. W. Bucheim and the staff of the Rand Corporation), 2d ed. New York: Random House, 1963.

Lathrop, P. A., and Rush, D. H. Photographic Instrumentation from Outer Space. Report of the Missile and Space Dept., General Electric Co., Philadelphia, Pa., 1959.

Lowman, P. D., Jr. A Review of Photography of the Earth from Sounding Rockets and Satellites. NASA Technical Note D-1868, 1964.

Malkus, J. The cloud patterns over tropical oceans. *Science* 141: 767, 1963.

Merifield, P. M. Some Aspects of Hyperaltitude Photogrammetry. Report No. 1, Contract No. NAS 5-3390, Lockheed-California Company, Burbank, California, 1964.

Miller, V. C. *Photogeology.* New York: McGraw-Hill, 1961.

Newell, H. E. *High-Altitude Rocket Research.* New York: Academic Press, 1953.

—. *Sounding Rockets.* New York: McGraw-Hill, 1959.

Rochlin, R. S. Observation Satellites for Arms Control Inspection. Rept. No. 62GL78, General Engineering Laboratory, General Electric Co., Schenectady, N.Y., 1962.

Schirra, W. M., Jr. Pilot's Flight Report. In: Results of the Third United States Manned Orbital Space Flight, October 3, 1962, NASA Special Report SP-12, pp. 49–55, 1962.

Singer, S. F., and Popham, R. W. Non-meteorological observations from weather satellites. *Astronautics and Aerospace Engineering,* 1:89, 1963.

Spooner, C. S., Jr. Requirements for a Satellite Cartographic Camera. Consultation Brief, Army Map Service, Washington, D.C., 1959.

Yaffee, M. L. Camera may aid terrain study. *Aviation Week* 79:69 and 71, November 4, 1963.

BEFORE THE *advent of air photographs, maps were made by teams of surveyors and cartographers who painstakingly measured their way over the landscape. Despite the dedication of these early mappers, their work was frequently inaccurate, but a greater problem was the length of time necessary to produce a complete map from ground-based surveys. The use of air photographs made it possible to speed up map production and produce maps with increased geometric accuracy and detail. With the successful orbiting of satellites in the early 1950s, the potential followed for a third step; cartographers viewed space flight as a part of the logical sequence in the development of mapping from plane-table to aircraft to satellite. However, they have not been able to exploit this new technology because space photographs generated to date have not been produced by mapping cameras in a systematic fashion. Some of the photographs taken of the surface of the moon from the Apollo command ships give indications of both the promise and problems of producing metric photographs from space. The Earth Resources Technology Satellite (ERTS) currently in orbit and the scheduled Earth Resources Experiment Package (EREP) of Skylab A will add to coverage of the earth's surface, but neither will produce the kind of imagery and coverage needed for systematic extensive mapping. A measure of the problem is provided by our own country. Although it is one of the best-mapped large countries in the world, over 30 percent of the United States has not been mapped at large scale. Even with technologic breakthroughs, such as the use of air photographs, it still takes approximately three years to produce a topographic map—these maps are three years out-of-date before the first one is printed! In rapidly growing areas such as Dallas or Los Angeles the variance between maps and land use can be startling. A method of obtaining more rapid, sequential photo coverage would be to generate systematic, metric photographs from a space platform. A mapping satellite placed in a near-polar orbit could photograph the United States with a sufficient degree of overlap to provide a geometrically sound base for mapping the entire country on a relatively large scale. In addition, repeated coverage of areas over a long time would make possible the rapid, accurate up-dating of the content on existing maps.*

19-Can Satellite Photography Contribute to Topographic Mapping?

FREDERICK J. DOYLE

AS ONE reviews the accomplishments of the last decade, there can be little doubt that the routine achievement of successful missions in space is the most technologically significant. Photogrammetrists and cartographers naturally look upon orbiting spacecraft as a logical step in the progression from planetable to aircraft to satellite. They can foresee the same kind of quantum jump in production, geometric accuracy, and content of topographic maps which occurred when aerial photogrammetry replaced ground surveys. At the same time they are frustrated because as of this date there has not been a single space photograph taken in which photogrammetric considerations were paramount in the selection and operation of the camera system.

Paper presented to United Nations Seminar on Photogrammetric Techniques, Zurich, Switzerland, March, 1971. Reprinted with permission of the author.

Hasselblad Photography

Throughout the manned space program, Hasselblad and Maurer cameras were used for still photography. These are both 70 mm film cameras with interchangeable lenses of various focal lengths. In the Mercury and Gemini programs over 2400 color photographs were obtained. Many of them are exceptionally beautiful and certainly historical. Scientifically they demonstrated the utility of the synoptic view for geologic, hydrologic, and geographic interpretation of the earth's surface. They also demonstrated that, apart from cloud cover, photographing through the earth's atmosphere did not impose as great a problem as had been anticipated.

Though cartography was not an objective for the Gemini missions, and photographs were not acquired in typical mapping sequences, some attempts were made to demonstrate the potential usefulness of space photography for map compilation. A single frame exposed from Gemini 7 over the Cape Kennedy area was compared with the existing 1:250,000-scale map, and new cultural features were located. Mosaics of Peru and the southwest United States were assembled from rectified prints of the Gemini photographs. Geologic and land use maps were compiled to demonstrate the utility for small-scale thematic mapping.

In April, 1968, the unmanned Apollo 6 earth-orbiting mission carried an intervalometer-controlled 70 mm film Maurer camera with 76 mm focal length. A sequence of near-vertical stereo photographs over southwest United States was obtained. One photograph at original scale 1:2,730,000 of the Dallas-Forth Worth area in Texas was compared with the existing 1:250,000 topographic map. By measuring the coordinates of identified points on an enlarged and rectified print and comparing them with coordinates scaled from large-scale 1:24,000 maps, it was found that the space photograph could be used to check the planimetric accuracy of the small-scale map. Furthermore, it was clearly demonstrated that linear cultural features, such as roads and railroads, whose actual dimensions were far less than the ground resolution of the photograph, could be detected and indeed recognized on the photograph. In addition, the limits of urbanized areas were easily visible on the photograph. The report on the study concluded that the space photograph was useful for map revision both in terms of planimetric position and content.

The Apollo 9 manned earth-orbiting mission in March, 1969, carried an experiment known as SO-65. The principal objective of the experiment was to simulate the multispectral photography which will be obtained by the Earth Resources Technology Satellite. Four 70 mm Hasselblad cameras with 80 mm lenses were mounted in a fixture which held their optical axes parallel and provided simultaneous exposures by all four shutters. The camera array was mounted in the hatch window of the Apollo Command Module. The spacecraft was oriented so that the camera axes were vertical, and five strips of pictures across the southern United States were obtained under exceptionally favorable weather conditions. Three black-and-white films were exposed through appropriate filters to give the red, green, and near-infrared spectral bands proposed for the ERTS television cameras. The fourth camera obtained color infrared photographs which simulate the effects of superimposing the three black-and-white pictures. These pictures have been used extensively to determine the procedures for automatic thematic mapping of earth resources.

Two frames of the red band covered approximately the area of the 1:250,000 Phoenix, Arizona, map sheet. These were rectified and enlarged nearly 10× by fitting to detail points identified on the map and the photographs. The mosaic was then printed as an image base for the standard line map. A number of different color schemes were tried and versions were produced both with and without contours (Fig. 1).[1] The maps have been widely circulated in the cartographic community and the reaction has been overwhelmingly, though not unanimously, favorable.

On the standard line map, wherever white paper appears, there is no information. The terrain morphology, the field patterns, and other characteristics of the area are more clearly shown by the photographic image. At the same time, all of the information from the conventional line map is also available.

Though the line map had been recently checked for accuracy and revised for content, the photo image base disclosed corrections to both accuracy and content. In several areas there was an obvious mismatch between the photo and the line data. Checking against large-scale maps disclosed that the error was in the line map and not in the image. In other areas, the photo image showed additional cultural features added since the line map was revised.

Though the Phoenix 1:250,000 space photomap may be considered a qualified success, it should be evaluated more as a forerunner of things to come rather than as an example of the limits of utility of space photography for cartography.

Lunar Camera Systems

Photogrammetrists will readily recognize that the Hasselblad cameras are not to be compared with the sophistication of conventional mapping cameras used in aircraft. When focal lengths of 150 to 600 mm and formats of 230 mm are used at altitudes up to 10,000 meters in aircraft, it is patently absurd to use focal lengths of 80 mm and formats of 70 mm from spacecraft at nearly 20 times the altitude.

For reasons difficult to rationalize, the lunar exploration program has been able to obtain better cameras than have been used in the Earth program. Through lunar Apollo mission 12, the Hasselblad camera was again used with interchangeable lenses of 60, 80, 250, and 500 mm focal length. The 60 mm lens was used primarily for documentary photographs within the spacecraft and for operations on the lunar surface. The 80 mm lens was used with the camera mounted on a bracket in the command module hatch window to obtain near-vertical stereo strips for lunar orbit much like the Apollo Earth photography. The 250 mm lens was used hand-held to take oblique photographs of sites of particular geologic interest. On Apollo 12 an attempt was made to photograph potential landing sites for future missions by using the 500 mm lens. To provide both forward-

[1] The illustrations in this article were not available for reproduction.

motion compensation and adequate stereo base, the camera was bracket mounted in the hatch window. The entire spacecraft was pitched in synchronism with its forward motion to keep the same scene in the field of view. This complicated operation restricted the number of targets that could be photographed, because of the necessity of completely reorienting the spacecraft as a unit between scenes, and it was only marginally successful in providing the hoped-for high resolution.

In October, 1969, the Apollo Orbital Science Photographic Team was formed to recommend the photographic equipment and operations which would get the maximum scientific return from the remaining missions. The team, however, was not completely free in selecting the camera systems to be employed. It was necessary to choose cameras that were either available or developed to the extent that they could meet the rigid timetable of qualification for the Apollo missions, which were at that time scheduled for approximately 4-month intervals. Furthermore, severe space and operating conditions had to be met. Consequently, the cameras selected were compromises adapted from existing equipment.

The Lunar Topographic Camera

The first new photographic system is a modified Hycon KA-74 reconnaissance camera of 460 mm focal length, exposing 430 frames in each magazine of 130 mm roll film (Fig. 2). The camera is mounted on a special fixture in the command module hatch window. Forward-motion compensation is provided by rocking the entire camera in its mount during the exposure. Adequate stereo base is obtained by taking pictures with the camera axis vertical on one revolution and directed 20° aft on the succeeding revolution.

This camera was carried aboard Apollo 13 and was actually mounted in position ready for lunar approach photography when the oxygen tanks exploded, with consequences which are now well known. It was carried again on Apollo 14.

The Lunar Panoramic Camera

With Apollo 15, presently planned for July, 1971,[2] lunar photography will take on new dimensions. This will be the first operation of the SIM, which is the acronym for Scientific Instrumentation Module. In this section of the service module, a variety of orbital science experiments will be mounted.

A major component of the SIM will be a panoramic camera to provide high-resolution photographs of large areas of the lunar surface. The camera will have a 610 mm focal length and an aperture of $f/3.5$. The 108° sweep angle will expose a frame 115 by 1270 mm on 130 mm roll film. A total of 1900 meters of thin-base film will provide 1650 exposures.

The camera, now under construction by Itek Corporation, is a modification of an existing rotating-optical-bar camera. Its principal elements (Fig. 3) are the main frame, which mounts rigidly to the spacecraft and carries the supply and takeup film rolls; the stereo gimbal assembly, which can rotate about a transverse axis to provide stereo convergence and forward-motion compensation; and the roll-frame assembly, which carries the rotating lens and moving film rollers.

Stereo coverage can be obtained by rocking the camera 12½° forward or aft between succeeding exposures. A V/H sensor will provide the signal for controlling the forward-motion compensation which is also accomplished by rocking the camera. A light sensor mounted on the stereo gimbal will adjust the exposure time by modifying the width of the slit. From the nominal operating altitude of 111 km, the camera will provide about 1.5 meter surface resolution at the spacecraft nadir. In the monoscopic mode each frame will cover 300 km across track by 20 km along track, and each vertical photograph will overlap the preceding one by 10 percent at the nadir. In the stereo mode, the camera will rock through the 25° convergence angle between each pair of exposures. The forward exposure from station 1 and the aft exposure from station 6 will overlap by 100 percent to form a stereomodel, and each succeeding stereomodel will overlap the preceding one by 10 percent.

For many applications it will be desirable to transform the panoramic photographs into equivalent vertical photographs and a special transforming printer is under construction for this purpose.

The rectifier (Fig. 4) is an analog of the camera system. The film is carried on a cylindrical platen whose radius is equal to the 610 mm equivalent focal length of the camera. Longitudinal displacement of the film with respect to the center of the arc compensates for roll of the spacecraft. A moving light source traverses the length of the film to make the exposure. The projection lens rotates at a rate controlled by an inversor so that the Scheimpflug condition is always fulfilled. The projected image is then reflected to the easel. The tilt of the easel can be adjusted to compensate for the 12½° stereo convergence as well as for spacecraft pitch. The longitudinal curvature of the easel can be adjusted to simulate the curvature of the lunar surface for orbital altitudes up to 150 km.

The rectified picture will have a nadir enlargement of approximately 1.8× but will be restricted to a scan range of ±35°. The resolution, measured at the input scale, will vary from 90 line pairs/mm on axis to 50 line pairs/mm at 35° maximum scan. The displacement of images from all causes is specified as 1,000 μm circular probable error. Thus the rectified pictures are not designed for photogrammetric use, but will have adequate geometry for photointerpretation.

The Metric Camera System

Mounted on the front shelf of the SIM on Apollo missions 15 through 17 will be a Metric Camera System (Fig. 5) composed of a terrain camera, a stellar camera, a laser altimeter, and a precise timing mechanism.

The terrain camera will have a 75 mm focal length, $f/4.5$ lens and will expose a 115 by 115 mm format on 130 mm role film. The glass plate focal plane will be engraved with a 10 mm reseau. Two sets of artificially illuminated fiducials and one set of naturally illuminated fiducials will be provided. A data block on each frame of the terrain camera will record the midpoint of each

[2] [Note that this paper was presented in March, 1971.]

exposure to ±1 millisecond. The flight height measured by the laser altimeter and the shutter open time to ±0.1 millisecond will also be recorded. A V/H sensor will control forward-motion compensation with an accuracy of 3 percent by moving the focal plane during the camera exposure. Automatic exposure control will set the between-the-lens shutter in five steps between 1/5 and 1/240 second.

The resolution on the lunar surface will depend upon solar altitude and the accuracy of forward-motion compensation. For the nominal altitude of 111 km and exposure on Kodak 3400 film, the expected high-contrast resolution at low solar altitude will be about 30 meters per line pair, decreasing to about 60 meters for low contrast at high solar altitude.

The forward overlap of successive frames can be adjusted preflight. For the missions now planned, the available 3600 exposures will permit the total overflown area to be photographed with 78 percent forward overlap and 55 percent side overlap. This means that every point on the ground will be seen on four consecutive photographs with a maximum base-height ratio of 1.0. In addition to providing geometric strength for triangulation, the four views will permit study of the moon's peculiar photometric characteristics.

The stellar camera will look to the side of the orbital plane and 4° above the horizon so that no part of the lunar surface will be included in the field of view. The lens will be 75 mm focal length, *f*/2.8, and will have a 25 by 32 mm format on 35 mm film. Four fiducials and a 5 mm reseau on the glass focal plane will be edge-illuminated to record on the film. This prefogging will have the additional advantage of hypersensitizing the film for stellar exposure. A fixed exposure time of 1.5 seconds should provide 25 to 75 star images per frame, depending upon the location of the field of view in the celestial sphere. This sensitivity takes into account the nominal spacecraft attitude rates and the orbital rate.

The laser altimeter will be mounted with its transmission and receiving optical axes parallel to those of the terrain camera. Its accuracy will be about ±2 meters.

The terrain and stellar cameras are being built by Fairchild Space and Defense Systems. The laser altimeter is being built by RCA, but Fairchild is responsible for integrating it into the metric camera package. The installation of all systems in the SIM is the responsibility of North American Rockwell.

Film Recovery

The photographic film from both the panoramic and the metric camera systems will be accumulated in special quick-release magazines. An astronaut will leave the command module, activate a single handle which will cut the film, close the light-tight compartment, and release the film-return magazine. The magazines will then be stowed in the command module for the journey back to earth.

The design and construction of these complicated camera systems, their qualification for operation in a manned spacecraft, and the extensive planning which has been done for the data analysis of the lunar photography inspire confidence that an adequate system could be devised for topographic mapping of the earth at scales compatible with existing cartographic standards.

Rationale for an Earth Mapping System

Before attempting to define a system which would perform useful topographic mapping from space, it is necessary to examine some of the factors involved.

A topographic map contains three kinds of information:

1. Content—the details represented on the map
2. Position—the reference graticule
3. Elevation—spot heights and contour lines

Any mapping system must be able to provide all of these.

Map *content* is obtained from photographic resolution and scale, or more directly from ground resolution. It is difficult to establish a linear relation between map scale and required ground resolution because some features, like roads and railroads, must be shown on a map regardless of its scale. However, by considering the photograph as an image base (photomosaic or orthophotograph) for the map, a useful criterion can be produced.

Under normal conditions the unaided human eye resolves about 5 line pairs per millimeter. In order to retain that resolution in the final map, the photography should have better resolution, say 10 line pairs per millimeter, in order to allow for some losses in processing. This would mean that the required ground resolution would be 1/10 mm times the map scale number.

$$R_g = 10^{-4} S_m \tag{1}$$

where R_g = required ground resolution (meters)
S_m = map scale number

Obviously not all map content is obtainable directly from photographs, regardless of scale or ground resolution, whether obtained from aircraft or space. Data such as political boundaries, place names, and detail obscured by vegetation must be compiled on the ground or from other sources. Studies with existing photography indicate that if this ground resolution criterion is followed, about 80 percent of the mapworthy detail can be identified from the photography. The fact that linear features, whose lateral dimension is well below the resolution limit, can be identified works in favor of small-scale photography.

In aircraft photography, external factors such as attitude rates, vibration, and atmospheric anomalies generally restrict photographic resolution to 25 to 30 line pairs per millimeter regardless of the laboratory-determined lens-film performance. In the space environment all these degrading factors are considerably reduced so that the actual photographs may have very near the resolution measured in the laboratory.

Large-scale maps ($S_m < 25{,}000$) are usually compiled with up to 5× enlargement between air photo and map scale. Medium-scale maps ($25{,}000 < S_m < 100{,}000$) are compiled at 1× to 2× enlargement between photo and map. Small-scale maps ($S_m > 100{,}000$) are usually assembled from existing larger-scale source materials rather than compiled directly from photographs. These restrictions, however, generally result from the operational limits of aircraft, cameras, and photogrammetric instruments, rather than from the information inherent in the photography. According to the criterion established, high-resolution space photography can be enlarged until its equivalent resolution is 10 line pairs per millimeter.

$$M_a = \frac{r_p}{10} \tag{2}$$

where M_a = allowable enlargement from photo to map scale

r_p = original photo resolution in line pairs per millimeter

The second kind of map information is *planimetric position.* Absolute position is generally provided by reference to ground control points. For a space system, however, orbital tracking data can provide spacecraft position to a high order of accuracy. When this is coupled with precise camera attitude data and time of exposure, an independent means of determining absolute position is available.

Within a stereomodel or a photogrammetric triangulation, the internal positional accuracy is equal to the product of the photograph scale and the error in measuring or identifying a point on the picture

$$\sigma_P = S_p \, \sigma_p \tag{3}$$

where σ_P = standard error in position of a ground point

σ_p = residual measuring error

S_p = photograph scale number

U.S. Map Accuracy Standards require that the standard error of point positions should not exceed 0.3 mm at the map scale

$$\sigma_P = 3 \times 10^{-4} S_m \tag{4}$$

The third kind of information on a topographic map is *terrain relief* depicted by contour lines, spot elevations, and sometimes shaded relief.

In conventional aerial mapping, vertical control points are established with an accuracy of 0.1 to 0.2 of the contour interval. A statistical interpretation of U.S. Map Accuracy Standards requires that

$$\sigma_h = 0.3 \, c.i. \tag{5}$$

where σ_h = standard error of elevation

$c.i.$ = contour interval

It may be noted in passing, however, that recent tests by the U.S. Geological Survey have demonstrated that if a stereomodel is properly adjusted to redundant vertical control points, contours at an interval equal to the standard error may be compiled and still meet National Accuracy Standards.

Elevation data is obtained from aerial photographs by means of image parallax which depends upon the camera configuration, the image scale, and the measuring accuracy.

$$\sigma_h = S_p \frac{H}{B} \sigma_p \tag{6}$$

where H/B is the reciprocal of the base/height ratio.

The critical requirements for maps at various scales may be derived by applying formulas 1, 4, and 5, with the results given in Table 1.

Table 1 Map Accuracy Requirements

Map Scale Number S_m	Ground Resolution R_g	Std. Error Position σ_P	Contour Interval $c.i.$	Std. Error Elevation σ_h
1,000,000	100 meters	300 meters	100 meters	30 meters
250,000	25	75	50	15
100,000	10	30	25	8
50,000	5	15	10	3
25,000	2.5	7.5	5	1.5

A fixed contour interval does not necessarily go with a given map scale. An interval fine enough to depict the character of the terrain will be chosen.

The values listed in Table 1 represent the performance objectives against which a space cartographic system should be evaluated.

Space Camera System for Earth Mapping

In the summer of 1967, NASA asked the National Academy of Sciences to perform a study on Useful Applications of Earth-Oriented Satellites. Thirteen panel reports were presented, on a variety of interests. Although differing in details, the Forestry, Agriculture, Geography, Geology, and Hydrology Panels essentially confirmed their requirements for a television system providing repetitive cover at relatively coarse resolution.

The Geodesy-Cartography Panel attacked the problems of providing data for geodetic control and topographic mapping on a worldwide basis. The Geodetic Satellite Program is now producing a worldwide network of some 40 stations in a unified geocentric coordinate system. Continental intensification networks, if carried to completion as recommended by the panel, will locate points with better than 3 meter accuracy at about 800 km spacing. These points, together with existing geodetic control, can form the foundation for map compilation at scales as large as 1:25,000.

The panel recommended two camera systems: a metric camera system for complete small-scale mapping and establishing control for large-scale mapping; a long focal length convergent camera system for providing the detail necessary for large-scale map compilation. The recommended metric camera system would comprise:

1. A 305 mm focal length, 230 by 365 mm format, vertical frame camera with 70 percent forward overlap between frames
2. A 150 mm focal length, 70 mm format, stellar-attitude camera synchronized with the terrain camera
3. A laser altimeter measuring the distance from camera to terrain in synchronism with each camera exposure
4. A timing device to record the midpoint of each exposure

Though the panel did not define the large-scale camera in detail, the panoramic camera from the lunar program would be an obvious candidate for this application.

The major parameters of these camera systems are summarized in Table 2.

Table 2 Parameters of Space Camera Systems

Parameter	Control Camera	Stellar Camera	Compilation Camera
Type	Frame	Frame	Panoramic
Focal length mm	305	150	610
Film width mm	240	70	125
Sweep angle	—	—	70°
Format mm	230 × 365	60 × 60	115 × 750
Convergent angle	—	—	30°
Overlap between frames	70%	—	100%
Base/height ratio	0.72	—	0.54
Resolution AWAR	50 ℓp/mm	—	135 ℓp/mm

To retain the geometric integrity of the photographs, the panel recommended physical recovery of the film rather than transmission of the images by some form of TV system. The general procedure would be to mount the cameras in the spacecraft and to feed the exposed film to the attached reentry capsule (Fig. 6). When the photographic mission has been completed, the recovery vehicle would be separated from the spacecraft and reenter the atmosphere. A parachute would be deployed, and the film package would be picked up by aircraft. An operating altitude of about 200 km was recommended in order to obtain an adequate lifetime for the satellite.

In evaluating the expected performance of the system, a critical factor is the residual image coordinate error, σ_p. In analytical triangulation with conventional aerial photographs, $\sigma_p = 7\ \mu\text{m}$ is regularly obtained in a single stereomodel, and $\sigma_p = 15\ \mu\text{m}$ in a strip having six to ten photographs between control points. With space photography there is good reason to believe that these values can be appreciably reduced. Contributing factors include:

1. Better photograph resolution
2. Reseau to control film deformations
3. Attitude constraints in each stereo pair obtained from the stellar photographs
4. Exposure station positions obtained from orbital tracking data
5. Scale restraint in each stereo pair obtained from the laser altimetry

It is anticipated that $\sigma_p = 5\ \mu\text{m}$ can be obtained in a single stereomodel, and $\sigma_p = 10\ \mu\text{m}$ in a triangulated block.

When the appropriate values are substituted in formulas 2, 3, and 6, the expected performance of the two camera systems is given in Table 3.

It should be noted that no absolute values are listed for the panoramic photography. This is because the panoramic camera has very poor inherent geometry. The concept is to establish the control for large-scale mapping using the frame camera, and then use the high-resolution panoramic photography to compile the planimetry and contours for the large-scale maps.

Comparing the performance values in Table 3 with the map requirement values in Table 1, it is apparent that the systems can satisfy mapping needs as shown in Table 4.

If the photography from the control camera were enlarged 2.6× to form a photo image base for a map at 1:250,000 scale, a single frame would more than cover a standard map sheet. Within the limits of a stereomodel—the central third of each frame—terrain relief of nearly 400 meters could be accommodated without exceeding the allowable planimetric displacement. One can imagine the improvement in the Phoenix 1:250,000 space photomap if the resolution of the photography were increased fourfold.

Similarly if a 1:25,000 map sheet is located near the center of a panoramic photograph, nearly 300 meters of relief can be accommodated without exceeding planimetric accuracy requirements.

Contours at the indicated intervals of 10 to 25 meters will not be easily obtained. Conventional photogrammetric instruments operate at a *c*-factor ($c = H/c.i.$) of 1000 to 2500. It would require $c = 8000$ to 20,000 to reach the values which are inherent in the photography. Analytical-type plotting instruments or computer derivation of contours from digitized elevations are possible solutions to this problem. One should not be surprised that new data-handling techniques may be required to exploit the capability of space photography. After all, the optical mechanical plotting instrument required for aerial photogrammetry represented quite a development from the plane table and alidade used in ground survey.

Obviously, one could argue that any of the assumed values should be a factor of 2 larger or smaller. What is needed is an experiment to determine how well actual performance will conform to that predicted. The U.S. Geological Survey has proposed to NASA that a system with the characteristics of the control camera be flown.

Table 3 Expected Performance of Space Cameras

Parameter	Control Camera	Compilation Camera
Photo scale S_p	655,000	328,000
Ground resolution R_g	13 m	2.4 m
Allowable enlargement M_a	5	13.5
Maximum map scale S_m	130,000	24,000
Coverage per frame	150 × 240 km	38 × 280 km
Relative accuracy, position σ_p	3.3 m	1.6 m
($\sigma_p = 5\ \mu\text{m}$) elevation σ_h	4.6 m	3.0 m
contour interval	13.7 m	9.0 m
Absolute accuracy, position σ_p	6.5 m	—
($\sigma_p = 10\ \mu\text{m}$) elevation σ_h	9.2 m	—
contour interval	27.5 m	—

Table 4 Mapping Capability

Parameter	Control Camera	Compilation Camera
Relative mapping		
Content for map scale	130,000	25,000
Position accuracy for map scale	25,000	25,000
Elevation accuracy for contour interval	15 m	10 m
Absolute mapping		
Content for map scale	130,000	25,000
Position accuracy for map scale	25,000	—
Elevation accuracy for contour interval	25 m	—

This is not yet a funded project, but topographic mappers look forward to it as the best approach to the application of space photography to their requirements.

What About Economics?

Several contractors have submitted proposals to NASA to build a photogrammetric satellite as described at a cost between $15 and $20 million per mission. The exact capability cannot be determined until the system is designed, but the National Academy of Sciences estimated that 90 kg of film could be carried—enough to photograph 20 million km^2. This is enough to cover the entire United States twice. The cost per square kilometer is directly competitive with the most economical aircraft photography. The obvious problem, however, is that 1000 km^2 would cost the same amount as 20 million km^2. Thus space photography is economical only if extremely large areas are required.

The economic analysis must include the photogrammetric and cartographic reduction as well as the cost of the photography. To cover the United States with the proposed space frame photographs would require less than 1000 stereomodels as opposed to over 100,000 with high-altitude aircraft photographs. A large part of this advantage might be lost if conventional line drafting methods were employed for map production. But as has been pointed out, the central part of each frame is orthographic within National Map Accuracy Standards. This leads one to conclude that annotated orthophotomaps might be the most practical way of producing useful maps of extensive areas on a timely basis.

The formats and focal lengths of space photography are, to a large extent, incompatible with the current photogrammetric instrumentation. Clearly, if satellite photography is to be useful on a production basis, detailed consideration and planning is required throughout the whole course of the mapping cycle. This extends as far as a re-education of map users who may find it necessary to revise their notions of what is an acceptable map.

What About Clouds?

Extensive studies of weather satellite pictures have been performed to analyze the probability of obtaining cloud-free photography from space. Without going into details, these studies seem to converge on the conclusion that a space system with a lifetime adequate to permit four passes over each area will achieve about 97 percent successful photography. This will require ground control of the satellite based upon weather predictions.

What Are the Immediate Prospects?

Apart from weather satellites, there are currently only two NASA funded projects for photographing the earth from space.

The first of these is the Earth Resources Technology Satellite, scheduled for flight in 1972. This will carry three television cameras and a multispectral scanner. It is designed fundamentally to provide repetitive coverage for monitoring time-variant phenomena. Thematic mapping will be an important part of the utilization of the records, and the pictures may be used as an image base for maps as large as 1:250,000. But topographic mapping is not an objective.

The other project is Skylab designed primarily to investigate the problems of men operating for extended periods in space. An empty S-IVB rocket will be converted into a laboratory capable of supporting three men for periods up to 56 days. The Apollo Command and Service Module will be used as a ferry to change crews in the Skylab on a 3-month rotation plan. It is planned to launch Skylab A in the fall of 1972. Its operating altitude will be 435 km and the orbit inclination will be 50°. It will carry two photographic systems.

The first of these is a multispectral photographic experiment designated S-190. A contract has recently been issued to Itek Corporation to develop an array of six cameras (Fig. 7). These will be of 150 mm focal length, $f/2.8$ aperture and record on 70 mm film. The optical axes of all cameras will be accurately aligned, and the focal length and lens distortions will be matched so that all pictures can be accurately registered. Forward-motion compensation will be provided by rocking the frame which carries the six cameras. Four black and white spectral bands of 0.1 μm bandwidth between 0.5 and 0.9 μm are planned. The other two cameras will carry color and color infrared film.

The second photographic experiment carried on Skylab A will be the 460 mm Hycon camera from the lunar program. At the moment a fixed mount is contemplated, with convergent stereo being accomplished by pitching the entire spacecraft as it passes over an experimental area. Photogrammetrists, however, have recommended to NASA that the camera be installed with a rocking mount so that convergent stereo photography can be acquired over extensive flight lines.

Both systems will resolve about 80 line pairs per millimeter for usual ground object contrasts. Neither system will have precise timing, attitude, or altimetry support data. Their primary application will be for interpretation of earth resources information. Their mapping potential, evaluated by formulas 2, 3, 5, and 6, is given in Table 5. These values obviously do not approach those expected from the proposed 305 mm frame camera system

Table 5 Cartographic Potential of S-190 and Hycon Cameras in Skylab A

Parameter		S-190	Hycon
Photo scale S_p		2,900,000	945,000
Ground resolution R_g		36 m	12 m
Allowable enlargement M_a		8	8
Maximum map scale S_m		360,000	118,000
Coverage per frame		174 × 174 km	109 × 109 km
Base/height ratio		0.16	0.47 (25° conv.)
Accuracy	position σ_p	30 m	10 m
($\sigma_p = 10\ \mu$m)	elevation σ_h	180 m	20 m
	contour interval	550 m	60 m

supplemented by the 610 mm panoramic camera. Nevertheless the photographs will be an appreciable improvement over those obtained from Gemini and Apollo.

Conclusion

The space program to date has produced some astonishing photographs. Many of these have been of compelling interest from the historical, pictorial, and interpretive points of view. So far as photogrammetrists are concerned, however, the pictures have presented far more problems than they have solved. It is easy to question why things have been done, and are being done in the ways described. Most of the reasons have nothing to do with science, but with finances and politics. However, there can be no doubt that, as the difficulties of operation for men and machines in space are overcome, the advantages of space will be applied to the routine production of both thematic and topographic mapping for the benefit of all mankind.

Two major *types of geologic information can be obtained from earth orbital photographs: structure and lithology. Actual quantitative measurements can be made of some geologic structures. Lithologic investigations employing photographs are confined to qualitative data. Earth orbital photographs can be used in the office or laboratory. For such purposes the geometry of the photograph is used in conjunction with viewing, measuring, and plotting instruments to provide detailed and accurate interpretation and measurement of remote areas. But the photograph also is a valuable base for field mapping and field data collection. Using the photograph as a mapping base, identification and interpretation can be constantly field-checked. Identification and interpretation can be extended into other areas by the use of stereo or overlapping photo pairs.*

All measurements and geologic interpretation of structure and lithology are limited, however, by the resolution of the photographs. Small-scale photos offer the interpreter a regional overview. Detail is lost but conformation and interrelationship of major structures are clearly discernible. For pinpoint assessment of minor structures and lithology, large-scale photos are needed. A combination of two photographic scales is invaluable for interpretation. Color photographs provide more advantages for geologic interpretation than do black and white. Speeding up the interpretive process and increasing confidence in interpretation are the main advantages of using color. For future photogeologic work, the use of color photographs at two different scales should become standard procedure.

20-Geological Comparison of Spacecraft and Aircraft Photographs of the Potrillo Mountains, New Mexico, and Franklin Mountains, Texas

DAVID L. AMSBURY

Before planning photographic missions for geologic interpretation (or before purchasing already existing photographs), the geologist must determine the scale and film type required for delineation of the features of interest. His choices may be limited by operational and budgetary considerations, but an increasingly varied selection of film is becoming available. Very few published data are available because geologic photography from high-altitude aircraft and from spacecraft is so recent a development, because the potential applications of many types of color film are relatively unknown, and because geologic requirements vary with each job and with each type of terrain to be studied. "This question, as to the best photography for geological purposes, would form an interesting and valuable subject for research" (Allum, 1966). One goal of the NASA Earth Resources Program is to provide qualitative and quantitative data for planning the optimum use of photographic and other types of sensors for many resource-mapping purposes. A portion of the work in progress at the Manned Spacecraft Center, designed to complement concurrent studies by the U.S. Geological Survey, several universities, and industry, will be discussed in this paper.

From *Proceedings of the Sixth International Symposium on Remote Sensing of Environment*. Ann Arbor, Mich.: University of Michigan, Institute of Technology, Willow Run Laboratories, 1969. Pp. 493–515. Reprinted with permission of the author and the publisher.

To provide data on resolution, scale, and film-type requirements for delineation, identification, and mapping of certain geologic features, a region of southern New Mexico and western Texas was chosen for study (Fig. 1, see color insert). Space photographs[1] and aircraft photographs from several missions provide a great variety of scales and film types (Table 1). The region chosen for study is typical of the southwestern United States in

[1] Space photographs in this paper refer to 70 mm film exposed in Hasselblad and Maurer cameras without vacuum backs, using a 38 mm or 80 mm lens unless otherwise specified.

Table 1 Spacecraft and Aircraft Photographs Used for Geological Comparisons[a]

Mission	Date of Photography	Film			Approximate Scale
		Kodak[b] No.	Type	Size	
Apollo 6	April 4, 1968	SO-121	High-resolution color	70 mm	1:2,800,000
Apollo 7	October 12, 1968	SO-121	High-resolution color	70 mm	1:3,000,000
Earth Resources Program Mission 981 (Convair 240)	October 16 and 19, 1968	2448	Color	70 mm	1:40,000
		SO-121	High-resolution color	70 mm	1:40,000
		3400	Black and white (25A filter)	70 mm	1:40,000
		3400	Black and white (58 filter)	70 mm	1:40,000
		8442	Color	9.5 in.	1:20,000
		8443	Color infrared	9.5 in.	1:20,000
Earth Resources Program Mission 981 (High-altitude aircraft)	October 18, 1968	SO-121	High-resolution color	70 mm	1:250,000
		SO-117	Color infrared	70 mm	1:250,000
		SO-136	Black and white (25A filter)	70 mm	1:250,000
		SO-136	Black and white (58 filter)	70 mm	1:250,000
		SO-117	Color infrared	70 mm (Panoramic)	1:62,500 (at nadir)
Apollo 9	March 9 to 12, 1969	SO-368	Color	70 mm	1:3,000,000
		SO-180	Color infrared	70 mm	1:3,000,000
		3400	Black and white (58B filter)	70 mm	1:3,000,000
		3400	Black and white (25A filter)	70 mm	1:3,000,000
		SO-246	Black and white infrared (89B filter)	70 mm	1:3,000,000
Earth Resources Program Mission 89 (Convair 240)	March 18, 1969	SO-180	Color infrared	70 mm	1:46,000
		3400	Color infrared (25A filter)	70 mm	1:46,000
		3400	Color infrared (58 filter)	70 mm	1:46,000
		SO-246	Black and white infrared (89B filter)	70 mm	1:46,000
		2448	Color	9.5 in.	1:23,000
Earth Resources Program Mission 89 (High-altitude aircraft)	March 12, 1969	8443	Color infrared	70 mm (Panoramic)	1:62,500 (at nadir)
		2402	Black and white (25A filter)	70 mm (Panoramic)	1:62,500 (at nadir)
		8443	Color infrared	70 mm	1:250,000
		3400	Black and white (25A filter)	70 mm	1:250,000
		3400	Black and white (58B filter)	70 mm	1:250,000
		5424	Black and white infrared (89B filter)	70 mm	1:250,000

[a]Copies of these photographs have been disseminated to the U.S. Geological Survey and to the U.S. Department of Agriculture, and are available for viewing at the Earth Resources Research Data Facility at the Manned Spacecraft Center.

[b]Kodak films were used in this study because the photographic laboratory at the Manned Spacecraft Center is presently able to process only Kodak films on a routine basis. All color films were of the color-reversal type.

that bedrock hills and mountains are surrounded by wide, brush-covered pediments and grassy plains. The region is semi-arid and has many closed drainage basins with streams that have not yet been integrated into the through-flowing Rio Grande. Many different rock types, geomorphic features, and structural features occur.

Three specific questions were considered during the study. What resolutions and scales are necessary to identify and map geologic features such as outcrop belts, faults, and volcanic fields? Is color photography superior in informational content to black and white photography? Does the informational content of different color films affect geologic interpretation significantly?

Scale and Resolution

A significant amount of confusion exists about the photographic scale and resolution required for geologic interpretation. Much of the confusion may be attributed to differences in objectives between investigators and to imprecise definition of the features of interest. Large geologic features (particularly linear features such as faults, joints, and dikes) that are clearly marked by differences in topography, vegetation, soil color, or land use may readily be discerned on space photographs having fairly low spatial resolution (Lowman, 1968; Wobber, 1968; Drewes and Morrison, 1966; Carter, 1968; Hemphill and Danilchik, 1968; Abdel-Gawad, 1968). Pecora (1968, 1969) illustrated geologic features visible on Mercury, Gemini, and Apollo photographs and stated that similar features will be mappable on black and white line-scan television imagery with a nominal ground resolution of 100 to 200 feet.

It has been stated that such geologic features as mountain ranges, bajadas, and playas are "consistently identifiable" on Apollo 9 color photographs (Colwell et al., 1969). It is further stated that igneous and metamorphic rock masses are "consistently identifiable" and that sedimentary rock masses are "usually identifiable." This statement may be true for the Gila Bend area of Arizona, but certainly not for southern New Mexico. For example, the bedded volcanic rocks of the Sierra de las Uvas (Fig. 1) cannot be distinguished from the bedded sedimentary rocks in the San Andres Mountains, and the igneous mass of the Organ Mountains cannot be disinguished from the folded limestones of the Sierra Juarez.

With respect to small geologic features, the authors of many photogeology texts (e.g., Ray, 1960; von Bandat, 1962; Lattman and Ray, 1965; Allum, 1966) do not discuss the minimum ground resolution necessary for geologic interpretation. The authors of most texts do recommend a contact print scale of 1:20,000 as a good compromise among the requirements of visibility, cost, and handling. The use of smaller-scale (1:60,000 to 1:100,000) mosaics and stereophotographs in conjunction with the larger-scale prints is also recommended.

Although a complex subject, resolution, for purposes of practical photointerpretation, may be defined as the size of the smallest object of interest that is visible on a photograph. Identification of such an object requires collateral information from field checking or larger-scale photographs. The author agrees with Melton (1953) that "Geologists will always be interested to see the features revealed by the greatest magnification that the grain of the photographs will permit . . . . The careful geologist will always work to the limit of visibility in photographs, though his map may not show all the features thus seen." For a given format, the better the resolution of the photograph, the better will be the geologic interpretation derived. The trouble in photographic planning is that, for a given film size and altitude, the increased resolution obtained with a longer focal length lens decreases areal coverage correspondingly. Because there is a practical limit to increasing film size, an urgent need for more information as to the resolution requirements of geologists and other photographic users has become evident. In evaluating these requirements, one must also consider the generation of photograph to be used and the process control necessary at each step in reproduction.

The geologic features that are mappable using different scales of photography in each of three regions are illustrated in the following sections, which also contain some information on the resolution required to identify sedimentary outcrops, faults, igneous rock bodies, and geomorphic features.

Southern Franklin Mountains

In the southern Franklin Mountains, west-dipping Paleozoic sedimentary rocks overlie Precambrian granite (McAnulty, 1968). The rocks are cut by a complex fault system. On space photographs (Fig. 1), the general strike and dip of the sedimentary rocks may be determined, but not as well as in the San Andres Mountains to the north where more color and tone contrast exists between rock units. On Apollo 6 photographs, the original scale of which was approximately 1:2,800,000 a 10-acre field (600 by 600 feet) is almost the smallest object the shape of which can be discerned, although many narrower, linear, high-contrast objects may be seen (Lowman, 1969). The mass of lower Paleozoic rocks that comprises the southernmost Franklin Mountains (Fig. 2, see color insert) is 2000 to 2500 feet thick and dips at about 30°; the inface and dip slope of the outcrop belt are each about 2500 feet wide. Using the space photograph, one can infer that the material underlying the eastern portion of the mountain is different from the bedded western portion because the eastern slope is smoother and redder. North-trending faults may be inferred from straight valley segments that parallel isolated hill blocks. Reference to a 1:250,000 geologic map of the region (Barnes, 1968) confirms the faulting and allows identification of the two types of material —Precambrian granite and lower Paleozoic sedimentary rocks.

A much more complete photointerpretation can be made from 1:250,000 photographs, as would be expected. Figure 2 was printed from 70 mm negatives; 9-inch film exposed with a lens of the same focal length at the same altitude would give coverage of 36 by 36 miles (1300 square miles) compared with the 100 by 100 miles (10,000 square miles) of the space photographs. Automobiles (10 by 20 feet) cannot be distinguished on Figure 2, but house (30 by 30 to 50 by 50

Figure 3 Southernmost Franklin Mountains taken at 10,000 feet with color film (but shown here in black and white). Bushes and outcrops 10 to 20 feet across can be distinguished and details of sampling traverses can be planned. (2448 film, mapping camera; original scale, 1:20,000.)

Figure 4 Southernmost Franklin Mountains taken at 10,000 feet with panchromatic film. The resolution of this photograph is about the same as Figure 3. (Panchromatic film, modified Hasselblad camera; original scale, 1:40,000.)

feet) can be distinguished and classified. The effective ground resolution thus seems to be about 30 feet. The increase in resolution shows 50- to 100-foot-high cliffs and slopes representing discrete rock units that can be traced along the mountain front. Narrow roads on the west side of the mountain also may be distinguished. A 50-foot-wide sill of white rhyolite (McAnulty, 1968) within the dark reddish-brown Bliss sandstone (Figs. 3 to 6) is particularly striking. [See color insert for Figs. 5 and 6.]

The increase in resolution from Figure 2 to Figure 1 allows the reddish-brown Bliss sandstone to be distinguished from the reddish-brown granite at many places along the mountain front. These two rock units may have a very similar outcrop appearance (Figs. 7 to 9)[2]; but, when the entire outcrop belt is viewed, the bedding of the sandstone clearly contrasts with the massive granite (compare Figures 2, 7, and 10). Gravel-covered terraces can be distinguished from bedrock outcrops on the

[2] Not reproduced here.

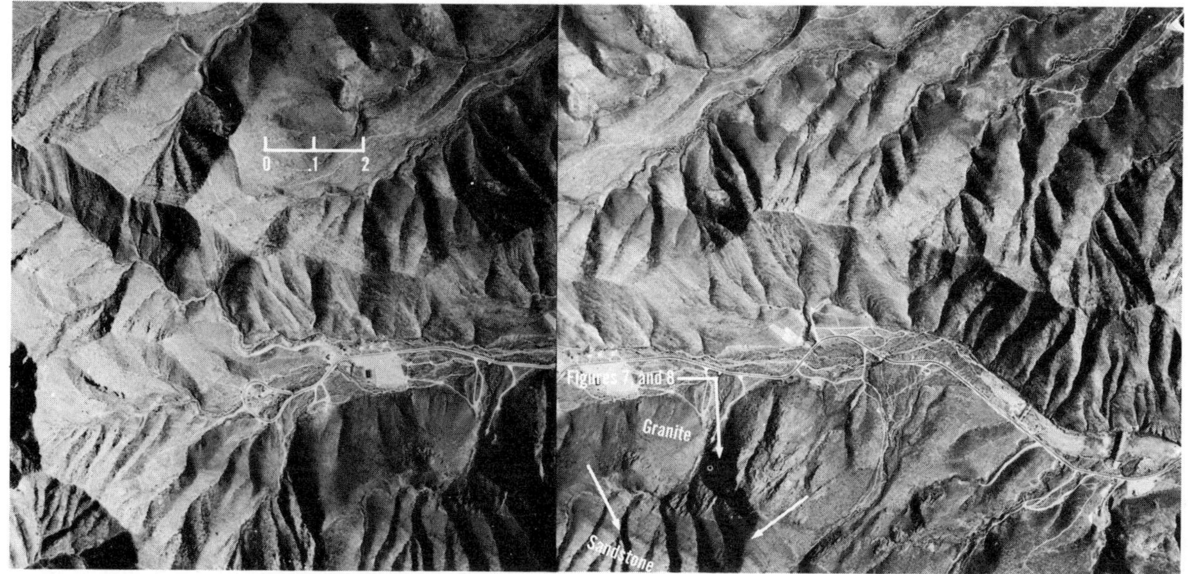

Figure 10 McKelligan Canyon. The sandstone outcrops clearly reflect bedding in contrast to the massive, but jointed, granite; however, without the highly reflective bed at the base of the sandstone, this contact would be much more difficult to map accurately. (Hasselblad camera; original scale, 1:40,000; compare with Figs. 2 and 7.)

black and white photographs, although to a lesser extent than on SO-121 (color) photographs taken at the same time. A much better rationale can be developed for mapping faults from Figure 2 than from Figure 1; for example, an apparent recent scarp bordering the west side of the mountain can be seen in Figure 2, but not in Figure 1.

Aircraft photographs of the southern Franklin Mountains at scales of 1:20,000 and 1:10,000 are shown in Figures 3, 4, and 10. The effective ground resolution of the mapping-camera photographs (Fig. 3) taken with a 6-inch-focal-length lens is similar to that of the Hasselblad photographs (Figs. 4 and 10) taken at the same time at the same altitude with a 3-inch lens—about 5 feet. In this instance, the 6-inch mapping-camera lens has compensated for the relatively poor resolution of the color film used compared to the black and white film used [32 line pairs per millimeter versus 65 line pairs per millimeter for a target object contrast ratio of 1.6:1 (Kodak, 1968)]. The resolution of these photographs is adequate for general geologic mapping. Ledges $\pm$ 10 feet high, covered slopes, and jeep trails show clearly, allowing accurate planning of traverses to investigate the important outcrops with minimal repetition and effort. Contacts between bedrock and alluvium are very clear, particularly on color photographs. The scale of 1:10,000 is more convenient for drawing contacts, small faults, and routes of measured sections, but the 1:20,000 photograph covers four times the area on the same size print. Neither scale is adequate for plotting sample locations, joint patterns, details of lateral changes in lithofacies, or depositional-current directional data.

Central Franklin Mountains

Roads in John Mays Memorial Park provide access to a block of Permian Hueco limestone and older rocks on the west side of the central Franklin Mountains (Fig. 11). The study of outcrops in this region provided a better understanding of what the photograph shows. The Hueco Formation is very evenly bedded, fairly resistant, and highly reflective (Fig. 12). Some experience in identifying outcrop characteristics in the region would make possible fairly certain photographic identification of the resistant beds as limestone or dolomite and tentative identification of the less resistant beds as shale or shaly limestone.

The banded-outcrop pattern is presumably related to differences in rock type, but the photographs show only differences in terrain color and tone. From 1:20,000 aerial photographs that have an effective ground resolution of approximately 5 feet, the outcrop bands could be interpreted as differences in rock or soil color. The resistant, highly reflective ledges are, in fact, bare rock,

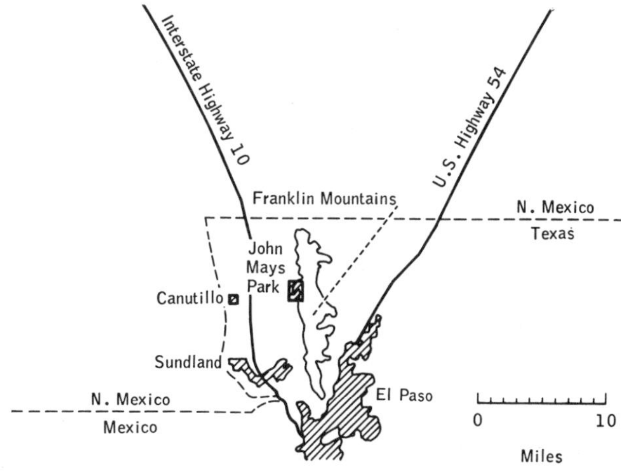

Figure 11 Index map locating John Mays Memorial Park.

Figure 12 Portion of John Mays Memorial Park. (Mapping camera; original scale, 1:18,000.)

but the less resistant, less reflective bands consist of lechuguilla, grass, sotol, dead vegetation, and some brown dirt (Fig. 13)[3]. The resolution of the aerial photographs does not permit differentiation among rock, soil, or vegetation, except for bushes larger than 5 feet across. This inability to differentiate objects was true for black and white, color, and color infrared photographs taken in October; presumably color infrared photographs taken during the growing season would help in distinguishing vegetation bands from rock or soil.

West Potrillo Mountains

The West Potrillo volcanic field consists of cinder cones, lava flows, and maare (volcanic-gas blowout craters) that cover approximately 600 square miles in southern New Mexico and northernmost Chihuahua (Fig. 1). Similar cones and flows bordering the Rio Grande Valley farther east are undoubtedly related (De Hon, 1965). The volcanic landforms are composed of black to reddish brown basaltic rock that is much less reflective than the surrounding sand and vegetation, so that the volcanic features are high-contrast targets that show up very clearly on space photographs (Fig. 1; and Lowman, 1964).

The spatter cone and lava flow of Little Black Peak (Fig. 14)[3] form a target about one-eighth by one-fourth miles (650 by 1300 feet) that can be distinguished but not identified on space photographs (Fig. 1); however, the target can be identified clearly on photographs taken at a scale of 1:250,000. The surface of the flow consists of cobbles and boulders surrounded by sand and brush (Fig. 15)[3]. Because of the vegetation, distinguishing the flow boundaries is much easier on vertical photographs than on closer oblique views. On the northern edge of the Potrillo Maar, a similar cone and lava flow of approximately the same size are barely distinguishable on space photographs, but can be mapped very well on photographs taken at 1:250,000 (Fig. 16)[3]. The cone and the flow form a very high-contrast target because the lava is surrounded by nearly white material (Fig. 17)[3]. For non-linear geologic features, the effective resolution of Apollo 6 photographs is evidently considerably poorer than the 200 feet often quoted.

Larger volcanic features can be distinguished on space photographs by shape, color, and tone. Large cones and the central craters are identified easily (Fig. 1). Dark streaks extend eastward from the cones, paralleling a pronounced east-west streaking of surficial deposits in much of the region. The first interpretations were that the volcanic field was very young and that the dark streaks were cinders spread eastward by prevailing winds during eruption. A closer look (Fig. 18, see color insert) revealed that the cones are deeply eroded, faulted, and probably older than originally supposed. Detailed interpretation disclosed a complex relationship among dark surfaces, erosional levels, probable alluvial material, and blown sand. Fieldwork—that is, making observations at a much higher resolution—showed that dark areas are underlain by volcanic rocks of various types, and that many of the light areas are covered by blown sand. The dark streaks apparent on space photographs may represent leeward protection of these areas, behind hills of any origin, from predominantly eastward sand transport.

The cinder cones that appear so smooth at relatively low resolution (Fig. 19)[3] are deeply gullied and tend to be covered with bombs and blocks up to 3 feet across (Fig. 20)[3]. The interior of the quarried cone investigated contains few such blocks, which is normal for basaltic cinder cones (MacDonald, 1967). Even with relatively high-resolution photographs of the region, one could seriously underestimate the difficulty of traversing these cones unless he had considerable interpretive experience and had done some field checking.

A similar misinterpretation of the flatter portion of Figure 19 could be easily made. Even at 5- to 10-foot resolution, the terrain appears smooth-surfaced except for small hummocky landforms that could be interpreted as stabilizing dunes. However, the surface consists mainly of lava blocks protruding through a few inches of sand (Fig. 21)[3], and the topography is formed by cooling features on lava flows. . . . Pockets of deeper sand are sparsely covered with large clumps of mesquite that are at the limit of identification on the 1:250,000 photographs . . . ; these clumps can be identified easily on 1:20,000 photographs. Of significance are the facts that the exclusive use of low-resolution photographs might lead to a serious misinterpretation of the engineering properties of the surficial material, and that underestimating the extent of the Potrillo lava field would be very easy, except where lava is bare or only slightly covered, as around Kilbourne Hole (Fig. 1).

The main value of small-scale photography, as has been stated many times, is the synoptic overview of large

[3] Not reproduced here.

features, particularly linear features. The West Potrillo Mountains illustrate this statement very well. In the western part of the volcanic field (Fig. 1), the cinder cones can be interpreted to occur along northeast-southwest trends. The three large maare on the east side of the field have a similar alignment, as does the chain of small volcanoes bordering the Rio Grande Valley (De Hon, 1965). De Hon postulated a relationship between the alignment of the maare and volcanoes, and the trends of known faults that are young enough to have displaced the surface of the desert. Two of these faults are diagramed on Figure 1. The west-northwest to east-southeast trending faultline scarp that bounds the eastern side of the East Potrillo Mountains may be seen in Figure 16. Many faults have been mapped along the Rio Grande north of Las Cruces by Kottlowski (1953). On photographs such as Figure 1, the north to northeast trending fault system mapped by Kottlowski can be seen to extend southward to the border of the West Potrillo volcanic field, perhaps tying into the interpreted alignment of cinder cones. These faults, and others in the Sierra de las Uvas, are not shown on the newest Geologic Map of New Mexico (Dane and Bachman, 1965), no doubt because a map of the faults has not been published formally. It should be remembered that the faults are not shown on the film, but must be inferred from straight valley segments and apparent vertical displacement of similar-appearing strata. The valley segments are approximately ½ mile wide by 5 to 10 miles long and consist of light-colored valley fill and dry vegetation bounded by darker bedrock hills. A line of dark evergreen phreatophyte vegetation occurs down the middle of many valleys, which heightens the contrast of these linear targets.

Comparison of Color and Black and White Films

Color photography has been acclaimed widely as a great improvement over black and white photography for geologic photointerpretation (e.g., Fischer, 1958; Pressman, 1968). Obviously, color provides a definite advantage if geologic features such as rock outcrops have different colors, and if these outcrops are recorded as the same gray tone on panchromatic film. Color photographs can be advantageous also if features are easier to discern and identify on color film than on panchromatic film. Very little published information exists on geologic interpretation of color versus black and white photography taken under controlled conditions; that is, of photographs exposed at the same time with the same camera system and at the same resolution and scale. Such a test is difficult to set up because, for example, the panchromatic films commonly used in aerial photography have significantly higher resolutions than the color films in general use (Kodak, 1968). Study of many different types of terrain under controlled conditions will be necessary before conclusions can be more than subjective and limited, because color differences between geologic features vary so markedly with changes of climate, physiographic history, weather, and lighting.

During the Apollo 7 and Apollo 9 missions, color and black and white photographs of the southern Franklin Mountains were obtained by the NASA Convair 240 aircraft (Missions 981 and 89, Table 1). The rocks in this region do not display wide variations in color, but range from light brownish gray to dark reddish brown. Comparison of prints made from aerial Ektachrome and panchromatic film of the region shown in Figures 3 and 4 indicates that few, if any, geologic features show up better on color film than on black and white film of the same resolution. Use of color film is advantageous, though, because many differences in soils and rocks can be seen more easily and more quickly, which not only speeds interpretation but also increases confidence in the interpretation.

Color photographs cannot be illustrated in this paper; however, with respect to this study of photographs of southern New Mexico taken by low-altitude aircraft, high-altitude aircraft, and spacecraft, it can safely be stated that color photographs are superior to black and white photographs for geologic interpretations. Color photographs portray the terrain more accurately and, therefore, are easier to interpret.

Comparison of Color Films

Controlled tests of the geologic value of different types of color film are even rarer than tests of color versus black and white, if the published literature is a valid indicator. Lowman (1969) stated that Kodak SO-121 film exposed in spacecraft over southern New Mexico is superior to both Kodak SO-217 film and Ansco D-50 film in rendition of ground color and resolution. Objectively distinguishing information derived by use of color tones from information derived by an increase in resolution is, of course, very difficult. Type SO-121 film has the highest resolution of any color film used in NASA spacecraft; in fact, SD-121 film has a higher resolution than most aerial panchromatic films (Kodak, 1968). The overall reddish cast of most duplicate transparencies and prints made from SO-121 originals tends to match the true color of the southwestern deserts better than the other films that have been used to date. After studying photographs taken on Gemini and Apollo missions over southern New Mexico, the author agrees with Lowman. The color infrared photographs taken on the Apollo 9 mission are better for studying forestry and surface hydrology because the distribution of growing vegetation (related to the presence of water) is much easier to infer from color infrared film. Interpretation of some geologic features should be enhanced by patterns of vegetation distribution; unfortunately, the color infrared coverage obtained on the Apollo 9 mission did not extend into the region of southern New Mexico for which comparable spacecraft and aircraft color photographs were obtained for this study.

During the Apollo 7 and Apollo 9 missions, both color and color infrared photographs were obtained from high- and low-altitude aircraft (Table 1). Each of the four types of color film contain useful geologic information when properly filtered, exposed, processed, and duplicated. Because the vegetation in the photographed region was dormant and because no haze was present, the Kodak 8443 (color infrared) photographs did not differ significantly in geologic information content from the Kodak 8442 and 2448 (color) photographs, which have about the same resolution. Type SO-121 film is preferable because of the better resolution and color rendition of the desert, as might be expected from study of the space photography.

The film from the four aircraft missions was processed and duplicated by two different laboratories at different times during 1968 and 1969. Although no reason exists to suppose that the photographic prints were not the best possible of the laboratories, it is evident that the differences in color balance and brightness between two emulsions exposed at the same time may be smaller than the differences between strips of the same film exposed at different times, or processed by different laboratories, or processed by the same laboratory at different times. Most geologists are not aware of these problems, nor are they aware that so-called duplicate transparencies are actually exposed on different emulsions from the original film. Such copy films have different properties from the original films. The results obtained depend on such variables as type of emulsion, the age of the film, the light source used, variations in line voltage during and between exposures, processing chemistry and methods, and the light source used for viewing. These variables make discussions of color balance and brightness academic except for extreme differences such as color infrared versus color.

On the basis of this study, it was concluded that the highest resolution film available should be used; of course, consideration must be given to the cost of the film, the necessity for special filters, the availability and cost of processing, and the ease and cost of duplication. For purposes of geologic interpretation, a film such as SO-121 would seem to be preferable; but when the geologist anticipates that a significant amount of growing vegetation will be imaged or that a severe haze problem will be encountered, color infrared film should be exposed concurrently.

Summary

During the Apollo 6, 7, and 9 missions, color, color infrared, and black and white photographs were obtained of southern New Mexico during April, 1968, October, 1968, and March, 1969 respectively. During October, 1968, and March, 1969, similar photographs of portions of the Potrillo and Franklin Mountains were obtained from high- and low-altitude aircraft. The photographic scale of the original films ranges from 1:3,000,000 to 1:250,000 and from 1:60,000 to 1:20,000; the effective ground resolution ranges from several hundred feet to about 5 feet. The photographs were compared by photo-interpretative methods, in addition to limited field checking, to determine resolution requirements and optimum scales for mapping the several types of geologic features that occur in the region.

Resolution is the major limiting factor in the geologic interpretation of these photographs. On the 1:300,000 transparencies, flows of black lava, ⅛ by ½ mile in size, surrounded by highly reflective sand and silt, can be resolved but not identified without larger scale photographs of field checking. Larger flows, outcrop belts ½ to 1 mile wide by 10 to 40 miles long, and straight valley segments along fractures can be identified. Strike and dip directions of large outcrop belts can be determined, but folded Cretaceous limestones in the Sierra Juarez cannot be distinguished from intrusive igneous rock in the Organ Mountains because the limestone outcrop belts are too small to be resolved. A major advantage of the small-scale photographs is the areal coverage; each 70 mm 1:3,000,000 photograph covers 10,000 square miles, whereas each 70 mm 1:40,000 scale photograph covers less than 2 square miles. The entire West Potrillo volcanic field can be studied on one small-scale photograph and the relationship of cinder cone alignments to a prominent northeast to southwest trending fracture system in the Sierra de las Uvas is evident.

Enlarged prints of the 1:250,000 photographs are ideal for reconnaissance fieldwork in southern New Mexico and west Texas. These prints, at an approximate scale of 1:60,000, cover about 60 square miles with an effective ground resolution of about 30 feet. Dirt roads, 50- to 100-foot-high cliffs and slopes, and many details of lava flow surfaces and sand dunes may be identified. The drainage net, young faultline scarps, and outcrop differences between granite and sandstone show clearly. Prints and transparencies made from 1:40,000 and 1:20,000 photographs have an effective ground resolution of approximately 5 feet, depending on the film type. Photographs at these scales were considered most useful for tracing individual beds, locating surface samples, and locating ground photographs. The resolution was adequate for most geologic purposes, but not satisfactory for distinguishing rock from soil or from the dead vegetation that covers most of the less resistant units in the region.

Comparison of Ektachrome SO-121, 2448, and 8442 films indicated that differences in lighting conditions, exposure, film handling, and processing have more effect on color saturation, brightness, and (to some extent) hue, than do differences between film types. With respect to saturation and brightness, the same is also true for color infrared (Ektachrome 8443) film. So-called duplicate transparencies and prints do not actually duplicate the brightness, saturation, or color balance of the original film. For geologic interpretation, this study indicated that the geologic information content of any usable color film, whatever the color balance, will be approximately the same at comparable resolutions. Therefore, the highest resolution film available (in this instance, SO-121) is the best film to use for geologic work. In more humid regions or during the growing season in arid regions, additional information will be gained by the concurrent use of color infrared film, which not only emphasizes healthy vegetation but has a better haze-penetrating capability.

Comparison of the four types of color film with black and white exposed at the same time and at the same scale indicates that in this region during the dry season the main advantages of using color film are speeding the interpretive process and increasing confidence in the interpretation. The normally reddish tint of most prints and transparencies made from SO-121 most closely matches the subjective visual impression of the terrain.

The use of two scales of photography should become routine for future geologic interpretation. A combination of photographs taken at 1:2,500,000 and 1:60,000 scales is excellent for regional mapping and reconnaissance work; a combination of 1:250,000 and 1:24,000 scales is better for detailed studies that require accurate spatial location of rock samples, fossil locations, or other ground data measurements.

REFERENCES

Abdel-Gawad, M. Geologic exploration and mapping from space: Paper submitted to the Proceedings of the American Astronautical Society, 1968. Abstract in Am. Astron. Soc. Science and Technology Series, Vol. 17, 1968, p. 41.

Allum, J. A. E. *Photogeology and Regional Mapping.* New York: Pergamon, 1966.

Barnes, V. E. Van Horn-El Paso sheet. *Geologic Atlas of Texas.* Austin, Tex.: Bureau of Economic Geology, University of Texas at Austin, 1968.

Carter, W. D. Some Uses of Space Photography in Earth Resources Surveys. Smith, J. T., Jr., and Anson, A. (eds.), *Manual of Color Aerial Photography.* Falls Church, Va.: American Society of Photogrammetry, 1968. Pp. 398–399.

Colwell, R. N., et al. A survey of earth resources on Apollo 9 photography. Special report for NASA contract no. NAS 9-9348. Forestry Remote Sensing Laboratory, School of Forestry and Conservation, University of California, 1969.

Dane, C. H., and Bachman, G. O. Geologic Map of New Mexico. Washington, D.C.: Dept. of Interior, U.S. Geological Survey, 1965.

De Hon, R. A. Maare of La Mesa. Fitzsimmons, J. P., and Loch-Man-Balk, C. (eds.), New Mexico Geological Society, *Guidebook of Southwestern New Mexico* II, 16th Field Conference, Oct. 15–17, 1965. Pp. 204–209.

Drewes, H., and Morrison, R. Extent of relict soils revealed by Gemini IV photographs. Washington, D.C.: U.S. Geological Survey Technical Letter NASA–60, 1966.

Fischer, W. A. Color aerial photography in photogeological interpretation. *Photogram. Engng.* 24:545, 1958.

Hemphill, W. R., and Danilchik, W. Geologic interpretation of a Gemini photo. *Photogram. Engng.* 34:150, 1968.

Kodak. *Characteristics of Kodak Aerial Films, July, 1968.* Rochester, N.Y.: Eastman Kodak Co., 1968.

Kottlowski, F. E. Road log from El Paso to Las Cruces. New Mexico Geological Society, *Guidebook of Southwestern New Mexico,* 4th Field Conference, Oct. 15–18, 1953. Pp. 18–63.

Lattman, L. H., and Ray, R. G. *Aerial Photographs in Field Geology.* New York: Holt, Rinehart, and Winston, 1965.

Lowman, P. D., Jr. A review of photography of the earth from sounding rockets and satellites. NASA TN D-1868, 1964.

— Geologic orbital photography: experience from the Gemini Program. NASA TM X-63244, 1968.

— Preliminary results from Apollo 6 70-millimeter photographs of North America taken on the Apollo mission: science report on the 70-millimeter photography of the Apollo 6 mission. NASA S-217, 1969.

MacDonald, G. A. Extrusive Basaltic Rocks. Hess, H. H., and Poldervaart, A., (eds.), *Basalts,* vol. 1. New York: Interscience, 1967. Pp. 1–61.

McAnulty, W. N. The Franklin Mountains and Mount Cristo del Rey, El Paso County, Texas: Guidebook to the Guadalupe Mountains, Hueco Mountains, Franklin Mountains, and geology of the Carlsbad Caverns, West Texas Geological Society, 1968. Pp. 50–58.

Melton, F. A. Geologic Exploration and Mapping with Aerial Photographs. Pp. 219–225, *Selected Papers on Photogeology and Photointerpretation.* Pub. GG 209/1. Washington, D.C.: Committee on Geophysics and Geography, Research and Development Board, 1953.

Pecora, W. T. Geologic applications of earth orbital satellites. United Nations conference on the exploration and peaceful uses of outer space. Vienna, Austria, August 14–27, 1968, Paper 68–95441.

— The earth resources observation satellite program. Statement to Congress, May 1, 1969.

Pressman, A. E. Geological comparison of Ektachrome and infrared Ektachrome photography. Pp. 396–397, Smith, J. T., Jr., and Anson, A. (eds.), *Manual of Color Aerial Photography.* Falls Church, Va.: American Society of Photogrammetry, 1968.

Ray, R. G. Aerial photographs in geologic interpretation and mapping. U.S. Geological Survey Prof. Paper 373, 1960.

von Bandat, H. F. *Aerogeology.* Houston: Gulf Pub. Co., 1962.

Wobber, F. J. Orbital photography: applied earth survey tool. *Photographic Applications in Science and Technology,* v. 2, 1968.

AFTER WORLD WAR II, pictures taken from high-altitude sounding rockets were found to contain considerable geographic and geologic information. Many of these were of excellent quality. Yet it was not until the early 1960s that a study of their geologic utility was published. Moving the camera platform farther from the earth by mounting it on a space vehicle has produced photographs of considerably smaller scale. As scale decreases, a larger surface area of the earth is imaged in each photograph, but there are more generalities produced due to reduced scale and the more limited resolving power of the camera system.

Some value is achieved from this reduction in scale and resolution. For example, mapping of photo-linears has become a valuable tool in regional and local stress analyses. Surface structural features such as changes in vegetation or land use patterns, straight sectors of streams, and linear arrangement of topographic forms are usually the result of fractures or lineaments. Because of their size these features may not be perceived by ground-based observers even when the observer is using suborbital photographs but they are frequently visible on earth orbital images. Photography from space provides synoptic coverage of large areas and under nearly uniform lighting and image exposure conditions. The regional view and the analysis of linear features provide insights into the regional tectonic history of an area. Photographs from space are particularly valuable when studying a very large geologic structure such as the San Andreas fault system, a regional geologic feature which extends over several hundred miles. An earth orbital photograph can provide a view of the entire fault system

Other advantages of earth orbital photographs include rapid coverage of remote or inaccessible world areas. World-wide coverage allows investigations of previously unknown relationships. It should also help eliminate the provincialism of focusing on local examples; at the same time, placing geologic features in a global context allows geologic observers to avoid the pitfall of treating them as special cases. There is the advantage of repetitive or sequential coverage. Some geologic processes occur with surprising speed. The changing nature of river point bars or deltas, coastal erosion or deposition, and volcanic eruptions are examples of surface features that can be modified rapidly with time. Earth orbital photography taken from a globe-circling satellite can monitor this change.

21-Geologic Uses of Earth Orbital Photography

PAUL D. LOWMAN, JR.

THE EARTH is in many ways as mysterious as any body in the solar system. One of the most promising ways to explore the geology of this mysterious planet is through orbital photography, in much the same way as the moon and Mars have been explored. Several thousand pictures of the earth's surface have been taken from space since the end of World War II, and their geologic uses are becoming fairly well defined. The purpose of this paper is to review the geologic applications of earth orbital photography, with examples from various manned and unmanned missions.

Invited paper, 22nd International Astronautical Congress, Brussels, Belgium, Sept., 1971. X-644-71-359, Goddard Space Flight Center, Greenbelt, Md. Reprinted with permission of the author.

History of Orbital Photography

Photography of the earth from orbital altitudes began shortly after World War II, when photographs were taken by automatic cameras carried by V-2, Aerobee,

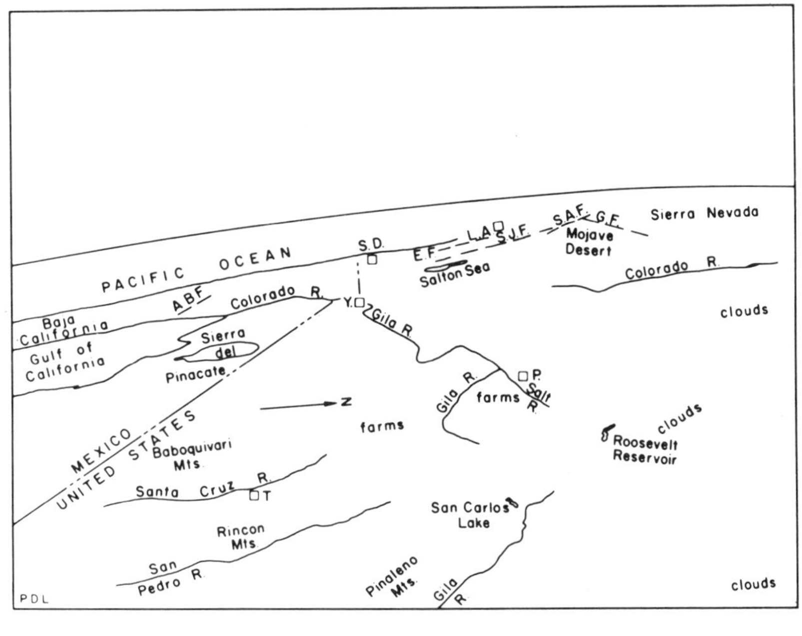

INDEX MAP

VIKING 12, PHOTOGRAPH 14

Figure 1 Viking 12 photograph taken in 1955 from altitude of about 140 miles (225 km) over White Sands Proving Ground, New Mexico, with black and white infrared film. Major faults visible include Agua Blanca (ABF), Elsinore (EF), San Jacinto (SJF), San Andreas (SAF), and Garlock (GF).

and Viking sounding rockets. These have been reviewed by Lowman (1964). Despite the excellent quality of many of these (Fig. 1), it was not until 1963 that a study of their geologic utility was published, by P. M. Merifield (1963). Shortly after, Morrison and Chown (1964) used the MA-4 Mercury photographs of North Africa in what is even today an outstanding piece of physical geography research. Merifield's work stimulated Lowman (1964) to propose synoptic terrain photography for the Mercury flights using hand-held cameras. A number of excellent pictures, chiefly of Tibet, were taken by the Mercury astronauts, leading to a similar terrain photography experiment for the Gemini flights. Because of the long duration of the Gemini missions and the availability of two men for in-flight experiments, the Synoptic Terrain Photography Experiment (S005) was extremely successful, with over 1100 photographs usable for geology, geography, or oceanography being returned. The experiment was summarized by Lowman and Tiedemann (1971).

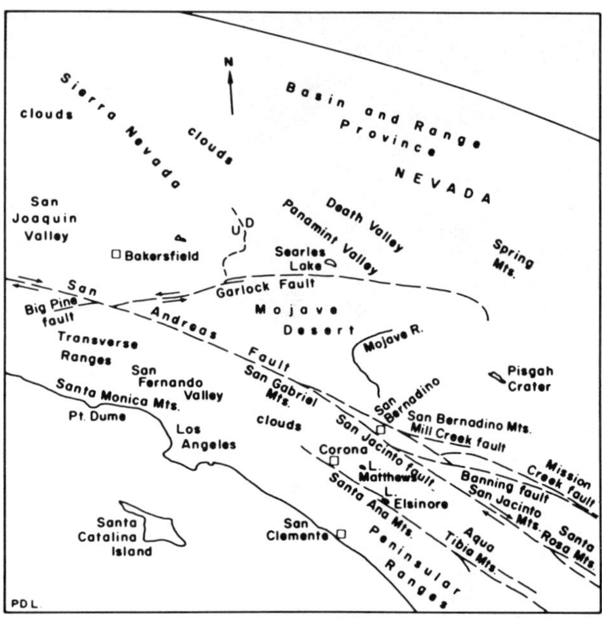

INDEX MAP
APOLLO 7 PHOTOGRAPH AS 7-11-2022
Note: Scale variable; Bakersfield-Searles Lake distance 95 miles.

Figure 2B Index map showing main physiographic and structural features of Apollo 7 photograph (see Fig. 2A, color insert).

Two terrain photography experiments were carried on the Apollo 7 and 9 missions. One was the same type of hand-held single camera photography (S005) that had been done on Gemini and Mercury missions, and was very successful, with several hundred high-quality color photographs (Fig. 2) being returned (Lowman, 1972). The other was a multispectral terrain photography experiment (S065) in which four 70 mm cameras, each covering a different spectral region, were used (Lowman, 1969). This too was successful, returning over 350 usable pictures, and demonstrating the advantage of multispectral over single-camera orbital photography. The S065 experiment also provided valuable experience in the acquisition and interpretation of multispectral photography that will prove useful in future earth resources satellites.

The unmanned Apollo 6 mission carried an automatic 70 mm camera that obtained a full revolution, minus dark areas, of high-quality color coverage, with overlapping vertical pictures. These have proven useful in a number of studies. Some geologic use has also been made of Nimbus and Tiros weather satellite images, but the low resolution of these systems makes them of interest chiefly for illustration of already-known geologic features (Fig. 3).

Four Geologic Case Histories

The geologic use of orbital photography can best be illustrated by presenting specific examples in which it has produced significant new geologic knowledge not previously uncovered by conventional methods (including aerial photography).

Structure of the Peninsular Ranges, California

The Peninsular Ranges of southern California are geologically the southern extension of the Sierra Nevada, consisting essentially of a block uplifted along faults on the east side and dipping westward into the Pacific Ocean. Lithologically they are also related to the Sierra Nevada in being composed largely of batholithic intrusions. Their regional setting is best shown by an Apollo 9 photograph (Fig. 4). Much of the Peninsular Ranges is in Baja California, whose remoteness is well-known and for which the published geologic mapping is of a reconnaissance nature only. However, the northern end of these mountains is in southern California, within an hour's drive of several major colleges and universities. It would therefore be expected that the geology of this part of the Peninsular Ranges would be well-known. However, photographs taken as part of the S065 multispectral photography experiment on Apollo 9 have uncovered a number of unmapped faults and thrown new light on the relation of the Peninsular Ranges to the regional geology.

One of the best published pre-Apollo geologic maps of the Peninsular Ranges in San Diego County is presented in Figure 5 (Johns, 1954). The predominance of northwest-trending wrench faults related to the San Andreas fault is apparent; examples include the San Jacinto and Elsinore faults. However, there are virtually no faults with a northeast trend in the block bounded by the Elsinore fault on the east. It was therefore a surprise to find, on the Apollo 9 photographs of this area, a number of northeast-trending lineaments (Fig. 6) several miles long. These lineaments are unusually straight valleys, for which structural control would be suspected. Field studies following the Apollo 9 mission has shown that some and possibly most of the lineaments are actually faults, which promote stream erosion along their traces. Other lineaments are the expression of flow structure and metamorphic screens in igneous rock, metamorphic foliation, or joints, although the last-named are generally below the resolution of the photographs.

Orbital photographs of this area also promise to clarify the nature of movement on the well-known northwest-trending faults of the San Andreas system, in particular the Elsinore fault. Although there are no numerical estimates of the amount of displacement on the Elsinore fault, its association with the San Andreas would suggest several miles of right lateral movement. (Estimates of total displacement along the San Andreas are about 250 kilometers.)

Photographs taken from Gemini 5, Apollo 7, and Apollo 9 missions show clearly that several lineaments cross the Elsinore fault without major lateral displacement (Fig. 7). One of these has been studied in some detail at its intersection with the Elsinore fault, and it has been found that the Elsinore fault must have an impossibly sinuous trace if it has had any appreciable lateral movement since the northeast-trending fault was formed. The age of the latter is not known, but it is clearly great enough to show that the Elsinore fault is not moving laterally at anything like the 1 to 5 cm per year rate of the San Andreas fault. It is quite possible, judging from

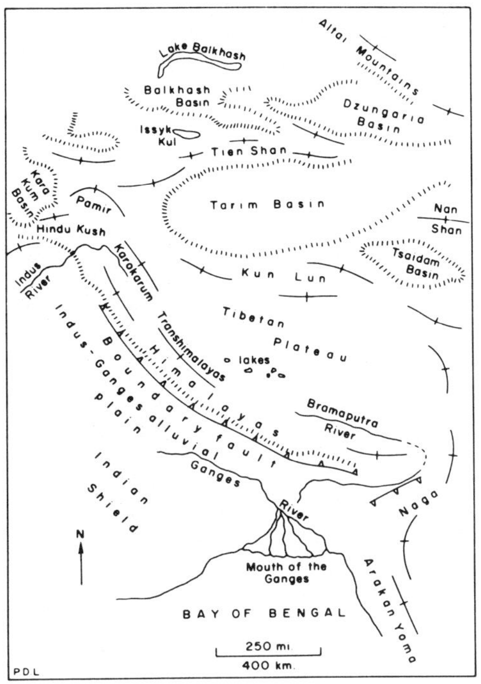

TECTONIC INDEX MAP OF SOUTH-CENTRAL ASIA BASED ON NIMBUS III IDCS PHOTOGRAPHS
Clouds not labeled

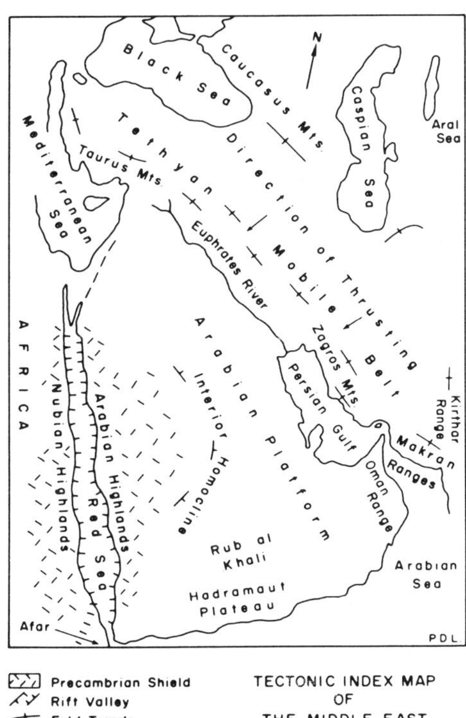

TECTONIC INDEX MAP OF THE MIDDLE EAST BASED ON NIMBUS III INFRARED RADIOMETER DAYTIME SCAN ORBIT 711, 6 JUNE 1969

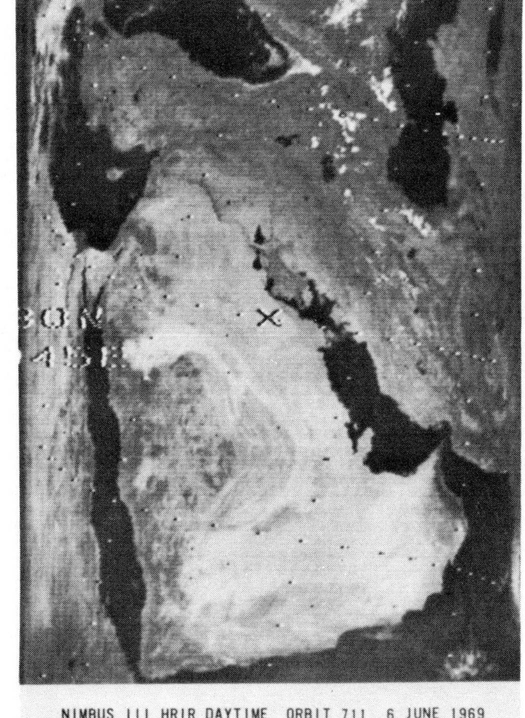

NIMBUS III HRIR DAYTIME ORBIT 711 6 JUNE 1969
ASIA MINOR

Figure 3 Nimbus IDCS photograph and HRIR scans, with maps illustrating main geologic features visible.

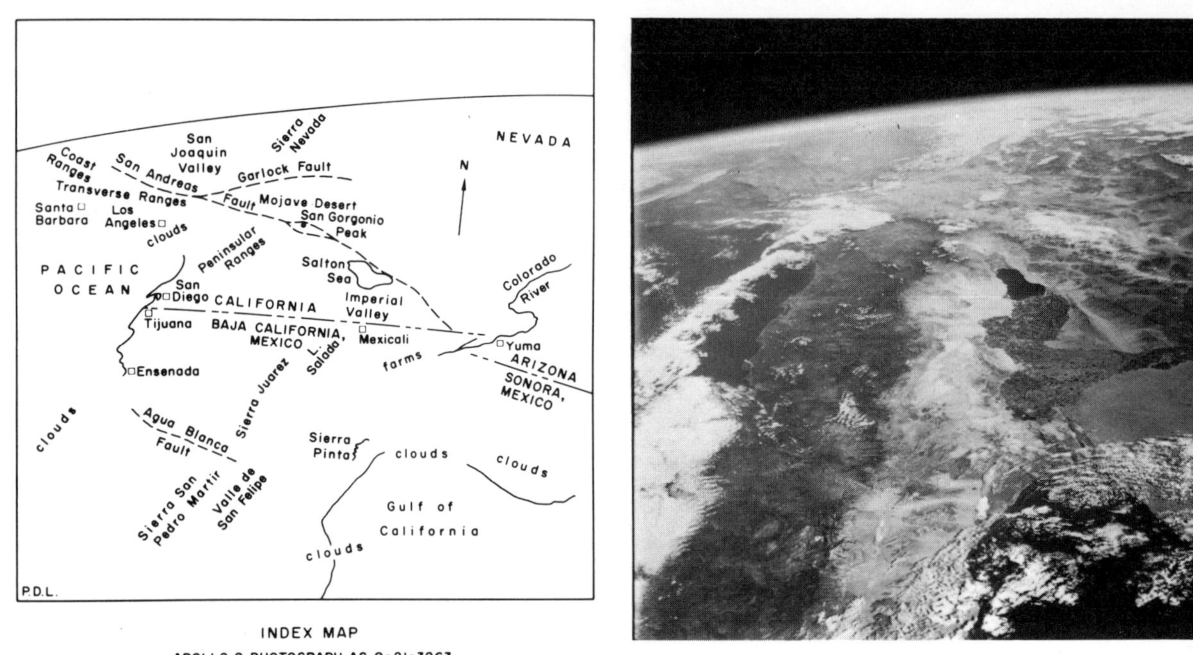

Figure 4 Apollo 9 hand-held photograph taken in 1969 with color film. (Shown in black and white.)

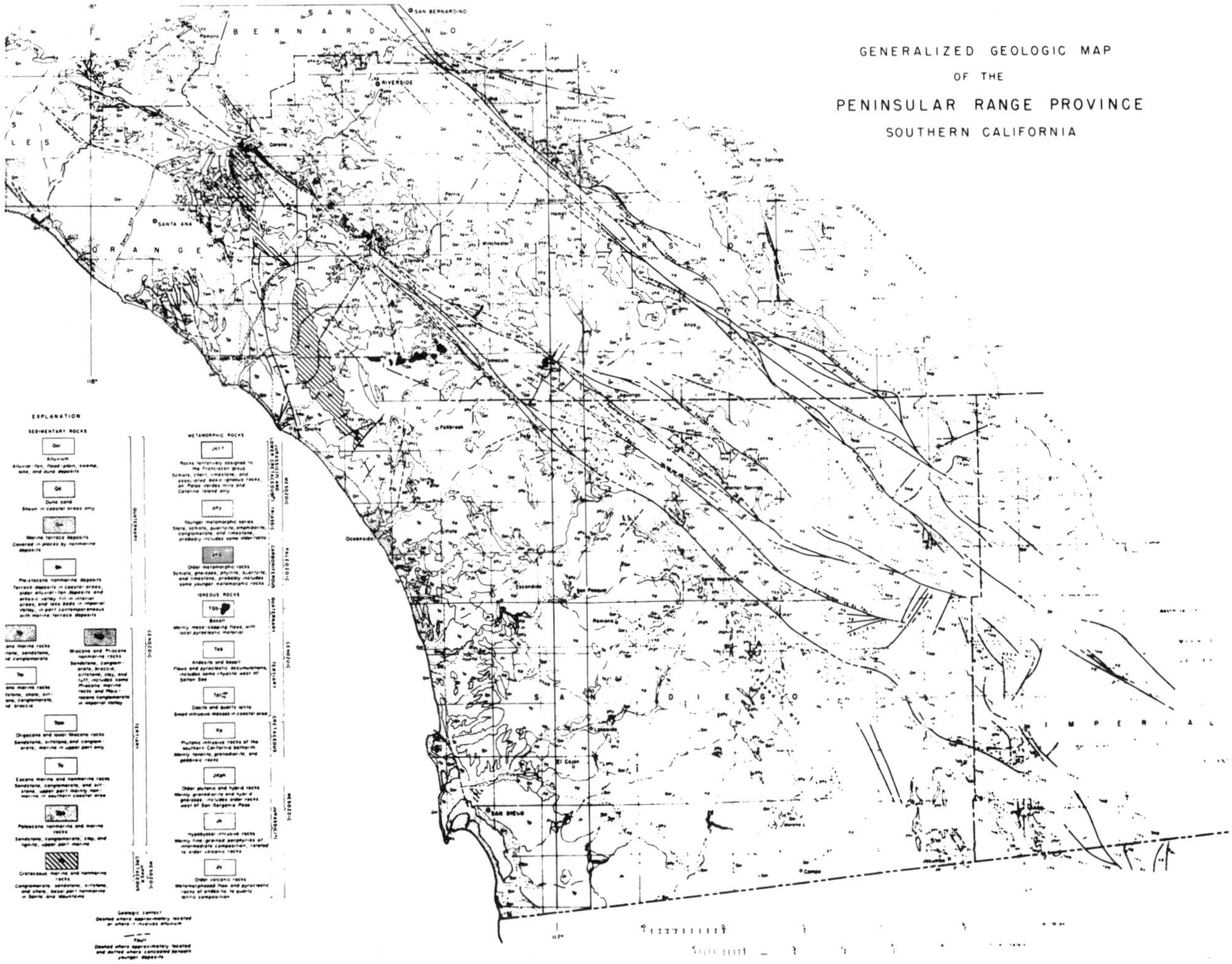

GENERALIZED GEOLOGIC MAP OF THE PENINSULAR RANGE PROVINCE SOUTHERN CALIFORNIA

Figure 5 Geologic map of San Diego County, California (Johns, 1954). Note virtual absence of north-east trending lineaments west of Elsinore fault.

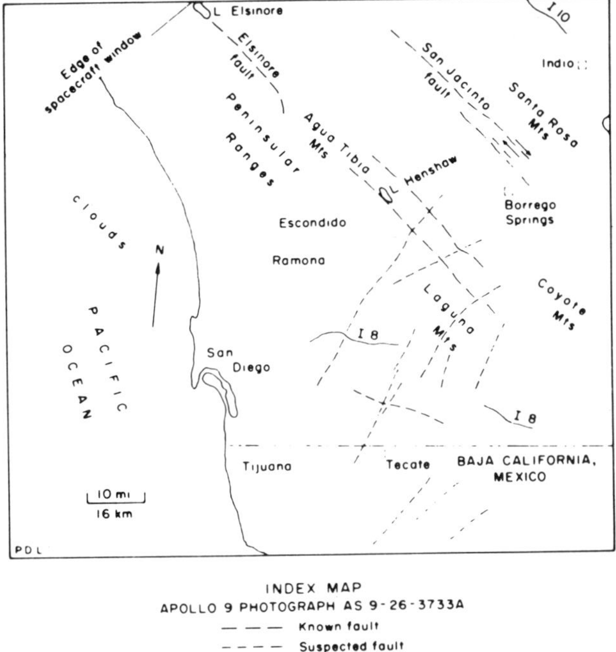

Figure 6B Index map of color infrared photograph of San Diego County, California (see Fig. 6A, color insert).

the photographs and the admittedly rapid field studies made to date, that the Elsinore fault is of primarily a dip-slip nature. Should this be proved, it would suggest re-examination of theories involving hundreds of kilometers of movement on the San Andreas system, and, more generally, of theories of continental drift depending on such movement.

Palomas Volcanic Field, Chihuahua

The Gemini 4 astronauts J. A. McDivitt and E. H. White, II, obtained a remarkably complete and cloud-free series of 39 overlapping pictures from Baja California to central Texas (Lowman, McDivitt, and White, 1967). Preliminary comparison of the photographs (Figs. 8 and 9[1]) with published geologic maps showed what appeared to be an unmapped, relatively young volcanic field in northern Chihuahua over 14 kilometers wide, with a large number of volcanoes. Further search of the Mexican and American literature revealed no mention of such a volcanic field. Accordingly, a short field reconnaissance was made by the author and H. A. Tiedemann in 1967.

It was found that the feature is indeed a volcanic field with over 30 individual volcanoes. Its age appears to be on the order of several score thousand years (Lowman and Tiedemann, 1971), judging from the degree of erosion of cones and flows; there were no signs of activity. The only volcanic rock type found was a uniform olivine basalt very similar to, and doubtless related to, scattered vents and flows just to the north, in Luna County, New Mexico.

The Palomas volcanic field has no obvious geologic importance by itself, being but one of many such Quaternary volcanic features in the southwest United States and northern Mexico. However, its "discovery" in an easily accessible area only 2 hours' drive from Ciudad Juarez and El Paso was an early demonstration of the use of orbital photography in correcting regional geologic maps. The value of such photography in initial mapping of remote areas is also strongly implied.

[1] Not reproduced here.

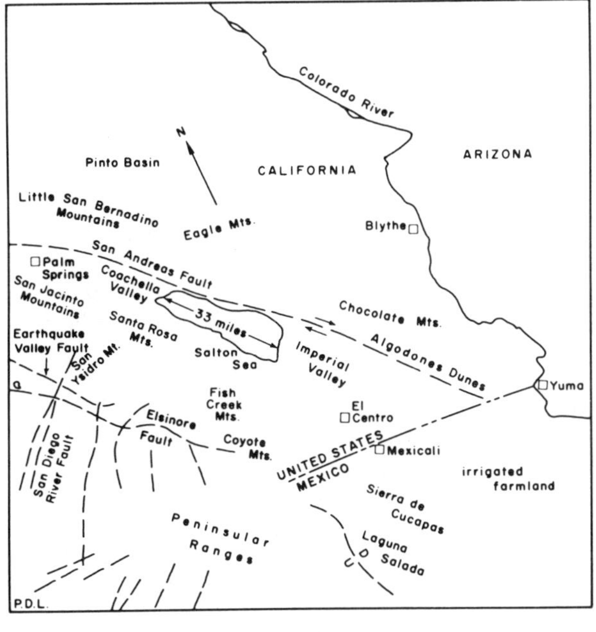

INDEX MAP
APOLLO 7 PHOTOGRAPH AS 7-11-2023

Figure 7 Apollo 7 hand-held photograph taken in 1968 with color film. (Shown in black and white.)

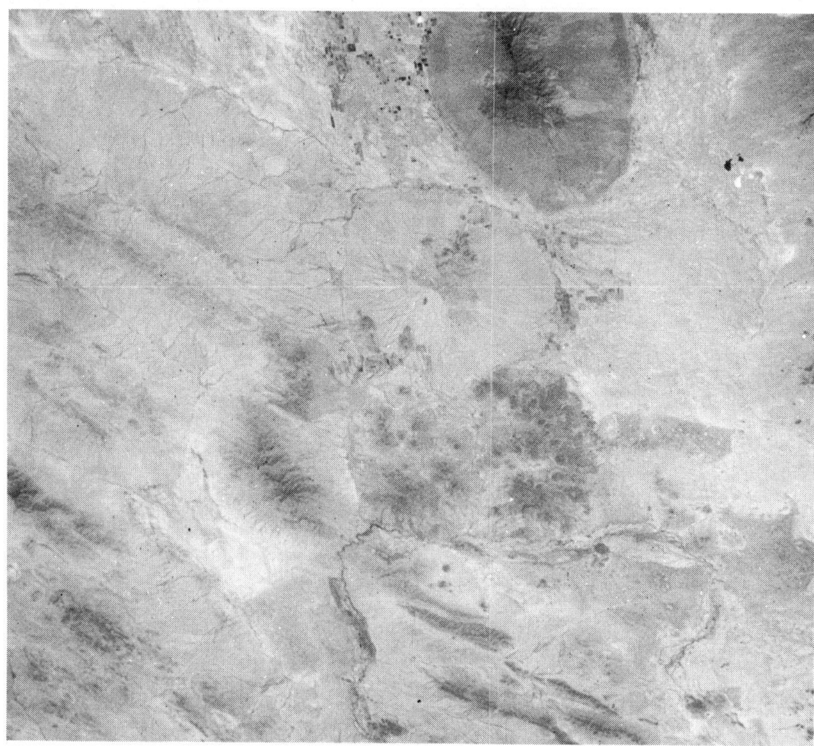

Figure 8 Gemini 4 hand-held photograph taken in 1965 with color film as part of S005 experiment. (Shown in black and white.)

The Texas Lineament

The Texas lineament is a hypothetical linear structure of subcontinental extent, extending northwest from west Texas through New Mexico, Arizona, and possibly California. It has been proposed, by various geologists since 1902, to control ore deposits, topography, ground water, and other geologic features in the southwest United States and northern Mexico. Several tectonic syntheses, reviewed by Lowman and Tiedemann (1971), have considered the lineament to be a transverse fracture zone of continental or even intercontinental significance. On the other hand, several geologists intimately familiar with the geology of the southwest United States, including the type locality 150 kilometers southwest of El Paso, completely ignore the lineament in their published reports on this area. Thus we see that while many geologists consider this structure to dominate the regional geology, others do not even think it exists.

A key area for the understanding of the Texas lineament problem is the southwest corner of New Mexico, since all treatments of the problem consider it to go through there. This area has been repeatedly photographed from sounding rockets and spacecraft, and the author and H. A. Tiedemann have therefore examined these photographs in some detail. An Apollo 9 oblique gives a good regional view (Fig. 10).

The results of this examination and related field checks cannot be presented here, but can be summarized as follows. First, it seems clear that no single fault goes through El Paso-Juarez to the northwest in the required direction; there is no hint of a fault trace, no alignment of Quaternary volcanoes in the necessary direction, and no major structural discordance between the Sierra Juarez and Franklin Mountains. Second, there is a broad zone of folding and dip-slip faulting in the N 60°W direction in southwest New Mexico, northern Mexico, and southeast Arizona, which grades into the structural continuation of the Sierra Madre Oriental in northern Chihuahua. This is shown in Figures 8 and 9. Finally, the zone commonly referred to as the Texas lineament is not a zone of transverse faulting, but one reflecting the extent of the former Mexican geosyncline, at least in the area discussed here.

If this interpretation is correct, it has definite economic implications. For example, there would be little point in basing mineral exploration in west Texas on the supposed Texas lineament, since the original structure can be shown to be a minor structure satellitic to the Sierra Madre Oriental. More generally, a better understanding of the problem can hardly fail to help in the application of tectonics to economic geology in this area.

Wind Erosion in Deserts

The importance of wind as a desert erosional agent is controversial, although the existence of this controversy is not obvious unless one examines many geology texts. One then finds that some geologists, such as Thornbury (1954), consider wind erosion relatively unimportant, while others, such as Holmes (1965), hold the opposite view. American geologists in general consider desert erosional landforms to be almost entirely the work of running water, rather than of wind erosion.

Orbital photographs have thrown considerable light on the question, since the combination of low-inclination

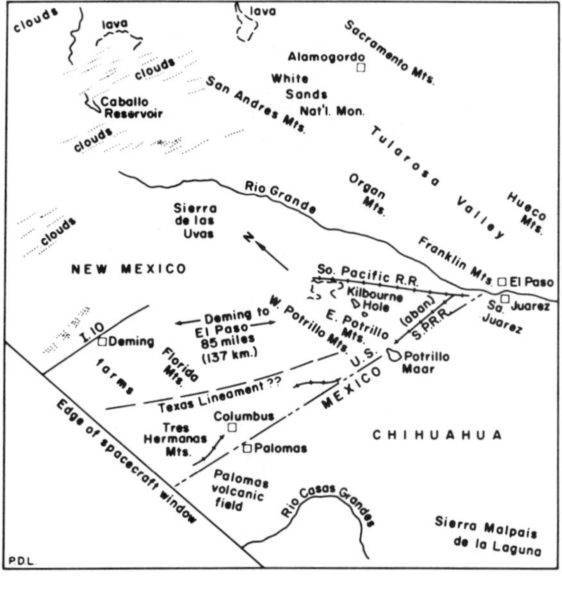

INDEX MAP
APOLLO 9 PHOTOGRAPH AS 9-21-3282
Oblique view to N.E.; scale variable.

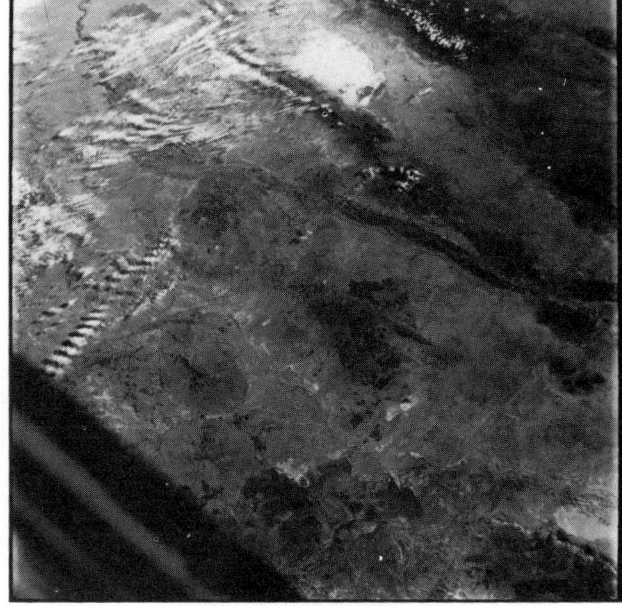

Figure 10 Apollo 9 hand-held photograph taken in 1969 with color film (shown here in black and white), showing location and geologic environment of supposed Texas lineament, interpreted as a single fault.

orbits and generally good weather in dry areas has permitted the accumulation of hundreds of excellent pictures of the world's great deserts. Study of these pictures has shown that wind erosion, by deflation and abrasion, has been a major though not necessarily the dominant land-sculpturing agent in North Africa and parts of the Middle East. In Figure 11, for example, we see an area of several thousand square kilometers in which most visible landforms are the result of deflation (removal of fine-grained material by wind). In Iran (Fig. 12)[2], a Gemini 5 picture has shown an immense field of yardangs—ridges and grooves formed by wind erosion in the soft sediments of the Dasht-i-Lut (salt desert). Numerous other examples of North African landforms produced chiefly by wind erosion have been found on the Gemini and Apollo photographs (see Lowman and Tiedemann, 1971).

On the other hand, it seems equally clear, from orbital photographs, that most of the erosion in North American deserts is the result of running water, not wind. The reason for this difference between North American and African deserts can be illustrated with still another space photograph, this one taken from the Apollo 11 spacecraft shortly after translunar injection at an altitude of several thousand kilometers (Fig. 13). Study of this picture shows that most North American deserts are part of the Basin and Range physiographic province, which consists largely of north-south trending fault block mountains. These are transverse to the prevailing winds, and thus the North American deserts are not subject to winds that can blow for hundreds of miles without obstruction, as in North Africa, which has only a few isolated massifs (Fig. 14). The North American deserts, furthermore, receive considerably more moisture than does North Africa, as shown by the heavier vegetation along mountain trends.

[2] Not available for reproduction.

Space photography, then, shows that the deserts of North America and North Africa are fundamentally different in tectonic structure and physiography, and that the relative unimportance of wind erosion in North America must not be extrapolated to the world's deserts as a group.

Advantages of Orbital Photography in Geology

The foregoing examples permit a summary of the advantages orbital photography offers the geologists. For a more general comparison of orbital and aerial photography, see Lowman (1969).

Large Area per Picture

A single photograph from orbital altitude covers several thousand square kilometers or, from deep space, an entire continent, with usable resolution, and in color or with multispectral arrays. This is of course the most striking characteristic of space photography, and cannot be duplicated by mosaics of aerial photography for several reasons. Color mosaics are impractical to produce; stereoscopic coverage is not provided by mosaics; dodging causes loss of tonal detail; constant sun angle over large areas is not provided; and mosaic coverage of large areas with similar film and scale is generally not available.

Personal Study of Large Areas

Important geologic structures, such as the Elsinore fault, are frequently so large that they can not be studied directly by individual geologists unless these geologists

INDEX MAP
APOLLO 9 PHOTOGRAPH AS 9-19-3034

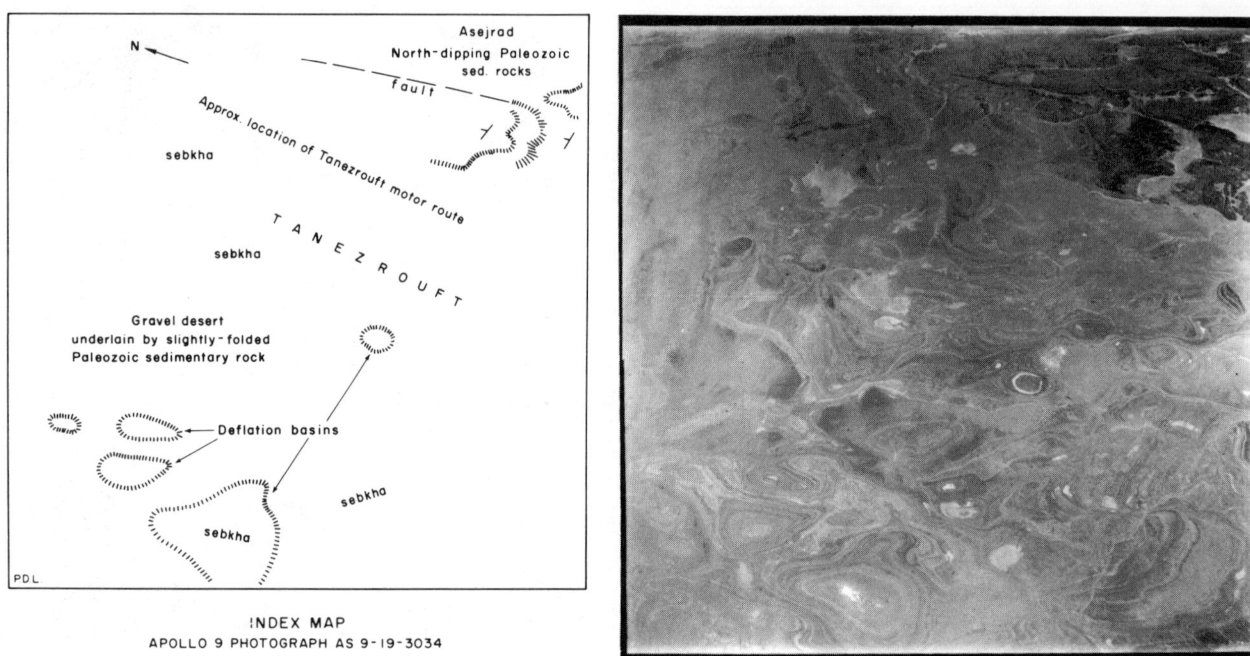

Figure 11 Apollo 9 hand-held photograph taken in 1969 with color film (shown here in black and white), showing deflation basins in southern Algeria.

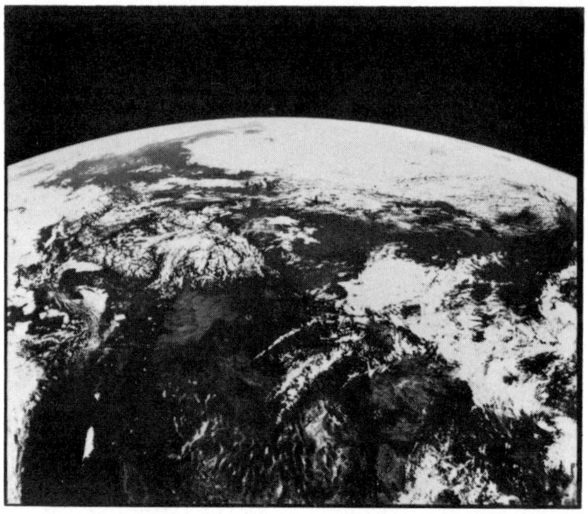

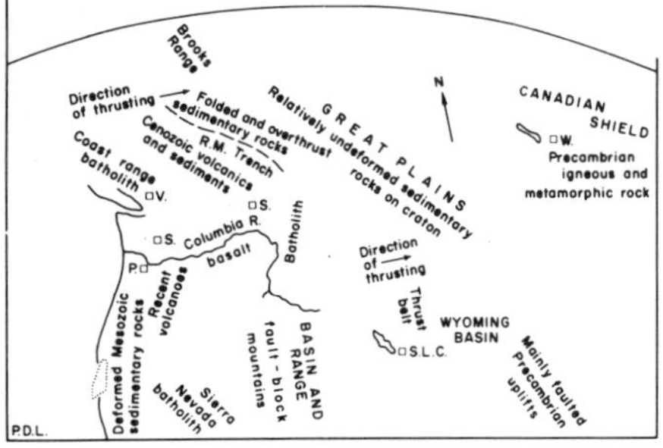

GEOLOGY
APOLLO II PHOTOGRAPH AS II-36-5302

GEOGRAPHY
APOLLO II PHOTOGRAPH AS II-36-5302

Figure 13 Apollo 11 hand-held photograph taken shortly after translunar injection at an altitude of about 6000 miles (9700 km).

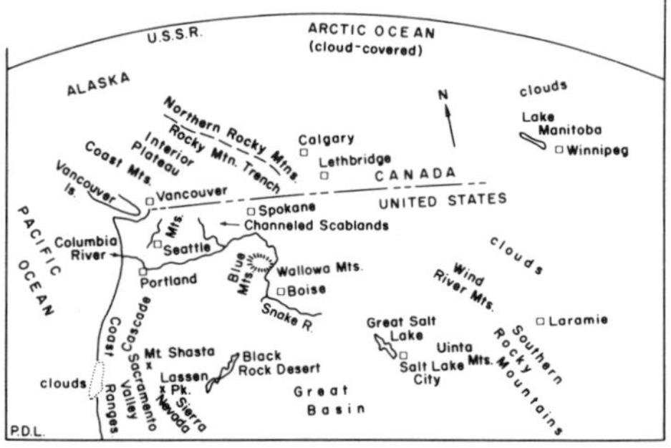

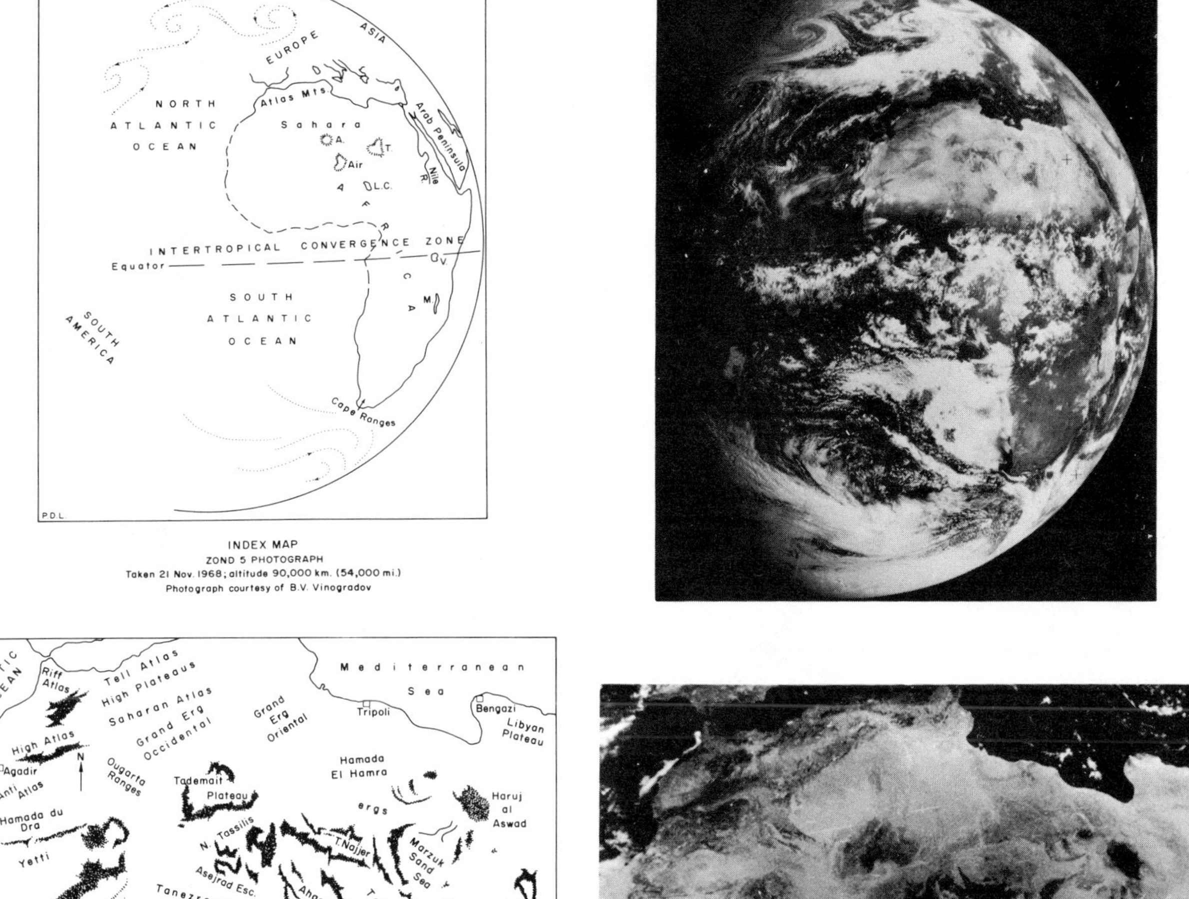

Figure 14 Zond 5 photographs showing Africa, supplied by Prof. B. V. Vinogradov. Maps from Lowman, 1971.

can devote several decades to the task. This is why the great syntheses of terrestrial geology are generally the work of men on the verge of retirement or death. Orbital photography, however, promises to remedy this unhappy situation. The large area per picture, just discussed, permits the geologist to actually see the entire length of, for example, the Elsinore fault at a glance; he can then map it, on photographs, as a whole in a relatively short time. Furthermore, the orbital photographs may show just what critical areas should be investigated in the field, thus sparing the geologist time-consuming (though enjoyable) mile-by-mile mapping on the ground.

Orbital photography, in summary, permits the direct personal investigation of extremely large structures or areas in a reasonably short time, providing continuity of thought and effort impossible with conventional methods.

Coverage of Inaccessible Areas

Darwin's observations during his world-circling voyage on H.M.S. Beagle transformed the sciences of biology and geology. And although Darwin was later to call on domestic pigeons for a proof of organic evolution, his key bird observations were made in the remote Galapagos Islands. It is often thus in geology, and orbital photography therefore presents a unique advantage by providing coverage of parts of the earth which are inaccessible for one reason or another and which can neither be visited nor photographed from the air. The Apollo 7 photograph presented in Figure 15, for example, covers a part of Tibet so poorly mapped that positive identification of several lakes scores of kilometers long was not possible. Yet the area is of great geologic interest, since it may throw light on the theory

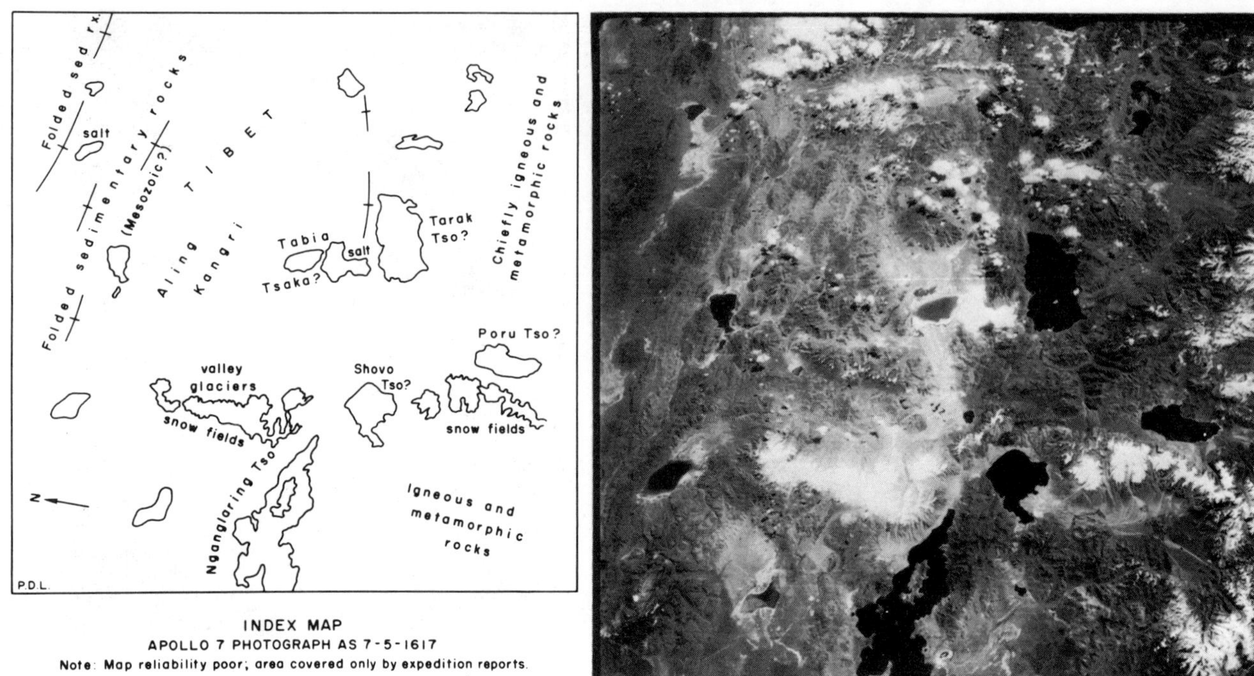

Figure 15 Apollo 7 hand-held photograph taken in 1968, showing south-central Tibet, with view to east.

that the Himalayas are the result of a continental collision (between India and Asia). Earth-circling spacecraft may permit any geologist to make his own Voyage of the Beagle, and to see such critical but remote areas.

Worldwide Coverage

Related to, but not synonomous with, the coverage of inaccessible areas just discussed is the global photographic coverage possible from earth-orbiting vehicles. Provincialism has long been a weakness of geology, and stems at least partly from the simple fact that the earth is simply too big and too poorly mapped for a valid global treatment of geologic problems. However, orbital photography reveals previously unknown relationships, such as the fact that the familiar Basin and Range province of North America (Fig. 13) is a tectonic freak. Nowhere else on earth is there such a broad region of block faulting, which appears to result from the intersection of the continent by the unusually broad East Pacific Rise. Furthermore, the south-central part of the North American Cordillera is now seen to be a tectonic montage in which the structures produced by block-faulting are superimposed on the earlier folded structures formed in the usual geosyncline-orogeny process.

Similarly, orbital photography will emphasize to every geologist facts which he may know but overlook. For example, the Ural Mountains have been ascribed by Hamilton (1970) to a continental collision between the European and Siberian cratonic blocks. Hamilton's evidence for this theory is impressive; but one familiar with orbital photography of the earth realizes that the Urals are but one of scores of similar folded mountain belts (Figs. 16, 17, and 18). So Hamilton's theory therefore either violates Occam's Razor, or implies scores of intercontinental collisions, both tending to weaken it.

Orbital photography may thus help geologists to avoid the error of treating geologic features as special cases, by seeing them in a global context.

Availability of Color and Multispectral Coverage

It is now generally realized that, for any area favorable for aerial photography in general, color coverage offers the geologist major advantages over black and white coverage. However, only small areas have yet been photographed from the air in color. Furthermore, even if the entire world could somehow be rephotographed with color film, the cost of reproducing and using color photography would be a major barrier to using it because of the great number of pictures required. Earth satellites, however, permit photography of entire continents in color with a relatively small number of prints, and substantial areas can be studied with even one print. . . .

The same argument applies to multispectral photography. As shown by the Apollo 9 S065 experiment (Lowman, 1969), multispectral coverage is not only desirable for orbital photography but necessary, because the great range of terrain, vegetative cover, and atmospheric conditions encountered in even single orbital photographs makes it impossible to get optimum rendition with one film/filter combination.

Obviously, the relatively low resolution of most orbital pictures must be taken into account in discussing the value of color and multispectral orbital photography. However, such photography could at least narrow down the areas for which the more expensive comparable aerial coverage would be justified.

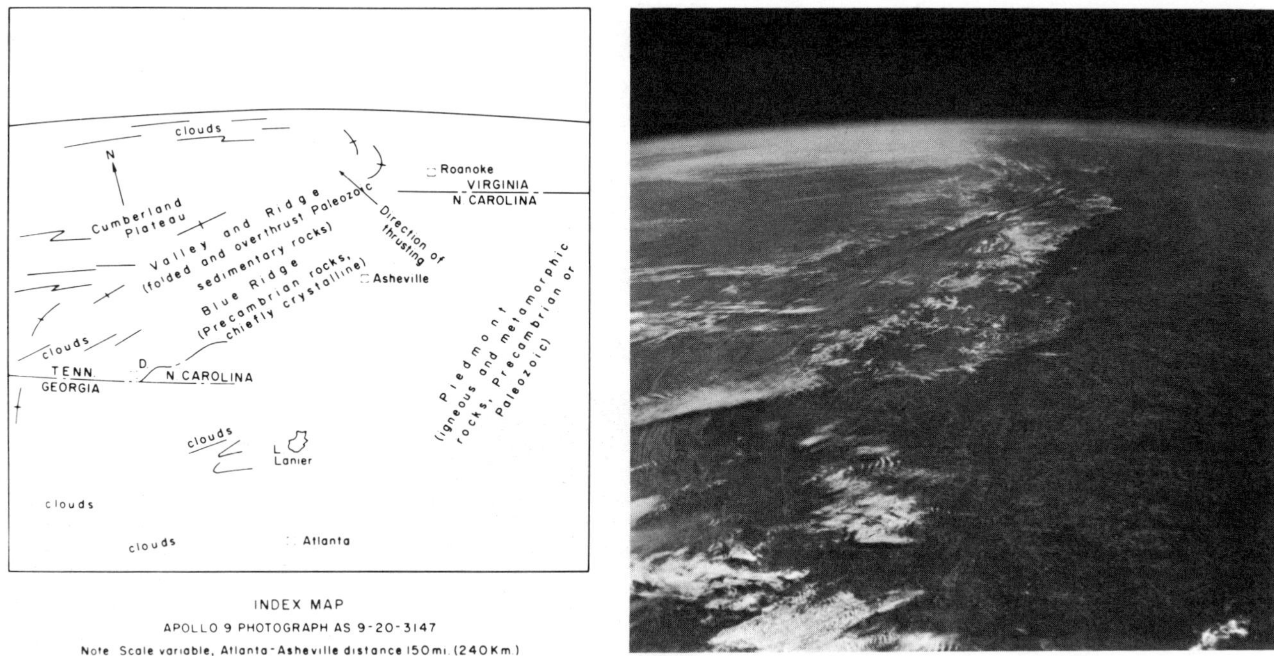

Figure 16 Apollo 9 hand-held photograph, original in color, looking north along Appalachian Mountains.

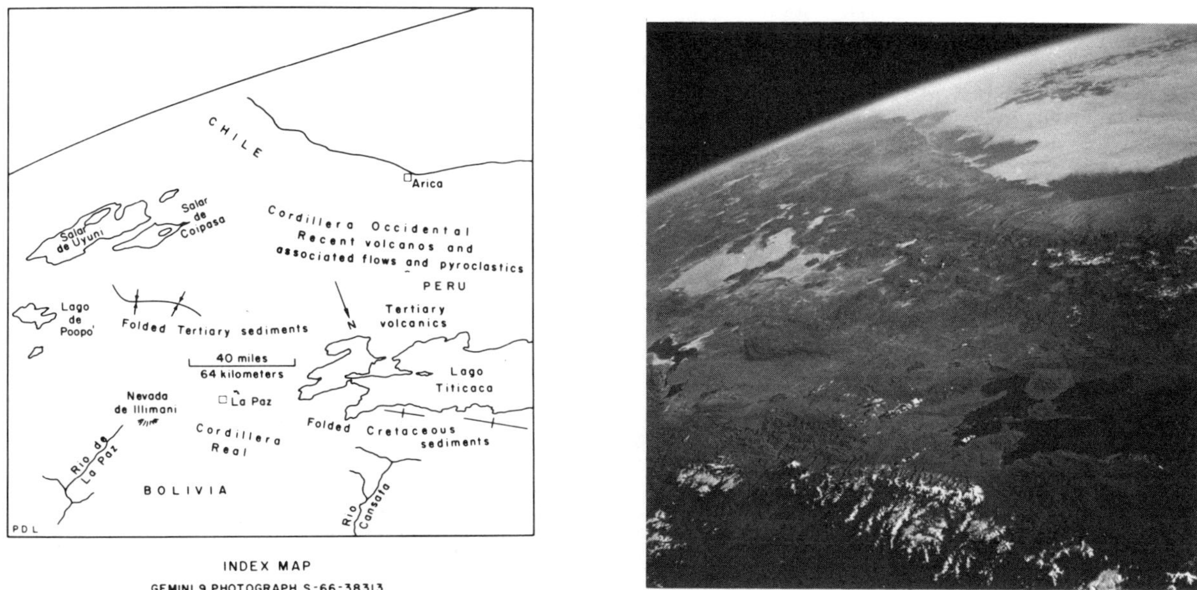

Figure 17 Gemini 9 photograph taken in 1966, showing structure of Andes Mountains.

Summary and Conclusions

The foregoing review has accentuated the positive side side of orbital photography in geology. There is of course another side. Most of the photography taken so far has low ground resolution, and many of the obliques are of little use for anything but general illustration. All orbital photographic schemes must contend with the earth's dense cloud cover, and of course with the fact that most of the earth is covered by water, ice, soil, and vegetation.

A question not explicitly discussed so far is that of just how orbital photographs can best be used by geologists. For areas in which no photography at all is available, the answer is obvious: they simply substitute for air photos. However, three other approaches have also proved valuable. The first of these is simply comparison of orbital photographs with existing geologic or topographic maps. In many areas, such gross discrepancies will be found that several lines of inquiry will immediately be opened; an example is the Palomas volcanic field

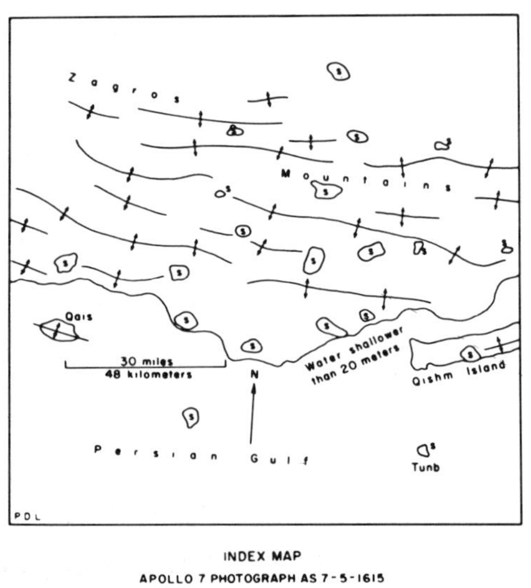

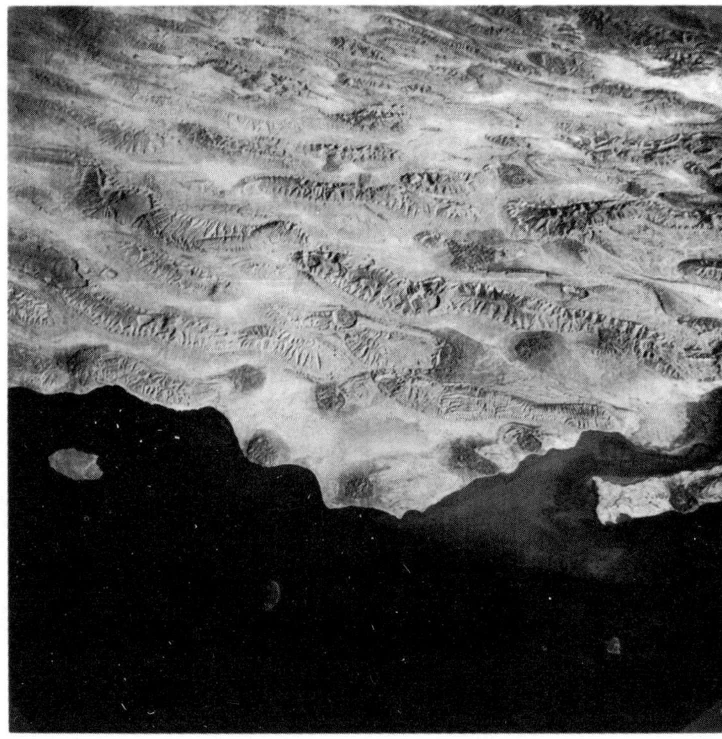

Figure 18 Apollo 7 hand-held photograph taken in 1968, showing Zagros Mountains in Iran; view to north.

(Fig. 8). A second approach is comparison of orbital photographs of widely separated though generally similar areas, such as the deserts of North America and North Africa. Explaining the differences and similarities will frequently throw new light on old fact or uncover new facts. Still a third approach is problem-oriented, in which orbital photography is applied to a particular geologic problem that may involve large areas, perhaps in different countries; the Texas lineament problem is a good example of this approach.

In summary, I wish to stress that orbital photography is not simply aerial photography from extremely high altitudes. Although it can serve as such, it is actually so different as to represent a qualitatively new geologic tool. Properly used, it can provide geologists with new approaches to old problems, with a new outlook on what is nominally a well-explored planet, and perhaps most important, with the recognition of some totally new problems.

REFERENCES

Holmes, A. *Principles of Physical Geology.* New York: Ronald Press, 1965.

Johns, R. H. Generalized Geologic Map of the Peninsular Range Province. Sacramento: Calif. Dept. Natural Resources, 1954.

Lowman, P. D., Jr. A review of photography of the earth from sounding rockets and satellites: NASA Technical Note D-1868, 1964.

Lowman, P. D., Jr. Geologic orbital photography: Experience from the Gemini program. *Photogrammetria* 24:77, 1969.

Lowman, P. D., Jr. Apollo 9 multispectral photography: Geologic analysis. X-644–69–423, Goddard Space Flight Center, Greenbelt, Md., 1969.

Lowman, P. D., Jr. The Third Planet: Terrestrial Geology in Orbital Photographs. *Weltflugbild.* Reinhold A. Muller, Feldmeilen/Zurich, Switzerland, 1972.

Lowman, P. D., Jr., et al. Terrain photography on the Gemini IV mission. NASA Technical Note D-3982, 1967.

Lowman, P. D., Jr., and Tiedemann, H. A. Terrain photography from Gemini spacecraft. Final geologic report: X-644–71–15, Goddard Space Flight Center, Greenbelt, Md., 1971.

Merifield, P. M., Geologic application of hyperaltitude photography. Ph.D. Thesis, Univ. of Colorado, Boulder, Colo., 1963.

Morrison, A., and Chown, M. C. Photography of the western Sahara desert from the Mercury MA-4 spacecraft. NASA Contractor Report CR-126, 1964.

Thornbury, W. D. *Principles of Geomorphology.* New York: Wiley, 1954.

SUGGESTED READINGS FOR PART FOUR, SECTION C

Alexander, R. H. Geographic data from space. *Professional Geographer* 16:1–5, 1964.

Atkinson, J. H., Jr. Atmosphere Limitations on Ground Resolution from Space Photography. *New York S.P.I.E. Journal,* Vol. 1, 7th Society of Photo-Optical Instrumentation Engineers Technical Symposium, 1963.

Badgley, P. C., and Childs, L. F. Earth Resources Survey from Space. The Ocean from Space Symposium for the American Society of Oceanography, 1967.

Badgley, P. C., Fisher, W. A., and Lyon, R. J. P. Geologic exploration from orbital altitudes. *Geotimes* 10:11–14, 1965.

Badgley, P. C., and Vest, W. L. Orbital remote sensing and natural resources. *Photogrammetric Engineering* 32:780–790, 1966.

Barringer, A. R., Newbury, B. C., and Moffatt, A. J. Surveillance of Air Pollution from Airborne and Space Platforms. *Proceedings of the Fifth Symposium on Remote Sensing of Environment.* Ann Arbor: University of Michigan, Institute of Science and Technology, Willow Run Laboratories, 1968.

Bird, J. B., and Morrison, A. Space photography and its geographical applications. *The Geographic Review* 54:463–486, 1964.

Data Staff. Is space photography the panacea? *Data Magazine* 10:33–37, 1965.

Drewes, H. An Evaluation of the Gemini IV Color Photos of the Gulf of California-Central Texas area. U.S. Geological Survey Technical Letter NASA-46, 1966.

Drewes, H., and Morrison, R. Extent of Relict Soil Revealed by Gemini IV Photographs. U.S. Geological Survey Technical Letter NASA-60, 1966.

Fischer, W. A. Orbital Surveys of the Earth. *Proceedings of the Symposium on the Peaceful Uses of Space.* Stanford, California, 1966.

Hahl, D. C., and Handy, A. H. Hydrologic Interpretation of Nimbus Vidicon Image—Great Salt Lake, Utah. U.S. Geological Survey Technical Letter NASA-61, 1966.

Hemphill, W. R., and Danilchik, W. Geologic interpretation of a Gemini photo. *Photogrammetric Engineering* 34: 150–154, 1968.

Lowman, P. D., Jr. The earth from orbit. *National Geographic* 130:645–671, Nov., 1966.

Tabor, R. Photogeologic Interpretation of Gemini IV Color Photography: Baja California. U.S. Geological Survey Technical Letter NASA-24, 1966.

Taggart, C. I. Satellite photography. *Canadian Surveyor* 18:105–112, 1964.

Wilson, R. C. Space photography for forestry. *Photogrammetric Engineering* 33:483–490, 1967.

Wobber, F. J. Environmental studies using earth orbital photography. *Photogrammetria* 24:107–165, 1969.

Wolfe, E. W. Gemini V Color Photography of Salton Sea Area, California. U.S.G.S. Technical Letter NASA-34, 1966.

PART FIVE
REMOTE SENSING IN THE REFLECTED OR NEAR INFRARED

Sensing the light
at wavelengths longer
than those we can see

As THE USE *of the airplane for military surveillance increased after World War I, military technicians became more adept at camouflaging ground installations. Carefully arranged cloth, painted the appropriate shade of green, made it almost impossible for an air observer to perceive a target against a background of surrounding vegetation. However, green-painted cloth is not highly reflective in the shorter wavelengths of the infrared; healthy vegetation is very strongly reflective in this part of the spectrum. A black and white film was designed with an emulsion layer that is sensitive to this reflected infrared. It was possible then to pinpoint camouflaged targets readily on this new film, and it became known as* camouflage-detection film. *Continued research led to the development of a color infrared film, on which there is a color shift; that is, objects do not appear on transparencies or prints in the same color as they do in nature. For this reason it has become known as* false-color film. *The only infrared film available in the United States is manufactured by the Eastman Kodak Company. It is sensitive to wavelengths from about 360 to 900 nanometers. The film is exposed with a yellow (minus blue) filter which attenuates wavelengths shorter than about 500–520 nanometers. Today in remote sensing, black and white infrared film is seldom used; most infrared photography employs color film, known specifically as Kodak, Ektachrome Infrared Aero Film, Type 8443. Other names that may appear in the literature for this film include color infrared film, infrared Aero Ektachrome, infrared-color film, and color camouflage-detection film. The scattered blue and other shorter wavelengths must be filtered out, and only the reflected green, red, and infrared wavelengths reach the emulsion layers of color infrared film. Because only these longer wavelengths are recorded on the film, it presents the interpreter with a better image of objects that otherwise might be obscured by haze. Longer wavelengths have greater haze-penetration capabilities.*

22-Optimum Methods for Using Infrared-Sensitive Color Films

NORMAN L. FRITZ

CONSIDERABLE INTEREST has recently been expressed in the use of KODAK EKTACHROME Infrared AERO Film, Type 8443, for applications which are as diverse as: military reconnaissance; the detection of disease and insect pests in forests, orchards, and grain crops; the identification of tree species; the study of soil conditions; geologic exploration; archeologic studies; and medical photography (American Society of Photogrammetry, 1960; Nelson, 1965; Norman and Fritz, 1965; Gibson, Buckley and Whitmore, 1965). It is thought that a comprehensive description of the film (Fritz, 1965; Tarkington and Sorem, 1965) and some of its unpublished characteristics would be useful to many whose field of specialization includes little or no technical photography. Often a knowledge of possible special techniques of use, and a better understanding of the capabilities and limitations of the film, will make the difference between obtaining optimum or unusable results. As this is a false-color film, different interpretive skills are required from those for either black and white or normal-color photography. For those potential applications where little concerted work has been done, it is often necessary to determine both the capabilities of the film and the optimum conditions of use for the specific application.

From *Photogrammetric Engineering* 33:1128–1138, 1967. Reprinted and edited with permission of the author and the American Society of Photogrammetry.

Film Characteristics

Figure 1 shows the arrangement of the principal layers of the film. The bottom layer is sensitive to the red spectral region and, on processing, forms a magenta positive image; the middle layer is green-sensitive and forms a yellow positive image; and the top layer is

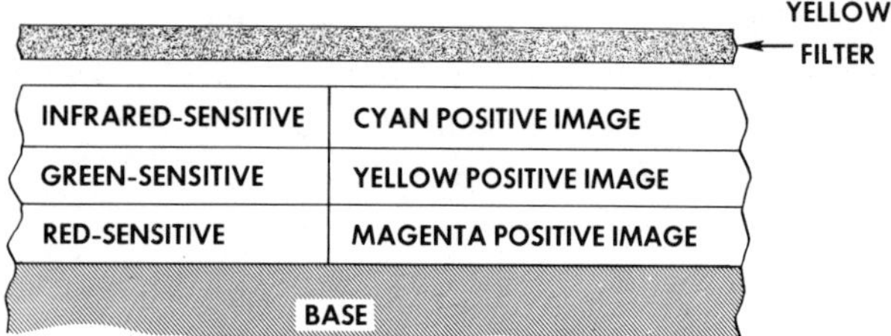

Figure 1 Arrangement of the layers and filter for KODAK EKTACHROME Infrared AERO Film, Type 8443.

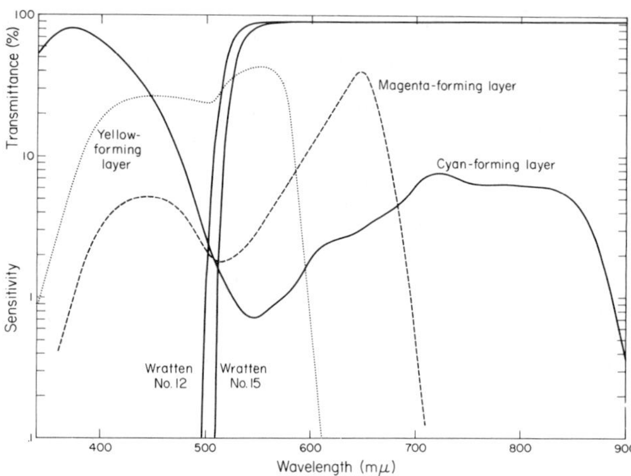

Figure 2 Spectral sensitivities of the three film layers. Spectral transmittance of KODAK WRATTEN Filters as indicated.

explained with the aid of Table 1. On the first line, the portion of the electromagnetic spectrum to which pho-infrared-sensitive and forms a cyan positive image. The relationship between the spectral regions of individual-layer sensitivities and the resulting dyes can be better tographic materials are sensitive is divided into five regions: the blue, green, and red regions of the visible spectrum, and the ultraviolet and infrared regions at shorter and longer wavelengths, respectively. A normal-color film has three principal layers, one of which is sensitive to blue light, one to green, and one to red (line 2). On processing, positive images of yellow, magenta, and cyan dyes are formed in the respective three layers (line 3).

Without going into the chemical and physical reasons of why the film behaves as it does, we know that the resulting images combine to form colors which closely match those of the original subject. That is, blues will be rendered blue, greens green, and so forth (Table 1, line 4). With EKTACHROME Infrared Film, one layer is sensitive to green, one to red, and the third to infrared. In addition, however, all three layers are sensitive to the blue region (line 5). To limit the exposure of each layer to only one spectral region, a yellow filter is always used over the camera lens to absorb this blue light before it reaches the film (line 6). The same three dyes are used as in normal color film, and in the same order—yellow, magenta, and cyan—for successively longer wavelength regions (line 7). Again, after processing, the colors blue, green, and red are formed, but the blue has resulted from green exposure, green from red exposure, and red from infrared exposure (line 8).

The relationship between the colors photographed and those resulting in the film can be remembered more easily by noting that the sequence of the reproduced colors is in the same order (blue, green, red) as it is in the spectrum, but the correspondence to the colors being photographed (green, red, and infrared) is one block (Table 1) toward longer wavelengths. With a knowledge of this relationship it is possible to predict readily the reproduction color of any colored object, provided its infrared reflectance is known. Likewise it is possible to predict reproduction color changes with color changes in the objects photographed. For example, if a tree loses infrared reflectance, the reproduction becomes less red or more cyan, which is the characteristic blue-green color of unhealthy trees, as recorded by this film.

So far, the individual layer sensitivities of this film have been presented as broad regions of the spectrum. The actual spectral sensitivities of each layer are given in Figure 2. Plotted on the same graph are the spectral transmittances of KODAK WRATTEN Filters No. 12 and No. 15. The WRATTEN Filter No. 15 was recommended for use with the forerunner of this film, KODAK EKTACHROME AERO Film (Camouflage Detection). When the present film was developed in 1962, it was decided to use a WRATTEN Filter No. 12, because most aerial photographers would already have this filter or a similar one for use with black and white photography. The basic color balance of the film was then changed to compensate for the change in filter.

Table 1 Principles of Operation of Normal Color Film and of KODAK EKTACHROME Infrared AERO Film, Type 8443

1. Spectral region	Ultraviolet	Blue	Green	Red	Infrared
2. Normal color film sensitivities		Blue	Green	Red	
3. Color of dye layers		Yellow	Magenta	Cyan	
4. Resulting color in photograph		Blue	Green	Red	
5. EKTACHROME Infrared sensitivities		Blue	Green	Red	Infrared
6. Sensitivities with yellow filter			Green	Red	Infrared
7. Color of dye layers			Yellow	Magenta	Cyan
8. Resulting color in photograph			Blue	Green	Red

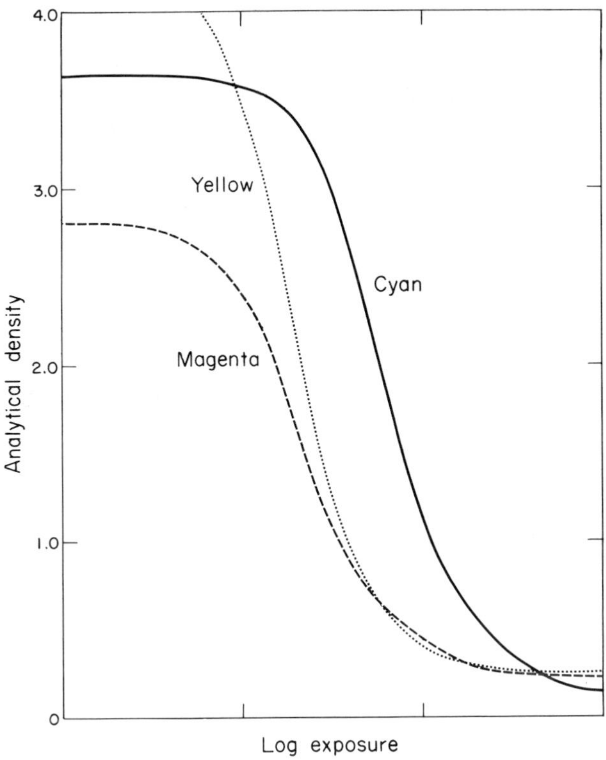

Figure 3 Sensitometric curves.

The sensitometric curves for the film are shown in Figure 3 and represent what is thought, at the present time, to provide the best color balance for the greatest number of applications. Special applications may require a slightly different balance, which can be obtained by means of filters. Note that, unlike normal color films in which the three layers are all of about the same speed, the cyan layer is considerably slower than the other two. The reason for this is explained later.

Scene Characteristics

The main applications of this film, at the present time, involve the photography of foliage of one form or another. Determining the characteristics of the foliage or plant itself may be the primary object of the photograph, or it may be to reveal the presence of disease or insect pests, or a means of determining soil or subsoil conditions. The detection of any condition which affects the growth of plants may be a potential application for the use of this film. As foliage is often the primary subject photographed, Figure 4 is included to show typical spectral reflectances of a number of types of foliage (Krinov, 1953). It should not be inferred that the types of foliage named always have the properties represented by these curves, because the actual values of reflectance are in many instances greatly affected by such conditions as the age of the leaf, the season, the water and mineral content of the soil, or the type of soil itself, to name just a few. An example of how soil conditions affect the color and growth of foliage is the familiar effect of fertilizer and water on lawn grass in the middle of summer. Most types of foliage are not very different from one another in spectral reflectance in the visible region of the spectrum. As shown in Figure 4, the small rise of each curve in the green region is all that is required to provide the characteristic green appearance. Although differences in foliage color certainly are visible, they are generally small compared to the gamut of greens possible with dyes. However, the generally high reflectance of foliage in the infrared region and the great differences in reflectance which can and do occur, explain the value of a film sensitive in this region for detecting differences in foliage conditions and between varieties of foliage. As these differences are recorded together with those which do occur in the visible region, the color film will provide more information than a black and white infrared-sensitive film.

In order to find the approximate relative exposure produced in each of the layers, measurements have been made of four types of foliage in about 50 aerial photographs. Figure 5 is a bar graph showing the average exposure of each layer of the film by grass, light and dark green deciduous foliage, and evergreens. Note that in all cases the exposure of the cyan layer is considerably greater than that of the other two layers. The grass was recorded almost identically to light green deciduous trees, whereas evergreen trees produced less exposure in all layers and their infrared exposure, relative to the green and red, was less than that for other types of foliage. These facts correspond well with what would be expected from the spectral reflectance curves shown in Figure 4. The exposure bar graphs are superimposed on the sensitometric curves of the film to show the average individual layer densities which result when the

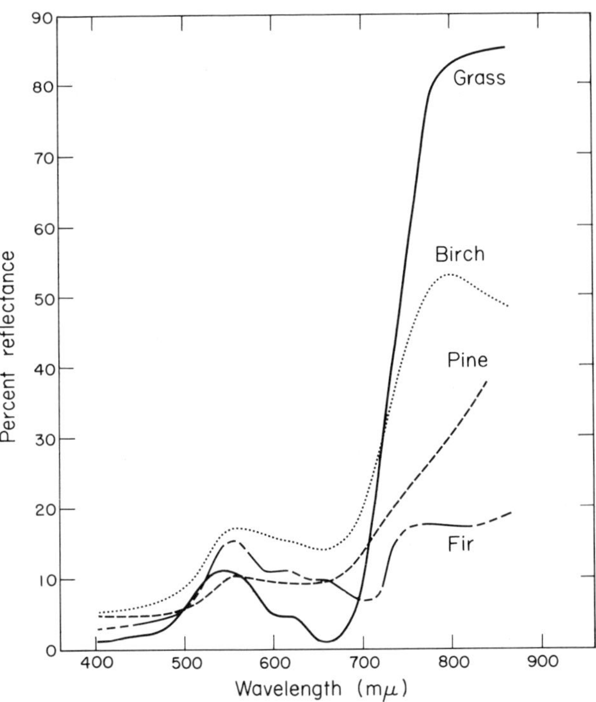

Figure 4 Spectral reflectance curves of various typical foliage types.

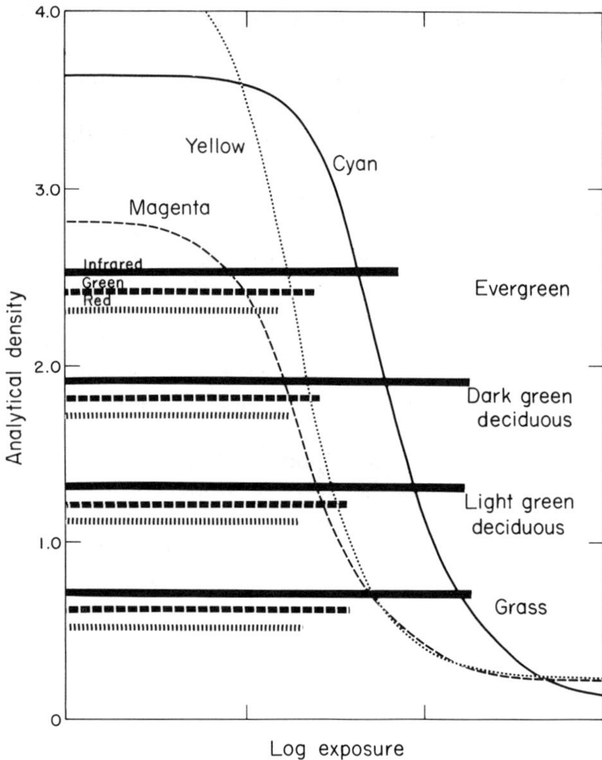

Figure 5 Bar graph showing the relative exposure produced in each layer of the film by four types of foliage, superimposed at an optimum exposure level on the sensitometric curves.

foliage is given what is considered to be an optimum exposure.

It can now be seen why the cyan layer has been deliberately made so much slower than the other two layers. If the cyan layer had been as fast as, say, the magenta layer, and the camera exposure was such that the other two layers were exposed as shown in Figure 5, the infrared exposure would be on the toe of the cyan curve, and any variation of infrared exposure would produce negligible cyan density change. Thus, all foliage, with the possible exception of evergreens, would be recorded excessively red, and small differences in infrared reflectance would not be detectable. With the properly balanced cyan layer, the infrared exposure is recorded well up on the sensitometric curve, and small variations in exposure result in significant differences in cyan density.

For many applications, the usefulness of this film will depend primarily on its ability to record slight changes in infrared reflectance. For example, the first indication of the loss of vigor in a plant due to the incidence of disease or infestation with pests may be a loss of infrared reflectance. This effect often occurs even before any visible symptoms can be detected. Where these conditions are suspected, testing the plants for loss of infrared reflectance, rather than waiting for visible symptoms to appear, will allow countermeasures to be started soon.

Photographic Characteristics

Ideally, if one wished to detect photographically the differences or changes in spectral reflectance, it would be desirable to have spectrophotometric curves from which to find the spectral regions of greatest difference. Thus, with suitable filters, photographs could be made in these regions, and the regions where no differences occurred could be eliminated. This method would provide the greatest degree of differentiation. Because suitable spectrophotometric equipment is not available to many people on low budgets, simple techniques exist which will assure that the best results possible have been attained. Without going to multiband black and white photographs which are quite difficult to analyze, we can choose the obvious first technique: to use color photography, which at least divides the spectrum into three discrete bands. If both normal-color and infrared-sensitive color films are used, there are four spectral bands with which to work.

Several characteristics of this film have a bearing on its ability to provide optimum differentiation. Two of them make it possible to obtain good results even on extremely hazy days. The first is the fact that a yellow filter is always used, thus eliminating the blue light, which contributes most to the degrading photographic effect of haze. The second is that the film has a high gamma, as can be seen from the characteristic curves. This high gamma tends to offset what effect there is of haze in the green and red spectral regions.

A consequence of the high gamma is that the film has a rather short exposure latitude. The range for optimum results is only about ½ stop on either side of the best exposure. Beyond that, some decrease in the information will occur, although some compensation can be made by adjustments in the processing if the exposure is known to be not optimum. Thus, for critical applications, or when one is studying the use of the film for a new application, the use of a short exposure series is recommended.

This film has an Aerial Exposure Index of 10 and an ASA rating of 100. Under normal conditions of use, an aerial exposure of about 1/500 second at *f*/5.6 has been found to be quite close to optimum. Thus a typical exposure series would include this value and ½ stop on either side. It is recommended that the camera be focused at the normal setting, rather than at the infrared setting which some cameras have. Two of the film layers, being sensitive in the visible region, require the normal setting. The infrared-sensitive cyan layer is usually well exposed, because of the high infrared reflectance of foliage, and thus its dye concentration is reduced. The effective image, therefore, is predominantly in the magenta and yellow layers.

Experiments have been made with this film for the detection of disease and damage by insects with close-up photography from the ground. These attempts have not always been successful, and some cases are on record where differentiation readily discernible in aerial photographs cannot be detected in photographs made on the ground. The reason for this effect is not entirely understood, but appreciable evidence indicates that at the large scale of a close-up photograph in which individual leaves are seen, considerable variation occurs in the rendition from leaf to leaf because of the high dependence of reflectivity on leaf orientation and on the direction of the light source. This variation is, in many cases, greater than the effect being sought. By increasing the distance from the subject, the variations in the individual leaves are integrated and the average effect can be detected.

Some attempts to record from the ground known abnormal conditions in trees within a forested area have failed because the effective scene luminance range is too great. This large range is due not only to the normal differences in illumination between sunlit areas and shade, but also to the fact that shaded areas are illuminated by skylight, which is largely eliminated by the yellow filter. As mentioned before, the exposure latitude of the film is short, and photographs which include both sunlit foliage and shaded areas under trees will not record both well. As the shaded areas often predominate, there is a tendency to overexpose, with the resultant loss of differentiation in the sunlit foliage. In these photographs, all such areas tend toward a white reproduction and the changes from red or magenta toward yellow and green cannot be obtained.

Several experimenters have attempted to study aquatic vegetation with this film. Knowing that infrared radiation is highly absorbed by water, we made a study to determine accurately the spectral absorption of pure water as recorded by the three layers of this film. One foot of water was found to have the effect of a density of 0.04 in reducing the exposure for the green-sensitive layer, 0.06 for the red-sensitive layer, and 0.22 for the infrared-sensitive layer. By including the surface reflectance of the water, and considering that the light must pass through the water twice, it can be calculated that a 1-foot depth of water will decrease the exposure of the green- and red-sensitive layers by about ½ stop and that of the infrared-sensitive layer by about 1½ stops. This film is not useful therefore for photography through any great depths of water.

There is a tendency to equate infrared radiation with heat, and to expect that this film can be used to record temperature differences. Tests have shown that, if no other source of radiation is present to affect the film, an object heated to 650°F will just be recorded by the infrared-sensitive layer of the film where the film is exposed for 15 minutes at *f*/2.0. The film will not record temperature differences at our environmental temperatures.

If exposures are made when the film is very cold, some loss in speed occurs, as well as a shift of color balance in the cyan direction. For normal low-altitude work, these effects are not significant, but for photography at high altitudes, with unheated or partially heated cameras, marked changes may result. As an example, a sensitometric test showed that where the exposures were made with the film at –40°F, the speeds of the green- and red-sensitive layers were decreased by 0.20 in log E, and that of the infrared-sensitive layer was decreased by 0.60.

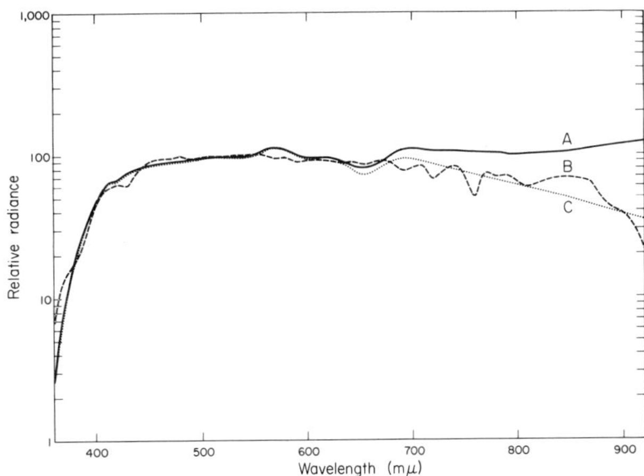

Figure 6 The spectral distribution of daylight *B*, of a sensitometric source suitable for films sensitive in the visible region *A*, and of the same source with the addition of a partial infrared-absorber *C*.

In exposing the film on a sensitometer, it is necessary to consider the spectral energy distribution of the light source in the infrared region. A commonly used daylight sensitometric source is composed of a tungsten-filament lamp with a suitable Corning 5900 glass filter. Although this source provides a satisfactory match for daylight in the visible region, the radiation in the infrared region is relatively too great. Figure 6 shows the current best estimate of the spectral distribution of daylight (Judd, MacAdam, and Wyszecki, 1964; Moon, 1940) at the camera film plane, the distribution of the filtered tungsten source described, and the improvement in the infrared region which can be attained by the addition to the filter of 1.0 mm of Pittsburgh 2043 glass. The lamp-filter combination provides a daylight source which is suitable for films which are sensitive either in the visible or the near-infrared regions of the spectrum.

Processing

This film was designed to be processed in KODAK EKTACHROME Film Process E-2 and E-3, and if fresh film is processed according to instructions, it will have the sensitometric characteristics shown in Figure 3. It is recommended that the user process his own film, but if this is impracticable, assurance should be obtained from the processor that Process E-2 and E-3 will be used, and that no infrared inspection will be made during processing. For critical applications, E-4 processing is not recommended. This process increases the speed of the cyan layer by about 0.20 in log E, causing the photographs to be more red than is desirable for optimum rendition.

If faster film is desired, or if a roll is known to be underexposed, an increase in speed can be produced by a change in processing. The speed can be increased by about one stop by increasing the first developer time by 4½ minutes. Sensitometric tests of this modified process indicate some decrease in the shoulder densities and a slight change in color balance. It is thought, though, that these changes are preferable to obtaining results which are more than ½ stop underexposed.

In the KODAK EKTACHROME RT Processor, Model 1411-M, the EA-4 process for EKTACHROME films increases the speed of the cyan layer by about 0.20 in log E. However, a process modification is available that will produce optimum results. It consists of an addition to the prehardener, and a change in the temperature of the first developer.

At present, no process exists which will produce a good color negative with this film. If it is processed in the KODAK Color Film Process C-22, the contrast of the cyan layer is very low. Prints from these negatives show not only poor color rendition but also a major loss of differentiation between foliage types having different infrared reflectances.

Color Balance

One of the characteristics of this film which must be understood in order to obtain consistently optimum results is the effect of age on its color balance. The infrared-sensitive cyan layer tends to decrease in speed as the film ages, and the speed of the green-sensitive yellow layer increases slightly. As the film ages, its color balance shifts toward cyan. If the film is kept at room temperature, its balance will eventually pass the point of optimum color discrimination. However, this effect can be reduced considerably by refrigeration, or almost eliminated by storage in a freezer.

As a time lag always occurs between when the film is manufactured and when it is marketed, it is coated with the cyan layer slightly fast. By the time the film reaches a customer, the speed of this layer will usually have slowed sufficiently so that the color balance is normal. Because the color of reproduction is important in the use of this product, it is sometimes desirable, for optimum results, first to determine the actual color balance of the film and then to make adjustments by means of filters.

Several tests have indicated that the ability of this film to discriminate between varieties of trees and between sick and healthy trees depends both on the exposure and on the color balance of the photograph (Norman and Fritz, 1965). Careful attention to these factors will assist greatly in determining the film's usefulness for particular applications. The color balance of normal color films is readily changed in any direction with gelatin KODAK Color Compensating (CC) Filters. Figure 7 shows the spectral transmittance of some typical CC filters superimposed on the spectral sensitivity curves of the film. As can be seen, not one of these filters absorbs in the infrared region. Thus, with no effective absorber for the infrared-sensitive cyan layer, it is impossible to obtain the entire gamut of colors with the convenient gelatin filters alone. The Corning 3966 glass filter does absorb infrared (Fig. 7), and suitable thicknesses can be made to obtain any desired decrease in the speed of the cyan layer. This filter is thus useful both for experimenting with color balance and for slowing the cyan layer of film having an overly red balance. Tests have shown that a standard thickness of this filter will change the color balance by about ½ stop in the cyan direction. Seldom is more than half this thickness needed.

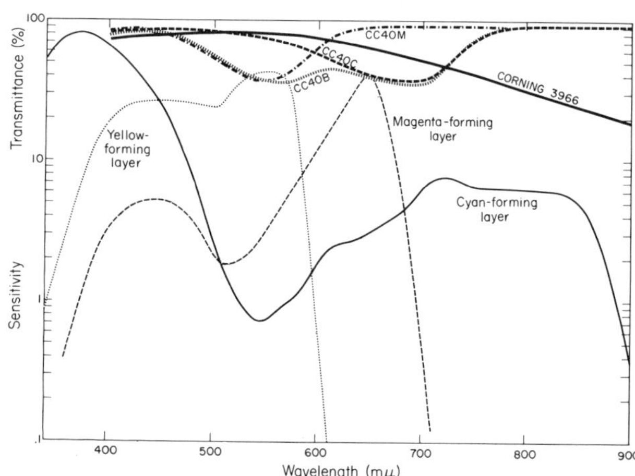

Figure 7 Transmittance of typical KODAK Color Compensating Filters superimposed on the spectral sensitivity curves.

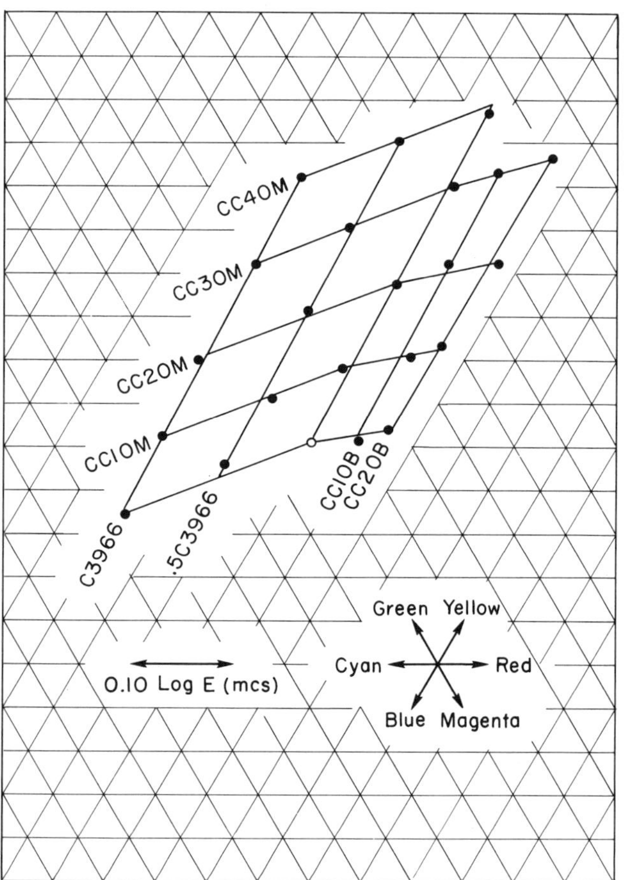

Figure 8 Color shifts produced by various combinations of KODAK Color Compensating and Corning No. 3966 filters.

To determine the optimum color balance of the film to provide maximum detectability for particular applications, it has been found desirable to make a series of photographs in which both the exposure and the color balance are varied systematically. Figure 8 is a convenient trilinear plot showing the results of a sensitometric test made to determine the effect of various filters on the color balance. The 0 point is the balance of the particular coating used, and the direction and distance from 0 indicate the color and magnitude of the shift produced by the filter or combination of filters shown. The CC-magenta filters shift the balance in the yellow direction, the CC-blue filters in the red direction, and the Corning 3966 toward the cyan. The balance can be shifted in the magenta direction with CC-cyan filters, or in the blue direction with combinations of CC-cyan and Corning 3966 filters, but these are seldom necessary. The CC-red filters produce the same results as the CC-magenta because the difference between them is primarily in their blue absorption, and this spectral region is eliminated with the WRATTEN Filter No. 12. For the same reason, the CC-green and CC-cyan filters produce similar results.

If, now, for a particular application, an exposure series is made at the color balance of each of the points shown in Figure 8, one will obtain an array of various colored photographs which can be studied to find optimum discrimination. Such arrays have been made to determine the optimum color balance to distinguish between varieties of deciduous trees in a northern forest area, and to indicate decline and disease in citrus trees in Florida.

In Figure 9 the results of these investigations are plotted and show the color-balance regions of maximum discrimination encompassed within the area of the inner black lines, some discrimination still being apparent out to the outer lines. The data for the second part of the figure are the same as those for Figure 1 in Norman and Fritz (1965), except that they are presented here in terms of analytical rather than integral densitometry, and an adjustment has been made because the original color balance of the film was different in the two tests. As can be seen, the optimum balance depends somewhat on the subject matter and, in general, lies within a rather small region. This illustrates that for those seriously interested in developing techniques for using this film for particular applications, the generation of such an array can be quite helpful. The results of these tests indicate that a rather purplish reproduction is preferable to the

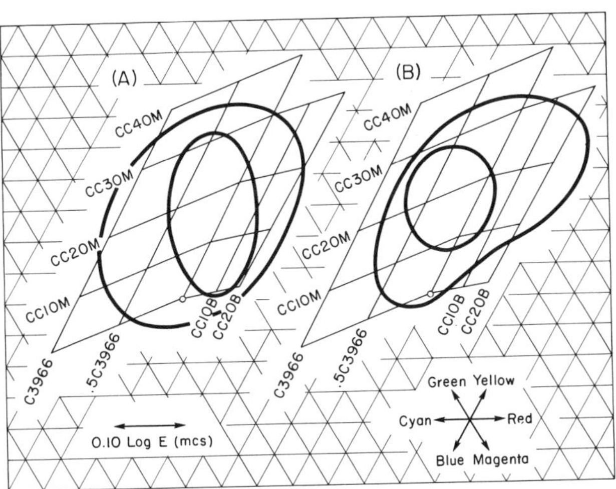

Figure 9 Color-balance regions for maximum discrimination in *A*, northern forest areas, and *B*, Florida citrus trees.

Figure 10 Paired camera arrangement for making simultaneous photographs.

more esthetically pleasing red preferred by some people. This conclusion has also been reached by others who have compared results obtained with fresh and somewhat older films.

To obtain the most consistent results, all photographs should be made of the same area from about the same point, which, unless a helicopter is used, requires quite a bit of flying around. It may also be found desirable to evaluate the effects of such variables as the plane altitude, the solar attitude, the angle between the direction of the sun's rays and that of the camera, etc.

An exposure series of three photographs spaced at ½-stop intervals is usually sufficient, so a complete color-balance-exposure-series array will consist of 69 photographs. A convenient arrangement is to use two identical 35 mm cameras mounted on a pair of handles (Fig. 10) so that both shutters can be tripped at once. By rapidly operating the film-advancing and aperture-setting mechanisms, it is possible from 8000 feet to make one exposure series on each camera on each pass past the target, and filters can be changed while the plane is going around for the next pass.

Because all filters have some absorption in their regions of highest spectral transmittance, and because the three layers of a color film have somewhat overlapping sensitivities, the effect of adding filters is not only to change the color balance, but also to decrease the exposure. Table 2 shows the complete set of filter combinations and exposures required to produce the above array on a clear day if the solar altitude is above 40 degrees.

Table 2 Exposures Required for Various Filter Combinations

Filter Combinations	Aperture (for 1/500 sec. exposure time)		
Wr. No. 12[a]	f/4.7	5.6	6.7
Wr. No. 12+CC10M[b]	f/4.7	5.6	6.7
Wr. No. 12+CC20M	f/4.7	5.6	6.7
Wr. No. 12+CC30M	f/4.7	5.6	6.7
Wr. No. 12+CC40M	f/4.0	4.7	5.6
Wr. No. 12+CC10B	f/4.0	4.7	5.6
Wr. No. 12+CC10B+CC10M	f/4.0	4.7	5.6
Wr. No. 12+CC10B+CC20M	f/4.0	4.7	5.6
Wr. No. 12+CC10B+CC30M	f/4.0	4.7	5.6
Wr. No. 12+CC20B	f/4.0	4.7	5.6
Wr. No. 12+CC20B+CC10M	f/4.0	4.7	5.6
Wr. No. 12+CC20B+CC20M	f/4.0	4.7	5.6
Wr. No. 12+CC20B+CC30M	f/4.0	4.7	5.6
Wr. No. 12+½C3966[c]	f/4.0	4.7	5.6
Wr. No. 12+½C3966+CC10M	f/4.0	4.7	5.6
Wr. No. 12+½C3966+CC20M	f/4.0	4.7	5.6
Wr. No. 12+½C3966+CC30M	f/4.0	4.7	5.6
Wr. No. 12+½C3966+CC40M	f/3.4	4.0	4.7
Wr. No. 12+C3966	f/3.4	4.0	4.7
Wr. No. 12+C3966+CC10M	f/3.4	4.0	4.7
Wr. No. 12+C3966+CC20M	f/3.4	4.0	4.7
Wr. No. 12+C3966+CC30M	f/3.4	4.0	4.7
Wr. No. 12+C3966+CC40M	f/2.8	3.4	4.0

[a] KODAK WRATTEN Filter Number.
[b] KODAK Color Compensating Filter.
[c] CORNING Filter Number.

Sharp-Cutting Filters

Considerable experimental work has been done to determine the effect of using various sharp-cutting filters, including band pass, band rejection, and high pass at longer wavelengths than that of the WRATTEN Filter No. 12. These filters have included the high-pass WRATTEN Filters Nos. 22, 25, and 70, and the band-rejection filters WRATTEN No. 55, Corning 9830, and a special dichroic one which cuts out the near-infrared sensitivity of the cyan layer. Combinations of these filters are essentially band-pass filters for isolating individual spectral regions. The effect of all these filters has been to change the color balance from optimum, with the resultant loss in discriminating ability. It has also confirmed what has been known for pictorial color photography for some time, namely, that one- and two-color photographs do not provide as much information as do three-color photographs.

The effect of using high-pass filters of shorter wavelength cutoff than that of WRATTEN Filter No. 12 has not been tested adequately. These filters include the WRATTEN Nos. 8, 3, 2E, and 2C. When these filters are used, the color balance can be brought back to optimum with the color-balance-adjusting filters described. As all three layers are sensitive in the blue region, some increase in speed should be obtained, but it is believed that the desaturation caused by exposing all three layers with blue light will decrease the sensitivity of discrimination.

Conclusions

From the facts presented, it can be seen that, like all advances in the state of the art, improved results do not necessarily come easily. It is considered that with a better understanding of the characteristics of the film, and a willingness to perform some experimentation, the effectiveness of this film for various remote sensing applications can be demonstrated. The experiments conducted to date indicate considerable promise of useful results, along with savings in cost over present ground-based methods.

REFERENCES

American Society of Photogrammetry. *Manual of Photographic Interpretation*. Washington, D.C., 1960.

Fritz, N. L. Film sees new world of color. *Citrus World Magazine* 2 (No. 2):11, July, 1965.

Gibson, H. L., Buckley, W. R., and Whitmore, K. E. New vistas in infrared photography for biological surveys. *J. Biol. Phot. Ass.* 33:1, 1965.

Judd, D. B., MacAdam, D. L., and Wyszecki, G. Spectral distribution of typical daylight as a function of correlated color temperature. *J. Opt. Soc. Am.* 54:1034, 1964.

Krinov, E. L. Spectral reflectance properties of natural formations. Aero Methods Laboratory, Academy of Sciences, USSR. Trans. by G. Belkov, National Research Council of Canada, *Technical Trans.* TT-439, 1953.

Moon, P. Proposed standard solar-radiation curves for engineering use. *J. Franklin Inst.* 230:583, 1940.

Nelson, B. S. Annotated bibliography of "Spectral Analysis and Aerial Photographic Application to Plant Species Indentification and Disease Detection." Part of 1965 Honors Program report to Dept. of Forestry, Iowa State University of Science and Technology.

Norman, G. G., and Fritz, N. L. Infrared photography as an indicator of disease and decline in citrus trees. *Proc. Florida State Hort. Soc.* 78:59, 1965.

Tarkington, R. G., and Sorem, A. L. Color and false-color films for aerial photography. *Photogram. Engng.* 29:88, 1965.

ARCHAEOLOGY IS the scientific study of a prehistoric culture, normally by means of excavation and description of the uncovered remains. In some environments—for example, the Middle East—locating an archaeologic site for excavation is relatively easy because of the paucity of masking vegetation. In more humid environments or areas with more agriculture, vegetation or man's activities may make it difficult to locate sites. Not only may cultural features be obscured by vegetation, but their ground scale may make them difficult to be perceived by an earthbound observer. Archaeologists early learned the value of airborne platforms not only in locating prehistoric sites but also in determining their spatial extent. These latter observations are quite valuable for they can be used to tell the excavators exactly where to dig and what plan to follow in exposing the structure of the site.

Recent investigations have shown that many crop and soil marks can be recorded by infrared photography, even though they are invisible to human eyes or fail to be recorded on panchromatic film. Quite often the culture traces in the soil result in subtle differences in the growth of vegetation. Most broad-leaf vegetation is highly reflective in the infrared, and minor differences in this reflectivity can be easily discerned on infrared photographs. Current research suggests that the greatest value to archaeology in this kind of remote sensing is detection of cultural features and delineation of environmental zones and microenvironments.

23-An Archaeological Survey of the Tehuacan Valley, Mexico: A Test of Color Infrared Photography

GEORGE J. GUMERMAN
JAMES A. NEELY

FOR THE last few years various remote sensing techniques, such as false-color infrared aerial photography and infrared scanner imagery, have been utilized as detection techniques with great success in such disciplines as geology, geography, and forestry. More recently these methods and devices have been applied to archaeological problems with varied degrees of success (Gumerman and Lyons, 1971). Indications are that the greatest value remotely sensed data has to archaeology is in the detection of cultural features and in the delineation of environmental zones and microenvironments. To date, however, no tests have been undertaken to obtain quantifiable data in an area characterized by differing microenvironments. Our tests indicate that the capabilities of black and white film as well as false-color infrared film were potentially most sensitive and useful for such a project. The effectiveness of panchromatic black and white film as a tool for archaeological survey has long been known and, as a result, color infrared (ir) was the medium selected for testing.

From *American Antiquity* 37:520–527, 1972. Reprinted with permission of the authors and the publisher.

The Tehuacan Valley in the State of Puebla, Mexico, was chosen as the area of study for four main reasons. Primary among these was the extensive coverage with various aerial remote sensing devices that was available for the Tehuacan Valley because NASA had photographed the area at the request of the Mexican Comisión de Espacio Exterior. This coverage was obtained to collect data on the hydrology, agriculture, and geology of the Valley. It is important to keep in mind that these data were collected for other than archaeological purposes. Since the cost of NASA missions are high, and obviously beyond the reach of an archaeologist's budget, one facet of the test, and admittedly an uncontrolled variable, was the determination of the applicability of remotely sensed data to archaeology even though that information was collected for other purposes.

A second major reason for the selection of the Tehuacan Valley as the test area was the extensive work undertaken there in the early 1960s by Richard S. MacNeish and his associates of the Tehuacan

Archaeological-Botanical Project. MacNeish (1964, 1967) has demonstrated a long and continuous occupation of the Valley from at least 7000 B.C. to the present. In addition, there is a great variety of cultural features, such as platform mounds, ballcourts, etc., associated with the occupation sites. Consequently, the aerial photographs can be checked for the visibility of different types of sites and sites of different ages.

Thirdly, the Tehuacan Valley seemed a good choice to test the usefulness of false-color infrared photography in the construction of environmental maps for the archaeologists. This region was the location of some of the earliest and still ongoing studies emphasizing the necessity for the archaeologist to consider microenvironmental implications in the analysis for economic and sociocultural systems and processes (Coe and Flannery, 1964; Flannery, 1967; Flannery and Coe, 1968; Flannery, 1968). As a result, the microenvironments have been delineated in considerable detail and provide a good test situation.

Finally, the Tehuacan Valley was chosen because the region was known to Gumerman only through the publications of the Tehuacan Archaeological-Botonical Project, thus insuring a less biased, more valid testing for the identification of cultural and natural features.

Methodology

Data were collected from the Tehuacan Valley by NASA in April of 1969, using a camera cluster with multiband black and white photography, color photography, side-looking radar imagery (SLAR), infrared scanner imagery, and false-color infrared photography. Our test was restricted to the use of false-color infrared photography due to two reasons. First, experience in other regions, such as the southwestern portion of the United States (Gumerman and Lyons, 1971), has indicated the best results might be obtained from this type of film. Second, was the fact that time and money limitations made the examination of the other photographic and imagery materials impossible. Unfortunately, a direct comparison of panchromatic black and white film with the color infrared film could not be made as no standard panchromatic photographs without filters were taken of the Valley by the NASA team.

Infrared scanner imagery, which records infrared radiation further into the invisible part of the spectrum than regular infrared films, has previously been used successfully to locate sites, some of which are invisible upon surface inspection (Schaber and Gumerman, 1969). However, the 25,000 foot altitude at which the infrared scanner imagery was taken in the Tehuacan Valley made it useless for archaeological reconnaissance.

Color infrared film, also called false-color infrared film, camouflage detection film, and infrared Aero Ektachrome film, is sensitized to green, red, and infrared, thus translating both cultural and natural features into dramatic "false" colors such as magenta trees and blue bedrock. The film records electromagnetic radiation in the visible portion of the spectrum as well as the near or reflected infrared portion of the spectrum that is not visible to the human eye. In principle, cultural and natural features should be visible on color infrared film because the features absorb and reflect radiation differentially, resulting in different colors or tonal contrasts on the photographs. Yet another objective of this project in Tehuacan was to determine why the various features visible on the film did absorb and reflect radiation differentially.

A total of 250 feet of color infrared photographs in a 9 inch by 9 inch format were taken at 5000 and 25,000 feet, providing our basic data. Since it is necessary to use the color infrared film as transparencies, unless extremely expensive color prints are made, we found it convenient to make two sets of black and white contact prints, one set for photo mosaics for field use, and the second set for use as stereoscopic pairs. It is important to note, however, that all of the interpretative work related herein was accomplished through the study of the color ir transparencies. The loss of detail and data through the use of black and white prints made from the color transparencies was recognized. As the color ir transparencies were provided by NASA on a large continuous roll and required special equipment for proper viewing, the black and white prints functioned better for transportation in the field and for field study with the pocket stereoscopic viewer we utilized. Selected use of the stereo pairs was made at several sites, but no well-controlled test of this technique was undertaken during this program. Although the resolution is not as great on the black and white contact prints as it is on the color infrared film, most cultural features visible on the transparencies are visible on the contact prints. This is because cultural features usually do not show as different colors, but rather as different tones of the same color. Different natural features often do show as distinct colors and as a result the black and white contact prints should not be used for delineating natural zones.

Before going to the field, a careful inspection of the low-altitude color infrared film was made by archaeologists without field experience in Mesoamerica in an attempt to locate all presumed prehistoric cultural features. It was soon apparent that so many sites were visible that representative types of features such as platform mounds, and canals, were selected for inspection on the ground. Furthermore, the area photographed was so large that only certain selected sections within the different microenvironments characterizing the entire valley were chosen for survey.

No attempt was made at delineating microenvironments until after a brief visual inspection was made of the valley. Coe and Flannery's 1964 *Science* article which briefly defines the microenvironments of the Tehuacan Valley was read, but Flannery's (1967) more detailed analysis in Volume 1 of *The Prehistory of the Tehuacan Valley* was purposely not consulted until after an attempt was made to determine ecologic subareas from the color infrared photographs.

Finally, a month was spent checking the features observable on the infrared photographs with the ground truth in an attempt to determine why different types of features were visible. After one week in the field, Neely, who had been a member of the Tehuacan Archaeological-Botanical Project survey team, joined the project to point out sites in different microenvironments which were not visible on the color infrared photographs. At this point the process of explanation was reversed; that is, we had to determine why these sites were not visible on the photographs.

Data and Results

We soon observed that a major factor in determining the visibility or nonvisibility of a site on the color infrared photographs is the microenvironment in which the site is located. Therefore, before discussing the value of color infrared film in site determination we will discuss its use in constructing a detailed picture of the landscape or natural maps.

The detection and outlining of all of Flannery's ecologic subareas and "specialized niches" from the color infrared photographs were done with virtually 100 percent accuracy before a study of Flannery's 1967 article. The ecologic subareas and specialized niches were conspicuous because of color or tonal differences. This was true of all of the microenvironments with the exception of the specialized niche Flannery has named the Short Grass Steppe West of Venta Salada. It was at first assumed that this area was photographically overexposed on the color infrared photographs because the transparencies are light and show relatively little color density and differentiation. The ground truth survey indicated that the reason for the lack of color on the film was not due to a photographic overexposure, but to a lack of vegetation. Flannery (1967, p. 135) describes the specialized niche as a barren plain, ". . . devoid of all but shortgrass cover and an occasional tree legume." If we had had more familiarity with color infrared photography in widely differing environmental situations, we would have realized that this apparent photographic "overexposure" indicated in reality a specialized ecologic niche. With this single exception all subareas of specialized niches were clearly visible and their boundaries, where distinct, could be sharply defined.

More difficult to distinguish were the various cultural features. Even though all specialized niches were easily distinguishable, it is most convenient when discussing site visibility to use Flannery's four main ecologic divisions or "subareas": (1) Alluvial Valley Center and Broad Travertine Terraces on the West Side of the Valley; (2) the Limestone-Travertine Slopes and Steppes on the West Edge of the Valley; (3) the Broader Alluvial Slopes on the East Edge of the Valley; and (4) Narrow Canyons and Dissected Alluvial Slopes on the Southeast Edge of the Valley.

Table 1 summarizes the success of distinguishing sites in the different ecologic subareas or microenvironments. Flannery's Narrow Canyons and Dissected Alluvial Slopes . . . subarea was considered as part of the Alluvial Slopes subarea due to the identical problems in seeing sites in both of the subareas and the fact that Flannery (1967, p. 139) himself observes that "Subarea 4 (Narrow Canyons and Dissected Alluvial Slopes . . .) might just as accurately have been considered a niche within Subarea 3 (the Broader Alluvial Slopes . . .), for it lies within and alternates with the latter."

Table 1 An Index of the Success of Discerning Archaeological Sites and Features Located within Three Ecologic Subareas of the Tehuacan Valley through the Study of False-Color Infrared Photographs

Factor	Alluvial Valley Center and Terraces	Travertine Slopes	Alluvial Slopes (Thorn Forest)
Total items	20	11	11
Percent visible	85%	100%	0%
Percent not visible	15%	0%	100%
Number correct	17	10	0
Number incorrect	0	1	0

Total items: the number of sites inspected on foot in each of the ecologic subareas. Percent visible: the percentage of sites and features visible on the photographs within each individual ecologic subarea. Percent not visible: the percentage of sites and features not visible on the photographs that were present within each of the ecologic subareas. Number correct: those sites or features which are visible on the photographs that were correctly identified as to their function. Number incorrect: those sites or features which are visible on the photographs that were not correctly identified as to their function.

Site discrimination was quite successful for the Alluvial Valley Center and Terrace subarea. Of the 20 sites visited in this subarea, 17, or 85 percent, had previously been discovered on the color infrared photographs. Only 3, or 15 percent, could not be seen. Furthermore, in all instances the types of features noted—platform mounds, canals, etc.—were correctly identified.

Mounds were especially pronounced in this subarea because modern farmers avoid the larger, more difficult to plow and irrigate raised land forms. The distinction between plowed fields and associated cultivated plant forms and the natural vegetation on the valley floor is the recognition signal. It seems very likely that mounds surrounded by plowed fields would be almost as easily discernible on regular panchromatic film as on color infrared film, except that the vegetation differences are enhanced with the latter medium.

More difficult to distinguish are the smaller sites. In such cases the surface features have been almost completely destroyed by modern farming and all that remains are scattered sherd and cobble areas. The three sites within this subarea that are not observable on the film are sherd areas. However, under certain circumstances sherd areas may show clearly in plowed or harvested fields. This was evidenced by a site that is visible on the photographs as the outline of a rectangular structure. Surface inspection revealed only a sherd area and no rectangular pattern. The pattern is visible on the film because of an absence of color, suggesting that where the structure stood, or the foundation presently exists below the surface, the corn did not grow as well, leaving the rectangular pattern not discernible on surface inspection.

Linear agricultural borders constructed of stone and brush are strikingly visible on the color infrared photographs of the broad alluvial slopes on the west edge of the valley (Figs. 1 and 2). Hundreds of linear borders of two types are visible, those registering as dark lines and those visible as light lines. The dark lines are rock borders, most of which are only one course wide and one or two courses, about 10 to 25 cm, in height. These features are visible as dark lines on color infrared film because the rock radiates or reflects solar radiation to a different extent than the surrounding soil and grass surfaces. The light lines represent long, narrow erosional surfaces characterized by almost a complete absence of vegetation. However, the long, narrow strips appear as light lines on the transparencies not because of the lack of vegetation per se, but due to their particular reflective qualities or characteristics which differ from immediately adjacent areas. There are several sources of evidence that indicate these erosional surfaces are the remnants of linear borders constructed of brush and other perishable materials. In fact, most of these narrow erosional surfaces appear to be parts of, or continuations of, the same water and soil control systems as the agricultural borders constructed of rock. The rock borders tend to characterize the upper slopes of the terrain while the erosional surfaces continue the evenly spaced pattern, closely following the contours of the lower, less steep portions of the slopes. The less steep portions of the slopes were apparently modified through the construction of very low terrace or linear border walls of brush and wood which have since disintegrated, leaving the slight alluvial terraces with an erosion surface at the break in slope where the border once stood. Although the stone borders, and only the stone borders, were noted during the Tehuacan Archaeological-Botanical Project survey, the extent and form of this particular water and soil control system were not, and could not be, discerned or mapped from the panchromatic black and white aerial photographs available to that project. The entire system could be mapped in several hours from the color infrared photographs.

Even better success was had spotting sites from the color infrared photographs of the Travertine Slope microenvironment, although individual cultural features were not usually as clear as they were in the Alluvial Valley Center subarea.

Table 1 indicates that 11, or 100%, of the sites recorded on the travertine slopes within the study area are observable on our photographs. The individual features include "fossil" and modern canals, platform mounds, courtyards, etc. The reason for the prominence of cultural features on the photographs is that the ground surface is almost barren limestone or travertine with little or no soil cover. Consequently, the dominant vegetation is comprised of short grasses which do not grow well on the cultural features, resulting in their definition on the photographs (Fig. 3). In some instances, such as the platform mounds at the large Post-Classic site of Venta Salada (Tehuacan Archaeological-Botanical Project site TR-57), grass covers the features, but not as densely as it does the surrounding lands, resulting in a poor but still recognizable definition of the features. Although this site and all associated features noted on the ground were visible on the photographs, one feature was misiden-

Figure 1 Black and white rendition of a color ir photograph. The upper portion of the photograph is the alluvial slope microenvironment. The dark lines are erosional surfaces where brush terraces were constructed. Scale: 1:10,000.

Figure 2 Erosional surfaces where prehistoric brush terraces were built. Each figure stands on an erosional surface and there are two erosional surfaces between the figures. The lack of vegetation where the terraces once existed emphasizes their existence on color ir film.

Figure 3 The figure stands on a "fossil" canal which is partially covered with grasses. Arrows point to the denser growth of grass on each side of the canal. The differential growth of the grasses permits easy recognition on color ir photographs.

tified. This was a partially filled-in, grass-covered "fossil" canal that was at first identified as a wall partially surrounding the site. This misidentification was a result of Gumerman's relative unfamiliarity with the valley's archaeology.

One entire "site" was incorrectly identified from the photographs of the travertine slopes. The object on the photograph, approximately 1 kilometer northwest of the Venta Salada site and in the same ecologic subarea, appeared similar to one of the Venta Salada platform mounds. After a considerable search for the site on foot, it was discovered to be a recently abandoned field which had been cleared of vegetation. The lack of relief was not apparent on the photograph, but after visiting the "site" and re-examining the transparencies with a hand lens, furrows were discerned and should have been noted before the field check.

The Alluvial Slopes or "thorn forest" ecologic subarea effectively concealed 100 percent of the sites. It is in this microenvironment that a number of important sites, ranging in date from the Santa Marie (Formative) Phase through the Venta Salada (Post-Classic) Phase, had been discovered by the Tehuacan Archaeological-Botanical Project survey team. Only after Neely pointed out some of the larger sites which he had visited previously did they become discernible on the transparencies. Recognition was with much difficulty and uncertain, and most of the individual cultural features within the sites could not be seen at all. For example, the Purron Dam (TR-435), a structure 400 meters in length, 125 meters wide, and 25 meters high, appears to be a natural topographic feature. It was discernible as a manmade structure only when, and if, it was considered in relation to the canyon bottom context across which it had been constructed. A nearby Palo Blanco (Classic) Phase hilltop site (TR-73) was extremely difficult to detect on the transparencies even after visiting the site and having knowledge of its exact location. This site is approximately 1.0 by 1.5 kilometers in size, and is characterized by architectural terraces upon which were constructed large platform mounds flanking plaza areas, a ball court, and other architectural features.

The reason for the concealment of the sites within this ecologic subarea is that although the climate is semi-arid, and the plant cover primarily consists of cactus and small-leaf shrubs and trees, the vegetation is forest-like with a broad canopy closing from 1 to 2 meters above the surface (Smith, 1965, p. 117; Flannery, 1967, pp. 138–139). The canopy is not uniformly dense nor does it consistently obscure the surface of the ground. Nevertheless, it does effectively "blur" the cultural features so that they are indistinguishable from natural features. High-quality panchromatic black and white aerial photographs might be more suitable than color ir for site identification in this microenvironmental situation.

Conclusion

The results of the Tehuacan Valley test of color infrared film for archaeological utilization indicate that the visibility of sites depends primarily on their environmental situation, and also that the delineation of environments and microenviroments is especially easy with this type of film. Furthermore, we can state that the age and size of the sites are not necessarily deciding factors in their discernment. A special effort was made to study sites and features ranging widely in age and size within each of the ecologic subareas represented. Unfortunately, one of the more important tests of this type was frustrated by the almost complete erosional destruction of the shallow Abejas Phase (ca. 3500 to 2300 B.C. pithouse hamlet discovered northwest of the modern village of San Gabriel Chilac in the early 1960s by R. S. MacNeish (1964). More important than age is the effect of the site on the vegetation and the resultant patterning in color and tonal variations on the film. For example, a single alignment of cobbles and small boulders one course in height or a bare spot 30 cm in diameter is conspicuous in an environment of grass cover because it interrupts the natural vegetative pattern.

Since the immediate environment is the most important factor in site discrimination, the data obtained in the Tehuacan Valley can be transposed to similar environments and microenvironments on a world-wide basis.

We conclude that:

1. Agricultural land which is under crop or has recently been harvested reveals sites on color ir film for several reasons. Color ir film is especially adapted for determining plant speciation and plant vigor (Colwell, 1970). Habitation areas which are now used for agriculture affect plant vigor either by increasing or decreasing production. As noted previously, large cultural features may be partially or completely outlined by cultivation and the difference in vegetation is readily apparent on color infrared film although regular panchromatic film may be almost as effective in this instance.

2. Sites are distinguishable in areas devoid of most vegetation in eroded, saline, or extremely arid zones if cultural features are constructed of materials different in consistency or type from the surrounding soil matrix because they absorb and reflect solar radiation differently.

3. Sites in a predominantly grassy environment are visible because in most cases either the grasses grow differently—that is, with greater or lesser vigor—or different types of grasses grow on these features. As noted above, color infrared film is capable of reflecting differences in the presence of plant species and varying degrees of plant vigor.

4. Finally, it is apparent that for archaeological reconnaissance color infrared photography is unsuitable for use in a canopy forest environment because the canopy visible to the cameras usually does not reflect the disturbances caused by the past human habitations it covers.

It is important to note that many of the positive characteristics of color in film are equally valid for panchromatic film and that the ideal archaeological situation is to complement panchromatic film with color ir.

REFERENCES

Coe, M. D., and Flannery, K. V. Microenvironments and mesoamerican prehistory. *Science* 143:650, 1964.

Colwell, R. N. Applications of Remote Sensing in Agriculture and Forestry. In *Remote Sensing with Special Reference to Agriculture and Forestry.* Washington, D.C.: National Academy of Sciences, 1970.
Flannery, K. V. Vertebrate Fauna and Hunting Patterns. In *The Prehistory of the Tehuacan Valley,* Vol. 1. D. S. Byers, ed. Austin, Tex.: University of Texas Press, for the R. S. Peabody Foundation, 1967. Pp. 132–177.
— Archaeological Systems Theory and Early Mesoamerica. In *Anthropological Archaeology in the Americas.* B. J. Meggers, ed. Washington, D.C.: The Anthropological Society of Washington, 1968. Pp. 67–87.
—, and Coe, M. D. Social and Economic Systems in Formative Mesoamerica. In *New Perspectives in Archaeology.* S. R. Binford and L. R. Binford, eds. Chicago: Aldine, 1968, Pp. 267–283.
Gumerman, G. J., and Lyons, T. R. Archaeological methodology and remote sensing. *Science* 172:126, 1971.
MacNeish, R. S. Ancient mesoamerican civilization. *Science* 143: 531, 1964.
— An Interdisciplinary Approach to an Archaeological Problem. In *The Prehistory of the Tehuacan Valley,* Vol. 1. D. S. Byers, ed. Austin, Tex.: University of Texas Press, for the R. S. Peabody Foundation, 1967. Pp. 14–24.
Schaber, G. G., and Gumerman, G. J. Infrared scanning images: An archaeological application. *Science* 164:712, 1969.
Smith, C. E., Jr. Flora, Tehuacan Valley. *Fieldiana: Botany* 31: 107, 1965. Chicago Natural History Museum.

USING AIR *photographs to detect plant diseases is not a new technique. As early as 1929, J. J. Taubenhaus and co-workers were able to detect cotton root rot on aerial photographs. The tremendous advances made in photographic techniques, cameras, and films have further improved the ability to detect plant disease and insect infestations on air photographs. Infrared aerial photography in particular has been used with remarkable success in detecting diseases in critus, vegetable, and cereal crops. Healthy vegetation is highly reflective in the near infrared but quickly loses this reflectivity with loss of plant vigor. Some investigators have reported that the high reflectivity is due to the chlorophyll in the leaves. More careful research has revealed that infrared reflectance is a complex interaction. Experimentally it seems to result from the interaction of light with plant cell wall interfaces. High infrared reflection in healthy vegetation then is the result of the physical structure of the leaves of the plant. As soon as this structure begins to change, the infrared reflectance is diminished, and its variation can be recorded on infrared photographs. Probably the best potential application for the use of air photographs in locating plant disease is identification of infected areas in perennial crops such as orchards, sugar cane, or other plantations. Annual crops are grown on vast areas and would require monitoring several times during the short growing season. This would be very expensive and too extensive a task for human photointerpreters. Perennial crops would need to be monitored once each year or less. The overall field response to disease in perennial crops would be slower and more limited spatially. Individual members of the plant community could be treated or removed before disease or insect infestation spread.*

24-Infrared Photography as an Indicator of Disease and Decline in Citrus Trees

GERALD G. NORMAN
NORMAN L. FRITZ

In the 50 years of the existence of the Division of Plant Industry, Florida Department of Agriculture, the number of citrus trees in the state has increased over 700 percent. During the same period an even greater expansion has taken place in the production of citrus nursery stock, tropical fruits, ornamental plants, and related horticultural crops. Because of this growth and the consequent multiplication of the Division's workload, the need for improved techniques in methods of inspection has long been apparent.

The word "inspection" has been defined as "to look, or to view closely or critically." This definition has actually been the general basis of agricultural inspection throughout the world since plant quarantine organizations were first established.

In recent years, recording-sensing instruments, other electronic devices, and photography have been applied to inspection in both industry and agriculture and have demonstrated their usefulness in producing permanent documentary records. In some applications these methods have proved superior to that of visual inspection by trained observers. Such records are permanent, objective, and reliable, and when properly interpreted, eliminate human error. Because of the Division's workload, and the availability of such instruments and techniques, exploratory trials were begun in 1964 with those devices which appeared to have possible applications for the detection of diseases in citrus trees (Norman, 1965).

Preliminary Experimentation

Initially, tests were run with a special Barnes bolometer, which is an extremely sensitive thermometer capable of measuring very small amounts of radiation, especially in the infrared region of the spectrum. Measurements with this instrument on Key Lime seedlings, *Citrus aurantifolia (Christm.) Swing.*, which were infected with severe strains of Tristeza virus, showed that the amount of infrared radiation from infected trees is different from that of healthy trees. Similar measurements indicated no differences in radiation between young trees grown in soil infested with nematodes [*Rodolphus similis (Cobb) Thorne*] and nematode-free trees, at least in the early stages of nematode invasion.

From *Proceedings of the Florida State Horticultural Society* 78:59–63, November, 1965. Reprinted with permission of the authors and the Florida State Horticultural Society.

In a second series of tests, a Barnes T-4 infrared scanning camera, which is normally used to make thermographs of the distribution of temperatures over wide areas, was used to scan both healthy and diseased trees. It was concluded that at the present time under Florida conditions, this instrument is unsuitable for detecting differences between sick and healthy trees.

Concurrent with the trials just described, detection of disease by means of black and white aerial infrared photography was also explored. The first flights were made by photographers from the Division's Information Section in the several types of aircraft available for charter in the area. Some of the resulting photographs, made for the most part with press cameras, were considered sufficiently promising to justify continuance of this phase of the study. Cost estimates supplied by two organizations engaged in commercial aerial photography indicated that charges even for preliminary disease surveys would be prohibitive. Neither company had had any previous experience with, or knowledge of, this type of special-purpose photography, and when the Division subsequently chartered a plane and pilot from one of them for a trial flight, it was found that the pilot had serious objections to flying at the 600-foot altitude required for this application. Because of the promise of this method, and because of the cost of having it done commercially, the Commissioner of Agriculture authorized the creation of a photographic unit to be established in Winter Haven to continue the photographic exploration on a systematic basis, and to assemble the necessary components of a laboratory capable of processing both black and white and color aerial infrared films.

Equipment

Two Fairchild K17B aerial cameras were contributed to the project by the United States Department of Agriculture, Plant Pest Control Branch, a cooperating agency. These cameras produce a 9-inch-square format from roll film available in lengths from 75 feet to 280 feet. Later, when infrared color film became available in the 35 mm size, two Exakta cameras with matching Biotar lenses graduated in half-stops were used. These cameras were mounted side by side on a metal bar, the vertical handles of which were equipped with dual cable releases so that paired exposures could be made more or less simultaneously. The use of this smaller film size reduced the cost of the experiments considerably.

Both fixed-wing aircraft and helicopters have been used in the project. The best results to date have been obtained from helicopters flying at altitudes between 200 and 600 feet.

Bases of Investigation

For the purposes of this investigation it was assumed that the principal diseases of citrus were footrot, Tristeza, Psorosis, Xyloporosis, and Exocortis. Spreading decline, caused by burrowing nematodes but generally referred to as a disease, was also included because of its economic importance. Five groves in the Winter Haven area were selected, each containing one or more of these diseases. In some, the exact number and location of each diseased tree were known; for example, one of the sites selected was the Division's own virus test plot. Other areas were platted tree by tree. Nematode-infested trees were platted first on the basis of visible symptoms to evaluate the size of the infested area and the degree of progress of decline in the grove. These same areas were later platted by taking root samples so that the precise extent of infestation could be determined for use in detailed interpretation and comparison of the photographs.

In order for the photographic method of detecting disease to be effective, it is necessary that there be some difference between the reflectances of healthy and infected trees. This difference in reflectance must occur within the spectral sensitivity range of the film.

Ideally, it would be desirable to have spectrophotometric measurements of healthy trees and of infected trees at various stages of progress of the disease. Such measurements would show in which spectral regions there is the greatest difference in reflectance, and it would then be necessary only to photograph in that region by means of suitable filters and film, to obtain the greatest possible detectability. Unfortunately, equipment which will measure the light reflected from trees as it would be recorded photographically from the air is unavailable at the present time. Measurement of individual leaves is not entirely satisfactory because an appreciable effect on the over-all photography is produced by the various orientations of the leaves on a tree with respect to the sun and the camera, and by the interreflectance between the leaves.

Without this method of approach, it is possible to infer the spectral characteristics from current knowledge of healthy and diseased foliage, and to try various combinations of filters and film sensitivities to distinguish between them.

One of the first changes that may occur in diseased or dying foliage is a decrease in infrared reflectance, and this effect may take place before any visible change can be noted. Also, most healthy foliage has a much higher reflectance in the infrared region than in the visible. Thus the greatest difference in reflectance between diseased and nondiseased foliage is apt to occur in the infrared region. Both black and white and color films which are sensitive in this region are available, and both are being tested. Present indications are that the color film may be more useful for this application since color differences are often more detectable than the changes in gray level produced by a black and white film.

Kodak Ektachrome Infrared Aero Film, Type 8443, is a false-color reversal film originally designed for detecting camouflage, but which is now more widely used for medical applications and to detect differences in and between plants of various types. The principal characteristics of this film have been described in the literature (Fritz, 1965; Tarkington and Sorem, 1963), so a brief description will suffice. A normal color film contains three layers which are sensitive to the blue, green, and red spectral regions, respectively, and when photographs are made, the images produced will be composed of colors which closely match those of the original scene photographed. The Ektachrome Infrared Film, however, is composed of three layers which are sensitive to the green, red, and infrared. Exposures are made through a yellow filter which cuts out the blue light to which all three layers are sensitive. The dye layers of the film are correlated with the sensitivities in such a way that green will be recorded as blue, red will be recorded as green, and infrared as red. Thus, because of its high infrared reflectance, healthy foliage will photograph as various shades of red, with perhaps a slightly bluish cast, depending on the intensity of its green reflectance. As the infrared reflectance is lost, this color will change toward magenta, purple, or green, depending on the magnitude of the loss.

The detectability of small color differences in color photography is primarily dependent on three characteristics: the saturation of the color being photographed, the over-all color balance of the film, and the darkness or lightness of the photographic record. The color characteristics of the objects which are being photographed, and thus the saturation, cannot be changed, but the color balance of a photograph can be changed by placing suitable filters over the camera lens, and the lightness or darkness can be controlled by the exposure. To determine the optimum color balance and exposure for this film for maximum detectability of disease in citrus trees, a number of series of photographs have been made in which both the color balance and the exposure have been systematically varied. A study of these photographs has indicated the optimum region of color balance and exposure.

An illustration of the differences in color balance which were produced by a single test is given in Figure 1, which shows the convenient trilinear method of plotting these differences. The 0 point represents the balance of the particular coating of the film used, and the direction and distance of other points from 0 indicate the color and magnitude of the shift caused by the filter or combination of filters indicated. Thus, it can be seen that increasing concentrations of Kodak Color Compensating Magenta Filters produce a progressively more yellow color balance. Similarly, Color Compensating Blue Filters shift the balance in the red direction, Corning Filter 3966 produces a cyan shift, and combinations of filters combine the effects of the individual ones. Since it is known from previous tests that the optimum color

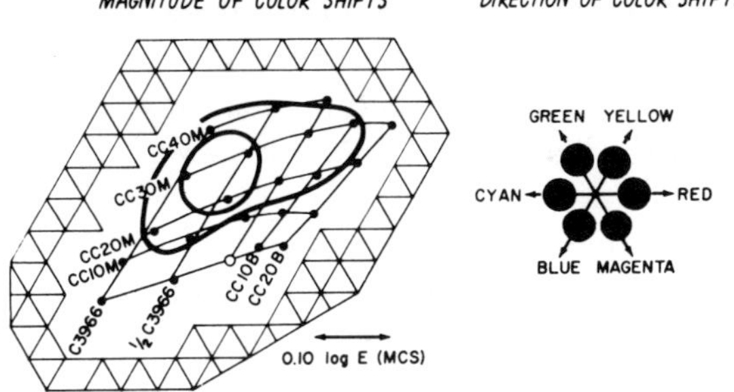

Figure 1 Color shifts produced by various combinations of KODAK Color Compensating and CORNING No. 3966 Filters.

balance is in the region encompassed by these points, no attempt was made to shift the color balance in the blue or magenta directions.

Procedures

In order to determine the most effective film color balance for detection of disease, three of the areas previously described were photographed at three exposure levels at each of the color balances indicated by a point on the graph. A single test thus consists of 207 individual photographs, and several tests have been made. All flights were made between 10:00 A.M. and 3:00 P.M. on bright, relatively cloudless days.

Some work has been done to evaluate the effect on the aerial photography of variables such as the plane altitude and the angle between the direction of the sun's rays and that of the camera. It is planned to evaluate further these variables and others such as solar altitude, haze, special bandpass, and sharp cutoff filters, moisture content of the air, soil, and trees, the stage of leaf growth, and perhaps other factors at present unknown.

Results

From the color-balance photographs it has been determined that the color-balance region enclosed within the outer black line in Figure 1 provides some differentiation between diseased and nondiseased trees, the best differentiation being provided by the balances within the inner line. All subsequent tests will be made using this balance.

From some 1000 photographs studied to date it is apparent that at certain times and under certain conditions infrared-sensitive color film will reveal the presence of sick and declining citrus trees, in some cases even before these conditions are apparent to the trained eye. It is also apparent that under other conditions, photography is of limited usefulness, or is ineffective. Further study may indicate the conditions necessary to produce optimum discrimination. The latent nature of some of the diseases of citrus create some interpretation problems. For example, photographs of one grove area indicated an infrared-reflectance loss in approximately 50 percent of the trees in a large block of the Temple variety, none of which at the present time show visible symptoms of disease. Later examination of the trees showing a reflectance loss may prove that they are infected with Psorosis, indexing may show they are infected with Tristeza virus, or it may be found to be due to some other cause. In photographs made of trees infected with Psorosis virus, where eruptive scaling has progressed to the point of limb-girdling, these limbs appeared diseased while the balance of the tree seemed healthy. In trees on sour orange rootstock, reaction to Tristeza virus has been found in the infrared photographs long before visible symptoms could be seen in the field. Burrowing nematodes are apparently not detectable by this method, in young trees which have been infested up to 3 or 4 years. However, they are easily detectable in older trees in areas where they have been present for longer periods of time. From somewhat inconclusive evidence to date, it appears that Xyloporosis and Exocortis virus can be detected where the disease has progressed beyond mild stages.

Conclusions

Very encouraging results have been attained with aerial infrared photography in the detection of what are assumed in this paper to be the principal diseases of citrus—footrot, Psorosis, Tristeza, Xyloporosis, and Excortis. (See Fig. 2 [in color insert].) Apparently, all of them can be detected. However, the effects on the photography of many variables, such as atmospheric and photographic conditions, must be evaluated before detection of disease by this means can be used with assurance of accuracy and dependability.

REFERENCES

Fritz, N. L. Film sees new world of color. *Citrus World Magazine* 2:11, 1965.

Norman, G. G. Exploring disease detection in citrus trees by means of aerial photography. *Proc. Am. Soc. Hort. Sci., Caribbean Region,* 23rd Annual Meeting, Kingston, Jamaica, 9:128, 1965.

Tarkington, R. G., and Sorem, A. L. Color and false-color films for aerial photography. *Photogram. Engng.* 39:88, 1963.

THE AERIAL *camera records on the emulsion of the photographic film all reflected energy collected by the lens in the wavelengths at which the emulsion is sensitive. The photograph is a record of all objects in the field of view that are larger than the minimum resolution limits of the camera system. Thus the camera is not selective, except by wavelength or object size. For this reason the air photograph offers more data about the earth's surface than any other cartographic device. Precise geographic relations also can be determined by photogrammetric techniques. Air photographs provide the geologist with a more comprehensive record than he can obtain from the ground. Collection of ground data in the field is slow and tedious; it is often difficult to get into some areas. Air photographs can be generated over these areas and quickly utilized by trained interpreters. The photo serves as a historic record which can be referred to frequently and which is a convenient measure of temporal change. Another advantage of photographs is their reduced scale. Although field work can never be entirely replaced, it is now evident that a field observer is sometimes overwhelmed by large-scale detail. High-altitude photography offers an overview which makes it possible to recognize the dominant structural units more easily. If these photos are produced from color or infrared color film, the interpreter has available over a thousand times more color tones than he could distinguish as shades of gray on black and white imagery. Infrared color film is particularly valuable because it clearly delineates the water/land interface better than any other film. Vegetation is frequently a surrogate of underlying geologic formations. It is possible to understand the underlying terrain conditions by interpretation of vegetative response and pattern. These are clearly visible on color infrared photographs.*

25-High-Altitude False-Color Photointerpretation in Prospecting

HEIKKI PAARMA
JOUKO TALVITIE

BLACK AND white aerial photographs taken at high altitudes (9 kilometers) have been available in Finland since 1963. Photomosaics made on the scale 1:100,000 have proved useful in the geologic interpretation of this material. It is easier to recognize the dominant tectonic structural units from a small-scale, large-area combination of aerial maps than from corresponding large-scale maps. In the field of black and white stereo work, the superiority of small-scale pictures has been emphasized by W. R. Hemphill (1958).

Since it was probable that false-color pictures could be taken with the available equipment also at higher altitudes, up to 9 kilometers, it seemed reasonable to try to lay photomosaics from the false-color material.

From Smith, J. T., Jr. (ed.), *Manual of Color Aerial Photography*. Falls Church, Va.: American Society of Photogrammetry, 1968. Pp. 431–438, 440. Reprinted with permission of the authors and the American Society of Photogrammetry.

This became possible through the considerable financial support obtained from the Finnish Foundation for the Research of Natural Resources (Suomen Luonnonvarian Tutkimussäätiö) for this research, which was incomplete at the time this article was written. Some of the results obtained in Northern Finland will be reported in this article.

Photographic Altitude and Season

It was proved rather quickly that dominant tectonic structures, significant for prospecting, could not be reliably located by more intensive investigation of a single picture or one pair of stereo pictures. A picture taken with a wide-angle camera (Zeiss pleogon RMKA 15/23 $\times$ 23 cm^2) at an altitude of 9 kilometers includes an area of 190 km^2 (picture scale 1:60,000). The 1:100,000 map sheet commonly used in Finland covers an area of 30 km $\times$ 40 km. For a normal stereo coverage 20 exposures are needed at an altitude of 9 km. At an operating altitude of 4.5 km, 100 pictures would be needed correspondingly. It was obvious that, because of several cost considerations, low altitudes were out of the question when taking false-color pictures of larger areas. On the other hand, the available aircraft did not allow an altitude above 9 km. During this research, a total area of about 8500 km² has been covered by high-altitude photography in different parts of Finland. Only local targets have been photographed at low altitudes. In order to make the interpretation of the aerial pictures simpler, false-color pictures were also taken on the ground, and geologic observations have been made.

The most important special features of the false-color film from the viewpoint of tectonic interpretation are: the distribution of moisture in the terrain is clearly shown and, because of the sensitivity of the film for the near infrared (Billings and Morris, 1951), a multitone picture is obtained of the distribution of the living vegetation. Both these features depend, however, on the season. The false-color pictures shown with this article were taken on June 20, 1966. The spring flood caused by the melting of snow was over at that time, but the even dryness of the midsummer soil had not had sufficient time to develop. The differences in the moist areas and the amount of area covered with water were quite distinctly visible at the time of photography. On the other hand, the infrared reflectivity of the vegetation had not yet reached its obvious maximum which occurs during the blossom time at midsummer. The largest tone differences for different geologic substrates can be obtained when the birch leaves are still growing. At that time, the reflection of the near infrared from barren substrates was most clearly distinguished in a terrain in which there was a fertilizing influence in the soil, caused by basic rocks, etc.

Making of the Photo Mosaics

The infrared color aerial film (8443) Eastman Kodak Co. was developed to a positive transparency in the usual manner. As such, the transparencies were not suitable for making the mosaic. Paper prints had to be made of individual pictures. A mosaic for a large area is naturally cheaper, if a reduction—e.g., to the scale 1:100,000—is made at printing of the original 1:60,000 diapositives. Hence, a color internegative on a 24 mm × 36 mm film was first made of each picture. Paper prints on the scale desired were then obtained from these for making the mosaic.

The false-color film can also be developed directly into a color negative. From this, it is simpler to make different kinds of copies (false-color and black and white prints and prints obtained by using filters).

On the other hand, the corresponding Russian films are of the negative type (Iordanskij, 1964; Smirnov, 1961). It was unfortunately impossible to arrange a comparison of the Russian and western films.

The aeromagnetic map sheet 1:100,000, with the most important landmarks of the region included, was used as the model for assembling the production of the mosaic, as well as the changes of scale.

Topography and Rock Types of the Investigated Area

The area is topographically a hilly region rich in lakes. The water surface of the lakes lies about 160–200 meters above the sea level. The greatest elevation difference is 112 meters, but the region is generally speaking relatively flat and covered with moraine. The bulk of the moraine is coarse, sandy, and devoid of clay. The stones are usually angular. Open bogs are common. The peat layers are mostly thin, about 1–3 meters thick. Underneath the peat there is usually some moraine. No clay has been found. The eskers are of gravel and their tops have been smoothed to the level of the ancient Yoldia sea, on the average to a height of about 205 meters (private communication by Professor E. Hyyppä). According to Hyyppä, the forming of the eskers is not primarily because of topographic, but rather tectonic reasons. By these he means fractured zones which were active at the postglacial time, rifts formed in the continental glacier, and eskers formed in the rifts (Paarma, 1963). The amount of glacial wear is about normal in this region. Smooth bedrock occurs, and weathered preglacial bedrock is also found. There are only few outcrops. The average thickness of the overburden is about 5 meters.

The predominant rocks of the region containing ores are early magmatic gabbros rich in Na and poor in K, amphibolites and albite gabbros, and rather commonly also albitites. Rocks formed out of sediments, such as quartzite and dolomites, etc., are found in smaller amounts. This ore province is surrounded in every direction by microline granite and migmatite. According to J. Nuutilainen, the igneous rocks containing the iron ores of the region are hypabyssic infracrustal rocks which are comparable with the spilitic-keratophyric association. The ores are formed out of a spilitic hydromagma, which has been formed by differentiation at a rather low temperature. Nuutilainen emphasizes that tectonic factors have been decisive at the localization of the intrusion of iron ores. Distinct fault surfaces, the most prominent of which lie in the E–W direction, govern the dimensions and the position of the ore bodies.

Tectonic Interpretation of the False-Color Films

The lakes, wet bogs and eskers have been picked up in Figure 1 from a photomosaic. The lakes are normally completely black in a false-color picture. Hence, even small ponds and the orientation of the lakes in general stand out for observation much more clearly than in black and white pictures. Wet bogs appear either black or with a blue tone (see Figs. 3–7). As explained above, the terrain conditions of the photographed region make it possible to understand that the precambrian tectonic structure is often projected through the overburden sufficiently well to be understood.

The results of an interpretation of the overall morphologic orientation are shown in Figure 2. The orientation of the terrain is more easily seen in the original mosaic, with its color tones, than from the interpretation element in Figure 1. The magnetic orientations obtained using an aeromagnetic map are also shown in Figure 2. Only the

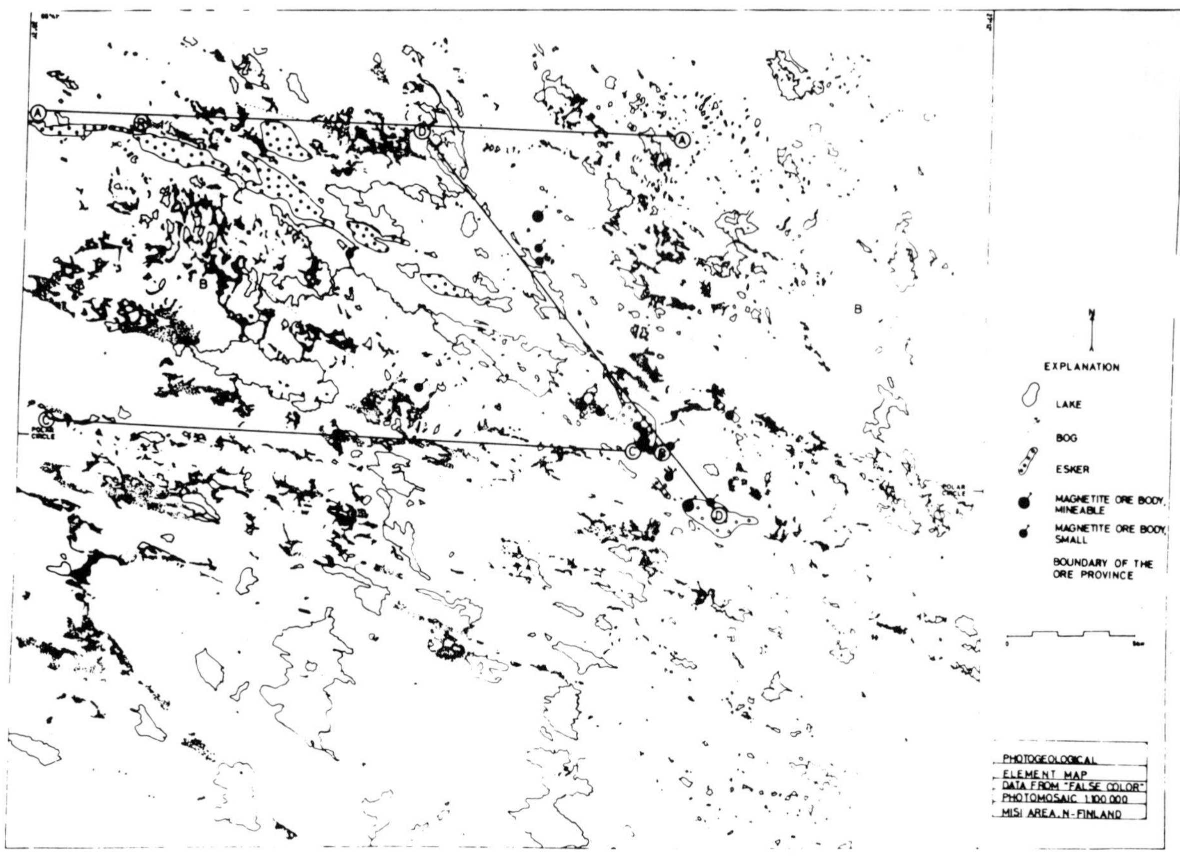

Figure 1 Photogeologic element map.

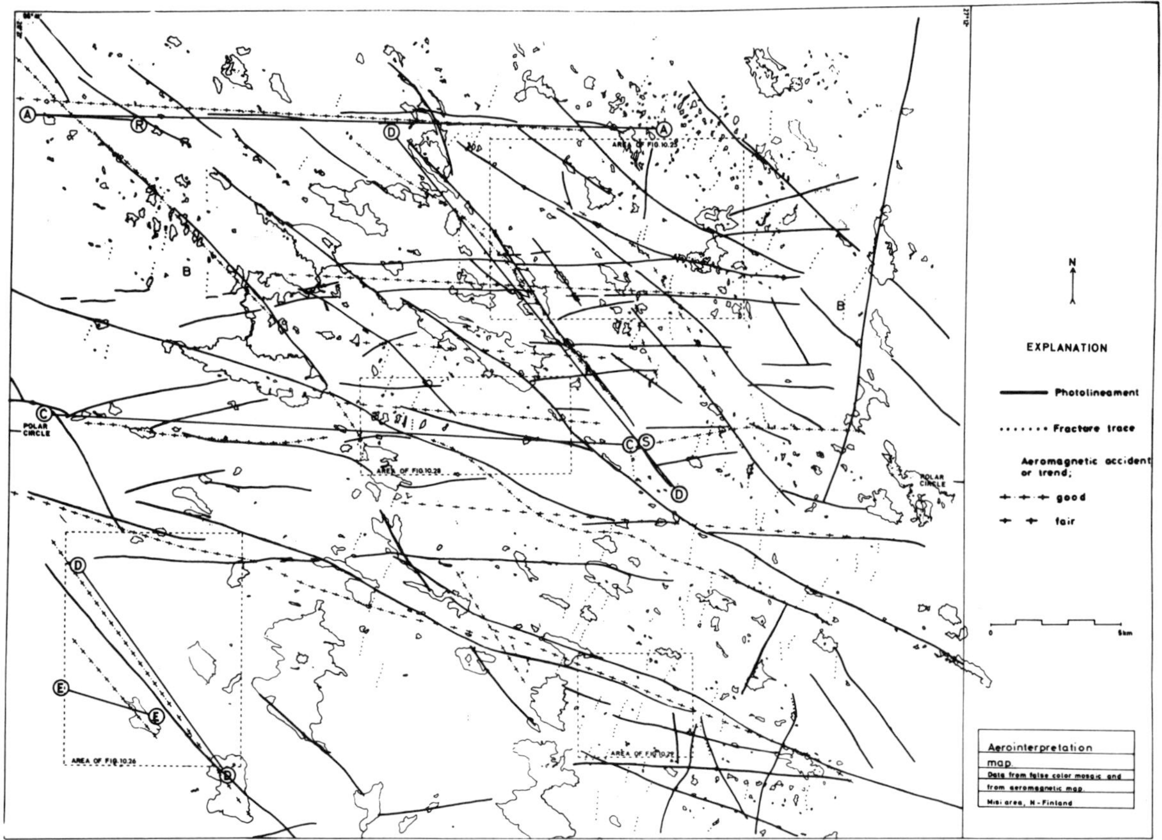

Figure 2 Aerointerpretation map.

most distinct and far-reaching orientations have been included. A good correlation is found between the morphologic and aeromagnetic orientations. There are several dominant fracture zones in the E–W direction in the map area. The line B-B shows up in the photomosaic, for instance, as the orientation of the bogs (Fig. 3)[1]. The line A-A coincides with the change in the direction of the eskers at point R (Figs. 1 and 2). The eskers system obviously follows the tectonic structure of the base rock of the foundation (Talvitie, 1965). The orientation D-D is particularly interesting from the prospecting viewpoint. An orientation parallel with it is found, for instance, in Figures 4 and 5. The intersection of the multiband morphologic and aeromagnetic orientation C-C (Fig. 6) with the direction D-D is obviously the most critical one for the formation of ores (cf. iron ore at point S) (Beyer, 1964; Cousins, 1959; Gorjersky, 1965; Mayo, 1958). The point S can, in fact, be considered as the junction of several tectonic orientations. The fracture trace orientation (terminology after Lattman (1958) in the approximate direction N 20°E shows up botanically (Figs. 3–7). It should be noted that it ends at a block boundary which passes through point S.

[1] See color insert for Figures 3, 4, 6, and 7.

Methods Used in the Field Work

Figure 7 shows a color print made from a part of a false-color mosaic. In Figure 8 the results of an electrogeophysical measurement have been reduced into the scale of Figure 7. The electric measurement covers only a small part of the area of the false-color picture. It is found that the photogeologic, botanic, and geophysical orientations all fit well together. It is clear that the many color tones of the false-color pictures cannot be explained by photointerpretation only. Nevertheless, it is very often possible to obtain aid for the interpretation of the false-color pictures from geophysical measurements. This can be best accomplished during the field work. Since the available 1:60,000 transparencies were used for field work, it was reasonable to include in the field equipment a light table, a large mirror stereoscope, and a small generator for the production of electric power. If making of paper prints is accomplished, one can omit the light table, which is difficult to use in field. A transparency is,

Figure 5 Aeromagnetic survey.

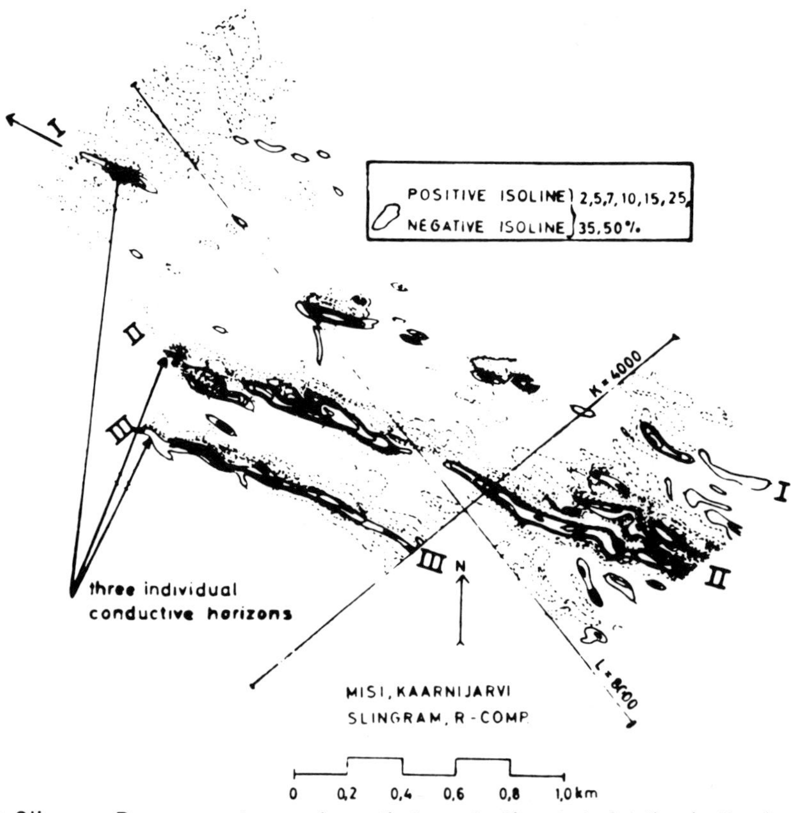

Figure 8 Slingram R-component map shows that conductive material lies in the fault zone. The northernmost horizon (I) may continue to the direction of the arrow. The thickness of the overburden varies between 7 and 11 meters in geophysically interpreted points. The thickest soil accumulation occurs over the horizon I. The appearance of Figure 7 suggests the same soil condition.

however, better for studying the details than copies made from it. Since the false-color picture permits an enlarging of nine times, even small details can be seen quite well from the 1:60,000 stereo pairs. The understanding of these details and the all-over structures to be seen from the photomosaic are aided in the field work by the parallel use of all kinds of geophysical material. It was found practical to make all geophysical and geochemical maps as soon as possible into transparencies (i.e., material with maximal optical transparency, as in glass). By placing these on aerial picture mosaics or individual pictures, one could understand much quicker those features which are important for prospecting and geologic research. It should be mentioned that by placing on the other stereo pair a transparent geophysical map, etc., made at the same scale, a false-color picture of the terrain can then be investigated in three dimensions simultaneously with the geophysical map. The authors will, however, explain later the methods of this continuing study with respect to the field work, copying, and making the mosaics. It seems suitable for field studies to form a work group consisting of a geologist who is familiar with the interpretation of aerial pictures and aided by a phototechnician also familiar with this field of work. An optical pantograph and a suitable developing apparatus have been used. This equipment made possible the parallel use of false-color information along with other prospecting information.

Summary

A false-color mosaic, which has been interpreted tectonically, was prepared in an area with only very few outcrops. The information obtained from conventional geologic mapping is usually very scarce under these conditions.

It is generally known that aerial pictures are useful for geologic and geophysical mapping. In conventional mapping work done in the field it is often impossible to map and to judge the pattern of those tectonic features and principal feature zones which are important for the location of rocks and, above all, of ores of economic importance. Working with aerial stereo pairs does not permit, however, the simultaneous judging of areas large enough for observing the dominant fractures. This is accomplished by the use of false-color mosaics.

It was of paramount importance for the work reported here, that high-quality pictures be obtained with false-color material from a height of at least 9 kilometers (29,528 feet).

When the overburden is thin, the faults and fractures are reflected in the form of the lakes. They are also often seen as bog regions, wet spots with abundant vegetation and eskers. These features show up strongly in the false-color picture by causing a strong contrast of the colors with respect to the surroundings. They are difficult to recognize immediately from a black and white mosaic. The advantage of a color picture is the generally known fact that the eye recognizes about 1000 times more color tones than shades of gray.

The usefulness and interpreting dependability of the false-color mosaic is considerably improved by parallel use of airborne geophysical equipment. This should be produced on transparent material at the scale of the mosaic. The geophysical information can then be accurately placed on the mosaic, without preventing the aerial picture from being seen clearly.

As a result of this work, the ores are concluded to be located in the investigated region in certain principal fractures or at their intersections (Gorjersky, 1965; Mayo, 1958). Fractures in the W–E direction are abundant in the region of the iron ore province. A correlation has been found between the aeromagnetic data and the photolineament. This shows that the recent crustal movements follow the orientation of the precambrian deformation zones (Indans et al., 1960; Paarma, 1963; Talvitie, 1965).

REFERENCES

Affleckt, J. *Geophysics* 28, No. 3, 1963.

Beyer, W. *Autorenkollektiv, Z. angew. Geol.* Bd. 10, Heft 12, 1964.

Billings, W. D., and Morris, R. J. *Am. J. Bot.* 38, May, 1951.

Cousins, C. A. *Trans. Geol. Soc. So. Afr.* 50A, 1959.

Gorjersky, D. J. *Geolog sbornik Lobskogo geolog abtsctsh* No. 9, 1965 (Russian).

Hemphill, W. R. *Photogram. Engng.* 24:1, 1958.

Indans, A., Kovalevsky, M., and Springis, E. *Akad. NAUK Estonskoi SSSR* Tallin, 1960 (English summary).

Iordanskij, A. N. *Zhurn. Nautshn. Prikt. Fotogr. Kinematogr.* 9, No. 3, 1964 (Russian).

Lattman, L. H. *Photogram. Engng.* 24:4, 1958.

Mayo, E. B. *Mining Engineer* 10, 1958.

Nuutilainen, J. *Bull. Comm. geol.*, Finland, in press.

Paarma, H. *Fennia* n:o 1, Helsinki, 1963.

Paarma, H., and Marmo, V. *Terra.* n:o 2, Helsinki, 1961 (English summary).

Smirnov, A. J. *Zhurn. Nautshn. Prikt. Fotogr. Kinematogr.* 6, No. 3, 1961 (Russian).

Talvitie, J. *Filosofinen Tiedekunta.* Oulun Yliopisto, Oulu, Finland, 1965 (Finnish).

Tarkington, R. C., and Sorem, A. L. *Photogram. Engng.* Jan., 1963.

SUGGESTED READINGS FOR PART FIVE

Antonini, G. A. Infrared Image Applications in Studies of the Marine Environment. *Proceedings of the Third Symposium on Remote Sensing of Environment.* Ann Arbor: University of Michigan, Institute of Science and Technology, Willow Run Laboratories, 1965.

Colwell, R. N. Some uses of infrared aerial photography in the management of wildland areas. *Photogrammetric Engineering* 26:774–785, 1960.

Clark, W. *Photography by Infrared: Its Principles and Applications.* New York: Wiley, 1946.

England, G., and Morgan, J. O. Quantitative Airborne Infrared Mapping. *Proceedings of the Third Symposium on Remote Sensing of Environment.* Ann Arbor: University of Michigan, Institute of Science and Technology, Willow Run Laboratories, 1965.

Estes, J. E. Some geographic applications of aerial infrared imagery. *Annals of the Association of American Geographers* 56:673–682, 1966.

Hoffer, R. M., and Johannsen, C. J. Ecological Potentials in Spectral Signature Analysis. *Remote Sensing in Ecology.* Athens: University of Georgia Press, 1969. Pp. 1–16.

Holter, M. R., et al. *Fundamentals of Infrared Technology.* New York: Macmillan, 1962.

Jamieson, J. A., McFee, R. H., Plass, G., Grube, R., and Richards, R. *Infrared Physics and Engineering.* New York: McGraw-Hill, 1963.

Knipling, E. B. Leaf Reflectance and Image Formation on Color Infrared Film. *Remote Sensing in Ecology.* Athens: University of Georgia Press, 1969. Pp. 17–29.

Martin, A. E. *Infrared Instrumentation and Techniques.* Amsterdam: Elsevier, 1966.

Olson, L. E. Accuracy of land-use interpretation from infrared imagery in the 4.5 to 5.5 micron band. *Annals of the Association of American Geographers* 57:382–388, 1967.

Peace, R. W., and Bowden, L. W. Making color infrared film a more effective high-altitude remote sensor. *Remote Sensing of Environment* 1:23–30, 1969.

Philpotts, L. E., and Wallen, V. R. Infrared color for crop identification. *Photogrammetric Engineering* 35:1116–1125, 1969.

Weiss, M. Infrared in meteorology. *Weatherwise* 20:156–161, 1967.

Wolfe, W. L. (ed.). *Handbook of Military Infrared Technology.* Washington, D.C.: Dept. of the Navy, Office of Naval Research, 1965.

PART SIX

REMOTE SENSING IN THE THERMAL OR EMITTED INFRARED

Radiation that results from molecular or atomic motion

The INFRARED *portion of the electromagnetic spectrum is generally considered to be from 0.7 to 1000 μ. A micron (μ) is a millionth of a meter. In this range, infrared radiation of a body is determined by two factors: (1) the nature of the object's surface—its emissivity—and (2) its temperature. Infrared energy is recorded in two ways. The first is with a* radiometer, *a device that records the radiation received and compares it to a fixed or known standard. The electronics of the system can then be used to generate an electrical signal based upon the differences between the standard reference and the object or scene being viewed. The other recording device is an* infrared scanner. *The scanner is normally a system of mirrors which rotate or oscillate and focus the incoming radiation on a detector. The infrared stimulation of the detector creates an electrical charge which can be amplified and recorded on a television tube, magnetic tape, video tape, or photographic film. Using this kind of system it is possible to get a "picture" of the thermal environment that we can experience with our normal human sensors. The capability of recording variations in infrared radiation has tremendous application in extending man's observation of many types of phenomena in which minor temperature variations could be extremely significant or valuable in understanding our environment.*

26-Research in Infrared Sensing

DONALD E. THACKREY

THE TENDENCY of research to open up entirely new fields of knowledge and technology has been noted often in recent decades. One of the best examples of this tendency is provided by the field of infrared sensing. Sensors of infrared radiation, which for several years have been used by the military—notably in heat-seeking missiles—have today become a promising means of extending man's observation of many kinds of phenomena. From analyzing forest fires to spotting crevasses in snow-covered ice, from locating blood constrictions within the human body to mapping diseased vegetation in an orchard, infrared sensors can produce amazingly detailed and exact information, some of which can be gathered by no other means. Moreover, research in infrared sensing has led to the rapid recent development of a new field called remote sensing of the environment in which infrared sensing is only one of several techniques using the radiation of various parts of the spectrum to obtain the maximum amount of information about remote environments. This issue of the *Research News* will first survey briefly the background of research for military sponsors that has placed the University's infrared scientists in the forefront of this field, and then will discuss some of the civilian uses for infrared sensors that have been developed or are being explored by these same scientists.

From *Research News* 18:1–12, February, 1968. Reprinted with permission of the author and the Office of Research Administration of the University of Michigan.

Infrared sensing exploits the fact that everything above absolute zero (−459°F) emits radiation in the infrared range of the electromagnetic spectrum, in quantities and at wavelengths that depend on the nature of the surface (its emissivity) and on its temperature. This radiation can be recorded (1) by a radiometer, a device that compares the radiation received from an object or scene with a fixed standard and generates an electrical signal in accordance with the differences between these radiations, or (2) by an infrared scanner, which produces an image on film that in certain respects resembles a conventional photograph, as some of the accompanying illustrations show. (See Figs. 1–3.) What infrared images show are differences in the received infrared radiation from various objects; and since these differences are frequently considerable, an infrared image can exhibit a wide range of contrasts as does a photograph, though when an infrared image is compared with a panchromatic photograph of the same object, the differences between them are quite apparent.

The study of infrared radiation had long been an interest of University physicists, but intensive, large-scale research into infrared sensing did not begin on this campus until the inception of Project Michigan at the Willow Run Laboratories in 1953. The interest of the sponsoring Department of Defense agencies was in the potentialities of this radiation as a means of acquiring military information—particularly surveillance and reconnaissance information. As a consequence of sustained government support in this area, the infrared specialists

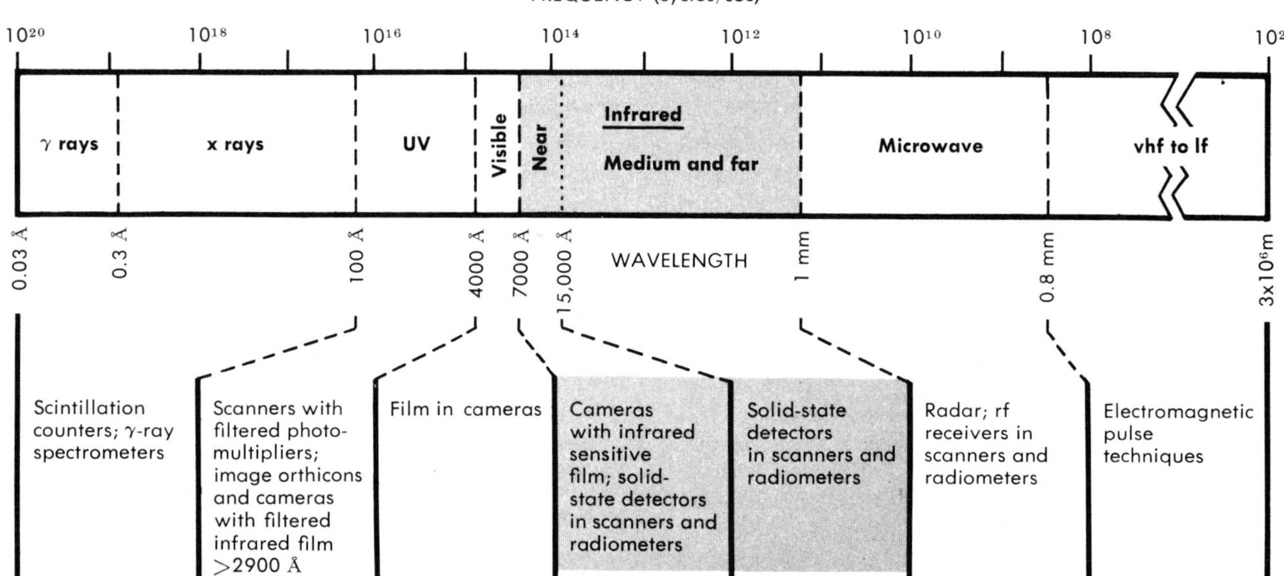

Figure 1 The electromagnetic spectrum and some sensors that use it. The infrared portion of the spectrum, though much wider than the band of visible radiation, is still a small part of the total. Recent research has focused on the integrated use of several sensors, each making use of a different portion of the spectrum.

at the Willow Run Laboratories are among the nation's major contributors to the field of infrared sensing, and two infrared laboratories have evolved to handle the total research requirements. The Infrared and Optical Sensor Laboratory, under the direction of Marvin R. Holter, and the Infrared Physics Laboratory, under the direction of George J. Zissis, are both active in the basic science of infrared as well as in developing new infrared sensing techniques. Each of these laboratories has been supported by various agencies of the Department of Defense and until recently has been concerned principally with military applications of infrared.

It is not surprising that the military would have been interested in the development of infrared sensors. Infrared images can be made at night and they can be made passively; unlike radar, infrared sensing does not require sending out energy to be reflected (and perhaps detected by an enemy). Furthermore, infrared images often show things that neither visual photographs nor radar techniques can show—for instance, the presence of a warm object, say a vehicle, that is completely concealed beneath camouflage. Infrared radiation can penetrate smoke or haze that is opaque to radiation at visual frequencies, although it cannot penetrate rain or heavy fog as well as radar. Infrared is particularly useful in providing data that can be used to locate an enemy's position, to estimate his numbers, and to detect his movements. Movement or activity, whether of bodies or of vehicles, generates extra heat; thus, an infrared scanner carried by an airplane can gather a surprisingly large amount of information on enemy maneuvers.

Another major, and well-publicized, military use of infrared techniques is missile guidance. The Sidewinder missile, an air-to-air weapon developed by the Naval Ordnance Laboratory, is equipped with an infrared detector and a tracking-and-homing device that fixes on a hot target such as the exhaust of a jet aircraft or a ballistic missile. The missile then homes in on the target and explodes when it reaches it.

The University's contribution to the development of military uses for infrared sensing has been indirect because the work has been primarily in basic infrared science. Before infrared sensing could become an effective tool, much research had to be done on basic problems. Detector cells, which sense the infrared radiation, had to be perfected; means for recording infrared imagery had to be improved; and criteria for interpreting the imagery had to be found and systematized. The physicists and engineers of the two infrared laboratories at Willow Run have worked in all these areas. They have experimented with various materials for possible use in infrared detectors; they have made important studies in techniques applicable to equipment for heat-seeking missiles; and they have made extensive contributions toward the establishment of guidelines for interpreting infrared imagery. Keys and signatures studies in which physicists attempt to learn how to distinguish between radiations from different objects and different backgrounds under various conditions, have been an active part of the infrared research at Willow Run for more than ten years.

While meeting various specific needs of sponsors, the University has over the years become an important repository of general knowledge in the field of infrared. Two national information centers are operated by the infrared laboratories. One is BAMIRAC, the Ballistic Missile Radiation Analysis Center, established under contract with the Department of Defense. This Center maintains a reference library, analyzes data from field measurements, does research on phenomena related to ballistic missiles, and works to establish predictive models for radiation from ballistic missiles. BAMIRAC

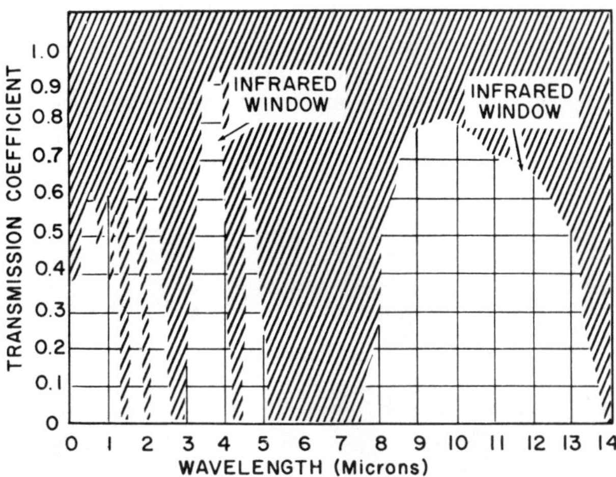

Figure 2 Pattern of infrared atmospheric "windows." Infrared radiation is transmitted through the atmosphere only in the spectral regions shown. Various solid-state detector materials sensitive to radiation at different wavelengths are used to take advantage of these windows.

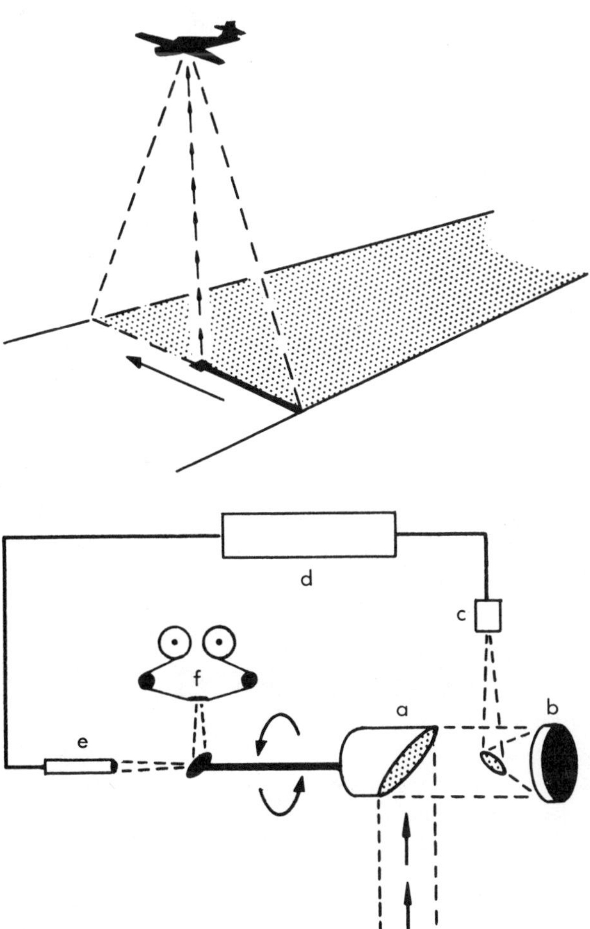

Figure 3 A typical infrared scanning system. This drawing shows the essential features of scanning systems. Radiation strikes the surface of a rotating mirror (a), and is reflected to the surface of a parabolic mirror (b), and then to a solid-state detector (c). The output of the detector is amplified (d) and modulates the output of a light source (e). The modulated light is recorded on film (f).

publishes bibliographies, reports, proceedings, and other technical documents; all its materials are made available to qualified requestors. The other information center is IRIA (Infrared Information and Analysis Center), which is supported by the Department of the Navy. It is a center for collecting, analyzing, and disseminating information on infrared research and technology; qualifications for its services are the same as for BAMIRAC's. IRIA publishes annotated bibliographies and state-of-the-art reports, publishes the *Proceedings of IRIS* (Infrared Information Symposia), and provides library and consultation services. IRIA collects all information that will advance infrared technology—classified and unclassified, published or unpublished.

The focus of infrared knowledge and expertise that has existed at the University of Michigan for many years has had the effect of stimulating programs in infrared science at other institutions. The University has conducted short summer courses in infrared for engineers annually for ten years. There are three such courses; altogether, over the years, they have been attended by some 1500 to 2000 men in the field—each of whom has taken new knowledge of infrared back to his own institution.

Two important books by staff members of the Willow Run Laboratories have also come out of the infrared studies. One, published by the Macmillan Company in 1962, *is Fundamentals of Infrared Technology* by Marvin R. Holter, Sol Nudelman, Gwynn H. Suits, William L. Wolfe, and George J. Zissis. This book was one of the first comprehensive and fundamental texts in infrared technology. The second, published by the Office of Naval Research in 1965, is *Handbook of Military Infrared Technology,* edited by William L. Wolfe.

The preceding discussion has suggested the variety of work done by the University's infrared scientists during the years when this science was being developed largely through the sponsorship of the federal defense agencies. In the course of this work, they have dealt with theoretical and developmental problems of many kinds. In the process of conducting feasibility studies, gathering samples of infrared imagery, and furnishing on-the-spot consulting services, they have traveled to the ends of the earth; and as special librarians and teachers, they have kept themselves acquainted with the larger problems and opportunities in their field. In all this work, they have accumulated a wealth of experience that has in recent years been turned—often again with the support of federal agencies—to the development of civilian uses for infrared.

A most interesting recent development in the exploration of civilian uses for infrared (and other) sensors has been a series of symposia on the remote sensing of the environment conducted on campus with the support of the Office of Naval Research, the Air Force Cambridge Research Laboratories, and the Army Research Office. The symposia, which began in 1962, have brought physicists and engineers who know the potentialities of the various sensors together with scientists who are working in areas where a sensing problem of some kind might be solved by one or a combination of the sensors. The spread of knowledge occasioned by these symposia has been impressive. Experts in radar, conventional photography, sonar, scintillation detectors, seismometers, and various other sensory devices, as well as infrared specialists, have put their knowledge to work in the consideration of ways in which these sensors can be used in space programs, oceanography, atmospheric research, meteorology, geologic and geographic research, studies in agriculture and forestry, and in many other fields. Although numbers alone cannot express the influence these symposia have had, it is interesting to note that attendance has risen from 72 at the first in 1962 to about 550 at the fourth in 1966. The proceedings from these symposia have increased in size from 110 to 870 pages and are distributed to approximately 2000 people.

Increasingly, the remote sensing symposia have led to related courses conducted by Willow Run scientists. This winter two intensive courses are being given on the remote sensing of the environment—one, a 12-week course for United States and foreign scientists in support of NASA's Earth Resources Program, and the other, a 2-week course supported by NSF to introduce college science teachers to the basic concepts of remote sensing.

Another way in which University scientists have contributed to the development of the field of remote sensing of environment has been through work on national committees. For example, James T. Wilson, Director of the Institute of Science and Technology, has since 1964 chaired the National Academy of Sciences' committee on remote sensing of environment. A subcommittee of this group responsible for working with the Department of Defense on matters of security classification of sensors is headed by Gwynn Suits of the Institute of Science and Technology. The University, working principally through this subcommittee, has made a sustained effort to have classified sensors and data reviewed when it appears that there is no longer a security need for the classification. University infrared scientists have continually emphasized the need to make the results of military sensor research as widely known as possible in the interests of developing and applying the new field of remote sensing of environment.

To illustrate the kinds of infrared applications that are being studied or developed in the field of remote sensing of environment, the rest of this article will briefly note several civilian uses for infrared that the two infrared laboratories have been instrumental in exploring.

In medicine, infrared technology is proving to have an increasing number of uses (Fig. 4). Recently, a way to detect the amount of carbon dioxide in human blood by infrared was devised. Oxygen is relatively easy to detect, but carbon dioxide is not because it escapes from solution rapidly. The infrared device found useful in detecting carbon dioxide is an infrared spectrophotometer, which can be positioned a little above the blood sample to sense the spectral radiation emitted by the carbon dioxide as it escapes from the blood. Another medical use for infrared, which was originated at the Albert Einstein Medical Center in Philadelphia but which has led to follow-up study at this University, is medical themography, a process of making infrared pictures of the human body. These thermographs reveal very slight differences in skin temperature, and one can thereby infer important information about any phenomenon that affects it, such as breast cancer, the constriction of blood vessels associated with varicose veins, and ailments in which blood flow to the brain is obstructed.

A homelier use for infrared, and one that has already been used for many years, is the detection of "hotboxes,"

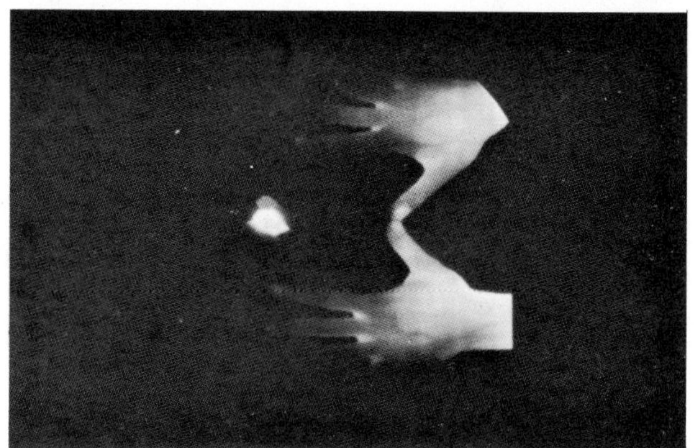

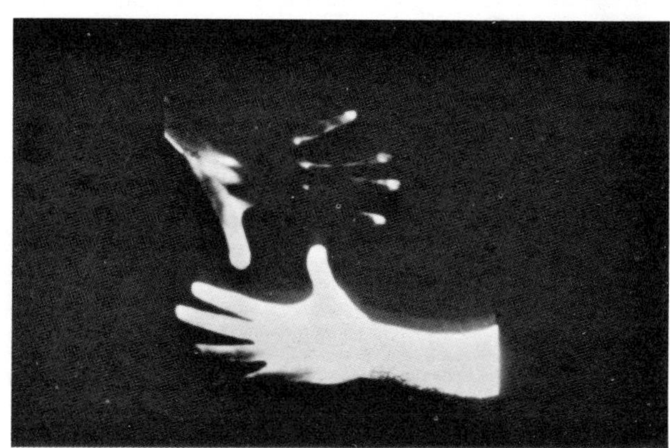

Figure 4 Infrared in medicine. These three infrared images, taken with a scanner being developed by Willow Run scientists for use in medical studies and diagnosis, show the kind of information that can be obtained by this technique. Note the cool dark areas over the sinus cavities in the man's face. The pair of hands touching at the thumbs show almost no radiation from the little fingers. The hot spot between the hands is a cigar lying in an ashtray. The third image is of the hands of two different persons with markedly different heat radiation.

or overheated journal boxes, on the wheels of railroad trains. The usual system is a "go or no-go" device, stationed along the tracks to survey the passing trains. The device passes any journal box that is below a certain safe temperature, but sets off an alarm when it detects a dangerously hot one.

It may seem surprising that much of the most fruitful research on the heat-sensing capability of infrared sensors has been done in the coldest regions of the world—the polar zones. If one recalls, however, that visual observation and conventional photography are impossible during the polar nights, the value of a passive heat-sensing device becomes clear. Although the normal temperatures in the polar regions are as much as 80°C lower than those in temperate climates, the temperature *differences* (what the sensor actually measures) that exist in the polar regions are often greater than those elsewhere. For example, an open lead in the Arctic ice pack is just below 0°C while adjacent snow surfaces may be −50°C or colder—a temperature difference much greater than any occuring in the tropics. The environmental scientists and infrared systems engineers at Willow Run Laboratories have collaborated with U.S. and Canadian government scientists during the past decade to demonstrate the utility of the infrared sensor in polar and sub-Arctic research. Infrared sensors allow scientists to identify and delimit various types of sea ice, even sea ice covered by snow (Fig. 5). Water, of course, is warmer than ice; water under thick ice transmits some of its warmth through the ice and through whatever snow there may be on top of it. The temperature differences here are slight, to be sure, but present-day infrared sensors can detect them, and from the infrared imagery obtained, one can determine the relative thickness of ice and the topographic characteristics that permit one to classify it. The present method of deciding, in the spring, whether ice on the St. Lawrence Seaway is thin enough for the icebreakers to start plowing through it is for a trained observer to fly over the seaway and estimate the ice thickness at various points by its appearance. Trained observers have developed this kind of estimation into an art, but infrared detection promises to make the evaluation of ice thickness more precise and more reliable. The routine use of infrared sensors might make it possible to extend by several weeks the shipping season of the St. Lawrence Seaway, the Upper Great Lakes, and other seaways that must be cleared of ice every year.

In polar studies, infrared sensors have demonstrated their value in detecting snow-bridge crevasses in glaciers

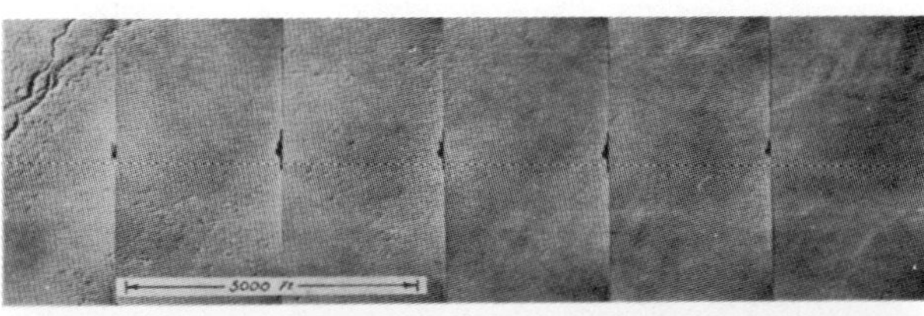

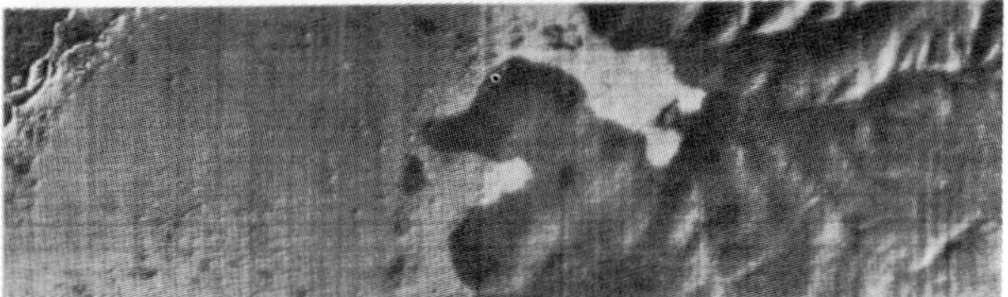

Figure 5 Infrared detection of a snow-covered icebound coastline. The coastline, completely hidden in the conventional aerial photographs (top), is clearly visible in the infrared imagery (bottom).

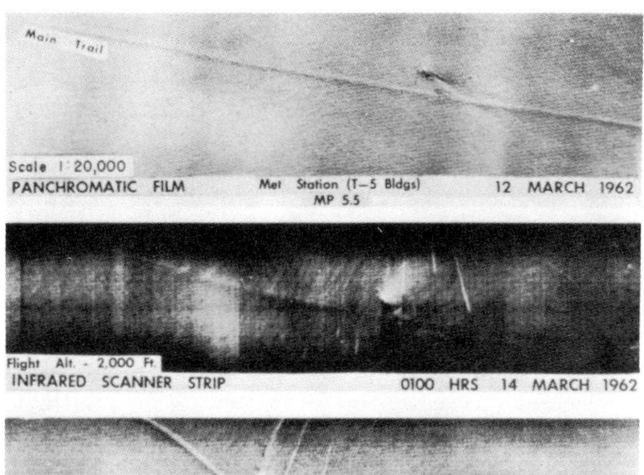

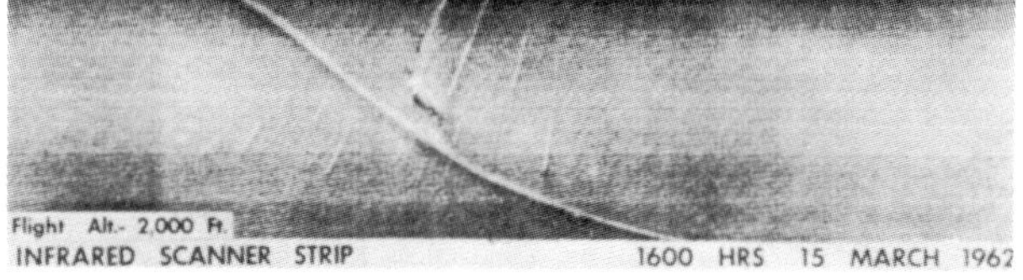

Figure 6 Crevasse detection by infrared. The top strip, a conventional aerial photograph, shows a trail passing by a meteorologic station near the edge of the Greenland Ice Cap. The lower two strips are infrared images of the same area. The light streaks show crevasses that are not visible either to the naked eye or to ordinary photography.

Figure 7 Infrared for forest fire detection and mapping. Infrared sensors are potent tools for detecting and mapping forest fires. In this imagery taken at night from 8000 feet over the Stinchfield Woods near the University of Michigan campus, the features contrasting most against the relatively cool terrain are the lake (at top) and the stream and road (at right). Also evident, however, are 15 scattered simulated fires—each a bucket of charcoal approximately a foot in diameter (lower left). The heat emission from these shows white against the light gray tone of the hardwood forest. Note the darker tones of rectangular clearings within the forest in the same vicinity. (Imagery by Northern Forest Fire Laboratory, U.S. Forest Service)

or large snowfields—crevasses that can suddenly swallow a man or a vehicle. Since the air in these crevasses is protected from environmental fluctuations, it is nearly always at a temperature different from that at the surface. The crevasse air, which is usually warmer than the ambient air, escapes through the snow plug covering the crevasse and is then revealed by infrared imagery taken from an airplane flying over the icecap margin, glacier, or snowfield (Fig 6). Such imagery can serve as lifesaving guides to those who must cross these surfaces.

In detecting and mapping forest fires (Figs. 7 to 9), infrared sensors are useful in two ways. First, they represent the best available method of spotting small fires before they have had time to get out of control. Campers often douse campfires with water and leave their campsites with consciences clear, unaware that they have left smoldering coals that can break into fire later. Lightning also frequently ignites tree trunks or logs that smolder

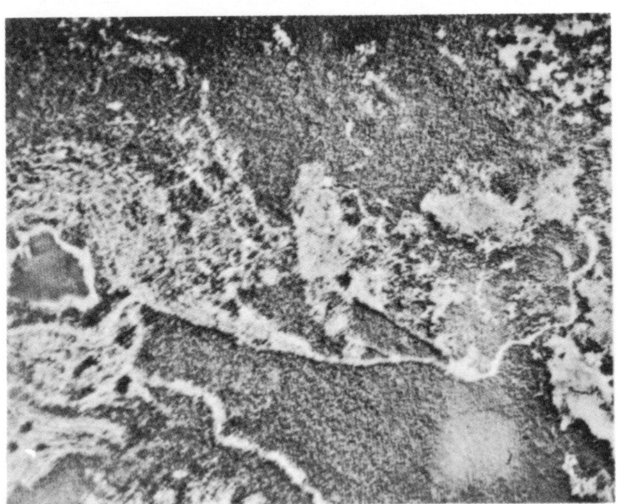

Figure 8 Infrared detection of fire. The infrared imagery on the bottom shows clusters of smoldering coals (lower right of picture) in a dense spruce stand. The black lines extending upward from the hot spots are optical effects. The hot spots are not at all visible in the conventional photograph on the top.

Figure 9 An actual forest fire mapped by infrared.

for many days before producing smoke or starting a visible fire. Constant infrared surveillance can detect those smoldering spots. Once a full-fledged forest fire has started, and is obscured by the smoke it produces, infrared sensors are again useful. They can pick up infrared radiation emitted from the fire itself, which penetrates all but the very densest smoke. In this way, the exact area of a big forest fire can be mapped, especially the critical downwind area where the fire is traveling and where the smoke obscures the rate of its progress, and firefighters can be deployed accordingly. Infrared sensors give immediate notice of "jump fires," or fires started away from the main fire. A jump fire can itself develop into a big fire capable of cutting off the firefighters' escape route. The U.S. Forest Service has established the feasibility of conducting forest fire surveillance in the intermountain west and routinely employs infrared sensors in fire mapping. In connection with these developments, scientists and engineers of the University's infrared laboratories have provided consulting services to the Northern Forest Fire Laboratories in Missoula, Montana, the organization that developed the systems and techniques.

In the science of geology, volcanologic and hydrothermal studies have already made considerable use of infrared sensing. For instance, infrared imagery taken from the air has revealed in dramatic detail the thermal characteristics of such volcanically active areas as Kilauea, Hawaii (Figs. 10 and 11), Taal Volcano in the Philippines, Surtsey Volcano in Iceland, and Erebus and Terror Volcanoes in Antarctica. From infrared imagery, one can map all of the surface manifestations of volcanism including areas that are abnormally warm from subterranean molten material. The United States Geological Survey, the Air Force Cambridge Research Laboratories, and the National Science Foundation have collaborated with or supported members of the infrared laboratories at Willow Run in investigating the

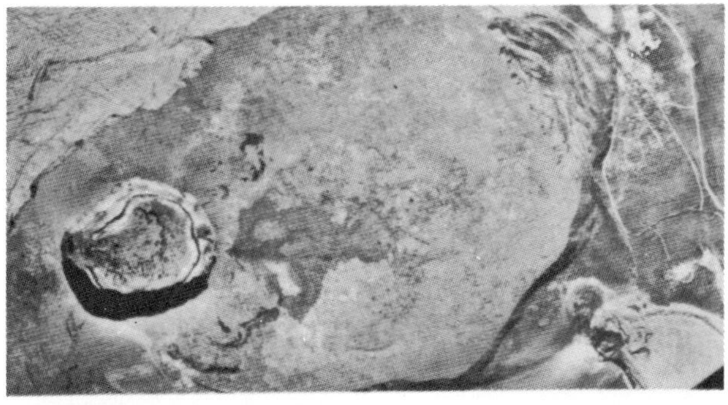

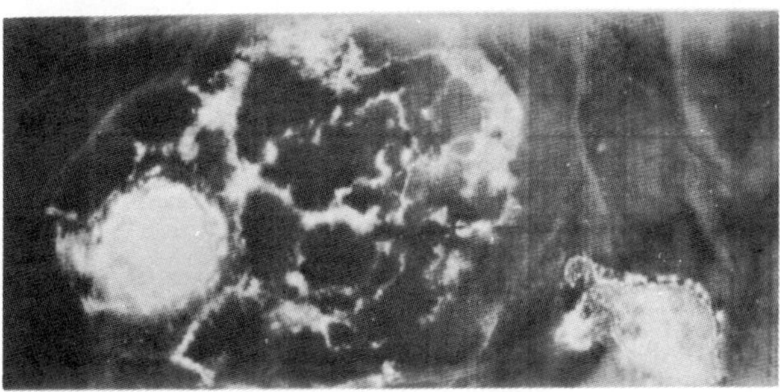

Figure 10 Summit of the volcano, Kilauea, in Hawaii. (Top) Conventional aerial photograph. (Bottom) Infrared image of the same area showing the distribution of large amounts of heat associated with the volcano. At the lower right, a lava lake formed during the 1959–1960 eruption and various hot fissures are delimited, and at the left the heat of the fire pit, Halemaumau, shows up dramatically.

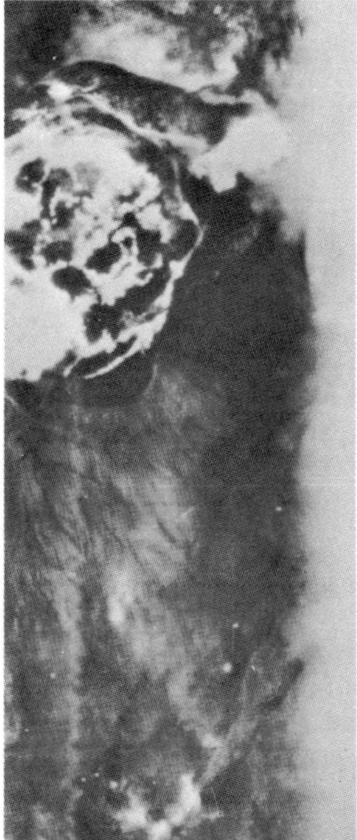

Figure 11 Simultaneous infrared images of Kilauea volcano area. The image on the right was filtered in order to show only the hottest areas. The spots visible indicate finger-size vents at temperatures in excess of 100°C.

capability of infrared scanners to produce useful information about areas of active volcanism.

Infrared surveys can also be used in physical oceanography (Figs. 12 and 13) to determine the temperature of water at various points, and thus to map currents, such as the Gulf Stream, that have well-defined thermal boundaries. A long-standing method of gathering such information has been to take up a bucket of water at each of various points and to take the temperature of each bucket. Infrared imagery provides a much faster and more accurate method and should contribute greatly to this science. (See also Fig. 14.)

The surveying capabilities of infrared sensors have led to a variety of other uses in "inventory." For example, in combatting air pollution, scientists need to pinpoint the source of pollution, and to be able to determine the extent of its influence. Visibility is often an inadequate criterion for evaluating the extent of air pollution. For example, smoke is quite visible, but sulfur dioxide, just as bad a pollutant, is not. Preliminary evidence has suggested that infrared sensors can be used to detect the presence of sulfur dioxide, and thus to secure the necessary information for monitoring smog-producing conditions, and eventually for making sure that smog does not exceed tolerable levels.

Conservationists have found infrared detectors useful in taking censuses of animal populations. Many animals range more freely by night than by day, and therefore are easier to observe at night, but flash photography or other means depending on visible light for observing would frighten them and would also be very inefficient. However, passive infrared detection that senses the radiation produced by the animals' body heat is proving invaluable in this application.

A potential use of infrared sensing in another kind of inventory recently arose when Willow Run scientists assisted the Director of Metropolitan Planning of Oakland County, Michigan, in planning an annual mapping of changes in the county's industrial and other resources. Maps made by ground survey methods, since they are time-consuming to produce, are typically years out of date and may well give little information about new additions to a community that will greatly affect its future development—a factory, for instance, that will alter traffic patterns, perhaps increase smog, produce residential and commercial changes, and alter other aspects of the county. An up-to-date map can alert urban planners more quickly to trouble spots requiring their attention and can suggest possibilties for the use of land resources that might otherwise be overlooked. Clearly, an up-to-date map is a prerequisite to urban planning, and infrared sensors and other remote sensing techniques, since they promise to simplify the problem of keeping maps current, should contribute substantially to land use studies and urban area analysis.

A very important potential inventory use of remote sensing techniques is in agriculture. A reliable means of obtaining a rapid large-scale inventory of crops would provide important information regarding their adequacy for preventing famine. Such an inventory would need to be able to determine the health of the crops, how

Figure 12 Infrared imagery of ocean features. Infrared imagery, taken near Santa Catalina Island off the coast of California, shows marked differences in water temperature.

Figure 13 Infrared in oceanography. Infrared imagery shows distinctly the demarcation of an oceanic front at a boundary of the Gulf Stream.

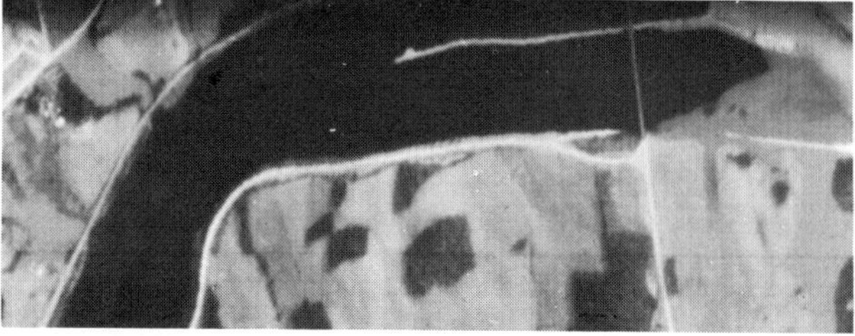

Figure 14 River monitoring. These infrared images of a river were made for the Tennessee Valley Authority, which wanted to determine the path and diffusion of cold water released into the river from a dam. Both images are of a portion of the river 20 miles downstream from the dam. The top image shows the river's normal appearance in the morning; the dark strip along the bank is cold water inside a retaining wall. The bottom image, made 23 hours after the dam was opened, shows that the cold water, which still retains a sharply defined front, has just reached this point.

much they can be expected to produce, and when they will be ready for harvest. Working from a satellite, remote sensors will, in a few years, if present work leads to success, be able to identify crops, assess their maturity, and detect disease among them. To make an inventory system of this type successful, present sensors will have to be improved, and much more knowledge about the imagery produced by different crops in different states of ripeness and health will have to be accumulated.

Illustrations of the potential uses of infrared and other remote sensing techniques could be extended almost indefinitely. Partly through the remote sensing symposia and other educational activities of the University in this field, the demand for these techniques in scientific and industrial applications has multiplied, and there is a continuing need today to improve remote sensing equipment. A powerful new aerial multispectral scanner just developed by Willow Run scientists is an important step in meeting this need (Fig. 15)[1]. This scanner, the most advanced now available for nonmilitary uses, is sensitive to 18 channels of radiation over the ultraviolet, visible, and infrared parts of the spectrum. The outputs from these channels produce continuous images somewhat resembling television pictures of the ground passed over. The electrical signals from the 18 outputs can be calibrated and fed into a computer that will "recognize" only a certain pattern of signals. Investigators can thus use the multispectral scanner to search for a specific ground feature or condition by determining in advance what pattern they want the computer to look for. Since the multispectral scanner is sensitive to a much wider range of radiation than the human eye, it has great potential particularly for large-scale investigations of natural resources. The challenge now is to learn how to use most effectively the vast amount of information that the scanner can produce. With satellite-borne scanners and computer processing of data, the technique offers very exciting prospects for biologists, geologists, geographers, geophysicists, ecologists, and agriculturalists.

A field test of the multispectral scanner was conducted in the fall of 1967 when a group of Willow Run scientists made an aerial survey around Point Barrow on Alaska's Arctic coast. The scanner gathered thousands of feet of exceptionally clear imagery which reveals information about the Arctic heat sink, ice packs and ice islands, and the contours of parts of the Arctic coast.

Another interesting development in scanning equipment at the Willow Run Laboratories is an inexpensive infrared scanner for remote sensing devised by Gwyn Suits (Fig. 16). Whereas previously available airborne scanners cost about $40,000 or more, the new scanner, designed for use on the ground, can be put together for about $1500. Such a scanner will allow smaller universities and other institutions that could not afford expensive airborne sensors to begin work in thermal sensing. To construct the inexpensive scanner, Dr. Suits simply used the parts of an amateur Newtonian telescope available in kit form from a number of suppliers and put an infrared detector cell where the eyepiece would normally be. The output of the scanner is recorded on an

[1] Fig. 15 appears on pp. 218–219.

Figure 16 Infrared image made with inexpensive scanner. This image, made in the dark with the inexpensive scanner newly designed by Dr. Gwynn Suits, clearly shows the heat distribution in a room. Persons across the room are clearly defined by their thermal radiation, as are the extinguished but still warm overhead light fixtures and a sheet of heated aluminum in the corner on the left. Above the heads of the persons, the heating pipes show a significant heat loss despite the thermal insulation covering the pipe.

FM tape recorder. In order to permit immediate and continuous viewing of the output as inexpensively as possible, Dr. Suits is now working to modify an ordinary television set so that, with a flip of a switch, it can be ready to present the images that the scanner sees.

While infrared and other remote sensing techniques will not solve such perennial problems as food shortage, dirty air, polluted water, eroded soil, and wasted natural resources, they may provide an improved means of assessing them. The role of the Willow Run infrared laboratories in these developments provides a good illustration of how a once-narrow field can be broadened enormously through the efforts of University scientists eager to extend the principles and explore the implications of their work.

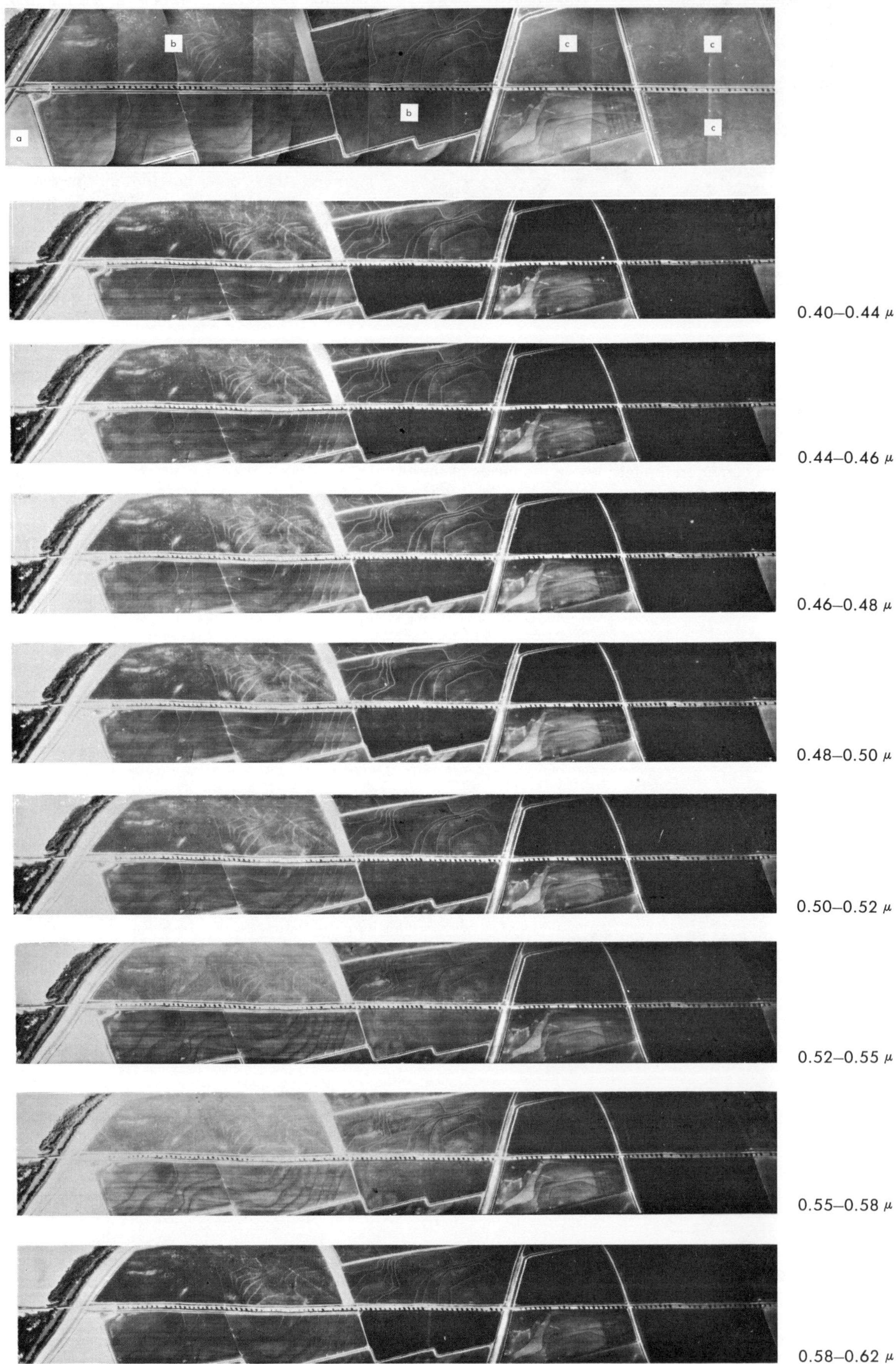
b
c
c
a
b
c
0.40–0.44 μ
0.44–0.46 μ
0.46–0.48 μ
0.48–0.50 μ
0.50–0.52 μ
0.52–0.55 μ
0.55–0.58 μ
0.58–0.62 μ

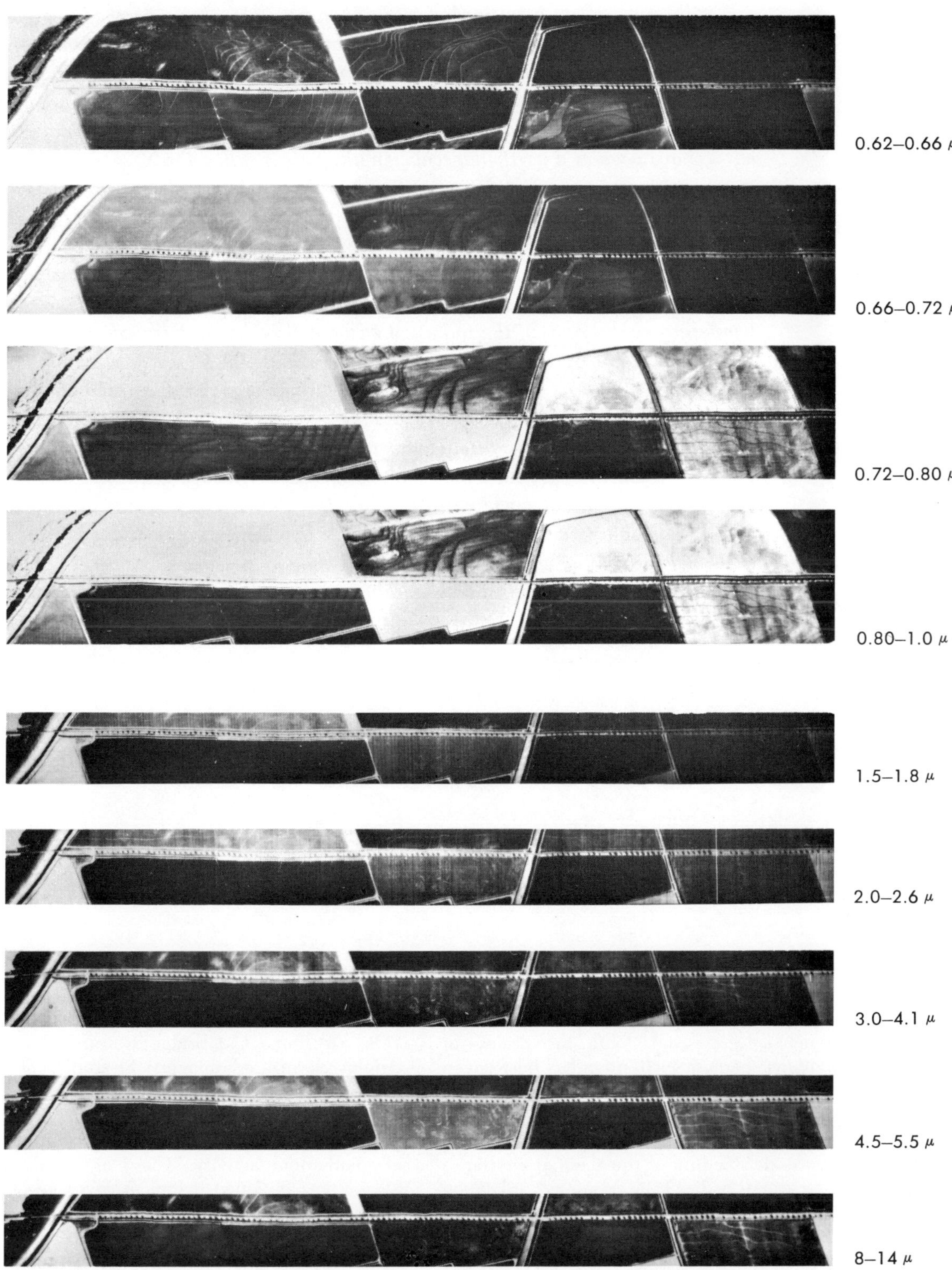

Figure 15 Multispectral sensing. These strips of imagery, made in California with Willow Run Laboratories' new multispectral scanner, show how different kinds of terrain will appear differently as viewed in the radiation of various wavelengths. At the top is an ordinary aerial photograph of the terrain shown in the strips. The triangular field on the far left (a) is bare earth. This and the several roads are light in most of the strips. In the center of the strips are two fields of safflower (b) and at the far right are three fields of green rice (c). The contrasts, which are stronger in all of the strips than they are in the aerial photograph and which change significantly from strip to strip, can be used as the basis for identifying specific material.

ANY OBJECT with a temperature above zero degrees Kelvin (–459°F) has molecular motion. Because of this motion—rotation, oscillation, vibration—heat radiation is emitted directly. In general, natural objects emit energy over a broad portion of the electromagnetic spectrum. But if the total radiation of any object is graphically plotted by wavelength, a curve develops which usually has a sharp peak at a particular wavelength. This is called the λ_{max} (wavelength at which maximum radiation occurs). From the work of such researchers as Planck, Wien, and others, it is possible to show both experimentally and empirically that the λ_{max} varies with the temperature of the object. The hotter the object, the shorter the wavelength at which λ_{max} radiation occurs. Conversely, as an object cools, its peak radiation shifts to longer wavelengths. At relatively low temperatures the peak of the radiation curve will be in the infrared. For example, the ambient temperature of the atmosphere is about 300°K; the λ_{max} at this temperature is at 9.7μ. This is well within the range of emitted infrared energy. If we stand near a bonfire or step out into the sun on a hot summer day, we have no difficulty sensing "heat" radiation. However, for cooler objects our body sensors prove too insensitive to record their emissions. What is needed is a sensor that will respond to the longer wavelengths. Further, if we know the peak radiation wavelength of an object, we can design the sensor to be most receptive to the narrow band of wavelengths at or near λ_{max}. This would make it possible to recognize or distinguish the object against considerable other background radiation or "noise."

27-Infrared Imaging for Water Resources Studies

JAMES I. TAYLOR
RONALD W. STINGELIN

SYSTEMS CAPABLE of producing operational infrared imagery have been available for approximately 15 years and are presently an accepted and integral tool in the military reconnaissance arsenal. With the easing of security restrictions, increasing attention has been given to the applicability of infrared remote sensing to the solution of civil problems. The high sensitivity to radiation emitted in the infrared spectrum and continuing progress in the understanding of the thermal emittance properties of terrestrial bodies have prompted utilization of these systems in wider and more diverse areas.

Since the infrared imaging system was commonly considered a "heat" sensor, it was natural that the early studies concentrated on pronounced surface thermal phenomena. The United States Geological Survey obtained infrared imagery over an active volcano in Hawaii in order to study the distribution of thermal energy in the immediate area (Fischer, 1964). (It was also noted that the various historically separated lava flows could be easily differentiated on the imagery.) More recently, U.S. Air Force Cambridge Research Laboratories has employed infrared imagery in studies of Surtsey, the new volcanic island off the coast of Iceland (Friedman and Williams, 1968).

Infrared imagery has also been used to delineate areas of geothermal activity. The United States Geological Survey has obtained complete coverage of the Yellowstone National Park and has been able to delineate the areas of subsurface geothermal activity, as well as the many surface "hot springs" (McLerran and Morgan, 1965). Geothermal steam represents a potential source of power in several areas of the world, but detection and mapping are prerequisites to development and utilization of these thermal energy sources.

It was soon noted, however, that features not visible on conventional aerial photographs, and not so obviously "thermal" in nature, could be interpreted from infrared imagery. Considerable attention was directed to the use of infrared imagery in geologic investigations. In particular, studies have shown that bedrock boundaries can

From *Journal of the Hydraulics Division, Proceedings of the American Society of Civil Engineers* 95 (HY1):175–189, 1969. Reprinted with permission of the authors and the American Society of Civil Engineers.

often be determined (Lattman, 1963); structural anomalies (faults, fractures, etc.) not detected on photography can be discerned; and regional geologic analysis can be aided by the planning and execution of infrared aerial surveys so as to produce infrared imagery strips amenable to mosaicking (Williams and Ory, 1967).

Application of thermal imaging techniques to ecologic problems is also found to be feasible and the thermal response of plant communities to changing air temperatures may be studied (Stingelin, 1968). Detection of plant diseases also appears as a promising technique in aiding plant pathologists fight the spread of disease.

The airborne infrared imaging system has, in recent years, been shown to possess an outstanding capability in the detection of small thermal differences in water bodies and is recognized today as a significant research tool in water resources work and studies. Much success has been achieved in detecting thermal pollution in fluvial, lacustrine, estuarine, and littoral environments (Stingelin and Fisher, 1967). Studies of currents and mixing conditions in estuarine areas have been greatly aided by this new imaging technique.

Nature of Infrared Imagery

It is necessary to understand the basic principles of operation of line scan equipments, their image-forming techniques, and the effect that such techniques have on the pictorial display of ground features and terrain details before the use of infrared systems in the collection of hydrologic data can be appreciated. These principles will be briefly reviewed here, as infrared technology has not received widespread attention in the civil engineering literature.

Figure 1 shows, in block diagram, the elements of the infrared (ir) imaging technique. In brief, there is primary radiating source, or target (e.g., a cultural object, terrain feature, effluent into a receiving body of water) which emits an electromagnetic signal. The intensity of the signal, released in the form of radiation, is a function of the temperature and emissivity of the target.

Adjacent to or surrounding the target, and likewise emitting radiation, is the target background. Detection of the target against this background depends upon the magnitude of the radiation difference between them and the ability of the system to discriminate such differences.

Both the target and the background radiation pass through and are attenuated by the atmosphere (the degree and distribution of attenuation depending upon geographic, meterologic, and operational conditions) and are delivered to the optical system, which focuses incoming radiation onto the surface of a detecting element sensitive to radiation in the infrared region of the electromagnetic spectrum. Detectors serve to translate radiation differences into weak electrical signals which are amplified, processed, and finally fed into display of some type.

A schematic drawing indicating the basic system operation is shown in Figure 2. Radiation from the ground is received at the system and reflected by a revolving scanner mirror onto a parabolic mirror, which focuses the rays on a sensitive element of an infrared

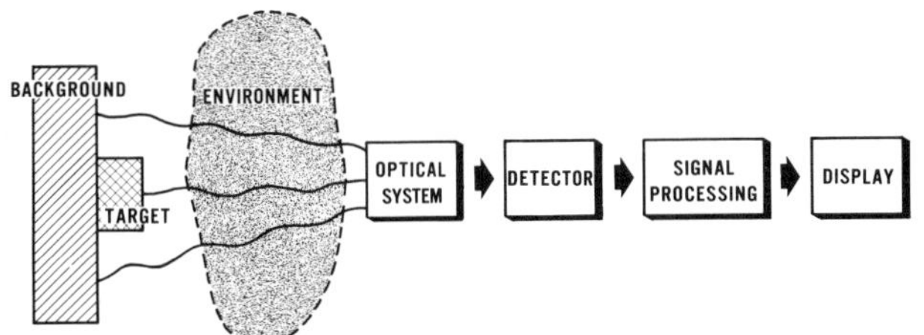

Figure 1 Generalized infrared (ir) imaging technique.

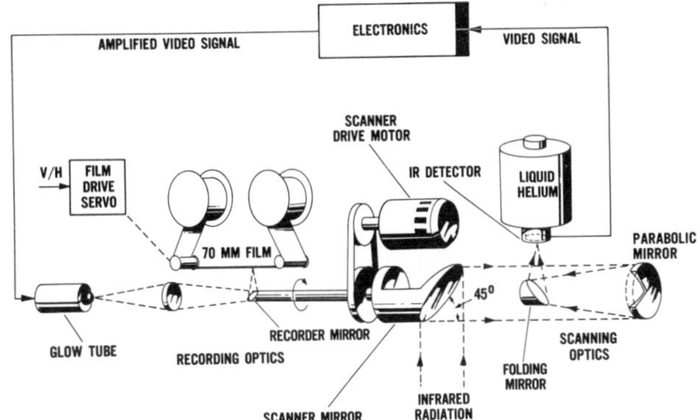

Figure 2 Infrared system.

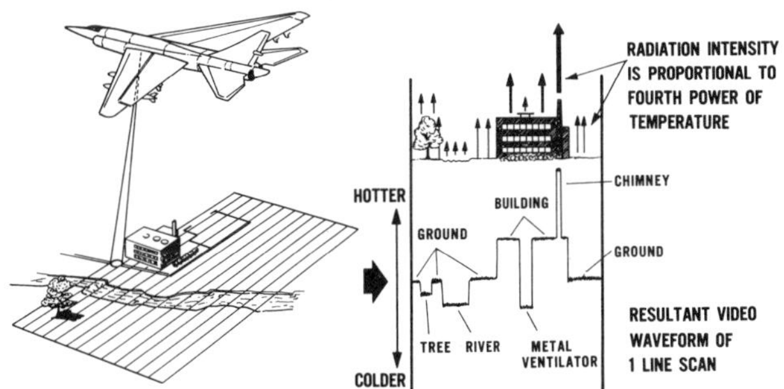

Figure 3 Infrared (ir) radiation illustration.

detector by way of a secondary folding mirror. The infrared detector, which is mounted in a dewar (a special double-walled vessel—like a thermos bottle—which is designed to prevent transmission of heat) filled with a coolant to increase sensitivity, converts the infrared radiation to an electrical signal in which variations are proportional to differences in the radiation received by the detector as the scanner mirror revolves. Liquid nitrogen and liquid helium are the two most commonly used coolants.

The electrical signals (called *video signals*) from the detertor are amplified and used to modulate a variable intensity light source, such as a glow tube or a cathode ray tube. In the case of the glow tube printer (the most widely used recorder), the output of the tube is focused to a small spot upon the emulsion side of a conventional photographic film by way of the recorder mirror, which may be fixed to the same shaft as the revolving scanner mirror.

The essential elements of detection and signal processing are summarized in Figure 3, which illustrates the scanning of a typical scene by the ir system. At any given instant, the system looks at a small elemental area, or scanning spot, on the ground and records a measure of the ir radiation from this elemental area as a spot or dot of light on the film.

The rotational axis of the scanner shift is mounted parallel to the longitudinal axis of the aircraft. As the scanner mirror revolves, the elemental area sweeps across the terrain and a strip of ground essentially perpendicular to the aircraft heading is observed by the system. The modulated spot of light from the glow tube is synchronously swept across the width of the film. As indicated in Figure 3, the forward motion of the aircraft results in scanning successive strips and recording them on the film, which is transported through the film magazine at a rate synchronized to the relative angular velocity of the aircraft.

The intensity of radiation from an object in the scene is proportional to the fourth power of the absolute temperature of the object and directly proportional to its emissivity. The resultant video wave form for a single scan is shown in Figure 3, with the trees, the river, and the metal ventilator (the emissivities of metal surfaces are characteristically low) generating signals "colder" than the ground, and the building and the stack producing signals "warmer" than the ground signal. "Colder" and "warmer" in this case do not correspond directly to the temperatures of the objects, but rather to the apparent radiation temperature, which includes the influence of the varying emissivities.

Note that infrared imaging systems are passive, utilizing energy radiated from the scene; thus, illumination of the target is not required. In fact, most infrared imagery is collected at night to eliminate the effects of any reflected energy in the infrared region.

There is a natural tendency to compare infrared imagery with aerial photography. It is important to remember, however, that even though both are pictorial in nature, the distribution of tonal contrast in each image represents entirely different information. In the photograph the gray scales represent variations in reflectivity of the elements of the image, whereas the gray scales in the infrared imagery represent the pattern of energy radiated from the scene due to its temperature and emissivity distribution.

Conventional photographic films are sensitive to electromagnetic energy in the visible region (roughly the wavelengths from 0.38 μ to 0.76 μ). Infrared photographic films with extended sensitivities (out to about 0.9 μ) are available but still depend on reflected energy. The generally accepted ir spectrum lies between the wavelengths of 0.76 μ and 1000 μ. However, most operational infrared imagery is collected in either the 3–5 μ or the 8–14 μ regions. These regions correspond to atmospheric windows—regions of minimum energy attenuation.

Geometry of Infrared Imagery

For a line scan system, the longitudinal scale on the imagery is determined by the rate at which the film is drawn through the film magazine, and by the velocity of the aircraft. The two velocities are related by the time span over which the ground feature is imaged. For example, suppose the aircraft flies parallel to a 5000-foot runway at a speed of 200 fps; then, the aircraft would be over the runway a total of 25 seconds, and the signal from the runway will be picked up on every scan within that 25 seconds. Now, if the film velocity is 0.2 inch per second, the image length of the runway will be 5.0 inches, as the runway will be imaged on each printing scan throughout a 25-second period. Within limits, then, it is possible to vary the film speed and generate imagery at some predetermined longitudinal scale (obviously, if the film is run too fast, gaps will occur between adjacent scan lines.)

In line scan imaging, the scanner mirror is rotated at a constant angular velocity. As a consequence the rate at which the scanning spot sweeps across the ground is not linear, but varies as a function of the tangent of the angle off nadir. In the printer, the spot of light which exposes the film is swept across the film at a uniform rate, determined by the angular rate of the scanner. This noncorrespondence of scanner spot and printer spot rates results in a lateral scale which is nonconstant, varying inversely with the distance from the nadir line.

Lateral scale of the line scan system may be formulated by referring to Figure 4. (The cupping of the film in Figure 4 suggests a glow tube printing technique commonly used in civilian applications.) It is apparent that at any given point which is at an angle θ off the nadir, the lateral scale is defined by $r/H \sec \theta$; r is the radius of the printer barrel over which the film is cupped, H = altitude over terrain, and $H \sec \theta$ = the slant range to the ground plane at angle θ off the nadir. Observe that $H \sec \theta$ varies with different values of θ, and r remains constant. Hence, the lateral scale varies with θ.

Because the lateral scale varies with the angle off nadir and the longitudinal scale is constant, it is not possible for these two scales to be equal except at one par-

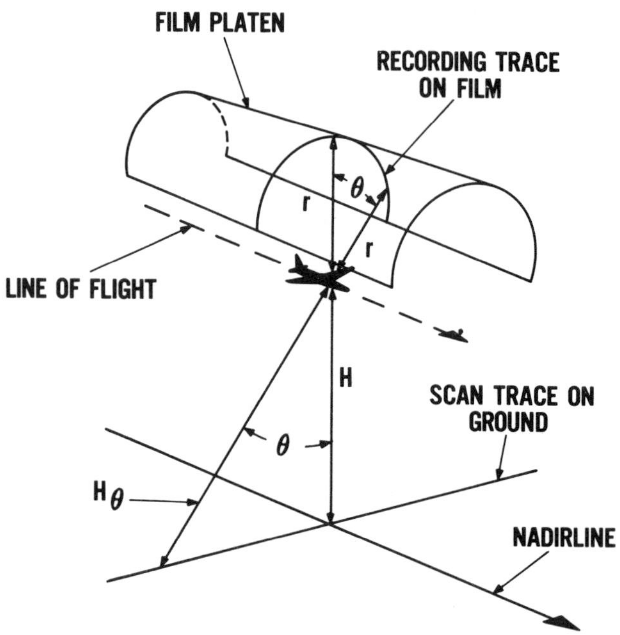

Figure 4 Single scan geometry.

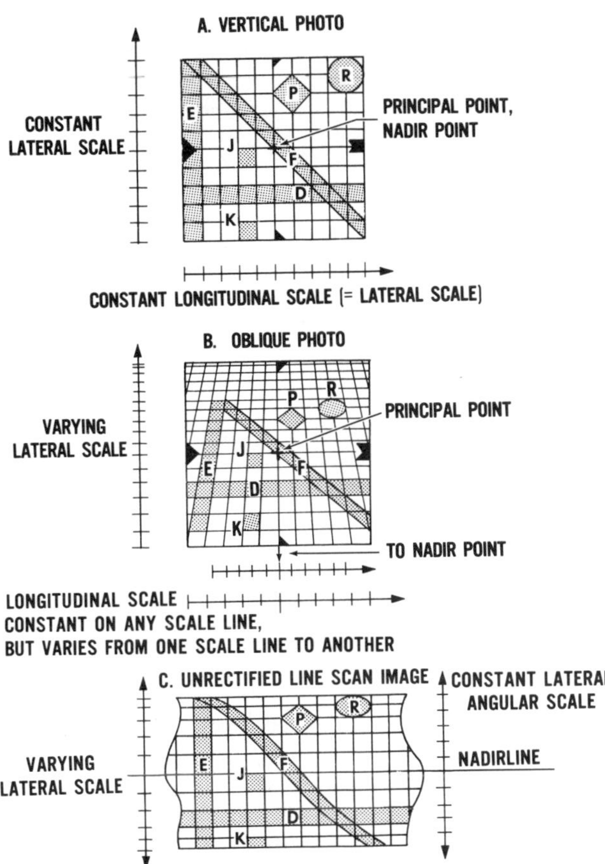

Figure 5 Comparison of images.

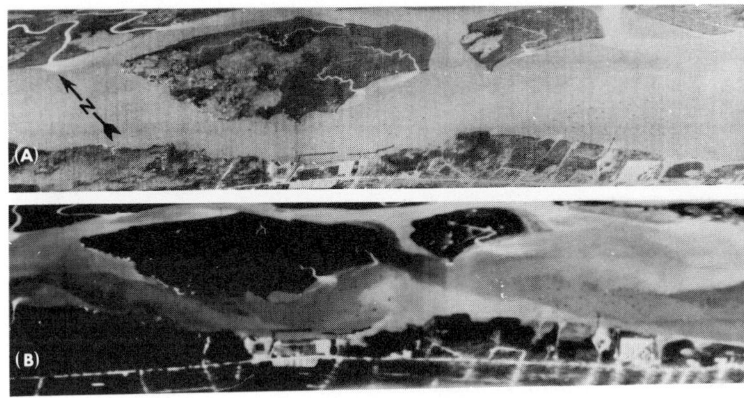

Figure 6 Delineation of tidal currents in the Merrimack Estuary, Massachusetts. (A) Small-area detector. (B) Large-area detector.

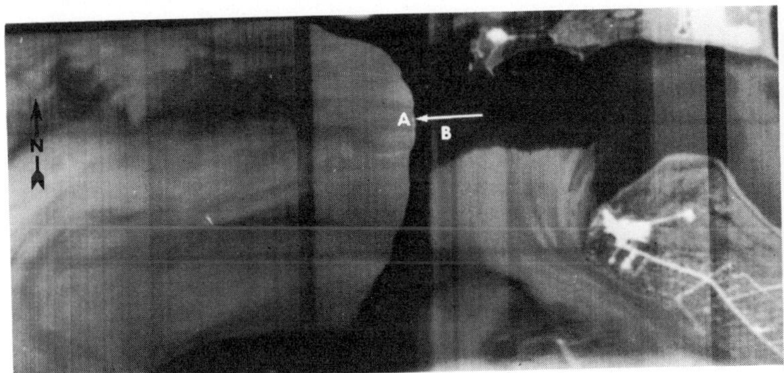

Figure 7 River water/sea water interface in the Merrimack River harbor mouth, Newburyport, Massachusetts (high tide). Arrow indicates interface. (A) Riverwater. (B) Seawater.

ticular value of θ. If it is desired to have the scales match at the nadir, this can be accomplished by adjusting the film speed so that the ratio of the film speed to the ground speed of the aircraft is equal to the ratio of the printing barrel radius to the altitude above terrain, or $v/V = r/H$.

Figure 5 compares the geometry of the line scan image with that of conventional vertical and oblique photographs.

Another significant difference between line scan imagery and conventional photography, from the geometric viewpoint, is the fact that line scan imagery is collected as a continuous process rather than in central projection pulses, or frames. Thus, variations in the flight trajectory or scanner system orientation affect the relative positions of image points on the film, whereas all image points on a single photograph have a fixed geometric relationship, as they are collected at the same instant with uniform camera exterior orientation and a known interior orientation.

In short, then, infrared imagery has some of the geometric properties (and advantages and disadvantages) of vertical, oblique, Sonne strip, and panoramic photography.

Recent Applications to Hydrologic Problems

The Water Resources Division of the United States Geological Survey (USGS) is actively engaged in research studies utilizing airborne thermal mapping systems. In cooperation with the State of Massachusetts, an intensive survey of the lower reaches of the Merrimack estuary, seaward from Haverhill, Massachusetts, was conducted. In this study, interest centers on the interactions of the river and sea waters, and the attendant effects on current patterns during a full tidal cycle. The position of the river water/sea water interface in the harbor at Newburyport, Massachusetts, during high tide is of particular concern in an investigation of the effects of the polluted river water on shellfish culture in the harbor.

Infrared imagery was collected during nighttime flights, scheduled to correspond with the four stages of tidal flow—flood, high, ebb, and low. The estuary was instrumented during these flights, and data pertaining to salinity, temperature, and current flow were collected by USGS and University of Massachusetts personnel. These data, correlated with the ir imagery, provided a synoptic description of currents in the estuary. Figure 6 shows the excellent delineation of tidal currents during flood tide. Figure 7 shows that the river water/sea water interface is held within the harbor during high tide and that most of the polluted river water is trapped along the southern shore of the harbor over existing shellfish

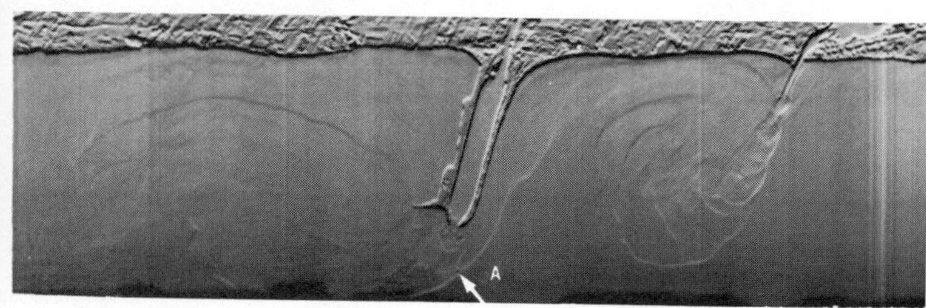

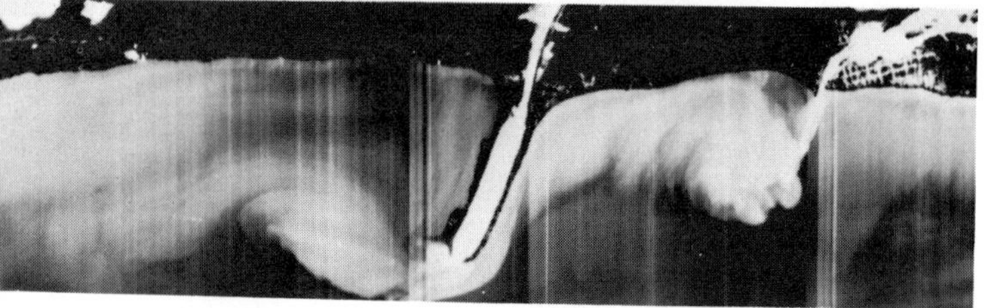

Figure 8 Delineation of current patterns of warm effluents introduced into Lake Ontario at the Welland Canal and at Port Dalhousie, Ontario. (Top) Differentiated image. (Bottom) Linear image. Arrow indicates typical representation of thermal boundary on differentiated imagery.

culture areas. This dramatic representation of thermal distribution in the harbor is a function of emissivity changes resulting from salinity variations between river and sea water, as well as temperature differentials.

The Geological Survey of Canada, Geophysics Division, pioneered infrared hydrologic studies for the Canadians in the Great Lakes. These studies are currently being continued by the Canada Centre for Inland Waters, Great Lakes Division, operating from Burlington, Ontario.

Of primary concern to the Canadians are the current patterns in Lakes Erie and Ontario and the influx and dispersal of thermal, industrial, and sewage pollution in these lakes. Figure 8 shows the current patterns of the warm effluents introduced into Lake Ontario at the Welland Canal and at Port Dalhousie, Ontario. Continued study of infrared imagery of the Great Lakes, collected under varying meteorologic and seasonal conditions, will ultimately lead to a better understanding of the dynamic processes active in these great bodies of water.

Potential Applications

There is an increasing interest in the effects of thermal effluents from large power plants on the ecology of the receiving waters. Dispersion and mixing patterns of these thermal effluents are strikingly displayed on infrared imagery. Since the emissivities of the effluents and the receiving bodies of water can be assumed to be essentially equal, the gray scale patterns can safely be interpreted as temperature dependent and, therefore, a pictorial display of relative temperatures is provided. Absolute temperature traces along the flight track can be obtained with an airborne radiometer and these data can then be used to "calibrate" the imagery gray scales to produce an absolute temperature map. In many cases, however, the qualitative delineation of dispersal and mixing patterns will suffice.

As almost all industrial or domestic waste discharges at temperatures somewhat higher (or possibly lower) than the receiving river or lake, these pollution sources can be readily located on infrared imagery. Temperature rises (or decreases) of less than 1°C in the receiving waters will produce a noticeable tonal change. Even submerged effluents have been located in several instances through the detection of the surfacing of the warmer effluents. Hence, it is possible to monitor large stretches of waterways during the day and/or night and provide an "effluent map" which will delineate all effluents with a combination of temperature differential and volume sufficient to alter the temperature of the receiving waters. The dispersal patterns—i.e., whether the effluent diffuses in the immediate area, follows one bank or the shoreline, travels upstream or downstream, etc.—will be recorded.

Figure 9 exhibits the capability of infrared scanning systems to detect and delineate dispersed patterns of industrial effluents. A comparison of the two images, separated in time by 38 min, illustrates the transient nature of most current and effluent dispersed patterns. Hence, the ability of aerial infrared imagery to provide synoptic data is especially advantageous in current-related studies.

Infrared imagery can be utilized effectively in some aspects of irrigation engineering. For instance, it would be possible to delineate the distribution of evaporation suppression films on large reservoirs. The temperature differential of 1°C to 2°C associated with areas covered by the film as opposed to areas not covered should be easily detected. Also, a probable difference in emissivity of the covered versus uncovered areas may enhance the differentiation.

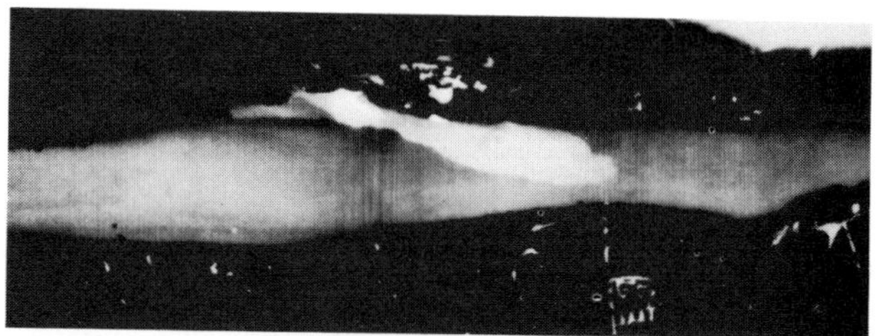

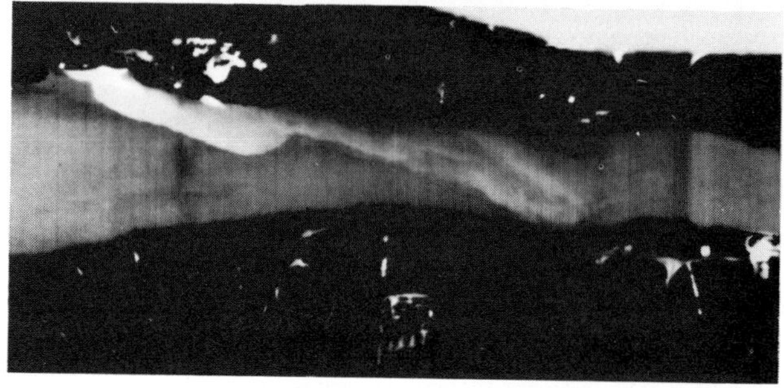

Figure 9 Industrial effluent being carried upstream and dispersed by flood-tide waters in the Merrimack River estuary. Lower image was recorded 38 minutes later than upper image.

The presence of moisture affects the thermal properties of soils—i.e., the ability to absorb energy during the day and re-emit it during the night—and, hence, even small amounts of moisture in relatively dry areas produce relatively strong signals. Consequently, it is to be expected that areas of seepage along irrigation canals could be discerned.

Infrared imagery will also be useful to the groundwater hydrologist as careful interpretation will reveal important information regarding the geologic structure of the area of interest—particularly in the detection and delineation of faults and fracture traces. Also, minor moisture differences are reflected in the imagery, and drainage features can usually be more easily distinguished through the interpretation of infrared imagery than through controversial photographic interpretation.

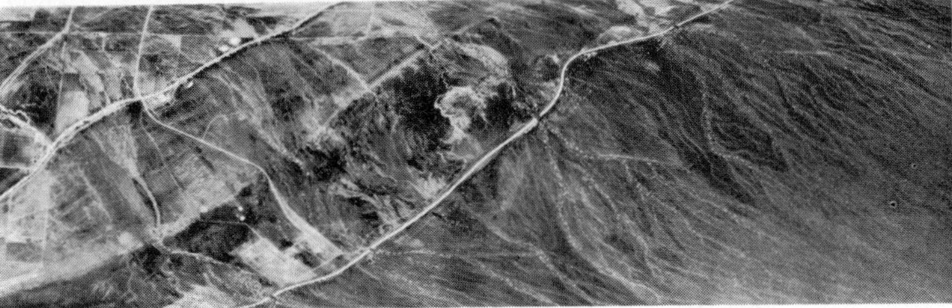

Figure 10 Two adjacent flight lines of infrared imagery taken with a germanium detector and illustrating the spatial resolution currently available for survey work. Locality: Imperial Valley, California.

Information Enhancement

In the course of conducting aerial infrared imaging surveys over diverse terrains, new criteria and techniques for obtaining optimum results have been developed at HRB-Singer, Inc.

Detector Parameters

In planning the collection of infrared imagery for any regional terrain study, consideration must be given to the type and size of detector to be used in the scanning system to produce the optimum desired results. Thermal sensitivity and spatial resolution are functions of the surface area of the detector exposed to incident radiation. The larger the surface area exposed, the greater the thermal sensitivity and the poorer the spatial resolution.

Spatial resolution is related to the size of the "scanning spot" shown in Figure 3—the larger the spot, the poorer the spatial resolution. The spot size, in turn, is a function of the angular field-of-view of the scanner system and the aircraft altitude above terrain. The angular field-of-view (determined by the collecting optics configuration and detector size) for most commercially available infrared scanning systems is between 1 and 8 milliradians—i.e., the spot (which need not be circular) will have a linear dimension of 1 to 8 feet per 1000 feet of aircraft altitude when the scanning spot is at nadir.

In imaging water bodies, which offer a background of relatively uniform emissivity and where spatial resolution is not critical, the potential exists for recording extremely small thermal differences at the water surface. Large-area detectors are generally utilized when imaging water bodies as the sacrifice of spatial resolution is more than compensated for by the increased thermal sensitivity of the system. Figure 6 illustrates this principle. The increased thermal sensitivity obtained in the lower image (large-area detector) has been gained at the expense of the fine spatial detail seen in the upper image (small-area detector). Such a sacrifice in detail, however, is preferable in that the increased thermal sensitivity greatly enhances the detection and interpretability of thermal phenomena in the river.

In the collection of imagery over geologic terrains, however, the need for fine spatial resolution usually outweighs the need for extreme sensitivity. Fine spatial resolution is especially valuable when the precise location and configuration of thermal phenomena detected on the imagery are to be transferred to an existing map base. The imagery in Figure 10 illustrates the excellent spatial resolution that can be obtained with airborne ir scanning systems currently in use for regional terrain surveys.

In thermal imaging, the atmospheric windows in the 3 to 5 μ and 8 to 14 μ region of the electromagnetic spectrum permit the use of two general types of detector crystals. In the middle infrared region (3 to 5 μ), indium antimonide crystals cooled with liquid nitrogen are most commonly employed; whereas in the far infrared (8 to 14), crystals of doped germanium, requiring a liquid helium coolant, are most common. Trimetallic detectors sensitive to the far infrared region which can be cooled adequately with liquid nitrogen are available, but the sensitivities of these detectors remain comparatively poor as compared to the germanium detectors.

The indium antimonide detectors are employed on most surveys because the logistics are simpler and the results comparable. Liquid nitrogen is less expensive and easier to store and handle than liquid helium (liquid nitrogen can be carried in ordinary vacuum flasks). In comparing the resulting imagery from indium and germanium detectors, little or no difference exists in the interpretability with respect to geologic and hydrologic determinations. Germanium detectors do produce superior imagery, however, when ground temperatures fall well below freezing. Figure 6 provides examples of imagery produced with indium antimonide detectors of different cross-sectional area; Figure 10 presents imagery produced with a germanium detector.

Daytime Versus Nighttime Data Collection

In addition to the selection of the proper detector parameters to be used on a mission, the time of overflight is a major concern. Nighttime data collection is found to be

Figure 11 Early evening flight over a corn nursery in Central Pennsylvania. Arrows indicate the cool (dark) tone associated with diseased plants. The contrast in the image is typical for an early evening flight.

superior to daytime imaging for almost all targets. During daytime data collection, some reflected solar radiation is recorded, even with the use of spectral bandpass filters. Cloud shadows are invariably recorded (if present) and mask considerable detail. Nighttime imaging, on the other hand, produces an end product free of extraneous solar interferences, and has proven to be the most suitable imagery for general thermal analysis. An exception is that late afternoon imagery usually displays the best enhancement of topography; and linear features, such as faults, often show best at this time.

Generally the best nighttime image contrast will be obtained on early evening flights. A gradual degradation of image contrast occurs throughout the night, with the poorest contrast obtained just before sunrise. Early evening flights on nights with temperatures above freezing appear to be optimum for ecologic studies where vegetational delineation is desired. Figure 11 represents an early evening flight over a corn nursery in central Pennsylvania. The dark (cool) areas, indicated by arrows, in the cornfield represent diseased corn plants. The long rows have been inoculated with a virus that is transmitted to healthy plants by aphids. The spread of the virus can be seen along the tiers in the cornfield. The sharp tonal contrasts displayed on this image are typical of early evening imagery.

System Modifications

Modifications to the basic infrared scanning unit permit several variations in terrain imaging techniques. In standard units an automatic gain and level control regulates the signal to a median value for each scan line. With this arrangement it is impossible to record fine thermal detail in areas of land/water contact due to the overriding signal of the warmer of the two bodies and the limited dynamic range of the recording film. Only when the two bodies are close to the same temperature can thermal detail in both bodies be imaged. In the modified system, the automatic signal control is replaced by a manual control, which permits the operator to clamp the signal level to the "average" of either the land or water. In this manner greater sensitivity is provided over water when conducting a hydrologic survey, for example, at the expense of the land signal. Figure 6 illustrates this principle. Note the excellent sensitivity obtained in the river in the lower image at the expense of the land signal, which, because of its much lower radiation temperature, was outside the dynamic recording range of the film.

In modifying the system to provide manual gain and level control, a signal differentation capability was also incorporated. With this capability low-frequency signals can be either partially or completely removed and the system made sensitive to high-frequency/high-resolution targets only. The background of the image is reduced to a uniform gray tone and the thermal boundaries are enhanced—giving a pseudo-relief to the image. Signal differentiation outlines all the significant thermal boundaries on an image and eliminates the loss of detail which occurs when the dynamic range of the film is exceeded in linear mode operation. However, the correspondence between radiation intensity and gray scale is lost. Figure 8 illustrates these points. The upper print is a differentiated image of the Welland Canal–Lake Ontario area. The lower print is a linear image of the same area. Note the distinct thermal boundary at *A* and the land and water detail available in the differentiated image.

The value of infrared imagery can be increased significantly through the development of a calibrated infrared line scan imaging system. A system of this sort would allow the assimilation of synoptic absolute temperature data—not available in the examples included in this paper. This system could also record the data on magnetic tape and provide a computerized data output. From these output data, thermal contour maps could be constructed, adding quantitative data to the pictorial infrared images. In principle, this calibrated system would provide a step waveform, with a known correspondence between signal intensity and radiation, during each dead-time between terrain scan lines. (See Figure 3 for an illustration of the waveform from a single terrain scan line.)

Battelle Northwest Laboratory (Eliason, Foote, and Taylor, 1968) have developed a false-color printing technique in which the signal intensity is expressed as color rather than as a shade of gray. They have also developed a printing technique termed as oblique three-dimensional display which provides visual impression of relief in the image—the third dimension being proportional to signal intensity. These techniques should aid in discrimination of small radiation intensity variations.

Conclusions

Infrared imaging systems are operational today. These systems have not been widely used to date due to security restrictions which hampered their distribution and application in civil works. However, with the recent declassification of several sophisticated infrared imaging systems and the imagery generated by these systems, previously obtained results and applications can be publicized. New applications will become apparent as knowledge of the basic characteristics and capabilities of this technology is disseminated.

Thermal expressions are associated with most hydrologic phenomena. Infrared imaging systems, with the ability to detect and record very small relative water temperature differences (less than 1°C), provide the hydraulic engineer with a type of data not previously available—a synoptic view of the thermal character of large expanses of water.

Various proven enhancement techniques can be employed to increase the value of the data derivable from thermal imagery. The simplest "enhancement" consists of knowledgeable selection of the optimal detector-filter configuration for the specific data collection program. Manual gain control permits the survey operator to maximize detail recorded on the imagery over the areas of interest at the expense of extraneous detail in other areas—e.g., maximize detail in water areas, and accept the consequent loss of detail in land areas. Signal differentiation provides spatial definition of thermal detail, but the relative temperature—gray scale relationship is sacrificed. Absolute temperatures over extensive areas can be obtained through the development of a calibrated infrared imaging system. (The present systems provide relative temperatures only.)

Infrared imaging systems have proven to be valuable aids in several recent water resources projects. The utilization of infrared systems is certain to grow with increased awareness of this relatively new technology now available to the water resources community.

REFERENCES

Eliason, J. R., Foote, H. P., and Taylor, G. R. Techniques for Qualitative and Quantitative Evaluation of Infrared Imagery. *BNWL-SA-1698,* Battelle Northwest Laboratories. Richland, Washington, Feb., 1968.

Fischer, W. A., et al. Infrared survey of Hawaiian volcanoes. *Science* 146:733, 1964.

Friedman, J. D., and Williams, R. S. Infrared Sensing of Active Geologic Processes. *Proceedings,* Fifth Michigan Symposium on Remote Sensing of Environment, Institute of Science and Technology Report No. 4164–18-x, April, 1968.

Lattman, L. H. Geological interpretation of airborne infrared imagery. *Photogrammetric Engineering* 29:83, 1963.

McLerran, J. H., and Morgan, J. O. Thermal Mapping of Yellowstone National Park. *Proceedings,* Third Michigan Symposium on Remote Sensing of Environment, Institute of Science and Technology Report No. 4864-9-x, Feb., 1965. Pp. 517–530.

Stingelin, R. W. An Application of Infrared Remote Sensing to Bear Meadows Bog, Pennsylvania. *Proceedings,* Fifth Michigan Symposium on Remote Sensing of Environment, Institute of Science and Technology Report No. 4164-18-x, April, 1968.

Stingelin, R. W., and Fisher, W. Advancements in Airborne Infrared Imaging Techniques in Hydrological Studies. *Proceedings,* Third Annual American Water Resources Conference, Nov., 1967. Pp. 466–471.

Williams, R. S., and Ory, T. R. Infrared imagery mosaics for geological investigations. *Photogrammetric Engineering* 33:1377, 1967.

ONE OF THE *early researchers in the field of infrared remote sensing began his lectures with the statement: "Think thermal." Any object or scene in the natural environment which is a strong source of infrared radiation is a good target for infrared sensing. If a phenomenon does not have a strong ir signature, a different kind of sensor should be used to study it remotely. Some strong sources of infrared are: volcanoes, thermal springs, forest fires, warm and cold ocean currents, steel mills, and thermoelectric plants. An infrared sensor should be used to study heat or a heat-related problem. These sensors probably have been most successfully applied to the problem of locating and combating forest fires. One of the most difficult tasks in fighting a forest conflagration is to determine the extent, size, shape, and movement direction of a large fire—information that can be a matter of life and death to the firefighters at the site. Yet these spatial and movement characteristics of the fire are almost impossible to determine on the ground. Even from the air, they can be obscured by smoke, haze, or darkness. An infrared image generated from an aircraft over the fire can determine the size and shape of the fire through haze or smoke or in darkness. Sequential images generated over time indicate the speed and direction of movement. Another application of infrared sensors is in early fire detection. A high-altitude airborne scanner can image hundreds of thousands of square miles of forest in remote areas each night. The airborne scanners have discovered slow-burning ground fires not visible to observers from a nearby road or in a fire tower.*

28-The Bispectral Forest Fire Detection System

S. N. HIRSCH
R. F. KRUCKEBERG
F. H. MADDEN

EIGHT YEARS of research and development by Project Fire Scan, USDA Forest Service, Northern Forest Fire Laboratory, Missoula, Montana, finally paid off during the summer of 1970 in the first successful operational test of an airborne infrared (ir) forest fire detection system.

To be effective, an ir forest fire detection system must be capable of differentiating 600°C, 1-square-foot targets from thermal background anomalies of up to 50°C. It must operate at a cost of approximately 6¢ per square mile to be economically competitive with existing visual, airborne systems. Infrared line scanning equipment must be operated at 15,000 feet above terrain, scan 120°, at speeds in excess of 200 knots to meet the cost limitation. At this altitude, a 2-mrd system has a ground resolution of 30 by 30 feet directly under the aircraft and 60 by 120 feet at 60° from the nadir. This resolution provides enough image detail for an interpreter to accurately locate targets.

Fire targets of the size we must detect (1 to 10 square feet) will occupy between 1.4×10^{-4} and 1.1×10^{-2} of the instantaneous field of view, depending upon the angle of observation and fire size. The effective target size may be further reduced by obscuration from timber overstory. Many of the targets detected in early measurement programs were 95 percent obscured (Wilson and Noste, 1966). The work we did in measuring the effect of timber characteristics on detection probability from 1962 to 1967 is well documented in previous symposia proceedings (Hirsch, 1963, 1965, 1968; Wilson, 1968).

We attempted an operational fire detection test during 1967 with an infrared line scanner sensing the 3 to 6 μm region. False alarms from hot springs, roads, and other miscellaneous thermal anomalies limited the effectiveness of that system. During 1968 and 1969, we developed a bispectral system using the difference between 3 to 4 μm and 8.5 to 11 μm signals to differentiate small hot targets from background thermal anomalies. This bispectral technique, along with scan-to-scan correlation to eliminate false alarms from aircraft electrical noise, and onboard rapid film processing made possible the operational test during the summer of 1970.

From *Proceedings of the Seventh International Symposium on Remote Sensing of Environment*, Vol. III. Ann Arbor: University of Michigan, Institute of Science and Technology, Willow Run Laboratories, 1971. Pp. 2253–2259. Reprinted with permission of the authors and the publisher.

Radiometric Considerations

The radiant energy from a forest fire (600°C) peaks in the 3 to 6 μm region. If a 600°C target fills less than 2×10^{-2} of the field of view of the scanner, the target's signal-to-background ratio in the 3 to 6 μm region is insufficient to distinguish fire targets from background thermal anomalies. In the 8 to 14 μm region, the target's signal-to-background ratio is less than it is in the 3 to 6 μm region. Because the background signal is coherent between these two spectral regions, we wondered if we could algebraically sum the signals to improve the signal-to-noise ratio.

Figure 1 shows the 3 to 4 μm (A) and 8.5 to 11 μm (B) signals from both terrain background and fire targets, as seen by a 2-mrd system from 15,000 feet, plotted in two-dimensional vector space. The lower curve contains all background temperatures from 0° to 60°C.[1] The area above the curve (shaded area) contains target signatures from fires ranging in temperatures from 600° to 800°C, areas ranging from 1 to 10 square feet, angles of view from 0° to 60°, and obscuration ranging from 0 to 95 percent. Amplitude discrimination in the A channel (horizontal line) would not provide effective separation of targets from background, nor would amplitude discrimination in the B channel (vertical line).

A sloping, straight line separating the lower curve from the target points does provide discrimination for most targets. The equation of that line is:

$$A = KB + C$$

where $A = {}_{3}\int^{4} W_\lambda T_A T_{F_A} R_{D_A}\, d\lambda$
$B = {}_{8.5}\int^{11} W_\lambda T_A T_{F_B} R_{D_B}\, d\lambda$
K = slope of discrimination line
C = displacement from origin
W_λ = Planck function
T_A = transmission of atmosphere
T_{F_A} = transmission of A channel filter
T_{F_B} = transmission of B channel filter
R_{D_A} = responsivity of A channel detector
R_{D_B} = responsivity of B channel detector

or, rewriting and redefining K:

$$K'A - B - C = 0$$

K is the slope of the discrimination line. It is determined by the range of background temperatures we expect to

[1] Variations in spectral emissivity will spread the background points, forming a band rather than the discrete curve shown in Figure 1.

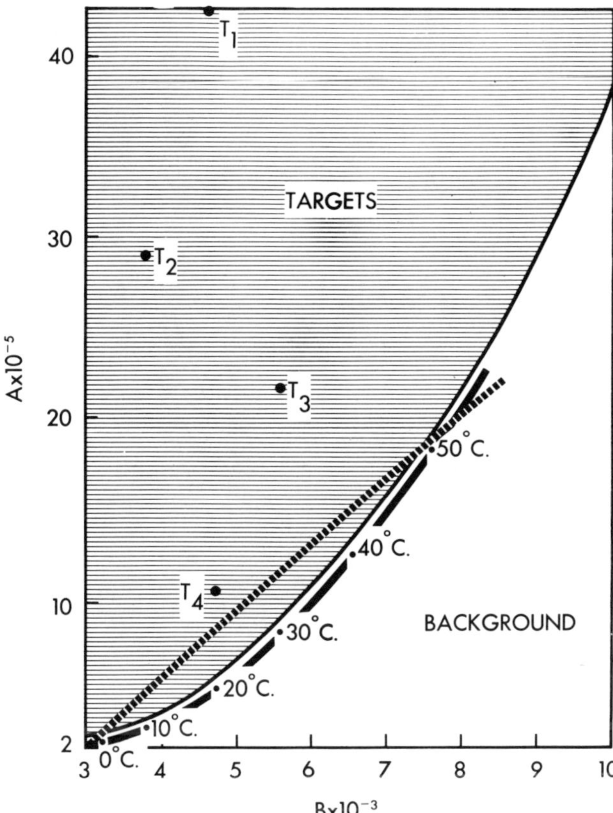

Figure 1 The signals for both targets and background, as seen by the 2-mrd ir detection system from 15,000 feet, are plotted in two-dimensional vector space, where A is the 3 to 4 μm and B is the 8.5 to 11 μm vector. Annotation of the target and background temperatures, areas, and angles of view for targets is as follows: T_1 = 1 ft.^{2}, 600°C, 20° background, 0° angle. T_2 = 1 ft.^{2}, 600°C, 10° background, 30° angle. T_3 = 1 ft.^{2}, 800°C, 30° background, 60° angle. T_4 = 1 ft.^{2}, 600°C, 20° background, 60° angle.

encounter. If the background temperature range can be predicted for a given flight, an optimum K can be selected. An effective decision rule and one that is easy to implement in an analog system is: A target exists if the signal is greater than $KA - B$. A nonlinear function that more closely matched the lower curve would be an even more efficient rule, but it would be more difficult to implement. We employed the $KA - B$ function and found it quite effective.

To minimize the effects of changes in atmospheric moisture, we selected the 3 to 4 μm and 8.5 to 11 μm bands rather than the total 3 to 6 μm and 8 to 14 μm windows. We empirically selected spectral regions near 3 μm and 8 μm, where the ratio of the power is relatively insensitive to changes in atmospheric moisture.

The Bispectral System

The way we implemented the bispectral system is shown in Figure 2. The 8.5 to 11 μm channel produces high quality ir imagery from which the position of targets can be accurately determined with respect to terrain features.

The $KA - B$ signal is pulse-height/pulse-width discriminated. A pulse above a preset threshold and within

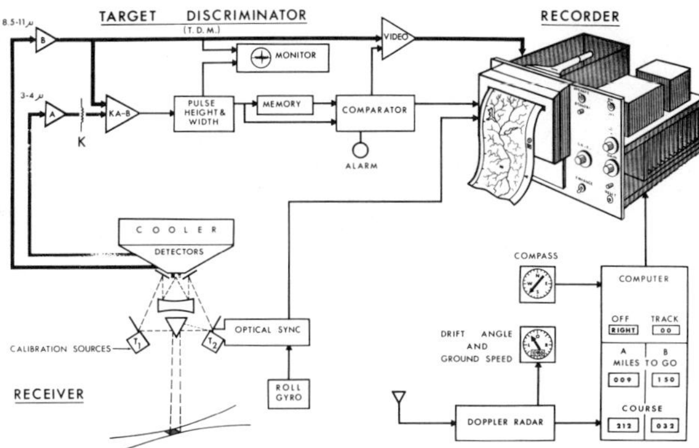

Figure 2 Block diagram of infrared fire surveillance system.

the pulse-width[2] limits produces a logic pulse that is stored in a digital memory for one scan line. To eliminate false alarms caused by electrical noise, we made use of the overscan available to do scan-to-scan comparison. If a logic pulse is produced at the same point in two successive scan lines, an output pulse is generated. The output pulse produces a mark on the edge of the film and reinserts a pulse in the video, which cues the operator (Fig. 3).

The curvature of the function shown in Figure 4 permits selection of two temperatures where the $KA - B$ signal amplitude is equal. To calibrate the system we select a pair of temperatures that will produce equal signals in the difference channel for the desired K value. We set the internal calibration sources at these temperatures,[3] and then adjust the gain of the 8.5 to 11 μm channel until the signal from the two sources is equal. This simple procedure assures that the ratio of total gain (amplifier, detector, and optics) in both channels is optimum for differentiating between targets and any preselected background range.

The receiver is a modified Texas Instruments, RS-7 line scanner.[4] The two detectors (Ge:Hg and InSb) are mounted side by side in the focal plane along a line perpendicular to the rotational axis of the scanning mirror. The signals from the two detectors are put in register by time delaying the InSb signal.

The Ge:Hg preamplifier is direct coupled and contains an automatic bias loop to maintain the detector at optimum bias. The 3-db bandwidth is 0 to 300 kHz. The InSb preamplifier[5] is ac coupled to 250 kHz. Extreme care was taken in the design of the 3 to 4 μm preamplifier to provide the best signal-to-noise ratio for the very low signal amplitudes available from the filtered InSb detector.

A synchronous clamp after one ac coupling in the printing video circuit maintains the dc reference required to eliminate overshoot, undershoot, and ringing that

[2] Pulse-width limit is set to equal the dwell time for one resolution element.

[3] Atmospheric transmission is not included in the integration for A and B in Figure 4 since the path between the detector and calibration sources is short.

[4] The use of trade and corporation names in this publication is for the information of the reader. Such use does not constitute an official endorsement by the U.S. Department of Agriculture of any product to the exclusion of others which may be suitable.

[5] The InSb preamplifier was designed and built by Texas Instruments, Incorporated.

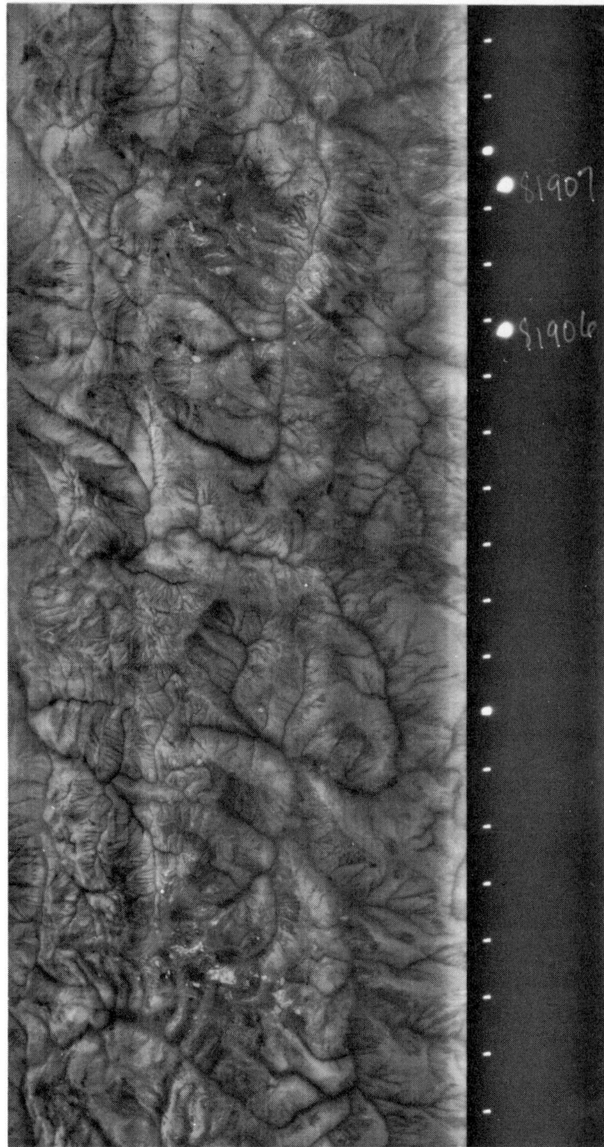

Figure 3 Imagery from 1970 patrol flight, Aug. 19, 1970 at flight altitude 23,000 feet m.s.l. Mileage marks, spaced 1 nautical mile with large mark every 10 miles, are recorded from Doppler navigation system. Target 81906 is believed to be a campfire located at the junction of a trail and target 81907 is a campfire at Mud Lake.

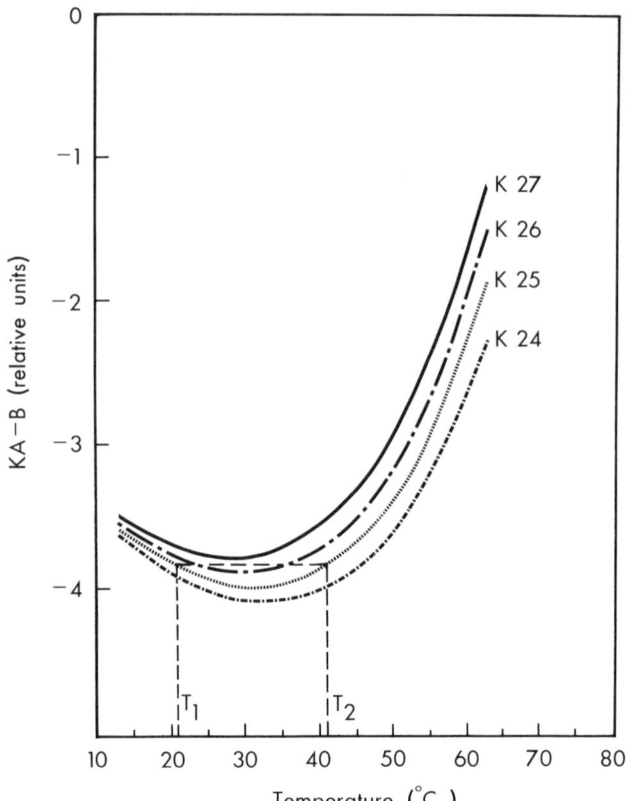

Figure 4 $KA - B$ signal plotted against temperatures for several values of K. To calibrate the system, T_1 and T_2 are selected to produce equal signals in the difference channel for the desired K value.

Table 1 Temperature Measurements for Specific Terrain Features

Terrain Feature	Temperature, °C
Roadway	−10
Snow (marked E, Fig. 5)	−13
Fire Hole River (marked D, Fig. 5)	+13
Geysers (marked B and C, Fig. 5)	>100
Hot Lake, west end (marked F, Fig. 5)	20
Hot Lake, center (marked F, Fig. 5)	50
Hot Lake, east end near the geyser (marked F, Fig. 5)	90

would otherwise obscure terrain detail when high-amplitude signals from large fires are encountered. The recorder contains a high-resolution CRT, a relay lens, and a two-stage, wet chemical film processor (Ansco KD-14).

The ir equipment was installed in a pressurized turboprop aircraft (Beechcraft B90 King Air), with a range of approximately 1200 miles, average groundspeed of 200 knots, and a service ceiling of 28,000 feet. In addition to the ir equipment, a Doppler radar and navigation computer were installed to permit accurately flying predetermined tracks.

Equipment Performance Test

We calculated the performance of both the single and bispectral systems and predicted that the bispectral system could produce a 12:1 improvement in detecting 1-square-foot, 600°C targets against backgrounds ranging from 0° to 50°C. To check our calculations, we performed a flight test using thermal anomalies in Yellowstone National Park to provide a high-contrast background.

Background temperatures encountered during the test in April, 1970, are shown in Table 1. The targets were buckets of glowing charcoal placed in a straight line and having surface areas of 1 and 2 square feet; the two 2-square-foot targets were on either end, the two 1-square-foot targets were in the center.

We overflew the area at 15,000 feet above terrain. In addition to the normal in-flight film recording and processing, we recorded the 3 to 4 μm, 8.5 to 11 μm, and $KA - B$ signals on magnetic tape.

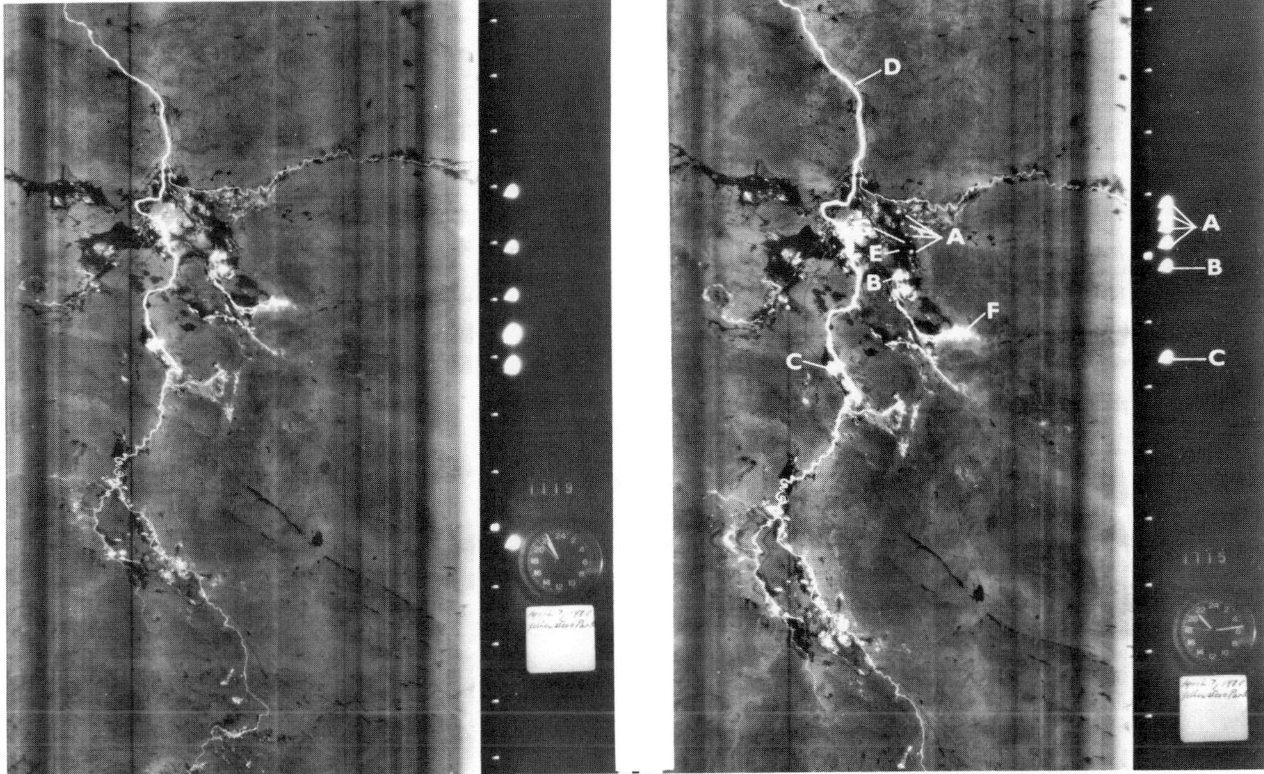

Figure 5 Imagery from the Yellowstone Park Test. (Left) A channel, TDM input. (Right) *KA* – *B* channel, TDM input. Annotation of the right-hand imagery is as follows: *A*, TDM trips on test targets (600 °C). *B*, Kaleidoscope Geyser. *C*, Excelsior Geyser. *D*, Fire Hole River (10 °C). *E*, Snowfield (–13.6 °C). *F*, Hot Lake (52 °C). *G*, asphalt road (–9 °C).

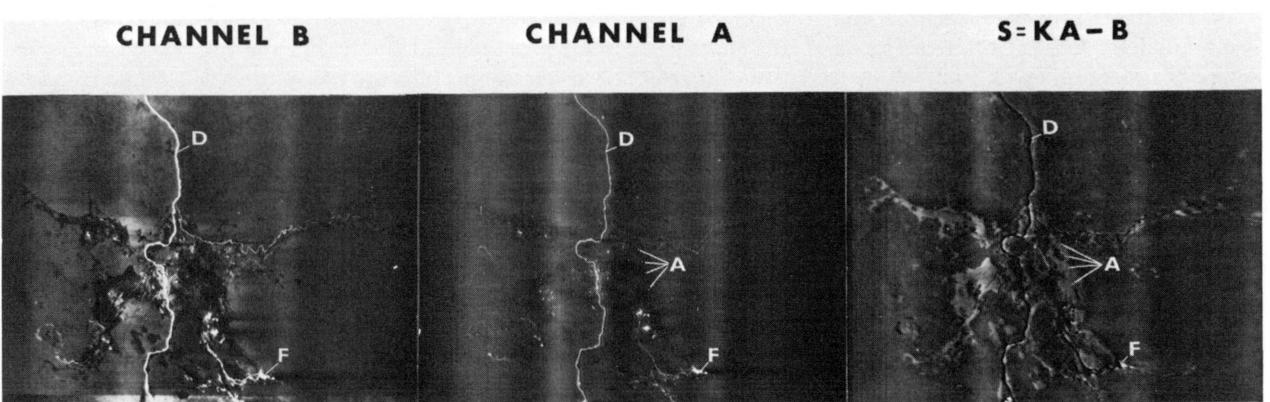

Figure 6 Composite imagery of Channel B, Channel A, and *KA* – *B*, recorded from magnetic tape, of the Yellowstone Park test. Annotation of this imagery is same as that in Figure 5.

All imagery (Fig. 5) was recorded using the 8.5 to 11 μm B channel. In the left-hand image of Figure 5, the 3 to 4 μm (A) signal was used as input to the target discrimination module (TDM). The 2-square-foot targets were marked by the TDM, but the 1-square-foot targets were missed. The system false-alarmed on geothermal activity in the area.

The right-hand image of Figure 5 is a similar pass with the $KA - B$ signal as input to the TDM. K was adjusted for a background range of 0° to 50°C. Both the 2-square-foot and the 1-square-foot targets were marked. The only nonfire alarms were on geysers whose temperatures were known to be above 90°C. Notice that the actual targets do not show up in either image. This further demonstrates the statement made in the section on Radiometric Considerations about the limitations of the 8 to 11 μm region for hot target detection.

After the flight, the images in Figure 6 were made directly from the magnetic tape with no TDM marks inserted. All four targets are visible in the Channel A print; however, the intensity of the signals from the Fire Hole River (D) and other geothermal activity at B, C, and F are greater than the signals from the targets at A. If pulse-height discrimination were used on the Channel

A signals, and if the threshold were set low enough to pick up the targets at A, false alarms would be produced by the geothermal activity.

None of the targets is visible in the direct print of Channel B. In the $KA - B$ image, all four targets are visible and they are whiter than anything else on the image, with the exception of the geysers at B and C. Notice that the river and parts of Hot Lake are black—a complete reversal from the A channel or B channel imagery. This clearly demonstrates the advantage of the bispectral system for enhancing hot targets.

In addition to making imagery from the magnetic tapes, we recorded individual scan lines as A-scan traces from an oscilloscope. Measurement of target-to-background ratio from the traces showed an improvement of 10:1 in the $KA - B$ channel as compared to A channel.

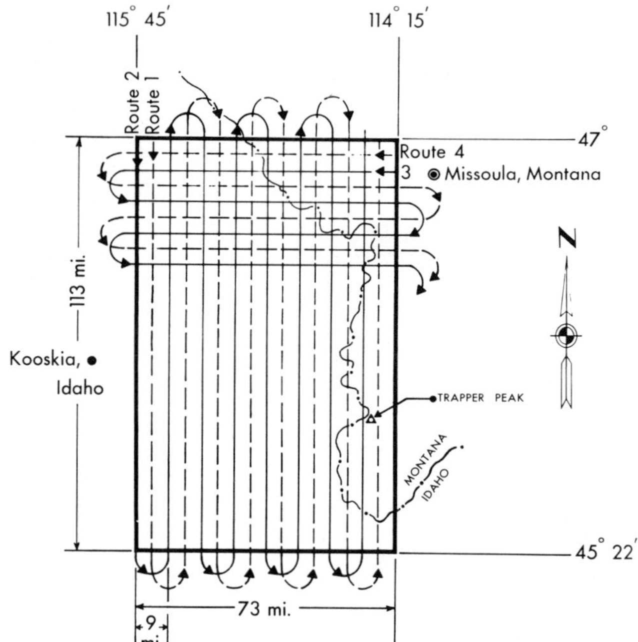

Figure 7 Layout of the 1970 patrol routes.

Evaluation of the System during the 1970 Fire Season

IR Fire Detection

After completing the Yellowstone Park test, we were convinced that the bispectral ir system had met all of our expectations in its ability to detect small hot targets. We then began the much more difficult task of determining its suitability for detecting latent forest fires. The strategy for employing this new tool in an overall fire detection system must consider fuel flammability, lightning storm occurrence, cloud cover, and the as yet unknown relationship between the radiant output from a fire and the smoke output used by the visual detection system.

To compare the performance of the ir system with visual fire detection, we selected a forested area containing 8200 square miles (5.2 million acres) in the Bitterroot, Clearwater, Lolo, and Nezperce National Forests in western Montana and northeastern Idaho. Elevations in the area range from less than 1300 feet near Kooskia, Idaho, on the Middlefork of the Clearwater River to 10,211 feet at Trapper Peak near Darby, Montana. Average elevation is 5400 feet. The area was selected because of high lightning fire occurrence (average 250 fires per year) and relatively little habitation.

Detection in the area is normally accomplished by 59 lookout towers, supplemented by seven light aircraft flying low-altitude visual patrol. The test we conducted in 1970 did not attempt to replace the existing detection system—ir flights were superimposed on the visual system.

Four ir patrol routes were established in an attempt to maximize detection probability. Two north-south routes, each containing eight strips, were laid out (Fig. 7). By alternating successive patrols, we assured that areas of low detection probability near the edge of the patrol route on one strip were given high detection probability by being near the center on the next patrol. These north-south flights were flown on the night patrols.

Two additional routes (Routes 3 and 4) on east-west headings were established for daytime patrols to minimize the effects of specular reflectance.

The flight crew consisted of a pilot and an equipment operator-interpreter. The operator continuously compared imagery to maps on which the desired flight tracks were marked. He immediately noted any deviations from the desired flight track and informed the pilot who updated the navigation computer and returned the aircraft to track.

The location of fires detected was recorded by the interpreter to the nearest 40-acre block (¼ mile) and the information radioed to the forests concerned during the turn-around at the end of the run. Radio frequency interference prohibited transmitting fire locations during runs.

Results of the IR Fire Detection Tests

From July 7 to September 3, we flew 41 patrol missions of approximately 5½ hours' duration each. Patrols were flown nightly at 23,000 feet m.s.l. (or 18,000 feet above terrain), with supplemental daylight patrols added when high fire occurrence was probable.

During this period, there were 418 forest fires in the test area that required suppression action; 12 of these fires were detected and extinguished during times when the aircraft was out for periodic inspection or involved in forest fire mapping missions. Two hundred and three of the fires were detected, manned, and extinguished before the ir aircraft had an opportunity to look at them, leaving 203 fires known to be burning when the ir aircraft passed over them. We detected 103 (approximately 50 percent) of the fires we flew over, or 25 percent of the total fires burning in the area.

Sixty-two of the 100 fires missed were only flown over once before they were seen visually, manned, and extinguished. A description of fires by detection category, size class, and elapsed time from visual to ir or ir to visual discovery is shown in Figure 8.

During the 41 patrol flights, the TDM marked 804 hot targets. While in flight, the interpreter identified these hot targets as: 169 wildfires and 635 miscellaneous

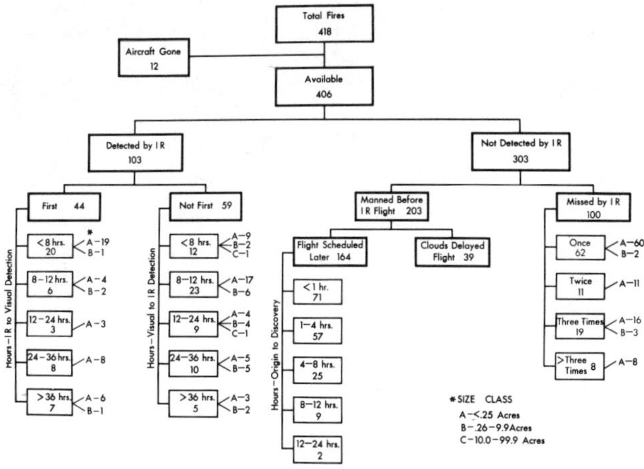

Figure 8 Summary of 1970 patrol test results.

targets (campfires, burners, etc.). Although these numbers do not agree exactly with those in Figure 8, they do indicate the interpreter's ability to sort out most wildfire targets from other TDM marks.

In the few cases where the location provided by the ir interpreter was different from that recorded on the fire suppression report, the ir location was the more accurate of the two.

From July 7 to September 25, over 265 flight hours were logged and nearly 500,000 square miles of terrain imaged with no equipment failures.

No firm conclusions about the relative effectiveness of ir versus visual detection can be made from the 1970 data. There were no large fires in the test area and no serious suppression problems were encountered. None of the fires missed by the ir system became larger than 1 acre. Ir detection could not be credited with substantial savings on the fires that were detected early.

We attempted to find criteria to determine which fires might have caused serious suppression problems if they had not been detected and manned when they were—we found none. We interviewed fire staff men, dispatchers, and fire control officers—all qualified and experienced men from the forests. They all felt any attempt at a subjective analysis would be meaningless. Conditions of fuel, weather, and topography were just too variable.

Our limited experience has shown that the ir equipment can find fires before they become visible, but many fires become visible quite rapidly—before a planned flight. Table 2 shows the number of fires occurring during each day of the fire season. From these data, it becomes readily apparent that the effectiveness of the fire detection system is determined by its performance on a very limited number of days. If the operational strategy is in error and flights are not made at the critical times, a large percentage of the total season's fires can be missed. We need to continue ir system tests under operational conditions in combination with strategically located lookouts and well-planned visual air patrols.

We did not fly the airplane over 50 percent of the fires that were burning in the test area. We made a nightly flight covering the *entire* patrol area. If we had reason to believe that two or more undetected fires remained, we scheduled another flight within the next

Table 2 Fires Originating by Day[a] in the Patrol Area During the 1970 Fire Detection Flights

July	No. of Fires	August	No. of Fires	September	No. of Fires
11	2	1	17	1	5
12	12	2	6	2	3
13	8	20	1		
16	54	23	1		
17	7	24	16		
18	3	28	1		
19	2	31	227		
20	8				
21	15				
22	2				
25	1				
27	15				
28	11				
29	1				
Total	141		269		8

Total fires in project area from July 7 to September 3 = 418

[a] No fires reported on days not shown.

6 hours. Cloud cover over all or part of the area delayed some flights and even caused us to cancel a few. By the time we did fly, many fires had smoked up, were seen, manned, and extinguished.

Daily fire occurrence patterns for the past season show that flying the entire area was not the optimal strategy. Lightning storms usually followed paths that included only small portions of the area. By waiting until conditions were right for a *full* flight, we lost the opportunity for early detection.

Forest dispatchers can provide up-to-the-minute weather data and storm pattern information. Areas of prime concern can be delineated prior to flight time and routes planned to provide overlapping coverage and high detection probability. Flights can be made at reduced altitude, below cloud cover, when the intent is to cover a specific part rather than the entire test area.

IR Fire Mapping

The bispectral equipment was used to map 13 large forest fires in Montana and Idaho during all phases of fire suppression; five large fires burning in the State of Washington were mapped shortly after control. Prints of the imagery were made as soon as the airplane returned to base, and delivered at dawn to the firecamps by light aircraft or helicopters.

Interpretation of the imagery at the firecamp was simple and anyone with even a minimal background in aerial photography was able to do a good job with little additional training.

The ability of the equipment to mark the small, hot target is invaluable during the final cleanup work. Fires too small to print on the image can be identified and located by ground crews. This provided the fireboss with information never before available to aid in making decisions on manpower needs and placement (Fig. 9).

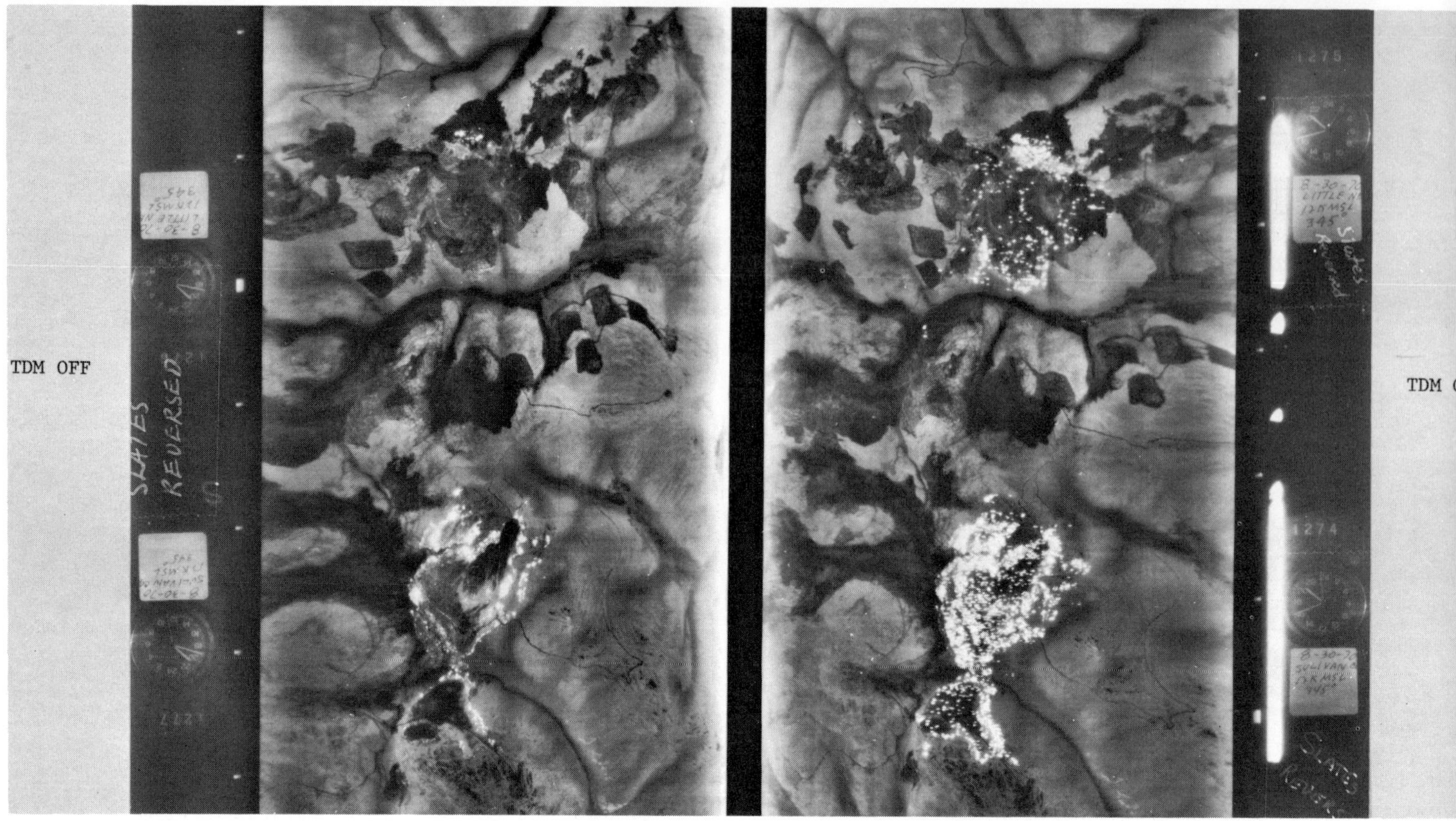

Figure 9 Imagery from fire mapping mission, Aug. 30, 1970.

Fire Spotter

During the summer of 1968 we became aware of the need for a device to pinpoint small fires that had previously put up enough smoke to be detected visually. The smoke output often dies down after initial detection, and suppression forces are unable to locate fires from the descriptions given by the lookout or air patrol who first saw them. We felt that a simple, inexpensive, airborne ir line scanner would be useful on light aircraft or helicopters to assist in locating these fires.

We built a feasibility model during the winter of 1968. It had a resolution of 10 mrds, scanned 120°, and used an uncooled PbSe detector filtered for the 3 to 4 μm region. This airborne "Fire Spotter" detected 1-square-foot, 600°C targets at altitudes up to 2000 feet above terrain against daytime thermal and solar backgrounds. After preliminary flight tests, we improved the design and produced ten prototypes for evaluation during 1970. A commercial version, produced by Barnes Engineering Company, became available during 1970.

These scanners employ a simple threshold detector to discriminate fire targets from background. The output from the threshold detector activates a horn and one of five lights to indicate which 24° segment of the total 120° scan contains the target. The spotters were mounted on forest air patrol planes for testing. An excellent example of the usefulness of these devices in conjunction with the high-altitude ir patrols occurred on July 16.

The Granite Creek Fire in the Nezperce National Forest started from a lightning storm on either July 10, 11, or 12. High-altitude ir patrols were flown on July 14 and 15 (patrol terminated at 0400 hours on 7/16). The fire was missed on the July 14 patrol but detected on the night of July 15. The ir image shows two fire targets—one of them a campfire at Twin Lakes and the other the latent fire (Fig. 10).

We flew over the fire at about 1000 hours on July 16 with the commercial fire spotter mounted on a Cessna 182. There was no smoke visible. The fire was missed on the first pass but picked up on each successive pass. After four or five passes, we were able to locate the fire. On one very low-altitude pass looking into the sun, we did see a small wisp of smoke. Smokejumpers from Missoula were unable to jump on the fire because they could not find it visually.

One of our project people walked into the fire, leaving Missoula late on the afternoon of July 16. At about 2200 hours that evening he camped about 2 miles from the fire. During a very heavy rain shower he saw a tree burst into flame at the fire location. He arrived at the fire on the morning of July 17 and reported that the fire was not producing enough smoke to be visible from the air. He measured approximately 5 square feet of burning material at a temperature between 600° to 700°C. He extinguished the fire late in the afternoon of July 17 when a burning log began to roll downslope toward heavy fuel where it could have started a major fire.

Results of the 1970 fire spotter testing were very encouraging (Kruckeberg, 1971). Fires that had been previously reported were located and manned even

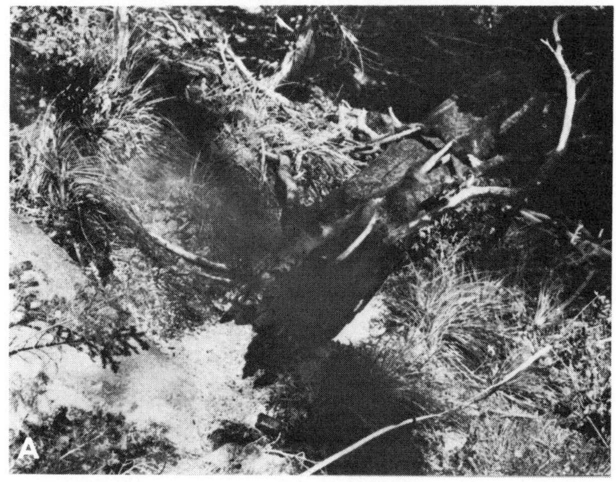

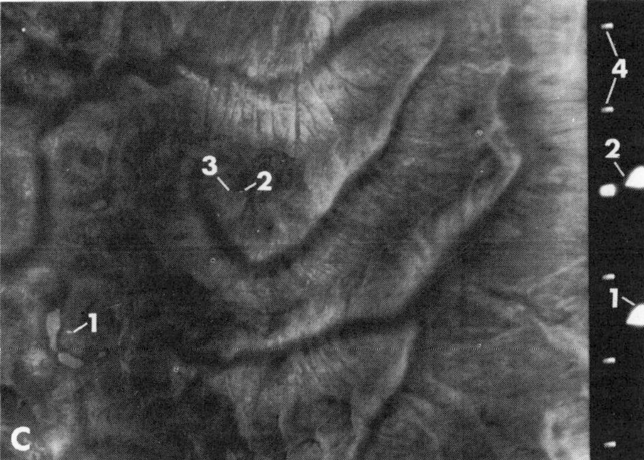

Figure 10 Granite Creek fire, Nezperce National Forest. This lightning-caused, latent or incipient fire (A) occurring in one tree in this area (B) was first spotted in this imagery (C) which was recorded at 0005 hours on July 16, 1970, in an aircraft flying at 21,000 feet. The fire was still burning when photo B was taken. The imagery is annotated as follows: (1) Target is a campfire at Twin Lakes. (2) TDM marks (double marks at the latent fire are due to aircraft pitching during the run). (3) Actual fire. (4) Navigation marks at 1-mile interval.

though not smoking. Others were found in areas where none were supposed to be, such as in controlled burns that had been extinguished and outside the control lines of large fires.

The equipment if properly used can perform a real service as part of the total ir detection system.

Future Plans

During 1971, an updated bispectral fire detection system will be operated by National Forest personnel, rather than our research group, in the same test area flown during 1970. Twenty-five of the 59 lookouts in the area will be left unmanned. The aircraft will become part of the operational fire detection system. The equipment has been improved by replacing much of the electronics used in 1970 with integrated circuits. The recorder has been replaced with a unit employing a fiber optics CRT and heat-processed, 9½-inch-wide dry silver recording paper. The recorder produces 8-inch-wide rectilinearized imagery with over 4000 line resolution and 12 gray shades. Our major effort will be concentrated on developing strategies to effectively integrate this new tool into the overall fire suppression system.

REFERENCES

Hirsch, S. N. Applications of remote sensing to forest fire detection and suppression. *Proceedings of the Second Symposium on Remote Sensing of Environment.* Ann Arbor: University of Michigan, Willow Run Laboratories, 1963. Pp. 295–308.

— Preliminary experimental results with infrared line scanners for forest fire surveillance. *Proceedings of the Third Symposium on Remote Sensing of Environment.* Ann Arbor: University of Michigan, Willow Run Laboratories, 1965. Pp. 623–648.

— Project Fire Scan—summary of 5 years' progress in airborne infrared fire detection. *Proceedings of the Fifth Symposium on Remote Sensing of Environment.* Ann Arbor: University of Michigan, Willow Run Laboratories, 1968. Pp. 447–457.

Kruckeberg, R. F. No smoke needed. USDA Forest Serv. Fire Control Notes. Washington, D.C.: Division of Fire Control, 1971.

Wilson, R. A. Fire detection feasibility tests and system development. *Proceedings of the Fifth Symposium on Remote Sensing of Environment.* Ann Arbor: University of Michigan, Willow Run Laboratories, 1968. Pp. 465–477.

—, and Noste, N. V. Project Fire Scan fire detection interim report, April, 1962, to December, 1964. USDA Forest Serv. Res. Pap. INT–25. Intermountain Forest and Range Exp. Sta., Ogden, Utah, 1966.

The human body is a complex system that is carefully shielded from the outside world by hair and skin; we do not assume the temperature of the surrounding environment. Despite extensive medical research over the centuries, we still know little about the overall heat budget of the human body. No normal or standard skin temperature has been established like the 98.6°F which is the accepted norm for oral temperature. In part this is due to slight variations in individuals and the skin itself, and in part to the very complex nature of the body's thermoregulatory mechanism. Man has known since the time of the early Greeks, before the birth of Christ, that the body surface has warm and cool regions. These thermal variations are determined largely by the heat conducted to the skin locally from underlying organs, or by variations in blood flow. A carefully calibrated infrared sensor can record the slight thermal variations of the skin's surface. From these images it is possible to then develop a normal thermal signature for each human being. Variations from this normal pattern could indicate the presence of disease or pathology. Malignant tumors can cause a rise of 1°C on the surface skin; this is well within the sensing capabilities of infrared sensors. Diagnostic work with infrared sensors offers promise in obstetrics and gynecology, peripheral vascular disease, cerebrovascular disease, trauma and wound healing, orthopedics, arthritis, and dermatology. It is not inconceivable that future medical check-ups will involve a thermographic scan as part of the regular examination.

29-Diagnostic Thermography

R. BOWLING BARNES

From a purely personal and selfish point of view, the one single area of nondestructive testing that we as individuals would most like to see become highly successful and universally accepted might well be that which involves the nondestructive testing for and the diagnosis of human ailments and disease. The potential of ir thermography in this connection was pointed out about 12 years ago (Lawson, 1956), and during the intervening period, considerable attention has been directed towards this technique. Half a dozen different ir scanners are commercially available and thermography is today being used in over 75 medical institutions in the United States and other countries. In general, the majority of these installations are located in larger institutions where they are being used by medical research teams to evaluate the over-all capabilities of and to establish helpful criteria and techniques for the successful use of this new diagnostic modality. As this exploratory work proceeds and as improved instrumentation designed specifically for clinical use becomes available, it is not difficult to imagine that medical thermography may soon become well established as a clinical tool and someday accepted routinely in the average hospital. Already, well over 175 scientific papers on various phases of thermography have appeared in medical journals. In these, the usefulness of thermography as a diagnostic aid in many specific areas of medicine seems to have been adequately documented.

From *Applied Optics* 7:1673–1685, 1968. Reprinted with permission of the author and *Applied Optics*.

From the earliest days of medical practice, it had been known that thermal abnormalities or unusual localized areas of skin temperature, either elevated or depressed, usually denoted the presence of some underlying pathology or disease. It was also recognized that, under normal conditions, the human body exhibited a high degree of right-left symmetry as far as skin temperature was concerned. Hippocrates, the "Father of Medicine," once recommended that, "The physician will examine to see whether one side is hotter than the other." Following this suggestion, medical students for centuries have been taught the basic truth of this concept of diagnosis. Unfortunately, the full utilization of this fundamental principle had to wait many centuries for the development of equipment suitable for accurate visualization of skin temperature.

Until relatively recently, the physician who wished to examine his patient's skin temperature for diagnostic purposes had to rely strictly upon his own hand, usually the back of his hand, as his only thermal sensor. While admittedly crude and far from accurate, such thermal measurements have nevertheless been invaluable in countless instances. Crude as they were, such determinations frequently proved to be important to the examining physician in his efforts to arrive at a correct diagnosis.

Shortly after the modern clinical thermometer was perfected in 1870, unsuccessful attempts were made to establish surface thermometry using a mercury thermometer whose bulb had been flattened to present a contact area of about 650 mm² (Seguin, 1876). No normal for skin temperature could be established such as the 98.6°F (37°C) that Wunderlich, in his monumental researches, had found for oral temperatures. The extreme complexity of the human body's thermoregulatory mechanism was not known at that time, including specifically the important role played by the skin itself (Montagna, 1965). These facts, combined with the over-all inaccuracies of the instruments and methods employed, prevented this diagnostic technique from becoming accepted. In more recent times, such modern devices as thermocouples, thermistors, and the like have been used with considerably more success. Although extremely useful in many research projects and in other special circumstances, and although they are still widely used today, these devices do not lend themselves to widespread clinical use. In addition, it has been demonstrated that they, like all contact methods, actually alter the temperature of the skin area which they would measure, and, accordingly, are prone to be unreliable and to yield results that may be in error by several degrees (Stoll, 1964). The fact that a noncontact method was needed, and that ir radiometry could fulfill this need was pointed out by Hardy in 1934. Together with his coworkers, he subsequently published many important papers on the subject of the accurate determination of skin temperature using a variety of ir radiometers.

Since Hardy's first radiometer for skin temperature measurements was announced, the science of ir has received a great deal of attention, both from the point of view of industry and also of the government. As a result, there are available today a number of greatly improved radiometers and scanners with which accurate determinations can be made. Not only is it possible to make point-by-point skin temperature measurements, but through the use of ir scanners, it is also feasible to make two-dimensional thermal maps of extended areas of human skin. From such maps, referred to as *thermograms*, it is readily possible to determine skin temperatures to the order of 0.1°C and to visualize temperature differences between adjacent or between contralateral areas of skin. The majority of the papers that have so far appeared on different phases of medical thermography have concerned themselves with the establishment of correlations between unusual or abnormal thermal patterns and the presence of pathology. Like crude or gross fingerprints, these thermal patterns have been found to vary from individual to individual; but unlike ordinary fingerprints, they have been shown to vary markedly from time to time in the case of any one individual with changes in his or her pathophysiological condition, thus giving them diagnostic capabilities. Because this technique is entirely passive, making use only of the radiant energy spontaneously emitted by the skin itself, it is entirely harmless and may be used freely by the diagnostician and repeated as often as desired, without the slightest danger or discomfort to the patient.

It must be stressed that thermography does not claim to be a self-sufficient diagnostic procedure, capable of providing specific and definitive answers when used to the exclusion of other techniques. This is by no means the case, and thermography should be looked upon not as a competitor to, but as a complement to other existing techniques. By providing information relative to one additional physical parameter, namely, skin temperature, it can, however, definitely contribute to the over-all accuracy of the total diagnostic procedure. Infrared instruments are capable of doing one thing and one thing only: they can, when designed carefully and used properly, provide accurate measurements of skin temperature. They have no judgment capability, nor can they tell the cause of any given thermal abnormality. They cannot, for example, distinguish between the local elevation of skin temperatures produced by an underlying malignancy, an abscess, or an innocent bruise. They can and do, however, provide pictorial evidence of the general size, shape, and localization of thermal abnormalities, and a quantitative measurement of their magnitude. When employed as a preliminary screening technique, thermography serves to direct attention to those patients and to those areas of the body where thermal abnormalities do exist and thus to indicate the need for further examination by all other available diagnostic tools.

The Basis for Diagnosis by Thermography

Like all other objects, human skin emits ir radiation as an exponential function of its absolute temperature at a rate given by the expression $W = \sigma \epsilon T^4$. In this equation, W is expressed as watts per square centimeter per second, σ is the Stefan-Boltzman constant, ϵ is the skin's emissivity, and T is the temperature of the skin in Kelvin. The exponent of T varies with the spectral range of each instrument's sensitivity and with the temperature of the object being measured. The emissivity ϵ is a wavelength-sensitive function of the chemical and physical nature of the surface of the object being measured, and describes the extent to which this surface absorbs and, therefore, emits like a true blackbody. Although this characteristic of human skin is difficult to measure, and although its exact value is still somewhat controversial, it is generally considered to approach unity throughout the spectral region from about 2.5 μ to about 15 μ (Hardy, 1939). If we assume it to be unity, we can then also assume that a well-designed ir instrument sensitive in the above spectral range and capable of measuring W will also be capable of measuring either T or differences (ΔT's) in skin temperatures. According to Wien's Displacement law, the skin's emission will show a maximum at a wavelength close to 10 μ, given by the expression:

$$\lambda_{max} = 2897/T, \text{ or } \lambda_{max} \approx 3000/T$$

The diagnostic capability of thermography is based upon the following facts:

1. All objects above absolute zero emit ir radiation as an exponential function of their temperatures.
2. Human skin is an excellent (almost perfect) emitter in the above-mentioned ir part of the electromagnetic spectrum.

3. This emitted radiant energy may, using modern instruments, be collected optically, transduced into electrical signals, amplified, and then processed and displayed as desired—either upon a cathode ray tube, which in turn may be photographed, or converted directly into visible light impulses that are proportional in intensity to the object's temperature and that may be photographed directly.

4. The photographic gradients in the final visible images or thermograms are to a very close approximation a direct function of the thermal variations that exist upon the skin at the time the measurements are made.

5. Pigmentations such as are found in and on the skin of different races, and in the nipple areas, blemishes, suntan, etc., do not alter the skin's emissivity.

6. The body, at rest in a constant cool ambient free of air currents and devoid of the proper clothing, will relatively quickly come to equilibrium with its surroundings, and the skin temperature patterns will approach a static condition.

7. In such an equilibrium condition in an ambient in the neighborhood of 21°C to 24°C, a major source of the heat loss experienced by the body takes place through the mechanism of the emission of ir radiation by the skin.

8. The thermal patterns that occur on the skin under the above conditions are occasioned largely by (a) vascular patterns, (b) localized conditions of the skin, such as necrosis, erythema, irritation, hyperemia, infection, and the like, or (c) heat brought to the surface through vascular channels or by thermal conduction through bones, fat, and other tissues from certain normal underlying organs of the body, or from discrete pathologic entities such, for example, as neoplasms, metastatic cancers, trauma, or deep-seated infections.

The human body continuously broadcasts thermal information, and, to an ir-sensitive instrument, it behaves exactly as if it were truly incandescent, and it is these ir signatures that form the basis for medical thermography (Barnes, 1963).

Thermographic Equipment

Although, as stated, there are today at least five commercially available ir scanners, the author is most familiar with the M1-A thermograph, an improved version of the scanner described some years ago by Astheimer and Wormser (1959). This instrument, shown in clinical use in Figure 1,[1] is actually a scanning radiometer that employs a thermistor bolometer, covers a field of view of 10° by 20°, has an instantaneous field of view of about 1.5 angular mils, has a basic frame time of 4 min, operates in the spectral range of about 2–16 μ, and has a thermal resolution better than 0.1°C. Operating at room temperature, the thermograph, as it scans the object plane, receives the net flux of energy emitted to it by the human skin, transduces this into proportional electrical impulses and converts these into proportional fluxes of visible light. These, in turn, are recorded on Polaroid film, with various temperatures of the skin revealed in different shades of gray, ranging from black (lower temperatures) to white (higher temperatures). In the resulting thermograms, as many as 60,000 spatial picture elements may be recorded, depending upon the actual frame time employed. The sensitivity of the scanner may be varied as desired over a tenfold range, so that any selected temperature difference (ΔT) from 1°C to 10°C will cause the film to go from black to white. A temperature offset, or brightness, control is also available, so that any desired temperature may be recorded in the center of the film's dynamic range.

In certain instances, the thermal information desired is purely qualitative in nature and may readily be obtained by visual inspection of the thermogram; however, most diagnostic situations require that this information be expressed quantitatively. To facilitate this, a thermal gray scale, Figure 2, consisting of ten accurately controlled blackbodies ranging from 38°C to 29°C, inclusive, is made a part of each thermogram. Each film is thus calibrated, and by use of the photocomparator, shown in Figure 3, the desired quantitative data may readily be obtained.

In the design of the M1–A thermograph, much emphasis was placed upon producing an image of the highest possible optical quality, commensurate with a reasonable and practical scan time and with a satisfactory thermal resolution. After examining literally thousands of medical thermograms, it is my personal belief that images of even higher optical quality are greatly to be desired. Particularly in the case of a relatively new diagnostic technique, such as thermography, where new correlations are sought between the existence of pathology and unusual values of the physical parameter being measured (skin temperature), it is of paramount importance that the instrumentation employed provide the best information possible within the limits of the existing state of the art. With certain of these correlations being based upon unusual qualitative thermal patterns and others being referenced quantitatively to the presence of temperature differences (ΔT's) greater than some specific amount, the above statement cannot be overemphasized. When decisions relative to the welfare of human beings depend directly upon the results obtained, the integrity of the instrument must never be compromised. To this end, all thermograms should, to the extent possible, be (a) true thermal maps of the skin's temperature, (b) produced entirely from the ir energy emitted by the skin itself, (c) independent of any artifacts produced by the skin's reflection of short wavelength ir emitted by hotter objects within the examination room, (d) of sufficiently high optical quality to enable the smallest pattern detail that exists on the skin to be recognized and identified, (e) accurate in portraying the smallest thermal detail that may be of diagnostic significance, (f) recorded within a time interval sufficiently short so that the skin, in partial equilibrium with its surroundings, will not alter its temperature during the time of the scan, and (g) be readily interpretable, both qualitatively and quantitatively, in terms of temperatures or differences in temperature.

As is the case with every other diagnostic technique, the total time per examination is of prime importance and

[1] Because of space limitation, Figures 1-3 are not reproduced here.

should obviously be held to a reasonable and practical minimum. Neither optical nor thermal resolution, however, should ever be prejudiced merely for the sake of obtaining an extremely short frame time, since it is only the total patient examination time that is significant. In this connection, it might be noted that, although a single x ray exposure of the type used in mammography may be made in a fraction of a second, the total time required for a mammography examination for possible breast cancer normally approaches 10 to 30 minutes.

Examination Technique

Ideally, thermography should be done in a closed room, held at a constant temperature somewhere in the vicinity of 21°C. Successful thermal diagnoses have, however, been made at bedside in conventional hospital rooms. It has been shown (Williams, Williams, and Handley, 1961) that the thermal contrast on the surface of the skin increases as the room temperature decreases, and vice versa, that this contrast will disappear if the ambient is held close to 37°C–38°C (Barnes, 1964). Drafts and air currents are to be avoided, for these, obviously, will produce artifacts. Clothing must be removed from those parts of the body being scanned, since clothing is opaque to the wavelengths in question, and since the desired information is actually the net flux of electromagnetic radiation between the ir instrument itself and that small area of skin at which it is aimed. It is also clear that the exposed skin area must remain in the above-described environment (not necessarily the same room) for a sufficient time (usually 5–10 minutes) to enable any insulation effects of the clothing just removed to disappear. Experience has shown that if reasonable precautions are taken in accordance with the above, thermograms of excellent quality may be produced that truly reveal the thermal distributions that actually exist upon the skin. Such thermograms are unique to the individual patient and may be reproduced with remarkable accuracy during subsequent examinations.

Since the skin is a barrier separating the internal parts of the body from the ambient, it is clear that its average temperature will seek an equilibrium somewhere between that of the body and that of the walls of the room. If, however, a given area of skin faces another area of skin instead of the walls of the room, a new and higher equilibrium temperature (see Fig. 4) will be established. The walls of the ear canal present an extreme case of this phenomenon and reach a temperature closely approaching that shown by the oral thermometer (Barnes, 1967). In producing thermograms of diagnostic quality, it is important that the patient be positioned with care, so as to minimize such skin-to-skin cross radiation. Thus, with the patient reclining, the arms should be extended with the hands above the head. If the lower extremities are to be examined, they should be spread apart, and if hands are being examined, the fingers should be spread. In view of the limited depth of focus of the scanner, it is also important that the patient be positioned so as to present to the scanner a full view of the area of skin under observation as nearly as possible in one plane. Frequently, where highly contoured surfaces are being examined, multiple thermograms should be made.

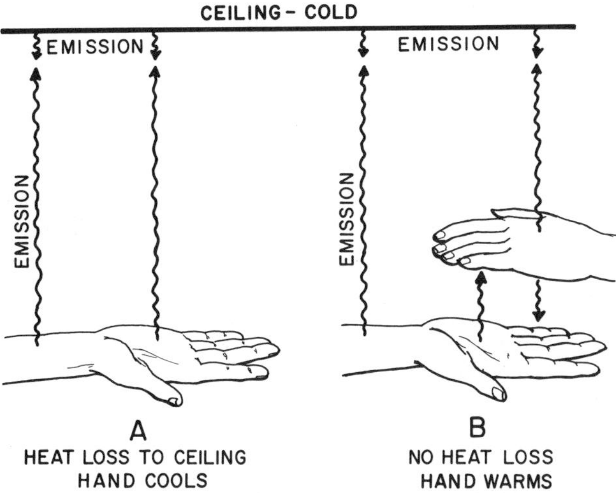

Figure 4 Radiation exchange between human skin and the surroundings.

The History of Medical Thermography

Modern medical thermography seems to have gotten its start in 1956 through the observation by Ray Lawson of Montreal that the nipple of a female breast containing a malignancy is hotter than that of the opposite breast. Using a thermograph of the Astheimer-Wormser type (1959), Lawson published in 1957 a thermogram of a metastatic cancer in the axilla of a woman who had previously had a breast cancer. In connection with the origin of the elevated skin temperatures apparently associated with the presence of underlying cancers, Lawson (1963) later showed that the temperature of the venous blood flowing away from the site of a malignancy was almost 1°C higher than that of the arterial blood flowing toward it. In 1961, Lloyd Williams and his associates in London published an important paper, in which they compared the skin temperatures over 100 palpable breast lesions with those of the same areas in the opposite breasts. Using a thermopile held by hand about 1 cm above the skin, the contralateral areas were examined, and where a $\Delta T \geqq 1°C$ was observed over a lesion, that lesion was recorded as hot. Since all lesions were biopsied, direct correlations between the thermal observations recorded and the resulting pathologic findings could be made. Out of the 100 lesions, 57 turned out to be malignant and of these, 54 were recorded as hot. Of 18 benign cysts, 17 were recorded as cold, and the 1 that was hot turned out upon operation to be abscessed. Eleven other benign lesions showed either no skin temperature elevation, or a ΔT's less than 1.0°C. Forty percent, or four out of ten, fibroadenomas (benign) showed positive (hot) thermal patterns. Even today, this same percentage of hot fibroadenomas seems to hold and to present somewhat of a puzzle.

The above results seemed to bear out Lawson's original findings that an increase of skin temperature appears to be associated with underlying malignancies, and gave impetus to further research in this general field (Barnes and Gershon-Cohen, 1963; Gershon-Cohen and Haberman, 1964; Gershon-Cohen et al., 1964). By December 1963, sufficient interest had been evidenced in this new diagnostic technique to justify a symposium held under the auspices of The New York Academy of Sciences (*Ann. N. Y. Acad. Sci.*, 1964). At that time, several ir scanners were described, as well as the diagnostic potential of this new technique in many fields of medicine. Since that time, several medical thermography symposia have been held, including one in Strasbourg, France (*J. Radiol.*, 1967) and another more recently in Leiden, Holland (Boerhave, 1968).

Specific Medical Applications

Unlike certain coldblooded animals, described as poikilothermic, man's body temperature does not assume that of the ambient. Actually, it is extremely well regulated and remarkably insensitive to wide changes in environmental temperature. The body's thermoregulatory system is complex, involving many separate organs and systems within the body, one of the most important of which is the skin (Montagna, 1965). Interlaced with capillary blood vessels under the control of the sympathetic nervous system, capable of opening or closing completely and of changing their caliber within wide ranges, the skin performs remarkably well as a heat exchanger and as regulator of body temperature. In the performance of this function, the skin itself changes widely with temperature changes in the ambient. In spite of this, however, the thermal contrast patterns of the skin are, under static ambient conditions, very indicative of underlying physiology and of pathologic conditions that may exist. Medical publications relative to the subject of thermography are appearing at a rapid rate and any attempt in this paper to present a complete picture of this very active subject would be out of place. At best, it will only be possible to outline certain of the broad areas of medical diagnosis that have so far been researched, and finally to discuss some recent developments designed to render this new diagnostic technique more practical and more readily acceptable to the medical profession.

Cancer

Since modern thermography had its start with the discovery that the skin overlying a malignant breast tumor exhibits an increased temperature as compared with the case of a benign tumor or with that of the contralateral area of the body, it is only right that we begin with this subject.

Lloyd Williams (1964) and his associates, as stated above, showed that some 95 percent of palpable breast lesions subsequently proven to be malignant were associated with elevated localized skin temperatures equal to or greater than 1°C. True positive findings of this same general order have been reported by the Royal Marsden Hospital in London (Harris et al., 1966), by Hoffman (1967), by Gershon-Cohen and his associates (1965), and by Freundlich and his associates (1968). Figures 5 and 6 show illustrations of two distinctive types of positive thermograms.[2]

[2] Figures 5–7 are not reproduced here.

Since the thermographic instrument possesses no decision-making capability, it is clear that it will inevitably be charged with calling out a certain percentage of false positives. Abscesses, bruises, and certain other benign diseases are known to produce elevated skin temperatures. Although if one is attempting to screen or diagnose for cancer alone these must be recorded statistically as false positives, it should be pointed out that the instrument did not per se produce a false result. It merely indicated that the skin temperature at one local area was abnormally high.

One of the remarkable features of this technique is the fact that extremely small breast cancers, frequently not detected by other means, have been indicated (Hoffman, 1967; Freundlich et al., 1968; Swearingen, 1965; Connell et al., 1966; Gershon-Cohen and Haberman, 1966). Although no conclusive evidence is at hand, it is widely believed that the early detection of cancer is of extreme importance, and some evidence has been presented that a correlation exists between the degree of skin temperature elevation (ΔT) and the prognosis of the case in question (Williams, 1964).

Soft tissue malignancies, both primary and secondary, often unrevealed by other diagnostic techniques, have also been shown to produce localized skin temperature elevations (Gershon-Cohen, Haberman-Brueschke, and Brueschke, 1965). Metastatic cancer of the bone has also given rise to positive thermographic findings, and, in some cases, even before their presence could be demonstrated by roentgenography or by isotope studies. Figure 7 shows just such a case.

As further improvement are made in this technology, thermography may prove to be an ideal diagnostic procedure for screening large populations of asymptomatic women. Hoffman (1967) has reported one series of 1924 women, 24 of whom had breast cancer. In 22 of these 24 cases, positive or suspicious thermographic evidence was obtained, and in 4 of the 22 cases, thermography was the only diagnostic modality to give positive indications. Of the 4 cases thus detected, 3 turned out to be cancers less than 4 mm in size.

Generally similar results have been obtained by others both in this country and abroad. In a recent paper, Freundlich et al. (1968) have reported favorable results in connection with the potential use of thermography as a screening method.

Obstetrics and Gynecology

During the early stages of pregnancy, the breasts become elevated in temperature. Characteristic thermal patterns result that have been used as an early test for pregnancy (Birnbaum, 1966). Somewhat similar patterns are produced by the taking of oral contraceptives (Gershon-Cohen et al., 1965).

During the later stages of pregnancy, thermography has proven helpful in locating the site of placental implants (Birnbaum and Kliot, 1965). The fact that it is an entirely passive procedure makes this method attractive. Dr. Haberman has indicated in a private communication (1968) an accuracy of 92 percent in a series of 85 patients delivered by caesarean section. In Figure 8 the

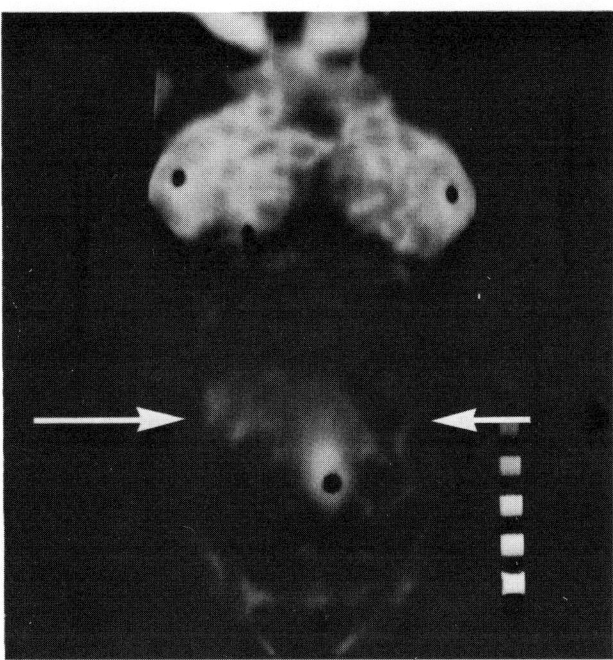

Figure 8 Thermogram showing an anterior placental implant during last trimester of pregnancy. Location confirmed at caesarean section. (Thermogram courtesy of JoAnn D. Haberman, M.D., Temple University Hospital)

location of the placenta as well as the increased skin temperature of the breasts is clearly shown.

Peripheral Vascular Disease

In a static condition such as that required for good-quality thermography, the temperatures of the extremities are largely determined by the flow of blood to and from the extremities and by the radiative losses of heat to the environment. Abnormalities in blood flow, therefore, should logically be expected to be detectable by thermography.

Occlusions, such as arteriosclerosis obliterans, thromboses, severe vasoconstrictions, etc., normally reveal themselves as areas of marked skin temperature reductions (Winsor and Bendezu, 1964; Winsor, 1968; Lane, 1967).

Vascular difficulties frequently associated with diabetes also produce well-defined thermal patterns, as shown in Figure 9. [See color insert.]

Vasoconstrictive effects associated with the inhalation of cigarette smoke are dramatically revealed by the thermograph. Figure 10 shows this phenomenon. [See color insert.]

Inflammatory processes, such as venous engorgement due to local constrictions, phlebitis, first-degree burns, trauma, etc., also reveal themselves as localized areas of increased skin temperature elevations. Figure 11[3] shows the influence of an abnormally tight wristwatch band on the blood flow to and from the hand.

The ability to visualize objectively skin temperature differences, which seem to parallel peripheral vascular diseases both pre- and post-operative, or before and after the administration of drugs, is of significant importance. Valuable information has been obtained relative to the severity of the damage in the case of accidents (Winsor and Bendezu, 1964), and to the course of healing.

[3] Figures 11–14 and 16 are not reproduced here.

Cerebrovascular Disease

What has been said above applies equally well to cases involving cerebral vascular problems with one major exception; hair, which is avascular, is an excellent thermal insulator and interferes. Although in the case of malignant brain tumors and other inflammatory conditions the skin temperature is elevated, this rise cannot be seen thermographically, unless the subject be bald or unless the hair be removed.

Wood (1965) has shown that if the internal carotid artery is occluded by roughly 50 percent or more, this fact is revealed by a marked lowering of skin temperature on the same side of the head just above the eyebrow. This results from the fact that one branch of the internal carotid artery becomes the ophthalmic artery, which then terminates in an area just above the eyebrow. Figure 12 shows the typical pattern associated with the occlusion of this artery.

Evidence of other diseases of the head has also been presented (Heinz, Goldberg, and Taveras, 1964).

Trauma and Wound Healing

Detailed studies of the skin temperature elevations associated with bruises can contribute much valuable information as to the extent and nature of the affected area and as to the course of healing. Figure 13 illustrates dramatically the appearance of a simple bruise.

Thermal changes appear to follow very closely the course of wound healing (Kliot and Birnbaum, 1965). Unfortunate postoperative infections have been revealed thermographically, even without the necessity of removing the dressings and bandages.

Orthopedics

Fractures, sprains, contusions, and herniated disks have revealed themselves thermographically (Albert, Glickman, and Kallish, 1964; Edeiken et al., 1968), and in some cases, elevated skin temperatures over these sites have indicated resulting soft tissue damage long after healing had apparently taken place.

Arthritis

Thermograms of patients suffering from rheumatoid arthritis are generally characterized by excessively hot joints. In view of the inflammatory nature of the disease, this is to be expected (Boas, 1964). Haberman, Ehrlich, and Levenson (1968) have shown that the various types of arthritis may be detected and discriminated thermographically. Figure 14 shows a typical case of rheumatoid arthritis.

Of significance is the fact that the skin temperature over an active arthritic joint may serve well as an objective test of the efficacy of chemotherapeutic treatment.

Dermatology

Since in thermography we are concerned entirely with the graphic portrayal of temperature patterns that

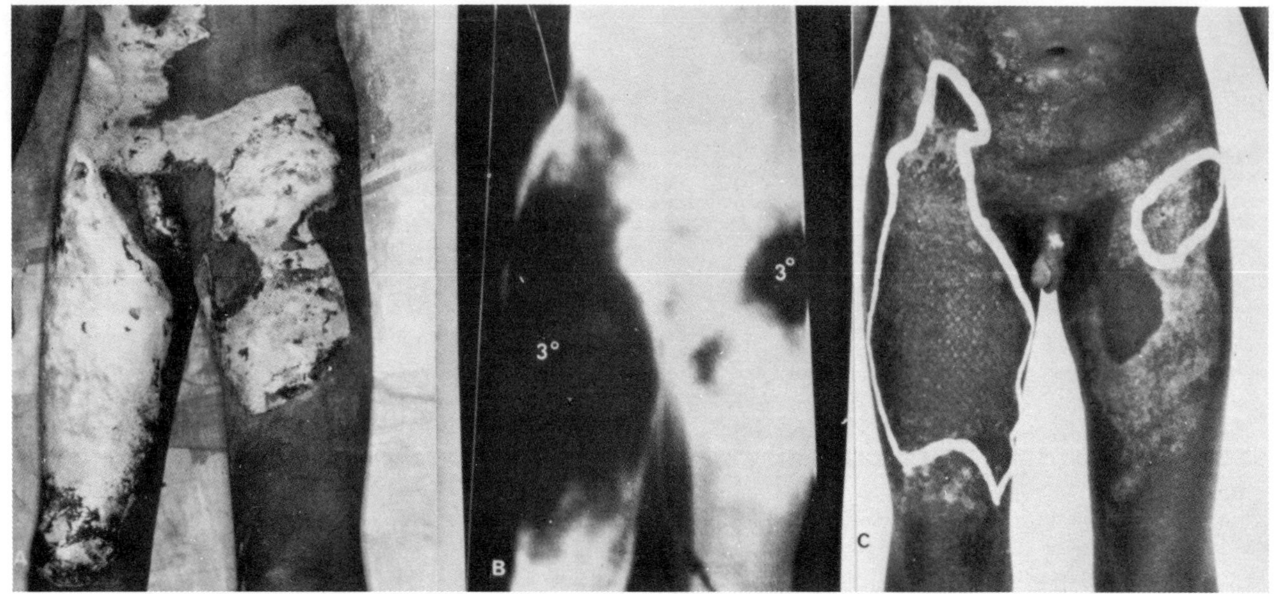

Figure 15 Illustration of thermography used in study of burned areas of skin. (A) Two days postflame burns of lower abdomen and thigh areas. (B) Thermograph of these burn areas showing large areas of third-degree loss (3°) with surrounding second-degree areas. (C) Two weeks following grafting of third-degree areas. The grafted areas correlated with areas outlined on the thermogram. Grafted areas outlined in white. (Mladick, Georgiade, and Thorne, 1966. © 1966. The Williams & Wilkins Co., Baltimore)

exist on the skin itself, it is obvious that dermatology should be a fruitful field for thermographic analysis and diagnosis.

Lawson, Wlodek, and Webster showed in 1961 that frostbitten skin areas are readily visualized with an accuracy approaching 100 percent. Such areas, being necrotic and, therefore, avascular, appear cold as contrasted with the surrounding skin.

Thermographic diagnosis of burned areas of skin is also very promising. First-degree burns being inflammatory show up as areas of increased temperatures, whereas third-degree burns are cold. Figure 15 shows areas of third-degree burns. The ability to appreciate areas of viable skin in the midst of what would appear to the eye to be extended areas of third-degree burns is of considerable importance to the surgeon. This is particularly true, since such determinations may frequently be made within a few hours after the burn accident has occurred (Mladick, Georgiade, and Thorne, 1966).

Results obtained by Lewin and Bloomenstein (Bloomenstein, 1968), using an ir radiometer, have indicated that the probable success or failure of various types of skin grafts including pedicle transfers may be predicted shortly after the graft has been made. Strong temperature gradients tending from normal values towards cold areas may indicate the onset of necrosis and thus forecast subsequent graft failure.

As stated above, pigmentation does not influence the normal radiation of ir. While an area of mild sunburn will show up as hot areas, deep suntan is not detected. The same also holds for pigmented areas such as birthmarks, freckles, etc., and also for the nonpigmented areas of vitiligo.

Recent Advances in Thermology

Until very recently, practically all thermograms have been presented as black and white images with warmer areas of the object appearing as white areas and colder areas as black. Thermal gray scales presenting blackbody targets of known temperatures have been simultaneously thermographed and thus made a part of the image for purposes of calibration. Because of the human eye's known limitations as a black and white photometer, precise quantitation of thermograms has only been possible through the use of some form of photocomparator or the thermoprofile scanner (Brueschke, Haberman-Brueschke, and Gershon-Cohen, 1965). Since each of these methods of obtaining quantitative data from thermograms is somewhat time consuming, an alternate method was sought. The fact that the human eye is far more sensitive to color differences than to shades of black and white was considered to be of prime importance. An alternate fact that was carefully considered was the ease with which the eye was capable of perceiving digital changes as contrasted with more subtle analog intensity changes. The results of these decisions are shown in the several colored thermograms presented in this paper.

The M1–A thermograph, the most extensively used medical ir scanner, is now capable of presenting its data in a multiplicity of forms, including, but not limited to, analog black and white or analog color, ten-digit black and white or ten-digit color, three-digit color or three-digit black and white, or black and white or color isotherms.

As an illustration of these modes of data presentation, let us consider the ten-digit color mode. Figure 16, taken at a sensitivity (ΔT) of 10°C, shows a thermogram of a young lady taken against an unevenly heated background. In this illustration, since there are ten color changes per 10°C, each color change represents a ΔT of 1°C. Note color changes along left edge of the background and throughout entire thermogram. Had this same scene been thermographed at a sensitivity of 5°C, each color change would be representative of 0.5°C. In each case, the standard gray scale (not shown on this thermogram) becomes a color scale and the eye is immediately able, without the aid of any auxiliary devices, to recognize both temperatures (T's) and temperature differences (ΔT's).

In the case of black and white thermograms recorded in ten digits, the above is also true, although to a somewhat lesser degree.

Conclusion

Under standard and normalized environmental conditions, the thermal contrasts or patterns that exist on the exposed human skin are determined largely by the heat conducted to the skin locally from underlying organs or variations in blood flow. These localized temperatures give rise to corresponding variations in the rates at which ir energy is radiated. Instruments are available that are capable of translating these differences in rates to photographic images or thermograms. In the absence of disease or pathology, there exists a normal thermal signature for each human. The presence of pathology grossly alters the thermogram and gives rise to a valuable diagnostic procedure—to a unique method of nondestructive testing or remote sensing.

REFERENCES

Albert, S. M., Glickman, M., and Kallish, M. *Ann. N.Y. Acad. Sci.* 121:157, 1964.

Ann. N.Y. Acad. Sci. 121. Thermography and its Clinical Applications, 1964.

Astheimer, R. W., and Wormser, E. M. *J. Opt. Soc. Am.* 49:179, 1959.

Barnes, R. B. *Science* 140:870, 1963.

Barnes, R. B. *Ann. N.Y. Acad. Sci.* 121:34, 1964.

Barnes, R. B. *J. Appl. Physiol.* 22:1143, 1967.

Barnes, R. B., and Gershon-Cohen, J. *J. Am. Med. Assoc.* 185:949, 1963.

Barnes, R. B., and Lane, W. Z. In *Proceedings of the Sixteenth Annual Conference on Engineering in Medicine and Biology* Vol. 5. Robinson, D. A. (ed.). Instrument Society of America, 1963.

Birnbaum, S. J. *Obstet. Gynec.* 27:378, 1966.

Birnbaum, S. J., and Kliot, D. *Obstet. Gynec.* 25:515, 1965.

Bloomenstein, R. *Plast. Reconstruct. Surg.* 42:252, 1968.

Boas, N. F. *Ann. N.Y. Acad. Sci.* 121:223, 1964.

Boerhaave Courses for Postgraduate Medical Education, Leiden, Holland, 1968. Medical Thermography.

Brueschke, E., Haberman-Brueschke, J. D., and Gershon-Cohen, J. *Am. J. Med. Electron.* 4:65, 1965.

Connell, J. F., Ruzicka, F. F., Grossi, C. E., Osborne, A. W., and Conte, A. J. *Cancer, J. Am. Cancer Soc.* 19:83, 1966.

Edeiken, J., Wallace, J. D., Curley, R. F., and Lee, S. *Am. J. Roentgenol.* 102:790, 1968.

Freundlich, I. M., Wallace, J. D., and Dodd, G. D. *Am. J. Roentgenol.* 102:927, 1968.

Gershon-Cohen, J., Berger, S. M., Haberman, J. D., and Barnes, R. B. *Am. J. Roentgenol.* 91:919, 1964.

Gershon-Cohen, J., and Haberman, J. D. *Radiology* 82:280, 1964.

Gershon-Cohen, J., and Haberman, J. D. *Ob/Gyn Digest* 8:55, 1966.

Gershon-Cohen, J., Haberman-Brueschke, J. D., and Brueschke, E. E. *Radiol. Clinics N. Am.* 3:403, 1965.

Haberman, J. D. Temple University Hospital, private communication, 1968.

Haberman, J. D., Ehrlich, G. E., and Levenson, C. *Arch. Phys. Med. Rehabil.* 49:187, 1968.

Hardy, J. D. *J. Clin. Invest.* 13:593, 1934.

Hardy, J. D. *Am. J. Physiol.* 127:454, 1939.

Harris, D. L., Greening, W. P., and Aichroth, P. M. *Brit. J. Cancer* 10:710, 1966.

Heinz, E. R., Goldberg, H. I., and Taveras, J. M. *Ann. N.Y. Acad. Sci.* 121:177, 1964.

Hoffman, R. L. *Am. J. Obstet. Gynec.* 98:681, 1967.

J. Radiol. 48, No. 1–2, 1967. Colloque International de Thermographie Médicale.

Kliot, D. A., and Birnbaum, S. J. *Am. J. Obstet. Gynec.* 93:515, 1965.

Lane, W. Z. *Hosp. Pract.* 2:36, 1967.

Lawson, R. N. *Can. Med. Assoc. J.* 75:309, 1956.

Lawson, R. N. *Can. Services Med. J.* 13:517, 1957.

Lawson, R. N., and Chughtai, M. S. *Can. Med. Assoc. J.* 88:68, 1963.

Lawson, R. N., Wlodek, G. D., and Webster, D. R. *Can. Med. Assoc. J.* 84:1129, 1961.

Mladick, R., Georgiade, N., and Thorne, F. *Plast. Reconstruct. Surg.* 38:512, 1966.

Montagna, W. *Sci. Amer.* 212 (No. 2):56, 1965.

Seguin, E. *Medical Thermometry and Human Temperature.* New York: William Wood, 1876.

Stoll, A. M. *Ann. N.Y. Acad. Sci.* 121:49, 1964.

Swearingen, A. G. *Radiology* 85:818, 1965.

Williams, K. Lloyd. *Ann. N.Y. Acad. Sci.* 121:272, 1964.

Williams, K. Lloyd, Williams, F. J. Lloyd, and Handley, R. S. *Lancet* 2:1378, 1961.

Winsor, T. In *Abbottempo Book I.* Richardson, R. G. (ed.). London: Abbott Universal, 1968.

Winsor, T., and Bendezu, J. *Ann. N.Y. Acad. Sci.* 121:135, 1964.

Wood, E. H. *Radiol.* 85:270, 1965.

SUGGESTED READINGS FOR PART SIX

Astheimer, R. W., and Wormser, E. M. Instruments for thermal photography. *Journal of the Optical Society of America* 49:184–187, 1959.

Bjornsen, R. L. Infrared Mapping of Large Fires. *Proceedings of the Fifth Symposium on Remote Sensing of Environment.* Ann Arbor: University of Michigan, Institute of Science and Technology, Willow Run Laboratories, 1968.

Colwell, R. N., and Olson, D. L. Thermal Infrared Imagery and Its Use in Vegetation Analysis by Remote Aerial Reconnaissance. *Proceedings of the Third Symposium on Remote Sensing of Environment.* Ann Arbor: University of Michigan, Institute of Science and Technology, Willow Run Laboratories, 1965.

Fischer, W. A., Moxham, R. M., Polcyn, F., and Landis, G. H. Infrared surveys of Hawaiian volcanoes. *Science* 146:733–742, 1964.

Fuchs, M., and Tanner, C. B. Infrared thermometry of vegetation. *Agronomy Journal* 58:597–601, 1966.

Gates, D. M. Infrared Measurement of Plant and Animal Surface Temperature and Their Interpretation. *Remote Sensing in Ecology.* Athens: University of Georgia Press, 1969. Pp. 95–107.

Hirsch, S. N. Project Fire Scan—Summary of Five Years Progress in Airborne Infrared Fire Detection. *Proceedings of the Fifth Symposium on Remote Sensing of Environment.* Ann Arbor: University of Michigan, Institute of Science and Technology, Willow Run Laboratories, 1968.

Komarov, V. B., Shilin, B. V., Miroshnikow, M. M., and Feoktislov, Ju. A. The Methods of Application of Infrared Photography when Studying the Volcanoes and Thermal Activities of Kamchatka Peninsula. *Proceedings of the Fifth Symposium on Remote Sensing of Environment.* Ann Arbor: University of Michigan, Institute of Science and Technology, Willow Run Laboratories, 1968.

Lee, K. Infrared Exploration for Shoreline Springs at Mono Lake, California, Test Site. *Proceedings of the Sixth International Symposium on Remote Sensing of Environment.* Ann Arbor: University of Michigan, Institute of Science and Technology, Willow Run Laboratories, 1969.

McCullough, D. R., Olson, C. E., Jr., and Queal, L. M. Progress in Large Animal Census by Thermal Mapping. *Remote Sensing in Ecology.* Athens: University of Georgia Press, 1969. Pp. 138–147.

Myers, V. I., and Heilman, M. D. Thermal ir for soil temperature studies. *Photogrammetric Engineering* 35:1024–1032, 1969.

Weaver, D. K., Butler, W. E., and Olson, C. E., Jr. Observations on Interpretation of Vegetation from Infrared Imagery. *Remote Sensing in Ecology.* Athens: University of Georgia Press, 1969. Pp. 132–137.

Wendland, W. M., and Bryson, R. A. Surface Temperature Patterns of Hudson Bay from Aerial Infrared Surveys. *Remote Sensing in Ecology.* Athens: University of Georgia Press, 1969. Pp. 185–193.

Wiegand, C. C., Myers, V. I., and Maxwell, N. Thermal patterns of solid fuel block-heated citrus trees. *Journal of the Rio Grande Valley Horticultural Society* 20:21–30, 1966.

PART SEVEN
PASSIVE MICROWAVE
Sensing naturally emitted or reflected radiometric wavelengths

THE INFRARED *portion of the electromagnetic spectrum grades imperceptibly into the spectral region known as microwave. The wavelength boundary between the two is generally placed at 1000 μ (with a frequency of 3×10^{11}). The name is somewhat of a misnomer. Microwaves are small when compared to radio or television waves but extremely long when contrasted with wavelengths in the visible portion of the spectrum. Microwaves are actually the longest wavelengths normally used in remote sensing.*

Remote sensors can be divided into two basic categories: passive and active sensors. Passive remote sensors are still undergoing considerable experimental development. Passive systems consist essentially of a receiving instrument which usually comprises a lens or antenna and a detector. The lens or antenna receives energy coming from an outside source and focuses it on the detector. The detector reacts to the introduced energy. If there is no energy incident upon the lens or antenna at the operational wavelength of the detector, the sensor cannot produce an image. This kind of sensor is passive because it is dependent upon exterior energy. In contrast, radar, which also utilizes the microwave spectral region, provides the best example of an operational active *sensor. Radar is composed of a transmitter and a receiver. The transmitter emits a wave, which strikes objects in the environmental field and is then reflected or echoed back to the receiver. Radar is power-source independent; it is not dependent upon incident energy in the environment. Active microwave radar is dealt with in Part Eight.*

From the discussion of infrared radiation we know that the hotter an object, the more energy it emits. But the object is also emitting energy in a broad band of wavelengths over much of the electromagnetic spectrum. If we study a perfect radiator at 300° K, which is a nominal temperature for a terrestrial object, we observe that from a peak at 9.6 μ power falls off rapidly toward the visible region and less rapidly but steadily toward the microwave region. All objects in the natural environment are emitting some microwave radiation. Even the atmosphere itself—through which this energy must travel to reach the antenna—emits radiation. Two properties of the atmosphere directly affect microwave detection. First, the atmosphere is a source of microwave energy and therefore illuminates all ground objects. Second, emission and transmission of the atmosphere place limits on the ability to detect ground objects. Luckily there are several atmospheric windows in the microwave region that are reasonably transparent and are "open" even during rather severe weather conditions. By using a sensor with a detector designed to receive wavelengths in one of these windows, it is possible to image ground objects remotely. There are also applications for microwave sensors which operate in the opaque regions of the atmosphere; these have proved valuable for measuring the amount of water vapor in the atmosphere.

30-Passive-Microwave Imaging

MARVIN R. HOLTER

PASSIVE-MICROWAVE sensors operate in the same spectral region as very-short-wavelength radar, 0.1 mm–3 cm, but they do not illuminate the scene artificially; instead, they make use of the natural radiation emanating from observed objects. The important properties of objects that determine the character of this radiation are emittance, transmittance, reflectance, and object temperature. The character of external natural radiation sources that contribute to the reflected and transmitted radiation are also important. These components are shown schematically in Figure 1. Treated in detail in the subsequent sections are the basic radiation principles and the instrumentation used in these sensors. Also discussed is the general applicability of passive microwaves to remote sensing for agriculture.

Basic Radiation Principles

Planck's fundamental radiation equation (1900) gives the spectral distribution of radiation flux from a perfect radiator (called a blackbody) with a uniform temperature T_B. This theoretical equation of Planck agrees with experimental results and applies to all regions of the electromagnetic spectrum. It states that the electromagnetic radiation given off by an object is a function of the object's absolute temperature and the wavelength (λ) of observation. In the microwave region, if we use the Rayleigh-Jeans[1] approximation to Planck's law, the emitted radiant power, $W_{B\lambda}$, from a blackbody is

$$W_{B\lambda} = T_{B\lambda}^{-4}$$

For real objects that are not perfect radiators, a radiation efficiency factor called spectral emittance, $\epsilon\lambda$, is used to describe the radiation efficiency of an object as a function of wavelength. This efficiency factor is defined as the ratio of observed radiant power, $W_{o\lambda}$, from an object to the radiant power from a blackbody at the same temperature and at the same wavelength:

$$\epsilon\lambda = \frac{W_{o\lambda}}{W_{B\lambda}} \qquad (T_o = T_B)$$

Effective target[2] temperature (or brightness temperaature), T_T, a commonly used term, evolves from these basic radiation concepts. The radiation emanating from an object is, in general, made up of three parts: a self-emitted component, a reflected component, and a transmitted component. The self-emitted component, for instance, is proportional to the object's spectral emittance $\epsilon\lambda$ and its temperature T_0. Thus, an effective temperature (T_{T_e}) due to self-emission can be defined as the product of the object temperature and its emittance; i.e.,

$$T_{Te} = \epsilon\lambda T_o \qquad (1)$$

Similarly, effective temperature contributions (T_{T_r} and T_{TT}) due to the object's spectral reflectance ($\rho\lambda$) and transmittance ($\tau\lambda$) can be written in the forms

$$T_{Tr} = \rho\lambda T_i \qquad (2)$$

and

$$T_{T\tau} = \tau\lambda T_i' \qquad (3)$$

where T_i and T_i' are terms proportional to the radiation incident on the object. The effective target temperature T_T is thus the sum of these three terms; i.e.

$$T_T = \epsilon\lambda T_o + \rho\lambda T_i + \tau\lambda T_i' \qquad (4)$$

This is the temperature of the object as one would measure it with a remote microwave sensor having an ideal antenna. It would be the same as the object's actual temperature only if the object were a blackbody, in which case $\rho\lambda = \tau\lambda = 0$, and $\epsilon\lambda = 1$. Thus for real objects, the remotely observed radiation intensity is dependent not only on the object temperature and the incident radiation but also on several other properties of the object. The emittance, reflectance, and transmittance are, in general, functions of the object material's absorption coefficient (Gardon, 1956; McMahon, 1950), its bulk configuration or shape, the aspect at which the object is viewed (Nicodemus, 1965), and the surface structure.

Object-Detection Parameters

Three basic parameters can be used for identification purposes: the polarization, and the temporal and spectral characteristics of the object's radiation. Each of them is discussed below.

From basic electromagnetic theory it is evident that the reflection properties of most surfaces are polarization-sensitive.

Examples of surfaces that exhibit pronounced polarization effects at microwave frequencies are water, concrete, asphalt, and ice. Results of theoretical computations (Lewis, Casey, and Vaccaro, 1954) of water reflectance and emittance as a function of angle (using

[1] These approximations, as well as a general discussion of the Planck law, can be found in most basic textbooks on modern physics.

[2] The word "target" is used here as a synonym for "object," and the identifying symbol is therefore the subscript T.

From *Remote Sensing with Special Reference to Agriculture and Forestry*. Washington, D.C.: National Academy of Sciences, 1970. Pp. 103–127, 161–162. Reproduced with permission of the author and the National Academy of Sciences.

Maxwell's field equations) are shown in Figures 2 and 3. From these values and a theoretical value for the angular dependence of the sky effective temperature T_i, the effective temperature of water $T_{T(H_2O)}$ can be computed as a function of angle:

$$T_{T(H_2O)} = \epsilon T_o + \rho T_i \tag{5}$$

The surface emittance shown in Figure 3 is obtained by

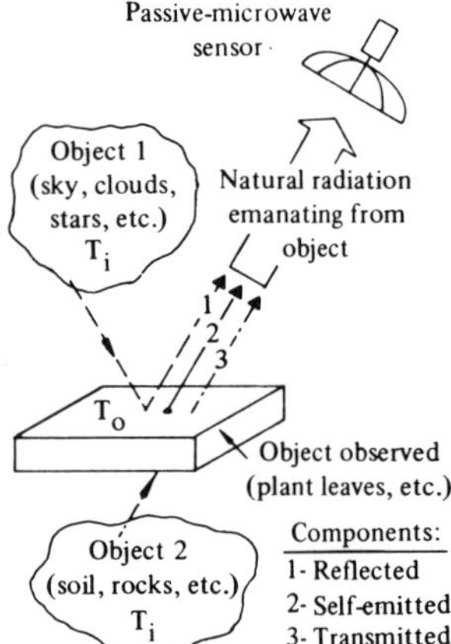

Figure 1 Natural microwave radiation emanating from an observed object.

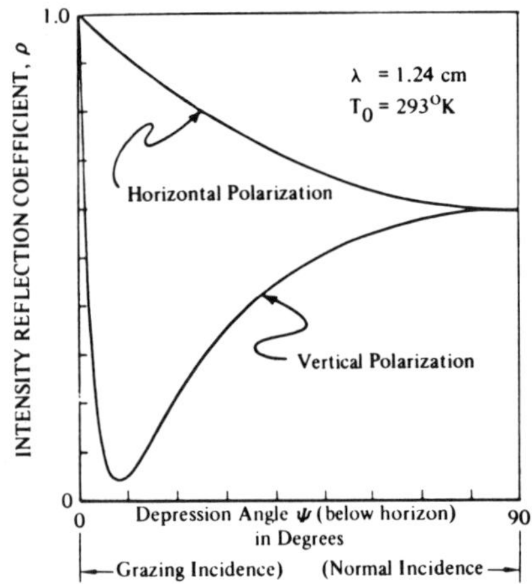

Figure 2 Calculated intensity reflection coefficients for a water surface. (Lewis, Casey, and Vaccaro, 1954)

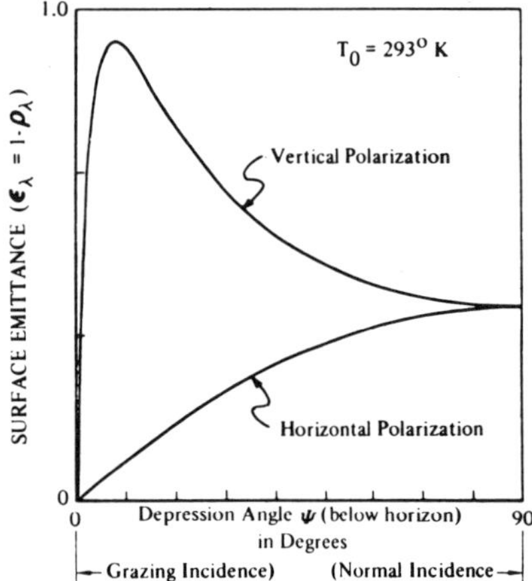

Figure 3 Calculated emittance for a water surface. (Lewis, Casey, and Vaccaro, 1954)

the relationship $\epsilon\lambda = 1 - \rho\lambda$, which is a form of Kirchhoff's law. It means that, for opaque objects where the transmittance is zero, the fraction of the incident radiation that is absorbed must equal 1 minus the fraction reflected; and for a body in thermal equilibrium with its surroundings and transferring power only by radiation, the power emitted must equal the power absorbed.

The results of computations of effective temperature for horizontal and vertical polarizations are shown in Figure 4. These calculations assume that the surface is perfectly smooth, and it should be kept in mind that surface roughness will affect the polarization characteristics of radiation from seawater. The use of this roughness relationship for monitoring sea state has been described in a recent publication (University of Michigan, 1966). The term "apparent" temperature is used in Figure 5. It applies to that temperature determined from the remote measurement. Apparent temperature is usually different from the effective target temperature T_T because the receiver's field of view is usually much

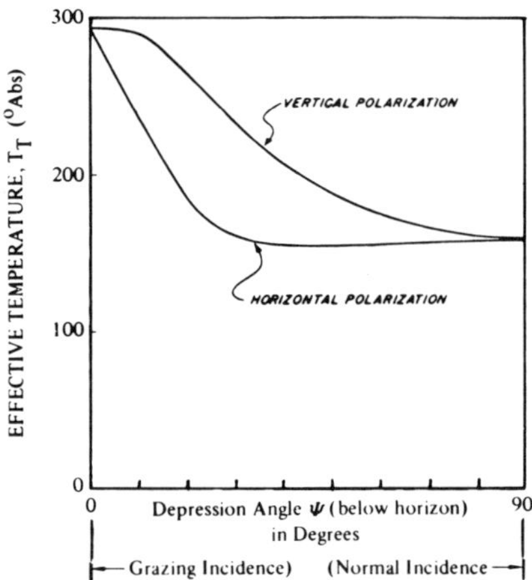

Figure 4 Calculated radiation from a water surface, including the reflected contribution from the sky. (Gardon, 1956)

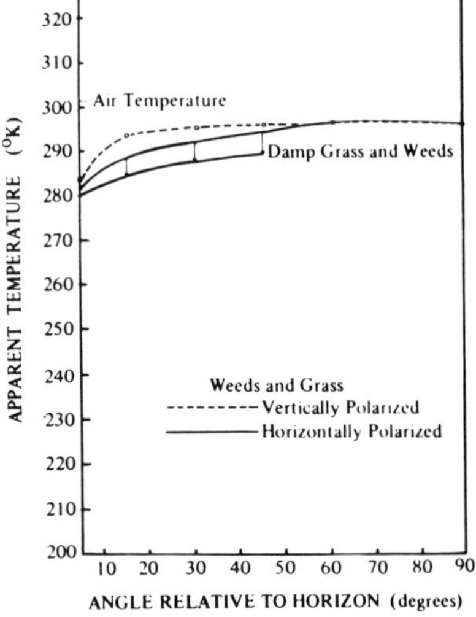

Figure 5 Temperature of grass and weeds. (Straiton, Tolbert, and Britt, 1958)

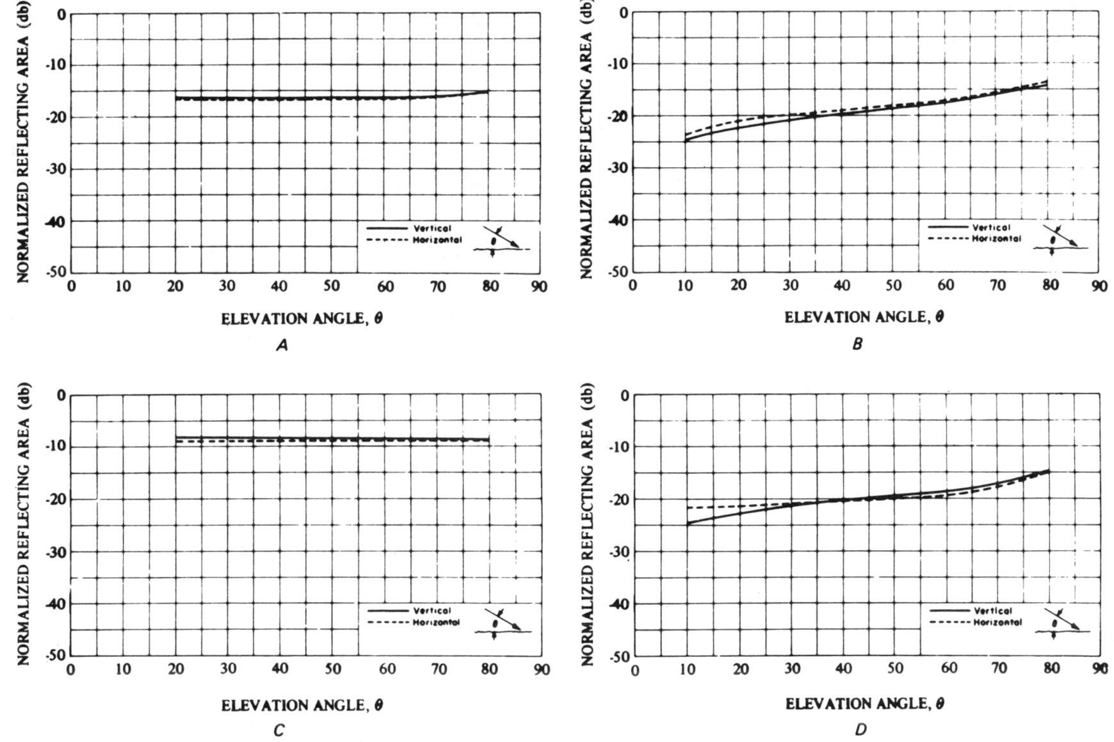

Figure 6 Normalized reflecting area of foliage. *A*, 8-inch grass flattened to a height of 3 inches at *K*-band (November). *B*, Soybeans at *K*-band (September). *C*, 6-inch alfalfa and grass at *X*-band (April). *D*, 2-inch wheat at *X*-band (April). (Adapted from Cosgriff, Peake, and Taylor, 1960)

larger than the size of the object. Also, the side lobes of the antenna response allow other radiation to "leak" through. Perfectly planar surfaces (which are mirror-like), such as the ones used in the calculations for Figures 2–5, are called *specular reflectors*. Surfaces that scatter radiation in every direction are said to be *diffuse*. Grass, bushes, and crops tend to be diffuse, and their polarization effects are not nearly so pronounced (Figs. 5 and 6), although a great deal more experimental work is required to understand their radiation characteristics more fully.

Some objects exhibit a time-dependent (temporal) change in their effective temperatures. These changes can be seasonal or diurnal and can result from fluctuations in the actual object temperature, changes in sky temperature, or changes in the reflection and emittance properties of the object. Shown in Figure 7 are typical thermodynamic temperature ranges for various objects. Thermodynamic temperatures vary over an interval of 40°C for some objects. Comparable changes in effective target temperature will occur only for those objects that are highly emissive (see Equation 5).

The reflection characteristics of vegetation can change with seasons because of geometry, shape, and changes in terrain-covering ability during its maturation. In Figure 8, the normalized reflecting area[3] for wheat is compared at two times during its maturation. A maximum

[3] In microwave work, reflectivity is frequently expressed in an alternative but equivalent manner as the cross-sectional area of a perfectly reflecting sphere that would reflect the same power as the object being measured.

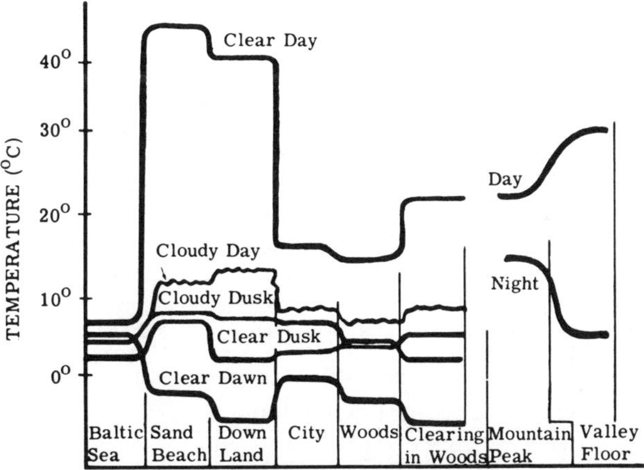

Figure 7 Typical gross temperature differences for various objects at different times of day. (Conway et al., 1963)

difference in echo occurs at an elevation angle of 10°; however, for the data shown, the differences at useful observation angles are much smaller. Unfortunately, a complete study of the temporal dependence of crop echoes on wavelength, polarization, and observation angles has not been made. Data of this kind are needed to make a judicious selection of the instrument parameters for obtaining maximum effects.

The spectral characteristics of an object are among the most useful parameters and have been exploited extensively in optical and infrared portions of the electromagnetic spectrum. This has not been the case in the microwave portion of the spectrum because of spectral bandpass constraints placed on conventional instrumentation. Some of the object data available show spectral characteristics (see Figs. 9 and 10). With the advent of certain new hybrid instrumentation that incorporates bolometer and Putley-type detectors (Chang and Lester, 1966; Breeden, Rivers, and Sheppard, 1966; Putley, 1965), thus combining the advantages of both microwave and infrared measurement techniques, submillimeter spectral data are becoming available. The "target" of most studies to date has been the atmosphere (Breeden, Rivers, and Sheppard, 1966; Williams and Chang, 1966). As more submillimeter data become available, the feasibility of multicolor and other discrimination schemes currently applied at optical and infrared wavelengths can be evaluated for use at submillimeter wavelengths.

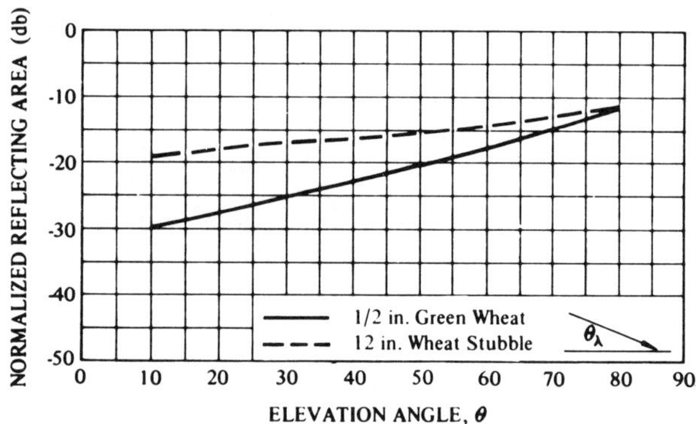

Figure 8 Seasonal variation of wheat, normalized reflecting area. (Adapted from Cosgriff, Peake, and Taylor, 1960)

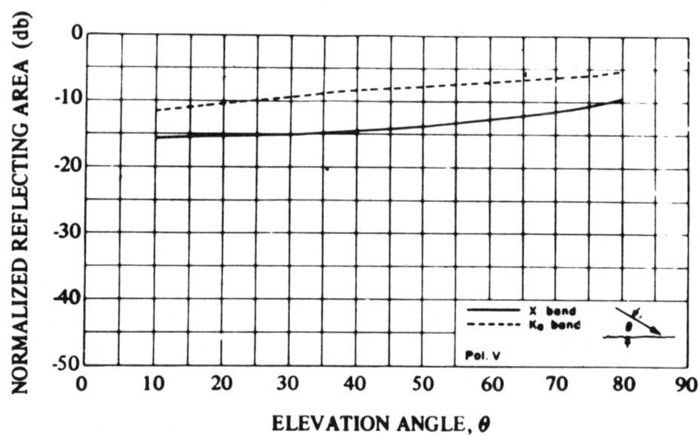

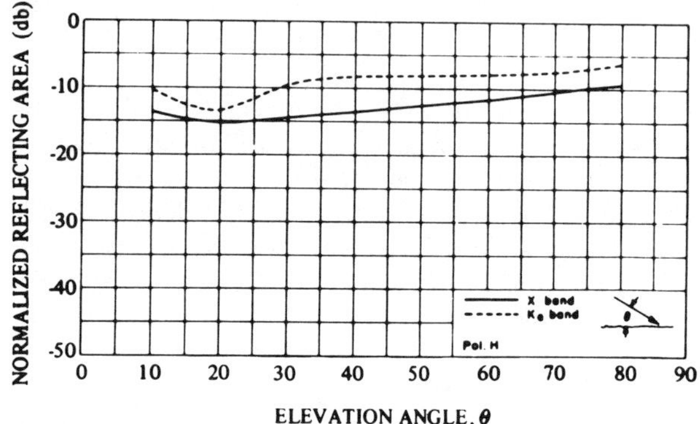

Figure 9 Normalized reflecting area of 4-inch wet soybean stubble. (Adapted from Cosgriff, Peake, and Taylor, 1960)

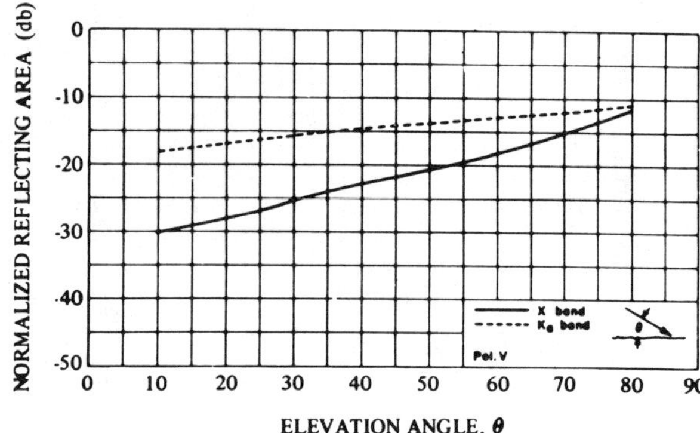

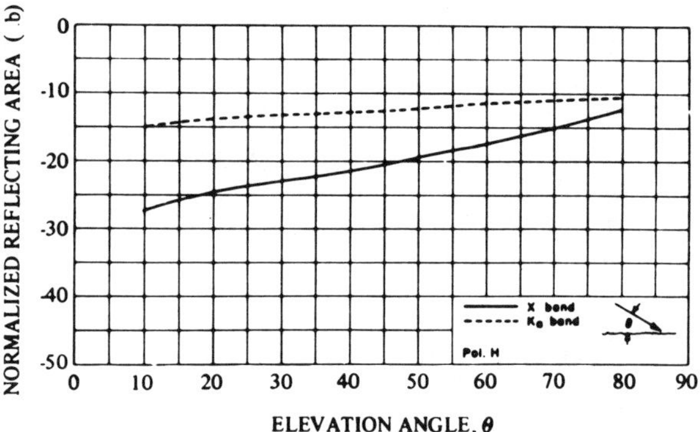

Figure 10 Normalized reflecting area of 0.5-inch green wheat. (Adapted from Cosgriff, Peake, and Taylor, 1960)

The spectral character of some materials can be expected to change because of the penetration capabilities of microwaves. For instance, the forest canopy, which is opaque at short wavelengths, becomes quite transparent at longer wavelengths. The depth of penetration in soils and snow is also a function of frequency.

Atmospheric Effects

Two parameters related to detection are affected by the properties of the atmosphere. First, radiation from the atmosphere itself is a source of illumination for all ground objects. Second, transmission of the atmosphere limits the ability to detect ground objects. Throughout all but the very-short-wavelength end of the region, molecular absorption by atmospheric constituents does not play a significant role. Here in the passive-microwave region, molecular absorption begins to be significant and, as we will see later, becomes very significant at infrared wavelengths. Several "windows" in the microwave region are reasonably transparent during moderately poor to severe weather conditions. Figure 11 shows at a glance the locations of these windows. The opaque regions are due to water vapor and atmospheric-oxygen-absorption bands. At those wavelengths where the atmospheric transmittance is low, the atmosphere appears hot[4] (close to ambient (Breeden, Rivers, and Sheppard,

[4] The illumination of the atmosphere, like the power received from any object at microwave frequencies, is expressed in terms of an effective temperature.

1966) i.e., the atmosphere has a high emittance at these frequencies and thus is a good emitter).

This correlation can readily be seen by comparing Figure 11 with Figure 12, which shows the atmospheric effective temperature due to clouds. It is therefore apparent that two reasons exist for operating in the windows. One is to minimize transmission losses, and the other is to maximize the effective temperature contrast. An example will illustrate this further. Suppose the object is a metal plate and the background is grass. The object at microwave frequencies is highly reflective, and the background is highly absorptive. Thus the effective temperature of the background is near its ambient temperature; however, the effective temperature of the highly reflecting object approaches the effective temperature of the illumination source, in this the sky. To achieve a high contrast, the object must appear either much hotter or much colder than the ambient temperature. Since the sky as a source of illumination (excluding the sun) has an effective temperature between ambient and 0°K, it is apparent that the contrast occurs when the sky appears "cold," i.e., when the atmosphere has a high transmissivity.

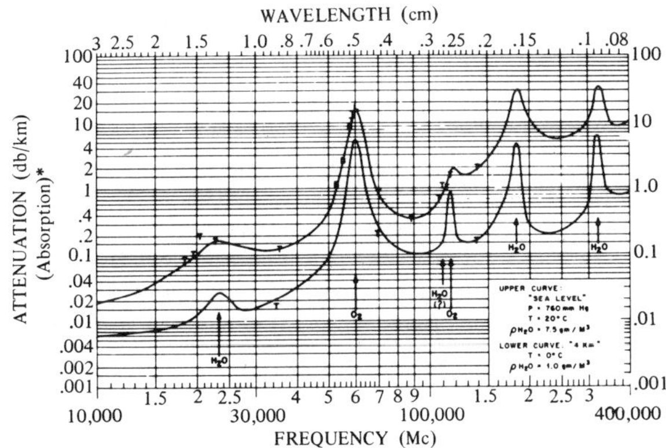

A

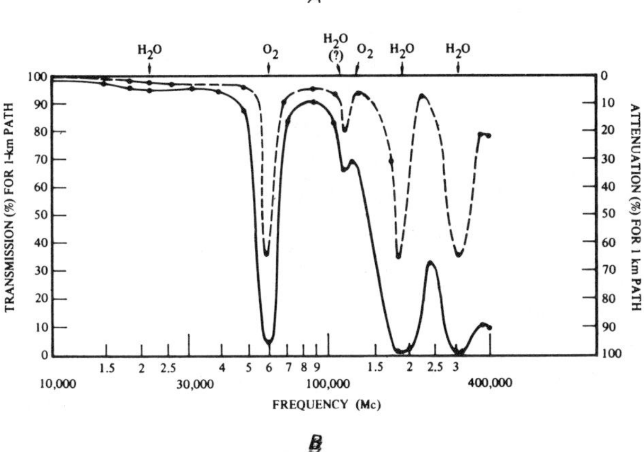

B

Figure 11 Attenuation (absorption) versus frequency. (*A*) Logarithmic plot. (*B*) Linear plot. (The term "attenuation" is construed to include losses from a beam of radiation by all possible mechanisms, principally absorption and scattering by any of several modes. Since at these wavelengths scattering is relatively insignificant, for normal atmospheres absorption can be read for attenuation with negligible error.) (Adapted from Meyer, 1966)

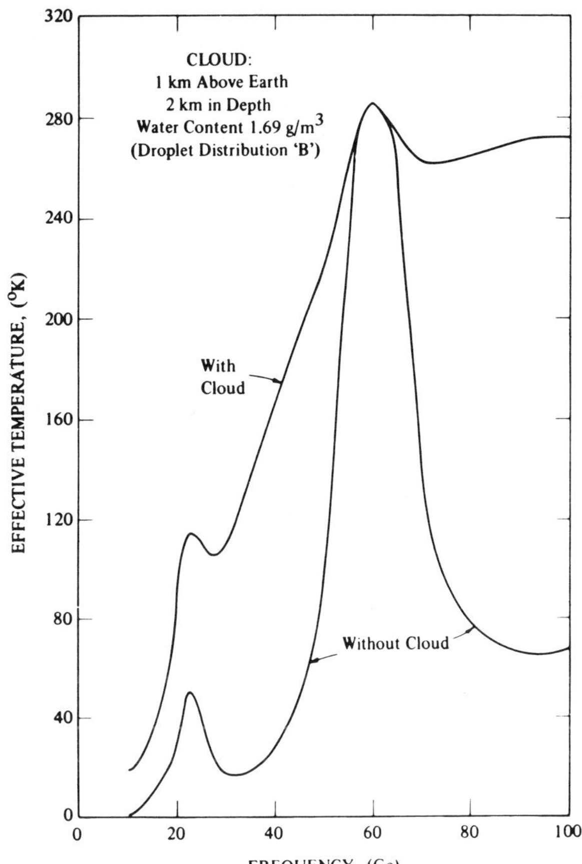

Figure 12 Zenith sky effective temperature with cloud and without cloud. (Edison, 1966)

It must be realized, however, that there are applications for which sensors operating in one of the nonwindow regions are possible solutions. For example, if it were the purpose of the instrument to monitor the amount of water content in the atmosphere, preferred operation would be in one of the water bands, e.g., at 23 GHz. The effective sky temperature at this frequency can be directly correlated with the amount of water vapor content (Cummings and Hull, 1966) (see Fig. 12).

Until after about 1958, atmospheric transmission data at frequencies greater than 140 GHz were not available. Recent developments in microwave-receiver technology have made it possible to make atmospheric propagation measurements at higher frequencies. As a result, some atmospheric transmission data are now available well into the submillimeter regions (Chang and Lester, 1966; Williams and Chang, 1966; Hoffman, 1966; Frenkel and Woods, 1966). In the centimeter and millimeter regions, atmospheric effects in the water- and oxygen-absorption bands at the lower frequencies have been studied in great detail (University of Texas, 1963; Tolbert, Stratton, and Simons, 1964).

Microwave Instrumentation

A microwave radiometer measures the radiant energy that emanates from a target within its field of view. The microwave instrumentation and the concepts developed to make these measurements are discussed in this section. Detection sensitivities of state-of-the-art instruments are summarized, the concept of apparent antenna temperature is reviewed, a basic radiometer is described, and finally, the scanning parameters are introduced, all to provide the reader with a basic understanding of microwave instrumentation.

The power received by the antenna of a radiometer and available at its output is commonly expressed in terms of an *apparent antenna temperature* (Williams and Chang, 1962; Ewen, 1967; McGillem and Seling, 1963). This temperature is the thermodynamic incident temperature, T_i, of a Johnson noise source, which will deliver an equivalent power to the remainder of the system, the power being proportional to $kT_i\Delta f$, where k is Boltzmann's constant and Δf is the electrical bandwidth of the electronics.

This can be justified by the following approach. Assume that an antenna is enclosed by a blackbody at a thermodynamic temperature T . . . and the antenna feed is terminated by a matched load, also at temperature T. From the second law of thermodynamics, it is known that the net power flow between the termination and the blackbody must be zero. Therefore, the conclusion can be drawn that the power delivered to the system is equal to that generated by the termination, i.e., the available Johnson noise power. These results are generally valid for wavelengths greater than 1 mm. For shorter wavelengths, the approximation errors become large and invalidate the results obtained (Williams and Chang, 1962; Ewen, 1967).

A number of radiometric system configurations have been developed in the past few years to measure accurately the apparent antenna temperature, T_A, and, more specifically, to measure the change in apparent temperature when a specific target enters the primary antenna pattern. Of these proposed configurations (McGillem and Seling, 1963) the input-modulated system proposed by Dicke in 1946 has since been widely and frequently used. The discussions presented stem from this system concept.

A functional block diagram of the input-signal-modulated system, a Dicke radiometer, is illustrated in Figure 13. The input signal proportional to T_A and a known reference signal proportional to T_R are alternately sampled at the input switch. The resulting radio-frequency

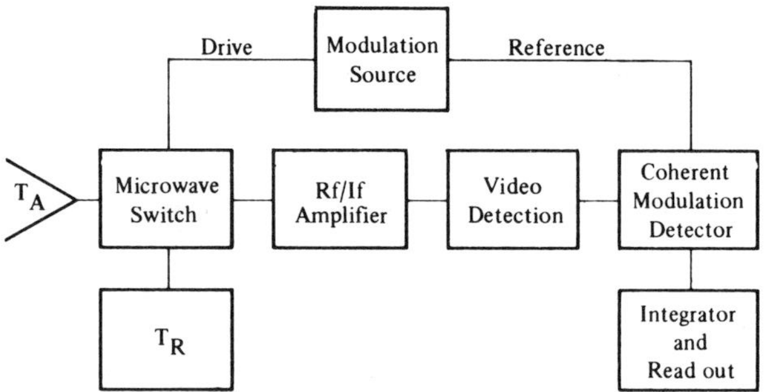

Figure 13 Functional block diagram of an input-modulated (Dicke) radiometer.

signal, which is amplitude modulated at the input switching rate, is amplified and detected. The modulation envelope is then sent to a coherent detection system that demodulates the signals. The filtered dc output signal then becomes a linear measure of the difference in signal between T_R and T_A. The sign of the signal and its relative magnitude are thus a measure of T_A.

The prime advantage of the Dicke system over others proposed is that output signal variations due to short-term gain instabilities of the receiver can be greatly reduced. Thus, considerable improvements in system performance can be realized without employment of elaborate gain-stabilizing devices.

To evaluate the performance of any radiometer, a signal-to-noise (S/N) analysis of the system must be performed by use of standard noise theories and definitions (Rand Corp., 1963; Edison, 1966). For $S/N = 1$, the minimum detectable root mean square signal is given by

$$\Delta T = \frac{T_A + (F-1)T_a}{(KB/B_o)^{1/2}} \tag{6}$$

or as

$$\Delta T = K_1 \frac{(T_A + T_{eff})}{(B\tau)^{1/2}} \tag{7}$$

where B = input signal (predetection) noise bandwidth

B_o = output signal (postdetection) noise bandwidth

K, K_1 = detection constants, determined by receiver configuration

F = receiver noise figure (McGillem and Seling, 1963; Steven, 1964)

ΔT = minimum detectable root mean square temperature change

T_A = apparent antenna temperature

T_a = noise figure normalization temperature (290°K) (McGillem and Seling, 1963)

T_{eff} = equivalent system input noise temperature

τ = integration time constant of the postdetection electronics (resistance-capacitance network is assumed)

Equations 6 and 7 are valid for any of the radiometer configurations described by McGillem and Seling (1963). The values of K_1 are given in Table 1, and the values of K can be computed, given that $\tau = \frac{1}{4} B_o$.

From Equation 7 it is evident that to obtain optimum system performance, i.e., a low value for ΔT, a large predetection bandwidth (B) and/or a long integration period (τ) is desired. In addition, a low receiver-noise figure (F) is desirable. As would be suspected, a trade-off between bandwidth (B) and noise figure (F) is necessary for optimum performance since, in general, receiver noise figures increase with increased bandwidths.

To evaluate various types of receivers, i.e., those using superheterodyne techniques, masers, parametric amplifiers, and/or tunnel diodes for the first stage following

Table 1 Radiometer Constants for Various Circuit Configurations

System Type	Modulation Waveform	Demodulation Waveform	Radiometer Constant (K_1)
Unmodulated	—	—	1/2
Postdetection modulated	Square wave	Sine wave	$2/\pi^2$
Signal modulated	Square wave	Square wave	1/8
	Square wave	Sine wave	$1/\pi^2$
	Square wave	([a])	$1/\pi^2$
	Sine wave	Sine wave	1/16
Cross correlation	—	—	1/4
Two-channel subtraction	Square wave	Square wave	1/4
	Square wave	Sine wave	$2/\pi^2$

From McGillem and Seling, 1963.
[a] Assuming the signal has passed through a narrowband filter before demodulation. If no filter is used, then K_1 is 1/8.

the input signal switch (see Fig. 14), a commonly used figure of merit, C, is given by

$$C = \Delta T(\tau)^{1/2} = \frac{K_1(290F)}{B^{1/2}} \tag{8}$$

This is obtained by substituting Equation 6 for ΔT and assuming T_A is equal to T_a. Receivers for the microwave spectrum have been evaluated at the University of Michigan[5] by use of Equation 8. . . .

From the results[6] of the evaluation one can predict what system improvements will be of greatest significance. In general, the receivers at long wavelengths ($\lambda > 3$ cm) are wideband and are background limited when one looks earthward; i.e., the apparent temperature (T_A) exceeds the system equivalent input noise temperature (T_{eff}). Improvements in receiver noise levels will not significantly increase the radiometric performance capabilities at these wavelengths. At shorter wavelengths ($\lambda < 3$ cm), however, this is not the case, and receiver improvements can be expected since the basic physical limitations have not been reached. Currently, efforts are being made to develop low-noise receivers for the shorter wavelengths.

Finally, when one considers basic concepts of the signal-modulated system, another noise term must be included to obtain the final value of ΔT. This term is the product of the short-term receiver-gain instabilities and the difference temperature ($T_A - T_R$) between the signal and reference channel; i.e.,

$$\Delta T \propto \frac{\Delta G}{G} \mid T_A - T_R \mid \tag{9}$$

where $\Delta G/G$ represents the short-term receiver-gain instabilities. This term (Hoffman, 1966) can become important when the temperature differential $\mid T_A - T_R \mid$ is large. This term can be negated by controlling the value of T_R such that

$$\frac{\Delta G}{G} \mid T_A - T_R \mid \ll \frac{K_1(T_{eff} + T_A)}{(B\tau)^{1/2}}$$

[5] Specifically, the Infrared and Optical Sensor Laboratory of Willow Run Laboratories, a unit of the University's Institute of Science and Technology.

[6] A graph that appeared in the original article is not included here.

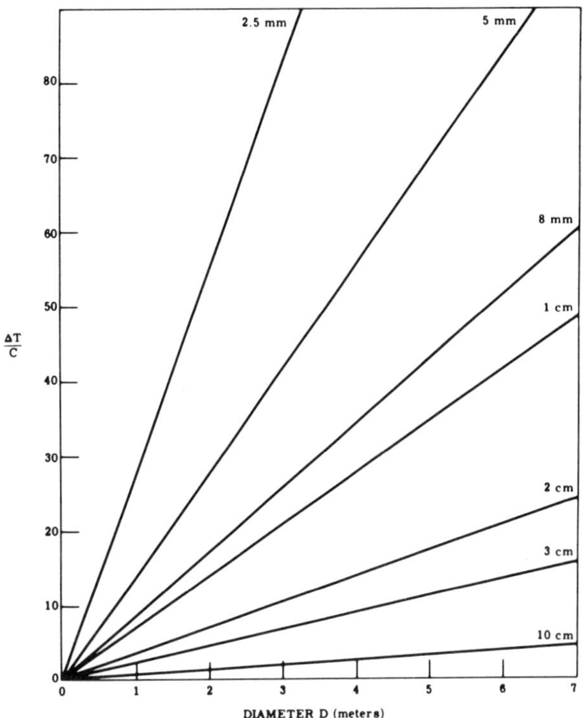

Figure 14 Plot of $\Delta T/C$ versus aperture diameter. This represents the operation that can be expected for a system at a velocity-altitude ratio of 0.02 and a scan angle of 10°. (University of Michigan, 1966)

As such, the control required, if any, is dependent upon the ultimate sensitivity required in ΔT.

To assess the scanning capabilities (University of Michigan, 1966) of the radiometer, the integration time τ or the postdetection bandwidth B_0 must be determined. Those parameters affecting the value of τ that must be evaluated include the antenna resolution β, the relative velocity V, the altitude H of the radiometer relative to the target, and the scan angle α of operation. It can be shown that, for an 87 percent rise in signal level while a point source traverses the main beam of the antenna, $\tau^{1/2}$ is expressed as

$$\tau^{1/2} = \frac{1.2\lambda}{D}\left(\frac{H}{2\alpha V}\right)^{1/2} \tag{10}$$

where $1.2\lambda / D = \beta$ and is the Rayleigh resolution limit of the antenna system. (D is the aperture diameter and λ the wavelength of measurement.) Through substitution it is found that

$$\Delta T = \frac{CD}{1.2\lambda}\left(\frac{2\alpha V}{H}\right)^{1/2} \tag{11}$$

With this information, an evaluation of the system capabilities becomes possible.

To evaluate the usefulness of remote microwave sensors for agricultural applications, system constraints and trade-offs are now considered. These factors (parameters) include the physical constraints, the functional interdependence of the previously considered parameters, and the choice of operational wavelength.

Spatial resolution of the microwave system must be considered in order to evaluate its capabilities as a remote sensor. Diffraction theory implies that it is necessary to use both large apertures and short wavelengths to obtain good angular resolution; however, other constraints such as receiver performance figures (figures of merit), atmospheric transmission factors, and mechanical limitation must also be considered. Equation 11 implies that system spatial resolution and temperature resolution are inverse functions, and, therefore, a compromise is necessary. The selection will be governed by both the spatial and temperature resolution desired as well as other physical limitations; i.e., for a given figure of merit and a given V/H, the temperature-detection ability can be improved only by degrading the spatial resolution or by reducing the scan angle. By assignment of values, then, a parametric plot of $\Delta T/C$ versus D, with wavelength as a parameter, can be established (see Fig. 14). From this [and other] information, . . . the system sensitivity (ΔT) can be obtained directly or by extrapolation if other parameters change.

The atmospheric losses encountered are a function of wavelength (see Fig. 11) and must be considered in the selection of an operational system. In general, the losses are low at wavelengths longer than 2 cm and are relatively low in a window occurring at $\sim$8 mm. The transmission losses have not been included in the equations and will degrade the operation to a varying extent. The degree of degradation will depend on the operating wavelengths chosen.

The mechanical constraints that must be considered apply primarily to the antenna configuration and size. For aircraft installations, maximum antenna apertures of 2 meters are considered practical, and antenna surface tolerances can be maintained into the millimeter region. Several possibilities exist for space antenna structures. They may be designed into the surface of the vehicle; an antenna may be assembled outside the vehicle by a man; or the antenna may be an unfurlable or inflatable type. It is generally believed that the physical size of the collecting aperture should be 20 meters or less. In general, the achievable size of the antenna is a function of wavelength; until 1970, practically achievable antenna diameters are expected to be as shown in Figure 15. The size limits for an antenna are imposed by the surface tolerances that can be maintained. The angular resolution versus diameter of an antenna is plotted in Figure 16 with wavelength as a parameter. Also shown are the practical limits on a space installation that can be assumed for the 1970 time frame.

Selections of the wavelength for operation and the aperture size would be determined primarily by the resolution needed to map particular sources. Thus, if one assumes the use of the maximum aperture attainable at each wavelength, the wavelength to be used is determined by the resolution requirements. Then the system's temperature-resolution capabilities can be established, and the value of the system will thus be determined for each particular usage accordingly; conversely, since optimum detectivity implies a minimum aperture size (see Equation 11), resolution requirements dictate operation at the shortest possible wavelength. Therefore, the selection would generally be in the 8-mm region, necessitating slightly increased atmospheric losses.

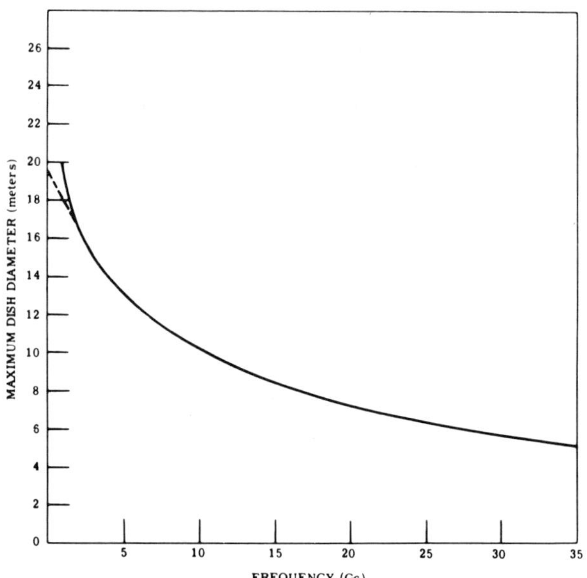

Figure 15 Attainable dish diameter versus frequency. For the 1970 time frame, the question of maximum dish diameter attainable in space with deployable antennas was explored at the University of Michigan in 1964 with these results. (University of Michigan, 1966)

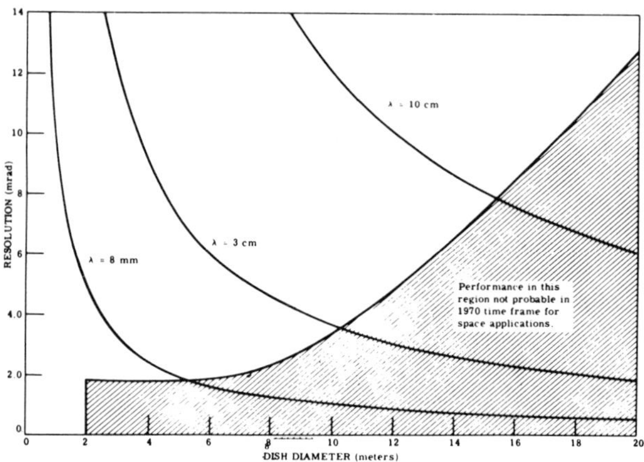

Figure 16 Plot of resolution versus antenna size. (University of Michigan, 1966)

Considering the system trade-offs discussed above, an airborne and a space platform for remote sensing of terrain features are examined. Typical airborne-system parameters for an antenna aperture diameter D of 1 meter are:

$$\lambda = 8\text{ mm}$$
$$C = 5 \times 10^{-2}$$
$$V/H = 0.02$$
$$\alpha = 10^\circ \ (0.175\text{ rad})$$

With these parameters, a temperature sensitivity of 0.5°K and a spatial resolution of 10 mrads are achievable. These system capabilities are adequate for remote sensing of terrain features. Among the features that should be detectable are field boundaries, roadways, water–land boundaries, snow cover, and buildings.

Typical space-system parameters are identical to those for an airborne system, except that the antenna aperture size can be larger. For this example the maximum available aperture (D=5.0 meters) was assumed, with a resulting spatial resolution of 2 mrads and a temperature sensitivity of approximately 2°K. At a satellite altitude of 200 miles, the system's angular resolution implies that a minimum ground-patch resolution of 0.4 mile is achievable. These capabilities would seem to be inadequate for many agricultural applications. However, a number of agricultural problems are amenable to study with only coarse resolutions; among them are obtaining information on soil temperatures and soil moisture over extended regions and hydrologic information such as area of snowpack.

In general, it would appear that, from airborne platforms, instrument capabilities are currently adequate for agricultural applications. However, for operation from space platforms, instrument capabilities must be improved before feasibility can be shown.

REFERENCES

Breeden, K. H., Rivers, W. K., and Sheppard, A. P. A submillimeter interference spectrometer: characteristics, performance and measurements. Georgia Inst. Technol., NASA CH-195, 1966.

Chang, S. Y., and Lester, J. D. Radiometric measurement of atmospheric adsorption at 600 Gc/s. *Proc. IEEE* 54:459, 1966.

Conway, W. H., et al. A gradient microwave radiometer flight test program. *Proc. 2nd Symp. Remote Sensing of Environment.* Ann Arbor: Univ. of Michigan, Inst. Sci. Technol., Rep. No. 4864-3-x, 1963.

Cosgriff, R. L., Peake, W. H., and Taylor, R. C. Terrain scattering properties for sensor system design (Terrain Handbook II). Eng Exp. Sta., Columbus: Ohio State University, 1960. Pp. 16–18.

Cummings, C. A., and Hull, J. W. Microwave radiometric meteorological observations. North American Aviation, Inc., Columbus Div. Rep. No. NA 66H-371, 1966.

Dicke, R. H. The measurement of thermal radiation at microwave frequencies. *Rev. Sci. Instr.* 17:268, 1946.

Edison, A. R. Calculated cloud contributions to sky temperature at millimeter-wave frequencies. U.S. Dept. Commerce, Natl. Bur. Standards. Boulder Labs., Boulder, Colo. NBS Rep. 9138, 1966.

Ewen, H. I. State of the art of microwave and millimeter wave radiometric sensors. *Int. Symp. on Electromagnetic Sensing of the Earth from Satellites.* Brooklyn, N.Y.: Polytechnic Press, Polytechnic Institute, 1967.

Frenkel, L., and Woods, D. The microwave absorption by H_2O vapor and its mixture with other gases between 100 and 300 Gc/s. *Proc. IEEE* 54:489, 1966. Gardon, R. The emissivity of transparent materials. *J. Am. Ceramic* Soc. 39:278, 1956.

Hoffman, L. A. Propagation observations at 3.2 millimeters. *Proc. IEEE.* 54:449, 1966.

Lewis, E. A., Casey, J. P., and Vaccaro, A. J. Polarized radiation from certain thermal emitters. Electronics Research Directorate, Air Force Cambridge Research Center, Air Force Research and Development Command. AFCRC Tech. Rep. No. 54-G, 1954.

McGillem, C. D., and Seling, T. V. Influence of system parameters on airborne microwave radiometer design. *IEEE Trans. on Mil. Electronics* MIL-7:(4), 1963.

McMahon, H. O. Thermal radiation from partially transparent reflecting bodies. *J. Opt. Soc. Am.* 40:376, 1950.

Meyer, J. W. Radio astronomy at millimeter and submillimeter wavelengths. *Proc. IEEE* 54:484, 1966.

Nicodemus, F. E. Directional reflectance and emissivity of an opaque surface. *Appl. Opt.* 4:767, 1965.

Planck, M. *Ann. Phys.* 4:553, 1900.

Putley, E. H. Indium antimonide submillimeter photoconductive detectors. *Appl. Opt.* 4:649, 1965.

Rand Corp. Atmospheres, clouds and spectra. *The Application of Passive Microwave Technology to Satellite Meteorology: A Symposium,* Sec. II. Santa Monica, Calif. Memorandum RM-3401-NASA, 1963.

Sleven, R. L. A guide to accurate noise measurement. *Microwaves.* 1964. Pp. 14–21.

Straiton, A. W., Tolbert, C. W., and Britt, C. O. Apparent temperatures of some terrestrial materials and the sun at 4.3 mm. *J. Appl. Phys.* 29:776, 1958.

Tolbert, C. W., Straiton, A. W., and Simons, R. A. The attenuation and emission in the earth's atmosphere of the complex of 60 Gc/s oxygen lines. Univ. of Texas, Elec. Eng. Res. Lab., Contract No. AF 33 (657)-7333, Project No. 1002, AD 39 245, 1964.

University of Michigan. Peaceful uses of earth-observation spacecraft. Vol. III: Sensor requirements and experiments. Univ. of Michigan, Willow Run Laboratories, Rep. No. 7219-1-F (III). NASA Contract NASw-1084, 1966.

University of Texas. Attenuation and emission of 58 to 62 kMc/s frequencies in the earth's atmosphere. Elec. Eng. Res. Lab. Contract No. AF 33 (657)-7333, Project 1002. AD 400 954, 1963.

Williams, R. A., and Chang, W. S. C. An analysis of the interferometric submillimeter radiometer. Ohio State Univ. Res. Found. Antenna Lab., Dep. Elec. Eng., Columbus. Rep. No. 1039–9. Grant No. NsG-74-60, 1962.

Williams, R. A., and Chang, W. C. Observation of solar radiation from 50 μ to 1 mm. *Proc. IEEE* 54:462, 1966.

Experiments in passive remote sensing at microwave frequencies have indicated useful applications in meteorology, oceanography, geology, and the study of soils and their moisture content. The applications in meteorology are best developed and most useful. Meteorologic observations of the atmosphere can be made from the ground or from space platforms. The observations include measurement of atmospheric temperature profiles and distribution of water vapor and ozone. The measurements are made in the shorter wavelength portion of the microwave spectral region near 1 cm; here the microwave resonances of water vapor, oxygen, and ozone occur. Research has shown that many other components of the atmosphere also emit measurable quantities of microwave energy but these are difficult to observe at present and are still considered only experimental.

Most meteorologic observations must be made near the ground. The atmosphere, however, is three-dimensional and the conditions at upper levels are significant in predicting weather developments. Observations of upper atmospheric levels have shown that microwave radiometric measurements are capable of sensing the direction of cloud mass movement, of measuring degrees of cloudiness, humidity, fog, and rain, and of delineating cloud mass.

We gather very little meteorologic data over the ocean areas of the earth—areas comprising over 70 percent of the earth's surface—and yet the oceans are the source areas for most of the moisture that falls on the land. Over the oceans humidity, heat fluxes, and surface winds are significant factors in measuring the energy released in the earth's atmosphere by thermal radiation and by moisture evaporation. This heat exchange between the earth's surface and the atmosphere over the oceans determines in large measure the global weather. Surface winds, humidity, and heat fluxes must be monitored in these inaccessible areas if accurate predictions and eventual control of weather is to be accomplished. It is possible that this kind of data could be collected by passive microwave radiometers in unattended buoys placed at carefully selected locations.

31-Passive Remote Sensing at Microwave Wavelengths

DAVID H. STAELIN

Perhaps the earliest passive microwave observations of the atmosphere were made in 1945 by Dicke (Dicke, 1946; Dicke et al., 1946) and his associates at the M.I.T. Radiation Laboratory. Their observations at 1.00-, 1.25-, and 1.50-cm wavelengths were generally consistent with the results expected on the basis of radiosonde measurements and the theoretical expressions of van Vleck (1947 a, b). Since that time many workers have measured atmospheric properties using wavelengths ranging from millimeters to meters.

In this paper ways in which the wavelength interval 0.1 to 100 cm can be utilized for passive ground-based or space-based meteorologic measurements are reviewed. Each region of the electromagnetic spectrum has unique properties which can be exploited for purposes of remote sensing, and in the microwave region these properties include an ability to penetrate clouds and to interact strongly with O_2, H_2O, and other atmospheric constituents. The interactions considered here are those of O_2, H_2O, O_3, clouds, precipitation, sea state, surface temperature, and surface emissivity. Barrett (1963) has listed many other constituents which also interact, such as OH, CO, N_2O, and SO_2, but these are presently difficult to observe and are not considered in this review.

The review has two major parts. The first part reviews the physics of the interactions, the mathematics of data

From *Proceedings of the Institute of Electrical and Electronics Engineers* 57:427–439, 1969. Reprinted with permission of the author and the Institute of Electrical and Electronics Engineers.

interpretation, and the state-of-the-art in instrumentation. The second part is applications-oriented and reviews the types, accuracy, and relevance of possible meteorologic measurements. It also contains several suggestions for further work.

Technical Review

Physics of the Interactions

The topics in this section include (1) the relevant equations of radiative transfer, (2) the absorption coefficient expressions for O_2, H_2O, O_3, clouds, and precipitation, and (3) the microwave properties of the terrestrial surface. Both the theoretical expressions and the corresponding experimental verifications are described and referenced. In general, only a fraction of all possible references are given.

The Equations of Radiative Transfer

At wavelengths longer than 1 mm the Planck function can be simplified because $h\nu << kT$ for all situations of meteorologic interest. The symbols h, ν, k, T are Planck's constant, frequency, Boltzmann's constant, and absolute temperature, respectively. In this paper rationalized MKS units are used unless stated otherwise. As a result of this low-frequency approximation the power P (watts) received by a microwave radiometer immersed in a blackbody enclosure at temperature T equals kTB, where B is the bandwidth (Hz). Since temperature has direct physical significance, most radiometers are calibrated in terms of antenna temperature T_A(°K), which is the temperature a blackbody at the antenna terminals must have to produce a signal of the observed power P. That is,

$$T_A = P/kB \tag{1}$$

The measured antenna temperature T_A is in turn related to the angular distribution of power incident upon the antenna. The appropriately polarized power incident upon the antenna from any given direction can be described by an equivalent brightness temperature $T_B(\nu, \theta, \phi)$ (°K). The antenna temperature is then an average of the brightness temperature weighted over 4π steradians according to the antenna gain function $G(\nu, \theta, \phi)$. This is,

$$T_A(\nu) = \frac{1}{4\pi}\int_{4\pi} T_B(\nu, \theta, \phi)G(\nu, \theta, \phi)d\Omega \tag{2}$$

The gain of any particular antenna can be calibrated on an appropriate test range.

The equation of radiative transfer can be used to relate the brightness temperature T_B to the atmospheric composition and temperature $T(z)$ along the line of sight, and to the brightness temperature $T_0(\theta, \phi)$ of the medium beyond the atmosphere. To an excellent approximation the equation of radiative transfer for a slightly lossy medium is

$$T_B(\nu) = T_0 e^{-\tau(\nu)} + \int_0^{z_{max}} T(z)\exp\left[-\int_0^z \alpha(z)dz\right]\alpha(z)dz \tag{3}$$

where scattering and variations in index of refraction have been neglected, and the variables are as defined in the following text. The equation loses validity primarily in the presence of relatively large particles like raindrops or snowflakes, and at 1-mm wavelength in the presence of some clouds. In Equation (3), $\alpha(z)$ (m^{-1}) is the absorption coefficient of the atmosphere, and τ (ν) is its integral; i.e., the total opacity of the atmosphere. Equation (3) expresses the fact that the brightness temperature in any given direction is the sum of the background radiation and the radiation emitted at each point along the ray trajectory, each component attenuated by the intervening atmosphere. The equations of radiative transfer with and without scattering are discussed by Chandrasekhar (1960) and are related to radio astronomy by Shklovsky (1960).

In certain cases the equation of radiative transfer is more appropriately expressed in matrix form. This form is desirable, for example, when describing the effects of Zeeman-split resonance lines upon the flow of polarized radiation. One such formulation was developed and applied by Lenoir (1967, 1968) to 5-mm wavelength measurements of the mesospheric temperature profile.

Opacity Expressions for Atmospheric Constituents

The absorption spectrum of the atmosphere is dominated by the H_2O resonance at 22.235 GHz and the O_2 complex near 60 GHz. Strong resonances also occur at frequencies above 100 GHz. Other resonant constituents contribute only weakly.

A general reference for microwave spectroscopy is Townes and Schawlow (1965) and good references for general atmospheric and surface effects include Stratton (1941) and Kerr (1951). Expressions for the absorption coefficients of atmospheric constituents are usually formed by empirically choosing constants in the quantum-mechanical formulation for opacity in order to provide the best agreement with experiment. If experimental verification of the absorption coefficient expressions under all possible conditions were practical, one could dispense with quantum mechanics and electromagnetism entirely. The degree to which the absorption coefficients are presently based upon theory rather than upon experiment varies considerably depending upon the constituents and the atmospheric conditions.

In the case of the oxygen complex near 5-mm wavelength the most widely used absorption coefficient is that of Meeks and Lilley (1963) who revised and updated the expressions developed by van Vleck (1947a). Gautier and Robert (1964) and Lenoir (1965) extended these expressions to include Zeeman splitting appropriate above 40-km altitude. The absorption coefficient was determined using the theoretical line-shape factor of van Vleck and Weisskopf (1945), and those linewidth parameters and line frequencies which were measured for the stronger lines. The linewidths have been measured at pressures near 1 atm and at low pressures. The dependence of linewidth on pressure and temperature has also been measured. A complete bibliography by Rosenblum (1961) references the literature on microwave absorption by oxygen and water vapor up to 1961.

Although much experimental work has been done, there is some need for still more precise laboratory measurements, particularly of atmospheric compositions over the full range of temperatures and pressures of meteorological interest.

The absorption coefficient of oxygen in the atmosphere can also be tested by direct measurement of antenna temperatures near 5-mm wavelength. Such measurements have been made from both ground-based and balloon-borne radiometers. The most recent such measurements are those of Lenoir et al. (1968). They examined the 9^+ resonance line at 61.151 GHz with a three-channel microwave radiometer mounted on balloons which flew to altitudes of 35 km. Observations were made at zenith angles of 60°, 75°, 120°, and 180°, and at frequencies separated from 61.151 GHz by ±20, ±60, and ±200 MHz. The most precise tests of the theoretical opacity expressions are those measurements made looking upward. In this case differences between the theoretical and experimental values of atmospheric opacity as small as 5 percent could be detected in certain regions of the atmosphere. No discrepancies this large were noted 60 and 200 MHz from resonance when the balloon was in the altitude region 25 to 32 km. One such data set from Lenoir et al. (1968) is shown in Figure 1. The experiments were less sensitive to opacity uncertainties in other regions of the atmosphere or spectrum, but in these regions the differences between the theoretical and experimental brightness temperatures were generally less than 5°K, the approximate experiment accuracy.

Sensitive solar absorption measurements at 53.4 to 56.4 GHz by Carter et al. (1968) have permitted van Vleck-Weisskopf line shape to be refined slightly.

Water vapor has transitions at 22.237 and 183.3 GHz, in addition to many transitions at higher frequencies. The opacity expressions for the 22-GHz line were developed by van Vleck (1947b; Kerr, 1951) and were refined by Barrett and Chung (1962) on the basis of the experimental findings of Becker and Autler (1946) and the theoretical calculations of Benedict and Kaplan (1959). Relatively good agreement between theory and

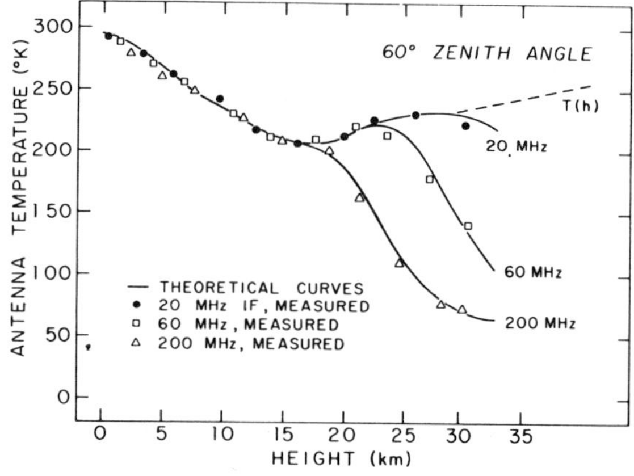

Figure 1 Experimental and theoretical values of antenna temperature at three IF frequencies for a 61.151 GHz balloon-borne radiometer. (After Lenoir et al., 1968)

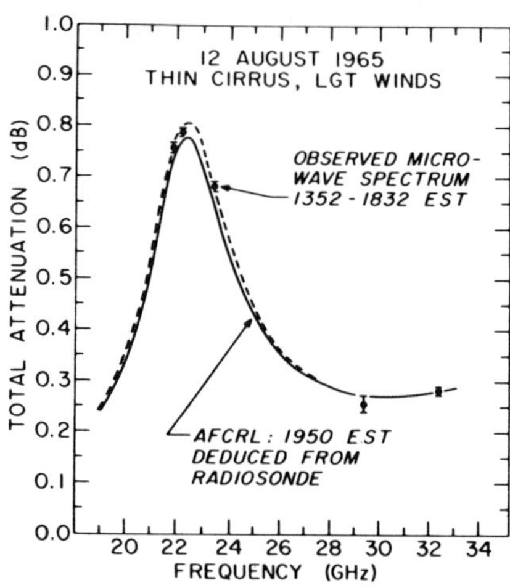

Figure 2 Zenith attenuation spectrum deduced from direct microwave measurements; and spectrum computed using temperature and humidity data from a radiosonde launched near the radiometer. (After Gaut, 1967)

experiment was obtained by combining the van Vleck and Weisskopf (1945) line shape with a nonresonant term which corresponds to contributions from the far wings of all the resonances at other frequencies. The magnitude of the nonresonant term was selected to provide the best agreement with the measurements of Becker and Autler, which were made at pressures near 1 atm. Gaut (1967) and Croom (1965a) have also developed expressions for the 22-GHz absorption coefficient.

Gaut (1967) has developed an expression for the absorption coefficient near the 183.310-GHz water vapor resonance. He included the effects of several neighboring lines by using in part the line strength calculations of King et al. (1947), the van Vleck-Weisskopf line shape, and measurements of Rusk (1965), Frenkel and Woods (1966), and Hemmi (1966). Croom (1965b) has done similar calculations. The review and bibliography by Rosenblum (1961) is a useful reference for work published prior to 1961.

Direct measurements of the atmosphere have been made primarily near 22 GHz. Most of these measurement programs have been handicapped because atmospheric water vapor varies somewhat more rapidly than the observing frequency can be changed, or more rapidly than accurate correlative measurements can be made. The remedy is improved instrumentation and stable meteorologic conditions. An extensive series of observations has been made with a five-frequency radiometer by Staelin (1966) and Gaut (1967). Five frequencies were observed simultaneously in absorption against the sun and permitted opacity measurements with an approximate accuracy of 0.02 dB or less than 5 percent of the total opacity. Meaningful comparison of the measured opacity spectrum and that spectrum expected on the basis of simultaneous radiosondes was possible only on days characterized by very stable meteorologic conditions. One such comparison made by Gaut (1967) is illustrated in Figure 2. Examination of nine spectra measured under these stable conditions

Table 1 Working Values of Attenuation in dB · km^{-1} (one way)

Absorber	Temp. (°C)	3.2 cm	1.8 cm	1.24 cm	0.9 cm
Rain[a]					
(R in mm · h^{-1})	18[b]	$0.0074R^{1.31}$	$0.045R^{1.14}$	$0.12R^{1.05}$	$0.22R^{1.00}$
Snow[a]	0	$3.3 \times 10^{-5}R^{1.6}$	$3.32 \times 10^{-4}R^{1.6}$	$1.48 \times 10^{-3}R^{1.6}$	$5.33 \times 10^{-3}R^{1.6}$
(R in mm · h^{-1} of		$68.6 \times 10^{-5}R$[c]	$12.2 \times 10^{-4}R$[c]	$1.78 \times 10^{-3}R$[c]	$2.44 \times 10^{-3}R$[c]
melted water)	–10	$3.3 \times 10^{-5}R^{1.6}$	$3.32 \times 10^{-4}R^{1.6}$	$1.48 \times 10^{-3}R^{1.6}$	$5.33 \times 10^{-3}R^{1.6}$
		$22.9 \times 10^{-5}R$[c]	$4.06 \times 10^{-4}R$[c]	$0.59 \times 10^{-3}R$[c]	$0.81 \times 10^{-3}R$[c]
	–20	$3.3 \times 10^{-5}R^{1.6}$	$3.32 \times 10^{-4}R^{1.6}$	$1.48 \times 10^{-3}R^{1.6}$	$5.3 \times 10^{-3}R^{1.6}$
		$15.7 \times 10^{-5}R$[c]	$2.80 \times 10^{-4}R$[c]	$0.41 \times 10^{-3}R$[c]	$0.56 \times 10^{-3}R$[c]
Water cloud	20	$0.0483\,M$	$0.128\,M$	$0.311\,M$	$0.647\,M$
(M in g · m^{-3})	10	$0.0630\,M$	$0.179\,M$	$0.406\,M$	$0.681\,M$
	0	$0.0858\,M$	$0.267\,M$	$0.532\,M$	$0.99\,M$
	–8	$0.112\,M$ (extrapolated)			
Ice cloud	0	$2.46 \times 10^{-3}\,M$	$4.36 \times 10^{-3}\,M$	$6.35 \times 10^{-3}\,M$	$8.7 \times 10^{-3}\,M$
(M in g · m^{-3})	–10	$8.19 \times 10^{-4}\,M$	$1.46 \times 10^{-3}\,M$	$2.11 \times 10^{-3}\,M$	$2.9 \times 10^{-3}\,M$
	–20	$5.63 \times 10^{-4}\,M$	$1 \times 10^{-3}\,M$	$1.45 \times 10^{-3}\,M$	$2 \times 10^{-3}\,M$

[a] These are empirical equations. For the more exact relations on which they are based, see Atlas et al. (1952).
[b] The effect of temperature is discussed in Atlas et al. (1952). The effect of nonsphericity on Q_s and Q_a has been neglected.
[c] As long as snowflakes are in the Rayleigh region. A value of R = 10 or $R^{1.6}$ = 39 is an upper limit for snowfall rates.

indicate that the theoretical expressions of Barrett and Chung (1962) are accurate to better than 5 percent for typical atmospheric conditions.

Ozone has many spectral lines at microwave frequencies, but because of the low abundance of ozone these lines are difficult to detect. Barrett (1963) listed resonances of ozone at frequencies (GHz) of 9.201, 10.226, 11.073, 15.116, 16.413, 23.861, 25.300, 25.511, 25.649, 27.862, 28.960, 30.052, 30.181, 30.525, 36.023, 37.832, 42.833, 43.654, 96.229, 101.737, and 118.364. Absorption coefficient expressions have been developed by Caton et al. (1967), Caton et al. (1968), and Weigand (1967). These theoretical expressions are based upon line frequencies and line strengths calculated by Gora (1959) and upon the van Vleck-Weisskopf line-shape assumption. At present five resonances in the terrestrial atmosphere have been observed. Mouw and Silver (1968) observed the 36.025-GHz line in absorption against the sun, Caton et al. (1967) observed the 37.836-GHz line in absorption and the 30.056-GHz line in emission, Barrett et al. (1967) observed the 23.861-GHz line in emission, and Caton et al. (1968) observed the 101.737-GHz line in both emission and absorption. For spectral resolution of a few megahertz, the line amplitudes in emission at zenith are generally a few tenths of a degree Kelvin except for the line at 101.737 GHz, which had an amplitude of a few degrees. The most precise line measurement was that made at 101.737 GHz by Caton et al. (1968), and their figure showing the line observed in absorption against the sun at 64° zenith angle is reproduced here in Figure 3. In order to match the theoretical and experimental curves they found it necessary to multiply the amplitude of the theoretical line by a factor of 1.5. It is uncertain whether this difference is due to errors in the opacity expression or errors in the assumed model atmosphere. More measurements of ozone would be very helpful.

Clouds exhibit no resonances in the microwave region of the spectrum but instead absorb by nonresonant processes. Mie (1908), Stratton (1941), and others have developed expressions for the absorption and scattering cross sections of dielectric spheres, and Ryde and Ryde (1945) and Haddock (1948) have then used the dielectric constants for water given by Saxton and Lane (1946) in order to compute the microwave absorption coefficients for clouds and rain. Goldstein (in Kerr, 1951) and Atlas et al. (1952) have reviewed and summarized much of this work. These equations can be extended to include snow and ice clouds through use of the refractive index of ice measured by Cumming (1952). These results were summarized by Atlas et al. (1952) and are reproduced here in Table 1. These studies show that scattering in clouds can generally be neglected if the wavelength is more than 30 times the droplet diameter.

Such theoretical calculations are approximate (Kerr, 1951) because they must be based upon some assumed distribution of particle sizes, shapes, etc. The expressions

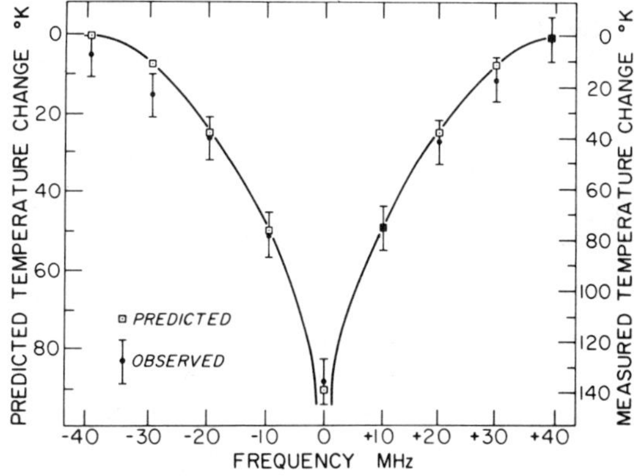

Figure 3 Solar absorption spectrum showing the 101.7 GHz resonance of atmospheric ozone. The predicted values were increased by a factor of 1.5 to match the measured values. (After Caton et al., 1968. © The University of Chicago)

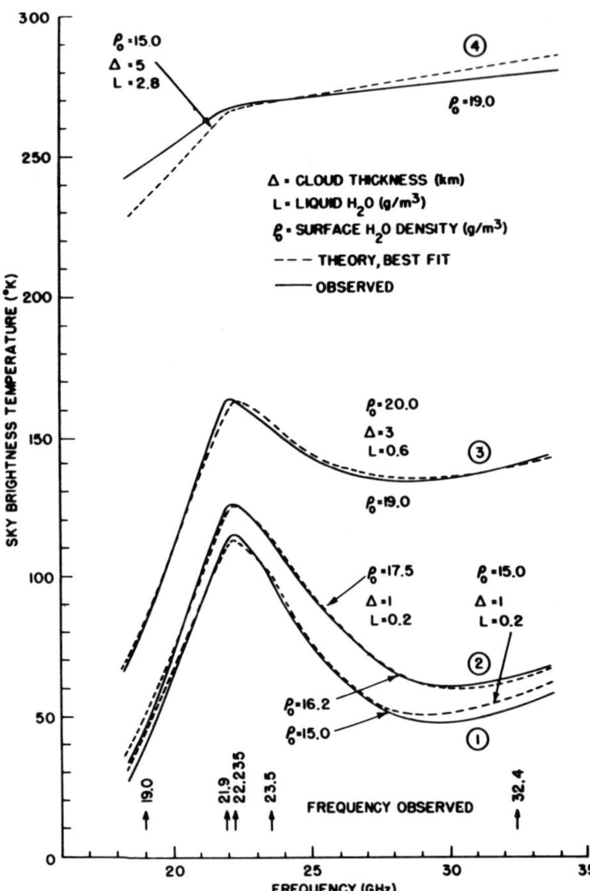

Figure 4 Theoretical and experimental atmospheric emission spectra obtained at 30° elevation angle during a frontal passage. (After Toong, 1967)

are less sensitive to drop size at the longer wavelengths. The expressions are best tested by field measurements. Goldstein (in Kerr, 1951) has reviewed several efforts to correlate rainfall rate with path loss measured along a short path. These measurements at wavelengths ranging from 0.62 to 3.2 cm are all within a factor of 2 of the predicted average attenuation. Emission spectra have been measured with a ground-based five-channel microwave radiometer observing at frequencies from 19.0 to 32.4 GHz (Toong, 1967). These spectra showed good agreement with the theoretical spectrum shapes, even through all but the heaviest rainfall. Data obtained at 30° elevation are compared with theoretical spectra in Figure 4. The cloud types inferred from these spectra were consistent with those characteristic of frontal passages.

The Terrestrial Surface

The terrestrial surface both emits and reflects radio waves, and the degree to which it does either is determined primarily by the surface dielectric constant and the detailed structure of the surface. In some cases the emission and reflection coefficients calculated for smooth surfaces are good approximations (Stratton, 1941). This is true for calm ocean and smooth homogeneous dirt surfaces, especially at longer wavelengths. Gaut (1967) and Marandino (1967) have calculated the emissivity of smooth water for a number of temperatures and wavelengths, and Marandino has made similar calculations for several solid surfaces.

Three interesting properties of the ocean surface are worthy of note: (1) the reflectivity is high, near 0.6, (2) the dielectric constant, and hence the emissivity, varies with temperature, and (3) the brightness temperature of the ocean surface is a function of surface roughness, which in turn is correlated with wind velocity. The high reflectivity results in a very low surface brightness temperature over most of the microwave spectrum, and thus spectral lines like H_2O or O_3 can be seen in emission from space. The temperature dependence of the emissivity results in certain combinations of sea surface temperature and frequency having a surface brightness temperature independent of surface kinetic temperature, and at other combinations the dependence can be moderately strong. For water temperatures near 285°K this temperature dependence is weak for wavelengths of 1 and 20 cm, and is stronger near wavelengths of 3 cm. Yap (1965) and Stogryn (1967) have calculated the dependence of surface brightness temperature upon equivalent wind speed, based upon the surface-slope statistics developed by Cox and Munk (1954). For example, Yap computed the effect at 3-cm wavelength at nadir to be approximately 0°K per knot equivalent wind speed, and 60° nadir angle to be 0.5°K for $E_\perp$ polarization and 0°K for $E_{||}$ polarization. At wind speeds over 25 knots whitecaps, foam, and spray become important and the brightness temperature of the sea may increase 15°K or more.

Since the terrestrial surface is difficult to analyze analytically, much emphasis must be placed on field measurement programs. Measurements by Porter (1966) of smooth asphalt and concrete are in reasonable agreement with theory, as are mapping measurements by Catoe et al. (1967). Measurements of rough surfaces have also been made, and qualitative agreement with intuition has been obtained. Very few well-calibrated measurements have been made.

Mathematics of Data Interpretation

One of the central problems in remote sensing is conversion of the measurements into estimates of the desired atmospheric parameters with the smallest possible error. It is often called the inversion problem because it can be interpreted as the problem of inverting the equations of radiative transfer. There are several major approaches which have been taken toward solving this problem. These approaches include (1) the trial and error or library technique, (2) linear analytical procedures, (3) nonlinear analytical procedures, (4) linear statistical methods, and (5) nonlinear statistical methods.

The trial and error or library procedures, as the names imply, simply involve calculation of a large number of theoretical values for the measurements, and selection of that model atmosphere which provides the best agreement with the measurements actually obtained. Although the method is useful because of its conceptual simplicity, there are few inversion problems for which this technique produces completely useful results. Examples include the inversion of Toong (1967) and most studies of the atmospheres of other planets. These are cases where a priori data is not available or where accurate correlative measurements were not feasible.

Linear analytical procedures include those of King (1961), Fow (1964), and others. These methods consist primarily of the solution of a set of simultaneous linear equations which relate N atmospheric parameters to N measured parameters. Writing these equations usually involves linearizing the applicable equations of radiative transfer. These equations are then solved simultaneously to yield estimates for the unknown parameters. Because this may involve the inversion of a nearly singular matrix, it is necessary to have data that is not noisy or to incorporate boundary conditions which lead to a less singular matrix. If the matrix to be inverted is not singular, then the method is often satisfactory.

Nonlinear analytical inversion procedures are similar to the linear procedures, except that nonlinear equations are to be solved. For example, King (1964) considered the problem of approximating the atmospheric temperature profile by a set of slabs of varying thickness, or by a set of ramps, and Chahine (1968) explored iteration procedures.

Linear statistical methods are similar to the linear analytical techniques, except that the final matrix which mutiplies the data vector to yield the parameter vector is selected so as to minimize some error criterion, such as the mean-square estimation error. There are several approaches to the problem. One approach is that of Twomey (1965, 1966), in which the mean-square error between the true and estimated parameter vectors is minimized by heuristic adjustment of a smoothing parameter which represents the degree to which the estimate is smoothed and biased toward the a priori mean. This bias toward the a priori mean avoids inversion of singular matrices.

Perhaps a more elegant technique is that applied by Rodgers (1966) to inversions of atmospheric temperature profiles at infrared wavelengths. In this procedure, also discussed by Deutsch (1965), Strand and Westwater (1968), and others, the mean-square error between the estimated and true parameter vector is minimized without any heuristic parameters. A complete set of a priori statistics is required to implement the procedure, however. The technique can be described simply as follows. Let the measured data and the desired parameters be represented by the column vectors d and p, respectively. In order to reduce the dimensionality of p, a statistically orthogonal set of basis functions for p-space is sometimes sought, as discussed by Rodgers (1966) and others. Let p^* be the estimate of p based on particular measured d, and let p^* be computed by multiplying d by the matrix D, where D is not necessarily a square matrix. The optimum matrix D may then be computed by finding that D which minimizes the mean-square error, $E[(p^* - p)^t (p^* - p)]$, where E is the expected-value operator and the superscript t means transpose. The optimum D can be shown to be given by

$$D^t = C_d^{-1} E[dp^t] \quad (4)$$

where C_d is the data correlation matrix, $E[dd^t]$. This procedure is equivalent to multidimensional regression analysis. In some cases the matrix C_d is sufficiently singular to degrade the inversion. In this case, as observed by Waters and Staelin (1968), the vector d may be transformed to d' such that $E[d'd'^t]$ is diagonal. It can be shown that d' is $R^t d$, where R^t is the transpose of the matrix formed from the normalized column eigenvectors of C_d. The estimate p^* then is given by $D'd'$. Those elements of d' corresponding to diagonal terms in $C_{d'}$ less than the computational noise level can then be discarded.

Linear statistical estimation is optimum in a least-square sense only in special circumstances. The procedure of Rodgers (1966) is optimum if the equations of radiative transfer are linear, and the parameter variations and the instrument errors are both jointly Gaussian random processes. In general these conditions do not apply, and there is room for improvement. For example, in the infrared region of the spectrum the Planck function is inherently a nonlinear function of atmospheric temperature. Even in the microwave region where the Planck function is linear, the equations of radiative transfer are somewhat nonlinear. In addition, the parameter statistics are seldom jointly Gaussian, although they often are approximately so. For nonlinear or non-Gaussian problems the optimum inversion procedures are generally nonlinear.

One quite general nonlinear statistical estimation procedure is Baye's estimate, as described for example by Helstrom (1960). The difficulty is that numerical evaluation of integrals over the parameter space can be impossibly time-consuming for estimation problems involving several data points and many parameters. A more practical technique is the nonlinear estimation procedure outlined by Staelin (1967) and Waters and Staelin (1968). This technique is similar to the linear statistical technique described above, except that the data vector d is replaced by the vector $\phi(d)$, where $\phi_i(d)$ is any arbitrary function of d. This technique permits most nonlinear relations between parameters and data to be inverted, and avoids the need for iteration which is sometimes employed with the linear procedures to obtain some of the benefits of nonlinearity.

Instrumentation

The sensors used for microwave meteorology are generally adaptations of systems developed for radio astronomy, a field in which receiver development is quite active. The major problem areas include sensitivity, absolute accuracy, spectral response, and directional response. In addition each application has constraints of cost, size, weight, power, reliability, and operational simplicity.

Receiver sensitivity may be characterized by the receiver noise temperature T_R, the bandwidth B, and integration time τ, a constant α which is usually in the range 1–3, and the equivalent rms fluctuations at the receiver input ΔT_{rms}. It can be shown (Kraus, 1966) that

$$\Delta T_{rms} = \frac{\alpha(T_A + T_R)}{\sqrt{B\tau}} \ {}^\circ\mathrm{K} \quad (5)$$

This equation can be used to estimate an ultimate limit to receiver sensitivity. For example, a radiometer in space viewing the earth will see an antenna temperature of approximately 300°K. For a noiseless receiver with 1-second integration, RF bandwidth 10^8 Hz, and α equal 2, then ΔT_{rms} is 0.06°K. If 1°K sensitivity is sufficient, then an integration time as short as 0.004 second may be used. These limits have a strong bearing on the ultimate spatial resolution and coverage which can be obtained by radiometers in space. For example, such a receiver in earth orbit at 1000-km altitude could have spatial coverage no better than 50 spots per mile of ground track, with 1°K-receiver sensitivity.

Although the receiver noise temperature T_R is not zero, the sensitivity of receivers is rapidly improving due to the development of improved solid-state components. At wavelengths longer than 3 cm receiver noise temperatures T_R of less than 150°K can be obtained with solid-state parametric amplifiers, and this performance is being extended to wavelengths near 1 cm. With masers receiver noise temperatures lower than 40°K have been obtained in operational systems near 20-cm wavelength. At wavelengths shorter than 3 cm the most common type of receiver is the superheterodyne. These systems have noise temperatures ranging from several hundred degrees at 3-cm wavelength to perhaps 20,000°K at 3-mm wavelength. Recent improvements in Schottky-barrier diodes show promise of reducing superheterodyne noise temperatures to a few hundred degrees for wavelengths as short as 5 mm. Progress in this area is so rapid but unpredictable that planning more than five years in the future is difficult.

The absolute accuracy of most radiometers is generally much less than could be obtained with care. Perhaps the most careful measurements ever made were those performed to measure the cosmic background radiation, as described by Penzias and Wilson (1965), Wilkinson (1967), and others. In these experiments absolute accuracies of approximately 0.1 to 0.2°K (rms) were obtained for antenna temperatures of approximately 6°K. Since the calibration problem becomes simpler as the antenna temperature approaches the physical temperature of the radiometer, such accuracies are obtainable in situations of meteorologic interest. Absolute accuracies of 1°K are readily obtainable without great care.

Almost any arbitrary spectral response can be obtained with a microwave radiometer, although most parametric or other low-noise amplifiers have instantaneous bandwidths less than approximately 200 MHz. Except for limitations imposed by the spectral response of low-noise amplifiers, bandwidths ranging from 1 to 1×10^{10} Hz could be obtained over most of the microwave region of the spectrum. Receivers capable of observing 1 to 400 spectral intervals simultaneously have been built, in addition to single-channel frequency scanning radiometers. For accurate observations of spectral lines the multichannel radiometers generally are more accurate and permit shorter time variations to be monitored.

The directional response of antennas varies considerably depending upon wavelength and antenna size. The halfpower beamwidth of most antennas is approximately $1.3\ \lambda/D$ radians, where λ is the wavelength and D is the antenna diameter. Thus a 5-mm wavelength radiometer with a 1-meter antenna could resolve 7-km spots from a 1000-km orbit. Still higher resolution could be obtained, but mechanical tolerances restrict most antennas to beamwidths greater than 1 to 5 minutes of arc. A second important property of an antenna is its sidelobe level, or the degree to which it is sensitivite to radiation incident upon the antenna from directions outside the main beam. The fraction of energy accepted from outside the main beam of an antenna is called the stray factor, which varies between 0.05 and 0.4 for most antennas. Low sidelobes and stray factors are obtained at the expense of antenna size, although the antenna diameter seldom needs to be more than doubled to obtain reasonable performance. Most such low sidelobe antennas must be custom-made, and such antennas are seldom used, although they could improve many meteorological experiments.

Some applications require scanned antennas. These antennas can be mechanically or electrically scanned. The advantages of mechanical scanning include a superior multifrequency capability lower antenna losses, and electronic simplicity. Electrically scanned antennas are more compact, need have no moving parts, and can scan more rapidly. A 19-GHz electrically scanned antenna system proposed for space flight has an antenna beamwidth of 2.6°, stray factor of 0.08, an insertion loss of 0.6 dB, a scan angle of ±50°, and is 18 by 3 inches. The instrument has operated in aircraft over various types of terrain and meteorologic conditions (Catoe et al., 1967).

Reduction of size, weight, and power is expensive, and is warranted primarily for space experiments. Examples include the two-frequency radiometer which successfully observed Venus from the Mariner-2 space probe (Barath et al., 1964), and a single-frequency 5-mm wavelength radiometer in earth orbit. The Mariner-2 radiometer weighed 20 lb and consumed 5 W average power and 10 W peak power. The 5-mm wavelength radiometer weighed 16 lb, consumed 42 W average power, and had a volume of 312 cubic inches. Substantial reductions in size, weight, and power are expected over the next ten years.

Meteorologic Applications of Passive Microwave Sensing

Microwave experiments can be categorized in several different ways. Here they have been divided into (1) temperature profile measurements, (2) composition measurements, and (3) surface measurements. A review of several possible meteorologic experiments from space is contained in a report edited by Ohring (1966).

Measurement of the Atmospheric Temperature Profile

The oxygen complex centered near 60 GHz offers opportunities to measure atmospheric temperature profiles from space or from the ground. This is so because the mixing ratio of oxygen in the terrestrial atmosphere is quite uniform and time invariant, and because the attenuation of the atmosphere varies with frequency from nearly zero to over 100 dB. The possibilities for such temperature profile measurements were first explored by Meeks and Lilley (1963) and have been extended by Lenoir (1968) to higher altitudes. Westwater and Strand (1967) have applied statistical estimation techniques to ground-based probing of the atmosphere and Waters and Staelin (1968) have applied similar techniques to space-based measurements.

Meeks and Lilley (1963) cast the expression for brightness temperature T_B in the form of a weighting function integral where $W(h, \nu)$ is the weighting function and $T(h)$ is the atmospheric temperature profile. That is,

$$T_B(\nu) = \int_0^\infty T(h)W(h, \nu)dh + T_0 e^{-\tau(\nu)} \qquad (6)$$

where $T_0 e^{-\tau}$ is the contribution of the background temperature. The weighting functions reveal the extent to which the brightness temperature measured at any particular frequency is sensitive to the kinetic temperature as a function of altitude. Lenoir (1965) has calculated many weighting functions appropriate to space-based microwave radiometers, and some of these are shown in Figure 5. The weighting functions above 50-km altitude are polarization dependent and vary in a predictable way with the terrestrial magnetic field. Parameters for several weighting functions computed by Lenoir (1965) are listed in Table 2. Examination of the half-power widths Δh of these weighting functions indicates that altitude resolution of several kilometers can be obtained, depending upon the receiver sensitivity.

Waters and Staelin (1968) have applied statistical estimation techniques to the problem of inverting such microwave data. An example of the inversion accuracy is shown in Figure 6, where the a priori standard deviation in the temperature is compared to the standard deviation obtainable with a seven-channel space-borne microwave radiometer of 1°K sensitivity. The six temperature weighting functions peaked at 4, 12, 18, 21, 27, and 31 km. One water vapor channel was also incorporated. When the satellite is over ocean, water vapor can be measured and used to improve the temperature estimates because temperature and humidity are correlated. This effect is shown in Figure 6. If the temperature

Table 2 Weighting Functions for Sounding the Atmospheric Temperature Profile

ν_0 (GHz)	W (MHz)	h_0 (km)	Δh (km)
64.47	200	12	11
60.82	200	18	7
58.388	30	27	9
60.4409	2.5	40	12
60.4365	1.0	50	20
60.5685 equator pole	1.5	60	21
		54	26
60.4348 equator pole	1.5	73	20
		66	26

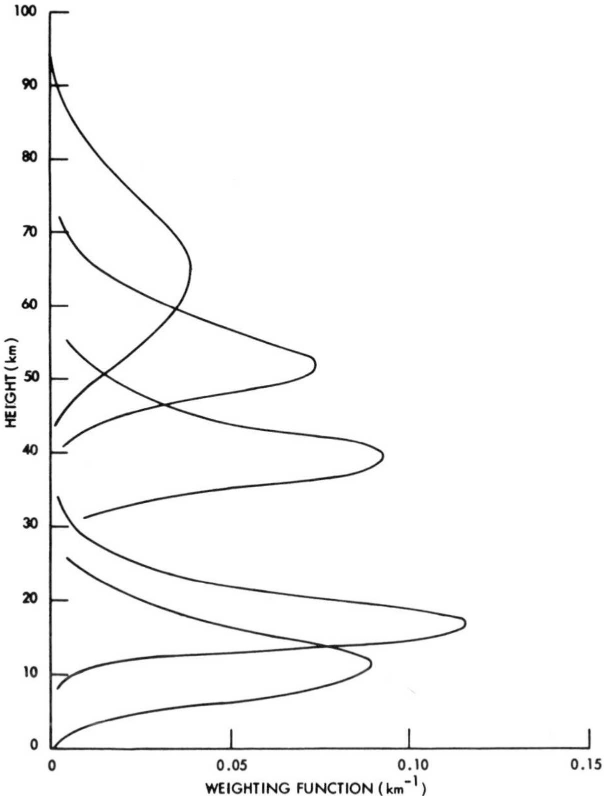

Figure 5 Temperature weighting functions for nadir observations from space. These weighting functions each correspond to different frequency bands near 60 GHz. (After Lenoir, 1965)

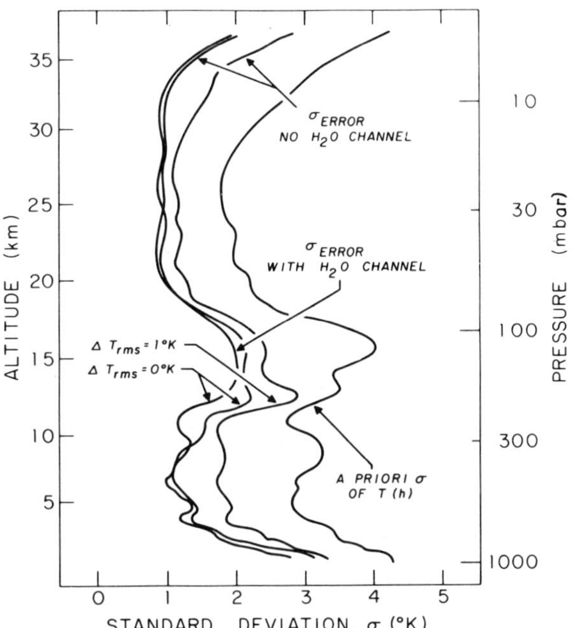

Figure 6 Results of 100 linear temperature profile inversions showing a priori and posteriori standard deviation for 0°K and 1°K rms receiver noise. The radiometer was assumed to be in space viewing nadir over ocean at 53.60, 58.39, 59.30, 60.82, 62.46, and 64.47 GHz in the oxygen band, and 22.235 GHz in the water vapor band. The inversion error was evaluated for 100 summer radiosondes from Peoria, Illinois, having a height resolution of 0.5 km.

profile can be measured with an accuracy of 2 to 3°K on a global scale, then such microwave systems would be very useful for the collection of synoptic meteorologic data.

The height resolution of the radiosondes was 0.5 km, much less than the width of the weighting functions. When the inferred $T(h)$ and the true $T(h)$ used for comparison were each smoothed by convolution with an 8-km Gaussian, then the error performance was improved, as shown in Figure 7. Thus average atmospheric temperatures over 5 to 10 km altitude can be inferred more accurately than can temperatures of narrower regions.

The effects of clouds and the terrestrial surface on such measurements can be estimated. In the preceding example only the weighting function which peaks at 4 km interacts appreciably with the terrestrial surface or with clouds. If the satellite is over land, then approximately 15 percent of the received radiation is received from the land, of which perhaps 95 percent represents the physical temperature of the land, and 5 percent is sky reflections, which have an effective temperature near 230°K. Thus a priori knowledge of the land temperature with 5°K rms uncertainty, plus knowledge of the emissivity within 1 percent rms reduces this contribution to the error in inferred atmospheric temperature to 1°K rms. Direct measurement of the surface brightness temperature at longer wavelengths can reduce this error still further. The effects of ice clouds are negligible, and even in the presence of heavy water clouds the error is small. For example, the heaviest nonraining cloud observed by Toong (1967) would have an approximate opacity of ½ at 5-mm wavelength, and thus would contribute an approximate error of less than 3°K, assuming the cloud were centered at 3-km altitude and that it were 20° cooler than the surface. This same heavy cloud over ocean would have an even smaller effect because the cloud temperature in this case would be even closer to the brightness temperature flux moving upward at 3-km altitude. Thus even this heavy-water cloud, approximately equivalent to a cumulus mediocris containing 0.18 g/cm^2 H_2O, would introduce no more than a few degrees error, even if it filled the entire antenna beam. Of course, if weighting functions were used which peaked nearer the surface, or if the bulk of the cloud were at much higher altitudes, then the effect could be larger.

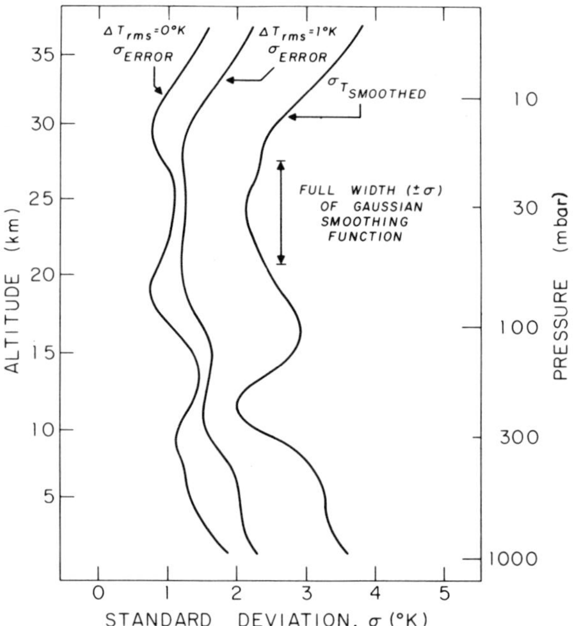

Figure 7 Results of 100 linear temperature profile inversions using the same data used in preparation of Figure 6. The inversion error was evaluated for a height resolution of 0.5 km after the inferred *T(h)* and the true *T(h)* were each convolved with a Gaussian of width 8 km.

Ground-based radiometers yield weighting functions which are quite different from those obtained for space-based measurements. These weighting functions are approximately exponentials with scale heights which depend upon the atmospheric opacity. This form of weighting function yields excellent height resolution near the observer, but the resolution is degraded at distances beyond 5 to 10 km. Westwater and Strand (1967) have calculated the errors expected for this example and find that for altitudes 0 to 10 km the rms errors range from 1.5°K to 4°K, respectively for 1°K receiver noise. Cloud effects for upward-looking radiometers are more severe than for those systems looking down because the equivalent temperature of space is much colder and therefore offers more contrast to clouds than does land. The fast response of microwave radiometers plus their ability to scan and to observe continuously may enable ground-based microwave radiometers to provide meteorologic information of a type not obtained before.

Measurement of Composition Profiles

Composition measurements are usually made in semi-transparent regions of the microwave spectrum where the spectrum is more sensitive to the distribution of absorbers in the atmosphere than to the temperature profile. For those constituents with resonances the problem of determining the distribution profile can also be expressed in terms of weighting functions. For example, measurements of a spectral line in absorption against the sun can yield $\tau(\nu)$. If those contributions to $\tau(\nu)$ from extraneous constituents are subtracted, then the remaining $\tau_r(\nu)$ for the constituent of interest can be expressed as

$$\tau_r(\nu) = \int_0^\infty \rho(h) W(\nu, h) dh \tag{7}$$

where

$$W(\nu, h) = \alpha(\nu, h)/\rho(h)$$

and where $\rho(h)$ is the constituent density profile, $W(\nu, h)$ is the weighting function, and $\alpha(\nu, h)$ is the absorption coefficient of the desired constituents. These weighting functions are weak functions of temperature, pressure, and $\rho(h)$, so a priori information about these parameters can determine the weighting function to a reasonable degree of accuracy. Staelin (1966) has computed such weighting functions for water

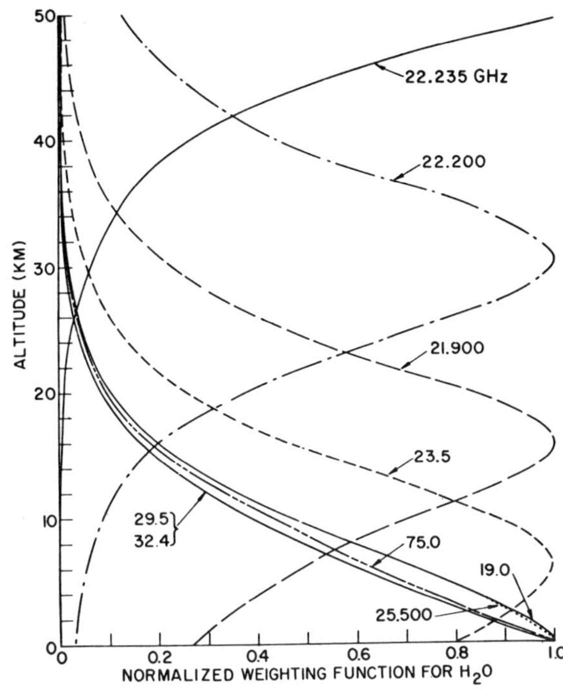

Figure 8 Normalized weighting functions for interpreting water vapor opacity measurements in the terrestrial atmosphere. (After Staelin, 1966)

vapor, as shown here in Figure 8. Very similar weighting functions result when the expressions are written for brightness temperature seen from the ground, or brightness temperature seen from space when over ocean. This is in contrast to the weighting functions found for the atmospheric temperature profile, which are quite different when observing from space or from the surface. Because the shape of the weighting function is determined almost exclusively by the variation of the linewidth parameter with altitude, almost identical weighting function shapes would apply for O_3, OH, or other trace constituents. The weighting function concept begins to break down, however, when the optical depth of the desired constituent is greater than 0.5, or when the background brightness temperature approaches the kinetic temperature of the atmosphere. Thus the weighting function concept can not readily be used to interpret the 183.3-GHz line of water vapor, nor any lines of trace constituents viewed from a spacecraft over land. Of course the development of weighting functions is not a prerequisite for use of the statistical estimation procedures described earlier. The existence of weighting functions does imply a certain degree of linearity in the estimation problem however.

Gaut (1967) has applied the method of Rodgers (1966) to inversion of the solar absorption measurements of water vapor at 22 GHz, and Waters and Staelin (1968) have considered the problem of estimating water vapor by measuring brightness temperatures from space or from the ground. Recent preliminary calculations of Waters and Staelin are presented in Tables 3 and 4. These calculations of estimated inversion errors were based on assumed perfect knowledge of the absorption coefficient and statistics computed from 100 radiosonde records. The results in Table 3 were computed for 20- and 22-GHz ground-based observations at zenith on the basis of 100 radiosondes, half from Tucson, Ariz., and half from Kwajalein Island. These calculations indicate that the integrated water vapor abundance can be determined to 0.1 g/cm². In Table 4 are presented similar results for observations from space over ocean. The atmosphere statistics were based upon 100 radiosondes from Huntington, W. Va., half in winter, and half in summer. The ability to resolve the water vapor profile with 2-km resolution results primarily from the a priori statistics, and secondarily from the resolution of the weighting functions. The weighting functions provide useful results with a resolution of 5 to 10 km, depending upon measurement accuracy.

Table 3 Inversion Performance for Ground-Based Radiometer Viewing Zenith at 20.0 and 22.0 GHz

Water Vapor Altitude	Water Vapor Statistics (g/cm²)		Water Vapor Inversion Errors (g/cm²)	
	Mean	σ	σ (ΔT_{rms} = 0°K)	σ (ΔT_{rms} = 1°K)
0–9.5 km	3.19	2.03	0.02	0.08
0–2 km	1.94	1.21	0.14	0.20
2–4 km	0.83	0.55	0.13	0.13
4–9.5 km	0.42	0.38	0.11	0.16

Table 4 Inversion Performance for Space-Based Radiometer Viewing Nadir at 22, 31, 53.6, 60.8, and 64.5 GHz, over Smooth Ocean

Water Vapor Altitude	Water Vapor Statistics (g/cm²)		Water Vapor Inversion Errors (g/cm²)	
	Mean	σ	σ (ΔT_{rms} = 0°K)	σ (ΔT_{rms} = 1°K)
0–9.5 km	2.09	1.34	0.09	0.11
0–2 km	1.35	0.92	0.22	0.22
2–4 km	0.51	0.35	0.14	0.14
4–9.5 km	0.23	0.15	0.09	0.09

With the improved radiometric systems which should become available over the next few years, similar performance might be sought for ozone and stratospheric water vapor. The measurements of ozone were reviewed earlier in this paper. Stratospheric water vapor has tentatively been detected at 22 GHz (Law et al., 1968). Since no such meteorologic data has been collected on the scale that passive microwaves should permit, the data will be quite unique. Spatial structure in three dimensions plus time variations should all be accessible.

Clouds and precipitation can also be measured. Because of the nonresonant nature of their absorption, however, it is difficult to measure altitude distribution or even to distinguish clouds from precipitation. Rain and clouds might be distinguished on the basis of form and intensity, and even on the basis of spectral shape, but the measurements of Toong (1967) indicate that such distinctions would be difficult to perform with any accuracy. Snow and ice might best be distinguished from water and rain on the basis of atmospheric temperature and climatology.

Nonresonant absorbers like clouds and precipitation can, however, be measured quantitatively and distinguished from water vapor or other resonant constituents. This was demonstrated theoretically by Staelin (1967) and experimentally by Toong (1967). Nonresonant constituents can best be measured from ground-based radiometers or from spacecraft over ocean. The accuracies which might be obtained can only be estimated. They are best expressed in terms of equivalent water cloud densities in g/cm^2 at some nominal temperature, like 283°K. A ground-based 0.9-cm receiver with sensitivity 1°K looking at zenith could detect a cloud with 0.005 g/cm^2, if the water vapor abundance were known exactly. Since the clouds must be distinguished from water vapor on the basis of spectral shape, the cloud sensitivity might be degraded to 0.01 g/cm^2. The measurement accuracy would be further degraded by uncertainties in cloud temperature, 5°K representing $\sim$ 15 percent in α.

Measurement of Surface Properties

Surface properties of interest include surface temperature, ground water, snow and ice cover, and sea state. Since the surface brightness temperature at long wavelengths is essentially the product of the surface emissivity and the surface temperature, the surface temperature can not be uniquely determined. If the surface brightness temperature is measured from space over a long period of time, then the surface emissivity may be averaged or calibrated out, and accurate measurements of temperature may be obtained. It is not known to what extent daily changes in emissivity may occur, but since the emissivity of most land surfaces is greater than 0.9, the variation is limited. The use of $E_\perp$ polarization (electric vector perpendicular to the plane of incidence) near the Brewster's angle may increase the emissivity further. Improvement may also be obtained by simultaneous monitoring of $E_\perp$ and $E_{||}$ polarization so as to detect changes in emissivity and perhaps permit corrections to the inferred temperature. The presence of ground water should decrease the brightness temperature of $E_\perp$ radiation, and snow or ice should normally increase it. This is an area where more analysis and experiments are needed before performance can be accurately predicted.

Sea state may be measured from space by observing $E_\perp$ and $E_{||}$ polarization at a nadir angle near 60°. Since the change in brightness temperature with equivalent wind speed for winds less than 30 knots has been calculated to be approximately 0.5°K per knot for $E_\perp$ polarization at 3-cm wavelength, and since the accuracy of the measurement should be approximately 1°K, the equivalent wind speed might be inferred with an accuracy of 2 knots. The analysis on which this was based neglected features like whitecaps and foam, and assumed that the ocean surface was composed of smooth facets, large compared to a wavelength. Furthermore, the sea surface properties depend not only upon wind speed, but also on fetch, current, surface pollution, etc. The true accuracy which might be obtained would best be determined by very carefully calibrated measurements at sea. Such measurements would be desirable not only at wavelengths where the temperature is nearly transparent, but also at wavelengths where the atmosphere absorbs up to one-half the radiation, because at these wavelengths that component of the radiation reflected from the sea surface into the antenna beam is quite sensitive to the surface slope probability distribution.

Meteorologic Relevance and Suggestions for Further Work

There are several meteorologic problem areas for which passive microwave sensors have unique capabilities.

1. Microwave sensors provide the only remote sensing technique capable of measuring atmospheric temperature profiles in the presence of clouds. This may be of crucial importance to global data collection for numerical weather prediction unless new techniques are developed which permit other remote sensors, super pressure balloons, etc., to operate more effectively in the 300 to 1000-mbar region than do microwaves. Above 300-mbar, microwave sensors are also a competitive technique.

2. Microwave sensors appear to be unique in their ability to measure the temperature profile above 50-km altitude. Synoptic mesopheric temperature data collected by satellite would be unparalleled as a tool for studying the mesospheric temperature structure.

3. Microwaves are unique in their ability to yield measurements of tropospheric water vapor in the presence of clouds. Although such sensors in space are effective only over ocean, the oceans cover over half the globe, and are very poorly monitored in contrast to most land masses. Even in the absence of clouds the great sensitivity of microwaves to water vapor and the ability of microwave sensors to average water vapor spatially permits measurements of integrated water vapor abundances which are competitive with and perhaps superior to radiosondes, which appear to be handicapped by aliasing errors (Gaut, 1967). This averaging ability and ability to operate through clouds may make such instruments valuable on the ground also.

4. Microwaves are unique in their ability to measure water vapor above the tropopause and ozone on a continuous basis. Although this is diffcult and has not yet been done, it is within the state-of-the art. Again, until these experiments are done, it is difficult to predict the meteorologic significance. One might certainly hope to learn more about the distribution and variability of these constituents, and perhaps to use them as tracers for circulation in the upper atmosphere. Such experiments could be done from the surface or from space.

5. Microwaves provide a powerful tool for measuring the total liquid water content of clouds, and even though there may be some ambiguity in the presence of large

particles or precipitation, the data are still quite unique. If such data could be taken with a high-resolution imaging system on board a satellite, storm cells, squall lines, etc., could be mapped with a precision not always available with optical sensors. Proper choice of wavelength would permit only very heavy clouds to be seen, or alternatively, perhaps almost all water clouds.

6. Microwaves offer promise of land temperature measurements from space. Such data are of interest in their own right, and also as an aid to determining the temperature profile of the troposphere.

7. Microwaves offer promise of yielding such surface characteristics as sea state, snow cover, ice cover, ground water, etc. More research is needed to determine the true potential of such experiments, although the sea state measurements appear quite promising.

Still other applications exist, and no doubt new ones will develop as the field of microwave meteorology grows.

Several suggestions for further work are obvious.

1. The expressions used for absorption coefficients of various atmospheric constituents should all be refined both experimentally and theoretically, particularly at millimeter wavelengths.

2. The microwave properties of the surface should be studied in a precise quantitative way in preparation for possible surface temperature measurements, sea state measurements, etc., from space.

3. The statistical inversion techniques should be improved and applied to a broader range of microwave problems where microwave sensors are accompanied by infrared or other types of sensors.

4. Efforts should continue to improve the sensitivity, accuracy, and antenna characteristics of radiometric systems, particularly at millimeter wavelengths, while reducing their size, weight, and cost.

5. Efforts to detect new spectral lines should continue as improved instruments become available.

6. Those which have been selected should become meteorologic tools and studied as such, in particular, the stratospheric water vapor and the ozone lines should be observed.

REFERENCES

Atlas, D., et al. Weather effects on radar. In *Air Force Surveys in Geophysics,* vol. 23, Geophys. Research Directorate, Air Force Cambridge Research Center, Cambridge, Mass, 1952.

Barath, A. T., Barrett, A. H., Copeland, J. Jones, D. E., and Lilley, A. E. Mariner 2 microwave radiometer experiments and results. *Astron. J.* 69:49, 1964.

Barrett, A. H. Microwave spectral lines as probes of planetary atmospheres. *Mem. Soc. Roy. Sci. Liège* 8:197, 1963.

Barrett, A. H., and Chung, V. K. A method for the determination of high-altitude water-vapor abundance from ground-based microwave observations. *J. Geophys. Res.* 67:4259, 1962.

Barrett, A. H., Neal, R. W., Staelin, D. H., and Weigand, R. M. Radiometric detection of atmospheric ozone. M.I.T. Research Lab. of Electronics, Cambridge, Mass., Quart. Progr. Rept. 86, 1967, p. 26.

Becker, G. E., and Autler, S. H. Water vapor absorption of electromagnetic radiation in the centimeter wavelength range. *Phys. Rev.* 70:300, 1946.

Benedict, W. S., and Kaplan, L. D. Calculations of line widths in H_2O–N_2 collisions. *J. Chem. Phys.* 30:388, 1959.

Carter, C. J., Mitchell, R. L., and Reber, E. E. Oxygen absorption measurements in the lower atmosphere. *J. Geophys. Res.* 73:3113, 1968.

Catoe, C., Nordberg, W., Thaddeus, P., and Ling, G. Preliminary results from aircraft flight tests of an electrically scanning microwave radiometer. NASA, Goddard Space Flight Center, Tech. Rept. X-622-67-352, 1967.

Caton, W. M., Mannella, G. G., Kalaghan, P. M., Barrington, A. E., and Ewen, H. Radio measurement of the atmospheric ozone transition at 101.7. GHz. *Astrophys. J. Letters* 151:L153, 1968.

Caton, W. M., Welch, W. J., and Silver, S. Absorption and emission in the 8-mm region by ozone in the upper atmosphere. Space Sci. Lab., ser. 8, issue 42, 1967.

Chahine, M. T. Determination of the temperature profile of an atmosphere from its outgoing radiation. *J. Opt. Soc. Am.* 58: 1968.

Chandrasekhar, S. *Radiative Transfer.* New York: Dover, 1960.

Cox, C. S., and Munk, W. H. Measurement of the roughness of the sea surface from photographs of the sun's glitter. *J. Opt. Soc. Am.* 44: 838, 1954.

Croom, D. L. Stratospheric thermal emission and absorption near the 22.235 GHz (1.35 cm) rotational line of water-vapour. *J. Atmospheric Terrest. Phys.* 27: 217, 1965a.

Croom, D. L. Stratospheric thermal emission and absorption near the 183.311 GHz (1.64 mm) rotational line of water-vapour. *J. Atmospheric Terrest. Phys.* 27:235, 1965b.

Cumming, W. A. The dielectric properties of ice and snow at 3.2 centimeters. *J. Appl. Phys.* 23:768, 1952.

Deutsch, R. *Estimation Theory.* Englewood Cliffs, N.J.: Prentice-Hall, 1965.

Dicke, R. H. The measurement of thermal radiation at microwave frequencies. *Rev. Sci. Instr.* 17:268, 1946.

Dicke, R. H., Beringer, R., Kyhl, R. L., and Vane, A. B., Atmospheric absorption measurements with a microwave radiometer. *Phys. Rev.* 70:340, 1946.

Fow, B. R. Atmospheric temperature structure from the microwave emission of oxygen. S. M. thesis, M.I.T., Cambridge, Mass., 1964.

Frenkel, L., and Woods, D. The microwave absorption by H_2O vapor and its mixtures with other gases between 100 and 300 Gc/s. *Proc. IEEE* 54:498, 1966.

Gaut, N. E. Studies of atmospheric water vapor by means of passive microwave techniques. Ph.D. dissertation, Dept. of Meteorology, M.I.T., Cambridge, Mass., 1967.

Gautier, D., and Robert, A. Calcal du coefficient d'absorption des ondes millimetriques dans l'oxygéne moléculaire en présence d'un champ magnétique faible. Application à l'atmosphere terrestre. *Ann. Geophys.* 20:480, 1964.

Gora, E. K. The rotational spectrum of ozone. *J. Mol. Spectr.* 3:78, 1959.

Haddock, F. T. Scattering and attenuation of microwave radiation through rain. Naval Research Laboratory, Washington, D.C., unpublished manuscript, 1948.

Helstrom, C. W. *Statistical Theory of Signal Detection.* New York: Pergamon, 1960.

Hemmi, C. Pressure broadening of the 1.63-mm water vapor absorption line. Elec. Engrg. Research Lab., University of Texas, Austin, Tech. Rept. 1, 1966.

Kerr, D. E. (ed.). *Propagation of Short Radio Waves.* New York: McGraw-Hill, 1951.

King, G. W., Hainer, R. M., and Cross, P. C. Effective microwave absorption coefficients of water and related molecules. *Phys. Rev.* 71:443, 1947.

King, J. I. F. Deduction of vertical thermal structure of a planetary atmosphere from a satellite. *Planetary Space Sci.* 7:423, 1961.

King, J. I. F. Inversion by slabs of varying thickness. *J. Atmospheric Sci.* 21:324, 1964.

Kraus, J. D. *Radio Astronomy.* New York: McGraw-Hill, 1966.

Law, S. E., Neal, R., and Staelin, D. H. K-band observations of stratospheric water vapor, M.I.T. Research Lab of Electronics, Cambridge, Mass., Quart. Progr. Rept. 89, 1968.

Lenoir, W. B. Remote sounding of the upper atmosphere by microwave measurements. Ph.D. dissertation, Dept. of Elec. Engrg., M.I.T., Cambridge, Mass., 1965.

— Propagation of partially polarized waves in a slightly anisotropic medium. *J. Appl. Phys.* 38:5283, 1967.

— Microwave spectrum of molecular oxygen in the mesophere. *J. Geophys. Res.* 73:361, 1968.

—, Barrett, J. W., and Papa, D. C. Observations of microwave emission by molecular oxygen in the stratosphere. *J. Geophys. Res.* 73:1119, 1968.

Marandino, G. E. Microwave signatures from various terrain. S.B. thesis, Dept. of Physics, M.I.T., Cambridge, Mass., 1967.

Meeks, M. L., and Lilley, A. E. The microwave spectrum of oxygen in the earth's atmosphere. *J. Geophys. Res.* 68:1683, 1963.

Mie, G. *Ann Phys.* 25:377, 1908.

Mouw, R. B., and Silver, S. Solar radiation and atmospheric absorption for the ozone line at 8.3 millimeters. Inst. Engrg. Research, University of California, Berkeley ser. 60, issue 277, 1960.

Ohring, G. (ed.) Meteorological experiments for manned earth orbiting missions. Geophysics Corp. of America, Tech. Rept. 66-10-N. Final Rept. NASW Contract NASW-1292, 1966.

Penzias, A. A., and Wilson, R. W. A measurement of excess antenna temperature at 4080 Mc/s. *Astrophys. J.* 142:419, 1965.

Porter, R. A. Microwave radiometric measurements of sea water, concrete and asphalt. Raytheon Co., Space and Information Sys. Div., Sudbury, Mass., Tech. Rept., June 20, 1966.

Rodgers, C. D. Satellite infrared radiometer, a discussion of inversion methods. Clarendon Lab., University of Oxford, England, Memo 66:13, 1966.

Rosenblum, E. S. Atmospheric absorption of 10–400 K Mcps radiation: summary and bibliography to 1961. *Microwave J.* 4:91, 1961.

Rusk, J. R. Line-breadth study of the 1.64-mm absorption in water vapor. *J. Chem. Phys.* 42:493, 1965.

Ryde, J. W., and Ryde, D. Attenuation of centimetre and millimetre waves by rain, hail, fogs, and clouds. British General Electric Co., Rept. 8670, 1945.

Saxton, J. A., and Lane, J. A. The anomalous dispersion of water at very high frequencies. In *Meteorological Factors in Radio Wave Propagation.* London: The Physical Society, 1946.

Shklovsky, I. S. *Cosmic Radio Waves.* Cambridge, Mass., Harvard University Press, 1960.

Staelin, D. H. Measurements and interpretation of the microwave spectrum of the terrestrial atmosphere near 1-centimeter wavelength. *J. Geophys. Res.* 71:2875, 1966.

— Interpretation of Spectral Data. M.I.T. Research Lab. of Electronics, Cambridge, Mass., Quart. Progr. Rep. 85, 1967.

Stogryn, A. The apparent temperature of the sea at microwave frequencies. *IEEE Trans. Antennas and Propagation,* AP-15:278. 1967.

Strand, O. N., and Westwater, E. R. The statistical estimation of the numerical solution of a Fredholm integral equation of the first kind. *J. ACM* 15:100, 1968.

Stratton, J. A. *Electromagnetic Theory.* New York: McGraw-Hill, 1941.

Toong, H. D. Interpretation of atmospheric emission spectra near 1-cm wavelength. *1967 NERM Rec., IEEE* no. 61-3749, 1967, p. 214.

Townes, C. H., and Schawlow, A. L. *Microwave Spectroscopy.* New York: McGraw-Hill, 1965.

Twomey, S. The application of numerical filtering to the solution of integral equations encountered in direct sensing measurements. *J. Franklin Inst.* 279:95, 1965.

— Indirect measurement of atmospheric temperature profiles from satellites: II. Mathematical aspects of the inversion problem. *Monthly Weather Rev.* 94:363, 1966.

Van Vleck, J. H. The absorption of microwaves by oxygen. *Phys. Rev.* 71:413, 1947a.

— Absorption of microwaves by water vapor. *Phys. Rev.* 71:425, 1947b.

— and Weisskopf, V. F. On the shape of collision broadened lines. *Rev. Mod. Phys.* 17:227, 1945.

Waters, J. W., and Staelin, D. H. Statistical inversion of radiometric data. M.I.T. Research Lab. of Electronics, Cambridge Mass., Quart. Progr. Rept. 89, 1968.

Weigand, R. M. Radiometric detection of atmospheric hydroxyl radical and ozone. S.M. thesis, Dept. of Elec. Eng., M.I.T., Cambridge, Mass., 1967.

Westwater, E. R., and Strand, O. N. Application of statistical estimation techniques to ground-based passive probing of the tropospheric temperature structure. Inst. for Telecommun. Sci. and Aeronomy, Boulder, Colo., ESSA Tech. Rept. IER 37-ITSA 37, 1967.

Wilkinson, D. T. Measurement of cosmic microwave background at 8.56-mm wavelength. *Phys. Rev. Letters* 19:1195, 1967.

Yap, B. K. Wind velocity and radio emission from the sea. S. M. thesis, Dept. of Elec. Engrg., M.I.T., Cambridge, 1965.

OCEANOGRAPHY IS *a field of investigation that offers some encouraging applications of remote sensing with passive microwave radiometry. The term "sea state" has been used widely for describing the characteristics of the wind-modified ocean surface. The measurement or profiling of ocean waves from aircraft or spacecraft is important to a number of oceanographic disciplines. A marked variation in microwave temperature brightness results from varying ocean surface conditions. The technique works effectively until whitecaps form or spray increases markedly. The presence of foam or spray can increase temperature brightness by 40° K or more. Higher water temperature increases microwave emission characteristics, but only slightly; a change of several degrees results in a very slight microwave brightness response. In the open ocean this minor response will probably be masked by the more striking variations caused by surface roughness. In this temperature realm of remote sensing investigation of ocean water, the far infrared portion of the spectrum offers more accurate results in measuring minor variations in surface water temperatures.*

The world's oceans provide a great transportation network on which most of the earth's international trade moves. Some of the main items in this trade are crude oil and associated petroleum products. Almost all of the ocean ships are powered by diesel engines, so an accident involves the risk of huge oil spillage. The wreck of the Torrey Canyon *off the coast of Great Britain and the more recent collision of two tankers in San Francisco Bay are prime examples of major oil spills. Oil pollution of the oceans has reached alarming proportions. Crossing the Atlantic in the* Ra,

Thor Heyerdahl noted extensive floating patches of oil in the open ocean far from major shipping lanes. Crude oil pollutants, even in almost trace amounts, are detectable on a relatively calm body of water with a microwave radiometer. This sensing system may provide a method of monitoring the world's oceans for oil pollution.

Sea ice has long been a problem in man's use of the ocean. The major part of the Arctic Ocean and its surrounding waters are usually covered by sea ice. Under the dynamic force of ocean currents and winds, this ice is continually in motion. The ice is constantly being deformed, and open water is exposed to freezing temperatures. Determining the thickness, condition, age, and movement of sea ice is one aspect of ice reconnaissance and ice forecasting, which are important to shipping in far northern waters. A typical arctic shipping season lasts only about three months. Ships can waste valuable time being trapped by ice flows or searching for "leads" of open water. Microwave radiometry can distinguish sea ice types. Perhaps more important is the sharp contrast such imagery exhibits between sea ice and water. An orbiting microwave system might prove a valuable tool in monitoring the drift of icebergs into the sea lanes of the world.

32-Oceanographic Applications of Remote Sensing with Passive Microwave Techniques

A. T. EDGERTON
D. T. TREXLER

SEVERAL OCEANOGRAPHIC applications for microwave radiometry are under investigation. Among these are the determination of sea state, water surface temperature, water salinity, oil pollution detection, and sea ice mapping. Interest in these applications stems both from the uniqueness of sensor signatures and the adverse weather capability afforded by microwave imaging systems.

Federal agencies sponsoring this research include the Office of Naval Research, NASA, the U.S. Coast Guard, the Naval Ordnance Laboratory, and the Naval Research Laboratory. Research summarized herein has been sponsored primarily by the Office of Naval Research and the U.S. Coast Guard. This work indicates that passive microwave techniques will be valuable for a variety of oceanographic studies. The following sections provide a review of each potential application.

Principle of Measurement

The subject of this presentation is radiometric mapping of ocean areas by sensing the thermal radiation at centimeter wavelengths. This wavelength region, extending from 0.3 cm to 30 cm, constitutes the microwave spectrum. At these wavelengths, which are very long compared to optical wavelengths, the radiation is relatively immune to scattering and attenuation by fog and clouds. It is the property of microwave radiation which is utilized in all-weather radar systems. An additional significant feature is that microwave radiation provides distinctive signatures not observable in the infrared and visual spectrum.

From *Proceedings of the Sixth International Symposium on Remote Sensing of Environment*. Ann Arbor: University of Michigan, Institute of Science and Technology, Willow Run Laboratories, 1969. Pp. 767–773. Reprinted with permission of the authors and the publisher.

The spectral intensity $I_\lambda(\theta)$ of the radiation emitted at wavelength λ in the direction of a nadir angle θ, is proportional to the absolute temperature T and the emissivity of the surface:

$$I_\lambda(\theta) \sim \epsilon(\theta)T \qquad (1)$$

The product $\epsilon(\theta)\ T$ is commonly referred to as brightness temperature, T_B. Measurements are generally expressed in terms of that product, and radiometers are calibrated with reference to a radiation source of known brightness temperature.

Depending on wavelength, the absorption and reemission of radiation by the atmosphere, due to the presence of clouds and gases such as water vapor and oxygen, must be considered. The brightness temperature measured from a platform at height h can then be expressed as

$$T'_B(\theta,h) = \left[\epsilon(\theta)T + r(\theta)T_S\right]\tau(h) + \int_0^h T_A(h)\frac{\partial\tau}{\partial h}\,dh \qquad (2)$$

where T_S is the brightness temperature of the sky integrated over the entire hemisphere at the surface, T_A is the ambient temperature of the atmosphere, r is the reflectivity of the surface, and τ is the transmissivity of the atmosphere.

The first term in Equation (2) relates to radiation from the surface; the second term relates to atmospheric radiation. Calm water has an emissivity of only about 0.4 and, therefore, always appears quite cold. For ice, on the other hand, the emissivity is near 1.0 so that its brightness temperature approaches the actual temperature of the ice surface. Water clouds vary in transmissivity virtually from 0.0 to 1.0 depending on thickness and liquid water content. Ice cloud transmittance is always very close to 1.0. Thus, over water, sea ice (high emissivity) or thick water clouds will appear much warmer than the water surface. Over land, brightness temperatures vary depending on vegetation, soil moisture, and composition. The emissivity of dry natural surfaces varies from about 0.8 to 0.95, depending on vegetation cover. The emissivity of moist soil can be much lower, depending on moisture content.

Considering that radiation emitted by most surfaces is polarized, Equation (2) can be written for both horizontal and vertical components. Horizontal and vertical polarizations are defined such that the electrical vectors lie in planes parallel and perpendicular to the earth's surface, respectively.

$$T'_{B,H}(\theta,h) = \left[\epsilon_H(\theta)T + r_H(\theta)T_S\right]\tau(h) + \int_0^h T_A\frac{\partial\tau}{\partial h}\,dh \qquad (3)$$

$$T'_{B,V}(\theta,h) = \left[\epsilon_V(\theta)T + r_V(\theta)T_S\right]\tau(h) + \int_0^h T_A\frac{\partial\tau}{\partial h}\,dh \qquad (4)$$

Subscripts H and V denote horizontal and vertical polarization, respectively. Only ϵ and r depend on polarization in Equations (3) and (4). Thus, polarization strictly relates to characteristics of the surface.

The difference between $T'_{B,V}$ and $T'_{B,H}$ varies strongly with nadir angle, θ. This difference is largest at the Brewster angle which, for most natural surfaces, occurs near $\theta = 70°$. $T'_{B,V}$ is never less than $T'_{B,H}$. By definition, $T'_{B,V}$ is equal to $T'_{B,H}$ *at nadir* $\theta = 0°$). For any given nadir angle, the difference $T'_{B,V} - T'_{B,H}$ decreases with increasing surface roughness. Thus, the roughness of surface indicating, for example, sea state or sea ice type may be inferred from the difference between $T'_{B,V}$ and $T'_{B,H}$. This effect is illustrated in Figure 1 which shows brightness temperatures at 0.81 cm observed with a ground-based radiometer over various surfaces.

Microwave radiation observed over ocean surfaces is a function of water temperature and salinity, surface roughness, and atmospheric radiation and transmission

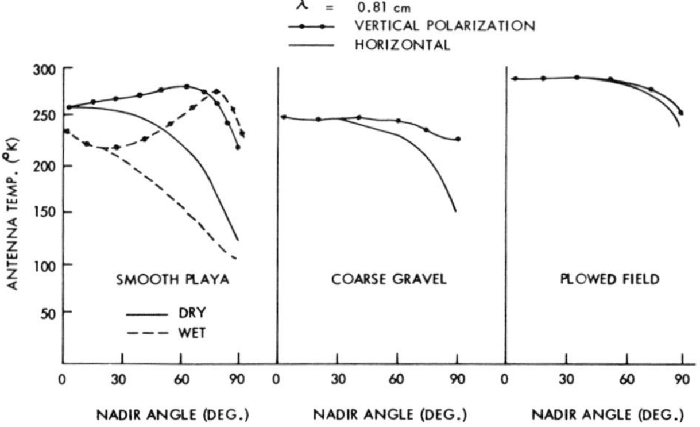

Figure 1 Microwave emission of several surfaces.

in the observational bandwidth. Other factors to be considered include observational wavelength, antenna polarization, and viewing angle. In the following discussion, each of these variables is reviewed and potential oceanographic applications are identified.

Sea State Determination

One of the first investigators to develop a theoretical relationship between ocean surface roughness and microwave emission was Stogryn (1967). He computed ocean brightness temperatures as a function of incidence angle and surface wind speed for both horizontal and vertical polarization by considering a theory based on the Kirchoff approximation for scattering from rough, finitely conducting surfaces. His model deals with emission resulting from large-scale roughness such as that associated with large waves and swells. Results of this analysis are shown in Figure 2. He concluded that large-scale sea roughness has a complex relation to the microwave brightness temperature, depending mainly on radiation wavelength and polarization, viewing angle, and orientation to the waves. He concludes that large-scale roughness should have little effect on T_B at viewing angles within about 20° of nadir and should have its greatest effect near a 50° viewing angle for horizontally polarized radiation. In his calculations he excluded contributions to T_B from small-scale roughness such as that produced by foam and spray which usually accompanies heavy seas.

The theoretical work of Stogryn and others precipitated several investigative studies including a series of aircraft flights by NASA/Goddard utilizing microwave instruments mounted in a Convair 990 (Nordberg, Conaway, and Thaddeus, 1968) and studies along the Southern California Coast sponsored by the Office of Naval Research (see ref.).

Ground-based measurements of the microwave emission characteristics of the littoral zone and near-shore environment were taken from a pier at Newport Beach, California. Measurements of the littoral zone were conducted to establish the microwave characteristics of breakers, foaming water, spray, the swash zone, etc. These experiments included stationary measurements taken from a pier while viewing selected areas of the

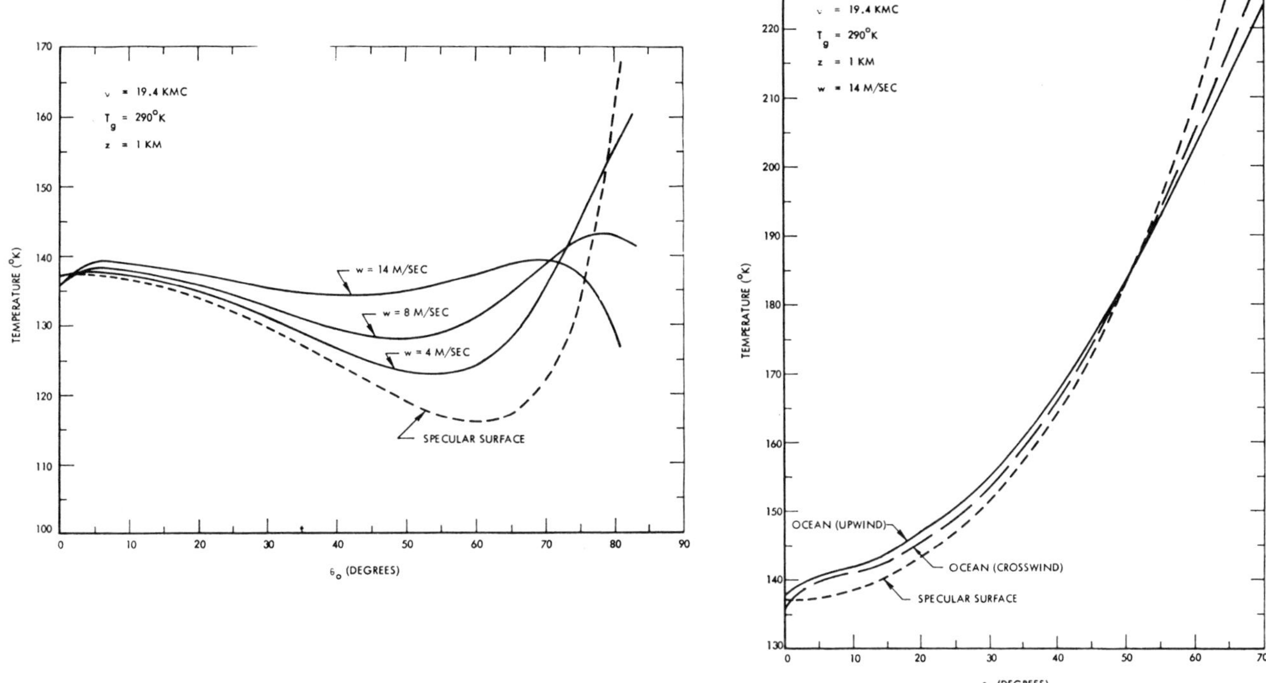

Figure 2 Ocean surface apparent temperature. (Left) Horizontally polarized radiation as a function of angle (upwind case). (Right) Temperature of vertically polarized radiation as a function of angle.

littoral zone, and continuous measurements or microwave profiles taken while the microwave system was moved outward from the shoreline along a pier. Near-shore investigations were concerned with establishing the microwave emission characteristics of the ocean surface as a function of wave height (sea state). These studies were conducted with the Aerojet multifrequency microwave field laboratory.

These investigations indicate that a marked variation in microwave brightness temperature occurs due to varying ocean surface conditions. Figure 3 exhibits the vertical and horizontally polarized 0.81 cm microwave temperatures measured during a traverse over the relatively dry beach sand and out into the near-shore environment. During these measurements the antenna was oriented with a 50 degree incidence angle, and an integration tome of 5.4 seconds was utilized. The projected field of view of the antennas on the ocean surface was relatively small, being estimated at approximately 7 × 4¼ feet. There are two curves shown for each antenna polarization. These represent the range of brightness temperatures observed during a sequence of five measurements at each station along the traverse.

The beach sand exhibited little variation in brightness temperature, and the temperatures were typically quite warm (260° to approximately 275°K). As the radiometer progressed down the beach face and encountered the swash zone, much greater variations were noted, corresponding to periodic inundation of the beach face. As the traverse continued out into the breaker region, where considerable foam was present, the temperatures

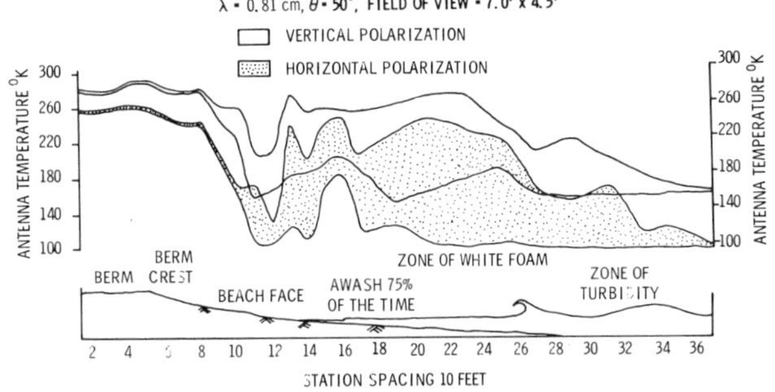

Figure 3 Traverses along Newport Pier showing maximum and minimum vertical and horizontal polarization antenna temperature.

become very warm—approaching that observed for the relatively dry beach sands. Beyond the breaker region, brightness temperature variations were much reduced. Near Station 37, which was well beyond the breaking waves, temperatures exhibited little variation and were typically quite cold as would be expected of an open ocean surface.

The significant feature of this curve is the very warm microwave temperature, or high emissivity, of foam in the breaker region. This behavior may result from diffuse scattering on the ocean surface associated with the fact that individual bubbles comprising the foam are not small in relation to the observational wavelength. The bubbles ranged from about 1 mm up to approximately 2 cm in diameter.

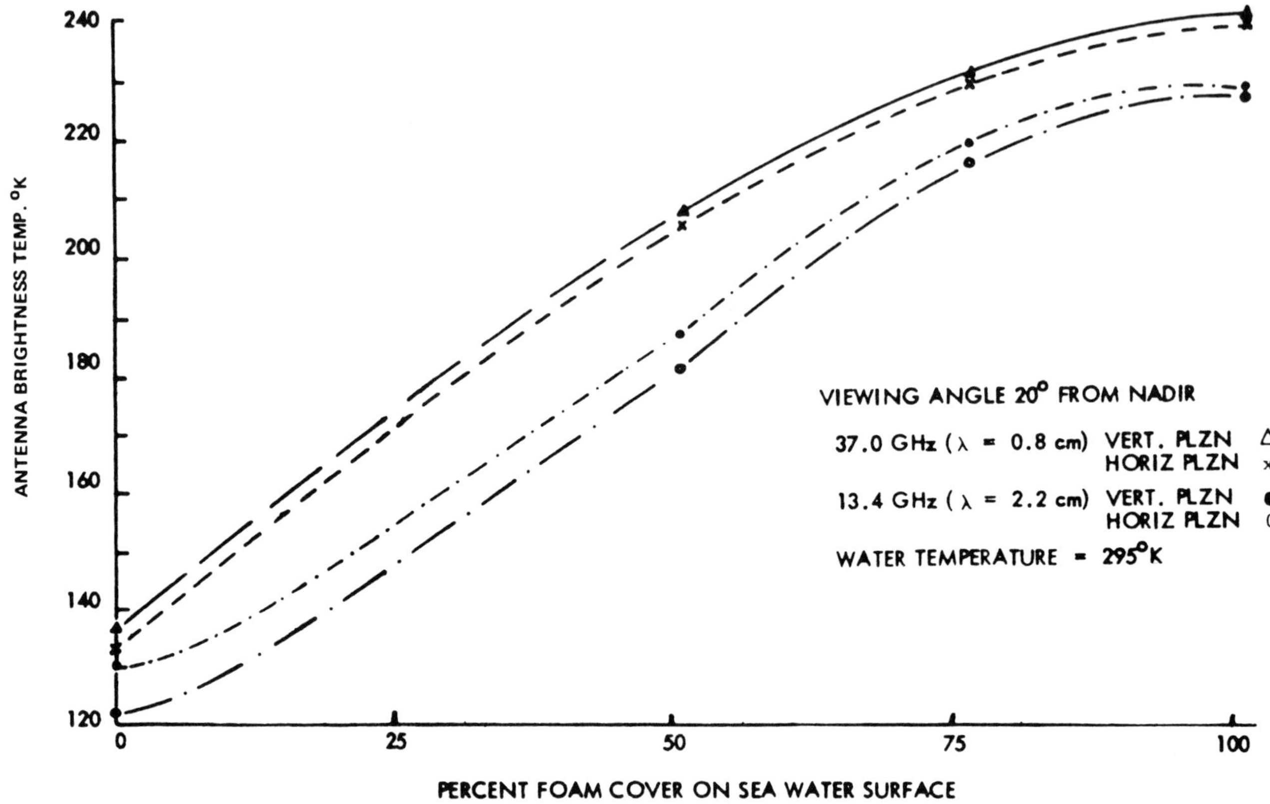

Figure 4 Effects of artificially induced foam on the radiometric temperature of salt water (tank experiment performed in Aerojet laboratory).

The anomalously high emissivities observed in the surf zone (area of foam) prompted an additional experiment. A foam generator was submerged in sea water, and simultaneous radiometric measurements at 0.81 cm and 2.2 cm were taken at a fixed viewing angle of 20° from nadir. Antenna temperatures for various percentages of foam on the water are presented in Figure 4. Note the increase in brightness temperature and emissivity of the water surface with increasing concentration of bubbles. An increase in brightness temperature ($T_{B,V}$ and $T_{B,H}$) of nearly 55°K was recorded for the 0.81 cm radiometer, while the 2.2 cm temperatures ($T'_{B,V}$ and $T'_{B,H}$) increased 50°K when 50 percent of the surface was covered with foam (bubbles). When the target was 100 percent foam, the total increase in brightness temperature above that of calm water was on the order of 100°K for both radiometers and emissivities approach 0.8. Note that a decrease in temperature difference ($T'_{B,V} - T'_{B,H}$) is apparent for the 2.2 cm sensor as the foam concentration increases.

Although this discussion only cursorily examines the effects of foam on sea surface brightness temperatures, it does substantiate the warm brightness temperatures recorded in the beach surf zone. Further investigations into the critical bubble size and concentrations of bubbles that appreciably affect brightness temperatures of the sea surface are planned for the future.

The high emissivities associated with foaming water imply that analytical models for predicting brightness temperatures of the open ocean under varying sea conditions must also consider the percentage of the surface covered by foam, spray, or whitecapping. Several investigators are now working on small-scale roughness models.

Figure 5 shows elevation scans taken in the surf zone and the open ocean beyond the breakers. These scans show the considerable differences that exist between the two regions. The scan of the surf zone is very irregular due to the constantly changing ocean surface. It also shows that temperatures in the surf zone are very warm in comparison with those of the open ocean. The scan of the open ocean is much smoother.

Several measurements were also taken of the open ocean beyond the breaker zone for various antenna viewing angles. This work is intended to clarify the relationship between microwave brightness temperatures and sea state conditions. To date these measurements have been concerned with relatively calm seas corresponding to low sea states.

The Convair 990 meterologic flights conducted by NASA/Goddard during 1968 and 1969 also provided microwave data concerning sea state (Nordberg, Conaway, and Thaddeus, 1968). In June, 1968, the aircraft overflew the Salton Sea at altitudes varying from 60 to 11,100 meters above the surface. Measurements of the brightness temperature of the sea surface were taken at 1.55 cm with an electronically scanned radiometer developed by Aerojet-General.

During these overflights calm conditions prevailed over the northern half of the sea (wind speed < 3 meters per second) while surface winds of about 15 meters per second appreciably increased the sea state over the southern half. The line of demarcation between the calm sea and the disturbed southern half was quite abrupt. Strong spray and frequent whitecaps were observed over the rough area while the northern portion was smooth with a slightly rippled appearance. Figure 6[1] shows a microwave image taken at a height of 11 km. Brightness temperatures are rendered as shades of gray. The image is produced automatically by displaying the radiometer scans as lines on a TV screen. Low brightness temperatures correspond to dark shades. The sea state patterns over the entire Salton Sea is readily apparent. Terrain surrounding the sea appears white because of the very high emissivities ($\epsilon > 0.9$), with brightness temperatures generally higher than 260°K. The much darker shades over the sea correspond to a brightness temperature range of 140 to 165°K, depending on sea state. The areal extent of variations in the sea state is clearly evident with low brightness temperatures (dark shades) corresponding to smooth seas and high brightness temperatures signifying rough seas. The water surface temperatures were 294°K for the rough sea and 300°K for the smooth sea. This is reasonable since one expects colder thermometric temperatures for the disturbed water. It is interesting to note that the 6°C difference in water temperature between the rough and smooth seas should have made the brightness temperature slightly colder over the rough sea. Due to the effect of roughness on the emissivity, however, brightness temperatures were actually 20°C warmer over the rough sea. These data provide additional evidence of a relationship between ocean surface roughness and emitted microwave radiation. The emitted radiation appears to be a function of the amount of foam and/or spray as well as of the larger-scale wave slopes, suggesting that emitted radiation may be related not only to sea but also to the magnitude of surface winds, since formation of foam and spray is intimately related to prevailing winds.

A sea state mission was subsequently conducted during the March, 1969, Convair 990 meteorologic flights, to further assess the feasibility of determining sea state by means of microwave radiometery (NASA, 1969; Nordberg, private communication). These experiments, conducted under the direction of Dr. W. Nordberg of NASA/Goddard, include measurements with the 1.55 cm microwave imager and with a 3.2 cm radiometer provided by JPL.

Six flights, each of about 5 hours duration, were conducted over water from Shannon, Ireland. In addition, data were obtained on the two flights crossing the Atlantic. During these flights, sea states were observed corresponding to surface winds ranging from 10 knots to 58 knots. Considerable data were obtained over very rough seas whose characteristics ranged from deep swells with long fetches to violent, wind-driven seas with very short fetches. Wave spectra of the sea surface were obtained with a laser profiler onboard the CV-990. The effects of clouds, which ranged from low-altitude, thin stratus to thick overcast with occasional cumulus buildups and rain cells, was also measured.

The strong dependence of 1.55 cm emission on sea

[1] See color insert.

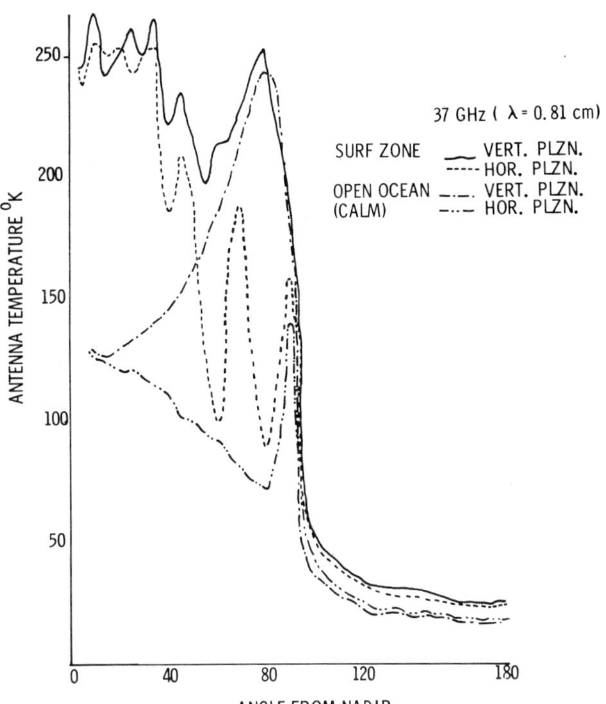

Figure 5 Antenna temperatures for near-shore sea conditions, Newport Beach, California.

surface roughness, originally observed over the Salton Sea, was confirmed. Over smooth seas (10–15 knot winds), brightness temperatures of about 115 to 120°K were measured. Over rough seas (35–40 knot winds), brightness temperatures of about 140°K were measured. This is in good agreement with the Salton Sea observations where a 1°K brightness temperature increase was associated with a 1–1.5 knot wind increase.

At windspeeds greater than 40 knots, an interesting situation arises. The same brightness temperatures were measured for windspeeds of 37 knots over a fully developed sea in the North Atlantic and for the North Sea at 55 knots. The laser indicated average wave heights of 18 and 25 feet, respectively, for these two conditions, and vastly different wave spectra for the two seas. This was also reflected in the instantaneous values of 1.55 cm microwave emission in the presence of high wind conditions. The fully developed sea with lower windspeeds showed fewer and much smaller fluctuations around the average brightness temperatures of 140°K. The wind-driven sea at 55 knots and 40 knots showed frequent excursions to brightness temperatures much higher than the average. In some instances, these excursions reach brightness temperatures of 200°K. From sea surface photographs and other simultaneous observations, these excursions can be correlated with large (50–100 feet) patches of foam on the sea surface. These foam patches exhibit very high emissivities or brightness temperatures which are observed only when the radiometer beam scans over their areal extent. Between the patches, the sea did not produce any higher emissivities than for the 37-knot case. When the aircraft was at higher altitudes, the individual foam patches could not be distinguished and the radiometer's instantaneous response assumed a characteristic and an average brightness temperature similar to that for the 37-knot wind case.

These data indicate that sea surface emission corresponds rather well to the wave heights, at least up to about 20 feet. However, it does not necessarily relate to windspeeds, especially for very strong winds. The 3.2 cm observations seem to indicate that, at that wavelength, the same effect exists qualitatively, but quantitatively it is somewhat smaller. Effects of clouds (negligible for thin stratus, but considerable for rain cells) at the 1.55 cm wavelength were confirmed in all flights. The cloud and rain effect is much less at 3.2 cm.

Additional experiments are in progress to clarify the physics of microwave emission associated with the sea surface roughness. These include analysis of 0.81 and 1.55 cm data gathered during recent Convair 990 flights in the Caribbean near Barbados, flights off the Southern California Coast and additional multiwavelength, dual-polarized measurements ($\lambda = 0.8$, 2.2, 5, and 21 cm) of near-shore wave conditions. Other experiments planned in the near future include stationary measurements from the ARGUS Island Tower (NRL) and multifrequency measurements over the Atlantic (NASA/MSC).

More data, preferably multiwavelength, are needed to clarify contributions from small- and large-scale roughness. By proper choice of sensor characteristics (wavelength, polarization, and viewing angle), it may be possible to distinguish between emission associated with the two types of roughness and to develop a practical means of determining sea state by passive microwave techniques. Multispectral methods may provide the solution.

Water Temperature and Salinity

Aside from roughness, microwave emission characteristics of water depend on water temperature and salinity. Investigators at Texas A & M University have computed the emissivity and absorption of sea water as functions of wavelength, salinity, water temperature, and dissolved gases (Geyer, 1968). These data are expressed as the product of the computed emissivities and physical temperatures. Figure 7 shows variations in apparent temperature as functions of water temperature for various values (essentially 0°/°° and 35°/°°)[2] of salinity. Changes in salinity have little effect above about 8 GHz ($\lambda = 3.75$ cm) and the higher frequencies exhibit a nonlinear water temperature dependence. A frequency of 4 GHz ($\lambda = 7.5$ cm) was deemed optimum for linear response to water temperature in areas having salinities of roughly 35°/°°. With salinity variations over a wider range, a frequency of 6 GHz ($\lambda = 5.0$ cm) was deemed optimum. Figure 7 also indicates that the very low frequencies are most responsive to salinity variations.

Laboratory data measured at 0.81 cm (37 GHz) and 2.2 cm (13.4 GHz) show good agreement with these results. Figures 8 and 9 show 0.81 and 2.2 cm antenna temperatures of fresh and sea water measured over a range of water temperatures. Measurements were conducted in an environmental chamber where the water was cooled at a slow rate. The antenna viewing angle

[2] °/°° = parts per thousand.

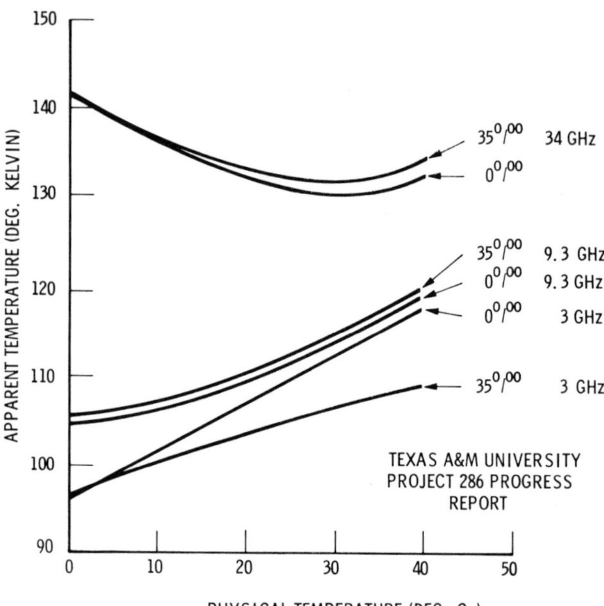

Figure 7 Theoretically predicted curves. Brightness temperature vs. salinity and temperature.

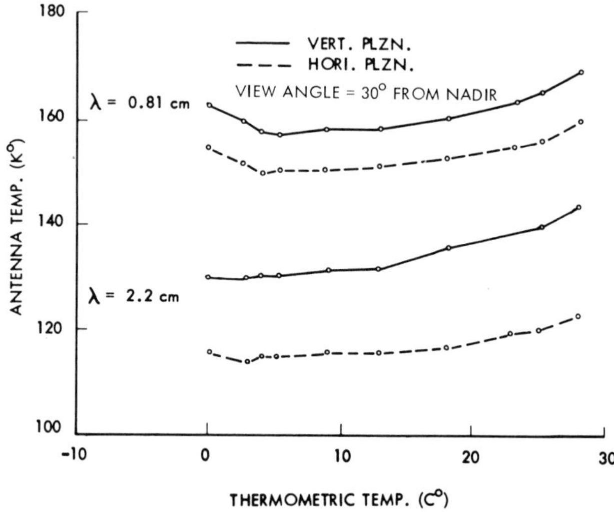

Figure 8 Antenna brightness temperature of fresh water as a function of thermometric temperature.

was 30°. The data indicate only slight changes in 0.81 cm (37 GHz) T_B for a relatively large range of water temperatures. Additional data will be taken soon using a wider selection of radiometers including observational wavelengths of 0.8 cm, 2.2 cm, 6 cm, and 21 cm.

Theoretical data indicate that microwave sensors should be most responsive to water temperature variations for observational wavelengths near 5 cm where the dependence is substantially linear. At this wavelength and for moderate viewing angles where atmospheric path lengths are minimal, ocean water exhibits an emissivity of about 0.4. Thus, a 20°C change in water temperature would result in a change in T_B of about 80°K and water temperature variations of the order of a few degrees will result in very slight brightness temperature changes. On the open ocean, these small changes are apt to be masked

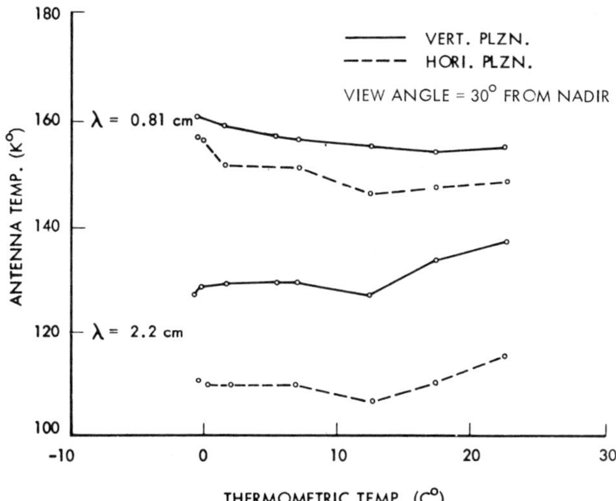

Figure 9 Antenna brightness temperature of sea water as a function of thermometric temperature.

by contributions associated with surface roughness and weather. For these reasons it does not appear practical to measure ocean surface temperature from spacecraft or high-altitude aircraft by means of microwave radiometry, as some have proposed.

Oil Pollution

A USCG-sponsored measurement program for detection and mapping of oil pollution with passive microwave techniques is under way (see ref.). The laboratory phase consisted of radiometric measurements at 0.8 and 2.2 cm of five petroleum pollutants, including gasoline, Bunker C fuel oil, 20, 30, and 40 API gravity crude oil. Measurements were performed as a function of viewing angle, oil film thickness, and age of pollutant. In addition, the dielectric properties of the pollutants were measured at 0.8-cm wavelength using an ellipsometer (precision reflectometer). The real part of the dielectric constant for the petroleum pollutants ranged from 1.85 to 2.41 as compared to a value of approximately 21 for sea water at 23°C. A slight increase in the real part of the dielectric constant and a large increase in the imaginary part of the dielectric constant was observed as pollutants are aged. As shown by Figure 10, ellipsometric measurements show an almost linear increase in the dielectric constant for crude oils with increasing viscosity, i.e., decreasing API gravity.

The low dielectric constants observed for petroleum products are reflected in the warm radiometric brightness temperatures of oil films on sea water. Horizontal 0.81 cm brightness temperatures were 7–70°K warmer than calm, unpolluted sea water. The wide range in brightness temperatures is a function of the pollutant type and viewing angle (see Figure 11). Note that higher API gravity crude oils have warmer radiometric temperatures and that gasoline has lower radiometric temperatures. The lower brightness temperatures for gasoline are due to its relative transparency at microwave frequencies. The somewhat irregular response of the 30 gravity crude and the Bunker C fuel oil stems from nonuniform distribution of these pollutants over the water

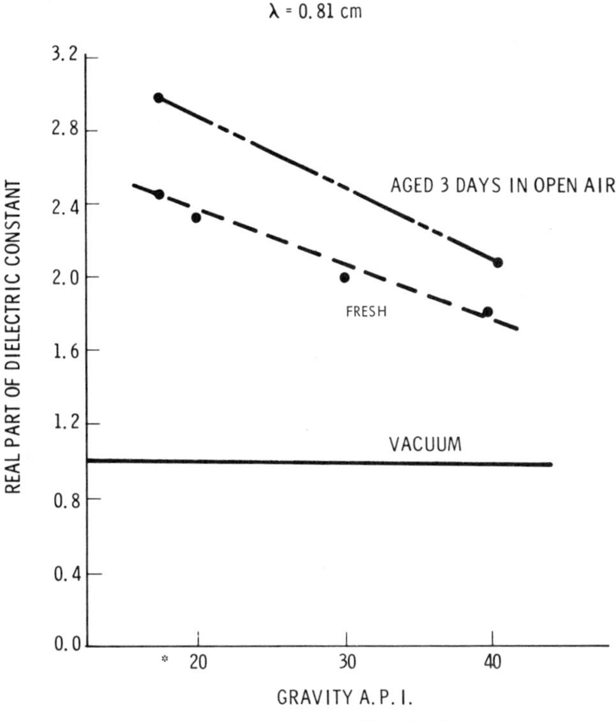

Figure 10 Dielectric constant of crude oils vs. gravity API.

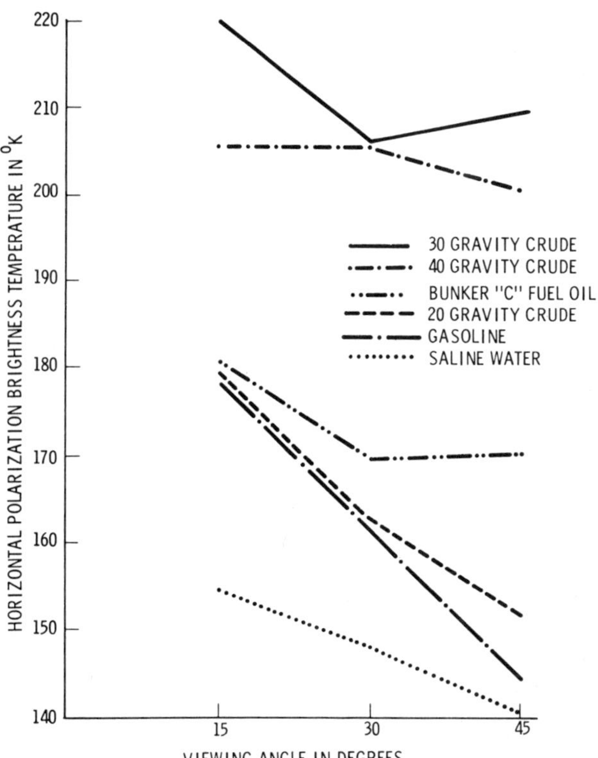

Figure 11 0.8 cm horizontal polarization brightness temperatures for pollutants used in laboratory measurement phase (pollutant thickness of 1 mm).

surface. Variations in radiometric temperatures for 20 gravity crude oil as a function of age (fresh and aged 3 days) and time of measurement are shown in Figure 12. During laboratory studies film thicknesses of less

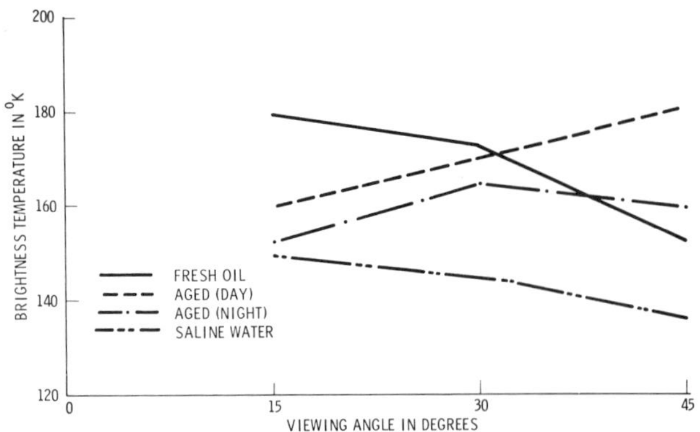

Figure 12 0.8 cm horizontal polarization brightness temperatures for 1 mm thickness of fresh 20 gravity crude oil, aged 20 gravity (day and night), and saline water.

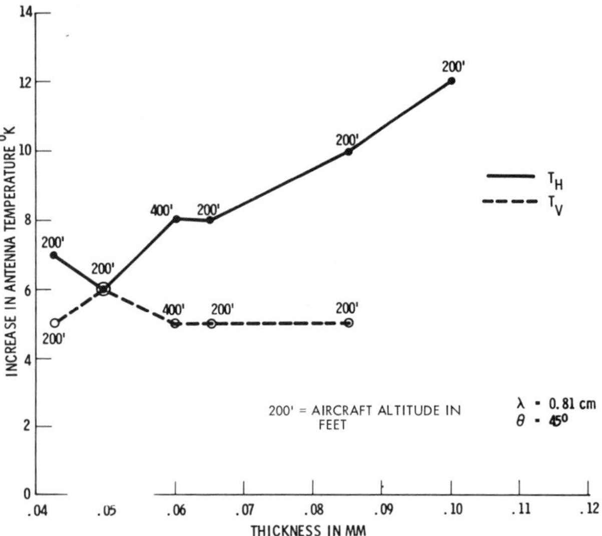

Figure 13 Diesel slick overflight, Aug. 28, 1969. Oil film thickness vs. antenna temperature increase.

than 0.1 mm had marginal detectability at 0.81 cm wavelength.

Airborne measurements of controlled spills of small quantities of pollutants off the Southern California Coast were performed at 0.81 cm wavelength for a view angle of 45°. Four petroleum pollutants were used for these tests, including marine diesel fuel oil and 20, 30, and 42 API gravity crude oil. Overflights were made at altitudes ranging from 200 feet to 800 feet. Antenna brightness temperatures of these small slicks (165 gallons of pollutant) showed increases on the order of 3–20°K. Radiometric temperature anomalies were dependent on the oil thickness and sea state conditions during the overflights. The brightness temperature anomaly for a given thickness of pollutant is less for higher sea states (warmer radiometrically) than for the same thickness on a calm sea. These measurements indicate thinner oil slicks are more detectable than was expected from analysis of the laboratory data. Figure 13 shows the increase in horizontal polarization antenna temperature as a function of oil slick thickness. The vertical polarization temperature does not vary appreciably for thicknesses less than 0.10 mm at a view angle of 45°.

From this study the following conclusions can be made: (1) the emissivity for petroleum products is significantly higher than that of a calm sea surface, (2) crude oil pollutants have decreasing dielectric constants (increasing emissivity) with increasing API gravity, (3) time of day and age of oil have only small effects on the radiometric response, (4) detection improves with decreasing sensor wavelengths, and (5) a reliable microwave oil pollution detection system can be configured for detecting oil slicks of the type examined during the described research program.

Sea Ice Mapping

Multispectral microwave measurements of sea ice and fresh water ice were performed utilizing an environmental chamber where ice formation could be monitored (Office of Naval Research ref.). The response of sea ice was distinctly different than that for fresh ice. Figure 14 shows the 0.81 and 2.2 cm radiometric response of fresh ice after initial ice formation. The oscillatory pattern of antenna brightness temperatures for both radiometers has an amplitude range of 80°K, although the 2.2 cm antenna temperatures are 30–40 °K, cooler than that of the 0.81 cm brightness temperatures. Oscillations are associated with interference phenomena. Note the wavelength dependence of the response from each radiometer. The period of the maximum 2.2 cm radiometer response is approximately 3 times that of the 0.8 cm radiometer. This relationship is a function of the physical ice thickness and sensor wavelength.

Sea ice was measured under the same environmental conditions. Figure 15 shows the radiometric response as a function of ice thickness. The 0.81 cm brightness temperature reached an equilibrium temperature of approximately 235°K for both vertical and horizontal polarizations. The ice thickness at equilibrium temperature was approximately 4 cm. Some minor oscillations are apparent in both the vertical and horizontal brightness temperatures. The amplitude of the oscillations is too small to determine if the horizontal polarization temperature $T_{B,H}$ is out of phase with the vertical polarization temperature ($T_{B,V}$). It is apparent, however. that the amplitude of $T_{B,V}$ is greater than $T_{B,H}$ and that both have diminishing amplitude as the ice thickness increases. These data indicate that microwave emission characteristics for fresh and sea ice are clearly different. This feature may be useful for distinguishing between various sea ice types.

Additional sea ice data were obtained during overflights conducted off Point Barrow, Alaska, by NASA/Goddard during May 1967. Observations at 1.55 cm indicate that sea ice emissivities were approximately 0.95 compared with sea water emissivities of 0.47. Figure 16[3] shows a radiometric image of arctic ice north of Pt. Barrow at approximately 76° 30′ North Latitude, and 157° 20′ West Longitude. The dark area across the center of the image is a lead of open water. Brightness temperatures range from black (cold) to white (warm). The thick solid sea ice shows the warmest brightness

[3] Not available for reproduction.

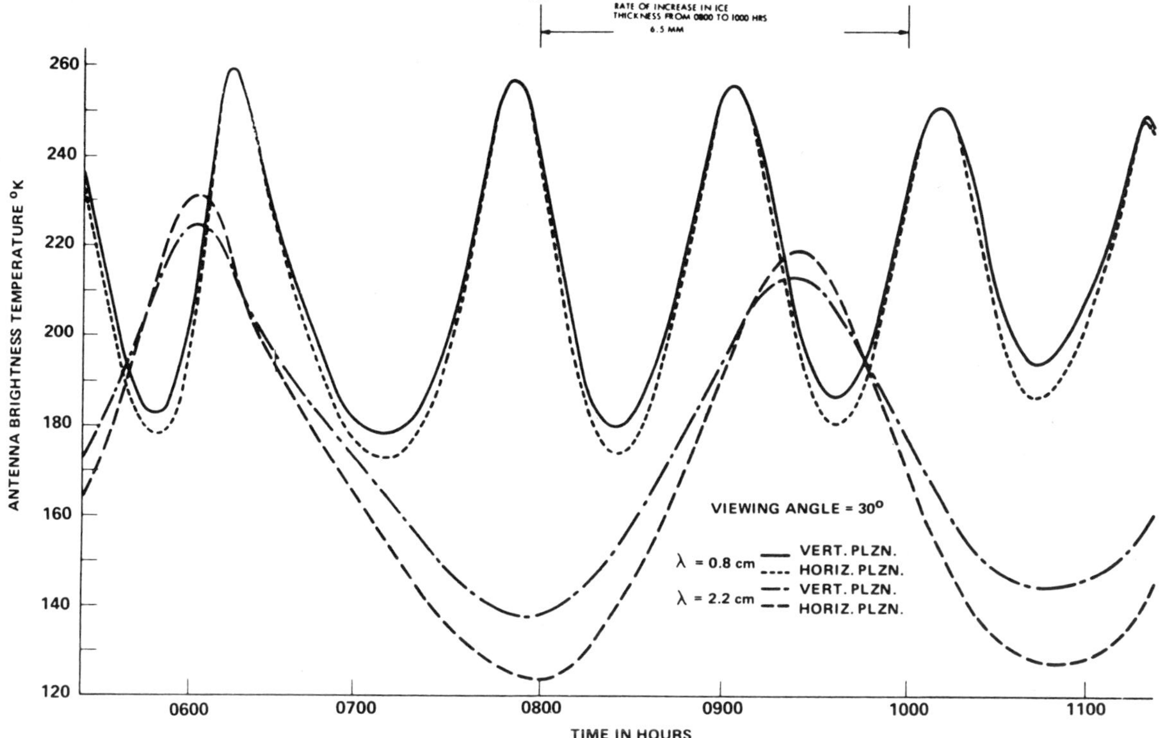

Figure 14 Radiometric response of 2.2 cm (13.4 GHz) and 0.8 cm (37 GHz) radiometers during initial stages of fresh ice formation.

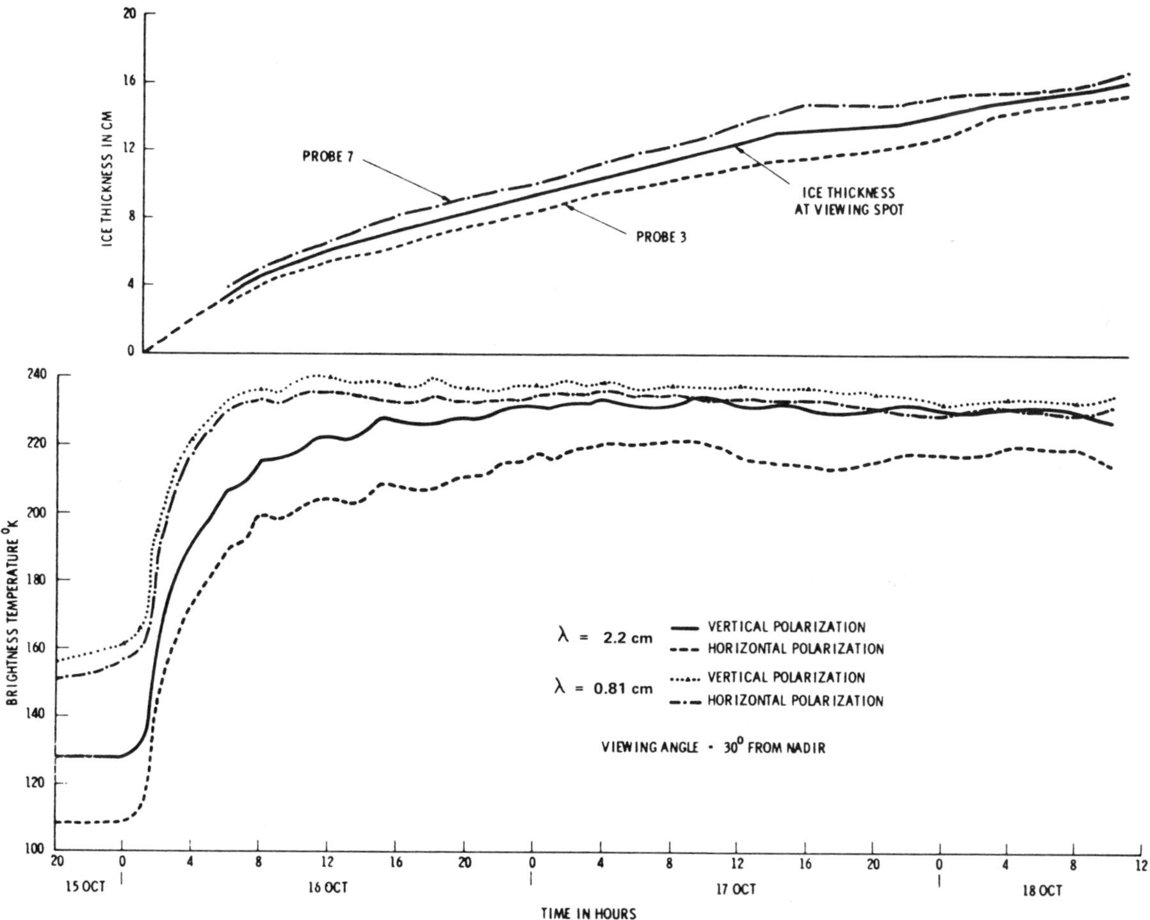

Figure 15 Brightness temperature, sea ice.

temperatures ($\sim$250°K) and the open water shows very cold ($\sim$100°K) brightness temperature.

These data demonstrate that microwave radiometry can be used as an adverse weather technique for mapping ice-water boundaries and sea ice distribution (i.e., leads, ice flows, etc.). More research on the use of multispectral microwave techniques for distinguishing ice types is needed. Identification of ice type on the basis of structure and surface roughness may prove feasible since the microwave emission characteristics are dependent on these features. Variations in brine content of various types of sea ice (discernible with long-wavelength sensors) may also be useful for discrimination.

REFERENCES

Edgerton, A. T., et al. *Passive Microwave Measurements of Snow, Soils, and Snow-Ice-Water Systems.* Tech. Rept. No. 4, SGD 829-6, Aerojet-General Corp., El Monte, Calif., Feb., 1968. Contract NOnr 4767(00) NR 387-033.

Geyer, R. A. A&M Project 286. *Oceanography of the Gulf of Mexico—Progress Report.* Texas A&M University, June, 1968.

NASA-Navy Microwave Review Meeting. Washington, D.C., June 11–12, 1969.

Nordberg, W., Conaway, J., and Thaddeus, P. *Microwave Observations of Sea State from Aircraft.* Greenbelt, Md.: Goddard Space Flight Center, Nov., 1968.

Office of Naval Research. Contract NOnr 4767 (00) Nr 387-033.

Stogryn, A. Apparent Temperatures of the Sea at Microwave Frequencies. *IEEE Transactions on Antennas and Propagation,* Vol. AP-15, No. 2, Mar., 1967.

U.S. Coast Guard. Contract DOT-CG-93, 228-A.

SUGGESTED READINGS FOR PART SEVEN

Aukland, J. C., Conway, W. H., and Sanders, N. K. Detection of Oil Slick Pollution on Water Surfaces with Microwave Radiometer Systems. *Proceedings of the Sixth International Symposium on Remote Sensing of Environment,* Vol. II. Ann Arbor: University of Michigan, Institute of Science and Technology, Willow Run Laboratories, 1969. Pp. 789–796.

Conway, W. H., Bacinski, R. R., Brugma, F. C., and Falco, C. U. A Gradient Microwave Radiometer Flight Test Program. *Proceedings of the Second Symposium on Remote Sensing of Environment.* Ann Arbor: University of Michigan, Institute of Science and Technology, Willow Run Laboratories, 1962. Pp. 145–174.

Dicke, R. H. The measurement of thermal radiation at microwave frequencies. *Review of Scientific Instruments* 17:268–275, 1946.

Edgerton, A. T., Engineering Applications of Microwave Radiometry. *Proceedings of the Fifth Symposium on Remote Sensing of Environment.* Ann Arbor: University of Michigan, Institute of Science and Technology, Willow Run Laboratories, 1968. Pp. 711–724.

Edgerton, A. T., Mandl, R. M., Poe, G. A., Jenkins, J. E., Soltis, F., and Sakamoto, S. *Passive Microwave Measurements of Snow, Soils, and Snow-Ice-Water Systems.* Washington, D.C.: Technical Report #4, Geography Branch, Earth Sciences Division, Office of Naval Research, 1968.

General Motors Corp., AC Spark Plug Division. *Iceberg Detection by Microwave Radiometry, Report on Experimental Test Flight Program.* U.S. Coast Guard, TCG-41456, 1960.

Hruby, R. J., Ragent, B., and Edgerton, A. T. An Experimental Evaluation of the Basic Assumptions Used in the Analysis of Microwave Radiometric Ground Truth Data. NASA Ames Research Center, Moffet Field, Calif. (revised), July, 1967.

Kennedy, J. M., and Sakamoto, R. T. Passive Microwave Determinations of Snow Wetness Factors. *Proceedings of the Fourth Symposium on Remote Sensing of Environment.* Ann Arbor: University of Michigan, Institute of Science and Technology, Willow Run Laboratories, 1966. Pp. 161–171.

Ohio State University. *Study of Thermal Microwave and Radar Reconnaissance and Applications.* Final Report, ASTIA Document #250363, July, 1960.

Sherman J. W. Passive Microwave Sensors for Satellites. *Proceedings of the Sixth International Symposium on Remote Sensing of Environment,* Vol. II. Ann Arbor: University of Michigan, Institute of Science and Technology, Willow Run Laboratories, 1969. Pp. 651–669.

Skiles, J. J., Grzelak, T. A., Dixon, R. S., Ragotzkie, R. A., and McFadden, J. D. An Airborne Instrumentation System for Microwave and Infrared Radiometry. *Proceedings of the Second Symposium on Remote Sensing of Environment.* Ann Arbor: University of Michigan, Institute of Science and Technology, Willow Run Laboratories, 1963. Pp. 175–185.

Vivian, W. E. Passive Microwave Technology and Remote Sensing. *Proceedings of the First Symposium on Remote Sensing of Environment.* Ann Arbor: University of Michigan, Institute of Science and Technology, Willow Run Laboratories, 1962. Pp. 37–45.

PART EIGHT

ACTIVE MICROWAVE

Radar, generating a wave and recording its reflection or echo

THE PRECEDING *discussions of remote sensing have dealt with passive sensors which produce imagery from energy coming to them from exterior sources. In contrast, radar is an active system. Among the commonly used sensors, radar is the only one which provides an independent source of energy or illumination. There are other active sensors, but to date these must be considered experimental, or they have been classified and utilized primarily for military purposes.*

The acronym RADAR is derived from the phrase RAdio Detection And Ranging. The shorter expression was probably first used by two U.S. Naval officers, F. R. Furth and S. M. Tucker. Implicit in this title is that radar is a radio apparatus which is capable of detecting a remote object, called a target, *and indicating its range and position (distance and direction).*

Radar sends out a radio wave whose properties are accurately known. Currently most systems operate at wavelengths between 0.86 and 3.3 cm, although some experimental systems utilize wavelengths up to 70 cm. The waves are emitted for brief periods of time, thousandths of seconds (milliseconds) for targets hundreds of miles away or millionths of seconds (microseconds) for close-range targets a mile or less away. These short, powerful bursts of energy are called pulses. *Radio pulses move through space, striking targets in the environmental field. Pulses are subsequently reflected or echoed back to the radar set as a much weaker pulse. In order to distinguish the weaker returning energy from the more powerful outgoing signal, the radar transmitter must be turned off. Thus, modern pulsed radars are switched on and off thousands or even millions of times each second. By comparing the wavelength of the returning wave with that of the transmitted wave it is possible to determine the Doppler shift, thereby measuring the relative speed of the target.*

Radar pulses are sent in a known direction with a directional antenna which is able to focus the radio wave into a carefully controlled beam. The more precise or sharp this beam of radio energy is, the better it can be used to measure direction. The size of the antenna, measured in units of the operative wavelength radio frequency, determines the sharpness of the beam. A great deal of current radar research is concerned with antenna design.

Because the system sends out the beam at a predetermined wavelength, the receiving antenna can be specifically designed to receive returning energy in the wavelength generated and only from the immediate direction in which the radar wave was generated. The time it takes for the wave to travel to a remote object and return can be measured. Electromagnetic waves travel at known speed (3×10^8 meters/sec in space, about 186,000 miles/sec). It is thus possible to compute the distance to the sensed object as a function of time. Therefore, the range or distance of the target is determined.

Modern radar systems consist of four major components. The first is the transmitter, *which generates and then emits the radio wave. The second is the* receiver, *which accepts the weak pulses of reflected energy. In most systems, the transmitter and receiver share the same antenna. The third major element of the system is the* amplifier, *which strengthens the echoed signal. The fourth is the method of display, the part of the system which turns the invisible radio wave into an image that can be seen or photographed.*

As early as the 1920s, experiments with radio waves indicated that passing ships or flying aircraft created a disturbance in the reception of the signal. A. H. Taylor and L. C. Young, experimenting with high-frequency radio communication in Washington, D.C., placed a transmitter on one bank of the Anacostia River and a receiver on the opposite bank. Their observations indicated that when a ship passed between the transmitter and receiver, radio tone suddenly nearly doubled in volume and then faded out before resuming its normal tone. This suggested the possibility of detecting a ship in darkness or—since the long waves of radio easily penetrate fog or cloud cover—under adverse weather conditions. Two researchers conducting a somewhat similar experiment discovered that aircraft flying over a receiver could be detected in much the same way. (For a more complete discussion of the early events that were so important in the development of radar, see Robert Morris Page, The Origin of Radar. *Garden City, N.Y.: Doubleday, 1962.)*

In 1934 Young and Page operated the first radar that received echoes from an aircraft at a range of 1 mile. The following year, Watson and Watt, working in England, successfully obtained echoes from a seaplane at a distance of 15 miles. A year later they were able to track aircraft beyond a range of 80 miles. In the same year Page and Guthrie in the United States received excellent response from a target beyond 25 miles with a second-generation radar set.

By the beginning of World War II, scientists from several countries had developed functional radar sets for military purposes, but these early systems were massive, crude, and unreliable. The generated images were hazy, imprecise, and sometimes difficult to interpret. Continued experimentation produced the small, portable, reliable radar systems now commonly in use. Begining about 1950, radars with better capabilities designed primarily for military reconnaissance were developed. This is the type of radar now employed for studying earth resources.

In the early use of radar, it was noted that objects in the surrounding environment, such as buildings, landforms, and forests, frequently produced a return or echo back to the receiver. To the radar engineers this was known as ground clutter. *They worked assiduously to suppress these returns. As radar was declassified and became more widely used, scholars from other disciplines quickly realized the value of ground clutter returns. They provided a fresh look at the earth's surface for the geoscientist. On board an aircraft, radar can rapidly cover large areas, day or night, and in almost any kind of weather. Only extremely heavy rainfall will attenuate these radio waves. Once the value of these ground returns was recognized, the complex problem of systematic interpretation was confronted. Interpretation of radar return requires an interdisciplinary approach. Engineers must explain the parameters of the radar system. Geoscientists must be able to interpret the diverse nature of terrain illuminated by the transmitted electromagnetic waves. The intricate relationship between radar returns and terrain is due in part to the complexity and variability of the earth's surface; material composition, moisture content, vegetation, surface texture, and temperature all modify radar return amplitude. In addition, the relationship is further complicated by characteristics of the generated radiowave; angle of incidence, wave polarization, and frequency of the emitted signal affect the image received. Despite this complexity, the radar image can be interpreted and it does provide valuable information about the natural environment over large areas and under adverse weather conditions.*

33-An Interpreter's Perspective on Modern Airborne Radar Imagery

RICHARD B. INNES

THE COMMON feature of remote sensors is that the energy which carries information from the subject to the sensor is in the form of a wave through some intervening medium. Radar senses remotely by making use of radio waves.

I said radio waves, not radar waves, to point out the fact that these waves are the same sort that are used for radio communications. Some of them have wavelengths of a foot or a few feet, like those used for FM and television broadcasting; others have wavelengths of a few inches, like those used between the relay stations that distribute radio and television network programs, or that carry long-distance telephone conversations. Certain of these wavelengths do become radar waves simply by being assigned to that kind of service, just as certain channels are assigned to broadcasting stations. Now that this point has been made, I will proceed to call them radar waves. It is from the characteristics of these waves that radar images derive their own distinctive characteristics.

The principal distinction between radar waves and the light waves that we are more accustomed to using for imaging purposes is the comparison of their wavelengths. The wavelength of light is, to most of us, unimaginably small—almost by definition, it is too small to be seen, and our ideas of tiny things are limited by our experience of seeing such things. But the radar wavelengths are comfortably man-sized. Distances like these are the ones that we can show to each other between thumb and forefinger, or by holding our hands a few feet apart. The shortest waves commonly used for radar are less than a centimeter long, yet these are about 10,000 times longer than the longest light wavelengths. The longest radar waves are about a meter long; these are about 3 million times longer than the shortest light waves.

On the other hand, when considered from the other end of the spectrum, radar waves are short, compared with other radio waves. Since the waves at the center of the AM broadcast dial have a length of 300 meters, radar waves are from 300 times to 30,000 times shorter. Radar waves stand, then, nearly midway between the most familiar kinds of optical and radio waves, combining some of the imaging capability of the one with some of the penetrating capability of the other. The most important characteristic of the waves chosen for radar use is their ability to retain both their form and their strength while propagating long distances through the earth's atmosphere.

From *Proceedings of the Fifth Symposium on Remote Sensing of Environment.* Ann Arbor: University of Michigan, Institute of Science and Technology, Willow Run Laboratories, 1968. Pp. 107–122. Revised by the author. Reprinted with permission of the author and the publisher.

The feature that I consider next in significance is that nature produces only small amounts of these waves, yet they can be manmade at high intensities. Therefore, we have the capability to illuminate remote objects with this kind of radiation at levels much higher than the sum of all natural sources of that same wavelength. Furthermore, we can, in the process of generating these waves, achieve close control of several of their other parameters, such as their timing, strength, phase, and polarization. Each of these parameters can be manipulated so as to produce or to enhance various imaging features. Consequently, we have available some modes of control over radar imagery that we do not have over optical imagery.

The opportunity to penetrate reliably a hundred or more miles of atmosphere in very orderly fashion, and to image through haze, cloud, smoke, heat shimmer, rain, and falling snow is the primary reason for sometimes turning from optical waves to radio ones, from photography to radar. The need to provide the radio-wave illumination artifically makes the factor of night-time or other time of day become also irrelevant to the possibility of obtaining imagery. In some areas of the world, a means for imaging through optically poor atmospheric conditions is essential for timeliness, economic practicality, or even ultimate possibility of imaging programs.

A second reason for choosing radar as the sensor for a particular application is that the reflection of its waves is governed by phenomena whose meanings, in terms of direct human experience with the objects imaged, are unlike those of the phenomena governing optical reflection. For example, color, as we think of it from visual experience, is inconsequential to radar. The processes by which radar waves are scattered are described, instead, by rather gross electrical properties of the scattering material. Roughly speaking, good electrical conductors, such as metals, scatter radar waves most strongly; moderately good conductors, such as wet natural materials, scatter more strongly than poorly conducting dry ones. Almost everything that forms part of the topmost surface of the earth sends detectable amounts of the incident energy back into the atmosphere through which the illuminating waves approached. But there are differences in the amounts of energy returned by stone, sand, grass, trees, water, ice, asphalt, cement, etc., and these differences offer hope for identifying some of these materials. (More will be said later about some of the ambiguities involved.) The value of radar in this respect is that it may make some distinctions which other sensors do not. Also, radars of different frequencies may offer some additional identifying comparisons. Finally, because of our direct control of the incident radiation, we have the option, not available in passive sensing systems, to choose the polarization characteristics of the incident waves as well as of the received re-radiation, and thus to gain yet another coordinate for possible identifiable variations.

A third reason why one might desire radar for an imaging task is that the geometric coverage available can be greater than is possible through photographic lenses, even in clear weather. This claim will be explained later, after some contributing facts have been stated. That explanation involves the manner in which radar determines the locations of image points. That manner is different from the type of object-space to image-space transformation used by lens-optics systems. The basis for the difference is in the relative ease and accuracy of controlling different parameters of the radar waves.

There are prices to be paid for the benefits that long wavelength gives to radar-wave imagery. One part of the price is the difficulty of defining the direction from which a particular echo is received, hence a difficulty of pinpointing the object point that reflected that bit of energy. This determining of direction is a matter of trilateration—of establishing the shape of a triangle by measuring its three sides. One side (it is convenient to think of it as the base of the triangle, or the baseline for the measurement of direction) is the opening (or aperture) of the antenna that emits or receives the waves. The other sides are, of course, the ranges from the two ends of the aperture to the object point.

The wavelength is important to this measurement because it turns out to represent the uncertainty in the comparison of the ranges. The most likely direction of the point from which a returning wave has spread is that direction for which the ranges from the scattering point to the opposite ends of the antenna are equal. But the best estimate that we can make of this direction is to say that it is somewhere between the direction where one range is half a wavelength longer than the other and the direction where it is half a wavelength shorter. The easy rule to use is this: if the aperture is N wavelengths wide, then the width of the beam that it uses is $1/N$th of a radian. For example, consider an antenna 5 feet across and using a 1-inch wavelength; its beam would be 1/60th of a radian wide (or about 1 degree, since 1 radian is about 57°). Objects that are 1 degree apart when seen from the radar would be barely resolvable, since, as the beam scans across them, one would be just leaving the beam as the other entered. In terms of distances, if the objects were 5 miles away (26,000 feet), the least object-separation that could be resolved would be over 450 feet.

The directions toward improved angular resolution are toward both smaller wavelengths and larger antennas. When wavelength reaches 1 cm, the loss of atmospheric penetration capability begins to become serious. Also, it is more difficult to produce and to control the power levels that are desired.

Before talking about antenna sizes, let me set some limits on my subject. This paper's title specifies airborne radar. It also implies imaging radar, and by that I mean devices used to image part of the ground that the airborne radar can look down upon. Furthermore, most of what I will say will apply best to side-looking radar. This means a radar whose antenna points its beam steadily toward one side of the aircraft path during a practically straight flight. This mode of operation is in contrast to the older practice of rotating the antenna so as to look all around the aircraft several times a minute. The side-looking radar defines positions of scatterers in cartesian coordinates, whereas the rotating-antenna radar uses polar coordinates.

When the antenna has to rotate, it must be kept short enough to fit inside a streamlined cover, or *radome,* on the aircraft. A radome big enough to contain a long rotating antenna has very serious effects on the flying abilities of an aircraft. But a nonrotating antenna intended to look only toward the side has its greatest dimension extended along the length of the aircraft. Long antennas can indeed be carried in this way; hence side-looking radars can use larger airborne antennas than can rotating radars.

So, now, suppose a side-looking antenna 20 feet long and using 1 cm waves. This means an aperture of about 600 wavelengths, hence a beamwidth of 0.1 degree, or less than 50 feet at 5 miles. Some of what my title calls "modern" radar imagery is of this sort.

Even when we use the side-looking scheme to get better beam resolution, we have improved the image resolution in only the along-track dimension. This is because we have been able to enlarge the antenna only along that direction. To narrow the beam as much in the other direction, the antenna would also become 20 feet tall. Besides, to scan this beam across the terrain, as is done with a mirror scanner, the whole antenna would have to spin at a rather terrific rate. Obviously, that is not the way to do it when using these large wavelengths. So we need a better way to locate the across-track positions of objects in radar-wave images.

I think it is fairly common knowledge that it is standard radar practice to determine distances to objects by timing the echoes received from them. This timing technique is possible because radars are active sensors, creating their own illumination. We are therefore able to "label" the outgoing energy, then detect the label on the returning energy, and thus measure the time it required to go and return. The commonest way to do the labelling is to emit short flashes of energy (radio engineers call them *pulses*) and then spread the successive bits of returning energy uniformly along a line to form an image of the radar reflectance of the terrain along the beam.

The success of this technique rests, in part, on the fact that precise timing is the thing that electronic

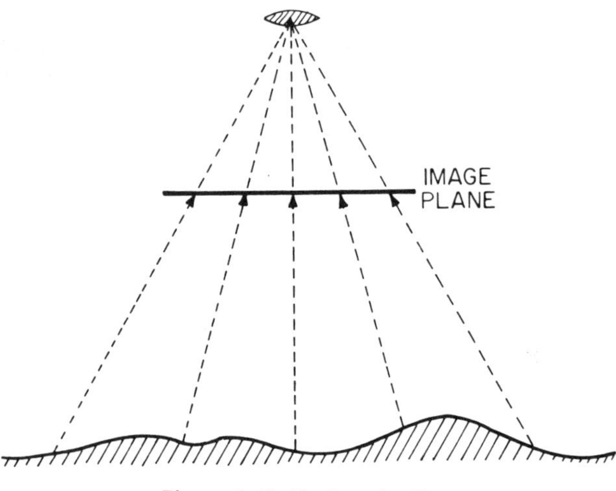

Figure 1 Optical projection.

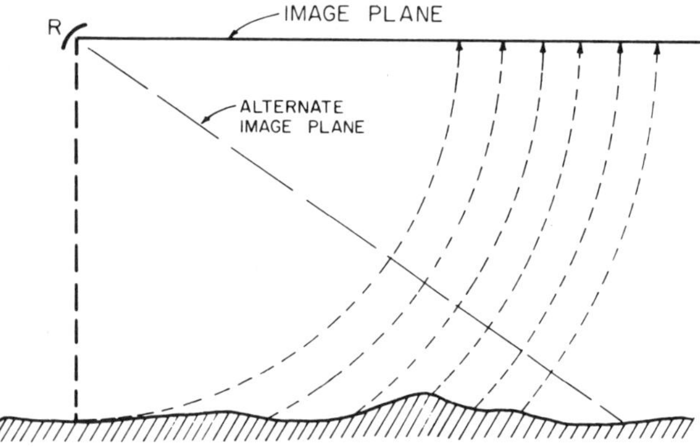

Figure 2 Radar projection.

circuits do best. Another element is the constancy of the speed of the radar waves through the atmosphere. However, distances found by timing the travel of waves are distances measured along straight lines from the radar to the objects. When an image-locating transformation, or projection, is accomplished in this way, the result is quite different from lens-imaging.

In the latter, points on the terrain are projected onto the imaging surface via straight lines through a common point (Fig. 1). But the radar projection (Fig. 2) transfers terrain points to its image plane along constant-range arcs about a common center. In a sense, these two kinds of projections are orthogonal to each other. Certainly the results look different—and they should be *looked at* differently.

Either image plane is viewed nearly perpendicularly. In the photographic case, the viewer who wishes to see a realistic perspective will place his eye at the lens point, so that his lines of sight to various image points match the original projection lines. This can be done because the eye uses the same projection scheme as the camera.

No such exact match is possible in viewing the radar image of Figure 2, because the original projection lines are arcs. There is no one eyepoint that can give an exact visual perspective. Furthermore, an "eyepoint disparity" exists: even the eyepoints offering the best pseudo-perspective never correspond to the radar position. One thing the radar-image interpreter can do to help put these images into an orderly and pseudo-familiar reference frame is to create for himself a mental viewpoint about what he is looking at, and to make his interpretations in the context of that viewpoint. One such viewpoint is developed below.

One of the visual cues to perspective is the foreshortening of shapes that are seen obliquely. Perspective foreshortening has two effects: (1) it distorts the image so that it is not geometrically similar to the terrain ground-plan, and (2) it causes the available linear resolution in the image plane to represent poorer ground resolution in the foreshortened areas than in the more accurately portrayed areas.

Radar images show similar foreshortening. Within any small part of the radar image (Fig. 3), short lengths of the arcs approximate parallel lines. Each local part of the image therefore approximates an orthographic projection of part of the terrain onto an oblique plane, such as RO. For terrain located at increasing depression angles from the radar's horizontal, the obliquity of the apparent orthographic projection plane increases. Since the apparent direction of viewing follows at 90° to this image plane, the foreshortening cue causes terrain far from the radar to appear as if seen from nearly above, while terrain closer to the radar appears to be seen at increasingly oblique aspects. That is, the viewer sees each part of the terrain as if from a different elevation angle (Fig. 4). The terrain nearly below

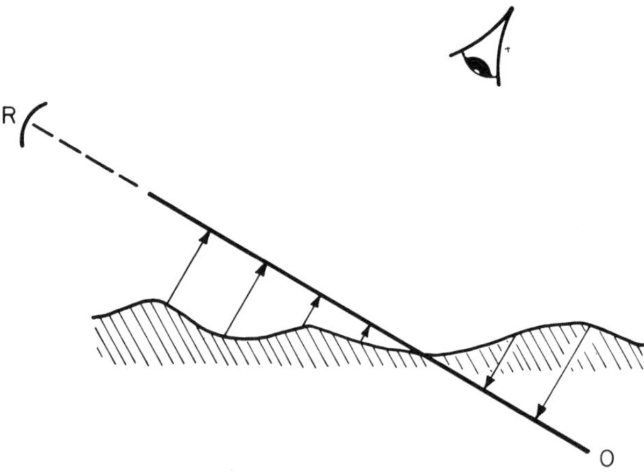

Figure 3 Local approximation to radar projection.

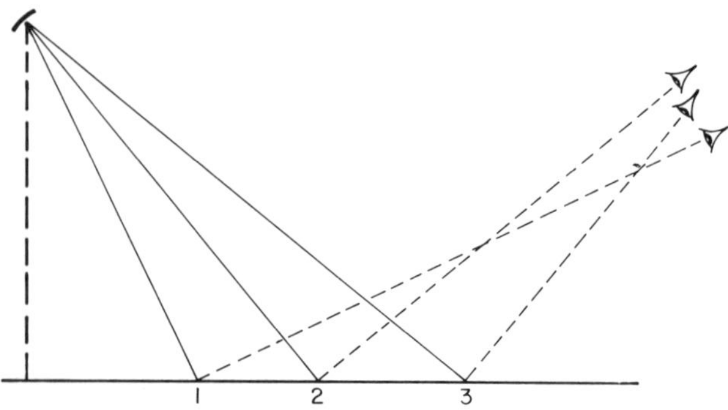

Figure 4 Apparent viewing directions vs. depression angles.

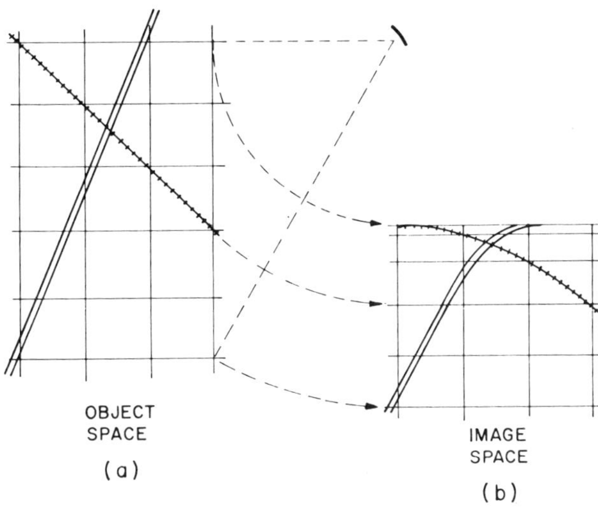

Figure 5 Ground-plan to radar-image transformation.

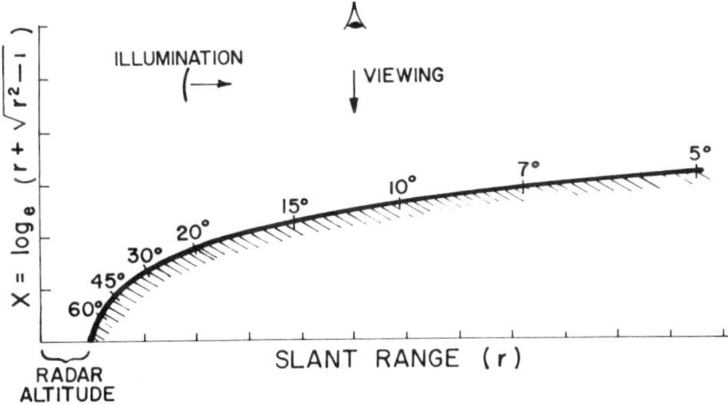

Figure 6 Orthographic model of terrain surface.

the radar appears as if viewed at nearly grazing incidence, and the nadir line (the ground track of a side-looking radar) looks like a horizon. In order to show this appearance, Figure 5a has been drawn to represent an area with a square road grid crossed by straight sections of a diagonal railroad and a divided expressway. Figure 5b shows how this terrain pattern would appear on a side-looking radar display covering depression angles from 90° (i.e., straight down) out to 30° from the horizontal.

Although this appearance may seem to represent the actual curved earth, it does not. The curvature seen is not related to that of the real earth, but to radii of curvature about the radar, which are on the order of the radar altitude. Also, the apparent surface is not a uniformly curved one. Figure 6 shows the form of the curved surface whose orthographic projection foreshortening matches the radar image; this curve is logarithmic. It is also expressible as the hyperbolic cosine curve, $r = \cosh x$. The curved lines in Figure 5 are simple hyperbolas. Radar ground-ranges are measured along the curve from $(r = 1, x = 0)$. The terrain's verticals are perpendiculars to this curve.

There is another perspective cue that matches (and thus re-enforces) the foreshortening one. This additional cue is the one given by shadow lengths and by the variations of image brightness with changes of slope. These characteristics appear as if the curved surface in Figure 6 were illuminated by parallel waves moving in the direction along the image surface (i.e., moving parallel to the r-axis).

When the above geometric model is understood and kept in mind, the radar-image interpreter can consider the radar image as a kind of oblique photograph. The fact that the radar image and the nearly matching oblique photo are made from sensor positions that are orthogonal (with respect to the object being viewed) does lead to a significant difference. Photography can be made to give a nearly undistorted plan view of flat ground out to somewhat beyond 45°, but encounters increasing amounts of foreshortening when used to form oblique images of terrain farther from the nadir. Radar, on the other hand, badly distorts the plan view of that part of the ground that is most nearly below, but loses this fault at lesser depression angles, hence at greater ranges. The distortion of a radar image at 45° depression angle equals that of a 45° oblique photograph. But, although there are equal angular regions over which each can be the victor in a distortion race, the ground area that can be contained in the radar's best-performance octant is several times larger than that in the photograph's. For example, Figure 7 shows some swath coverages available within different angular coverage regions. Here, at last, is the specific statement of the earlier claim that radar can satisfy certain image-quality criteria over greater areas than can optical-wave imaging devices.

My point is not to claim, via the argument above, that radar is a better sensor than photography. That would be a foolish claim. What I do want to point out is that there is a large geometric domain in which radar has certain advantages. But it is equally true that photography can far surpass radar (under the proper weather conditions) in the near-vertical domain, where scanning-radar images are at their poorest. What is most important in the comparison that I have made is that the domains of each are complementary. Aerial photography produces the best ground-plan image and makes the best use of its own resolution in the region nearly below the aircraft; airborne radar best meets these same requirements far to the side of aircraft's ground-track. Hence these two sensors complement each other's coverage when operated simultaneously from a single airborne remote sensing vehicle. As for mirror-scanning devices, since their domain is about the same as photography's (due to their dependence upon angular resolution), radar's geometric coverage is complementary to the coverage of these devices also.

Another part of the price paid for long wavelengths is that, when the wavelengths are as long as these,

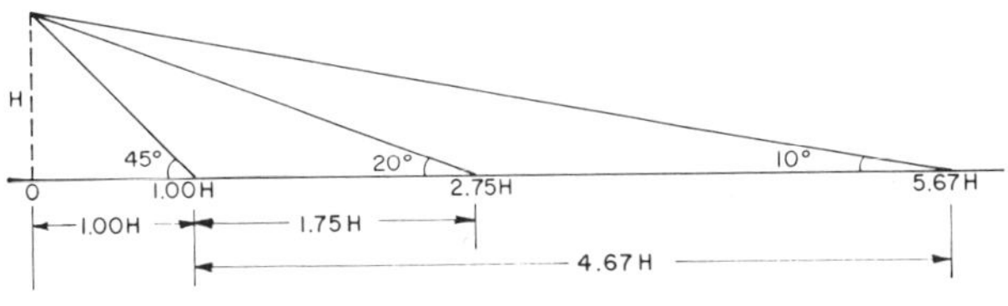

Figure 7 Terrain coverages.

many surfaces which appear rough to the much tinier optical waves will appear mirror-smooth to radar waves. When a radar attempts to form an image of such a surface, the surface will indeed act like a mirror. Even a brick wall acts in this way, and so do concrete and asphalt pavements. When illuminated by a radar beam, they reflect that beam in some well-defined direction. Usually, most specular reflections travel away from the radar, since not many flat surfaces will happen to be oriented perpendicular to the incident beam. However, where walls and pavements meet, successive specular reflections occur. In the particular case of three plane surfaces that are perpendicular to each other, the three successive reflections that can occur will combine to return the beam back along its original path, hence directly toward the radar antenna. Cultural areas, such as cities, are commonly created of plane surfaces, with numerous right angles. Radar echoes from such areas are therefore a matter of feast or famine—bright glints from interior corners and from surfaces fortuitously perpendicular to the beam, but little or no return from most isolated surfaces such as streets, roofs, parking lots, closely trimmed lawns, ponds, etc. The manmade world, as seen through a radar antenna, is a glittery, sparkling world, not so easy on the eyes as the optical one we are accustomed to seeing. It can present, to a degree, some of the confusion of a hall of mirrors and some of the mystery of a sculpture made of transparent material. But there is plenty of reality, too—patterns of streets, outlines of open areas and shore lines, alignments of fences and bridges, the continuity of rail lines, and numerous "signatures" of specialized arrangements of construction. These features and many others do prove to be detectable and interpretable and useful, especially to the practiced interpreter.

The photographic equivalent of this highly specular situation would be a world in which most cultural objects were chrome-plated—and this is certainly not the world we know today. But from an interpreter's viewpoint, this is not too bad a way to build the world he must look at. After all, reduction of optical visibility is most of the reason for putting so little chrome on military vehicles. Even everyday objects which don't mind being seen (but which make no particular attempt to show up, either) seldom contrast strongly with their surroundings in aerial photographs. In fact, weak contrast is a rather common thing in this imagery. Radar at least has enough contrast, and it is reasonable to expect that we will find ways to control an excess. The situation would be less hopeful if radar images showed less-than-useful amounts of contrast.

Quite a few sentences ago, I promised to say something about some ambiguities that affect the radar location and identification of surface materials. A fairly obvious one is the ambiguity between the ground-ranges and the elevations of objects whose positions in the radar image have been accurately timed. A time-position really specifies only that the object is located on a particular arc, but this amounts to specifying a whole family of elevation/ground-range locations, rather than any specific position. This is a well-known effect, and it is not unique

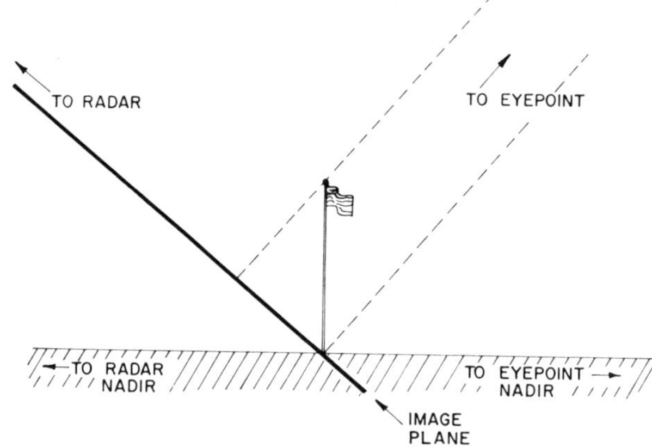

Figure 8 Relief displacement.

to radar. This is relief displacement, and is also encountered in aerial photography, where it sometimes is a nuisance but at other times is a positive value, as when used for stereoscopic effects.

You may read or hear comments that the relief displacement in radar images is in the direction opposite to that in the familiar photo images. This bit of truth does not mean that the relief features will have erroneous appearances in radar images. On the contrary, when the disparity between the radar position and the proper interpretation viewpoint is recognized, the relief appearance presented to the interpreter is found to be in the familiar direction, relative to that eyepoint (Fig. 8).

Unfortunately, taking the proper viewpoint (and thus getting the proper perspective) does not eliminate the ambiguity. It is not possible, with either radar images or photographs, for a single two-dimensional image to indicate terrain slopes unambiguously. The additional information needed to resolve this ambiguity might come from an outside source, such as a map, or even from the knowledge that regular cultivation patterns imply nearly flat terrain. Another well-known way to get the information is to use the stereoscopic approach, combining two images made from different locations of the same sensor. This method is applicable to radar images as well as to photographs. (I will say more about that subject later.)

This ambiguity not only leaves us with a lack of desirable topographic information, but also leaves the door open for still another kind of ambiguity. If we know the direction of the illumination, but do not know the slopes of the terrain, then we cannot be sure of the illumination level on a particular area. Measures of image brightness, however accurate, are then uncertain indicators of the reflecting properties of surface materials until the slope question has been answered.

It would be convenient if the rules of "radiometry" for radar images were like those of photometry for optical images. But, as I have said so often already, the rules are not the same. An optical image contains models of those projected areas of the various scattering surfaces which contribute optical-wave energy to the camera aperture. The brightnesses of those surfaces are therefore reproduced substantially proportionally in the photographic image. In a radar image, the projected area displayed in

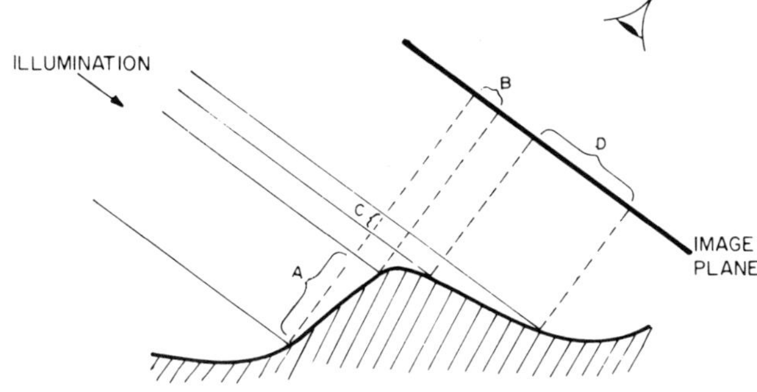

Figure 9 Projected areas toward radar and eyepoint.

the interpreter's eyepoint direction is not a projected area facing the radar. Thus the energy returned to the radar, and presented in the image, is governed by an area different from the model area over which its analog is distributed in the image. There is no proportionality between the energy per unit area in the object and image spaces, as there is in a photograph.

To give just one example, consider a steep slope that is nearly perpendicular to the radar beam, followed by another slope nearly parallel to the beam, as in Figure 9. Echo energy contributed from large projected area A becomes "piled up" in small area B of the image, while that from C is spread over D. The difference between the brightness of the two slopes is even greater than it would be if the image areas were models of the terrain areas facing the radar. This fact leads to a radar "exaggeration" of relief, which has been appreciated by some investigators, particularly those who have sought geologic information in radar images.

Incidentally, the increased terrain brightness at B is not related to the specular effects mentioned before, even though surface A is nearly normal to the beam. The pile-up effect occurs even when the slope facing the radar is rough natural terrain, giving no specular reflection.

To persons working in the field of radar, references to ambiguity usually imply two particular problems, neither being the ones described above. Both of these standard problems result from the fact that imaging radars illuminate their object fields with short pulses of radiation, as I have mentioned previously. Things that happen between these blinks may be misinterpreted if the spacing of the pulses does not happen to fit the situation being viewed. One of these standard ambiguities due to pulsing has to do with measuring the speeds of moving objects. If the pulse sampling is not done fast enough, the measured speeds can be wrong. The basis for this effect is exactly the same as the motion picture effect in which the wheels of a vehicle seem to be going at some fraction of their proper speed, or even rotating backwards. The other standard ambiguity comes from pulsing too fast, so that echoes from the newest pulse begin returning before the last ones from the previous pulse have all come home. This results in range ambiguity, and looks exactly like a photographic double exposure, the images of near and far terrain being superposed on each other.

Choosing the pulsing intervals properly is the way to avoid each of these ambiguities, but if the speed of the radar aircraft should become very great, we could get caught by these dual requirements for fast enough and slow enough pulsing, being unable to satisfy both at once. Fortunately, only when remote sensing aircraft travel much faster than they do today will this last problem become a real one. It is interesting to note that one of the basic reasons for this difficulty is that the speed of light, which we so commonly consider "out of this world," becomes inadequate to our wants, in that it cannot provide us with one range-direction strip of information before a very fast vehicle will have moved up to the point where another look is needed.

I have one more image disturbance to discuss. But again let me set the stage by going through some other considerations, picking up the earlier topic of improving resolution by increasing antenna size. Suppose we want to use our 1 cm waves to get resolution four times better than 50 feet, or about 13 feet, and are willing to hold our range requirement down to only 3 miles. We then calculate that we need an antenna about 1200 wavelengths long, hence 12 meters long, or about 40 feet. Although this might fit on the aircraft (but it would be no easy job), something quite new has turned up:—now the antenna is larger than the resolution distance. This means that its beam must converge upon the objects that are being imaged. In effect, we are requiring our radar to become focused. A problem that arises immediately is that a focused beam will be in focus only at one range. About a mile closer to the radar, and also a mile farther away, the beam would be about 22 feet wide. In order to focus the beam everywhere in a range interval, we need some new kind of trickery.

The trick we have learned is called *synthetic-aperture radar*. It consists of using a small antenna, with a wide diverging beam operated in the side-looking mode, to collect data from a large area. By measuring the phases as well as the amplitudes of the echo waves, we eventually have on hand enough information so that a specialized computer (which does exist) can sort out the thousands of overlapping signals and determine the location of each separate source with great precision. We use the term "synthetic aperture" because what the computer does is to assemble the signals into sets of the sort that a very large antenna would have received if it had been carried along the same flight path. Several unexpected benefits, in addition to the expected ones, have been found to be offered by synthetic-aperture systems. Not only can they be focused at all ranges, but they can get just as good resolution at long ranges as at short ones. (There are maximum limits, of course, but these are a satisfyingly long way out.) Also, we need not restrict our design of this kind of radar to very short wavelengths, for their resolution no longer depends on wavelength. And the simulated antenna can be hundreds, or even thousands, of feet long, since it is no longer constrained by the aircraft size. These are only a few of the benefits. There are also some undesirable features.

The attribute I am trying to bring up is not a feature of synthetic-aperture radars alone. However, it appears most noticeably when radars are focused, and, in practice, this nearly always means synthetic-aperture radars. When a radar that is not a focusing radar uses its diverging beam, it looks at each object over a variety of aspects during the time that the object is in the beam. But a focused radar effectively looks at each object only while the object passes through a very small region near the axis of the antenna. Over this short distance, the phase of the wave that indicates the object varies relatively little. The waves from adjacent objects have unrelated phases, and can interfere with each other in unpredictable ways. The result is that a focused image is criss-crossed with dark bands which are a kind of interference fringes. The energy removed from these dark fringes accumulates rather randomly in islands of brightness, some strong, some weak. The size of many of these islands is about the same as the resolution of the system. They can be thought of as a sort of half-tone dot structure which carries the picture information. But both the pattern and the strength of these dots is irregular. As a result, some of the desired image information becomes harder to make out, especially those details whose sizes are close to the dot size.

The general appearance, and the reasons behind it, are exactly the same as in images seen by the scattered light of a laser. In both cases, the effect is due to the nearly "perfect" coherence of the illuminating waves. But this "perfection" has some associated imperfection. Our eyes are neither adapted nor accustomed to using unvarying coherent light. The kind of illumination they are suited for is a fast succession of large numbers of individually coherent waves. In this mode of operation, many successive coherent images are formed, each having its own interference pattern. The eye averages these images, and normally has so many available that the combination has smooth tones while retaining the occasional sharp details. The individual coherent images, on the other hand, have no smooth tones, but an excess of fine detail, most of it spurious.

When I call it an excess, I do not mean that it does not have a meaning—it really tells about the wave phases in the object field. Technically, this can be useful information. Subjectively, though, it is useless. The trouble is that our eyes do not care about phase. They want to know about wave intensity, or the corresponding wave-amplitude. What the coherent image presents to them, instead, is a lot of *products* of wave-amplitudes multiplied by sines (or cosines) of phase angles—but no key to permit separating the amplitudes from the phases.

In general, the way to remove this ambiguity and leave only an amplitude pattern (hence an intensity pattern) is to get more information about the phase. A specific way is to obtain samples having enough different phases to approximate "all possible phases." We can do this optically, using laser illumination, and thus create images that show direct comparisons of the appearances of coherent and noncoherent images of the same object field. Figure 10 shows such a comparison. The coherent image is on the left, the noncoherent one on the right. The most important fact about these images is that *the resolutions* are the *same*. Both were made with the same lens aperture and wavelength (helium-neon laser light), and the detail shown here is diffraction-limited by those quantities. These images show rather well that resolution is not the only criterion of interpretability. Smoothness of tone is also highly important.

Let me emphasize again that the images in Figure 10 are *not* radar images, but optical images. Completely coherent radar images would have structures resembling the image on the left. An image as smooth as the one on the right would require so many separate radar contributions as to be impractical if not impossible. Diverging beam radars form images with irregularities of tone somewhere between these extremes. Let me also point out that even focused radars, while they have nearly fully coherent features, are generally less disturbed by this effect than are optical images like the ones in Figure 10. This is a benefit of the very high contrast that I previously said was a feature of radar imagery. Only waves of nearly equal strengths produce strong interference patterns. The very fine, very strong, cultural-object returns in radar images seldom have similarly strong neighbors located close enough to disturb each other. Natural terrain, on the other hand, usually has considerable fine structure already, so that sizable areas have to be observed for interpretive purposes. Over such areas, the coherence structure becomes a minor irritant, as is film grain in photography.

My final topic is radar stereoscopy. This is a nearly abstract subject at this time, because we don't know much about it. We are sure that it is desirable, partly because of its use in gaining topographic information, partly because it can aid in specifying radar-wave incidence angles and thus give more specific meaning to measures of returned-energy levels, and partly because it would aid the recognition of objects by giving them three-dimensional signatures instead of two-dimensional ones.

The presence of relief displacements in radar imagery means that the basis for stereoscopy is present. But the nonvisual perspective of the radar projection raises its head again, this time in a disturbing way. In aerial photographic practice, both of the image coordinates are perpendicular to the direction of wave-propagation. Either of these image coordinate directions can be used for the stereo baseline. The one ordinarily chosen is the along-track one, so that the pairs of exposures can be made at different times along a single flight path.

Because of the difference between the photo and radar projections, it turns out that the best stereo baseline for the radar does not lie in the flight path. It is then necessary for stereo pairs of side-looking radar imagery to be made in two parallel flights, in order that the differences between the two images be such as to produce the needed third coordinate. This dual-pass requirement is a very undesirable one, not only because of time and cost considerations, but because of the difficulty of precisely determining, even after-the-fact, the amount and direction of the path separations which can vary continually along the paths.

Because radar carries its own illumination, the direction of illumination is different in the two passes. The resultant differences between the image tones representing the same piece of terrain in the two radar images

Figure 10 Coherent and noncoherent optical images.

interfere with binocular fusion of the two images, since they are no longer just separate views of identical subject matter. Much of the reason for the seriousness of this change comes from the fact that the distance between the two flight paths is no small thing. In order that convergence angles of two views of any single object can be considerable (as in photographic practice, convergences of tens of degrees are wanted), the paths will be miles apart when slant ranges are also several miles.

Because radar appearances of many objects (especially specular ones) are more sensitive to viewing aspect than is the common case in photography, the admissible convergence angles between radar stereo image pairs may be less than those useful in stereo photography. A particular possibility is that we may be constrained to keep both flight lines on the same side of the image swath, since radar returns from sharply defined objects come largely only from the vertical side nearest the radar. For example, suppose a radar with 50-foot range resolution shows returns from a factory building 100 feet wide, but shows opposite sides in the two images. Attempts to combine these returns so as to learn the terrain elevation would lead to a large error, if not to a complete failure to associate the two unlike elements for, in this case, to mis-associate them).

Most worrisome of all is the probable effect of the coherent breakup that is present to some degree in all practical radar imagery. Since the speckly patterns in two independently formed members of a stereo pair will superpose unrelated fine structures on the real image content, these dissimilarities will tend to obstruct binocular fusion. Visual stereo succeeds as well as it does because it can actually match images with a precision several times finer than the resolution within either image. Such performance is possible because the image-to-image similarity of the response of an optical sensor (either a camera or a pair of eyes) normally exceeds the object-to-object dissimilarity that permits resolution. This is something like saying that densitometer traces over two point images can be matched to about the width of the line that drew them, whereas the point-image resolution possible when two such images are merged is much coarser. But radar images containing speckles of point-image size and larger can be matched to precisions only somewhat worse than the resolution, rather than somewhat better. The effect is like that of half-tone dots upon the quality of press-published stereograms. (It is also a feature of binocular observation of all optical images made with coherent light, including holographic images.)

Now that I have seemed to list all of the things that possibly could be wrong with radar imagery, I must point out that I do like radar imagery. It does show much excellent detail, it does locate detected objects precisely, it does have readily interpretable appearances, it does detect objects that other sensors do not, and it does do these things even under some optically hopeless conditions. The interpreter, on the other hand, does have to apply himself to learn the ways of this still-novel sensor, in order to be able to accomplish truly useful things with it. Vendors and purchasers of radar-imaging services also must learn its unique features and its differences from other sensors, in order to assign to radar those tasks that belong to it more than to other sensors, and to leave to the other sensors those tasks that radar does less well.

The range of topics I have tried to touch upon in this paper certainly does not represent one person's conclusions, although I do accept all responsibility for the manner in which they are stated here. Production of the radar imagery that has led to recognition of these features has been done by several organizations in addition to the Radar and Optics Laboratory of The University of Michigan. The writer must acknowledge the value of this material, and also the interpretive efforts of numerous other persons in several organizations, to the evolution of the ideas so briefly "explained" above. Particular credit is due to Charles Liskow of our Laboratory, who has been relentless in pointing out to me and others numerous features, faults, and follies affecting radar image interpretation.

SOME AREAS *on the surface of the earth have an almost perennial cloud cover. Periods of clear weather occur sporadically and unpredictably, and usually are of short duration. If such an area is to be photographed, expensive equipment (aircraft and camera systems) and valuable personnel might be on stand-by for weeks waiting for the weather to clear. If clearing occurs at night, the photography mission is still delayed. Obviously, what is needed is a sensing system that is not dependent upon solar illumination and is unaffected by clouds and moderate rainfall. Further, the system must be able to present a detailed image of the terrain under the cloud cover. Radar is such a system. Modern side-looking airborne radar (SLAR) models terrain sharply with an illuminated and shadow effect. Because the radio waves are generated at an oblique angle from above the land surface, they impact on landforms with about the same angle of incidence as early morning or late afternoon sunlight. Land surfaces that are higher, steeper, and facing toward the radar antenna produce a strong return. Surfaces that are lower, less steeply sloping, and facing away from the antenna produce a weaker return or perhaps none at all. A steep-sided, conical hill or ridge line appears with a brightly illuminated side—facing toward the antenna; this is a strong return. The same hill or ridge shows a heavily shaded, dark side—sloping away from the antenna; this is a weak return. The over-all effect produces an image remarkably similar to a carefully shaded relief map. The major ridge lines, surface structures, and lineament trends are visible as if mapped by conventional terrain shading methods. Drainage basins are clearly delineated. The radar image thus lends itself to regional geologic studies and geologic reconnaissance surveys. Careful examination of the image can reveal the general rock type in the area.*

34-SLR Reconnaissance of Panama

ANDRIS VIKSNE
THOMAS C. LISTON
CECIL D. SAPP

SIDE-LOOKING RADAR (SLR) was first developed for military targeting purposes in the early 1950s. It soon became obvious that such radar imagery would be capable of providing significant military geographic intelligence. As radar systems became more sophisticated and imagery resolutions improved, it was natural that SLR should also be suggested for nonmilitary applications in terrain reconnaissance surveys.

SLR can be considered a geophysical tool in that it provides a measure of the reflectance characteristics of the overflown terrain, the results of which are displayed as an image resembling an optical photograph but having lower resolution. Basically, radar imagery represents a record of the interaction of transmitted electromagnetic waves with nonuniform natural surfaces of the terrain. Soil and plant moisture content, dielectric properties of rock and soil, vegetative cover, surface roughness, geometry, and penetration characteristics of the particular frequency at which the radar is operated are the most significant parameters affecting the radar return.

In addition to its capability of providing wide areal coverage in a minimum amount of flight time, operation of a SLR system is almost independent of diurnal and atmospheric conditions owing to its being an "active" sensor (i.e., it transmits electromagnetic waves, thereby providing its own illuminating source), and to the fact that these waves are sufficiently long to penetrate cloud cover of all but the cumulonimbus type. Thus, the system has a day/night and all-weather capability.

Project RAMP (*RA*dar *M*apping of *P*anama) was initiated in 1965 by the U.S. Army Engineer Topographic Laboratories (formerly USAE-GIMRADA) to demonstrate the capability of the AN/APQ-97 SLR system to produce high-resolution radar imagery to be used in lieu of optical photography. The Darien area of Panama chosen for this operational test consists of about 6600 square miles (Fig. 1). Once-over coverage, with overlap, was obtained in approximately 4 hours flying time.

A primary consideration in the selection of Darien Province as a test area is that it has a near-perennial cloud cover that severely limits the acquisition of conventional, optical, aerial photography. The results obtained by this SLR experiment provided the first complete overview of the Darien. Prior to this experiment, persistent and unsuccessful attempts had been made for nearly 20 years by the U.S. Air Force and others to acquire optical photography of the area.

From *Geophysics* 34:54–64, 1969. Reprinted with permission of the authors and *Geophysics*.

Sensor Characteristics

The AN/APQ-97 is a real aperture radar system operating in the K band. It has better than a 99 percent cloud-penetration capability, but does not, however, penetrate heavy rain or vegetative cover. The high-resolution imagery is recorded in a slant-range presentation in which the scale increases with distance from the aircraft ground track. Nomographs are available for conversion to ground range, if this is required. The sweep is down and to one side of the aircraft (Fig. 2).

Procedures and Results

In order to acquire stereoscopic and cross-flight coverage, six missions were flown over the subject area from various directions. After eight of the best parallel flight strips were selected, an uncontrolled mosaic (Fig. 3) was constructed to the scale of 1:250,000. An uncontrolled mosaic is one which is not based upon ground control. The mosaic was subsequently reduced to a scale of about 1:850,000 for inclusion in this paper.

A number of overlays to the mosaic were prepared from interpretations of the SLR imagery, and four of these have been selected for discussion in this paper. These include:

1. Surface drainage
2. Surface configuration
3. Vegetation
4. Engineering geology

Surface Drainage

A valuable tool in reconnaissance terrain studies, particularly in areas of low-to-moderate relief, is the regional and local expression of the surface drainage. The recognition of variations in the distribution patterns, differences in drainage density and texture, depth of channel incision, etc., permit inferences to be made concerning structural and compositional conditions of the underlying terrain.

Drainage features are particularly well-defined in SLR records. Bodies of water give the appearance of black, or no-return, areas because of their smooth, horizontal surfaces. Black areas on imagery are due either to nonreflectance of materials or to reflectance away from the receiving antenna. Any surface whose roughness is less

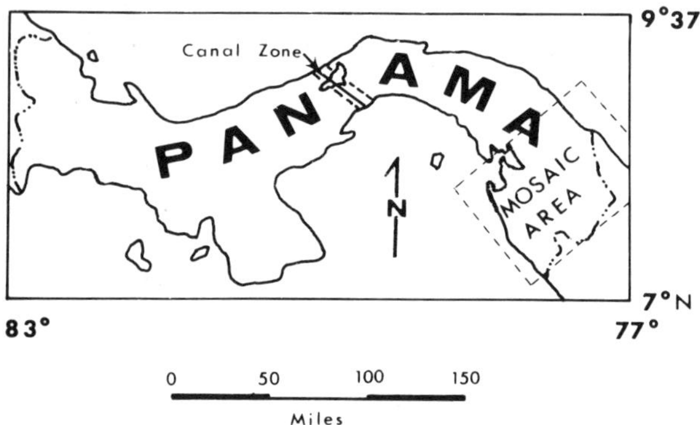

Figure 1 Index map.

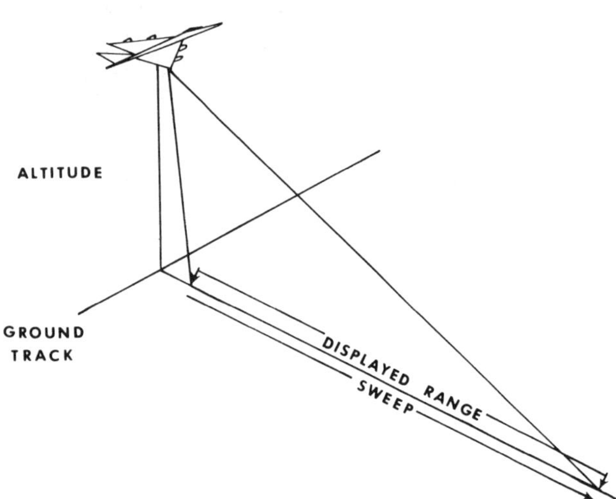

Figure 2 SLR sweep characteristics.

than half the wavelength of the SLR system acts as a specular reflector, while surfaces whose roughness is greater than half the wave-length act as diffuse reflectors. For example, a smooth water surface acts as a specular reflector, whereas rough or choppy water is a diffuse reflector. Drainage patterns are further defined both by their continuity and the fact that the radar energy rebounds from the banks and marginal vegetation of drainage channels, producing a brightly defined edge. The resulting contrast facilitates the delineation of such areas.

Drainage patterns are diagnostic of specific terrain conditions. To varying degrees they are capable of suggesting the nature and depth of soil cover lithology, morphology, and the structural and tectonic influences prevailing in the area. The patterns present themselves in an infinite variation of density and habit (or form), the density being mainly determined by the lithologic character of the rock traversed, i.e., hardness, porosity, solubility, and consolidation, while the habit is determined primarily by structure, i.e., faults, fractures, and the attitude of bedding. Thus, the drainage pattern overlay, as shown in Figure 4, was the first to be developed from the imagery and was thereafter used as a base for other overlays.

Some differences were noted between the SLR imagery and existing maps of the area. Most notably, the course and direction of flow of one of the rivers were incorrect on all maps examined. Many other minor examples, mostly changes in character of shoreline or drainage features, were noted. Although some of these differences might possibly be attributed to recent changes in the stream pattern, the existing coverage maps being quite old, most differences were the result of limited photo coverage and lack of detailed ground information concerning the area.

Surface Configuration

SLR records are an excellent medium for the portrayal of regional landforms owing to the continuity afforded by the wide, long-image segments. The "generalizing" ability of SLR, which permits the elimination of superfluous detail (such as individual tree crowns, etc.), provides the interpreter with a synoptic view without obscuring the presentation of major landforms.

The elements most useful in the interpretation and analysis of landforms on AN/APO-97 records are topographic expression and drainage patterns. The characteristics of a landform depend primarily upon its composition, morphology, and structure, the climate in which it was formed, and the various erosive processes.

As outlined in Figure 5, the regional geomorphic features of Darien consist of Plains, Low and High Hills, and Mountains. This classification of landforms was made on the basis of local relief, which is defined as the "difference (in elevation) between (the) highest parts of interstream areas (or with mountains, the crests) and the adjacent valley bottoms."

As was the case with the drainage and coastal features, substantial differences were observed between the SLR presentation of landforms and their portrayal on existing maps of the area. One such difference is the location and orientation of one of a series of hills east of the Tuira River and inland from a major swamp area (near the center of the mosaic). The major axis of this anticlinal hill is shown on earlier maps to be aligned approximately northwest-southeast instead of northeast-southwest as shown in the radar imagery. Previous maps of this specific area were made by ground surveys with no aerial photographic coverage available to serve as a check.

Some other outstanding differences observed were the character of coastlines, the alignments of minor ridges, and the alignment of the Continental Divide. In addition, numerous minor topographic expressions on the plains appeared on SLR but were not found on any maps.

Vegetation

As stated previously, under Sensor Characteristics, the K-band SLR does not penetrate vegetative cover, thus making it possible to evaluate the various vegetation forms on the basis of their reflectance characteristics. Of course, a general knowledge of the vegetation associations existing in the particular geographic location of the imagery, and the knowledge of the time of year the

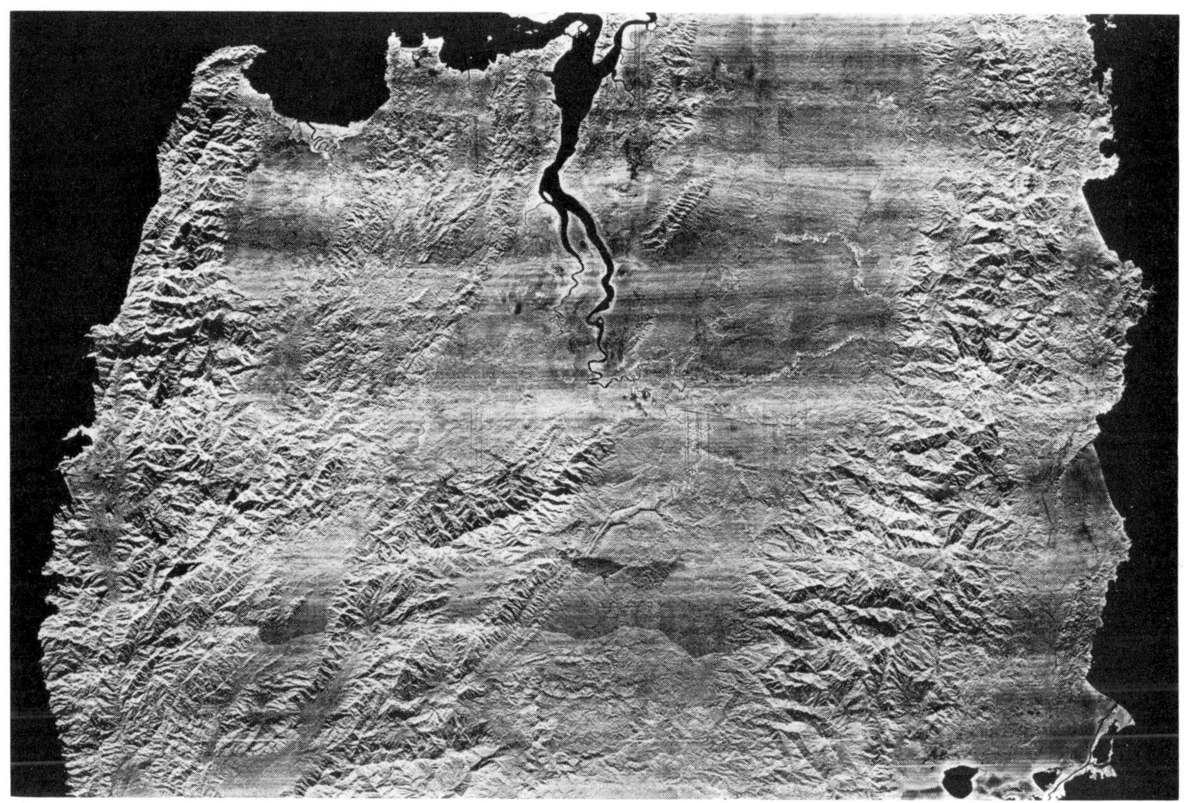

Figure 3 Radar mosaic of Darien and northwest Colombia.

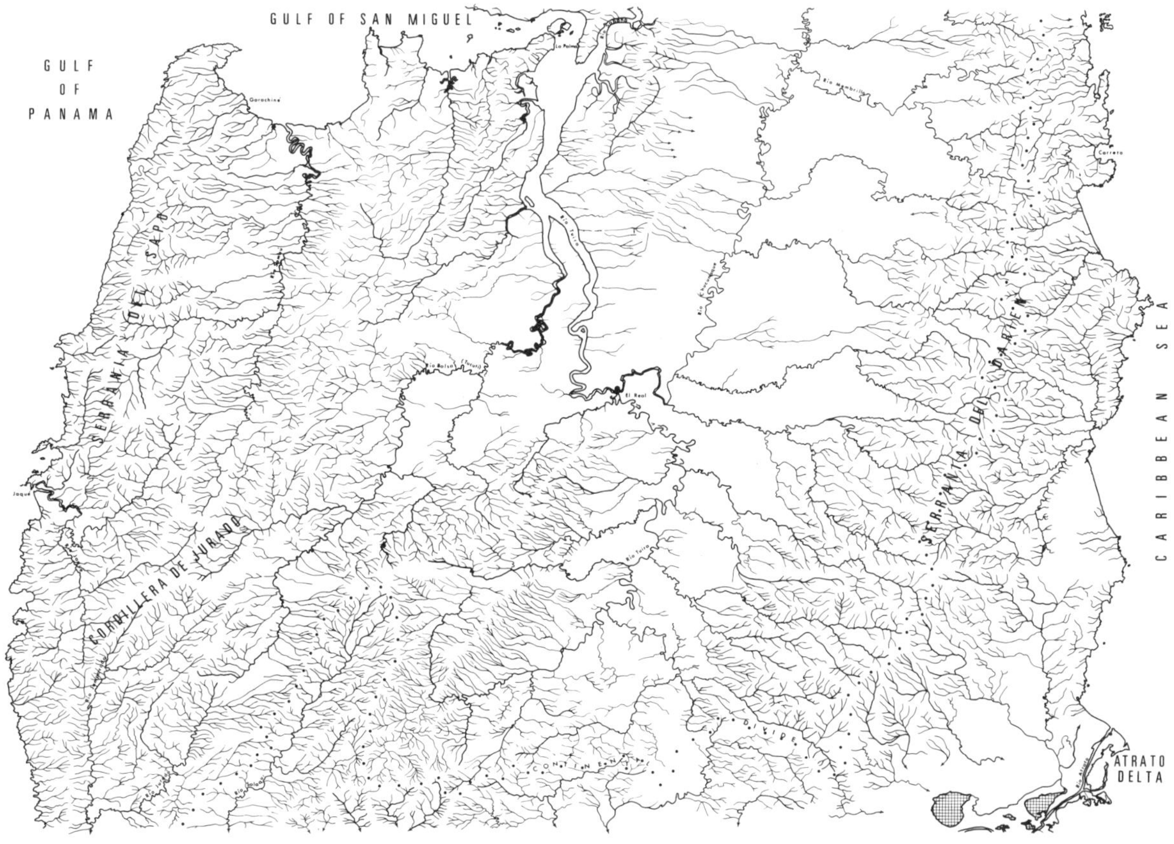

Figure 4 Surface drainage.

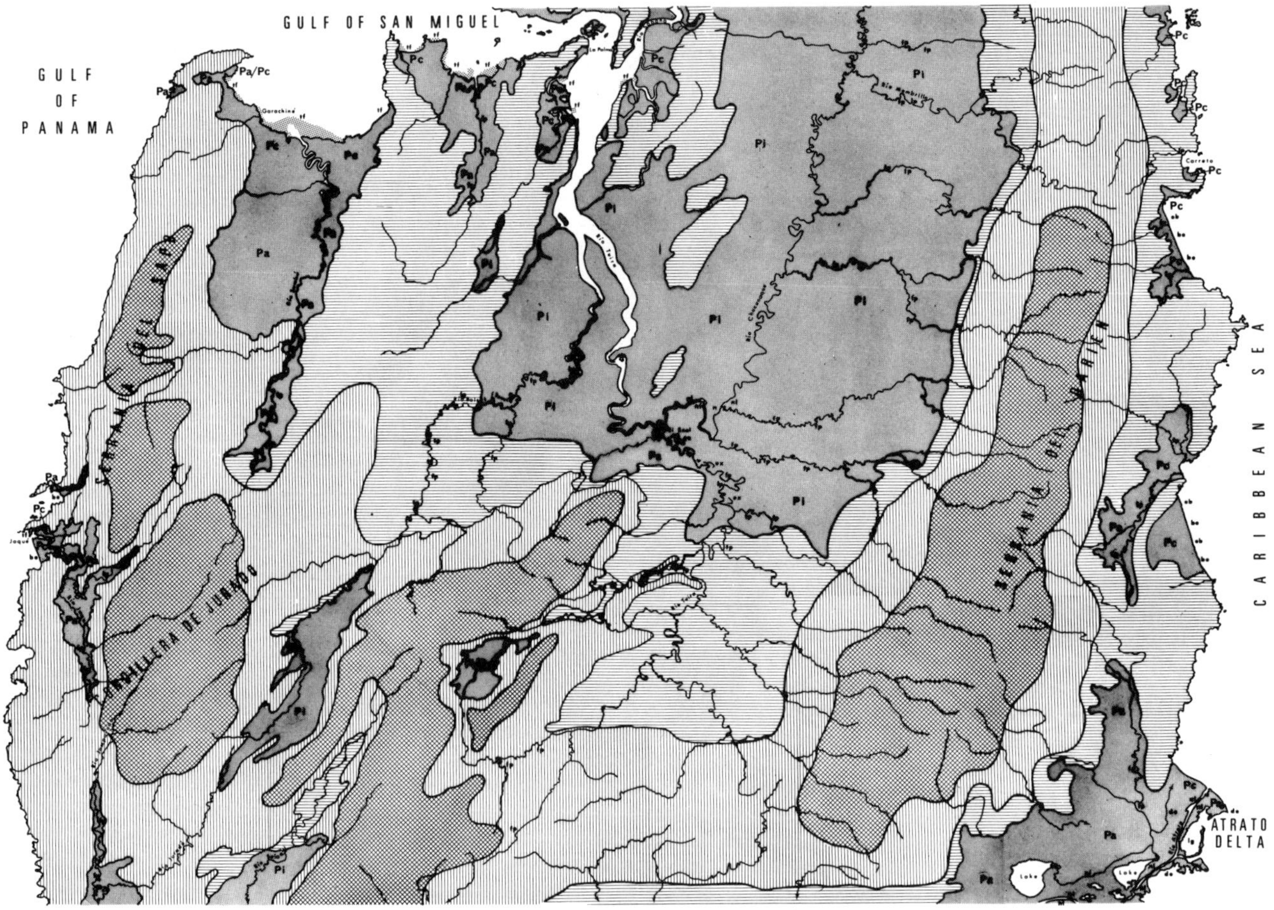

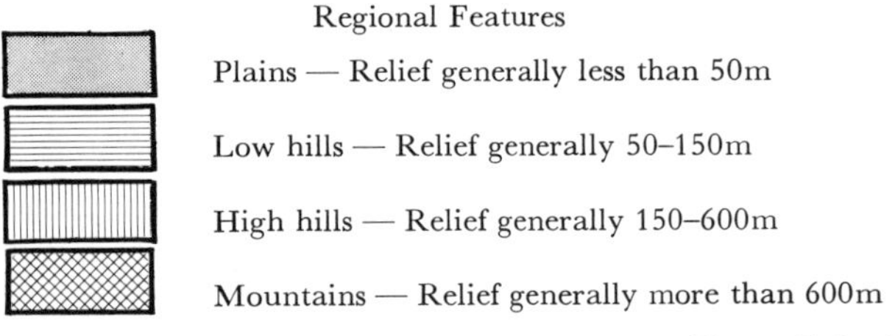

Regional Features

Plains — Relief generally less than 50m

Low hills — Relief generally 50–150m

High hills — Relief generally 150–600m

Mountains — Relief generally more than 600m

Local Geomorphic Features

Pu—Upland plain	lg—Lagoon
Pc—Coastal plain	be—Beach
Pi—Interior plain	ob—Offshore bar
Pa—Alluvial plain	nl—Natural levee
ox—Oxbow lake	tf—Tidal flats
de—Delta	fp—Flood plain

Figure 5 Surface configuration.

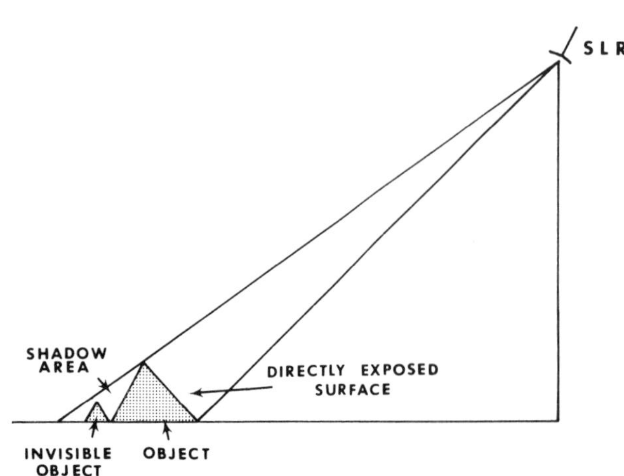

Figure 6 Radar-produced shadow effect.

imagery was obtained, are necessary before any assessment of specific gray tones and textures, in terms of particular vegetation types, is possible. In the high-relief zones of the subject area, the varying tones of gray that reflect distinct changes in vegetation were sometimes obscured by the "shadow effect" inherent in the SLR system. That is, the steep surfaces directly exposed to the sensor show up much brighter than the surrounding area while the opposite side of such protrusions lie in a shadow area (Fig. 6). An additional difficulty in delineating vegetation strata in the high-relief areas was due, not to the radar system, but to the heterogeneous nature of the vegetation types found.[1] Distinct boundaries between the evergreen rain forest and the mixed semi-deciduous and evergreen forest (jungle), for example, could not be discerned on the imagery since these two zones (and other forest zones) blend into each other

[1] Figure 7, not reproduced here.

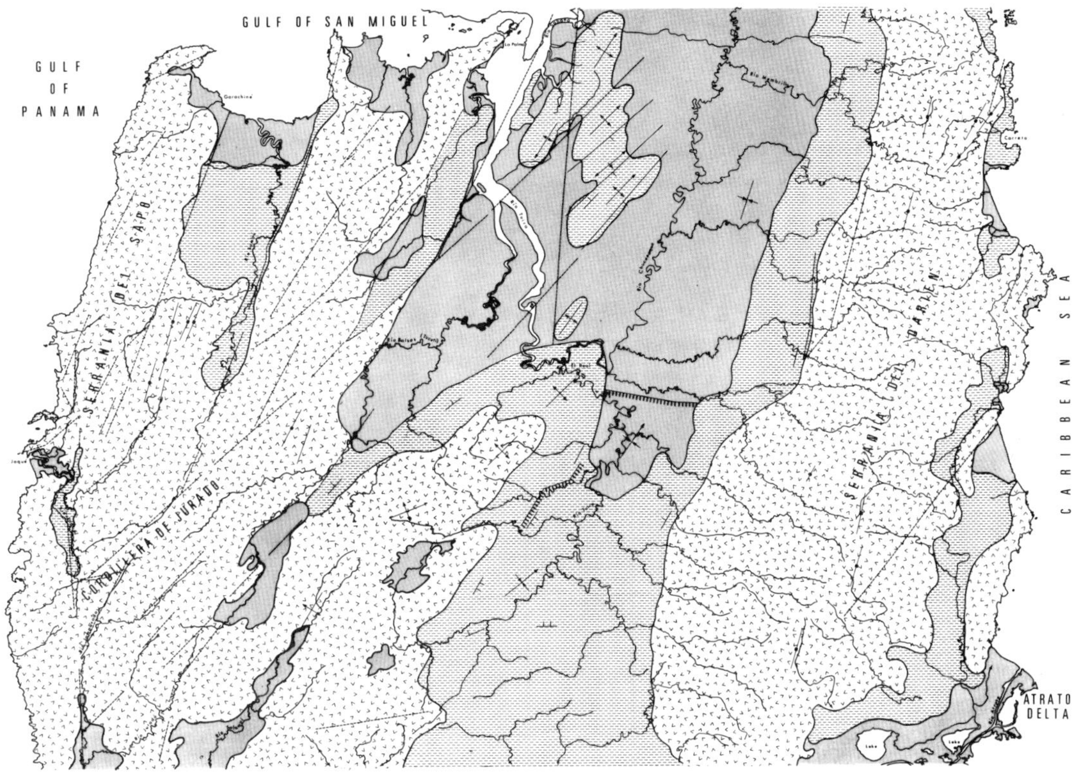

Igneous Rocks: Extrusive (Basaltic) — Andesite dominant, basalts common; medium to fine grained; gray, pink; massive; local irregular fractures common. Generally in mountains and hills; quarry sites hard to reach; heavy blasting required; suitable for building stone, aggregate, riprap, base course, and surface course; high bearing capacity; stable in steep slope; overburden to 6 m.

Sedimentary Rocks: Undifferentiated — Stratified chert, crystalline limestone (with interbedded tuffs), sandy shale, sandstone and conglomerates; in low hills and broad valleys; quarry site access difficult; light blasting required. Only hard sandstone, conglomerate and limestone suitable for building stone, riprap, aggregate, base/surface course; moderate bearing capacity. Others suitable for fill only. Moderately stable in steep slope.

Unconsolidated Deposits: Stream alluvium — Unconsolidated materials more than 6 m thick continuously covering bedrock. In plains, beaches, river beds, swamps, marshes. Generally unsuited for construction use except as fill.

Geologic Structure

Anticlinal Axis

Synclinal Axis

Plunging Anticline

Plunging Syncline

Strike and Dip

0°–30°

30°–60°

60°–90°

Fault (showing dip)

Fault (concealed or inferred)

Fractures and lineaments (dips not determined)

Dike

Escarpment (over 60% slopes; over 5 m high)

Figure 8 Engineering geology.

gradually over a distance of from several hundred meters to several kilometers. Thus, no distinct boundary exists.

In such areas, the best judgment of the interpreter is based on an "averaging" of the transitional zone for boundary placement. Some larger-scale aerial photos and local spot photography aided somewhat in the delineation of these difficult areas.

Engineering Geology

SLR imagery is also useful in the regional interpretation and mapping of folded mountain, plain, and plateau areas. The broad coverage accomplished through side-scanning and strip recording provides a view of large areas and permits continuous interpretation of the radar image. Aerial photography requires extensive mosaicking to provide a similar facility. Regional features are easily recognized and associated with other significant units necessary for an accurate areal evaluation.

Hard, resistant beds of sedimentary or igneous origin are easy to follow on the SLR records. Coarse clastic materials (thick sandstones, conglomerates, quartzites, etc.), with uniform image tonality and constant resistance, can be mapped continuously for many miles. Ridges produced by resistant beds are usually apparent and can be easily traced.

The major structural features of the area, and the structural attitudes of specific rocks in localized areas, were determined from the imagery by evaluation of topography, outcrop, drainage patterns, and vegetation variations.

The Engineering Geology Overlay (Fig. 8) includes major rock types categorized by gross engineering characteristics and shows their areal extent.

Conclusions

Although SLR has not eliminated the need for conventional aerial photography, radar imagery can be used in lieu of photography:

1. To provide imagery coverage required for the preparation of reconnaissance geoscience products in remote, unmapped or poorly mapped areas, especially those having nearly year-round cloud cover or low-angle illumination.
2. For the frequent and rapid updating of maps and charts, and for acquiring general information about an area.
3. As an effective tool for the mapping of regional geology.

As an agent *on the earth, man has greatly modified the surface characteristics of the landscape. These modifications are most apparent in the expression of agricultural land and physical structures such as cities, transportation routes, dams, and reservoirs. Less apparent are the subtle changes induced by man—such as speeding up of drainage or erosion, introduction of exotic fauna and flora, and spread or concentration of synthetic chemicals and pollutants. Monitoring the cultural landscape by remote sensing may offer solutions to some of man's most annoying problems. Generally, cultural features, especially physical structures, produce a strong return on radar imagery. Exactly why the strong returns are produced is not yet fully understood, but linearity, flatness, vertical and horizontal surfaces, sharpness of corners, and material composition and extent of flat surfaces are important factors in producing good returns from manmade concentrations.*

All electromagnetic waves are polarized; that is, once propagated, they continue to move at a given angle measured against a standard plane of reference unless some outside force or object changes that angle. If a radar wave is transmitted horizontally or vertically and received at the same angle or polarization, it is termed like-polarized. *If a wave is transmitted horizontally or vertically and received at a different angle or polarization, it is called* cross-polarized. *Generally, a cross-polarized signal produces a grainier and therefore less-sharp image than does a like-polarized wave. Surprisingly, cultural objects in the rural environment show up well on this cross-polarized imagery. This is due to the factors mentioned above and the fact that cultural objects reflect a relatively stronger orthogonally depolarized radar signal.*

35-Evaluation of Multiple-Polarized Radar Imagery for the Detection of Selected Cultural Features

ANTHONY J. LEWIS

The use of radar as a remote sensor is a relatively new innovation for the geoscientist. The capabilities which render radar an especially useful tool to geoscience are as follows: (1) independence from solar illumination and most atmospheric conditions, (2) ability to scan wide swaths of terrain, (3) presentation of collected data on a continuous strip of imagery, (4) resolution characteristics even at orbital altitudes to resolve a cell 15 meters by 15 meters. Multifrequency and multipolarization radar reconnaissance has been initiated recently to increase the information-collecting capabilities of imaging radar. By utilizing various frequency bands between 0.20 and 40 gigacycles and the total polarization matrix, more data are obtained, and as a result interpretations can be made with a higher level of confidence than with a single-frequency, single-polarized radar image.

A NASA-sponsored study is being conducted at the University of Kansas Center for Research in Engineering Science (CRES) to evaluate the use of multiple-polarized K-band radar imagery for geoscience purposes. This evaluation uses four types of polarization, namely:

1. Horizontal transmit, horizontal receive (HH)
2. Horizontal transmit, vertical receive (HV)
3. Vertical transmit, vertical receive (VV)
4. Vertical transmit, horizontal receive (VH)

Radar imagery produced by transmitting and receiving in the same polarization mode (HH and VV) is also referred to as like-polarized, whereas when two modes (HV and VH) are used, the imagery is termed cross-polarized or orthogonally depolarized.

Theoretical studies (Fung, 1965) have substantiated the possibility of differences of received signal amplitudes between the two like-polarizations (HH and VV) and between the cross-polarizations (HV or VH) and the like-polarizations (HH or VV). Due to reciprocity amplitude differences between the two cross components (HV and VH) should not exist. The degree of depolarization of the return signal has formulated to be a function of (1) object orientation (polarization) in both the azimuthal and range direction, and (2) the Fresnel reflection coefficient, which is in turn a function of the complex dielectric constant and the angle of incidence. Therefore, scanning with multiple-polarized radar provides the geoscientist with information concerning the complex physical properties of the target not available from one type of polarization alone.

From Technical Letter NASA 130, October, 1968. Reprinted with permission of the author.

Previous evaluations of radar imagery by geoscientists have been concerned primarily with the like-polarized component. Comparatively few studies are available which evaluate both cross- and like-polarized components. One such study by L. F. Dellwig and R. K. Moore (1966) showed a use of multiple-polarization radar imagery in the field of geology. They were able to distinguish alluvial material derived from various sources, and to differentiate rock types in areas of apparent similarity by comparing cross- and like-polarized imagery of the Pisgah Crater area in California. Their preliminary investigations also indicate that the absolute identification of each rock type on the basis of contrasts in return from various combinations of polarized radar imagery may be feasible. Dellwig and Moore suggest that differences between cross- and like-polarized return may be lithology dependent but that they are more probably a function of surface roughness. Cooper (1966) and Gillerman (1967) reported a striking difference in radar return between like- and cross-polarized images in several areas dominated by silicon-rich outcrops. More specifically, the silicon-rich (volcanic glass) areas produced a lower return on the cross- than the like-polarized image in relation to the surrounding environment. Field checking revealed the variation in relative return to be a complex function of surface roughness, topography, vegetation, and rock composition and not a simple relationship with the percentage of silicon in the outcrop as previously expected (Gillerman, 1967).

Papers by Morain and Simonett (1966) and Ellermeier, Fung, and Simonett (1966) presented at the Fourth Symposium on Remote Sensing of Environment at the University of Michigan respectively involved the interpretation of multiple-polarized imagery for vegetation analysis and the general applications of multiple-polarized imagery in interrelated fields of geoscience. Morain (1967) later reported visually detected variations in relative tonal signatures on the like- and cross-polarized radar imagery from two vegetation types, chaparral shrub and sagebrush, in the vicinity of Horsefly Mountain, Oregon. A follow-up study by Morain and Simonett (1967) utilized electronic techniques to determine the radar backscatter and probability density function on the cross- and like-polarized imagery from natural plant communities. They concluded that detection was enhanced and mapping facilitated by the use of both electronic techniques and multiple-polarized imagery.

Studies are also presently being conducted at the Center for Research, University of Kansas, using multiple-polarized radar imagery for the detection and discrimination of crop types. It has been found that depolarization is dependent in part upon the crop type and its stage of development. For example, depolarization of the radar signal is greater with headed sorghum than it is with sorghum prior to heading. Similar results were also found with alfalfa as it progressed to maturity. Other parameters are possible but to date have not been tested.

The purpose of this study is to evaluate empirically and statistically the like-polarized (HH and VV) and orthogonally depolarized (HV and VH) components of K-band radar imagery for detection of cultural features. Only selected cultural features have been investigated. These are: (1) rural, urban, and agricultural patterns, and (2) transportation and communication nets. Subsequent related reports will cover the use of multiple-polarized imagery in the sensing of physical features of the environment as well as aspects of the cultural landscape not dealt with in this report.

Wherever possible, interpretations presented in this report are based on field investigations and correlated with published maps. The differences between like- and cross-polarized images at first are few. Consequently it was decided to test statistically the nature of any differences between the several polarizations.

The statistical observations of this report are based in part on an interpretation exercise utilizing like-polarized and orthogonally depolarized imagery of K-band radar presented to 68 student observers with little or no previous experience interpreting radar imagery. Two groups at the University of Kansas, a Physical Geography lab section and the Geography Institute of Elementary Teachers, were selected to participate in an attempt to evaluate the two polarization components for the detection of cultural features. Each interpreter was supplied with four radar images, each of a different geographic area, two of which were like-polarized (HH) images and two of which were cross-polarized (HV) images. The images were distributed so that no one interpreter would receive two polarizations of the same geographic area and so that, although all would be working on the same geographic area, approximately half would have the same polarized image. Instructions accompanied the radar imagery and designated, for each area, the alloted time for interpretation and cultural features expected to be detected. Brief descriptions of the cultural features and several of their identifying signatures on radar imagery were also included. Results of the experiment were compiled and an analysis of variance computed to test for significant difference between the two polarization schemes.

Other statistical measures were applied in evaluating multiple-polarized imagery for the detection of spots of high-intensity return in Maryland. The procedures varied from study area to study area. Three of the studies involved the counting of spots of high-intensity return in selected geographic areas by several experienced radar interpreters. The total and average count of high-intensity spots for each polarization was used for tentative evaluation of the four polarizations under study, whereas standard deviations were computed to test variation around the mean. The fourth study encompassed correlation of field data with radar imagery in an attempt to evaluate the influence of roof orientation and material on the four polarization schemes available. Buildings detected by one or more of the polarizations were compared with maps of roof orientation and material based on intensive field work. Findings were then presented in tabular form.

Observations

In general, like-polarized K-band imagery is of better quality than the simultaneously recorded cross-polarized imagery, due in part to the greater dynamic range exhibited on the like-polarized imagery. Since the orthogonally depolarized (HV and VH) received signal is several decibels lower (approximately 10) than the like-polarized signal, it is necessary to increase the gain setting of the depolarized signal to an energy level comparable to the like-polarized signal. The increase in gain required to record the cross-polarized signal raises the noise level sufficiently to produce grainy appearance on the depolarized image.

Rural and Urban Patterns

In rural environments cross-polarized imagery is generally better for defining cultural objects such as farmsteads and transportation arteries (see Fig. 1)[1]. Certain cultural objects, such as bridges, stand out more than natural features on cross-polarized imagery because of their ability to return a relatively stronger orthogonally depolarized radar signal. The data from a number of preliminary studies in 1965–1966 reveal that spots of high-intensity return, interpreted as cultural objects, were more discernible on cross-polarized imagery. Results of one report (Lewis, 1966b) are shown graphically in Figure 2, indicating the predominance of high-intensity spots on the cross-polarized imagery. In Figure 2, the near-perfect alignment of spots between Laytonsville and Sunshine, Maryland, on the cross-polarized image represents the location of a power transmission line not visible on the like-polarized image. A second unpublished study by Morain and Lewis (1966) evaluated the ability of four polarization modes for detecting cultural targets by tabulating the number of spots of high-energy return in identical geographic locations. The areas were selected from unclassified K-band radar imagery of Frederick County, Maryland. The results help substantiate that cultural features, such as buildings, are more easily detected on the cross-polarized imagery than the like-polarized. The capability of the multiple-polarized imagery for detecting buildings was tentatively categorized in the following order:

1. HV — Highest capability for detection
2. VH — Intermediate capability for detection
3. HH — Intermediate capability for detection
4. VV — Lowest capability for detection

[1] Not available for reproduction.

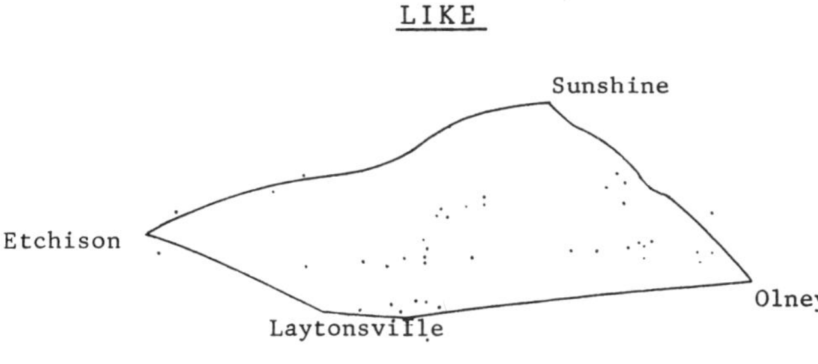

CROSS

Sunshine

Etchison

Olney

Laytonsville

Figure 2 Point cultural features detected on two radar polarizations, Montgomery County, Maryland. (CRES, University of Kansas)

The classification of the VH in the same category with the HH and not with the HV is seemingly contradictory to reciprocity. However, reciprocity assumes parameters, such as quality of the imagery, direction and altitude of flight path, and viewing angle, which are constant, a condition not satisfied in the above experiment. A true test of reciprocity would require the four differently polarized images to be recorded simultaneously on the same flight, an experiment which to date has not been carried out and in part accounts for the apparent contradiction.

A third unpublished report (Lewis, 1966a) tested six interpreters' ability to detect spots of high-intensity return on four geographic areas in Frederick County, Maryland, two of which were scanned by HH and HV polarizations and two by VV and VH polarizations. In all four geographic areas the average number of spots detected was higher on the cross-polarized image (HV and VH) than the like-polarized image (HH and VV) ranging from 1.25 to 1.9 times as great (see Table 1). Even with the larger number detected on the cross-polarizations, the standard deviations in three of the four areas were lower indicating a greater reliability in the detection of spots of high-intensity return for the cross-polarized images.

Other related but independent studies by CRES personnel have involved the effect of building materials and roof orientation on the radar return signal in the Woodsboro-Walkersville, Maryland, area. The results are tabulated in Tables 2 and 3 and further substantiate the increased detectability of certain cultural features on cross-polarized imagery and the advantage of scanning with more than one polarization. In all cases, regardless of roof orientation and material, the orthogonally depolarized image was equal to or better than the simultaneously received like-polarized image for the detection of buildings. These preliminary studies led to the adoption of a systematic test procedure to document in a structurally acceptable fashion the character of the differences between polarizations for detecting cultural objects.

In urbanized areas the cross-polarized mode enhances the interpreter's ability to discriminate large shopping centers, institutional complexes, and industrial areas such as fertilizer plants, oil refineries, cement plants, rail stations or yards, and grain storage bins. All of these areas characteristically have a dearth of natural vegetation. It is interesting to note that the Central Business District (CBD), also characterized by a lack of natural vegetation, does not seem to be more accurately delineated on the cross-polarized image than the like-polarized image. The results of the interpretation exercise presented to the physical geography lab section and Geography Institute for Elementary Teachers indicate that the CBD was better delineated on the HH polarization (see Table 4). A ratio of variance of 4.00 allows the null hypothesis to be rejected at the .05 level of confidence (see Table 5). A visual observation of the radar imagery of Lawrence, Kansas (Fig. 3), also helps to verify that the CBD is more easily delimited on the like-polarized radar image than it is on the cross-polarized image.

Discrimination between residential sections that differ by age, building material, and/or roof shape is also more feasible on multiple polarized imagery because each target exhibits variations in ability to depolarize the signal. K-band radar imagery of Lawrence, Kansas (Fig. 3), and San Diego, California (Fig. 4), illustrate the additional information acquired by obtaining both cross- and like-polarized imagery. Several shopping centers of Lawrence, Kansas, are more distinguishable on the cross-polarized imagery as is the University of Kansas campus and Lawrence High School. The slight increase in return on the cross-polarized image west of Route 59 indicates the location of recently developed residential sections. East of Lawrence on 23rd Street a high-intensity area appears on the like-polarized image but not on the cross-polarized image. Field investigation revealed this area to be a trailer park. The reason for this phenomenon is not clearly understood; however, the alignment of the individual trailers parallel to the flight path may in part be an explanation. The effect of target alignment to the flight path is considered later in more detail.

The effect of building materials and the amount of natural vegetation can be readily seen on both like- and cross-polarized radar images of a portion of San Diego (Fig. 4). For example, Balboa Park is easily distinguished from the surrounding residential area on the like-polarized image; however, the two areas are hardly discernible from each other on the cross-polarized image. This is illustrated by the higher positive detection of the park on the like-polarized image, 79.0 percent detection for interpreters viewing the HH image to 15.7 percent detection for the HV image, based on the observations of 68 students. (See Table 6.) The variance ratio tested to be significant at the .001 level (see Table 7), and

Table 1 Individual Results of Comparison of Like- and Cross-Polarized Radar Imagery by Counting Spots of High-Intensity Return

Interpreter	Area A Polarization		Area B Polarization		Area C Polarization		Area D Polarization	
	HH	HV	HH	HV	VV	VH	VV	VH
1	43	70	154	128	98	118	41	65
2	26	67	88	97	64	83	19	47
3	27	54	65	83	115	122	36	72
4	29	66	58	86	69	107	29	63
5	48	70	58	92	74	98	34	53
6	65	84	58	104	82	120	41	78
Total	238	411	481	590	502	648	200	378
Average	39	68	80	98	83	109	33	63
Standard deviation	11.3	8.8	34.2	14.5	17.2	13.9	7.6	10.5
Ratio of HV/HH or VH/VV	1.75		1.25		1.34		1.9	

Table 2 Percentage of Buildings According to Building Material and Roof Direction That Are Detectable by Multipolarized Radar Imagery[a]

Polarization	Building Material				Roof Direction in Relation to North				
	Metal	Composition	Slate	Asphalt	↕	⤢	↔	⤡	✥
HH only	1.4	0.0	0.0	0.0	0.0	0.0	2.0	0.0	1.1
HV only	5.6	3.2	0.0	0.0	3.0	40.0	6.0	0.0	3.4
VV only	4.6	8.1	0.0	0.0	4.1	0.0	8.0	0.0	3.4
VH only	10.2	9.7	0.0	0.0	5.1	20.0	15.0	0.0	9.1
Total on single polarization	21.8	21.0	0.0	0.0	12.2	60.0	34.0	0.0	17.0
Detected on more than one polarization	44.7	22.6	40.0	100.0	51.0	0.0	35.0	0.0	36.4
Total detected	66.7	43.5	40.0	100.0	63.3	60.0	69.0	0.0	53.4
Not detected on any polarization	33.3	56.5	60.0	0.0	36.7	40.0	31.0	100.0	46.6
Total targets	100.0	100.0	100.0	100.0	100.0	100.0	100.0	100.0	100.0

[a] Flight path was NW-SE.

therefore the null hypothesis can be rejected at a high level of confidence. Since the residential section surrounding Balboa Park constitutes some of the older residential sections in San Diego, the inability to discriminate the two on the cross-polarized image is in part explained by the abundance of vegetation in both areas. The effect of vegetation can also be noted by comparing the newer, less tree-sheltered residential areas north of Mission Valley with the older residential area near Balboa Park (see Fig. 4). The former area can be subdivided into two distinct categories on the cross-polarized image, each of which is characterized by roof tops of a particular material. Roofs covered with crushed dolomite predominate in the northwest residential section, whereas shingled roofs and larger buildings are prevalent in the northeast residential section.

The detection of airports, oil tank farms, an oil refinery, an oil field, and a small town on the two polarizations was also tested by the 68 student interpreters. Their findings indicate that except for the detection of a single town outside of Wichita, Kansas (see Fig. 7), neither polarization was better on a statistical basis for the delineation of the targets in question. The percentage of airports detected on radar imagery of San Diego, California, was 45.6 percent on the like-polarized imagery and 44.0 percent on the cross-polarized imagery (see Table 8). The low F value (see Table 9) from the data is indicative of the absence of additional distinguishing characteristics on either polarization. In Superior, Wisconsin, 47 percent of the interpreters with like-polarized imagery detected the small rural one-runway airport, whereas only 25 percent of those with cross-polarized

Table 3 Percentage of Buildings Detected on Each Polarization According to Roof Material–Roof Direction Combinations[a,b]

Roof Material and Direction	Polarization				Total % Detected At Least One Polarization	Total Possible Number
	HH	HV	VV	VH		
M-N	34	35	37	50	63	86
M-NE	0	50	0	0	50	4
M-E	35	42	31	58	82	71
M-NW	0	0	0	0	0	2
M-C	25	28	28	52	59	69
Co-N	6	31	31	31	44	16
Co-NE	0	0	0	0	0	0
Co-E	9	9	19	16	44	32
Co-NW	0	0	0	0	0	0
Co-C	8	8	15	23	39	13
SL-N	0	0	0	0	0	0
SL-NE	0	0	0	0	0	0
SL-E	0	0	0	0	0	2
SL-NW	0	0	0	0	0	0
SL-C	22	22	22	33	55	9

[a] Flight path was NW-SE.
[b] Legend: M = Metal, Co = Composition, SL = Slate, N = North-South, NE = Northeast-Southwest, E = East-West, NW = Northwest-Southwest, C = Complex.

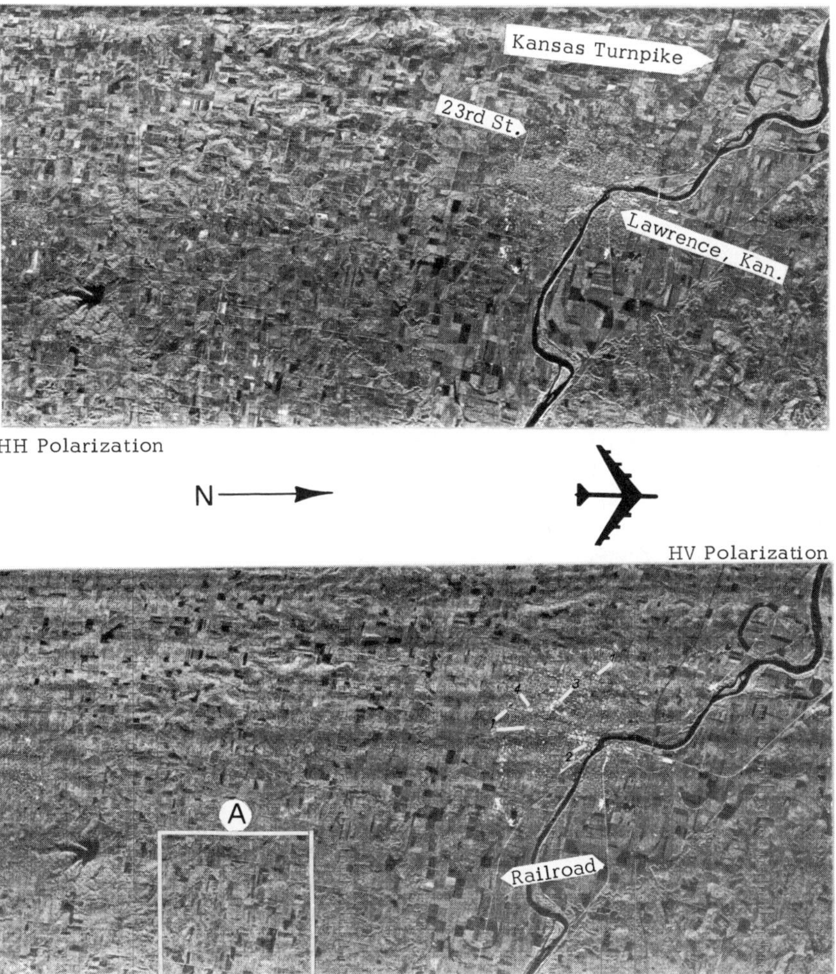

Figure 3 Multiple-polarization, K-band positive radar imagery of Lawrence, Kansas. 1, shopping centers; 2, Central Business District (CBD); 3, University of Kansas; 4, Lawrence High School. (CRES, University of Kansas)

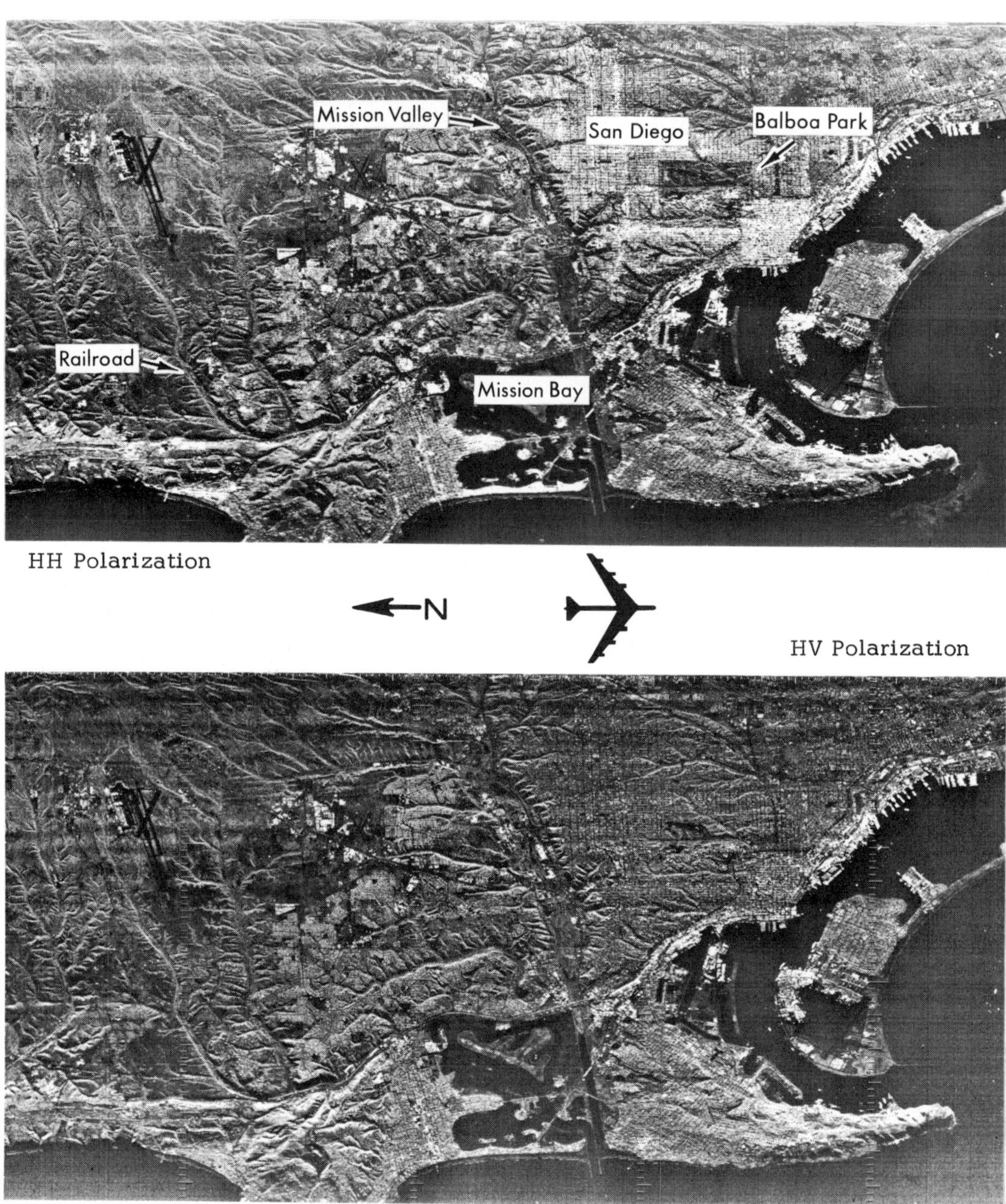

Figure 4 Separation of residential, park, and business areas using two radar polarizations, San Diego, California. (CRES, University of Kansas)

imagery detected the runway (see Table 10). The F value in this case is high enough to suspect a reasonable variation between polarizations although the difference is not statistically significant (see Table 11). The conclusion drawn from the data on detection of airports suggests that certain sizes and types of airports may be more easily delineated on a specific polarization. More testing needs to be done before this hypothesis is verified. An F value of 5.15 for the positive detection of a single town is not interpreted to be significant on the basis of the premises that detection was largely by chance as indicated by the extremely low positive to false-positive detection ratio. It is suggested, therefore, that a larger sample be accrued before any conclusions are made (see Tables 12 and 13). Lack of statistical significance between polarizations in the delineating of tank farms, oil refineries, and oil fields is interpreted as indicating that neither polarization was better for interpreting the above cultural features (see Tables 14 through 19). However, the possibility exists that a major source of variation was not included in the error term or that more testing needs to be done.

Transportation and Communication Nets

Detecting the tracing communication nets is performed more easily, completely, and accurately on cross-polarized imagery when the communication net traverses land and is at an angle to the flight path; communication nets either parallel to the flight path or crossing water bodies are more easily observed on like-polarized

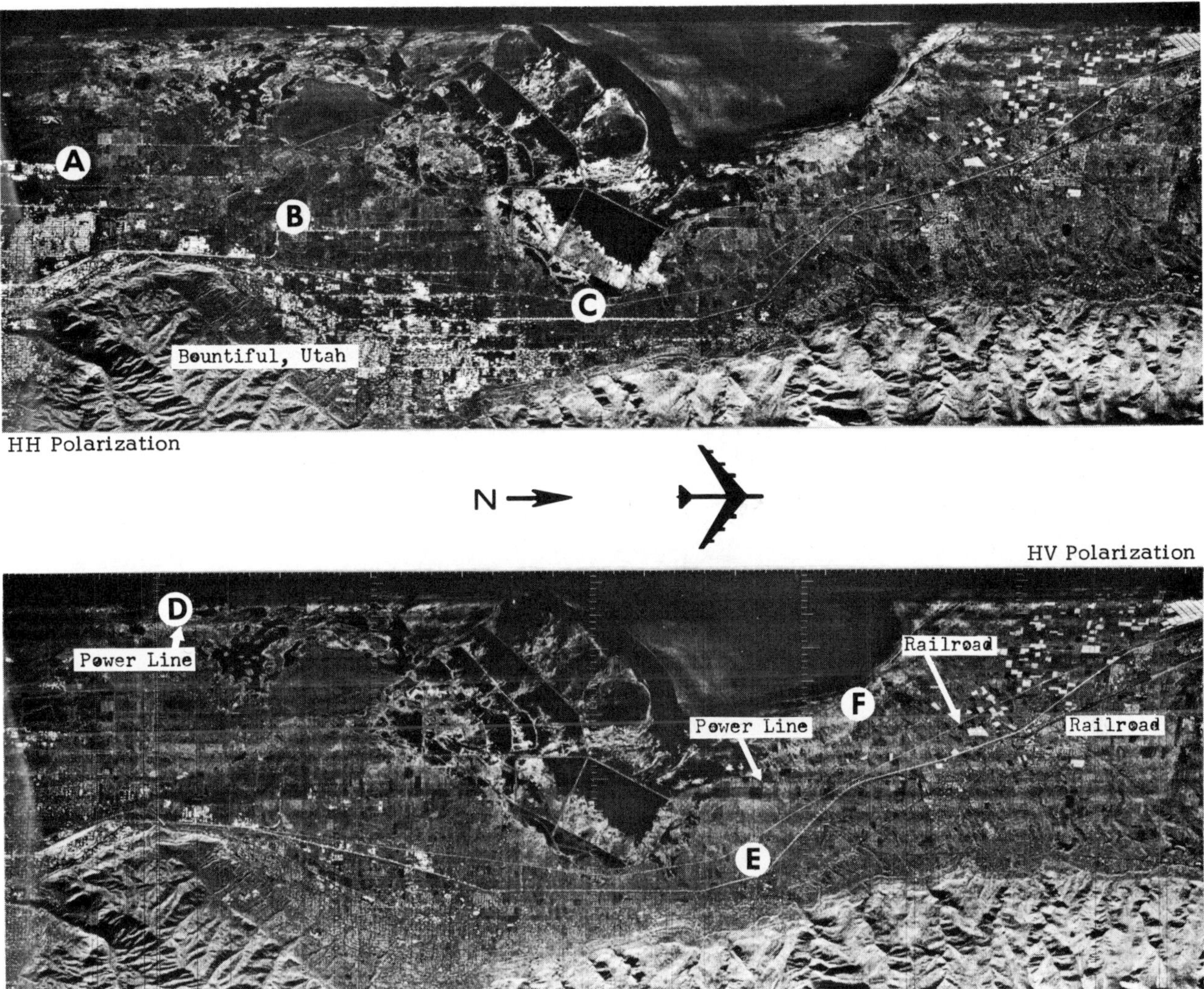

Figure 5 Detection of rail and power lines using two radar polarizations near Bountiful, Utah. (CRES, University of Kansas)

imagery. Multiple-polarized radar imagery of Bountiful, Utah (Fig. 5), has been included solely to demonstrate the ability of cross-polarized imagery to detect communication nets.

The effect of alignment in relation to the flight path is demonstrated on the HH and HV radar images near Bountiful, Utah (Fig. 5), where A, B, and C indicate increased return from power lines and railroad tracks parallel to the flight path on like-polarized imagery and D, E, and F illustrate the increased detectability on cross-polarized imagery of transportation and communication nets at an angle to the flight path. Evaluation of the Bountiful, Utah, imagery by 68 student interpreters showed that 76.0 percent of the power lines and railroad tracks parallel to the flight path were detected on the like-polarized image, whereas only 18.2 percent were detected on the cross-polarized image (see Table 20). The 18.2 percent seems high until one considers that the entire percentage detected by the interpreters with cross-polarized imagery represents one parallel segment of the transportation-communication net connecting two other segments oriented at an angle to the flight path and therefore the parallel segment may have been inferred by the appearance of the two unconnected communication lines (see power line E on Fig. 5). The ratio of variance, F, demonstrates the difference between polarizations in detecting transportation and communication lines parallel to the flight path to be extremely significant, $P < .001$ (see Table 21). The delineation of power lines oriented at an angle to the flight in Bountiful, Utah, was better on the cross-polarized image, 50.5 percent, than on the like-polarized image, 3.5 percent, as shown in Table 22; the difference between the two polarizations in detecting power lines at an angle to the flight path tested to be extremely significant on a statistical basis (see Table 23).

The delineation of transportation arteries, railroads or roads, or both, appears to be influenced by the quality of the image, the geographic area, and the interpreter. This is indicated by the results of testing inexperienced interpreters in detection of transportation arteries on multiple-polarized radar imagery of Superior, Wisconsin (Fig. 6), and Wichita, Kansas (Fig. 7). The cross-polarization image was judged better for detecting total

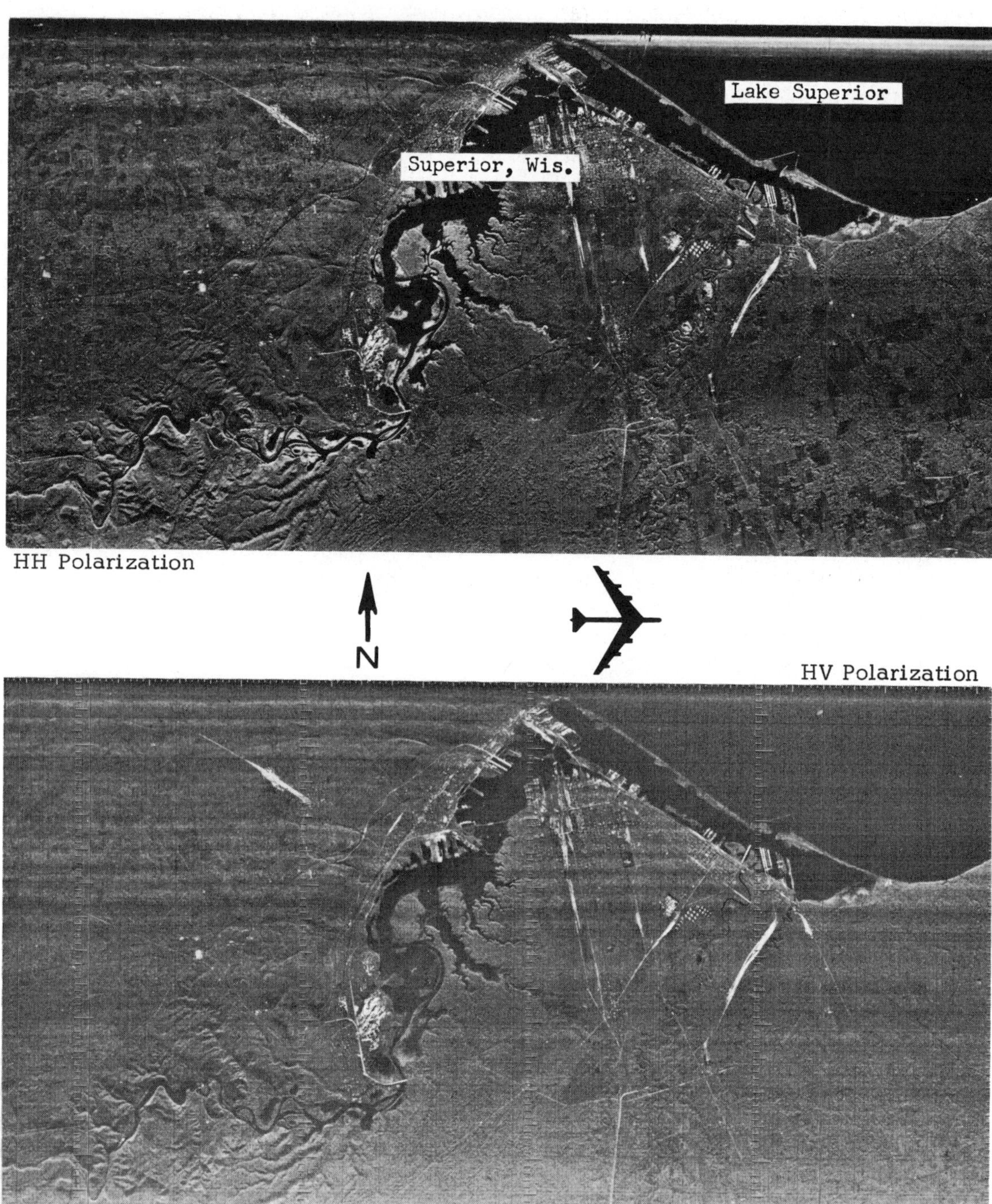

Figure 6 Detection of transportation routes and industrial sites using two radar polarizations, Superior, Wisconsin. (CRES, University of Kansas)

transportation in the former geographic area (see Table 24), whereas the like-polarized image was judged better at Superior, Wisconsin (see Table 26). Both of the above were tested to be significant at the .05 level of confidence (see Tables 25 and 27). Detection of road and railroads on both the like- and cross-polarized radar images of Wichita, Kansas, and Superior, Wisconsin, range from 40 to 60 percent. Both A and B on Figure 8 demonstrate the variation between polarizations in the detection of transportation arteries near Banida, Idaho. Although this report is concerned with cultural features, it is interesting to point out the difference in return from the dry stream bed at C on the like- and cross-polarized image as well as the signature from agricultural land north of A.

Identification of roads varied by only 1 percent between polarizations of Wichita, Kansas, 28.0 percent on the like-polarized image to 27.0 percent on the cross-polarized image (see Table 28). The ratio of variance, .03, was not large enough to be statistically significant (see Table 29). These results suggest that in Wichita, Kansas, the difference between polarizations is not significant for detecting roads, although it is possible that a major source of variation was omitted in the error team. In

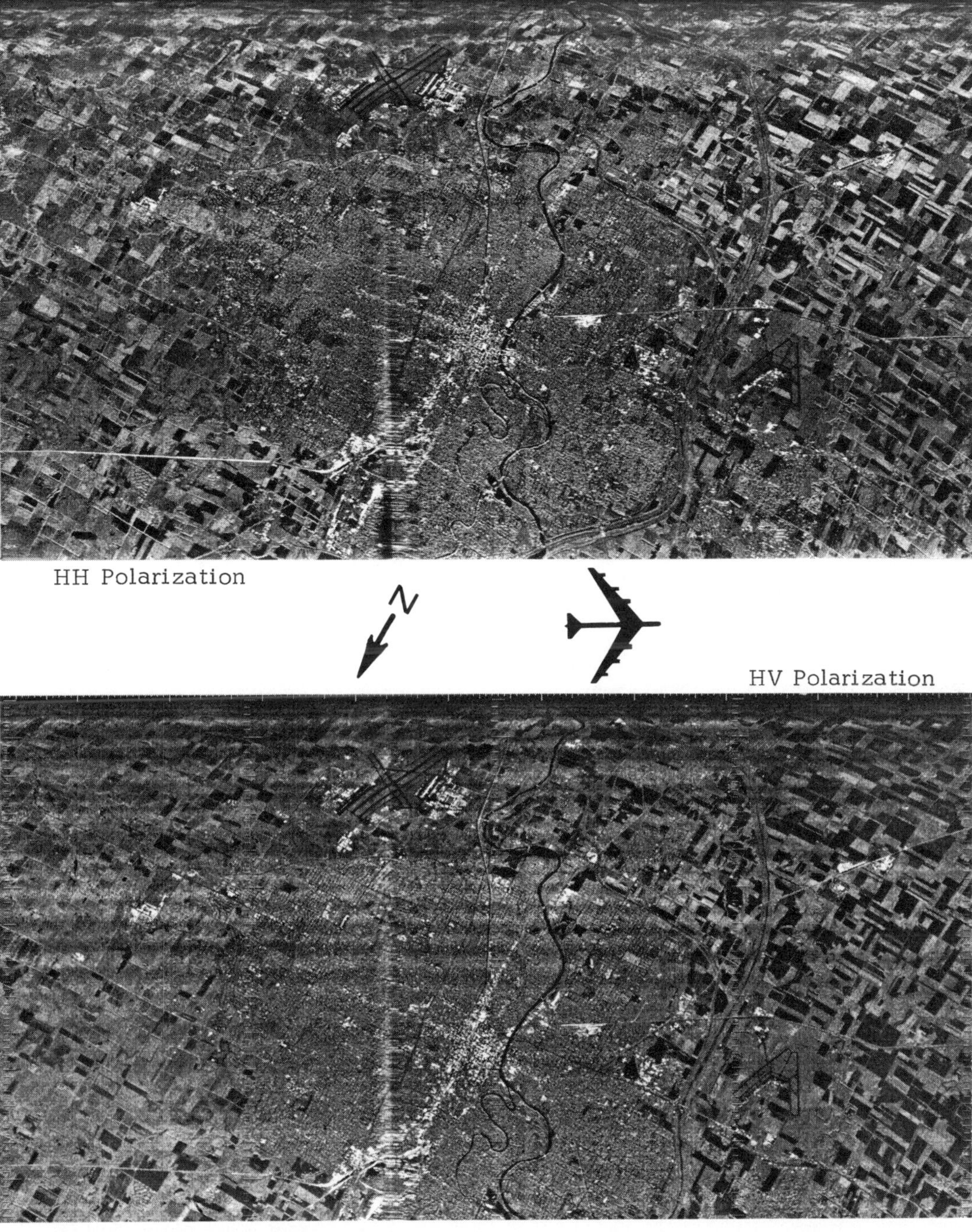

Figure 7 Detection of bridges across the Arkansas River with two radar polarizations, Wichita, Kansas. (CRES, University of Kansas)

Superior, the difference in the ability of the polarizations for detecting roads was tested to be extremely significant, .001 level of confidence (see Table 31), as the percent detected varied from 50.5 on the HH or like-polarization to 22.4 on the HV or cross-polarization (see Table 30). The low radar return characteristic of roads in relation to railroads does not lend itself to a high number of false-positive detections. The total detection of roads, including those detected falsely as railroads, followed the same pattern exemplified by positive detection of roads, i.e., the percentage detected in Wichita was higher on the like-polarization but the data did not prove to be significant at the .05 level of confidence, whereas, in Superior, Wisconsin, the data proved to be extremely significant with the like-polarized image interpreters detecting 54 percent on the like-polarized and only 20.3 percent on the cross-polarized imagery (see Tables 32–35).

Railroads were more completely delineated on radar than were roads, although false-positive detection (railroads marked as roads) was greater. False-positive detection of roads as railroads, mentioned previously, is very low which in part is accounted for because railroads give more radar return than roads. Positive detection of

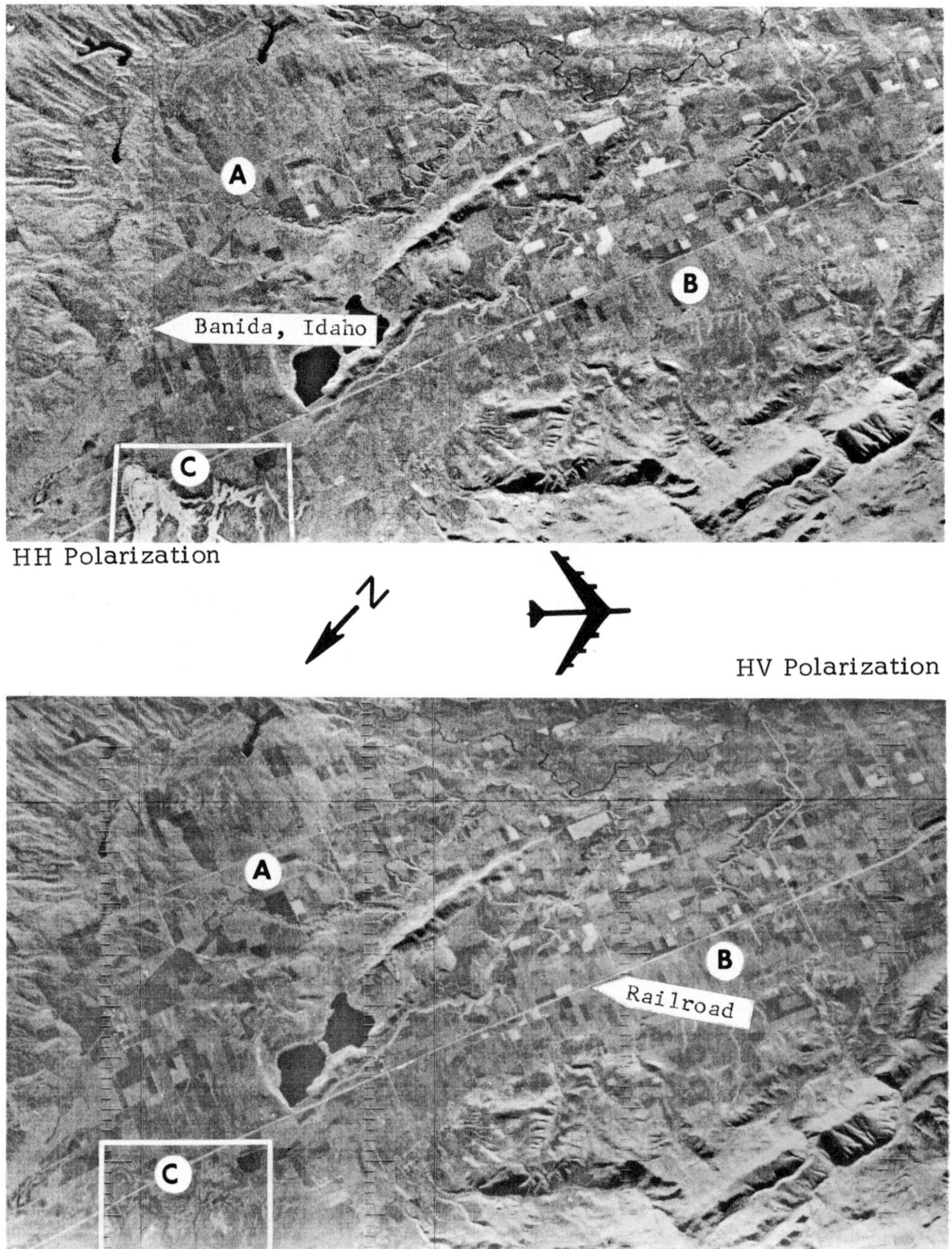

Figure 8 Detection of transportation routes using two polarizations near Banida, Idaho. (CRES, University of Kansas)

railroads in Wichita, Kansas, and Superior, Wisconsin, deviated less than ± 5 percent around 50 percent for either polarization. In Wichita, Kansas, the greater percentage was detected on the cross-polarized image, 54.2 percent in relation to 45.4 percent on the like-polarized image (see Table 36). The F test showed that this data was significant at a .001 level of confidence with a $F = 12.22$ (see Table 37). It can be concluded, therefore, that in Wichita the HV, cross-polarized radar image, was better for the detection of railroads than the HH, like-polarized radar image. The conclusion cannot be carried over and applied to the radar imagery of Superior, Wisconsin, where the interpretation on the like and cross-polarization did not differ significantly (see Tables 38 and 39). The F of .32 was far below the level needed to reject the null hypothesis at the .05 level of confidence. The percentage of railroads detected, whether marked as railroads or roads, was greater on the cross-polarized images of both geographic areas (see Tables 40 and 42) although only the data from Wichita, Kansas, tested to be extremely significant, .001 level of confidence (see Tables 41 and 43).

Detection of transportation and communication nets that traverse water bodies (bridges and powerlines) was more complete with like-polarized imagery in Wichita, Kansas (see Fig. 7), and in San Diego, California (see Fig. 4). Of the 22 bridges across the Arkansas River in the Wichita, Kansas, area encompassed

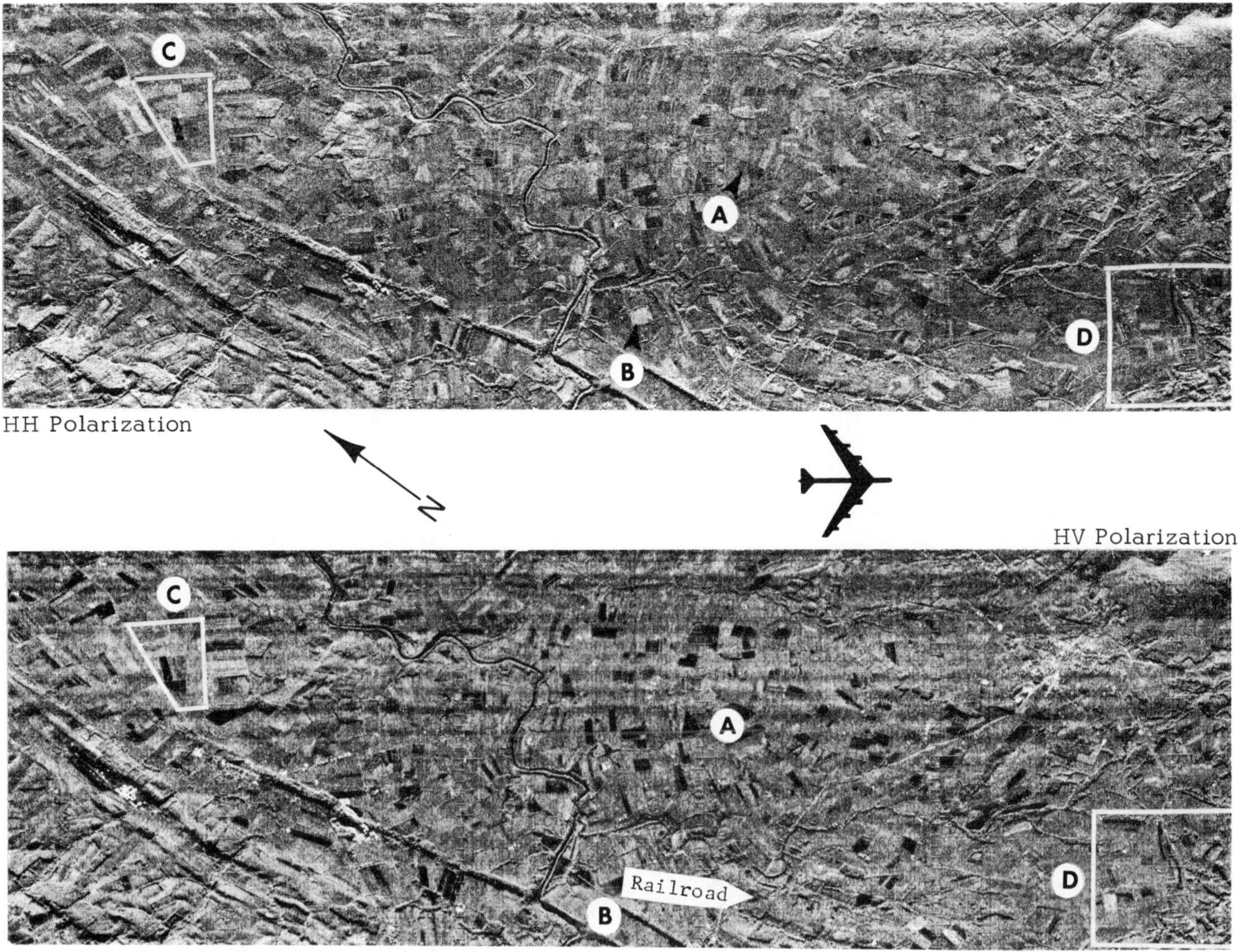

Figure 9 Agricultural patterns detected by two radar polarizations in the vicinity of Thurmont, Maryland. (CRES, University of Kansas)

by radar imagery, an average of 15.7 (78.8 percent), were detected by the 36 student interpreters viewing those with like-polarized imagery while only 6.2 (31.0 percent) were detected by those with cross-polarized imagery (see Table 44). The difference in the detection of bridges is significant at P = .001 (see Table 45). A positive to false-positive ratio of 566 to 1 on like-polarized imagery indicates the high degree of positive identification associated with bridge detection. The ratio on cross-polarized imagery was lower, 18 to 1. Several of the false-positive detections on both images were range marks on the image that had been mistaken for bridges, a mistake not likely to be made by an experienced interpreter. The Kansas Turnpike bridge over the Kansas River (see Fig. 3) illustrates the more pronounced radar return on the like-polarized imagery of a transportation artery traversing a water body.

Forty-eight percent of the channel markers traversing San Diego Bay were detected by the interpreters using the like-polarization, and only 5.8 percent by those with the cross-polarized image (see Table 46). As with the detection of bridges, the statistics proved to be significant at P = .001 (see Table 47). The ratio of positive to false-positive detection, though low, on the like-polarized (3.3 to 1) was at least positive, whereas on the cross-polarized image there were more false than correct identifications (0.35 to 1) (see Table 46).

Agricultural Patterns

Variations in intensity of return are also found on the like- and cross-polarized imagery of agricultural areas. These variations, which reflect differences in crop types and/or field conditions, provide the interpreter with additional information not available prior to the use of multiple-polarized radar imagery. Near Thurmont, Maryland (Fig. 9), dark fields on the cross-image immediately attract one's attention; however, more subtle gray value differences between the like- and cross-polarized images are visible at B, C, and D. Detection of field A was enhanced on cross-polarized imagery, whereas field B is more predominant on the like-polarized image. Areas C and D illustrate some of the changes in relative gray tone values associated with like- and cross-polarized imagery and should dispel the belief that the additional information on the cross-polarized image is only a result of lifting the noise level. In some areas, such as near

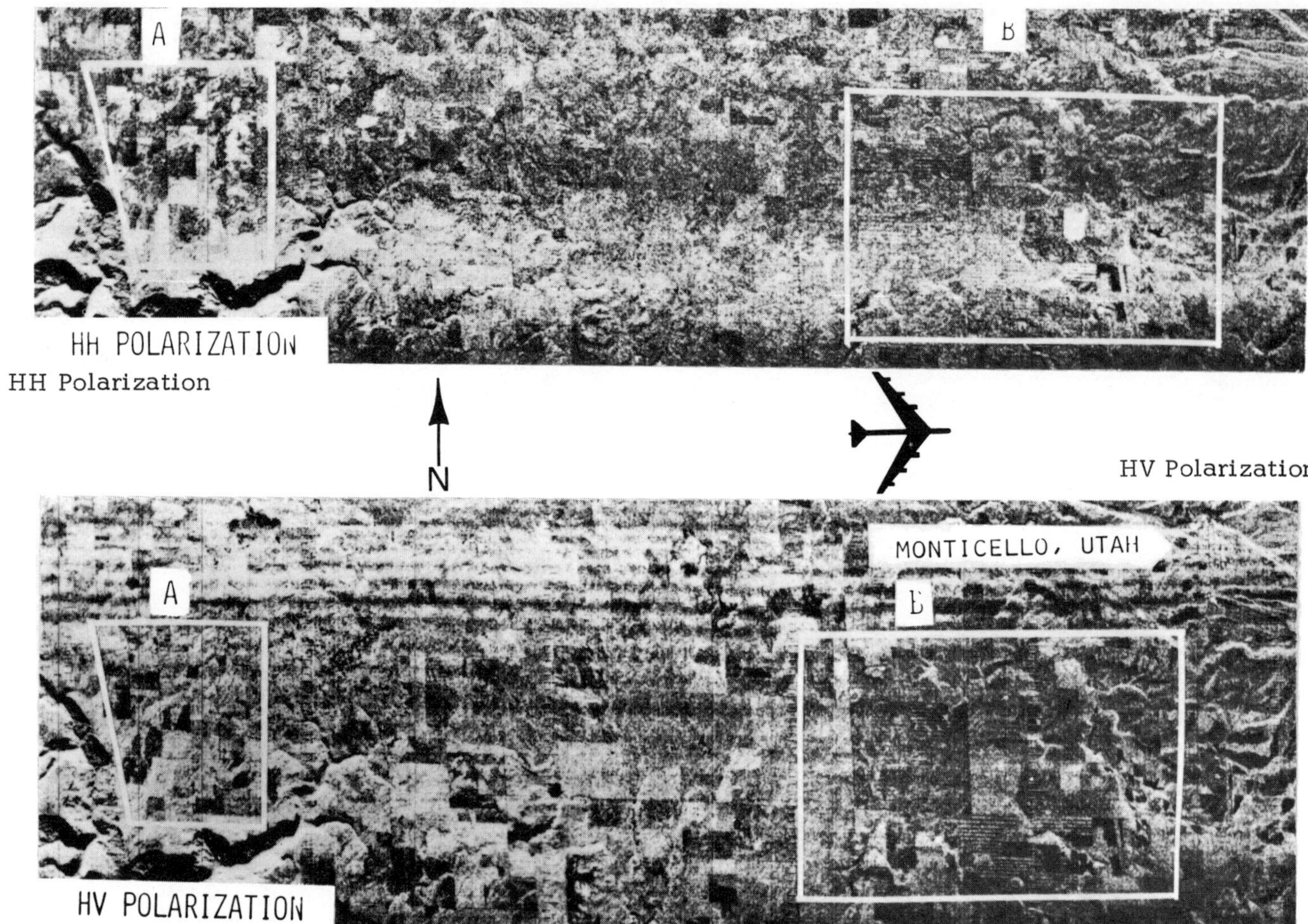

Figure 10 Detection of pasture and cultivated lands using two radar polarizations near Monticello, Utah. (CRES, University of Kansas)

Monticello, Utah (see Fig. 10), where land-use is limited primarily to grazing, the cross-polarized image appears to be better for the delimitation of field boundaries. Areas A and B on Figure 10 illustrate the greater detectability of field boundaries on the cross-polarized image than on the like-polarized image.

Statistical Summary

The studies involving detection of spots of high intensity provide the following ranking of polarizations: 1. HV—highest detection capability. 2. HH and VH—intermediate detection capability. 3. VV—lowest detection capability.

Statistical analysis of the student interpretation employing HH and HV imagery revealed the following:

1. At the 99.9 percent confidence level

   a. The like-polarized (HH) imagery was better for detecting vegetated residential areas and parks versus nonvegetated urban areas; power lines and railroads when aligned parallel to the flight path or crossing water bodies; and bridges and channel markers.

   b. The cross-polarized (HV) imagery was better for detecting powerlines and railroads when at an angle (other than parallel) to the flight path or traversing land.

2. At the 95 percent confidence level the like-polarized imagery was better for detecting the central business district (CBD).

3. Detection of airports and roads did not prove to be significantly different on either the HH or HV polarizations.

4. Cross-polarized radar imagery was better for detecting railroad nets; however, the degree of confidence varied from 99.9 percent to less than 95 percent depending on the area studied.

Summary

A summary of the results comparing two polarizations received simultaneously on one pass, HH to HV or VV to VH, is as follows:

1. The HH polarization in general proved better for

   a. delineation of vegetated areas within an urban complex, such as Balboa Park in San Diego or the Golden Gate Park in San Francisco,

   b. detection of transportation and communication arteries that traverse water bodies, and

   c. detection of communication lines oriented parallel to the flight path.

2. The HV polarization in general was better for

   a. detection of buildings in a rural setting, as well as shopping centers, industrial and manufacturing plants, and other cultural conglomerations that produce a high orthogonally depolarized signal, and

b. detection of communication lines oriented at an angle other than parallel.

3. Variations between VV and VH imagery were only tested in the detection of buildings where it was found that VH imagery was better than VV.

A decisive conclusion is not warranted for evaluating polarization schemes in the detection of transportation lines without further investigation. Other parameters—quality of image, the geographic area, and experience of the interpreter—along with polarization variations should be included in future testing. Visual observations, however, suggest that railroads are more completely detectable on the cross-polarized image whereas no one polarization is better for detection of roads.

Detection of airports larger than one runway was not enhanced by any polarization scheme; however, the variation experienced in the interpretation of a single runway airport in a rural setting was large enough to suggest increased detectability on the like-polarized imagery.

Variations between like- and cross-polarized radar imagery were visible in agricultural areas, suggesting differences in crop types and/or field conditions. The lack of ground data prohibited determination of the cause-effect relationship and will until such data are collected at the time of overpass.

Conclusion

Multiple-polarized radar imagery provides the geoscientist with information relating to the complex physical properties of the target not otherwise available with a single polarized system. More specifically several of the applications in the sensing of cultural features are:

1. Helping to discriminate between residential and business or industrial districts.

2. Increasing the ability to plot more complete transportation and communication nets.

3. Providing the observer with additional information concerning the rural setting or location and number of farmsteads.

4. Separating fields of different crops or crop states.

Table 4 Detection of Central Business District in Wichita, Kansas

	Polarization	
	HH	HV
Number of interpreters	36	32
Total positive detection	35	27
Average positive detection	0.97	0.84
Percent positive detection	97.20	84.50
Total false-positive detection	7	7
Ratio positive/false-positive detection	5 to 1	3.86 to 1

Table 5 Analysis of Variance of Total Positive Detection of Central Business District, Wichita, Kansas

Source of Variation	Sums of Squares	Degrees of Freedom	Mean Sq. Variance	Ratio of Variances, F
Total	548	67	—	—
Between region	28	1	.28	4.00
Within region	520	66	.07	—

$P(F_{1,67} = 4.00) = .05$

Table 6 Detection of Balboa Park, San Diego, California

	Polarization	
	HH	HV
Number of interpreters	34	34
Total positive detection	27	5
Average positive detection	0.79	0.15
Percent positive detection	79.00	15.70
Total false-positive detection	67	65
Ratio positive/false-positive detection	.40 to 1	.077 to 1

Table 7 Analysis of Variance of Total Positive Detection of Balboa Park, San Diego, California

Source of Variation	Sums of Squares	Degrees of Freedom	Mean Sq. Variance	Ratio of Variances, F
Total	17.18	67	—	—
Between region	6.18	1	6.18	38.62
Within region	11.00	66	.16	—

$P(F_{1,67} = 38.62) < .001$

Table 8 Detection of Airports, San Diego, California

	Polarization	
	HH	HV
Number of interpreters	34	34
Total positive detection	64	60
Average positive detection	1.84	1.76
Percent positive detection	45.60	44.00
Total false-positive detection	9	6
Ratio positive/false-positive detection	7.1 to 1	10 to 1

Table 9 Analysis of Variance of Total Positive Detection of Airports, San Diego, California

Source of Variation	Sums of Squares	Degrees of Freedom	Mean Sq. Variance	Ratio of Variances, F
Total	58.29	67	—	—
Between region	.10	1	.10	.12
Within region	56.19	66	.83	—

$P(F_{1,67} = .12) > .05$

Table 10 Detection of an Airport, Superior, Wisconsin

	Polarization	
	HH	HV
Number of interpreters	32	36
Total positive detection	15	9
Average positive detection	0.47	0.25
Percent positive detection	47	47
Total false-positive detection	11	11
Ratio positive/false-positive detection	1.4 to 1	.82 to 1

Table 11 Analysis of Variance of Total Detection of an Airport, Superior, Wisconsin

Source of Variation	Sums of Squares	Degrees of Freedom	Mean Sq. Variance	Ratio of Variances, F
Total	15.53	67	—	—
Between region	.81	1	.81	3.68[a]
Within region	14.72	66	.22	—

$P(F_{1,67} = 3.68) > .05$

[a] Close to .05 level and is probably significant at .10.

Table 12 Detection of a Town, Wichita, Kansas

	Polarization	
	HH	HV
Number of interpreters	36	32
Total positive detection	6	13
Average positive detection	0.16	0.41
Percent positive detection	16	41
Total false-positive detection	75	85
Ratio positive/false-positive detection	.08 to 1	.15 to 1

Table 13 Analysis of Variance of Total Positive Detection of a Town, Wichita, Kansas

Source of Variation	Sums of Squares	Degrees of Freedom	Mean Sq. Variance	Ratio of Variances, F
Total	13.70	67	—	—
Between region	.98	1	.98	5.15
Within region	12.72	66	.19	—

$P(F_{1,67} = 5.15) < .05$

Table 14 Detection of Tank Farms, Superior, Wisconsin

	Polarization	
	HH	HV
Number of interpreters	32	36
Total positive detection	35	44
Average positive detection	0.91	1.20
Percent positive detection	46	61
Total false-positive detection	5	9
Ratio positive/false-positive detection	1.4 to 1	2 to 1

Table 15 Analysis of Variance of Total Detection of Tank Farms, Superior, Wisconsin

Source of Variation	Sums of Squares	Degrees of Freedom	Mean Sq. Variance	Ratio of Variances, F
Total	37.23	67	—	—
Between region	.28	1	.28	.50
Within region	36.95	66	.55	—

$P(F_{1,67} = .50) > .05$

Table 16 Detection of Oil Refineries, Superior, Wisconsin

	Polarization	
	HH	HV
Number of interpreters	32	36
Total positive detection	23	28
Average positive detection	0.72	0.77
Percent positive detection	72	77
Total false-positive detection	7	4
Ratio positive/false-positive detection	.32 to 1	.36 to 1

Table 17 Analysis of Variance of Total Positive Detection of Oil Refineries, Superior, Wisconsin

Source of Variation	Sums of Squares	Degrees of Freedom	Mean Sq. Variance	Ratio of Variances, F
Total	13.31	67	—	—
Between region	.06	1	.06	.30
Within region	13.25	66	.20	—

$P(F_{1,67} = .30) > .05$

Table 18 Detection of Oil Fields in Wichita, Kansas

	Polarization	
	HH	HV
Number of interpreters	36	32
Total positive detection	3	8
Average positive detection	0.08	0.25
Percent positive detection	8.3	25.0
Total false-positive detection	42	46
Ratio positive/false-positive detection	.07 to 1	.17 to 1

Table 19 Analysis of Variance of Total Positive Detection of Oil Fields, Wichita, Kansas

Source of Variation	Sums of Squares	Degrees of Freedom	Mean Sq. Variance	Ratio of Variances, F
Total	9.23	67	—	—
Between region	.28	1	.28	2.15
Within region	8.95	66	.13	—

$P(F_{1,67} = 2.15) > .05$

Table 20 Detection of Power Lines Parallel to Flight Path, Bountiful, Utah

	Polarization	
	HH	HV
Number of interpreters	35	33
Total positive detection in inches	72.72	16.7
Average positive detection in inches	1.98	0.50
Percent positive detection	76.00	18.20[a]
Total false-positive detection	80	49
Ratio positive/false-positive detection	.91 to 1	.34 to 1

[a] Entire percent is accounted for by detection of one power line that connected power lines at an angle to the flight path.

Table 21 Analysis of Variance of Total Positive Detections of Power Lines in Inches, Bountiful, Utah

Source of Variation	Sums of Squares	Degrees of Freedom	Mean Sq. Variance	Ratio of Variances, F
Total	129.31	67	—	—
Between region	86.68	1	86.68	135.43
Within region	42.63	66	.64	—

$P(F_{1,67} = 135.43) < .001$

Table 22 Detection of Power Lines Oriented at an Angle to Flight Path, Bountiful, Utah

	Polarization	
	HH	HV
Number of interpreters	35	33
Total positive detection in inches	9.98	133.9
Average positive detection in inches	0.28	4.04
Percent positive detection	3.50	50.50
Total false-positive detection	Totaled with the	
Ratio positive/false-positive detection	power lines parallel	

Table 23 Analysis of Variance of Total Positive Detections of Power Lines in Inches, Bountiful, Utah

Source of Variation	Sums of Squares	Degrees of Freedom	Mean Sq. Variance	Ratio of Variances, F
Total	29.19	67	—	—
Between region	19.78	1	19.78	141.2
Within region	9.41	66	.14	—

$P(F_{1,67} = 141.2) < .001$

Table 24 Detection of Total Transportation Lines, Wichita, Kansas

	Polarization	
	HH	HV
Number of interpreters	36	32
Total positive detection	582.25	625.75
Average positive detection	16.20	19.57
Percent positive detection	48.70	58.70
Total false-positive detection	0	0

Table 25 Analysis of Variance of Total Transportation Lines, Wichita, Kansas

Source of Variation	Sums of Squares	Degrees of Freedom	Mean Sq. Variance	Ratio of Variances, F
Total	2125.97	67	—	—
Between region	192.07	1	192.07	6.55
Within region	1933.90	66	29.30	—

$P(F_{1,67} = 6.55) < .05$

Table 26 Detection of Total Transportation Lines, Superior, Wisconsin

	Polarization	
	HH	HV
Number of interpreters	32	36
Total positive detection in inches	644.25	579.50
Average positive detection in inches	20.00	16.00
Percent positive detection	53.00	42.40
Total false-positive detection	0	0

Table 27 Analysis of Variance of Total Transportation Lines, Superior, Wisconsin

Source of Variation	Sums of Squares	Degrees of Freedom	Mean Sq. Variance	Ratio of Variances, F
Total	3146.24	67	—	—
Between region	267.42	1	267.42	6.13
Within region	2878.82	66	43.61	—

$P(F_{1,67} = 6.13) < .05$

Table 28 Detection of Roads, Wichita, Kansas

	Polarization	
	HH	HV
Number of interpreters	36	32
Total positive detection in inches	146	125.75
Average positive detection in inches	4.04	3.92
Percent positive detection	28.00	27.00
Total false-positive detection in inches[a]	5	4

[a] Roads indicated as railroads.

Table 29 Analysis of Variance of Total Positive Detection of Roads, Wichita, Kansas

Source of Variation	Sums of Squares	Degrees of Freedom	Mean Sq. Variance	Ratio of Variances, F
Total	643.40	67	—	—
Between region	.38	1	.38	.03
Within region	643.02	66	9.74	—

$P(F_{1,67} = .03) > .05$

Table 30 Detection of Roads, Superior, Wisconsin

	Polarization	
	HH	HV
Number of interpreters	32	36
Total positive detection in inches	254	113.75
Average positive detection in inches	7.95	3.15
Percent positive detection	50.50	22.40
Total false-positive detection in inches[a]	17.75	2.25

[a] Roads indicated as railroads.

Table 31 Analysis of Variance of Total Positive Detection of Roads, Superior, Wisconsin

Source of Variation	Sums of Squares	Degrees of Freedom	Mean Sq. Variance	Ratio of Variances, F
Total	1102.30	67	—	—
Between region	390.27	1	390.27	36.20
Within region	712.03	66	10.78	—

$P(F_{1,67} = 36.20) < .001$

Table 32 Total Positive and False-Positive Detection of Roads, Wichita, Kansas

	Polarization	
	HH	HV
Number of interpreters	36	32
Total detection in inches	151.00	129.75
Average detection in inches	4.20	4.05
Percent detection	29.00	27.90

Table 33 Analysis of Variance of Positive and False-Positive Roads Detected in Wichita, Kansas

Source of Variation	Sums of Squares	Degrees of Freedom	Mean Sq. Variance	Ratio of Variances, F
Total	651.12	67	—	—
Between region	.16	1	.16	.01
Within region	650.96	66	9.86	—

$P(F_{1,67} = .01) > .05$

Table 34 Total Positive and False-Positive Detection of Roads, Superior, Wisconsin

	Polarization	
	HH	HV
Number of interpreters	32	36
Total detection in inches	271.75	116.00
Average detection in inches	8.50	3.20
Percent detection	54.00	20.30

Table 35 Analysis of Variance of Positive and False-Positive Roads Detected in Superior, Wisconsin

Source of Variation	Sums of Squares	Degrees of Freedom	Mean Sq. Variance	Ratio of Variances, F
Total	1138.60	67	—	—
Between region	431.65	1	431.65	40.68
Within region	706.95	66	10.61	—

$P(F_{1,67} = 40.68) < .001$

Table 36 Detection of Railroads, Wichita, Kansas

	Polarization	
	HH	HV
Number of interpreters	36	32
Total positive detection in inches	306.75	325.50
Average positive detection in inches	8.52	10.17
Percent positive detection	45.40	54.20
Total false-positive detection in inches[a]	135.00	168.50

[a] Railroads indicated as roads.

Table 37 Analysis of Variance of Total Positive Detection of Railroads, Wichita, Kansas

Source of Variation	Sums of Squares	Degrees of Freedom	Mean Sq. Variance	Ratio of Variances, F
Total	1317.51	67	—	—
Between region	205.85	1	205.85	12.22
Within region	1111.66	66	16.84	—

$P(F_{1,67} = 12.22) < .001$

Table 38 Detection of Railroads, Superior, Wisconsin

	Polarization	
	HH	HV
Number of interpreters	32	36
Total positive detection in inches	270.25	282.75
Average positive detection in inches	8.50	7.80
Percent positive detection	54.00	49.50
Total false-positive detection[a]	101.25	180.75

[a] Railroads indicated as roads.

Table 39 Analysis of Variance of Total Positive Detection of Railroads, Superior, Wisconsin

Source of Variation	Sums of Squares	Degrees of Freedom	Mean Sq. Variance	Ratio of Variances, F
Total	1216.15	67	—	—
Between region	6.00	1	6.00	.32
Within region	1210.15	66	18.33	—

$P(F_{1,67} = .32) > .05$

Table 40 Total Positive and False-Positive Detection of Railroads, Wichita, Kansas

	Polarization	
	HH	HV
Number of interpreters	36	32
Total detection in inches	441.75	494.00
Average detection in inches	12.25	15.45
Percent detection in inches	65.40	82.50

Table 41 Analysis of Variance of Positive and False-Positive Railroads Detected in Wichita, Kansas

Source of Variation	Sums of Squares	Degrees of Freedom	Mean Sq. Variance	Ratio of Variances, F
Total	1583.56	67	—	—
Between region	277.55	1	277.50	14.03
Within region	1306.01	66	19.78	—

$P(F_{1,67} = 14.03) < .001$

Table 42 Total Positive and False-Positive Detection of Railroads, Superior, Wisconsin

	Polarization	
	HH	HV
Number of interpreters	32	36
Total detection in inches	371.50	463.50
Average detection in inches	11.60	12.88
Percent detection in inches	52.80	58.20

Table 43 Analysis of Variance of Positive and False-Positive Railroads Detected in Superior, Wisconsin

Source of Variation	Sums of Squares	Degrees of Freedom	Mean Sq. Variance	Ratio of Variances, F
Total	1482.69	67	—	—
Between region	31.33	1	31.33	1.42
Within region	1451.36	66	21.99	—

$P(F_{1,67} = 1.42) > .05$

Table 44 Detection of Bridges, Wichita, Kansas

	Polarization	
	HH	HV
Number of interpreters	36	32
Total positive detection	566	199
Average positive detection	15.70	6.20
Percent positive detection	78.80	31.00
Total false-positive detection	1	11
Ratio positive/false-positive detection	566 to 1	18 to 1

Table 45 Analysis of Variance of Total Positive Bridges, Wichita, Kansas

Source of Variation	Sums of Squares	Degrees of Freedom	Mean Sq. Variance	Ratio of Variances, F
Total	2116.75	67	—	—
Between region	1530.05	1	1530.05	172.30
Within region	586.70	66	8.88	—

$P(F_{1,67} = 172.30) < .001$

Table 46 Detection of Channel Marker Traversing Water, San Diego, California

	Polarization	
	HH	HV
Number of interpreters	34	34
Total positive detection	82	10
Average positive detection	2.40	0.29
Percent positive detection	48.00	5.80
Total false-positive detection	25	28
Ratio positive/false-positive detection	3.3 to 1	0.35 to 1

Table 47 Analysis of Variance of Total Positive Detection at Channel Marker Traversing Water, San Diego, California

Source of Variation	Sums of Squares	Degrees of Freedom	Mean Sq. Variance	Ratio of Variances, F
Total	165.34	67	—	—
Between region	67.14	1	67.14	45.99
Within region	98.20	66	1.46	—

$P(F_{1,67} = 45.99) < .001$

REFERENCES

Dellwig, L. F., and Moore, R. K. The geologic value of simultaneously produced like- and cross-polarized radar imagery. *J. Geophys. Res.* 71:3597, 1966. Also published as CRES Report 61-13, Univ. of Kansas.

Ellermeier, R. D., Fung, A. K., and Simonett, D. S. Empirical and Theoretical Interpretation of Multiple-Polarization Radar Data in the Geosciences. Fourth Symposium on Remote Sensing of the Environment, Univ. of Michigan, Ann Arbor, Michigan, April, 1966.

Fung, A. K. Scattering and Depolarization of Electromagnetic Waves by Rough Surfaces. CRES Report 48–5, Univ. of Kansas, 1965.

Gillerman, E. Investigation of Cross-Polarized Radar on Volcanic Rocks. CRES Report 61-25, Univ. of Kansas, 1967.

Lewis, A. J. Comparison of Like- and Cross-Polarized Radar Imagery by Counting Spots of High-Intensity Return. Unpublished CRES Memorandum, 1966a.

Lewis, A. J. Summary of Observations on Like, Cross, and Difference Radar Imagery. Unpublished CRES Memorandum, 1966b.

Morain, S. A. Field Studies on Vegetation at Horsefly Mountain, Oregon and Its Relation to Radar Imagery. CRES Report 61-22, Univ. of Kansas, 1967.

Morain, S. A., and Lewis, A. J. Interpretation of Multiple-Polarized K-Band Radar Imagery. Unpublished CRES Memorandum, 1966.

Morain, S. A., and Simonett, D. S. Vegetation Analysis with Radar Imagery. Fourth Symposium on Remote Sensing of the Environment, Univ. of Michigan, Ann Arbor, Michigan, April, 1966. Also published as CRES Report No. 61-9, Univ. of Kansas, 1966.

Morain, S. A., and Simonett, D. S. (1967): K-Band radar in vegetation mapping. *Photogram. Engng.* 33:730, 1967. Also published as CRES Report 61-23, Univ. of Kansas.

How do we recognize phenomena displayed on remote sensing imagery? This is not an easy question to answer. The answer involves the psychophysical response of light interacting with the eye and the brain. It also involves memory, training, and correlation with other sources of information and data. We do know that some image variations are discernible and measurable; the more important of these are size, shape, tone, texture, pattern, linearity, frequency, spacing, and spatial position. A radar system illuminating the earth from above presents a blend of the physical and cultural landscape which combines elements of all these image variations. This integration of the landscape on the radar image must be interpreted. To do this, the patterns delimited on the image have to be correlated with known observable variations in the landscape. This will allow delineation of integrated spatial associations of phenomena. It has been established that general regions can be delimited on radar imagery quickly and relatively accurately. This technique offers some advantages over other approaches to regional synthesis in terms of time, cost, accuracy, and comparability.

36-Integrated Landscape Analysis with Radar Imagery

NELSON R. NUNNALLY

Remote sensor systems are now receiving much attention to determine how they may most effectively and economically be used in earth resource studies. One system which appears to have a wide range of applications is radar. Several published reports have demonstrated the feasibility of identifying such diverse phenomena as geologic structures, physiographic units, vegetative communities, and types of polar pack ice from radar imagery (Rydstrom, 1967; Leighty, 1966; Morain and Simonett, 1966; Anderson, 1966; and Simpson, 1966).

While geographers are interested in specific earth phenomena, one of their major concerns is with the total landscape—the associations of phenomena as they exist on the earth's surface and how these associations vary with location. Since radar imagers are capable of distinguishing a wide variety of phenomena, it was felt that the question of whether they could detect varying associations of phenomena should be investigated.

The imagery which was used for this study covers a portion of the Asheville Basin in North Carolina and extends from Hot Springs to Hendersonville, an area some 50 miles long and 12 miles wide. The images are from a K-band multiple-polarization radar system (HH, horizontal transmit, horizontal receive; HV, horizontal transmit, vertical receive), and were obtained in September, 1965. The scale of the original contact print varies from approximately 1:170,000 to 1:205,000. These scale figures are based on measurements between recognizable points on the radar print.

From *Remote Sensing of Environment* 1:1–6, 1969. Reprinted with permission of the author and American Elsevier Publishing Company, Inc.

Techniques Used in the Analysis

The hypothesis of this paper is that radar provides a means for delimiting varying associations of physical and cultural phenomena through outlining image variation in tone, texture, pattern, and shape. The term "integrated landscape" has been chosen as a name for the types of units delimited.

It is the nature of the scientific method that hypotheses are investigated to see if empirical data or empirical results lead to their support or rejection. Support of a hypothesis in this manner does not prove its validity—rather it may then be developed into a theory. One cannot test the validity of a theory with empirical observations, since by its very nature a theory must fit the empirical data. The only true test of a theory is whether it enables one to make predictions and develop other hypotheses and theories and ultimately leads to the expansion of the general body of knowledge.

If it can be shown that patterns delimited on a radar image are correlated with known, observable variations of physical and economic phenomena, then this empirical observation could be said to support an hypothesis that radar images may be used to delineate integrated spatial associations of phenomena. There are, however, varying degrees of support. It is admitted that statistically valid sampling techniques and analysis are more meaningful than visual means of correlation. While statistical approaches will be used eventually, they are beyond the scope of this preliminary report. Instead, supporting evidence cited has come from aerial photos, the findings of other investigators, and field investigations.

As will be shown in this report, enough information can be interpreted from the radar image to basically characterize a number of relatively distinct regions in the Asheville Basin. However, it must be stressed that the very thing which permits the rapid identification of regions—namely, the small scale and limited resolution —prohibits interpretation of detailed regional variations. Nevertheless, the technique has demonstrable value, for other methods of regional analysis demand both vast arrays of data collected on a large scale, and time-consuming analysis of the data to establish regional categories and permit generalizations of the findings. With radar regions one needs to sample only when detailed data about regional characteristics are required. And, in those cases the sample size need be only a fraction of that required by other approaches.

If it can be established that general regions can be delimited on radar quickly and accurately, the technique will represent a considerable improvement over other approaches from the standpoint of time, cost, comparability, and accuracy.

Delimitation and Description of Landscape Types

On the HH polarization imagery, it is possible to identify areas which exhibit a more or less homogeneous pattern throughout and which are distinctly different in appearance from other areas. The difference in appearance is a result of the patterns produced on the image by the tone, size, shape, and arrangement of the objects which were imaged. Examples include both cultural and physical items such as field patterns, vegetation, roads, water bodies, and landforms. The identifications are based entirely on qualitative assessments of the imagery. It is possible to offer tentative explanations of some patterns with only limited prior knowledge of the area if a person has photointerpretation experience and some general knowledge of radar systems.

Eleven distinct areal patterns were delimited on the Asheville image, and some could be subdivided on the basis of variations within the main pattern (Fig. 1). In this paper examples taken from six of the eleven areas, hereafter referred to as regions,[1] are described, analyzed, and shown to agree with empirical variations.

Region 1

The most noticeable pattern and the easiest to delimit occurs in the mountainous areas of the imagery. The distinctive pattern in these areas is the result of two conditions: (1) gross tone changes from very light to very dark, which cover large areas and occur with regularity; and (2) the lack of any noticeable secondary patterns within the lighter-toned areas. The enlargement and the density trace in Figure 2 illustrate the nature of the pattern. The pattern is the result of the rugged terrain in the first case, and a nearly continuous forest cover in the latter situation. Some nonforest areas are visible, but most are apparently in grass. The topographic effect is related to the active nature of the radar system which results in a high return from the illuminated slope, and little or none from the shadowed slope.

It can be inferred that there is sparse settlement and little cultivated land in region 1. In fact some of the areas are largely within national forests and contain few farms. In most of the region there are two farm types—mountain commercial and mountain marginal (Figs. 3 and 4). Spatially both are composed largely of woodlands, with the bulk of the remaining acreage used for

[1] The term "region" is used in both text and illustrations in lieu of the bulkier expression "integrated landscape type." I recognize that geographers normally do not consider widely separated areas to belong to the same region even though the areas may exhibit similar characteristics. For further reference see Bunge (1962).

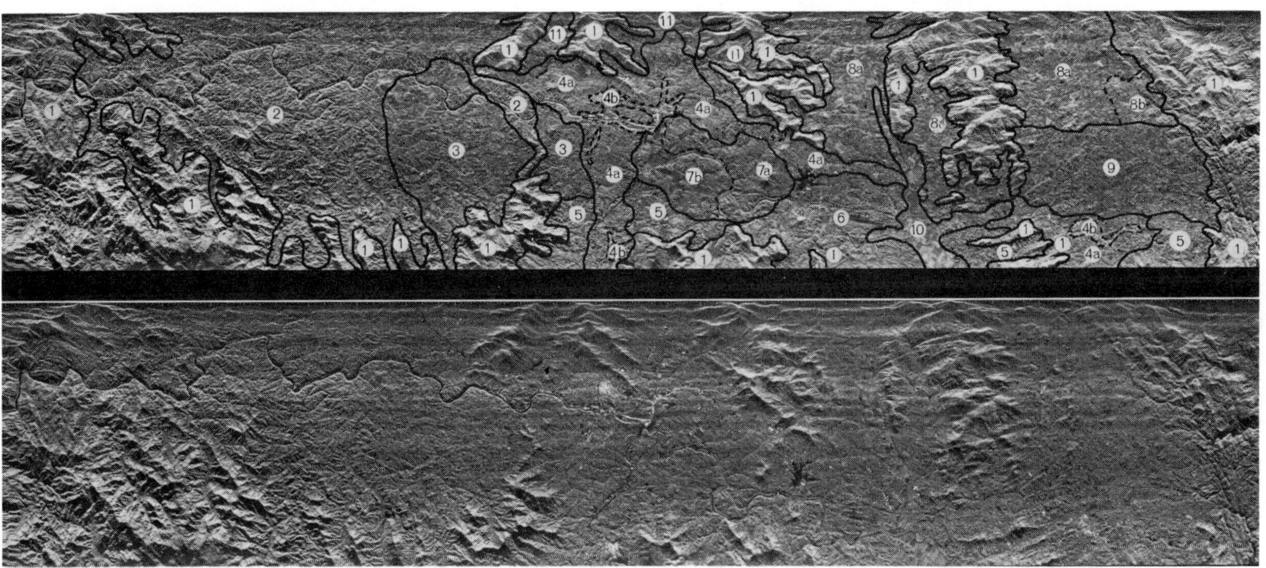

Figure 1 Integrated landscape types distinguished on K-band radar imagery. The numbers on the HH polarization refer to the regions discussed in this paper. The HV image with the names of features is intended for orientation.

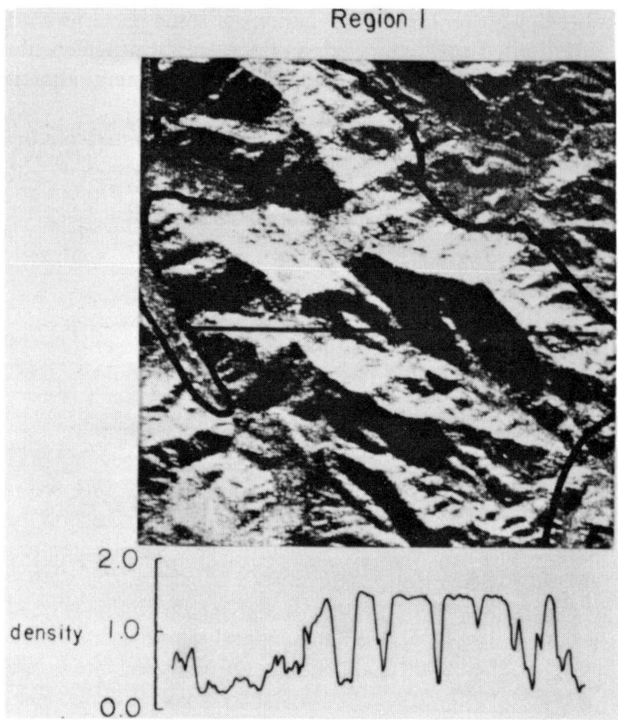

Figure 2 An enlarged portion of radar region 1 with microdensitometer trace. The trace was made along the horizontal black line in the center of the photograph. Note the size and density of the illuminated and shadowed slopes of the ridges.

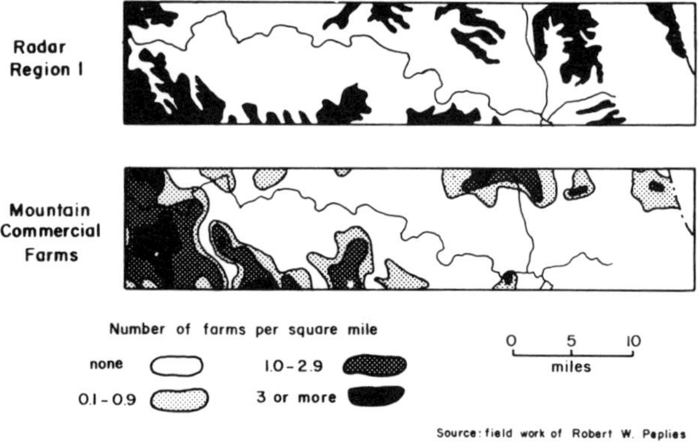

Figure 3 Radar-delimited region 1 compared to the density of mountain commercial farms. The main concentrations of mountain commercial farms are in the same general locations as the segments of region 1. The boundaries do not coincide, partly because of the sizes of the farms and the generalizing effect of isopleth mapping.

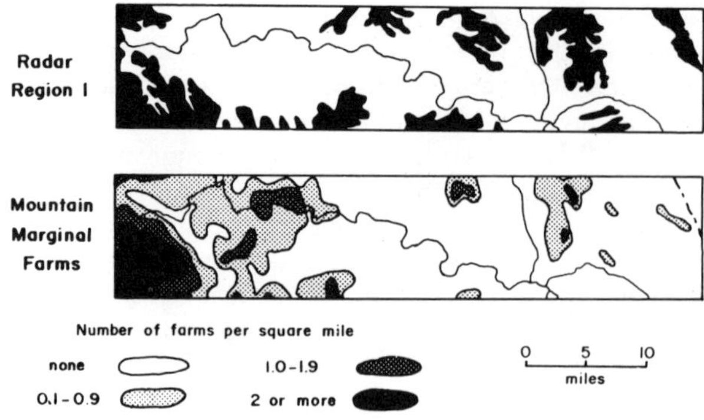

Figure 4 Radar-delimited region 1 compared to the density of mountain marginal farms. See legend to Figure 3.

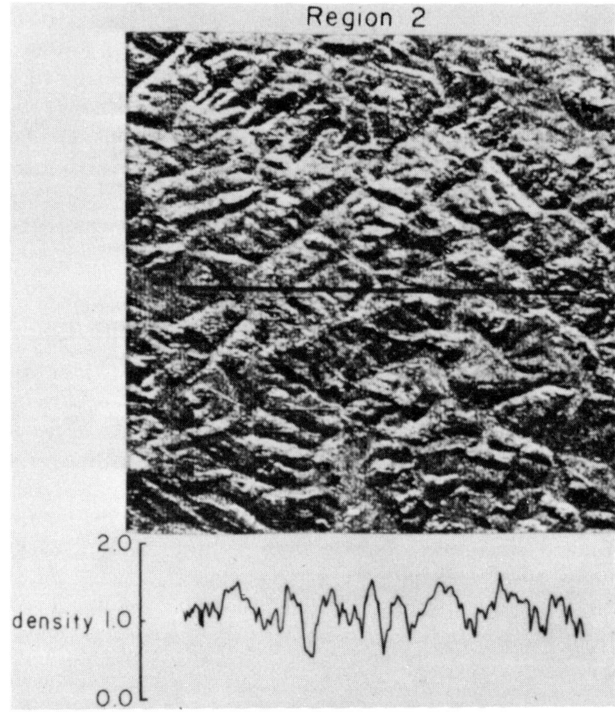

Figure 5 Enlarged portion of radar region 2 with microdensitometer trace. The stream valleys orthogonal to the radar scan show up well. The moderate topographic texture can be inferred from the frequency and magnitude of change in the density trace.

pasture and hay owing to the emphasis on grazing. The amount of cropland is small, usually limited to 3 acres or less (Peplies, 1968).

Region 2

Adjacent to the mountains in the northwestern portion of the Basin a second pattern may be identified. The density trace and the enlargement in Figure 5 show the pattern—many regularly spaced, linear light and dark areas and other irregularly shaped, but uniformly distributed, tonal patterns.

There is little difficulty in interpreting the major pattern. The linear effect is caused by the tone change associated with stream valleys, with one side giving a bright return and the other being in a shadow. Since the high and low returns are narrow and linear, the relief can be interpreted as moderate.

The nontopographic tone changes are associated largely with vegetation and crop patterns. The lighter tones on the interfluves—continuous on some, fragmented on others—are due to the higher returns which wooded areas give.

The appearance of the region 2 landscape suggests that a significant portion of the area is cleared (60 to 80

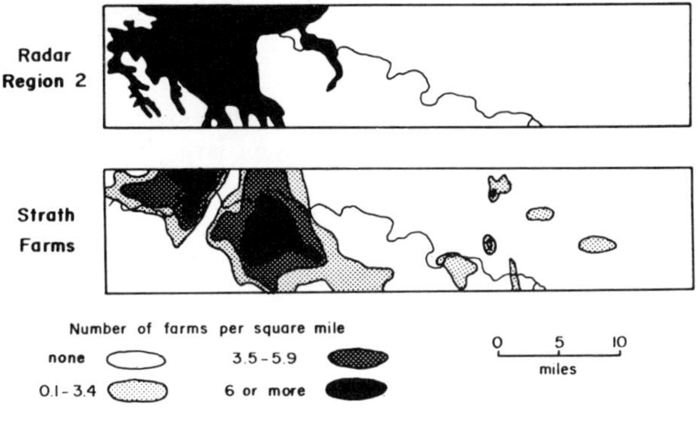

Figure 6 Radar-delimited region 2 compared to the density of strath farms. Although the boundaries do not coincide, the correlation of region 2 with areas containing strath farms is evident.

percent) and perhaps cultivated. Cleared land seems to occupy most of the stream valleys, much of the sloping land, and some of the interfluves. The lack of tonal variation in the nonwooded areas probably means that most of the cleared land is put to one use (such as pasture or abandoned land) or that field sizes are so small that individual fields are not resolved, yielding an integrated signal.

The dominant farm type in region 2 is a family operation which has been given the name "strath farm" (Fig. 6) (Peplies, 1968). Land use descriptions of both regions 2 and 3 apparently relate to these family farm characteristics. Woodland, pasture, and cropland portions are located in relation to slope. Forests are located on interfluves and in some cases along streams, while crops are largely on uplands and pastures are found on slopes. Crops include tobacco, hay, corn, and vegetables, with hay occupying the most area. The main noncropland use is pasture, and woodland is generally limited to less than 25 percent of the farm area.

Region 3

The tonal pattern in the third landscape type varies more in size, shape, and variety of tones and, overall, is of a more subtle texture than regions 1 and 2 (Fig. 7). All of this indicates subdued topographic expression and a more intense occupance with more agriculture, less forest, and a greater variety of phenomena. The lack of any large areas of uniform tone indicates generally small field patterns and woodland area (but large enough to be resolved).

The differences between regions 2 and 3 are, in part, related to topography and level of intensity. But there are additional and different elements in region 3. It, like region 2, contains a large number of family farms. But, in addition, it contains a large concentration of nonfarm residences (Fig. 8).

Region 4

The most intricate and finest textured pattern is found in region 4. In addition to its fine texture and wide tonal variation there are numerous lineaments arranged into rectangular and other regular geometric patterns. Also, some of the highest returns present on the imagery are found in the region 4 areas. This is particularly evident in the HV polarization.

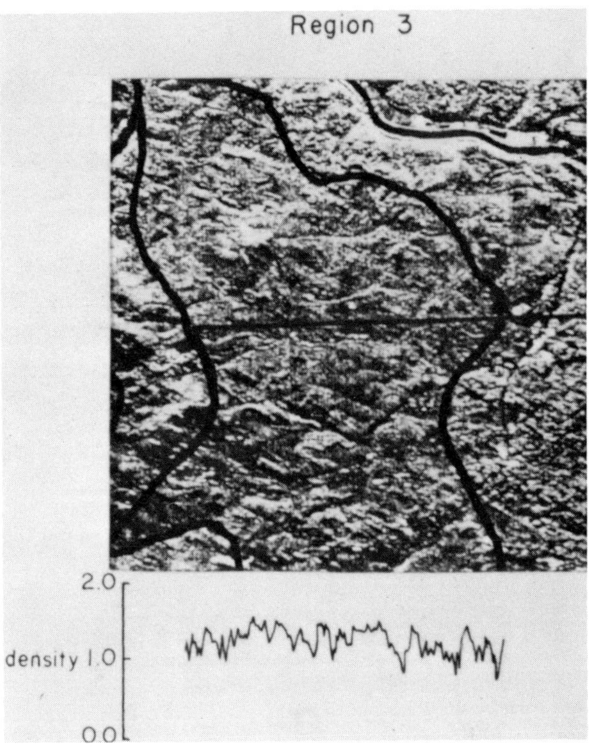

Figure 7 Enlarged portion of radar region 3 with microdensitometer trace. The density range of this trace is less than that of previous regions, reflecting more uniform topography. The photographic texture and the form of the density trace in this region are largely the product of cultural practices and phenomena rather than land forms or other physical phenomena.

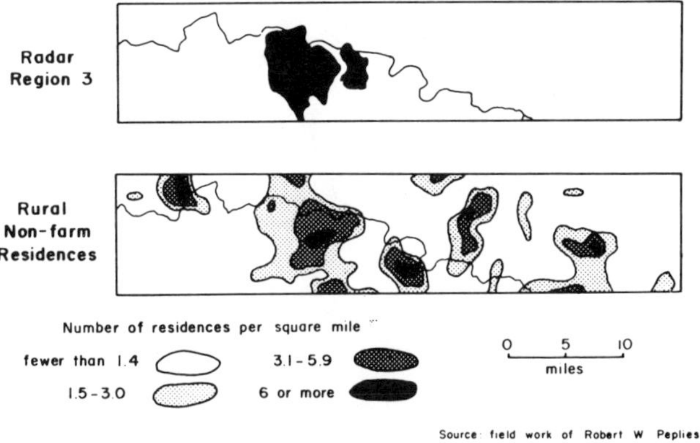

Figure 8 Radar-delimited region 3 compared to the density of rural nonfarm residences. While there are rural nonfarm residences elsewhere in the radar area, the concentration in region 3 is the largest and most dense. This emphasizes the significance of rural nonfarm elements to the landscapes of region 3.

Region 4 areas represent the urbanized areas of the Asheville Basin, with Asheville, Enka, and Hendersonville being particularly noticeable. On the basis of the

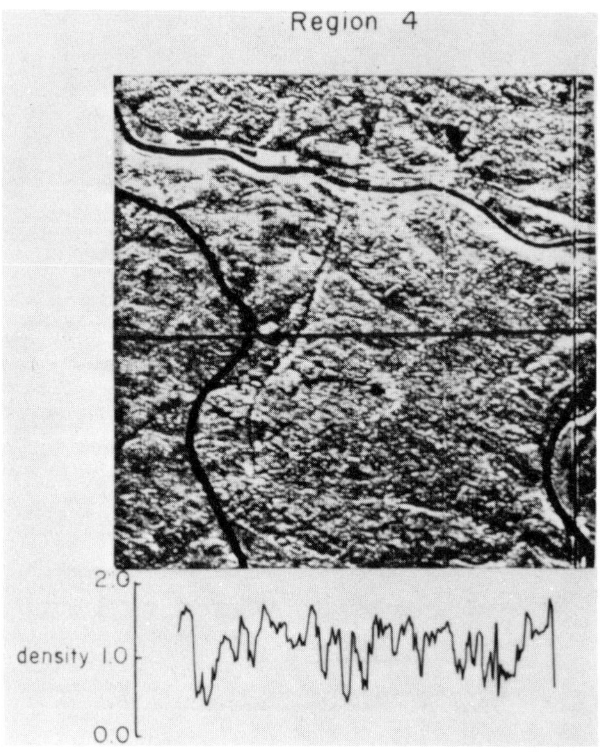

Figure 9 Enlarged portion of radar region 4 with microdensitometer trace. The unique texture and the corresponding density trace are the consequence of the urban pattern, with its streets and buildings. The most intense tones are reflected from large buildings in the central business district, warehouse district, and apartment complexes.

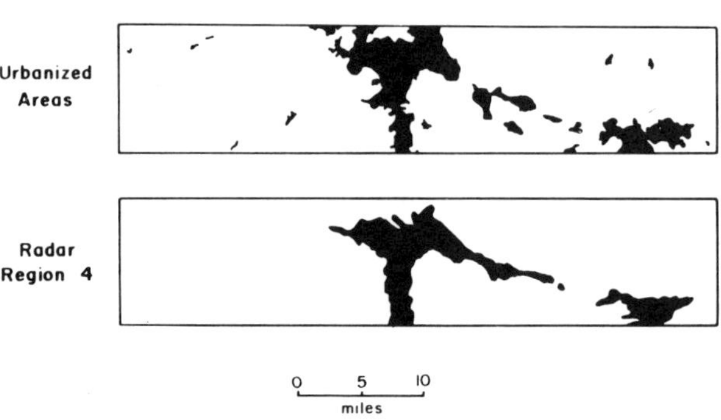

Figure 10 Radar-delimited region 4 compared to the distribution of urbanized areas. The good correlation between radar region 4 and the urbanized areas of the Asheville Basin is evident from this figure.

dense, light tones it is possible to subdivide these urban areas into "core areas (not necessarily limited to the central business district) and less-developed "fringe" areas.

The fine texture, the lineaments, and the regular geometric patterns evident in the type 4 areas are the result of the urban street patterns and block developments (Fig. 9). A definite change in texture can be observed as the eye scans outward from the urban centers—the texture becomes coarser and the rectangular patterns become less noticeable.

The intense reflections of the "core" areas (subtype b) which are not confined to one central location probably result from the high density of large buildings and transportation facilities which provide numerous corner reflectors for the incident radar energy.

Region 4 areas correlated strikingly with the urbanized areas as identified by the detailed field work of Robert W. Peplies (1968) (Fig. 10). The light-toned areas, called "core" areas earlier, are commercial centers (the central business district, shopping centers, and the larger neighborhood centers and strip commercial developments), areas with high concentration of industrial and warehouse functions, and railroad switchyards.

Region 7

Immediately southeast of Asheville is an area which yields one of the least complex returns other than the mountain regions (Fig. 11). Tonal contrast within the area is limited, and tones occur in large tracts which display little internal variation. This suggests that field sizes are much larger than average and that there is little variety in land use. The dark tones imply that there are no crops which have heavy, broadleaf foliage such as corn and beans, but rather that the fields are used largely for hay and pastures or were planted to crops such as grain which would have been harvested before Septem-

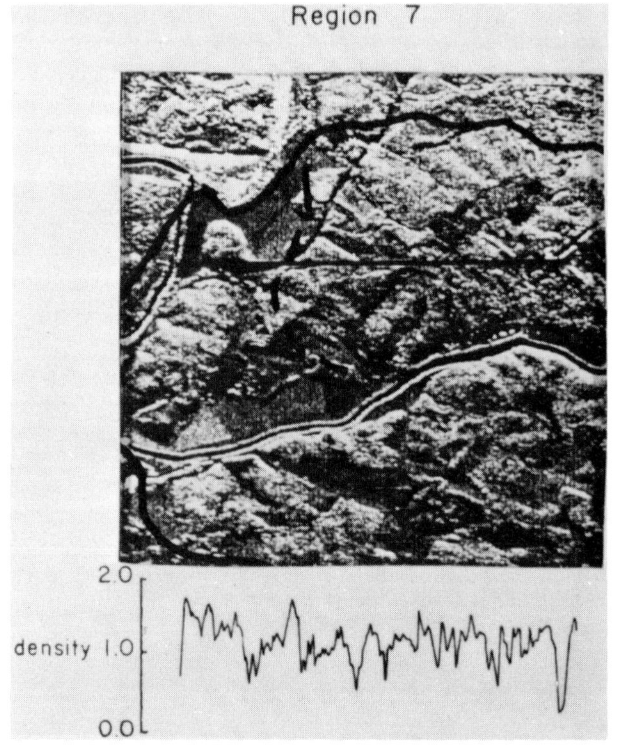

Figure 11 Enlarged portion of radar region 7 with microdensitometer trace. The main features of the region are the large field of uniform tone and the large areas of continuous forest cover. The dashed line separates the area occupied by the fields from the predominantly forested portion.

ber. This leads one to believe that some form of operation involving cows or cattle dominates the agriculture here.

Investigation revealed that region 7 consists almost entirely of one large landholding—the Biltmore Estate. The estate is functional as a tourist attraction as well as an agricultural operation. Dairying is the primary agricultural enterprise but horticultural and forestry activities also are important. Dairying is centered in the subregion 7b, while 7a contains the bulk of the horticulture and forestry operations.

Region 10

The area designated as region 10 contains several recognizable elements (Fig. 12): (1) wide tonal variety is evident; (2) definite rectangular shapes occupy part of the area; (3) tones are relatively uniform within the rectangles and within other large, but more irregularly shaped areas; and (4) there is no tonal evidence of relief. All of these factors indicate that the area is one of intensive agriculture which is conducive to large fields and a crop variety suggestive of a "corn-belt" type of farming.

Region 10 is unique in terms of its description as well as its appearance. The large rectangular fields are associated with cash corn and corn-livestock operations and with large dairying operations (Figs. 13 and 14). In neither case are the entire operations included within the limits of region 10. Crops, which cover an average 27 percent of the area of dairy farms, are limited to the alluvial floodplains, while pastures, woodlands, and farmsteads are located on the sloping land outside the floodplain. The cash corn and the corn-livestock operations are located mostly on the alluvial floodplains. They are rectangular or square in shape, 25 to 200 acres in size, and have about 80 percent of the total area devoted to crops. Farmsteads and woodlands which account for less than 20 percent of the farm area may be located outside the alluvial valleys (Peplies, 1968).

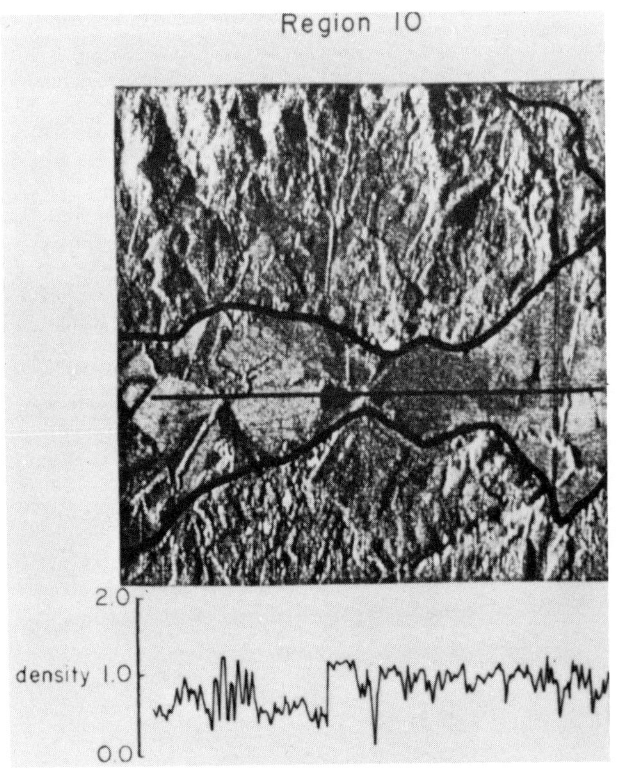

Figure 12 Enlarged portion of radar region 10 with microdensitometer trace. The areas of uniform texture and density are large fields of standing corn, harvested fields, and pasture.

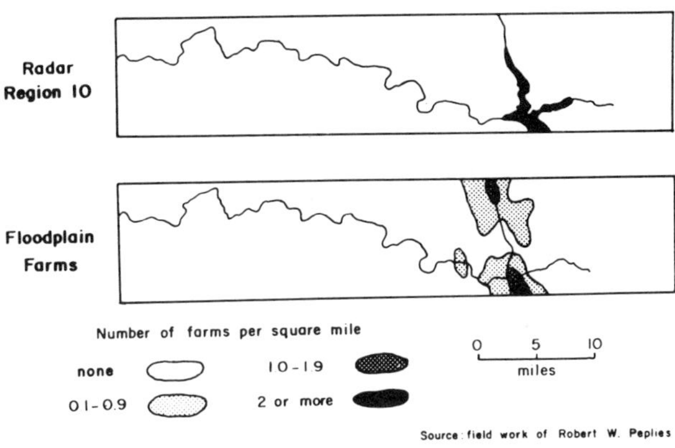

Figure 13 Radar-derived region 10 compared to the distribution and density of floodplain farms. While some floodplain farms are found along upper Cane Creek, the largest concentration is at the confluence of Cane Creek and the French Broad River where region 10 is centered.

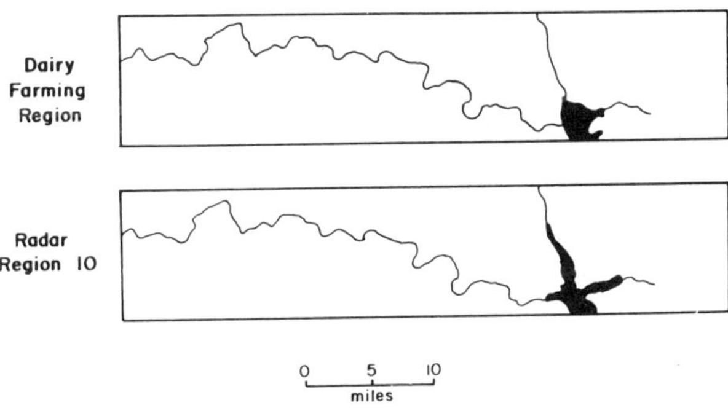

Figure 14 Radar-derived region 10 compared to the dairy farming region. A good correlation exists between radar region 10 and the dairy farming region identified by Robert Peplies (see text).

Conclusions

The radar regions delimited in this study do appear to correlate well with distinctive integrated landscape types, and further study of radar derived regions is warranted.

Several comments are pertinent concerning the approach used here; (1) The region outlined may not all belong to the same hierarchial category; (2) in some cases it is difficult to say whether an area is sufficiently different from others to represent a separate region; and (3) the technique is basically subjective. However, the fact that these questions cannot be resolved does not invalidate the method of study or the usefulness of the results. For example, even if quantifiable data can be obtained from radar, there is no agreement among geographers as to the best analytic techniques for using such data to delimit regions. Furthermore, there is virtually an infinite number of measurable phenomena and someone must make a subjective evaluation regarding the best ones for analysis.

The interpretations in this report are not complete enough to demonstrate the eventual value of radar in regional analysis. Certainly there is a very real possibility that radar, and perhaps small-scale, very-high-altitude photography, can be used for rapid, reliable regional delimitation. Also, some interpretation is possible at the same scale. Not only can broad land use categories be identified, but it may be possible to identify specific uses (including crop types, vegetative communities, and urban land uses) where the size of the pattern is large enough to be resolved. Assuming that regional delimitation and even limited interpretation are possible, then field interpretation and field checking of data could be held to a minimum—thus lowering costs associated with data collection while covering larger areas with better reliability than is presently possible over a major portion of the earth.

REFERENCES

Anderson, V. H. High-altitude, side-looking radar images of sea ice in the Arctic. In *Proceedings of the 4th Symposium on Remote Sensing of Environment.* Willow Run Laboratories of the Institute of Science and Technology, The University of Michigan, Ann Arbor, Michigan, 1966. Pp. 845–857.

Bunge, W. Theoretical geography. *Lund Studies in Geography, Series C. General and Mathematical Geography, No. 1.* Lund. Gleerup, 1962.

Leighty, R. D. Terrain information from high-altitude side-looking radar imagery of an Arctic area. In *Proceedings of the 4th Symposium on Remote Sensing of Environment,* pp. 575–585. Willow Run Laboratories of Science and Technology, The University of Michigan, Ann Arbor, Michigan, 1966. Pp. 575–585.

Morain, S. A., and Simonett, D. S., (1966), Vegetation analysis with radar imagery. In *Proceedings of the 4th Symposium on Remote Sensing of Environment.* Willow Run Laboratories of Science and Technology. The University of Michigan, Ann Arbor, Michigan, 1966. Pp. 605–622.

Peplies, R. W. Occupance Formation Theory: A Case Study of the Asheville Basin. Unpublished Ph.D. dissertation, University of Georgia, Athens, Georgia, 1968.

Rydstrom, H. P. Interpreting local geology from radar imagery. *Geol. Soc. Am. Bull.* 73:429, 1967.

Simpson, R. W. Radar: geographic tool. *Ann. Assoc. Am. Geographers* 56:80, 1966.

SUGGESTED READINGS FOR PART EIGHT

Barringer, A. B. The Use of Audio and Radio Frequency Pulses for Terrain Sensing. *Proceedings of the Second Symposium on Remote Sensing of Environment.* Ann Arbor: University of Michigan, Institute of Science and Technology, Willow Run Laboratories, 1963. Pp. 201–212.

Barton, D. K. *Radar Systems Analysis.* Englewood Cliffs, N.J.: Prentice Hall, 1964.

Berkowitz, R. S. *Modern Radar—Analysis, Evaluation and System Design.* New York: Wiley, 1965.

Cummings, W. A. Radiation measurements at radio frequencies: A survey of current techniques. *Proceedings of the Institute of Radio Engineers* 14:705–735, 1959.

Ellermeier, R. D., and Simonett, D. S. Imaging Radars on Spacecraft as a Tool for Studying the Earth. *International Symposium on Electromagnetic Sensing of the Earth from Satellites. Miami Beach, Fla.,* November 22–24, 1965.

Feder, A. M. Interpreting natural terrain from radar displays. *Photogrammetric Engineering* 26:618–630, 1960.

Lack, D. L. Watching migration by radar. *British Birds* 52:258–267, 1959.

Lyytikainen, H. E. An analysis of radar profiles over mountainous terrain. *Photogrammetric Engineering* 26.403–412, 1960.

MacDonald, H. C. Geologic evaluation of radar imagery from Darien Province, Panama. *Modern Geology* 1:1–63, 1969.

Marchinton, R. L. Portable Radios in Determination of Ecological Parameters of Large Vertebrates with Reference to Deer. *Remote Sensing in Ecology.* Athens: University of Georgia Press, 1969. Pp. 148–163.

Moore, R. K. *Radar as a Remote Sensor.* CRES Report 61–7, Center for Research in Engineering Science, University of Kansas, 1966.

Moore, R. K., and Ulaby, F. T. The radar radiometer. *Proceedings of the Institute of Electrical and Electronic Engineers* 57:587–589, 1969.

Morain, S. A., and Simonett, D. S. K-band radar in vegetation mapping. *Photogrammetric Engineering* 33:730–740, 1967.

Page, R. M. The Discovery of Radar and the Radar Idea. *The Origin of Radar.* New York: Doubleday, 1962.

Poejsil, D. J., Raven, R. S., and Waterman, P. *Airborne Radar.* Cambridge, Mass.: Boston Technical, 1965.

Rouse, J. W., Jr. Arctic ice type identification by radar. *Proceedings of the Institute of Electrical and Electronic Engineers* 57:605–611, 1969.

Simpson, R. B. Radar, geographic tool. *Annals of the Association of American Geographers* 56:80–96, 1966.

PART NINE
MULTISPECTRAL REMOTE SENSING

Imaging a target at more than one wavelength

SOME REMOTE *sensors operate at a single, rather narrow—almost specific—wavelength. This might be the case for an infrared scanner using a single detector. Other sensors are not specific but integrate energy from many wavelengths into a composite image. A normal photograph in the visual portion of the spectrum records the many wavelengths reflected to the surface of the film within the upper and lower wavelength limits of the film emulsion. From previous discussion we know that most objects reflect or emit energy over a broad spectrum of wavelengths, but normally there is a peak wavelength at which the maximum amount of energy is being reflected or emitted. We could best isolate this object's reflection or emission from all the other energy in the background by utilizing a sensor that operates at the point of peak transmission. It would give the greatest contrast between target and background. In effect, this is the essence of multispectral imaging. A sensor or series of sensors is designed so as to operate at a very narrow band width; objects with peak responses at or near this point in the spectrum will produce a strong signature on the image. In practical application we might choose a sensor operating in the 720–890 nanometer band to distinguish broadleaf trees from conifers. At this band broadleaf trees image very light; the conifers' signature is much darker. In the 580–720 nanometer band, clear water appears dark and turbid water light. The more specific a target response is, the easier it is to design a sensor that will locate it against other background radiation.*

Actually we know very little about the spectral response of targets in the natural environment. Many factors affect the response, including temporal variations such as time of day and season of the year. Multispectral images of the same earth scene have been generated by sensors with as many as 24 channels, but it is difficult to correlate target signatures among so many wavelength responses unless some type of mechanical or electrical device is used. For most purposes a sensor with four carefully selected channels will provide maximum contrast and easier correlation.

37-Remote Sensing of Natural Resources

ROBERT N. COLWELL

AS PRESSURES on natural resources increase, because of growing populations and rising standards of living, it becomes steadily more important to manage the available resources effectively. The task requires that accurate inventories of resources be periodically taken. Until as recently as a generation ago such inventories were made almost entirely on the ground. Geologists traveled widely in exploring for minerals; foresters and agronomists examined trees and crops at close hand in order to assess their condition; surveyors walked the countryside in the course of preparing the necessary maps. The advent of aerial photography represented a big step forward. Within the past few years the making of aerial photographs has been augmented by a new technique, in which sensing is done simultaneously in several bands of the electromagnetic spectrum. The name often given the technique is remote sensing. In its fullest form the technique ranges through the spectrum from the very short wavelengths at which gamma rays are emitted to the comparatively long wavelengths at which radar operates. In this way one can secure far more information about an area than can be obtained with conventional photography, which is limited to the visible-light portion of the spectrum.

From *Scientific American* 218:54–69, 1968. Reprinted with permission of the author and *Scientific American*.

Remote sensing can be done from aircraft or spacecraft, including unmanned satellites. It employs cameras and a number of other sensing devices. To some extent the data obtained by the sensing devices can be processed and interpreted automatically, so that a large volume of information can be dealt with rapidly.

The information thus obtained is useful to investigators in many disciplines. Geologists use remote sensing to find deposits of minerals and petroleum, to improve their understanding of the distribution and origin of major geological features and to study the exchanges of energy associated with such crustal disturbances as earthquakes and volcanic eruptions. Soil scientists can take inventory of the important physical and chemical characteristics of soil by relating these characteristics to the geological features and the types of vegetation found on images obtained by remote sensing. Foresters and agriculturists can determine what kinds of trees and plants are growing in an area, can assess the health of the forest or crop and can estimate harvests. Similar information can be obtained by workers interested in populations of livestock, wildlife, and fish.

By means of remote sensing hydrologists can locate useful aquifers and can estimate the volume of surface and subsurface flow in watersheds. Oceanographers can map the movements of ocean currents, marine organisms, and water pollutants. They can study in detail the daily and seasonal changes in tides, shorelines, and the state of the sea. Geographers can analyze land-use patterns over broad areas and can study the interplay of climate, topography, plant life, animal life, and human activity in a particular area. Engineers planning large construction projects such as highways, airports, and dams can obtain data on landforms, rock materials, soils, types of vegetation, and conditions of drainage. It goes almost without saying that remote sensing in various parts of the spectrum is invaluable to map makers in their efforts to identify ground features and to position them accurately.[1]

The earliest aerial photographs, made somewhat more than a century ago, suffered from the deficiencies of the cameras and emulsions and from the necessity of using such unsteady vehicles as balloons and kites for platforms. Today the array of equipment available for remote sensing can be matched to almost any requirement. Whatever the platform—helicopter, airplane, or satellite—the camera can be mounted so that it is stabilized against roll, pitch, and yaw and insulated against vibration. The aberrations of the lenses in cameras have been greatly reduced so that sharp images can usually be obtained. Roll film of high dimensional stability has almost entirely replaced emulsion-coated glass plates. Several kinds of color film are available to augment or replace black-and-white film in both the visible and the infrared portion of the spectrum.[2]

Remote-Sensing Equipment

Among the many types of equipment developed for remote sensing, six show the most value or promise for the inventory of natural resources. They are the conventional aerial camera, the panoramic camera, the multiband camera, the optical-mechanical scanner, side-looking airborne radar, and the gamma ray spectrometer.

A *conventional aerial camera* has four basic components: a magazine, a drive mechanism, a cone, and a lens (see Fig. 5). The magazine is the light-tight box that holds the film. Usually it can be detached from the rest of the camera. The film is ordinarily in the form of a continuous roll 9½ inches wide and 200 feet long. Such a roll will accommodate about 250 exposures, each 9 inches square.

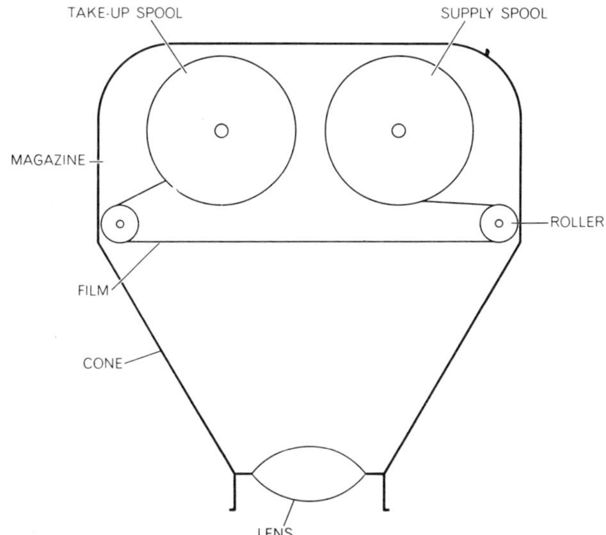

Figure 5 Conventional aerial camera uses film in roll form and can make about 250 exposures, each 9 inches square. Magazine holds the film. Cone positions lens with respect to the film, at a distance governed by the focal length of the lens. Film being exposed is held flat against a locating plate to minimize distortions and provide uniformly sharp images. (Thomas Prentiss)

The drive mechanism is a series of cams, gears, and shafts designed to move the film from the supply spool to the take-up spool. As the film moves, rollers guide it over the front surface of a locating plate. One of the rollers is designed to meter the amount of film passing from the supply spool to the take-up spool between exposures, thereby providing a correct and uniform spacing of exposures on the roll of film.

During an exposure, suction is created behind the locating plate by means of a venturi tube or a special vacuum-cylinder-and-piston apparatus built into the magazine. The suction, transmitted to the film through small perforations and grooves in the locating plate, holds the film in a flat plane against the locating plate at the instant of exposure. In this way distortions that would be caused by wrinkles in the film at the moment of exposure are minimized.

The cone is a light-tight unit that holds the lens in the correct relation to the film. The length of the cone is governed by the focal length of the lens, which is essentially the distance from the center of the lens to the film. It is not unusual for a magazine to have interchangeable cones to accommodate lenses of differing focal lengths. Most of the aerial photography done for the inventory of natural resources uses focal lengths of 6, 8¼, or 12 inches.

The lens is a compound one that is carefully designed to cast an undistorted image on the large area of the film. Aerial cameras usually have fixed-focus lenses with the focus at infinity; the camera is used so far above the ground that such a focus will provide a sharp image of all objects on the ground. In most aerial cameras the shutter is between the front and the rear elements of

[1]Figures 1, 3, and 4 not available for reproduction.
[2] See Figure 2 in color insert.

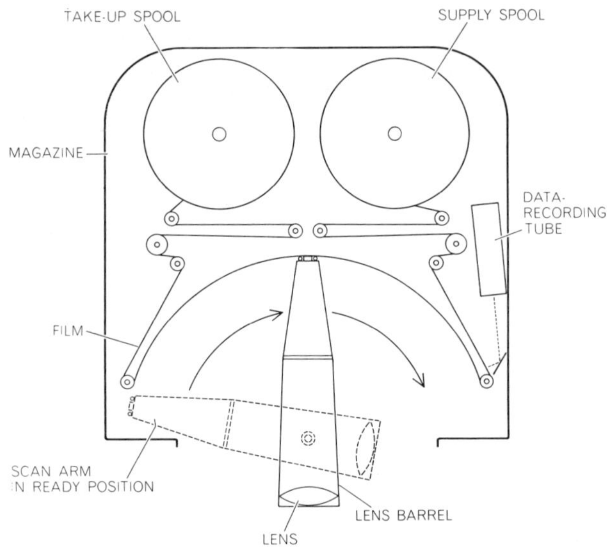

Figure 6 Panoramic camera was developed to provide sharp aerial photographs of large scale. Since the camera must have a long focal length, it must also have a narrow angular field. As a result it requires a scanning mechanism that moves the lens barrel to left and right. At any instant during the course of a scan only light passing through a narrow slit falls on the film.

the lens. The drive mechanism of the camera recocks the shutter automatically after an exposure.

The *panoramic camera* (see Fig. 6) makes it possible to photograph a large area in a single exposure at very high resolution, meaning with a high degree of sharpness of image in every part of the photograph. The camera meets a need but creates some special problems. In order to get a sharp image when photographing large areas, one paradoxically needs a narrow angular field so as to minimize aberrations of the lens. Such a field is provided in the panoramic camera by a narrow slit in an opaque partition near the focal plane of the camera. The slit is parallel to the camera platform's line of flight. With such a slit, however, one will be able to photograph only a narrow swath of terrain unless the optical train of the camera is equipped to pan, or move from side to side, as the aircraft advances. The optical train of the panoramic camera is designed to make such movements.

On the other hand, for the panoramic camera to maintain a uniformly clear focus as the optical train moves, the frame of film being exposed must be held in the form of an arc instead of being kept flat as in a conventional camera. With the film in an arc the photographic scale becomes progressively smaller as the distance of objects on the ground increases to the left and right of the flight path. In some applications the scale problems outweigh the advantage of a panoramic field of view, so that it is preferable to use a conventional camera.

Related Devices

The *multiband camera* makes photographs simultaneously in several bands of the spectrum. In essence it provides a variety of lens, filter, and film combinations, each designed to obtain maximum information from a particular band. A typical camera might have nine such combinations (see Figs. 7 and 8)[3]. Together the lenses give the camera a capacity to sense in the range of wavelengths from .4 to .9 micron, which is to say throughout the visible spectrum and into the very near infrared. All nine shutters click simultaneously, thus yielding nine photographs, each with tonal values that are distinctive for its portion of the spectrum. Study of distinctive tonal values in nine photographs of an area enables the interpreter to determine a "tone signature" for each type of object. As a result he obtains much more information about the area's natural resources than he could obtain from any one of the photographs.

The *optical-mechanical scanner* meets the need for a device that will sense farther in the infrared—in what is commonly called the thermal infrared region. Ordinary photographic film is not sensitive to wavelengths in the thermal infrared region. It would be possible to coat a film with a material sensitive to such wavelengths, but then the problem would arise of protecting the film from the thermal energy being emitted by the camera. Just as the conventional camera must be a light-tight box to keep light-sensitive film from fogging, so a thermal infrared camera would have to have a heat-tight box to keep heat-sensitive film from fogging. In fact, the box would have to be continuously cooled almost to absolute zero, which is a practical impossibility for a large airborne sensing device.

Thus a "camera" that translates thermal energy directly onto film is out of the question. It is possible, however, to obtain photographic images of thermal energy indirectly, and that is what the optical-mechanical scanner does. The device uses a detector that consists of a coating of some infrared-sensitive material such as copper-doped or gold-doped germanium on the end of an electrical conductor. The material occupies an area no bigger than a pinhead. It is entirely feasible, even in an airborne system, to cool this small detector with liquid nitrogen for sensing at wavelengths of from 3 to 6 microns and with liquid helium for longer wavelengths.

A rotating mirror directs to the detector energy emanating from the terrain. At any instant the mirror views only a small segment of terrain. Infrared photons striking the detector generate an electrical signal that varies in intensity according to the amount of thermal energy coming from the part of the terrain then being viewed by the mirror. The signal, by being converted to a beam of electrons, can generate visible light, such as the moving luminous spot on the face of a cathode-ray tube. The spot grows brighter or dimmer in direct proportion to the strength of the electron beam. An image of the light is recorded on photographic film, and the analyst obtains what is in effect a thermal map of the ground.

The scanner is not limited to sensing in the thermal infrared region of the spectrum. It can provide multiband imagery in any band from the near ultraviolet through the visible and photographic infrared regions and into the thermal infrared. Moreover, in "photographs" made by the instrument the general shape of

[3] Figures 7–10 not available for reproduction.

ground features is essentially the same in every band, so that the images can be superposed or otherwise compared readily.

Side-looking airborne radar, commonly called SLAR, brings to remote sensing such valuable attributes as all-weather and around-the-clock usefulness and the ability to penetrate a cover of vegetation. Because radar operates at much longer wavelengths than the other equipment I have described, however, it does present difficulties in obtaining high resolution. Recent developments such as SLAR equipped with a synthetic aperture system have brought about large improvements in the quality of radar imagery.

In the SLAR system a transmitting antenna in the airplane sends a short pulse of microwave energy out one side of the plane. The energy strikes a roughly circular area on the ground, and a receiving antenna collects the energy reflected back to the plane. The greater the distance from the aircraft to any portion of the target, the greater the time delay in the return of the reflected signal. By accurately measuring the time delay, SLAR differentiates the echoes that return to it from various small concentric rings. Each ring represents the locus of all points within the large circle that are roughly equidistant from the plane.

Within any ring there is a spot just opposite the aircraft that moves along at the same speed as the aircraft. At any given time the distance from the aircraft to all other points on the ring is either increasing or decreasing. Here the Doppler effect comes into play: the frequency of the reflected signal changes according to whether the plane is approaching a given point or receding from it. As a result the microwave energy reflected back to the aircraft from such points differs in frequency from the energy transmitted to them. The radar receiver is designed to accept energy of approximately the same frequency as the initial pulse and to reject significantly different frequencies.

Because of the two discriminating effects—one depending on time delay and the other on the Doppler effect—the radar receiver accepts at any given instant only the energy that meets two conditions: that it be from the narrow ring within which the time delay is such that the energy is at that instant striking the receiving antenna, and that it be in the particular part of the ring that is directly opposite the aircraft—the part having almost no relative velocity with respect to the aircraft and thus exhibiting no Doppler effect. Together the two discriminating features provide the synthetic aperture. The technique greatly improves the spatial resolution of the system.

Figure 11 Thermal infrared image of a site in Yosemite Valley shows several campfires better than sensing in other bands would. Thermograph, which senses infrared wavelengths and uses them to govern a source of visible light that is recorded on film, was about a mile above the valley. Smallest fire detected was one charcoal briquette less than a cubic inch in size. (Edwin Roberts)

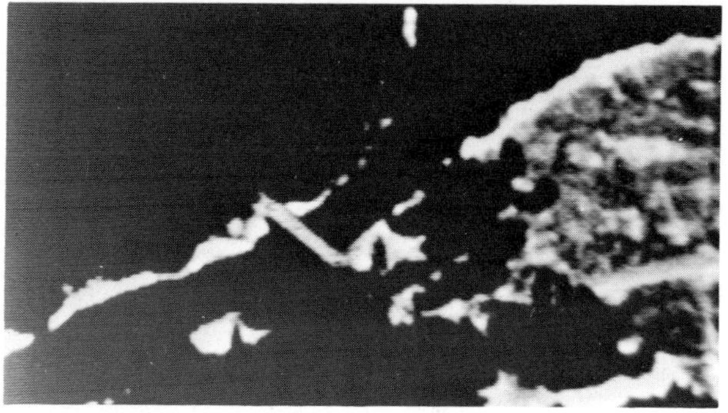

Figure 12 Additional view of the Yosemite Valley site from the same station was made with the thermograph set to function at wavelengths of 8 to 14 μ and so brought out vegetation in meadows (right). The fire-sensing thermograph functioned at 3 to 5 μ. (Edwin Roberts)

Figure 13 Timber resources stand out in this thermal infrared view of same site. Thermograph was set for 8 to 14 μ but the image was obtained by day rather than by night. (Edwin Roberts)

Radar images are transformed into photographic images in the same way that photographic images are produced by the optical-mechanical scanner. The microwave energy is converted to an electron beam that operates a cathode-ray tube, and the light is recorded on film. The density on each portion of exposed film is in proportion to the brightness of the radar signal coming from the corresponding spot on the terrain.

The *gamma ray spectrometer* functions at very short wavelengths—a millionth of a micron or less, compared with the billion microns or more at which radar and other microwave sensors operate. The spectrometer is excellent for locating radioactive substances, even when it is operated at altitudes of several thousand feet above the ground. It is therefore useful in prospecting for minerals. Moreover, a gamma ray spectrometer can be designed to operate in as many as 400 different channels, or wavelength bands, so that the instrument has considerable ability to differentiate each of several radioactive minerals.[4]

Analysis of Data

Remote sensing of natural resources rests on the fact that every feature of the terrain emits or reflects electromagnetic energy at specific and distinctive wavelengths. The analyst cannot hope to accomplish much in the way of interpreting the data, however, until he takes the time to determine what spectral response pattern—what multiband tone signature—to expect from a given feature. The best means of accomplishing this is to set up a test site in which each type of feature that is to be identified by remote sensing is exhibited. By studying multiband images obtained from the test site, the analyst will equip himself to recognize, by their unique spectral response patterns, the features that are of interest to him in a sensing mission. Ideally at least one such test site should be included in each sensing flight for calibration purposes.

Eventually it may become possible to identify every feature in a given area. The technique of sensing in a variety of wavelengths promises to speed progress toward that objective. As the number of spectral bands used in remote sensing is increased, the identifying response pattern for each natural resource becomes more complete and more reliable.

At the same time the increase in spectral bands sensed means that the task of analyzing data grows larger. It can become unmanageable unless the analyst has equipment that helps him to correlate the multiband images. The problem is that he confronts several black-and-white images, each with distinctive tone values for particular features. He can find himself in confusion if he interprets one image, goes on to another, refers back to the first, and so on for a number of images.

One way to deal with the problem is to reconstitute the various multispectral black-and-white images into a single, composite color image. The usual technique is to project each black-and-white image through a colored filter. A battery of projectors is used so that all the images can be superposed simultaneously on the screen.

In such a composite image the tone or brightness of a ground feature as recorded in any given spectral band is used to govern the intensity of one of the colors used in the composite. By varying the selection of colored filters, the analyst can change the color contrast of the composite. Often by this means he finds that one combination of filters provides the best interpretability of one kind of feature, whereas other filters provide better interpretability of other features. The bottom two illustrations in Figure 20 are composites made in this way.

A second way of correlating multiband images is to use a battery of photoelectric sensors to scan all the black-and-white images simultaneously. The sensors record degrees of brightness. For each spot scanned, the sensors automatically determine a tone signature, which

[4] Figures 14, 15, 18–20 not available for reproduction; see color insert for Figures 16 and 17.

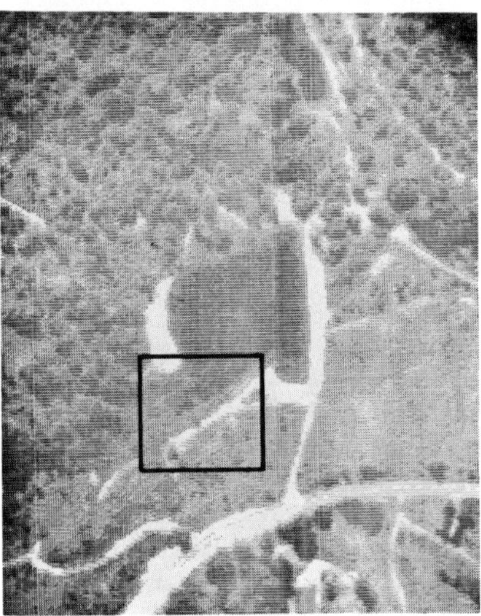

Figure 21 Automatic analysis of tonal qualities can be done with a photoelectric scanner. At top is a photograph made with a multiband camera. At center is a scanner's print in which "N" shows darkest tones and other symbols represent lighter tones. At bottom is an enlargement of the outlined area. Since each natural resource tends to have a unique multiband tone signature, automatic encoding of tones may lead to automatic inventories. (Philip G. Langley, Pacific Southwest Forest and Range Experiment Station, U.S. Forest Service)

in theory will be identifiable with some signature established from the test site. By this means the analyst can identify what features the remote-sensing equipment detected on the ground.

In its ultimate form the technique will result in a tape printout indicating the objects and conditions encountered at every spot in the multiband imagery. The method has not been developed to that stage, but even at its present stage of development it is able to provide enough automatic analysis of images to reduce considerably the amount of work done by the analyst. The illustrations in Figure 21 show the results of photoelectric scanning of an aerial photograph.

In a third technique the multiband sensing system records on magnetic tape, rather than on photographic film, the signal strength from each object in each spectral band. Thereafter the procedure is essentially the same as it is in the photoelectric scanning technique. The third method provides a complete inventory only moments after the remote sensors have been flown over the areas of interest. It also makes possible an analysis of the signal strengths emanating directly from the sensed objects, whereas in the second method the analysis is of signals that may have been degraded in the process of forming multiband images of the objects.

Some Applications

Against the background of sensing equipment and analytical techniques that I have described it is possible to consider in more detail some of the ways in which remote sensing can contribute to the management of natural resources. Several of the posssibilities are illustrated in Figure 2, which was made from the spacecraft *Gemini V* and shows a large area of central Australia. The principal features of the area are identified by letters in the black-and-white reproduction of the photograph in Figure 4.

The southern part of the MacDonnell Range (*A* in Fig. 4) has steeply dipping parallel beds that the geologist would recognize as indicating the presence of folded sedimentary rocks varying in hardness and in susceptibility to erosion. The characteristics of the northern part of the range would suggest to the geologist that the rocks there are igneous or metamorphic. Evidence of faulting appears in the linear ridge that runs through the northern part of the range. The characteristics of the Waterhouse Range (*B* in the illustration) suggest sedimentary rocks that long ago were folded into an anticline, or upfolded structure, and have since been eroded to varying degrees. Careful study of shadow detail in the vicinity of the circular structure known as Gosse's Bluff (*C*) reveals that it is a hollow outcropping of rocks that probably resulted from the impact of a large meteorite.

From even this crude interpretation of the photograph a mineralogical prospector would be able to deduce that some of the best prospects for metallic minerals are to be found along the discernible fault lines in the MacDonnell Range. The petroleum geologist would be interested in the folded anticline of the Waterhouse Range. It is equally significant that the searchers for both metals and petroleum often can eliminate nearly 90 percent of the vast area shown in a small-scale photograph as being unworthy of detailed mineralogical or petroleum surveys. Important deposits of either kind are rarely found in areas that photographically show little geologic evidence of their presence.

The *Gemini* photograph is also helpful in assessing the soil resources of the area. For example, it can be assumed that most of the central region [*D*] has deep alluvial soils because there are nearby mountain ranges from which alluvial deposits are likely to have come, because the pattern of streams indicates that a considerable amount of outwashing activity has taken place even though the area now seems arid, and because in the outwash plains geologic features have become so deeply buried, presumably by deposited soil, as to be indiscernible. In the top left portion of the photograph [*E*] the presence of sandy soil is suggested by the dunelike patterns, which continue appreciably beyond the edge of the photograph. A dry lake bed [*F*] is likely to contain heavy clay soils.

The photograph is of further usefulness in determining the vegetational resources of the area. Even though the photographic scale is small, several vegetational boundaries can be seen. One of considerable significance [*H*] shows two types of grass: Mitchell grass on the left and spinifex on the right. Areas of Mitchell grass are far better than other grassland for maintaining livestock. Moreover, they normally are indicative of the most fertile soils in an area, a point of great importance if the objective is to find new land to put to the plow.

In mapping vegetational boundaries the lack of fine detail in a photograph such as the *Gemini* one may actually be helpful. The fact is evident if one looks at the area marked *G* in the *Gemini* photograph and at the corresponding oblique aerial photograph at the bottom left in Figure 3. The boundary is between mulga (a type of acacia tree) and spinifex. In the *Gemini* photograph the boundary is clear; in the oblique photograph it is difficult to follow even though more detail is discernible there.

Recently I accompanied Ray Perry of the Commonwealth Scientific and Industrial Research Organization in a check of the ground shown in the *Gemini* photograph. We made the oblique aerial photographs that appear in Figure 3. The check showed that the interpretations previously made from the *Gemini* photograph were correct in all respects.

I have dwelt at length on this single *Gemini* photograph in order to suggest the capabilities of spacecraft photography in the remote sensing of natural resources. Since the whole of an area as big as Australia can be depicted in a short time with a few photographs from a spacecraft, the possibilities of the technique are enormous, particularly for the vast areas of the world that are yet to be developed. Australia is a case in point. According to Australian scientists, virtually all the significant geologic, soil, and vegetational features found in approximately 70 percent of the continent's arid regions are represented in the *Gemini* photograph that I have described. It seems evident that one of the best ways to produce suitable reconnaissance maps for the remainder of underdeveloped Australia and for other

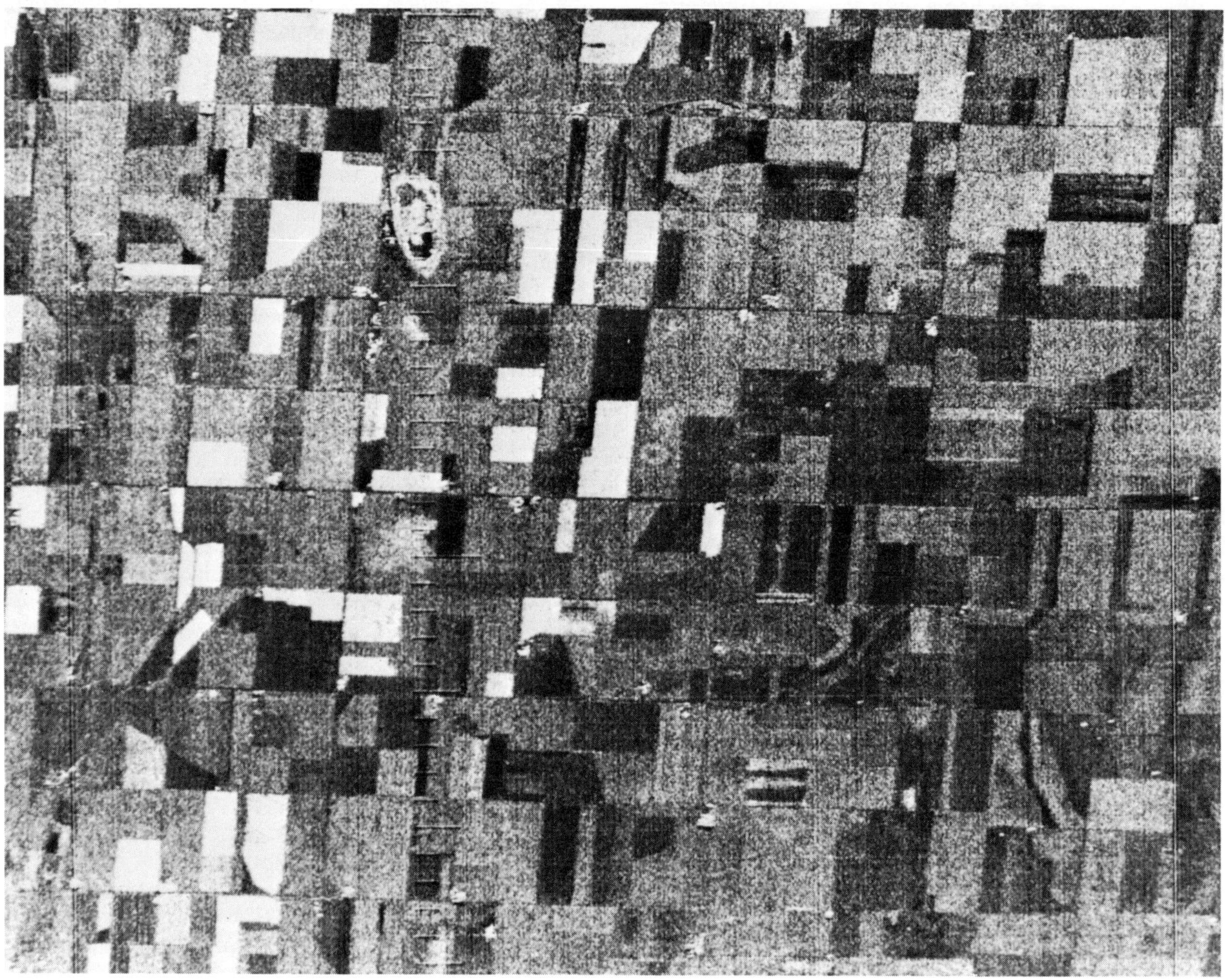

Figure 23 Radar image of farmland in Kansas shows certain types of crops more clearly than images from other bands of the spectrum could. Lightest fields, for example, contain sugar beets. Clear radar images can be obtained day and night and through clouds. (Westinghouse Electric Corporation)

undeveloped areas of the world would be through the use of space photography, supplemented as necessary with large-scale aerial photographs and field checks.[5]

Additional Applications

The catalogue of uses for remote sensing is extensive. In forestry, for example, it is possible in small-scale photographs, such as those from spacecraft, to delineate the timberland, brushland, and grassland in a wild area. With proper film and filter it is possible to differentiate the three major types of timber—hardwood, softwood, and mixed wood. In larger-scale photographs one can determine the size of trees, the density of growth, and the volume of timber. Foresters also use aerial photographs to detect trees that are diseased or infested with insects. Aerial photographs can be used to help in the planning of forest roads and of means for fighting forest fires.

The management of rangelands is assisted by remote sensing. From photographs one can learn the species of vegetation in an area, together with their volumes and their forage value. Photographs also reveal other data pertinent to range management, such as watering places, salt ground, plants that are poisonous to livestock, highly erodible sites, and areas that need reseeding.

Wildlife managers can use aerial photographs for censuses of various kinds of animals and fish. The information is important in determining the impact of hunting, fishing, and the works of man on fish and wildlife populations.

Administrators of agricultural programs need information on the type of crop growing in each field of a large area, the vigor of each crop, and the probable yield. Where crops lack vigor, the agriculturist wants to know what is wrong. All such information can often be obtained through the interpretation of aerial photographs if the photographs have been made under appropriate conditions, including the scale, the type of filter and film, and the seasonal state of development of the crops.

Work already done along these lines has indicated that the classification of crops and land use in six categories will suffice for the preliminary assessment of

[5] Figures 22 and 24 not available for reproduction.

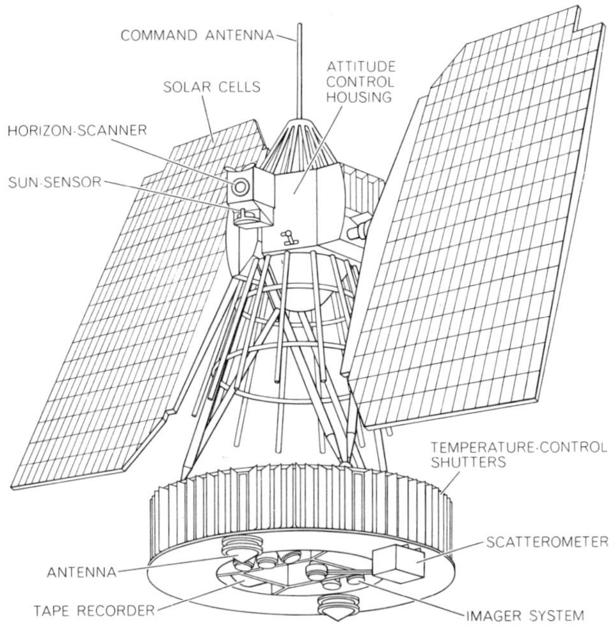

Figure 25 Unmanned satellite that will sense data in several wave bands and transmit the information to earth by television may be in operation by 1970 [Note that this article was published in 1968]. The U.S. Department of the Interior and the National Aeronautics and Space Administration have been working on plans for such a spacecraft, to be known as EROS for Earth Resources Observation Satellite. (Thomas Prentiss)

almost any agricultural area. The categories are orchard crops, vine and bush crops, row crops, continuous cover (such as alfalfa and cereal crops), irrigated pasture crops, and fallow ground. Each of these categories can be recognized by an experienced interpreter of photographs; usually he can also make further identifications of specific crop types within each of the six categories.

Let us consider the matter of crop vigor a little further. The first photographic evidence of loss of vigor due to black stem rust in wheat and oats or to blight on potatoes is to be found in the near infrared part of the spectrum, where reflectance rather than emission phenomena are of primary importance. On positive prints made from infrared photography the unhealthy plants register in abnormally dark tones. The technique is successful even in photographs made from spacecraft. Moreover, haze does not interfere appreciably with the technique because haze is easily penetrated by the long wavelengths used in making infrared photographs.

Water resources are susceptible to a degree of management through remote sensing. Aerial photography can show the area and depth of snowpacks on important watersheds at various times of the year. By following seasonal changes in the snowpack hydrologists can more intelligently regulate the impounding and release of water in reservoirs. Watershed managers also need to keep track of vegetation in order to estimate the loss of water to plants.

Vast ocean areas, about which a great deal remains to be learned, can be surveyed by remote sensing, particularly from spacecraft. Typically a camera in a satellite orbiting the earth can photograph a strip 3000 miles long in 10 minutes, so that it is easily possible to keep track of changes over huge reaches of ocean. Among the phenomena that can be followed are the flow of currents, the course of tidal waves, and the movements of marine animals, kelp beds, and icebergs.

Many other applications of remote sensing come to mind; I can only touch on them. Numerous archaeological sites have been discovered through conventional aerial photographs; it is probable that spacecraft photographs will reveal still more sites. Tax authorities can use aerial photographs to update maps showing land use and to spot efforts to change a land use without detection, such as by turning timberland into farmland while leaving a strip of forest along the road that a ground-based tax assessor might be expected to travel. Violations of law often show up in photographs; examples are illegal mining or logging in remote areas, pollution of waters by illegal dumping of chemicals, release through industrial smokestacks of materials that contribute to smog, and fishing in waters where fishing is prohibited. The analysis of such disasters as floods, fires, and hurricanes can be assisted by the study of remote-sensing data, and the information so obtained can be used in making emergency decisions and in combating future catastrophes of a similar nature.

Techniques of remote sensing are in a fairly early stage of development. Many of the applications I have suggested are therefore yet to be realized in practice. Their success, and the achievement of still other applications, will depend heavily on further research into the kinds of data that can be obtained from remote sensing—in learning, for example, where in the spectrum a certain plant disease will appear most distinctly. My colleagues and I have found it helpful to set up arrays of various natural resources and photograph them from high but stationary places, such as water towers and the tops of cliffs. The work helps to determine, economically and under controllable conditions, the bands of the spectrum that might best be used in remote sensing directed at finding the same resources.

A Prospect

I can foresee the possibility that the techniques for remote sensing will evolve into a highly automatic operation, in which an unmanned satellite orbiting the earth will carry multiband sensing equipment together with a computer. Thus equipped, the satellite could, for any particular area, take inventory of the resources and produce a printout that would amount to a resource map of the area. The computer could then use the inventory data in conjunction with preprogrammed factors (such as what ratio of costs to benefits would be likely to result from various resource management practices) and could reach a decision for the optimum management of the resources in the area. The decision would be telemetered to the ground for whatever action seemed necessary.

As a simple example, the satellite's sensors might spot a fire in a large forest. Its computer might then derive information on the location and extent of the fire and

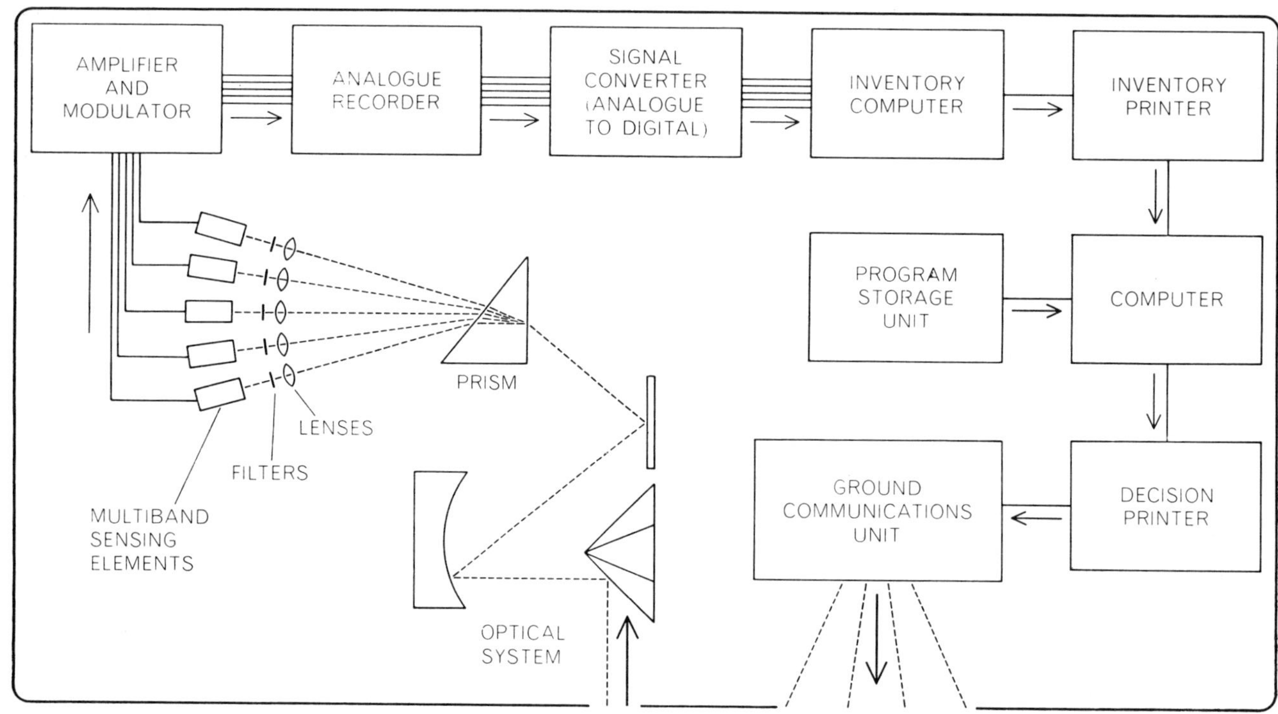

Figure 26 Computerized satellite is a prospect for the future. It would sense resources in several wave bands, automatically identify them, weigh them against previously programmed data on the cost effectiveness of various management possibilities, and send to the ground a decision on what should be done. It also could be used to monitor developing situations, such as a forest fire, suggesting how ground crews might fight it, and to perform automatically such tasks as turning irrigation valves on and off as required. (Thomas Prentiss)

could assess such factors as the type and value of the timber, the direction and speed of the wind, and the means of access to the fire. On the basis of the assessment the computer would send to the ground a recommendation for combating the fire.

Capabilities of this kind need not be limited to emergencies. Many routine housekeeping chores now done manually by the resource manager could be made automatic by electronic command signals. Examples might include turning on an irrigation valve when remote sensing shows that a field is becoming too dry and turning off the valve when, a few orbits later, the satellite ascertains that the field has been sufficiently watered.

A satellite of such capabilities may seem now to be a rather distant prospect. After a few more years of developing the techniques for remote sensing, the prospect may well have become a reality.

Acknowledgments

This article has sought to trace some of the progress that has been made recently in one of the most vital of human endeavors—remote sensing of the earth's natural resources. This progress has not stemmed from the work of any one individual. Rather, it has resulted from the well-integrated efforts of scientists and engineers in a wide variety of disciplines. Among these have been the instrument designers and manufacturers through whose efforts various ingenious kinds of equipment have been produced for the acquiring of remote sensing data. Special acknowledgment is hereby given to the Barnes Engineering Company, whose equipment was used in taking the thermograms appearing in this paper. Also of importance has been the development of vehicles (aircraft and spacecraft) from which these remote sensors might be operated. Consequently the acknowledgments for this article are properly extended to include thousands of highly competent scientists who have been employed, whether in federal, state, or private agencies, on various projects sponsored by the National Aeronautics and Space Administration. These are the scientists who have designed our space vehicles and developed the complex propulsion, guidance and tracking systems necessary for their effective use. But raw data as acquired by the remote sensing process ordinarily cannot be regarded as an end in itself. There also is a need for data analysis by both human and mechanical means. Consequently, still another large group of scientists is hereby acknowledged—the group concerned with developing suitable equipment for viewing, measuring, identifying and plotting remotely sensed data.

Hopefully, some individuals will be stimulated through this article to seek further information about certain aspects of remote sensing. It is with this possibility in mind, and also with a desire to acknowledge here the excellent remote sensing research that is being conducted in many agencies throughout the United States that the following list is provided. The list is limited to activities in the United States, and apologies are hereby extended for any important errors of omission or commission. The various research groups are listed under three major headings: universities, private industry, and government-related institutions. Because of space limitations, only a few of the leading agencies are listed under each heading. The sequence in which the various groups are listed below is semi-alphabetical and bears no necessary relation to the quantity or importance of work done by them. More specific evidence of work done in the field of remote sensing by personnel of these and other institutions can be obtained from a study of the section entitled References with which this article will conclude.

a. Universities

1. Cornell University—geologic and land use interpretation; cost effectiveness studies for remote sensing techniques.
2. Long Island University—multiband cameras and color projection equipment.
3. Northwestern University—use of statistics in ground truth data collection.
4. Oregon State University—remote sensing in range management; forestry applications.
5. Purdue University—automatic interpretation of agricultural crops, primarily by "pattern recognition" based on multiband tone signatures; basic research on spectral signatures and spectrophotometry.
6. Stanford University—applications of multiband reconnaissance (especially in the thermal infrared region) to geologic inventory.
7. Texas A&M University—remote sensing in oceanographic and arid land use studies.
8. University of Arizona—long focal length tracking cameras; multiband photography.
9. University of California (primarily at Berkeley, Davis, Los Angeles, Riverside, and La Jolla)—applications of remote sensing in forestry, range management, geography, agriculture, oceanography studies of air and water pollution, and extra-terrestrial applications.
10. University of Kansas—remote sensing applications of radar; image enhancement techniques for the study of agricultural and other resources.
11. University of Michigan—development of remote sensing equipment; multiband image analysis; compilation of spectrometric data.
12. University of Minnesota—forestry applications for both black-and-white and color aerial photographs; detection of tree diseases.
13. University of Nevada—geologic and wildland resource applications of remote sensing.
14. University of Washington—applications of remote sensing in engineering, life sciences, and earth sciences.
15. University of Illinois—compilation and distribution of airphoto stereograms; engineering and forestry applications.
16. Utah State University—applications of aerial photography to wildland recreation, forest, and range problems.

b. Private Industry

1. Abrams Aerial Survey Corp.—viewing and plotting equipment; aerial photographic equipment.
2. Aero Service Corp.—mosaics; terrain models and maps; image analysis for civil and military purposes.
3. Barnes Engineering Co.—thermal infrared cameras and radiometer.
4. Bausch and Lomb Co.—viewing equipment.
5. Bendix Corp.—thermal infrared scanning equipment.
6. Boeing Scientific Research Laboratories—terrestrial and extraterrestrial applications of thermal infrared and other remote sensing data.
7. Eastman Kodak Co.—photographic films and cameras; image quality research.
8. Fairchild Camera and Instrument Corp.—aerial photographic equipment; viewing equipment.
9. General Aniline and Film Corp.—aerial photographic films.
10. General Electric Corp.—recoverable capsule systems for space reconnaissance.
11. HRB Singer Corp.—thermal infrared scanning equipment and its applications.
12. Lockheed Corp.—reconnaissance vehicles and remote sensing systems with electric readout.
13. Itek Corp.—cameras and viewing equipment; multiband image analysis.
14. Radio Corporation of America—earth resource satellite systems.
15. Texas Instruments, Inc.—multiband remote sensing equipment; military and geoscience applications of remote sensing.
16. Westinghouse Corp.—image forming, side-looking radar equipment.

c. Government-Related Institutions (nonmilitary research)

1. Agricultural Research Service, USDA—agricultural applications of remote sensing.
2. Economic Research Service, USDA—cost effectiveness studies for various applications of remote sensing in agriculture.
3. Forest Service, USDA—the inventory of timber, forage, and other wildland resources by remote sensing; thermal infrared for fire detection.
4. Statistical Reporting Service, USDA—the inventory of crops and livestock by means of remote sensing.
5. Environmental Science Services Administration—applications of remote sensing in meteorology and other environmental sciences.
6. Geological Survey, USDI—remote sensing in geology, hydrology, and geography; development of an earth resources technology satellite.
7. Bureau of Land Management, USDI—applications of remote sensing in range management and wildland resource inventory.
8. National Park Service, USDI—study of the population dynamics of timber-damaging insects; wildland recreation applications.
9. Naval Oceanographic Office—remote sensing of oceanographic phenomena.
10. Fish and Wildlife Service, USDI—applications of remote sensing in the inventory of big game animals, waterfowl, and fish.
11. Office of Space Sciences and Applications, NASA—remote sensing of earth resources and meteorological phenomena.
12. Earth Sciences Division and Division of Biology and Agriculture, National Research Council—applications of remote sensing in the earth sciences and life sciences.

REFERENCES (LITERATURE SURVEY)

Alexander, R. H. 1964. Geographical data from space. *The Professional Geographer* 16 (6):1–5.

American Society of Photogrammetry. 1960. *Manual of Photographic Interpretation.* George Banta Co. Menasha, Wisconsin. 868 p.

Astheimer, R. W., and Wormser, E. M. 1959. Instrument for thermal photography. *Journal of the Optical Society of America.* Vol. 49(2): 184–187.

Badgley, P. C. 1966. Orbital remote sensing and natural resources. *Photogrammetric Engineering* 32(5):780–790.

Badgley, P. C., and L. F. Childs. 1967. Earth resources survey from space. Presented at the Ocean From Space Symposium of American Society of Oceanography.

Badgley, P. C., Fischer, W. A., and Lyon, R. J. P. 1965. Geologic exploration from orbital altitudes. *Geotimes,* 10(2).

Badgley, P. C., and Vest, W. L. 1966. Orbital remote sensing and natural resources. NASA Headquarters Special Publication.

Beran, D. W., and Merritt, E. S. 1967. Satellite observed cloud cover in southeast Asia. U.S. Army Electronics Command, Fort Monmouth, N.J. Technical Report ECOM-02308-F, June, 1967.

Bird, J. B., and Morrison, A. 1964. Space photography and its geographical applications. *Geographic Review* 54(4):463–486.

Brock, G. C., Harvey, D. I., Kohler, R. J., and Myskowske, E. P. 1965. Photographic considerations from aerospace. Itek Corp., Lexington, Mass.

Carneggie, D. M., Draeger, W. C., and Lauer, D. T. 1966. The use of high altitude color and spectrozonal imagery for the inventory of wildland resources, Vol. I of III: *The Timber Resource.* Annual Progress Report to NASA by Forestry Remote Sensing Laboratory, University of California, Berkeley, California. 41 pages, illustrated.

Colwell, R. N. 1952. Report from the President of Commission VII, to the International Society of Photogrammetry. *Photogrammetric Engineering* 18(3):375–400.

Colwell, R. N. 1966. Uses and limitations of multispeed remote sensing. Proceedings of Fourth Symposium on Remote Sensing. University of Michigan, 71–100 p.

Combs, A. C., Weickmann, H. K. et al. 1965. Applications of infrared radiometers to meteorology. *Journal of Applied Meteorology.* Vol. 4(2):253–262.

Conti, M. A. 1966. Evaluation of Nimbus I high resolution infrared radiometer (HRIR) imagery. U.S.G.S. Technical Letter NASA-35.

Cronin, J. F. 1966. Terrestrial features of the United States as viewed by TIROS. Aracon Geophysics Co., Concord, Mass. Science Report No. 2.

CSIRO 1962. General report on lands of the Alice Springs area. Commonwealth Scientific and Industrial Research Organization, Australia. Land Research Report No. 6. (Includes maps of land systems, pasture lands, geology and ground water provinces.)

Data Staff. 1965. Is space photography the panacea? *Data Magazine* 10(4):33–37.

Drewes, H. 1966. An evaluation of the Gemini IV color photos of the Gulf of California–central Texas area. U.S.G.S. Technical Letter NASA-46.

Drewes, H., and Morrison, R. 1966. Extent of relict soils revealed by Gemini IV photographs. U.S.G.S. Technical Letter NASA-60.

Ellermeier, R. D. and Simonett, D. S. 1965. Imaging radars on spacecraft as a tool for studying the earth. International Symposium on Electromagnetic Sensing of the Earth from Satellites. Miami Beach, Florida. Nov. 22–24.

Fink, D. E. 1965. Gemini photos advance AES experiments. *Aviation Week and Space Technology.* August 9. 61–62 p.

Fischer, W. A. 1966. Orbital surveys of the earth. Proceedings of Symposium on Peaceful Uses of Space. Stanford, California. Unpublished.

Fischer, W. A., and Robinove, C. J. 1968: A rationale for a general purpose earth resources observation satellite. *In* Proceedings of University of Washington Remote Sensing Symposium. Feb. 15–16.

Gawarecki, S. J., Lyon, R. J. P., and Nordberg, W. 1965. Infrared spectral returns and imagery of the earth from space and their application to geologic problems. *American Astronautical Society* Vol. 4:13–33.

Gillis, J. E., and Leestma, R. A. 1965. Research for the development of an earth atlas based on viewing the earth from space with various remote sensors and the applications for global planning. Paper presented at American Astronautical Society, March.

Hahl, D. C., and Handy, A. H. 1966. Hydrologic interpretation of Nimbus vidicon image—Great Salt Lake, Utah. U.S.G.S. Technical Letter NASA-61.

Heller, R. C., Bean, J. L., and Marsh, J. W. 1952. Aerial survey of spruce budworm damage in Maine in 1950. *Journal of Forestry.* Vol. 50(1):8–11.

Hemphill, W. B., and Danilchik, W. 1968. Geologic Interpretation of a Gemini photo. *Photogrammetric Engineering* 24(2):150–154.

Hoffer, R. M. 1967. Interpretation of remote multispectral imagery of agricultural crops. Purdue University. Agricultural Research Bull. No. 831–33 p.

Holter, M. R. 1967. Infrared and multispectral sensing. *Bioscience,* 17(6):376–383.

Katz, A. H. 1960. Observation satellites: problems and prospects. *Astronautics,* Vol. 2.

Katz, A. H. 1967. Reflections on satellites for earth resource surveys. Rand Corporation P-3753 28 p.

Keller, F. L. 1953. Resources inventory—a basic step in economic development. *Economic Geography* 29(1):39–47.

Langley, P. G. 1965. Automating aerial photo interpretation in forestry—how it works and what it will do for you. *Proceedings, Society of American Foresters.* 172–177 p.

Lauer, D. T. 1967. The feasibility of identifying forest species and delineating major timber types in California by means of high-altitude multispectral imagery. Annual Progress Report to NASA on Remote Sensing Applications in Forestry. 95 p.

Lent, J. D. 1966. Cloud cover interference with remote sensing of forested areas from earth-orbital and lower altitudes. Report to NASA on Remote Sensing Applications in Forestry. 55 p.

Lewis, C. R., and Davis, W. E. 1966. Geological evaluation of Nimbus vidicon photography, Cheseapeake Bay–Blue Ridge. U.S.G.S. Technical Letter NASA-64.

Lowe, D. S. 1968. Future trends in optical sensing and R & D requirements. Remote Sensing Symposium, University of Washington.

Lowe, D. S., Braithwaite, J., and Larrowe, V. L. 1966. An investigative study of a spectrum-matching system. Final Report under NASA Contract 8-21000. 105p.

Lowman, P. D., Jr. 1966. The earth from orbit. *National Geographic* 130(5):645–671. November.

Lowman, P. D., Jr., and Chang, Te Lou. 1964. Hyperaltitude photography and its applications. Proceedings of Symposium on Remote Sensing of Environment, University of Michigan. 153–169 p.

Lyons, E. H. 1967. Forest sampling with 70mm fixed air-base photography from helicopters. *Photogrammetria* 22(6):213–231. September.

McKallor, J. 1967. A photomosaic of western Peru from Gemini photography. U.S.G.S. Technical Letter NASA-87.

Meyers, V. I., Weigand, C. L., Heilman, M. D., and Thomas, J. R. 1966. Remote sensing in soil and water conservation research. Proceedings, 4th Symposium on Remote Sensing of Environment, University of Michigan.

Moore, R. K. 1966. Radar as a remote sensor. CRES Report No. 61–7. Center for Research in Engineering Science, University of Kansas.

Morrison, A., and Bird, J. B. 1964. Photography of the earth from space and its non-meteorological applications. Proceedings of Symposium on Remote Sensing of Environment, University of Michigan.

National Aeronautics and Space Administration. 1967. Earth photographs from Gemini III, IV and V. NASA SP-129.

National Aeronautics and Space Administration. 1968. Meteorological Data Catalog for the Applications Technology Satellite—a Users Guide for ATS-1. Goddard Space Flight Center. 17 pp. 1968.

National Research Council. 1965. Spacecraft in geographic research. NAS-NRC Publication 1353.

Nugent, R. H., and Starr, L. E. 1967. Gemini photography evaluation. U.S.G.S. Technical Letter NASA-69.

Olson, C. E., Jr. 1960. Elements of photographic interpretation common to several sensors. *Photogram. Engineering.* 26(3): 630–637.

Parker, D. C., and Wolff, M. F. 1965. Remote sensing. *International Science and Technology,* July 30–31 p.

Pecora, W. T. 1967. Surveying the earth's resources from space. Proceedings of 27th Annual Meeting, Congress on Surveying and Mapping.

Perry, R. A. 1962. Pasture Lands of the Alice Springs area. *In* CSIRO, Australia, Land Research Report No. 6.

Pope, R. B., McLean, C. D., and Bernstein, D. A. 1961. Forestry uses of aerial photographs. Pacific Northwest Forest and Range Experiment Station, U.S.F.S.

Research and Development Board. 1953. Procedure for the Construction of Photo Interpretation Keys Report to Committee on Geophysics and Geography by R. N. Colwell. Publication No. CG 209/1 pp. 135–154. April, 1953.

Risley, E. M. 1967. Satellites for earth survey. *Earth Sciences Newsletter* No. 2.

Robinove, C. J. 1966. A preliminary evaluation of airborne and spaceborne remote sensing data for hydrologic uses. U.S.G.S. Technical Letter NASA-50.

Rouse, J. W., Jr., Waite, W. P., and Walters, R. L. 1966. Use of orbital radars for geoscience investigations. CRES, University of Kansas, Report No. 61–8.

Sattinger, I. J., and Polcyn, F. C. 1966. Peaceful uses of earth-observation spacecraft. Vol. II. Survey of Applications and Benefits. Report No. 7219-1-F(II):159 p. Institute of Science and Technology, University of Michigan.

Schwartz, A. L., and Zeidner, J. 1961. Comparison of photo interpretation under stereo and non-stereo viewing conditions. *Photogrammetric Engineering* 27(5):720–724.

Simonett, D. S., and Morain, S. A. 1965. Remote sensing from spacecraft as a tool for investigating arctic environments. CRES, University of Kansas. Report No. 61-5.

Starr, L. E., and Sibert, W. 1966. Potential time-cost benefits from use of orbital height photographic data in cartographic programs. U.S.G.S. Technical Letter NASA-54.

Stroud, W. G. 1960. Our earth as a satellite sees it. *National Geographic Magazine.* 118(2):292–302.

Tabor, R. 1966. Photogeologic interpretation of Gemini IV color photography: Baja California. U.S.G.S. Technical Letter NASA-24.

Taggart, C. I. 1964. Satellite photography. *The Canadian Surveyor* 18(2):105–112.

Thaman, R. R. 1967. A study of the feasibility of mapping vegetation on a world scale using satellite imagery. Masters Degree Thesis, Department of Geography, University of California. 214 p.

U.S. Army. 1966. Earth resource surveys from spacecraft. Prepared by U.S. Army Engineers for the Earth Resources Program, Space Applications Program Office, NASA.

U.S. Dept. of Commerce. 1968. Probability of vertical penetrable optical path for high intensity, high contrast optical targets. National Weather Records Center, Asheville, North Carolina.

Wear, J. F. 1966. The development of spectro-signature indicators of root disease on large forest areas. Pacific Southwest Forest & Range Experiment Station Annual Report to NASA. 30 Sept.

Weiss, M. 1967. Infrared in meteorology. *Weatherwise,* Vol. 20(4):156–161.

Wilson, R. C. 1967. Space photography for forestry. *Photogrammetric Engineering* 23(5):483–490.

Wobber, F. J. 1967. Put geology surveys in orbit to find oil. *Oil and Gas Jour.* December.

Wobber, F. J. 1968. Space photography—a new analytical tool for the sedimentologist. *Sedimentology.* (In press)

Wolfe, E. W. 1966. Gemini V color photography of Salton Sea area, California. U.S.G.S. Technical Letter NASA-34.

Wormser, E. M. 1962. Infrared meteorological instrumentation for space vehicles. *Electrical Engineering.* Vol. 81(4):285–289.

Yost, E. F., and Wenderoth, S. 1967. Multispectral color and aerial photography. *Photogrammetric Engineering* 33(9): 1020–1033.

ALTHOUGH MULTISPECTRAL *sensing could be accomplished at any point or points along the continuum of the electromagnetic spectrum, it is done mostly in the visible and near and far infrared portions. In the visible, a number of multilens cameras have been developed, with from two to as many as nine lenses. It is extremely difficult for a human interpreter to compare even a few multispectral images, correlate image signatures, and then interpret the results. A single color photograph is a special kind of multispectral image. At the time of exposure, variations in wavelength intensities interact with the dye layers of the film emulsion to produce a surface which when developed will subtract all undesired colors from the incident light. The wavelengths reflected from the photograph will interact with the observer's eye to produce the sensation of color. Depending upon many factors, this film color response may or may not approach the color fidelity of the target response in the natural environment. Some disadvantages in using a single color photograph for remote sensing include: limited spectral sensitivity of the dye layers, fixed relative exposure for each emulsion layer, limited exposure range, difficulty and complexity of processing the exposed film, and a possible lack of true color response or the inability to produce observable color differences between targets with minor wavelength reflectance differences.*

Black and white film overcomes most of the disadvantages of color film. Moreover, by superimposing two or more black and white images generated at different wavelengths, a true color fidelity image can be created. This technique makes use of the additive color theory. This method of combining wavelengths to create color sensation enables detection of subtle reflectance differences between objects on the images.

38-Multispectral Color Aerial Photography

EDWARD F. YOST
SONDRA WENDEROTH

MULTISPECTRAL AERIAL photography is conventionally taken by using a number of film-filter combinations to obtain a set of photographs in different bands of the spectrum. The spectrum covered by this photographic technique can include the near ultraviolet, the visible, and the near infrared. The lower bound of spectral sensitivity is believed to exist at about 260 nm (nanometer; 10^{-9} meter) due to ozone absorption in the atmosphere. An upper bound exists at 980 nm which is the current upper limit of the spectral sensitivity of available practical photographic emulsions.

Ground objects usually exhibit a variation in the percentage of radiant energy they reflect. This difference in reflectance in the visible part of the spectrum is what causes the apparent color of an object. A difference in spectral reflectance of an object can be detected as images of different density on a set of multispectral photographs.

To be certain that this density difference is in fact caused by the difference in spectral reflectance of the object on the ground, it is essential that: (1) the camera system be spectrophotometrically calibrated, (2) the spectral distribution of the illumination be known, (3) the spectral bands covered by each photograph be correctly chosen, and (4) the photographic processing be precisely controlled. Under such controlled conditions it *may* be possible to obtain image densities which can be accurately related to the spectral reflectance of the object repeatably and with reasonable precision flight after flight.

Numerous applications of multispectral techniques using a multiplicity of cameras or using one camera to take successive exposures at different times have been reported (Colwell, 1963). A special nine-lens camera was constructed in the USSR as a research tool to establish optimum film-filter combinations to use in photographing selected ground objects (Zaitov and Tspurun, 1962). A similar camera has been constructed in the USA originally as part of a multispectral system for detection of surface indications of underground nuclear explosions (Molineux, 1965).

Experiments using multispectral cameras have shown that more information concerning physical features of our environment can be obtained with sensors which operate in spectral bands compared to conventional panchromatic photography (Holter and Legault, 1965). However, many such studies have clearly demonstrated that the human interpreter possesses very low data input and output rates. Herein lies the basic difficulty in using conventional multi-spectral photography: *it does not discriminate between information collected and, therefore, provides the interpreter with much more nonrelevant than relevant data.* The inherent complexity in attempting to compare tonal values on even a few multispectral images and to interpret the results with confidence has been well documented (Legault and Polcyn, 1965).

A color photograph can be considered as a special type of multispectral photo. In conventional color films the yellow, magenta, and cyan dye layers respond to the blue, green, and red spectral regions of an equal-energy visible spectrum in a proportion fixed by the chemistry of the emulsion. In infrared color film the dye layers respond to the green, red, and infrared parts of the spectrum in a similar manner. When viewed under white light, color photographs *subtract* from the viewing light the undesired colors, and the remaining spectral components of light fall upon the observer's eye to produce the sensation of color. The considerable advantages of this subtractive technique of color in aerial photography have been explored in several papers (Smith, 1963; Swanson, 1960).

A number of authors have also pointed out the disadvantages of color aerial photography, which are primarily: fixed spectral sensitivity; fixed relative exposure for each dye layer; inadequate exposure latitude; relative processing complexity compared to black and white films; and either (1) a lack of *true* color fidelity to what is seen by a human observer or, conversely, (2) inability to produce significant color differences between objects which have slight spectral reflectance differences.

Color can also be produced by the addition of colored lights rather than by the subtraction of *unwanted* colors from white light. This so-called *additive color theory* can be used to create a composite color image from photographs taken in different parts of the spectrum under certain conditions (Evans, 1948).

If blue, green, and red primary colors are used to illuminate three positive transparencies taken in these respective regions of the spectrum, and these spectral positives have images in identical spatial locations rela-

From *Photogrammetric Engineering* 33:1020–1033, 1967. Reprinted and edited with permission of the authors and the American Society of Photogrammetry.

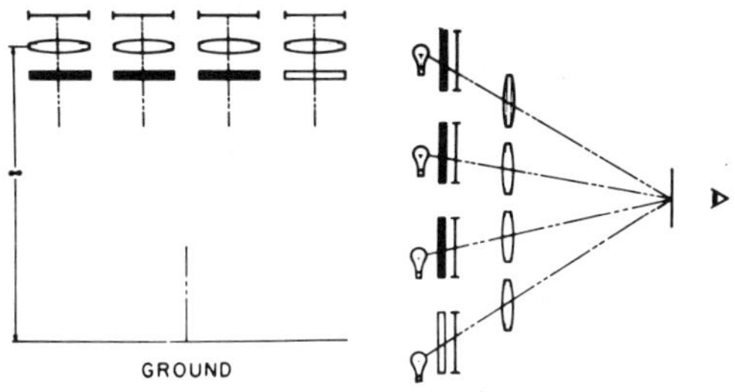

Figure 1 Schematic representation of the taking and viewing situations for additive color presentation.

tive to their respective principal points, and if these photographs are optically projected one upon the other so that no misregistration exists, a composite color rendition of the scene will be produced (Fig. 1) (Yost and Wenderoth, 1965; Winterberg and Wulfeck, 1961). Not only will the primary projection colors be reproduced, but *every* hue will be seen in varying degrees of saturation, and the colors of most natural objects will be recreated rather well. Every image which exhibits a density difference on the individual black and white spectral positives will be seen as a color. If no density difference exists, the composite image will be achromatic (a shade of gray). If the minimum perceivable density difference is about 0.02, not more than 200 shades of gray can be differentiated on a black and white photograph, whereas under certain conditions over 7,500,000 *color* differences can be perceived (Committee on Colorimetry, 1953).

The Initial System Breadboard Model

The research described herein was oriented toward developing a Spectral Zonal Color Reconnaissance System[1] which would provide an interpreter with an image presentation having the following features (Wenderoth and Yost, 1966):

1. A "true" color presentation of the ground scene as would be seen by a "standard" human observer.
2. False-color presentations which would allow very small density differences between multispectral photographs to be seen as color differences.
3. A dynamic color presentation which would permit the interpreter to correct for variations in color caused by the spectral distribution of the sun at different times of the day and for different atmospheric conditions.
4. A black and white presentation of one or more of the individual multi-spectral photographs.
5. An acceptable level of spatial resolution in composite presentation along with the spectral discrimination achieved by color.
6. A minimum amount of time between taking the photograph and subsequent viewing of the composite color image.

A breadboard model of the camera system was constructed using four Fairchild KA-56 panoramic cameras which were used to take photographs in different parts of the spectrum. The four spectral bands initially chosen were: blue (385 to 520 nm), green (480 to 610 nm), red (590 to 700 nm), and infrared (700 to 960 nm). The choice of filter bands was made to cover completely the visible spectrum, to approximately the standard observer color sensitivity mechanism of the human eye, and to permit comparison with conventional color and infrared color films.

The four cameras were aligned in an *inner* rack in such a manner that all the lens optical axes were geometrically normal to a plane common to each exposure slit. The cameras also had the same azimuth orientation to insure that the film of all cameras was transported parallel with respect to each other. Care was taken to assure that differential distortion and focal-length differences between each of the four cameras was minimized for the spectral bands used.

A control unit was constructed which permitted control of both the exposure of all four cameras simultaneously, as well as each camera individually. This allowed for the variation of exposure with changes in brightness as well as changes in the relative spectral reflectance of the scene.

The objective in using this camera arrangement was to obtain four photographs which contained images in identical spatial locations with respect to the principal point of each individual spectral negative except that the density of similar images would differ from negative to negative. Extreme care was taken to assure that the density of similar images on the four spectral negatives was proportional to the intensity of reflected radiation in the particular interval of radiation *sensed* by each camera. Plus X (EK 8401) film was used with the blue, green, and red filters, and Infrared Aerographic (EK 5424) film with the infrared filter.

The sets of spectral negatives were developed so that the density of any image on each individual spectral negative was a correct representation of the brightness of the object. In addition to compensation for differences in exposure, it was necessary to correct for reduced gamma, particularly in the blue negative, in order to produce an identical relationship of exposure-to-density on all four negatives. A gamma was chosen to produce medium contrast without excessively reducing the exposure range. Positive transparencies were then made which were also medium contrast with low minimum density (Fig. 2).

The composite color rendition of the spectral positives was accomplished by construction of a rear projection viewer using additive color principles. The source of illumination for each of the four spectral positives was controlled in illuminance, dominant wavelength, and purity, which in turn controlled the color sensations of brightness, hue, and saturation. Illuminating each spectral positive by a different color of controlled brightness, hue, and saturation, and at the same time superimposing each spectral positive, one upon the other, on a screen by optical projection, produced a composite color presentation.

The understanding of the phenomena of additive color requires a concept of color in a geometric sense if not in a detailed analytic context. Color can be conceptualized as a cone standing on its apex (Fig. 3). The axis of

[1] U.S. and foreign patents pending.

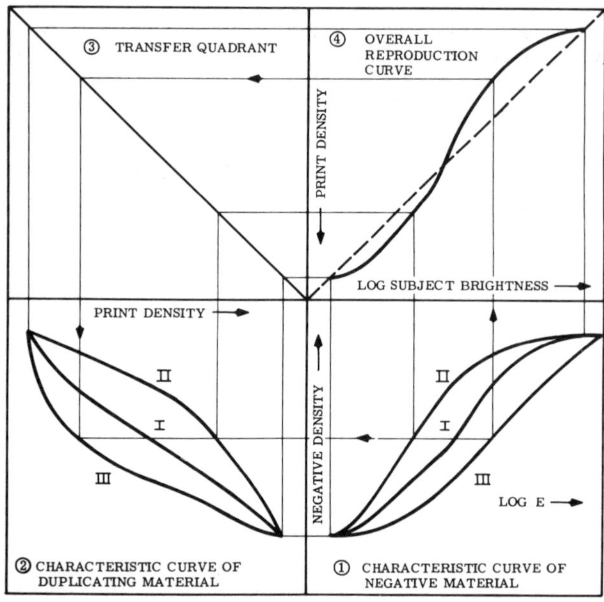

Figure 2 A tone-reproduction curve for three spectral photographs showing the relationship between regional scene and the final positive reproduction.

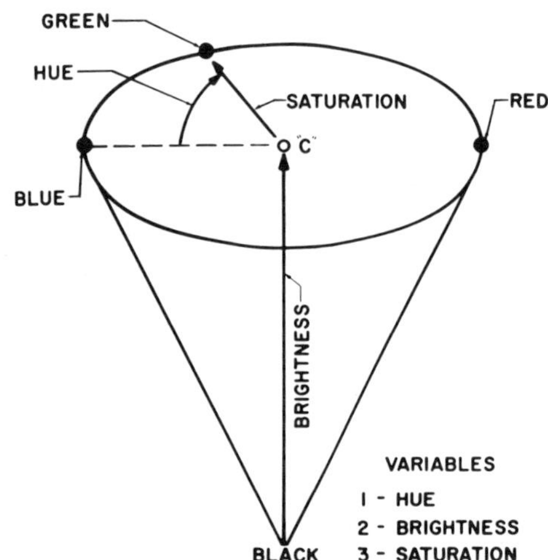

Figure 3 Color solid indicating the geometric interpretation of color.

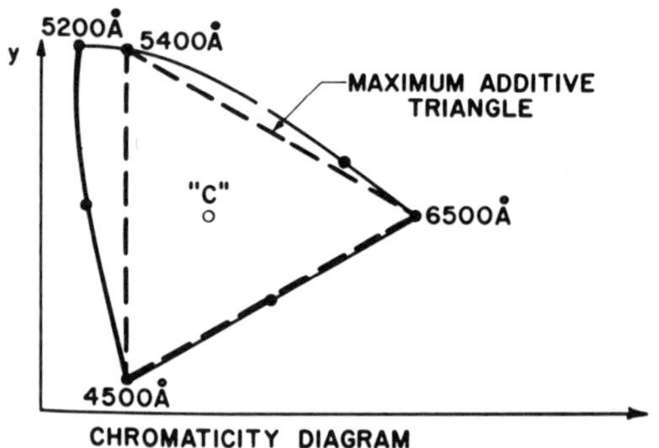

Figure 4 Chromaticity coordinates of viewing filters.

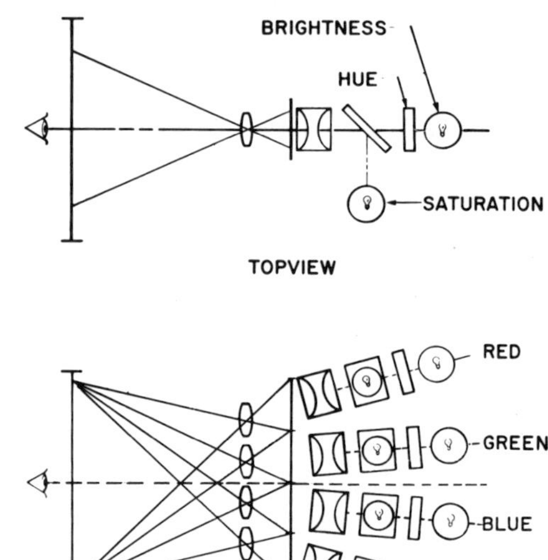

Figure 5 Schematic diagram of the additive color viewer.

revolution represents the brightness scale. All *black and white* photographic images are confined to this one-dimensional brightness axis and thus exhibit a shade of gray. In addition, color has two additional psychophysical variables, hue and saturation, represented in the figure as an angle and distance from the brightness axis.

The image created by the additive color projection of a set of multispectral photos can be established as a position in this color solid by a mathematical treatment. To facilitate color analysis, a mathematical manipulation can further transform the color description to a two-dimensional coordinate position in the chromaticity diagram (Hardy, 1936). In the chromaticity diagram form of presentation (Fig. 4), all colors are treated as having the same brightness. The filters used to establish the hue of the viewing colors have chromaticity coordinates shown in the accompanying figure. The triangle connecting these filters is of particular significance. Any color lying within this triangle can be reproduced accurately by additive color techniques.[2] The filters chosen have the two properties that they provide the maximum area within the chromaticity diagram, and the green-red side of the triangle lies close to the locus of pure spectrum colors. This is particularly important if accurate reproduction of color is desired because saturated yellow appears in nature but saturated cyan does not.

A question naturally arises as to how the position of an image in the color solid can be manipulated. A schematic view of the viewer breadboard model shows how this is accomplished (Fig. 5). The four spectral

[2] A detailed mathematical analysis is presented in: Miller, *Principles of Photographic Reproduction,* New York: Macmillan, 1943.

positives are contained in a single film plane. Each is illuminated by a *brightness* lamp, which contains a filter in the optical path, and a *saturation* lamp. The light from the *brightness* lamp controls the brightness of the illumination of each spectral positive. The filter establishes the hue, and the illumination of the *saturation* lamp controls the amount of desaturation. Manipulation of these color variables controls the apparent color of the composite image on the screen.

Experimental Results Using the Initial Breadboard

The camera breadboard model of four KA-56 panoramic cameras was flown in September, 1964, on Long Island, New York. The solar angle at the time the photography was taken was approximately 30° with scattered cloud cover and moderate haze.

More than 30 experiments were conducted in which the spectral negatives and positives were processed with various characteristics. Each set of experimental photography was placed in the viewer, aligned, and analyzed for registration and color characteristics.

The best evaluation of the experimental results can be achieved by looking at the viewer screen. However, in order to convey to the reader an indication of some of the results of these tests, several experiments were recorded by taking color photographs of the composite image on the viewer screen.

The set of photographs in Figure 6 shows the individual black and white spectral positives as they appeared on the viewer screen. These four positives are indicative of conventional multispectral photography discussed previously, in which the interpretation procedure is to compare the density differences of selected objects on the individual photographs.

If the blue, green, and red spectral positives are projected each with its respective primary to form a composite image on the viewer screen, a *true* color rendition is seen. The color characteristics of the composite rendition can be varied by adjustment of the brightness, hue, and saturation of the source of illumination for each spectral positive.

If the green spectral positive is projected as blue, the red as green, and the infrared as red, the standard *camouflage* false-color rendition is observed. Living deciduous foliage (a mixture of oak and maple) appears as red, dying foliage as magenta, and dead foliage as a green-brown color.

On examining the reproductions of these results, the reader should note that both composite color renditions show all density differences between the individual spectral positives as colors. Images which have the same density on all spectral positives, such as shadows, are a shade of gray in the composite presentation.

Plate 3[3] shows two multispectral color renditions of a stand of oak, some of which have been cut. Cutting was timed to occur from 2 weeks to 2 hours prior to the time of photography. The chromaticity coordinates of the standard camouflage color rendition of living (uncut), dead (cut more than 2 days), and dying (cut 2 hours to 2 days) foliage are shown. The second composite presentation shows a false-color space achieved by combining all four spectral positives. In this latter rendition, the dead trees are shown as red, the dying as pink, and the live as blue. The more important fact is that the color-signature overlap between the chromaticity coordinates of the live and dying categories which occurred in the standard camouflage rendition has been eliminated in the second false-color space.

A capability of the viewer to allow control of color in order to achieve a target detection was also demonstrated as a result of this experiment. As can be seen in Plate

[3] Plates 3 and 4 and some other illustrations in this article are not reproduced here.

Figure 6 Four multispectral photographs, each having images in identical spatial location with respect to the principal point. (A) Spectral photograph in the blue band (325 to 520 nm). (B) Spectral photograph in the green band (480 to 610 nm). (C) Spectral photograph in the red band (590 to 700 nm). (D) Spectral photograph in the near infrared band (700 to 980 nm).

4 a and b, it was possible to *wash out* the chromatic noise due to variations in reflectivity of natural objects while at the same time to enhance the apparent color of a manmade object, in this case a green vehicle.

Another flight test of significant interest was performed in May, 1965. The targets consisted of equipment in a typical military deployment camouflaged by various types of military camouflage nets. These nets were embedded in foliage cover and generally in deep shadow. Two reproductions (Plate 4 c and d) show a comparison of a panchromatic photograph (EK 8401 film with Wratten 25 filter) taken at 750 feet altitude, and a color-multispectral rendition of the same scene. Note the greater target detection capability of the composite color presentation, particularly the two nets embedded in the trees.

More recent flight tests have also demonstrated the possible usefulness of this technique for water pollution studies. Plate 4 e and f show both a multispectral color and an Ektachrome color photograph of Indian Creek in Miami just after hurricane Inez in October, 1966. The presence of salt water can be detected as green in the multispectral color rendition where the presence of color shift is not discernible in the conventional color photograph.

Recent Equipment for Research

A four-lens camera has been designed by the authors and manufactured by Fairchild Space and Defense Systems to take multispectral photos in four bands from 360 nm to 980 nm. The spectral region covered by the camera includes part of the near ultraviolet, the visible, and part of the near infrared. Fairchild has also manufactured a companion additive color viewer which projects the set of four spectral photographs on a screen to form a single composite color presentation for interpretation.

The camera takes a set of four spectral negatives at exactly the same time and records all of them on one piece of film.[4] The spectral regions recorded may be any three bands in the ultraviolet-visible, and one band in the infrared, without making any adjustment to the camera. Accommodation of more than one infrared band, if desired, can be made by an optical adjustment. If no prior knowledge exists of the spectral reflectance of a target, a set of broadband filters which overlap throughout the spectrum are used. In those instances where spectrophotometric analysis has isolated wavelength regions and where the phenomena may be spectrally detected, appropriate filter-film combinations can be used which, together with the transmission characteristics of the camera lenses, will permit accurate multispectral photography in the particular wavelength bands. As each band is photographed through its own lens, it is possible to obtain the correct exposure of each negative for the spectral radiance of any scene. This control of exposure for all four bands permits repeatable accuracy under a wide range of illumination and ground reflectance conditions.

[4] The camera also has a capability of using four separate rolls of film by changing the magazine.

Each one of the four spectral negatives, which together comprise a set of multispectral photographs, is taken at exactly the same time by four matched lenses. As the optical axes of all the lenses are normal to the film plane, four spatially identical negatives are produced. All the images appear in identical coordinate positions as measured from the principal point of each photograph.

The viewer illumination system is designed to give the interpreter control of the dominant wavelength (hue), purity (saturation), and brightness of the illumination *source* for each spectral positive. He can adjust any one of these three variables for each of the four spectral bands to achieve the desired color space for viewing. In any color space, the composite additive color photograph shows all density *differences* between the individual spectral positives as colors. Those images that have the same density on all spectral positives will be achromatic (colorless) in the composite additive color presentation. Images which may escape detection because the density differences may be too slight if the film is viewed on a light table are almost always easily identified by difference in hue, brightness, and saturation of the image in relation to its background in the composite additive color presentation.

This method of abridged spectroradiometric sensing of radiant energy reflected by ground objects in the 360 nm to 960 nm spectrum and subsequent colorimetric analysis of the composite additive color image is designed for rapid and accurate photointerpretation. The rapidity is obtained by using a unitary piece of film upon which the principal points of the four photographs have been precisely located with respect to the film edge. This permits automatic registration of the composite additive color image after the interpreter has checked the registration of the first frame in the roll.

Requirements for Accurate Multispectral Photography

All of the experiments performed to date have demonstrated the absolute necessity for precise photographic technique in order to obtain repeatable results in different geographic areas under different photographic conditions.

The first requirement for accurate multispectral photography is to establish the spectral reflectance of the ground object. When subtle differences are to be determined, object-to-background spectral reflectivity must be established, and the spectral bands where relative differences in reflectance occur must be located. A spectrophotometer with a diffuse reflectance attachment has been used for this purpose. However, as all such instruments integrate the energy reflected by the sample, preferential directional reflectance at particular wavelengths is not recorded. The airborne camera is a perspective sensor which records the directional reflectance characteristics of a ground object. In many instances considerable difference has been found to exist between total diffuse reflectance of an object as recorded by a spectrophotometer and the reflected energy measured at a specific angle of incidence and reflection. The spectral distribution of the solar irradiance falling on the ground at the time of exposure can be established by using a portable recording spectroradiometer.

The camera must be spectrally calibrated in order to relate the energy falling on the film plane to that entering the lens. This requires calibration of the spectral distribution and magnitude of radiant energy as a function of field angle of the camera lens. The camera system must be designed to obtain spatially identical photographs. This means that the focal lengths and distortions of the lenses must be identical for each wavelength band. All photographs must be taken at the same instant of time to avoid shift in relative position of the image on the spectral negatives due to aircraft angular motion.

Film processing is critical for accurate results. So-called panchromatic films have different characteristic curves as a function of wavelength. When the film is processed, each spectral band can be expected to have a different relationship of log exposure (or radiometric equivalent) to density. This difference in the characteristic curves of multispectral photos can be corrected by differential processing, or eliminated by printing on poly-contrast materials. These gamma differences cannot be incorporated in a *standard* definition of chromaticity of a ground object because density variation will occur in the particular spectral bands due to exposure differences, and will not be necessarily caused by a difference in spectral reflectance.

Using positive transparencies for additive color projection creates a dilemma between contrast and exposure latitude. The higher the contrast the more saturated the color, but the more compressed the scene brightness range, and vice versa. In general, good color reproduction is achieved using positives with low base density and moderate to high contrast.

If the spectral positives are to be viewed in additive color, the viewer optical design must allow accurate registration and be free of color errors. Manipulation of the color variables of hue, brightness, and saturation must be calibrated and not be distorted by shifts in color temperature of the illuminant as the brightness of the projection lamps is varied.

The above precautions will not guarantee that all interpreters will identify the color of an image as being the same under different viewing conditions. This can be explained by the fact that a color photograph does not conform to colorimetric specification of a two-degree circular field of color in an otherwise dark field. It is known that spatial relationships within the visual field can affect hue, brightness, and saturation of colors. Variation in the angular size of an image is known to cause a change in apparent color. Isolated images of the very smallest angular sizes produce no hue response at all. The position of an image in relation to other images can also cause apparent color changes. There is a tendency for the visual mechanism to accentuate the color difference in objects juxtaposed in spatial position (simultaneous contrast enhancement), but an image-directed attitude on the part of the interpreter tends to reduce contrast enhancement effects.

These psychological aspects of color viewing make the final specification of the color of an image in objective mathematical terms necessary. While the color of the presented image can be manipulated to suit the desires of the interpreter, unambiguous color specification independent of a particular interpreter-photograph relationship is considered necessary to achieve reproducibility of results under a variety of photographic conditions.

REFERENCES

Colwell, R. Some practical applications of multiband spectral reconnaissance. *Am. Sci.* 49, No. 1, 1963.

Committee on Colorimetry, Optical Society of America. *Science of Color.* New York: Crowell, 1953.

Evans, R. *Introduction to Color.* New York: Wiley, 1948.

Hardy, J. *Textbook of Colorimetry.* Cambridge: MIT Press, 1936.

Holter and Legault. The Motivation for Multispectral Sensing. *Proceedings of the Third Symposium on Remote Sensing of Environment.* Ann Arbor: Univ. of Michigan, 1965.

Legault and Polcyn. Investigations of Multispectral Interpretation. *Proceedings of the Third Symposium on Remote Sensing of Environment.* Ann Arbor: Univ. of Michigan, 1965.

Molineux, C. E. Aerial Reconnaissance of Surface Features with the Multiband Spectral System. *Proceedings of the Third Symposium on Remote Sensing of Environment.* Ann Arbor: Univ. of Michigan, 1965.

Smith, J. Color—A new dimension in photogrammetry. *Photogram. Engng.* 29, 1963.

Swanson. Photogrammetric surveys for nautical charting—use of color and infrared photography. *Photogram. Engng.* 26, 1960.

Wenderoth, S., and Yost, E. F. A Rapid Access Color Reconnaissance System. Society of Photographic Instrumentation Engineers, Symposium on Human Engineering in Photo Data Reduction, 1966.

Winterberg and Wulfeck. Additive color photography and projection of military photointerpretation. *Photogram. Engng.* 27, No. 3, 1961.

Yost, E. F., and Wenderoth, S. The chromatic characteristics of additive color aerial photography. *Photographic Sci. Engng.* 9, No. 3, 1965.

Zaitov and Tsupurun. A camera for the selection of film type. *Geodesy and Aerophotography* No. 3, 1962, Moscow.

The geologist is concerned with the surface of the earth and the composition of the earth's crustal materials. He must be aware also of the forces that affect or modify the crust. Although there will never be a complete substitute for actual field observation and mapping, certain kinds of geologic data can be obtained by the use of remote sensors. Geology has a long history of utilizing sensing devices; the magnetometer and seismometer are examples of reliable, well-established geologic tools.

To what kinds of geologic problem can modern remote sensors be applied? Studies by a number of researchers suggest that remote sensing is useful in this discipline if one or more of the following conditions are met. First, direct samples of materials sensed are not required to solve the problem. Second, an image of the spatial characteristics of the phenomenon being sensed will help in the solution of the problem. Third, the interpretation of the remote sensing imagery has no uncertainty of meaning or less uncertainty than any other known method of data gathering.

Aircraft or space platforms provide the geologist a unique position on which to station a remote sensor. Imagery generated from such a source is especially valuable if the problem is regional or global in scale, particularly if it is time dependent. The return of a satellite over the same earth scene on a periodic basis allows the sensors to monitor temporal phenomena. Collecting data by remote sensors may in some cases be more economical or convenient, or may be the only way of collecting data from remote areas or those obscured by cloud cover or political obfuscation. Before we can make full use of satellite imagery, much more information is needed on how electromagnetic energy interacts with the earth's surface. Until some of these fundamental questions are answered, the most important question answerable by the use of satellite imagery may be where *a phenomenon is located!*

39-The Multiband Approach to Geological Mapping from Orbiting Satellites: Is It Redundant or Vital?

R. J. P. LYON

In assessing the geological utilization of multiband spectral data from an orbiting satellite system, we must first ask what we want to do with the data. Let us assume the need for geological mapping has been established, and hence our task is "how to satisfy" that rather than "how to justify" it. This is logical, as it is feasible to make simple reconnaissance-type, geological maps from space, which requires only the most simple of instrumentation —a camera. From this then let us analyze the detailed requirements imposed by this task.

From *Remote Sensing of Environment* 1:237–244, 1970. Reprinted with permission of the author and American Elsevier Publishing Company, Inc.

Basically, geological mapping is a method to express, at a manageable scale, the spatial distribution of rock materials. If rocks can be further distinguished by their shape or attitudes, then a structural map can be produced. If chemical information can be added, then the map achieves its maximum significance, as now the rock and soil types can be differentiated.

How can remote sensing aid in this process? Do we add enough information to cover the increased cost? What extra information does multichannel (spectral) imagery bring to the geological analysis? Have we evidence that we can use spectral data when we get it? What are the atmospheric limitations on such techniques? How do these variable characteristics of the atmosphere reduce the quality of the data? What are the trade-offs

Table 1 Typical Data from Aircraft and Laboratory Experience[a]

ERTS A[b] (1) one 4-channel Multispectral Sensor (MSS) and (2) three Return Beam Vidicon Cameras (RBV)

ERTS A, MSS Bands	**RBV Cameras**
Band 1 0.5–0.6 μm	No. 1 0.475–0.575 μm
Band 2 0.6–0.7 μm	No. 2 0.580–0.68 μm
Band 3 0.7–0.8 μm	No. 3 0.69–0.83 μm
Band 4 0.8–1.1 μm	
Scan angle ±5.8°. Inst. F.O.V. 230 × 230 ft.	
ERTS B[c] Band 5 10.4–12.6 μm added	

	Band				
	1	2	3	4	5
	Reflectance, %				Temperature, °K
Rock and soil materials and covers					
Sand	5.19	4.32	3.46	6.71	—
Loam 1% H_2O	6.70	6.79	6.10	14.01	—
Loam 20% H_2O	4.21	4.02	3.38	7.57	—
Ice	18.30	16.10	12.20	11.00	225/283
Snow	19.10	15.00	10.90	9.20	265
Cultivated land	3.27	2.39	1.58	(not given)	—
Clay	14.34	14.40	11.99	(not given)	—
Gneiss	7.02	6.54	5.37	10.70	—
Loose soil	7.40	6.91	5.68	(not given)	301
Fallow	—	—	—	—	302
Soil	—	—	—	—	315
Vegetation					
Grass meadows	—	—	—	—	289
Wheat (low fertilizers)	3.44	2.27	3.56	8.95	—
Wheat (high fertilizers)	3.69	2.58	3.67	9.29	—
Water	3.75	2.24	1.20	1.89	—
Barley (healthy)	3.96	4.07	4.47	9.29	—
Barley (mildewed)	4.42	4.07	5.16	11.60	—
Oats	4.02	2.25	3.50	9.64	—
Oats	3.21	2.20	3.27	9.46	—
Soybean (high H_2O)	3.29	2.782	4.11	8.67	—
Soybean (low H_2O)	3.35	2.60	3.92	11.01	—

[a] Lars, 1958.
[b] ERTS A and B Design Study Specs., April 1969.
[c] *Aviation Week and Space Technology*, November 1969.

between spatial resolution and the spectral resolutions necessary to define rock compositions? As yet, we can answer only some of these valid queries, but we are faced with satellite systems whose data channel characteristics are already fixed in design.

Laboratory Spectral Evidence: What Is It? What Does It Mean to Us?

To answer these really basic questions we must establish (1) that there *are* significant differences between rock and soil materials, and (2) that these differences vary with wavelength, i.e., that there are spectral differences between materials, for on this is based the whole concept of the multiband "spectral" approach. If rocks do *not* show differences between chosen bands (and this seems to be the case in the visible and near infrared out to several microns), then any band is as good as any other, and hence we may as well choose the visible photographic band because of its well-established technology and wide interpretability. But let us examine the evidence we have by wavelength intervals, for this is a convenient mechanism. For comparison with this evidence, Table 1 lists the bands chosen for ERTS A and B, and the characteristics of the first instruments which will provide new data from space.

Photographic Band (0.3–0.9 μm)

This wavelength region may be divided into three main areas: ultraviolet, from 0.3 to 0.4 μm; visible, from 0.4 to 0.7 μm; and the so-called "photographic" infrared (that portion of the near infrared in which readily available films are sensitive), from 0.7 to 0.9 μm.

Ultraviolet (0.3–0.4 μm)

This band extends from the blue limit of our eyes down

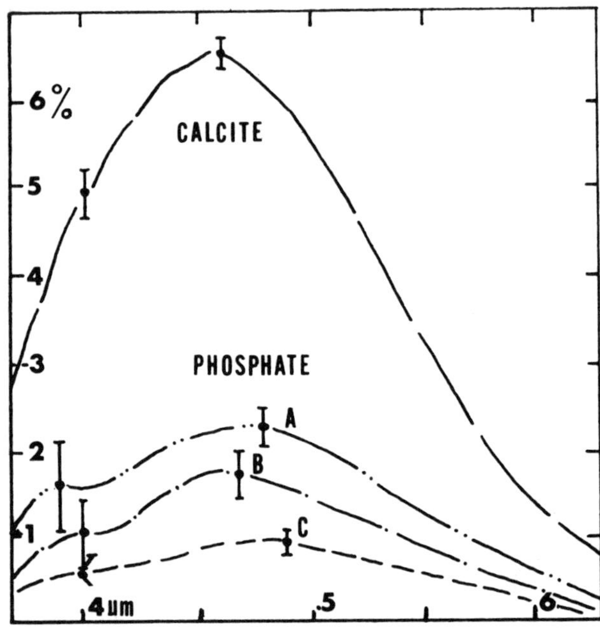

Figure 1 Curves showing ultraviolet-excited luminescence emission from three phosphate rocks and a calcite reference. (Data from Watts, 1967)

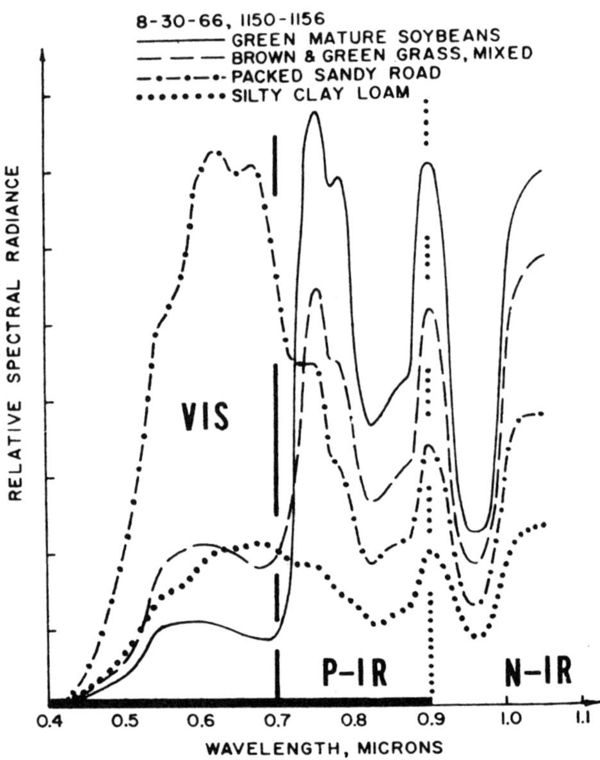

Figure 2 Relative spectral radiance of soils and vegetation showing the marked rise in vegetation reflectance at 0.72 μm. (Laboratory for Application of Remote Sensing, Purdue University, 1968)

to a vaguely defined edge near 0.3 μm, marked by a loss of transmission of energy through the atmosphere. Hemphill (1968) and (Watts (1967) have explored this region to determine if geological mapping of the earth is possible. The data are suggestive, but not clear-cut. Some rocks show clear spectral departures, but other samples, even of the same rock type, conform closely with chemically dissimilar rocks. Figure 1 shows data after Watts (1967). More significant spectral departures appear in the "vacuum" ultraviolet at wavelengths below 0.3 toward 0.1 μm, but terrestrial usage is impractical. Fluorite shows anomalous reflectances around 0.35 μm, and exposures have been detected by aerial photography.

Visible (0.4–0.7 μm)

In this band we already are all personally aware of the colors of various rocks, grading from blue to yellow, green, and red. But no single color is confined to a single rock or soil type. One can say that "fresh rocks" are usually bluish green (when their iron content is in the reduced Fe^{2+}state) and the weathered equivalents are reddish orange (when the iron is in the oxidized Fe^{3+} state), but these are vague terms.

Photographic Infrared (0.7–0.9 μm)

Unlike vegetation, rocks do not show the marked reflectance increase at wavelengths beyond the "chlorophyll-red" absorption band at 0.7 μm (Fig. 2). Rock spectra show only a general low gradient and, while they would not be confused with vegetation, no marked changes are present by which individual rock or soil types can be distinguished. Again the greatest variability in this band is from the water content in the sample. The H—OH stretching band of water occurs at 1.4 and 1.9

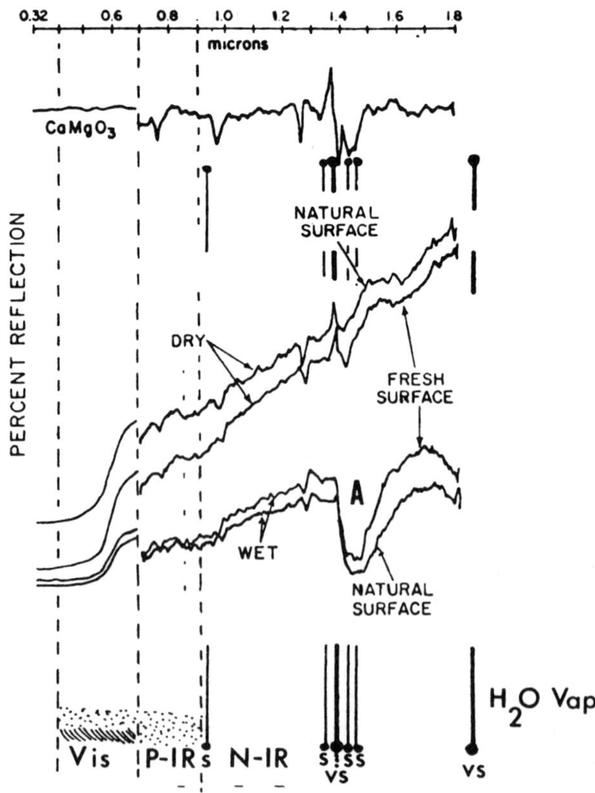

Figure 3 Reflectance spectra of wet and dry, fresh and weathered (salt-encrusted) surfaces of a red sandstone. (Brown et al., 1967)

μm in all "wet" samples. But even the published laboratory spectra often show other artifacts related to residual CO_2 or water vapor or both (Figs. 2 and 3) from incorrect purging of the equipment. Obviously, with field, airborne, and space-borne spectra, these effects will be markedly increased.

The biggest difference between rocks and soils, in the visible region, is related to their moisture content. This is clear in Fig. 3 (Brown et al., 1967), which represents duplicate runs on the same red rock sample in wet and dry states. Other spectra showing the same moisture-content effect are in the excellent compilation by Condit (1970).

Nonphotographic, Scanner-Dependent Bands (0.9–15.0 μm)

Near Infrared (0.9–5.5 μm)

Again this region may be divided into two bands, determined principally by the wavelength range of equipment used to obtain the spectra. Most data cover out to 2.5 μm, representing the capability of the Beckman DK-2 unit. This also represents the beginning of a very pronounced water vapor absorption band around 2.0 μm. Water appears in the air path and in the prism materials of the instrument, producing a low-energy region for measurement (Figs. 2 and 3).

Rock and soil spectra here are relatively featureless, except for the (superimposed) atmospheric bands for H_2O and CO_2, and for absorbed water (in several molecular forms) on the sample surfaces. Figure 4 (see also Table 2) shows diffuse reflectance spectra from three weathered and fresh rocks with two broad bands centered at 1.43 and 1.93 μm, which are probably due to CO_2 left in the instrument. Figure 3 shows several peaks in both samples, and in the CaMgO white-reflectance standard. These peaks are probably due to H_2O vapor left in the air path, but the marked absorbence (at *A*), which increases in depth on the wet sample spectrum, is one of

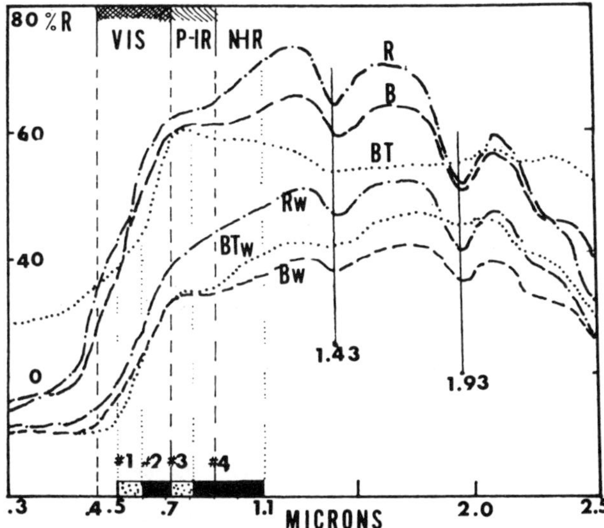

Figure 4 Typical reflectance spectra, from the Casa Diablo altered zone, Long Valley (near Bishop), California. (Total spectral reflectance using a DK-2A with reflectance sphere; after Hodder, 1969)

Table 2 Samples Used for Reflectance Data in Figure 4[a]

No. 10–24	Rhyolite (Tertiary), weathered and unweathered
No. 2.23A	Altered basalt (Quaternary), weathered and unweathered
No. B–24	Bishop Tuff (crystal-lithic, rhyolite tuff), weathered and unweathered

[a] Samples from Long Valley, California (between Mono Craters and Bishop). Total reflectance spectra were prepared on the Beckman DK 2-A, with an integrating sphere reflectance attachment (type 24500). The spectra are normalized in that slit width was varied to maintain constant energy. Data are relative to a CaMgO coating on the sphere (Hodder, 1969).

Table 3 Frequently Observed Spectral Peaks Which Are Anomalous to Rock Spectra

Wavelength (μm)	Absorber	Strength (s = strong, vs = very strong)
0.94	H_2O vapor	s
1.14	H_2O vapor	s
1.38	CO_2 vapor	vs
1.39	H_2O vapor	s
1.43	H_2O liquid	vs
1.46	H_2O vapor	vs
1.96	CO_2	s
2.2	H_2O liquid	vs
2.66 (2.5–2.8)	H_2O vapor	vvs
2.7	H_2O liquid	vs
4.26 (4.2–4.4)	CO_2	vvs
6.27	H_2O vapor	vvs
6.35	H_2O liquid	vs
9.6	O_3	s
15.0	CO_2	(above 120,000 ft) vvs

the liquid-water peaks from the water on the sample. Table 3 lists these bands, which often appear on the spectra, and which must be avoided in discussions of rock spectra (Hodder, 1969).

Hovis (1966a, 1966b) shows extensive and diagnostic spectral data from a variety of common rocks and minerals, and notes how these spectra are modified when using finer-grained materials.

Thermal Infrared (7–15 μm)

In these wavelengths many more data of direct chemical significance are available.

All materials above absolute zero radiate energy in proportion to their temperature, and earth surface materials at 300°K emit in the infrared region, peaking near 10 μm. These radiations may be observed through the most transparent windows in the atmosphere, from near 8 to about 13 μm wavelengths.

Most rock-forming minerals (silicates, carbonates, sulfates, etc.) show very strong absorption peaks in the infrared from about 7 to 12 μm, when the frequencies (or wavelengths) of the fundamental lattice vibrations of the functional anion group in the crystal are matched with the incoming infrared frequencies (Lyon, 1965). These are the "reststrahlen" bands.

For the dominant rock mineral group of the earth (silicates) and most planetary surfaces, the Si—O band absorption occurs between 8.5 and 9.0 μm and ranges out to 11.5 μm as the effect of Fe^{2+} and Mg^{2+} in secondary coordination with the Si–O groups lowers the frequencies of the bond vibration. Thus quartz (SiO_2) absorbs sharply at 9.0 μm, but olivine (Fe_2SiO_4) absorbs at 11.5 μm (Lyon, 1965). Similarly, granite rocks (which contain quartz and feldspar) absorb broadly from 8.5 to 9.5 μm and basalts (feldspar and pyroxene) absorb at 10.1 μm. The same phenomenon of marked changes in absorptive behavior may be observed in reflection of infrared energies from rock surfaces or, as an exact parallel, as minima in the envelope of emitted energies from rock and mineral surfaces.

These facts are geological statements of experimental studies traceable back to Coblentz (1905) and beyond, to Rubens in 1897. With further reflection and emission studies it became possible to differentiate rough rock specimens in the laboratory by sole use of the spectrum of infrared energies emitted by the rock samples (Lyon, 1965). In the field and in the air, this method has been found equally applicable (Lyon and Patterson, 1966), and rock surfaces may be mapped at 200 miles per hour.

A simplified version of this experiment, using a two-band radiometer with filters in the 8.0–9.5 μm and the 10.0–12.0 μm bands, has been flown in an aircraft by Hovis (personal communication, 1969). Similar effects were seen and rock types were discerned in a broadly diagnostic manner. The use of narrower bandpass filters would markedly enhance the selectivity in the data.

Results of Airborne and Space-Borne Rock Discrimination Experiments

Visible and Photographic Infrared (0.4–0.9 μm)

Figure 5 shows the four bands of the multiband photographic experiment carried on Apollo 9 (S-065 experiment) in which the shutters of a set of four Hasselblad cameras are fired simultaneously. The films and the filters used are as follows:

Band A. Color infrared film (CIR) type SO108 with #15 filter, now shown as a black and white copy.

Band B. Black and white film, type 3400 with #58B (green) filter.

Band C. Black and white infrared film, type SO246, with a #89B filter.

Band D. Black and white film, type 3400, with a #25A (red) filter.

The bandpasses for these filters are shown on Figure 5. The prints are arranged from left to right, and the filter numbers are indicated for reference.

This experiment is designed primarily for the agriculturist, in order to detect the vigor of the vegetative growth (Colwell, 1968; Myers and Allen, 1968). For this purpose the experiment is a great success, as may readily be seen by the red-colored contrast (on the colored original) along the U.S.A.-Mexico border at the lower part of each print marking the vigorous growth in the intensively irrigated Imperial Valley, California. Two further patches are in the center top of the areas (*B*, Blythe, California) and in the bottom right-hand corner along the Gila River (*G*) passing upstream into Arizona, again marking irrigated fields.

Geologically the results are not so useful. Admittedly, the extreme clarity of regional views of the structure is present, as in most space photography, but one does not find readily geological advantages from splitting up the visible and near infrared into the four bands. As a specific example, let us take the region of the Cargo Muchacho Mountains (*C*). They appear on the photographs as two distinctly different rock masses to the north and south, respectively. Yet on the San Diego-El Centro sheet of the California Geological Map (California Division of Mines, 1962), these are shown as essentially the same rock units, a series of granite outcrops, with interlayered metamorphics. Granites (Gr_1) and (Gr_2) are on the south, thus must be more highly colored and darker than the granites (Gr_3) and (Gr_4) and the metamorphics to the north. Significantly, however, this brightness contrast between the northern and southern groups is the same on all four bands, again illustrating the lack of spectral color in rocks.

A comparable example may be taken from the areas immediately to the northeast, marked *E* and *F*, respectively. Again a brightness contour can be drawn around the two rock outcrops, and they may be differentiated. But they cannot be indentified, as again they do not show in different manners on the several bands of photography.

Area *E* in the Chocolate Mountains is shown on the geological map as Tertiary volcanics (undif.). The next area *F*, the darker of the two, is shown as Quaternary Cenozoic volcanics-basalts. Area *E* is lighter than *F* in all bands, negating any simple scheme of positive identification. In area *C* they are both granites, and now in *E* and *F* both are volcanics, but of differing types.

That the three or four bands of information really do not give the geologist any new data will be the greatest shortcoming of the early ERTS photographs. It may be too soon really to make such a statement, as Ballew (1969) and others have been able to differentiate rocks by using their spectral response in the visible in a discriminant function program on the computer (BMDO-7M, UCLA series). The spectral differences appear to be small, and sophisticated analysis accompanied by comparison with much more newly gathered spectral data from those rocks in that area may be a necessary prerequisite before analysis can be achieved.

This is not to say that the photos are useless. Far from it. But their main use is as spatial information, clearly showing the geographic distribution of rock types and structures. One can readily trace the major faults, and the CIR print gives a reasonable representation of their true colors. Desert highways (Nos. 60 and 95) may be followed, and the location of the cities and towns are defined on the CIR print. For the best geological use, however, it seems likely that the red (25A filter) photograph will be of maximum use.

Analysis. Spatial (not spectral) information seems to

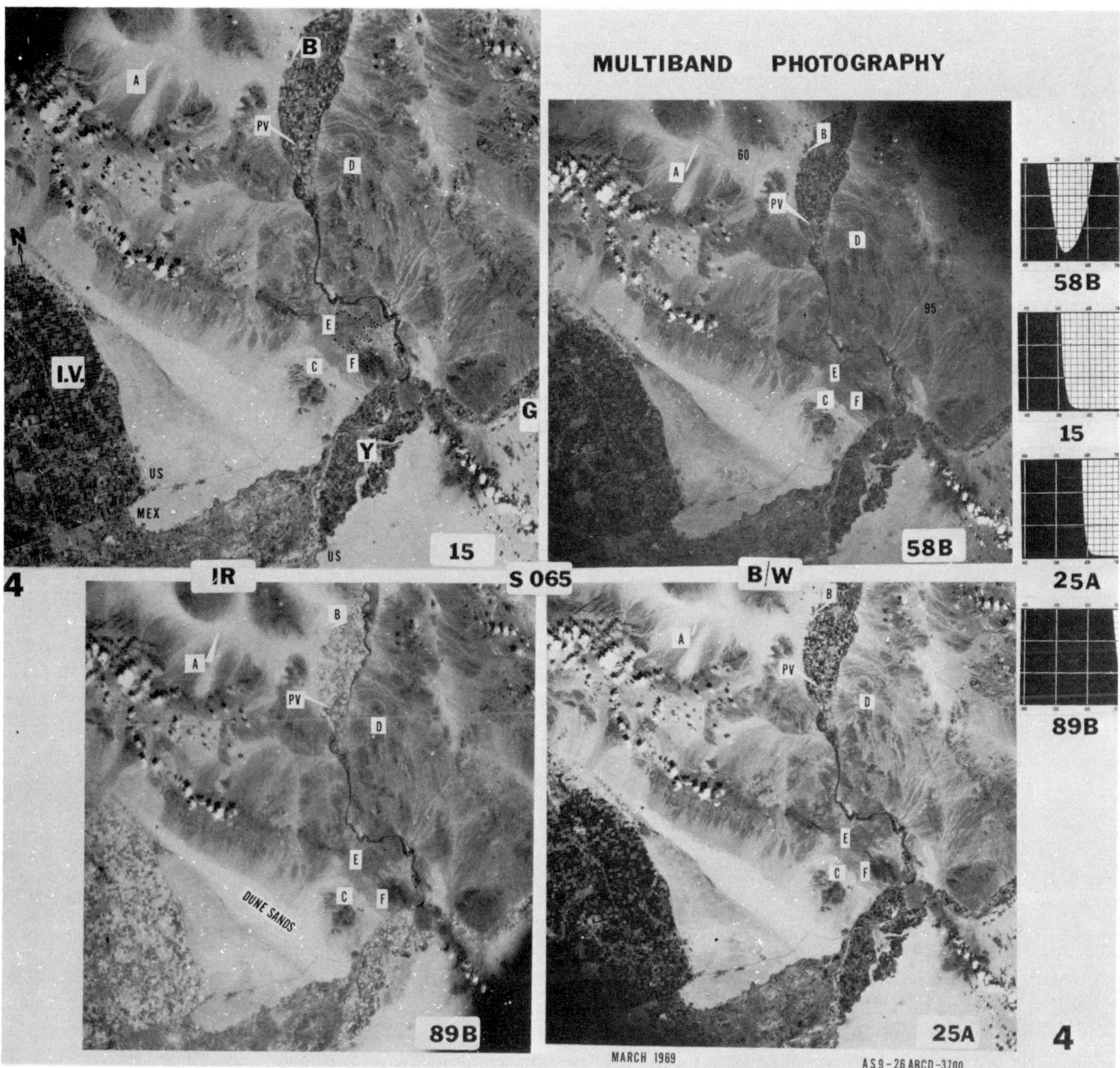

Figure 5 Four photographs in the visible and near infrared, taken in the S-065 experiment, on Apollo 9, Mar., 1969. Bands are shown by their filter designations. Bandpass details are shown on the associated graphs. Frames are AS9-26-A, B, C, D-3700 (NASA), all black and white copies. Annotated localities of interest are: *A*, Playa (Ford Lake) on Highway 60, seen more clearly on 58B band. *B*, township of Blythe, California, on the Colorado River. Arrow points toward the airfield to the left (west), seen best on 58B and 25A bands. *C*, Cargo Muchacho Mountains (see text). *D*, northern end of Trigo Mountains in Yuma Test Station. Mojave Peak is just to SE of *D*. *E*, *F*, two areas in the Chocolate Mountains, which run up to and off left top of picture (see text). *G*, Gila River, Arizona. *IV*, Imperial Valley, California, a rich farming area, heavily irrigated from the Colorado River. *PV*, Palo Verde, California, town and irrigation area. *Y*, Yuma, Arizona, town and irrigation area. East bank of Colorado River here is in the United States, west bank in Mexico.

be most useful so far in geological analyses. CIR is not especially useful, unless there is some geobotanical association.

Recommendation. Black and white with a red filter appears to be the best. Distinct advantages accrue if the sun illumination is at a low angle, between 20 and 35 degrees.

Problems. The low resolution from orbital altitudes, suitable only for regional studies, will continue to plague the geologist. There is no real substitute for high resolution.

Near Infrared (0.9–5.5 μm)

Only a few imaging experiments have been performed in this band, and really none of them was for geological purposes. The most dramatic is the NIMBUS HRIR (High Resolution Infrared Scanner in the 3.4–4.3 μm band. Although the instantaneous field of view (equiv-

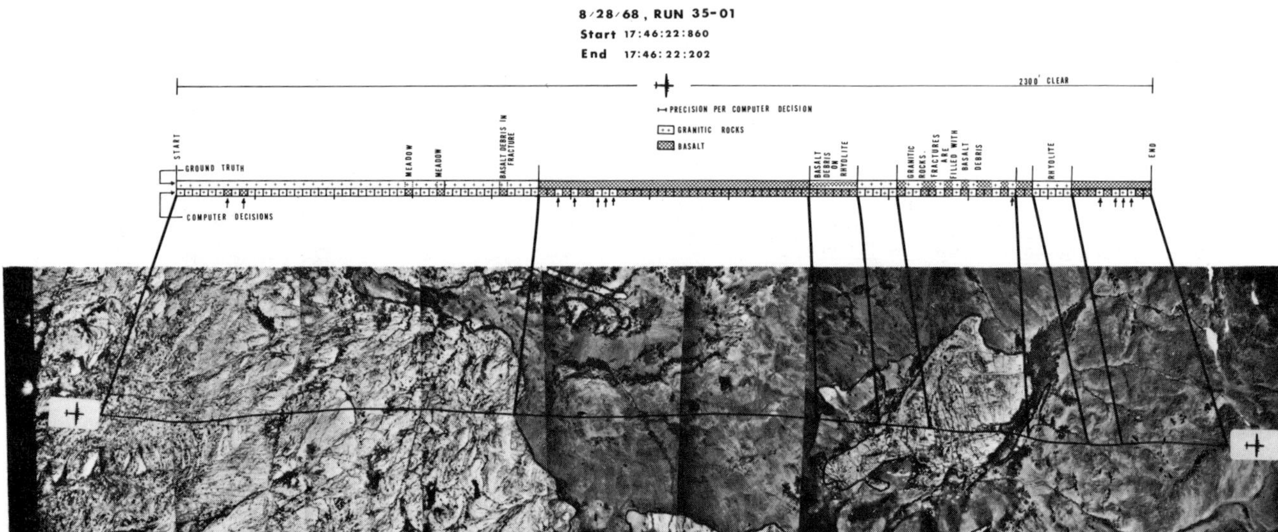

Figure 6 Aircraft ground track and computed rock types, using spectral infrared emission data. This figure consists of a series of RC-8 B/W photographs, used as a mosaic, to show the ground trace of an airborne (8–13 μm) ir spectrometer experiment (Lyon, 1965; Lyon and Patterson, 1966, 1969). At 2300 feet above terrain this is 25 feet wide, spectra being taken at a rate of 6 per second. The upper bar shows the geology (called ground truth) as quartz monzonites ("granite"), debris occurs in patches, and some shadows are shown. Area is Sonora Pass, Central California, in the Sierra Nevadas, at about 10,000 feet. Lower part of the bar graph shows the decisions made by the computer program which analyzes the spectra, using a discriminant analysis (BMD07M). Arrows indicate the 12 errors in the 121 decisions, a 90 percent success, entirely with machine analysis.

alent to spatial resolution, or "grain size") is 5 km on a side, truly major structural features like mountain ranges and islands can be seen in these thermal "maps." Volcanoes like Mauna Loa in Hawaii and Surtsey in Iceland can be discerned, but no one would feel that these were precise identifications of rock or soil types (Gawarecki et al., 1965).

Hovis (personal communication, 1969) has made several aircraft flights with near infrared spectrometer systems over a variety of desertic and vegetated terrain. He feels that field rock and soil spectra in this wavelength (1.4–5.5 μm) are not very characteristic (in contrast to his laboratory data) and that the subtle contrasts with vegetation and between soils are obliterated at jet aircraft altitudes and speeds. Thus another promising laboratory study has probably failed in the real-world applicability.

Thermal Infrared (7–15 μm)

Geological and mineralogical analysis of rock samples using their infrared emission spectra has become a well-developed and reasonably successful laboratory technique (Lyon, 1965). To develop the practical uses of these analytical techniques for field geological mapping, the instrumentation was made mobile, in a truck system (Lyon and Patterson, 1966). With it, rock-type determinations were made of targets up to 17,000 feet horizontally from the spectrometer. In 1967 a newly developed airborne infrared spectrometer was built, scanning the octave from 6.7 to 13.4 μm, and test-flown over simple geological targets (NASA Mission 56). Spectra were produced and recorded at a rate of 6 per second, or roughly every 40 feet of forward movement.

In 1968 a more sophisticated packaging of the three instruments was prepared, and two successful flights over quite different geological terrains gave initial feasibility to the airborne geological mapping experiment. Useful operations are marginal, however, where vegetation is a dominant feature of the target.

For geological uses, one is restricted to rapidly scanned spectra and narrow fields of view (6 spectra per second and 7-milliradian field of view, for example). Under typical flight conditions (2000 feet/terrain, 180–200 miles/hour), each spectrum represents patches of terrain 14 feet wide, smeared out to a rectangle for 40 feet while the spectrum is being recorded. Each and every spectrum (40 $\times$ 14 feet) has to be treated as a single entity and computed to an end result. This is a prime prerequisite for geological use: spectra must represent the minimum ground areas possible and, hence, averaging (or "co-adding") to increase S/N ratios is not allowable (Lyon and Patterson, 1969).

Already the techniques can show contacts between rock types and differentiate rock bodies, thus greatly facilitating the tasks of the field geologist entering a new area (Fig. 6). It is also possible to map rapidly and in detail, from low aircraft altitudes, a large outcrop (like an open pit of a mine) and to delineate the areal distribution of the mineral assemblages, or alteration zones in the ore. Perhaps then the greatest use of the techniques will be in detailed study of relatively small areas to pinpoint critical localities for even more detailed study by the field team.

Epilogue

Analysis of the Nature of Geological Mapping

What is geological mapping? What can we do from aircraft or from space to help? What is the nature of the geological experiment being performed? Just what is being mapped, and why, especially when over a terrestrial target area? What will it add to that provided by the traditional geologist? Several answers are possible, but they depend somewhat on the nature of the terrain.

Let us presume that the multiband techniques really work. Let us paint the rosiest picture we can imagine. The resultant data might show (1) rock contacts, i.e., that no significant change took place between times t_0 and t_1, and again between t_1 and t_2, but that the two materials were statistically similar within (and different between) those areas. Progressing further in a more sophisticated analysis, the data may show (2) that rock types in these two areas specifically were materials *A* and *B;* or even that *A* and *B* are identifiable as granite and andesite, for example. Both of these facts are of significance to the geologist, even if they provide only a calibrated line of rock variations beneath the aircraft. But this is still only one-dimensional, line-trace data. The second dimension, perhaps, could be laterally extended by the use of conventional photo-geological methods. Or, eventually a modification of the optical-mechanical scanner will be built to make images concurrently in several predefined wavelength slices in the 8–10 μm region. This would then allow both rock composition and their x-y areal extents to be defined simultaneously, providing true mapping of the geological targets below the aircraft.

Differentiation versus Identification

To pose the major questions again: as geologists, do we want a mapping system which differentiates dissimilar materials, or do we want one which gives us a unique answer by which each material can be identified, categorized, and labeled?

The techniques of field geological mapping consist in defining "mappable units" at a given spot, and tracing them in two or three dimensions until they change enough to warrant another name. If this is correct, then perhaps the aerial photograph (or its cousin, radar) is all a geologist needs. It may be redundant to give him more, as he may not be able to utilize this new information in his field task. It may even redefine his mappable units in a totally different and conflicting way. (Compare this with time boundaries versus lithologic boundaries in stratigraphic mapping.)

During the summer periods of 1967 and 1968 several groups of scientists and engineers met at Woods Hole, Massachusetts, under the auspices of the Division of Engineering of the National Academy of Science, to discuss the role and peaceful uses of earth-orbital spacecraft (National Academy of Science, 1969). The Geology Panel further explored this concept of differentiation and identification in the context of space flight aids to the economic (field) geologist. The panel felt that this man's job was primarily structural, and that the early Gemini-type photography and the (airborne) K-band radar had already contributed a vast amount for his immediate use. The photographic-like geometry was paramount. Such geometric fidelity was preferred to preserve the true relationships of linear elements of structure which he was seeking. It is clear, however, that at best this is only a request for differentiation and not a requirement for identification.

Perhaps we are asking too much, both of the sensor systems and of the field geologist. We are plagued on one side by poorly defined spectra from rocks and soils and the problems of seeing these after degradation by the atmosphere. On the other side, the recipients of the data, the field geologists, appear only vaguely interested in the detailed data—they want raw, unprocessed images. In all fairness, they have not had such detailed data commonly available before. Perhaps the greatest value from ERTS A and B will be the increased awareness of what a spectral approach can yield. The onus is on the researchers to provide the successful links.

To this end, we must have equipment as closely similar to ERTS A and B as possible, available now for airborne (and ground) use in a variety of terrains and seasonal times. Filter bandpasses and detector responses should be identical with those in ERTS A and B. It is not possible to simulate these subtle differences between rocks, and only practical operation over known terrains will do. Without this eleventh-hour research program, ERTS will be just another camera system for geologists.

Conclusions

In the foregoing analysis an attempt has been made to establish what spectral evidence exists by which rocks may be differentiated, and where in the photographic and infrared regions this evidence lies. The several experimental attempts to use spectral information for rock and soil mapping have been reviewed. From this discussion we can formulate the following questions and give some conclusions.

1. The simplest (and cheapest) form of spectral information can be obtained by using color films in which the preservation of the geometrical relationships between objects markedly assists the mapping process. Although geological interpretation of aerial color films is really still in its infancy, it does offer the maximum advantage to the field geologist of any remote sensing techniques.

What lies beyond this first step? Photographic film is an excellent medium for data storage, but a very poor one for data retrieval and particularly for use in a digital computer. Accordingly, only limited manipulation techniques are available for film-based data. To achieve the maximum from the computer using all the statistical and pattern-recognition programs, data must be in a more readily processible form, such as on magnetic tape.

2. To recognize the shape of an object, one must have high spatial resolution and this means immense volumes of data. To determine the nature of an areally extensive

material, however, one needs far less spatial resolution. Physical and chemical composition thus can be added to the image analysis with the processing of far fewer bits of spectral data than for spatial data at high resolutions.

3. We do not yet have the experimental evidence that spectral data can give us the required differentiation between materials of geological interest. At Stanford, in the 8–13 μm region, we can show an almost 1:1 correlation between the ir emission spectrum and rock chemistry. In other wavelength bands, and especially in the visible, the distinction is not so well made.

4. Physical composition such as soil particle size, sand grain size, and rock densities can be assessed with reasonable success, only if *repetitive* thermal-band measurements can be made. In this way the thermal inertia of the materials can be assessed by determining the rate of change of their radiometric temperatures with solar phase.

5. From orbit, the problem of rock and soil definition is complicated by the varied transmission of the atmosphere, which only allows viewing of the ground through small and restricted "windows." In the most diagnostic window for rock and soil composition (8–13 μm), ozone has a strong absorption doublet at 9.6 and 9.8 μm, precisely in the middle of the key data band. Likewise, one cannot make studies of thermal inertia of soils from orbit, as repeated passes over short-time frames are not practical.

6. Geologically, coarse resolution (100–200 feet) will still be useful for broad structural studies as for showing *where* it is, but the questions of composition, of *what* it is, will require much more aircraft-altitude study before they can be successfully asked from space.

REFERENCES

Ballew, G. L. Quantitative geologic analysis of multiband and photography from Mono Craters area, California. *M.S. Rept., Stanford University* (reprinted as *Stanford RSL Tech. Rept.* 69–2), 1969.

Brown, G. D., et al. Multispectral studies of a red-bed facies, Minas Basin, Nova Scotia, *AFCRL Document* 67–0603, 1967.

California Division of Mines. *Geologic Atlas of California, 1:250,000* scale, Kingman Sheet (1961); San Diego-El Centro sheet (1962); Trona sheet (1962); Salton Sea sheet (1967).

Coblentz, W. W. Investigations of infrared spectra, Carnegie Institution of Washington, D.C., *Publ.* 35, 1905.

Colwell, R. N. Remote sensing of natural resources, *Sci. Amer.* 218:54, 1968.

Condit, H. R. The spectral reflectance of American soils, *Photogr. Eng.*, Oct. 1970.

Gawarecki, S., Lyon, R. J. P., and Nordberg, W. Infrared spectral returns and imagery of the earth from space and their application to geologic problems, *Scientific Experiments for Manned Orbital Flight*, Amer. Astronaut. Soc., 1965.

Hemphill, W. R. Application of ultraviolet reflectance and stimulated luminescence to the remote detection of natural materials. Presented at the Annual Convention of the Amer. Soc. Photogram., 1968.

Hodder, D. T. Multisensor investigation of selected California geothermal areas. A.G.U. Fall Meetings, San Francisco (submitted to J.G.R.), 1969.

Hovis, W. A., Jr. Infrared spectral reflectance of some common minerals. *Appl. Optics* 5(2):245, 1966a.

Hovis, W. A., Jr. Infrared reflectance spectra of igneous rock, tuffs, and red sandstone from 0.5 to 22 μm. *J. Opt. Soc. Amer.* 56(5):639, 1966b.

Laboratory for Agricultural Remote Sensing (LARS). Purdue, University. Remote multispectral sensing in agriculture. *Ann. Rept.* 3, 1968.

Lyon, R. J. P. Analysis of rocks by spectral infrared emission (8 to 25 microns). *Economic Geol.* 60, 715, 1965.

Lyon, R. J. P., and Patterson, J. W. Infrared spectral signatures —A field geological tool, *Proc. Fourth Symp. Remote Sens. Environ.*, V. A., Michigan, 1966.

Lyon, R. J. P. and Patterson, J. W. Airborne geological mapping using infrared emission spectra. *Proc. Sixth Symp. Remote Sens. Environ.*, 1969.

Myers, V. I., and Allen, W. A. Electro-optical remote sensing methods as non-destructive testing and measuring techniques in agriculture. *Appl. Optics* 7 (9):1819, 1968.

National Academy of Science. Useful applications of earth-oriented satellites to geology. National Academy of Sciences, Washington, D.C., 1969.

Smedes, H. W., Pierce, K. L., Tanguay, M. G., and Hoffer, R. M. Digital computer terrain mapping from multispectral data, and evaluation of proposed earth resources technology satellite (ERTS) data channels. Yellowstone National Park, Preliminary Rep., U.S.G.S. (open file report), Denver, Colorado, 1970.

Watts, H. *AGI Remote Sensing Short Course Lecture Notes*. Fig. 3–8–2, 1967.

Ecology has *suddenly become a popular word in the American vocabulary. There are various kinds of ecology, but our concern here is primarily with that branch of biology that deals with the relationships between organisms and their environment. The burgeoning population and uncontrolled expansion of technology have produced stresses in our environment that were unimaginable in the past. It is hoped that the discipline of ecology can provide information to relieve these stresses or enable us to modify conditions before stresses develop.*

Most ecologic studies must be based upon an understanding of plant communities. It is the plant which is the ultimate source of energy-to-food conversion on which all life depends. Many parameters of plant communities lend themselves to the utilization of remote sensing. A multispectral sensor, in particular, is capable of answering many questions about vegetative associations which are significant to the study of flora. Little of the earth's natural vegetation (or cultivated vegetation, for that matter) has been adequately mapped in detail. But inventory is a first step toward rational use of cultivated and noncultivated vegetative associations. For example, we are never sure how many acres of wheat, corn, or oats are planted in a given year, yet this data is vital to strategic planning of food resources. Analyses of multispectral imagery offer the best evidence to date for discriminating crop types and distinguishing natural vegetation plant communities by a large-scale automated method. Instantaneous monitoring of several electromagnetic spectral regions frequently yields information about the physical characteristics of vegetation which otherwise could be obtained only by widespread and expensive ground sampling. Ecologic studies are not confined to vegetation alone, and multispectral imagery is also a valuable tool in the study of soil, water, pollution, geology, and the cultural works of man.

40-How Multispectral Sensing Can Help the Ecologist

FABIAN C. POLCYN
NORMA A. SPANSAIL
WILLIAM A. MALIDA

Since World War II, many techniques for producing imagery at wavelengths outside the photographic region have been developed. Imagery can now be generated over the entire optical range from 0.3 to 14.0 micrometers, thus spanning the ultraviolet, visible, and infrared wavelengths. At microwave (millimeter and centimeter) wavelengths, passive and active (radar) systems have been developed to produce imagery under all weather conditions. While much effort has been spent to develop and use the sensing instrumentation, less effort has been directed toward understanding the relationships between the remote sensor outputs, the dynamic parameters of the vegetation and the terrain surfaces, and the illuminating conditions available for observations. Furthermore, much of the experimentation has been carried out with sensors and camera systems confined to one or another relatively broad spectral band in or near the visible region, and often there has been relatively little "ground truth" collected at the time the images were made. In addition, interpretation of the imagery is always subject to the limitations of the photographic process and the skill and training of available interpreters.

At the University of Michigan, a unique approach is under development, centered around the use of a truly simultaneous multichannel optical sensor. This sensor detects more of the information about an object's radiation than has been possible previously and permits a

From Johnson, P. L. (ed.), *Remote Sensing in Ecology*. Athens, Ga.: University of Georgia Press, 1969. Pp. 194–218. Reprinted with permission of the authors, the editor, and the publisher.

new form of interpretation by providing data not available to the human eye with the added potential for automatic discrimination of objects from among a large number of observations based on their spectral properties. The key component of the first prototype multispectral sensor consists of a spectrometer mated to a conventional optical-mechanical scanner. With 12 channels between 0.4 and 1.0 μ, the system obtains data with automatic registration in both time and position at a rate of 50,000 spectra/sec. (Polcyn, 1967). Additional channels are obtained with detector arrays optically filtered to operate in selected bands between 1 and 5.5 μ and individual detectors at 8–14 μ and between 0.3 and 0.4 μ. Imagery can be produced in each individual band or from combinations of bands. Thus, for each instantaneous look of the scanner, a spectrum can be derived and a decision made by an automatic processor as to the identification of the object without using any spatial shape information. The design for a 30-channel system has been completed (Lowe et al., 1966), which will provide for perfect registration of all bands between 0.3 and 14.0 μ. Such an instrument would be the first device able to provide the full benefits of multispectral remote sensing to the study of ecologic and resource problems. Presently, research is centered around the existing prototype system, i.e., up to 18 channels (12 of which observe a given resolution element simultaneously) when two or more scanners are operated simultaneously.

For the ecologist who wants to understand the capability of multispectral sensing, an appreciation of the types of information that this new sensor provides and its problem areas is an essential first step. Consider the sensor-user chain suggested in Table 1. Sensor designs can provide outputs from which quantitative measurements can be derived and inferences can be drawn or deductions made to answer specific questions about ecologic parameters of interest. Table 2 gives a specific example of the types of information that are potentially derivable from a multispectral sensor to answer questions about plant productivity. The concept of plant productivity must be subdivided into those aspects which are measurable by multispectral techniques. This step is important no matter which phenomenon is under investigation. The important point is that one should not expect something of the sensor which it cannot provide, even though each added sophistication of the sensory outputs (left side of Table 2) provides increasingly useful information. A given set of sensor requirements, e.g., good signal-to-noise ratio, synchronous multichannel registration, voltage references, and radiance or reflectance calibration, will provide a set of sensor outputs which contain information about a specific parameter. The user must bring to the problem an a priori knowledge as to

Table 1 Sensor-User Chain

Sensor Requirements	→	Sensor Outputs	→	Derivables	→	Inferences or Deductions	→	Parameter of Interest

Table 2 Sensor-User Chain to Provide Information Concerning Plant Productivity

Sensor Requirements		Sensor Outputs		Potential Derivables		Possible Inferences or Deductions
				Spatial pattern	→	Plant community structure and stratification
			→	Time variation of relative image contrast	→	Maturation or wilting
Single channel data	→	Voltage		Relative apparent temperature (ir)	→	Effects due to change in energy balance
Synchronous multi-channel data and reference	→	↓ Raw spectrum (voltage)		Species identification if one has training set	→	Distribution of plant species over a site
			→	Soil type identification if one has training set	→	Probable soil fertility levels
Calibration	→	↓ Spectral radiance watt/cm²/ster.		Enhanced image contrast	→	Change in species: maturity, disease, nutrient status, or moisture status
			→	Outgoing energy over large areas	→	Energy budget calculations
Ground references (or sun sensor)	→	↓ Percent reflectance		Spectral shape, day-to-day basis	→	Distribution of plant species or soil types, day-to-day basis
			→	Absolute apparent temperature (ir)	→	Quantitative knowledge of thermal processes
				Moisture content of soil	→	Possible moisture availability for plant growth
				Species identification using signature library	→	Distribution of plant species or soil types, over a region with minimum ground truth

the meaning of the parameter in a given environmental context. A good example of this approach occurs when the sensor generates a spectrum which leads to an identification of a plant species which in turn permits an inference about the subsurface conditions.

Unfortunately, much of the research to date has been limited to the collection of multispectral data with little collection of the "ground truth" information necessary to properly interpret the various contrasts observed in each channel. It is hoped that in future research programs the necessary ground work will be completed and that much progress will be realized in obtaining a fuller understanding of the potential of the multispectral approach.

Applications of Multispectral Sensing Techniques to Ecologic Investigations

A better understanding of multispectral sensor capabilities in relation to the objectives of ecologic research may be gained by consideration of the following areas: (1) how multispectral signatures are used to provide information concerning objects or processes of interest; (2) the characteristic features of multispectral outputs; (3) the advantages of using multichannel scanners over photographic systems; and (4) the sources of variation in scanner video data which influence its interpretability.

All objects within a natural landscape absorb, transmit, reflect, and emit radiation selectively. The radiation collected by the optical-mechanical scanners is converted to electrical signals which are recorded on magnetic tape. The characteristics of the scanners in operation at Michigan have been described in detail (Lowe et al., 1966). The taped video data may be converted from magnetic tape to one or more of several types of signature data or imagery displays. The pattern of voltage signals, or image gray tone levels obtained for a given object at a given time in each of the different spectral channels, represents the spectral characteristics of the object and is called its multispectral signature. Multispectral signatures are used to differentiate objects of interest and to provide information concerning the energy budget of the objects.

An example of how gray tone levels from a film strip display of several channels of data may be used to differentiate objects of interest is given in Figure 1. Figure 1A is a photo mosaic of an agricultural area near Davis, California. Note that the gray tones registered for fields of rice and safflower are all similar. Now consider Figure 1B and 1C which are displays of scanner imagery of the same scene in 18 wavelength bands from 0.32–0.38 μ to 8.0–14.0 μ. By comparing the gray tone levels in the different images for pairs of fields, it is possible to differentiate each field type.

The sensor outputs from a multispectral system can be tape-recorded and a number of display formats can be obtained for analysis. Among them are (1) multispectral signatures as a set of voltage levels in each channel, spectral radiance graphs, and percent reflectance as a function of wavelength, (2) in each channel, a voltage vs. time presentation for point-to-point comparison, (3) imagery in the form of a "moving window" on a C-scope or converted to density variations on film. The main types of image outputs are: (1) standard imagery in each spectral channel; (2) single-channel data that have been processed in some way before being put in the form of an image (e.g., quantization and contouring); (3) single images with contrasts that have been enhanced by combining data in two or more channels by electronic means or by simultaneous display of these data (e.g., a color display); and (4) decision maps that present the results of automatic recognition or classification by computing devices and provide area estimates of particular species. For research purposes, the ecologist will likely want to study all types of image outputs plus calibrated statistical multispectral data from areas of interest, whereas once an operational system has been devised, he may be satisfied with only decision maps and related computations.

A very desirable feature of data obtained from synchronous multichannel, optical mechanical scanners is its amenability to automatic processing. For example,

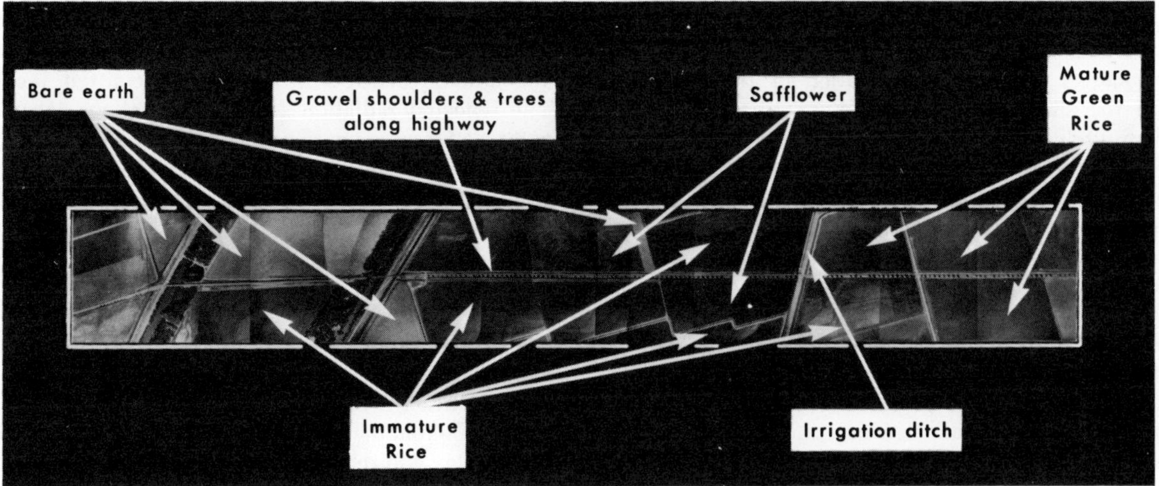

Figure 1A Panchromatic mosaic of Davis, California, agricultural area. May 26, 1966; 1600 hours; altitude 2000 feet; sky condition, clear and bright, 10 percent cloud cover at 30,000 feet; surface temperature 27°C. (University of Michigan)

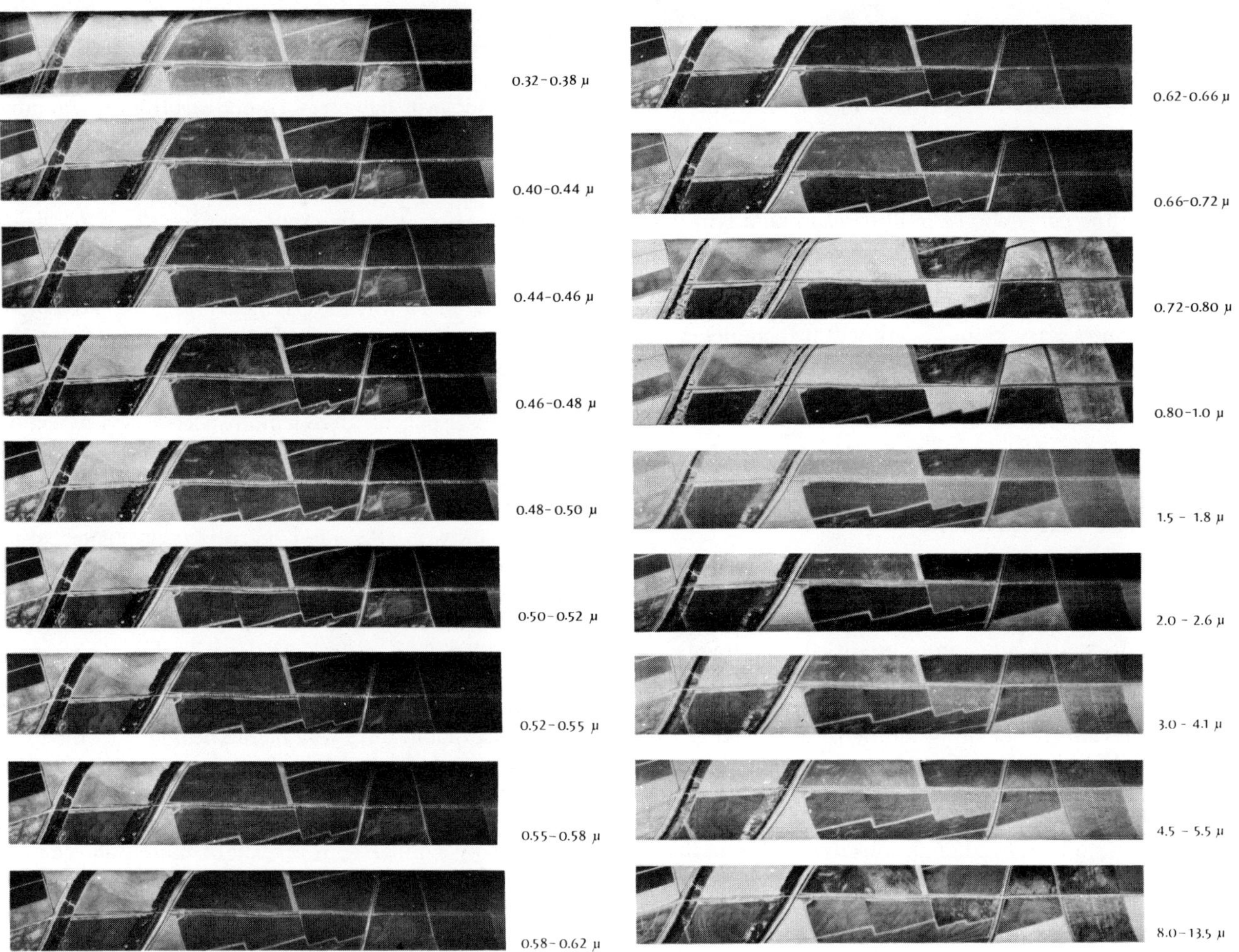

Figure 1B and **1C** Multispectral imagery of Davis, California, agricultural area. May 26, 1966; 1600 hours; altitude 2000 feet; sky condition, clear and bright, 10 percent cloud cover at 30,000 feet; surface temperature 27°C. (University of Michigan)

selected spectral channels of the data illustrated in Figure 1B and C have been processed electronically to achieve automatic recognition and mapping of the safflower, bare soil, and rice fields (Figure 1). Flexibility of signal processing also permits the removal of the backscatter component of the signals resulting in higher contrast imagery, and the optimization of the signals to reduce the effects of cloud shadows. Other advantages of the multichannel scanner are: (1) the extended spectral coverage which makes possible the establishment of a more unique signature for objects of interest; (2) use of thermal and reflective reference standards permitting calibration of the signals; (3) provision for extrapolation of detailed reflective and thermal measurements (made at a small site) to surrounding areas; and (4) the increased altitude of observation made possible by the greater dependence upon spectral information rather than spatial information. These are important advantages of the multichannel scanner over a multilens camera or several singlelens cameras with multilayer films such as conventional color film and Ektachrome infrared film.

Before discussing major types of contributions of multispectral sensing to ecologic investigations and presenting detailed examples of processing that can be performed on multispectral data, it is of interest to consider the sources of variation in the signals which produce the image tones and affect the interpretability of the data. A summary of the sources of variation in multispectral signatures of vegetation is presented in Table 3. There are two types of radiation that reach the sensor from the vegetation, reflected radiation and emitted radiation. The reflected radiation (from 0.3 to about 4 μ), originates at the sun, is modified in traversing the atmosphere path, and is reflected by the plant into the hemisphere around the plant according to a reflectance distribution function that depends on the physical properties of the plant. Certain wavelengths of the light are reflected more than others, thereby producing the color of the object in the visible portion of the spectrum (0.4 to 0.7 μ) and producing the spectral differences between various materials: differences on which multispectral discrimination is based.

The illumination from the sun depends on both the time of day and the time of year, as well as on any clouds or atmospheric particles that may act as intermediate reflectors or absorbers. The reflectance distribution function of the plant is determined by its geometry which in turn is dependent upon special maturity, and physiologic

Table 3 Sources of Variation in Multispectral Signatures of Vegetation

Illumination Conditions
Illumination geometry (sun angle, cloud distribution)
Spectral distribution of radiation
Site Environmental Conditions
Meteorologic
Micrometeorologic
Hydrologic
Edaphic
Geomorphologic
Reflective and Emissive Properties
Spatial properties (geometrical form, density of plants, and pattern of distribution)
Spectral properties (e.g., reflectance or color)
Thermal properties (emittance and temperature)
Plant Conditions
Maturity
Variety
Physiological condition
  Turgidity
  Nutrient levels
  Disease
  Heat-exchange processes
Atmospheric Conditions
Water vapor, aerosols, etc. (absorption, scattering, emission)
Viewing Conditions
Observation geometry (scan angle, heading relative to sun)
Time of observation
Altitude
Multichannel Sensor Parameters
Electronic noise, drift, gain change
Accuracy and precision of measurements on calibration references and standards
Differences in spectral responses of systems

factors. The plant's condition depends in part on the environmental conditions of the habitat. For most plants, their density and distribution on the ground also affects the radiation received by the sensor. These parameters may change from site to site, from day to day, or even from hour to hour. The constituents of the atmosphere can also change hourly or daily and may scatter extraneous sunlight into the sensor's field of view in addition to selectively attenuating radiation from the plant.

The observation geometry is important because vegetation reflects different amounts of radiation in various directions, e.g., it tends to reflect more radiation back toward the source than away from it. These effects have been observed in photographic data (Steiner and Haefner, 1965) and multichannel scanner data (Malila, 1968). The time of observation and the flight direction determine the relationship of the line of scan to the direction of the sun's angle of incidence. The flight altitude affects the amount of atmosphere that must be traversed by the radiation and also affects the size of the resolved area. The electronics and tape recorder of a multichannel scanner introduce noise, drift, and gain changes which cause additional signal variations. When the data are calibrated, there is always some uncertainty about the precise values of the calibration references and standards. This is particularly true when large reflectance standards are used on the ground. Finally, if more than one sensing instrument is used, there are usually differences in their spectral responses which can produce different output values for the same input radiation because the sensor performs a weighted integration of incident spectral power.

The process for emitted radiation (0.4 to 14 μ) is somewhat different than that for reflected radiation since the plant itself is the main source of radiation. The spectral emittance and temperature of the plant depend entirely on the condition of the plant and the environmental conditions. Since thermal changes occur slowly, the prior thermal history has a strong influence on the temperature of a material at any given time. Thermal data are important, since diseases and abnormalities in plants may be manifested by changes in their heat-exchange processes before their reflectance properties are affected.

Two primary objectives of current ecologic research are to understand the basis for biologic productivity of terrestrial and aquatic ecosystems, and to develop resource management practices to maintain or increase the productivity and quality of the ecosystems. Some of the types of ecologic research activities concerned with achieving these objectives which may be investigated by multispectral sensing techniques include: (1) the inventory of biotic and abiotic components of natural and modified ecosystems, (2) the assessment of spatial and temporal variation of selected biotic and abiotic components of ecosystems, and (3) the description of physical and biotic factors controlling biologic productivity.

Investigations of the three types given above are currently in progress or have recently been completed at [the University of] Michigan's Infrared and Optical Sensor Laboratory in cooperation with various users. The use of multispectral techniques for the inventory of biotic components of ecosystems includes the automatic recognition and mapping of plant species and plant communities. Fields of rice, safflower, and bare soil have been differentiated from one another and from other field types by their spectral (color) and spatial (gross crop geometry) differences, and their distributions have been mapped using automatic signal processing (Fig. 2). The automatic signal processing techniques used to map these field types are discussed later. Data have been collected for plant communities of sawgrass, willow brush, Australian pine, cocoplum, cattail, and submerged algal mats within the Everglades National Park and studies will be conducted to determine the feasibility of mapping these communities by automatic processing techniques. The information concerning rates and directions of plant succession derived from this type of inventory should make it especially useful for management purposes in areas of rapid ecologic change.

Investigations using multispectral techniques for inventory of abiotic components include the automatic recognition and mapping of rock outcrops and surface water bodies. Lava flows of different ages at Mono Lake, California, have been differentiated and mapped by their reflective spectral signatures including two specially filtered portions of the uv region. Multispectral scanner data of surface water bodies near Jamestown, North Dakota, were collected to determine the feasibility of

Figure 2 Recognition maps of rice, safflower, and soil for Davis, California, agricultural area. May 26, 1966; 1400 hours, 2000 feet. (University of Michigan)

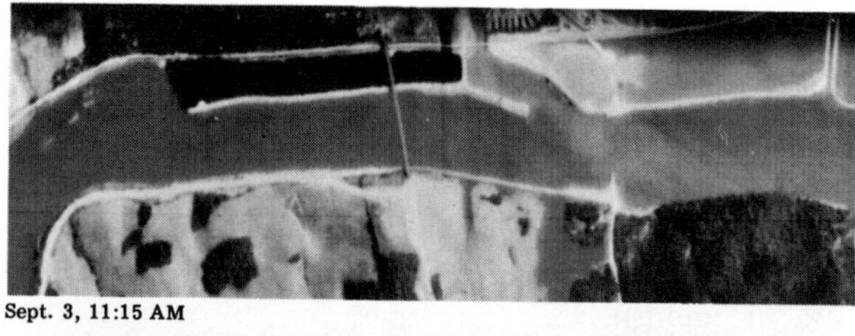

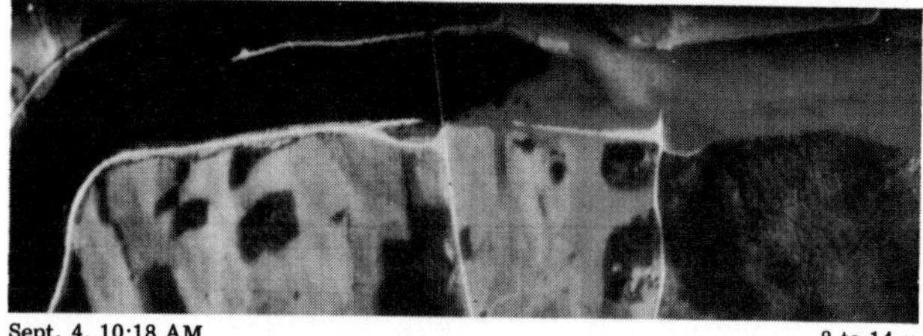

Figure 3 Monitoring of water temperature changes on the Clinch River, Tennessee. (University of Michigan)

conducting more rapid and accurate surveys of the number and sizes of suitable wildfowl breeding sites.

Studies concerned with spatial and temporal variations of ecosystem components include the mapping of surface water temperature changes on a river and the mapping of cooling rates of different Arctic terrain features at Point Barrow, Alaska. Calibrated thermal ir detectors were used to measure the rates of travel of cold water discharges from a dam on the Clinch River in Tennessee, and to measure the rates of surface water cooling. Figure 3 shows two views of a portion of the Clinch River near a power plant 20 miles downstream from a dam. In the top image, the warmest water entering the river from the plant appears as a light gray tone; the coldest water inside a retaining wall appears as a black tone; and the remainder of the river water which is intermediate in temperature appears as a medium gray tone. The bottom image was taken 23 hours after cold water was released at the dam 20 miles upstream (Kauth, 1968). A surface temperature map of this area has been produced by enhancement techniques which will be discussed in a later section.

Differences in the rates of cooling of selected Arctic terrain features is evident from examination of Figure 4.[1] The imagery was acquired during the transition period between summer and winter which is characterized by unstable thermal conditions. The tundra surface was frozen and snow-covered at this time, and ice had formed on the inland lakes. The warmest features in the scene appear as light gray tones; the coldest features appear dark gray in tone. The light gray tones of the inland lakes and the low center polygons indicate the presence of water beneath the surface of the snow and ice which serves as a heat source (Horvath, 1968).

Examples of investigations concerning biotic or physical factors which control biologic productivity include the mapping of moisture-stressed apple trees and ponderosa pine trees. Figure 5[1] illustrate the comparative capability of the 0.32–0.38 μ and the 2.0–2.6 μ spectral bands for the detection of a moisture-stressed condition of apple trees induced by rodent girdling at Ann Arbor, Michigan. In the 0.32–0.38 μ image the damaged trees appear lighter in tone than the healthy trees, whereas no difference in tonal values of the damaged and healthy trees is noted in the 2.0–2.6 μ image. Ground inspection of the orchard revealed that the damaged apple trees had smaller and fewer leaves and consequently more exposed branchwood. Laboratory spectrometer measurements showed that in the 0.32–0.38 μ region the difference in mean reflectance values for abaxial surfaces of healthy and damaged leaf samples was 0.5 percent, and 1 percent for adaxial surfaces. The reflectance values of bark samples ranged from 3 to 7 percent higher than the reflectance values of the leaf samples in the 0.32–0.38 μ region. The damaged apple trees were therefore differentiated from the healthy trees in the 0.32–0.38 μ imagery by their highly reflective branchwood. The lack of image contrast between healthy and damaged trees in the 2.0–2.6 μ region demonstrated that the reflectance values for both leaf and bark materials were nearly equal.

A typical procedure for the coordinated planning and research efforts of ecologists and remote sensing investigators involving nine separate tasks or steps is illustrated below. The procedure is described by the following example of an investigation underway at a forestry test site near Deadwood, South Dakota, for the purpose of detecting the intensity and extent of beetle-infested ponderosa pine trees (Heller, 1968). The investigation is incomplete, but the procedure used is typical.

Step 1: Definition of Objectives of Investigation

1. To determine if a temperature difference or nonvisual reflectance difference between healthy and beetle-infested ponderosa pine trees could be detected by

[1]Not reproduced here.

multispectral scanning techniques before any visual differences are observed.

2. To determine the accuracy of detection of beetle-infested trees by comparison of scanner results with ground surveys.

Step 2: Physical Manifestations of Parameters of Interest

1. Temperature: ground measurements of foliage temperatures indicated that the infested trees were 3°–8° warmer than the healthy trees due to decreased rates of transpiration.

2. Color: foliage of infested trees showed discoloration within 1 year after beetle attack.

Step 3: Sensor Characteristics Required to Solve Problem

1. Spectral regions: calibrated thermal ir scanner data in the 4.5–5.5 μ and 8.0–13.5 μ spectral bands were required, and additional spectral coverage in the 0.4–2.6 μ region.

2. Temperature resolution: 1°C temperature resolution was required in the thermal ir bands.

Step 4: Capability of Existing Sensors to Solve Problem

Existing sensors could provide calibrated thermal ir scanner data in the desired spectral bands with a temperature resolution less than 1°C.

Step 5: Sensor or Modification

Modification of existing sensors or design of new sensors are made in response to user needs for the solution of specific problems. This step was not considered necessary for the present investigation.

Step 6: Definition of Data Collection Program

Aerial-based operations

1. Flight schedule: May 29 and 30, 1968, at four time periods (early morning, mid-morning, noon, early afternoon) to obtain data under different levels of transportation.

2. Weather conditions: bright sunshine with less than 30 percent cloud cover.

3. Position and lengths of flight paths: the flight paths had to cover ground instrumented sites and extend for 2 miles.

4. Number and order of data runs over each target area: six to eight data runs were requested for each target area.

5. Altitude of flights: 800 feet above terrain was requested for all data flights.

6. Scanner configuration: 12-channel spectrometer (0.40–1.0 μ), a three-element detector (1.0–1.4, 1.5–1.8, and 2.0–2.6 μ), and a single-element detector (8.0–13.5 μ).

7. Calibration of thermal data: the thermal reference plates were set to register a 5 to 10°C temperature range without signal clipping; thermal ground resolution panels were overflown.

8. Calibration of reflective data: a "sun sensor" was used to record the intensity of total solar irradiance.

9. Radio communication: handie-talkie *Airnet* radios were used to coordinate timing and alignment of each run of the flight mission.

Ground-based operations

1. Temperature measurements: (a) five healthy and five insect-infested trees were instrumented for foliage absolute temperature; (b) apparent emitted temperature measurements were made of the thermal ground resolution panels: (c) radiometric surface temperatures of soil and other backgrounds of the trees were made from a stationary tower before, during, and after each flight.

2. Foliage moisture measurements: needle moisture tension measurements were made from samples of foliage collected from three trees of both healthy and dying classes.

3. Soil moisture measurements: soil pits were dug to determine water-holding capacity and current field capacity.

4. Micrometeorologic measurements: weather data from anemometers, pyreheliometers, hygrothermographs, and thermometers were collected in order to define the radiation environment.

5. Insect infestation measurements: type, extent, and intensity of insect infestation data were collected on a sample basis.

6. Foliage color measurements: Munsell color notations were made for all infested trees on five target sites.

7. Flight path marking devices: ground panels and lights were established for airborne orientation.

Step 7: Definition of the Data Reduction Program

The data reduction program has not been completely defined for this investigation. Typical considerations for this type of investigation include the following:

1. Evaluation of quality of the taped scanner data.

2. Evaluation of the "ground truth" data.

3. Definition of the purpose of the automatic signal processing and analysis (e.g., the differentiation and mapping of insect-infested pine trees).

4. Selection of scanner data for automatic signal processing and analysis (i.e., selection of data runs and times, selection of spectral channels).

5. Analysis of scanner data to determine the most promising processing technique(s).

6. Selection of the most promising automatic signal processing technique(s) and form(s) of display of processed data.

7. Processing of selected scanner data to discriminate objects of interest.

Step 8: Processor Design or Modification

This step would be taken if the existing processing facilities are found to be inadequate for the analysis of a particular user problem either before or after Step 7 had been taken, or after Step 9 had been taken.

Step 9: Analysis and Evaluation of Sensor and Processor Outputs

Again for this investigation the data analysis tasks have not been completely defined, but will likely include the following:

1. Correlation of physical characteristics of objects sensed as indicated by ground truth data with observed differences in reflectance or temperature recorded in the scanner data.

2. Determination of the statistical reliability of the results of processed scanner data, and the need for modification of existing processing equipment or design of new processors.

3. Determination of the influence of scanner system variables (e.g., electronic noise, accuracy of calibration references) upon the quality of the sensor and processor outputs, and the need for scanner system modification or design.

4. Determination of the influence of illumination conditions (e.g., sun angle, cloud cover) on the quality of the sensor outputs.

5. Determination of modifications needed for future aerial-based and ground-based data collection programs to solve this particular ecologic problem.

6. Reporting of results of the analysis of sensor outputs and processed data.

An important feature of the nine-step procedure described above is its provision for redefinition of tasks and reevaluation of processing and analysis techniques. We believe that such feedback originating from both ecologists and remote sensor investigations is necessary for the improvement of remote sensing techniques and the more effective solution of ecologic problems.

Examples of Results Using Multispectral Processing Techniques

Electronic processors have been used at [the University of] Michigan to perform recognition and classification of signals from several areas for which multichannel scanner data have been collected. Since the data are produced and recorded in analog form, the first processing of the data was performed by using a simple two-channel analog technique. Based on experience with this technique and on theoretical analyses, more complex decision rules were implemented, rules which make use of the likelihood ratio of statistical decision theory. A special-purpose analog computer which makes decisions at a real-time rate was designed and fabricated and is now in operation. Equivalent decision operations have also been performed at a much slower rate on a digital computer from digitized samples of multichannel data, and research is being directed toward hybrid computers which will combine the speed of analog devices with the flexibility and control of digital computers.

Selected examples of processed multispectral data are discussed to illustrate the capability of multispectral systems to recognize crop species, detect differences in the maturity and vigor of crops, detect differences in water depths in shallow waters, and detect temperature differences in river water near a power plant. A variety of ways of presenting the data was employed.

The first example illustrates the recognition of crop species with the simple two-channel processor. Shown in Figure 2 (Thomson, 1967) with a conventional image for the spectral band 0.4 to 1.0 μ are three black and white recognition maps for an agricultural area in California. The recognition map for rice (produced using data in the channels 0.72–0.80 μ) exhibits a good percentage of detection in the two rice fields, except for areas in the lower left-hand corner of each which may have had different soil moisture levels or different amounts of plant cover than the rest of the fields. The recognition map for safflowers (produced using data in the channels 0.48–0.50 μ and 0.62–0.66 μ) also exhibits a high detection percentage and correctly rejects the irrigation canals of dikes which have spectral characteristics that differ from those of the crop; portions of the lower field appear different from the remainder and were rejected for unknown reasons. The bare-soil recognition map (produced using data in the channels 0.48–0.50 μ and 0.58–0.62 μ) shows detection of gravel road shoulders along the highway and dirt roads between some of the fields in addition to the field of bare soil. It has been found very helpful to produce color-coded recognition maps which simultaneously present recognition data for several different crops. For example, the areas classified in Figure 2 as bare-soil could be brown, those for rice green, and those for safflowers red on a single color image. The remaining four fields were relatively immature rice which also have been differentiated by signal processing techniques.

It is also possible to produce a symbol on a digital computer printout to signify detection of a crop species (Hasell et al., 1968). Figure 6A is such an output decision map for wheat. As can be seen from the ground-truth tables and the ground-truth map of Figure 6B, all wheat fields were correctly identified. The skewness of the computer printout is due to crabbing of the aircraft during overflight, and the horizontal compression of the fields is due to the sampling interval used in producing the digital data. It is possible to correct electronically for

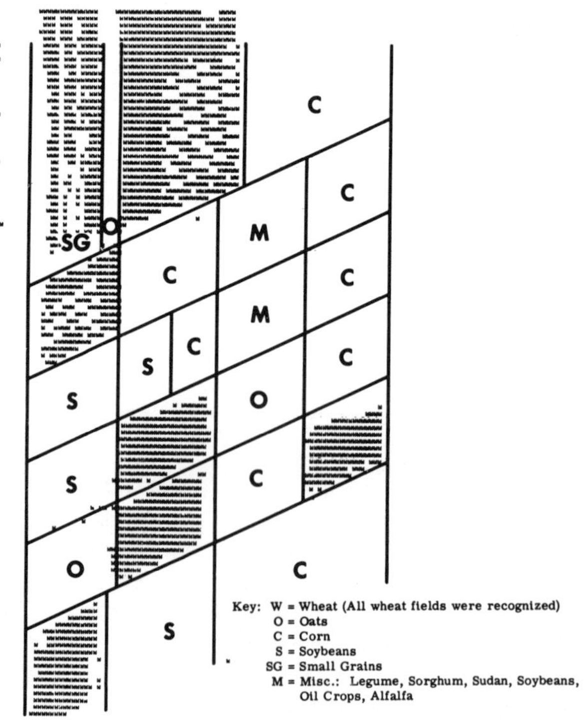

Figure 6A Recognition map for wheat. Maximum likelihood processing of digitized 12-channel scanner data. (Skewness due to crabbing of aircraft and sampling distortion.) (University of Michigan)

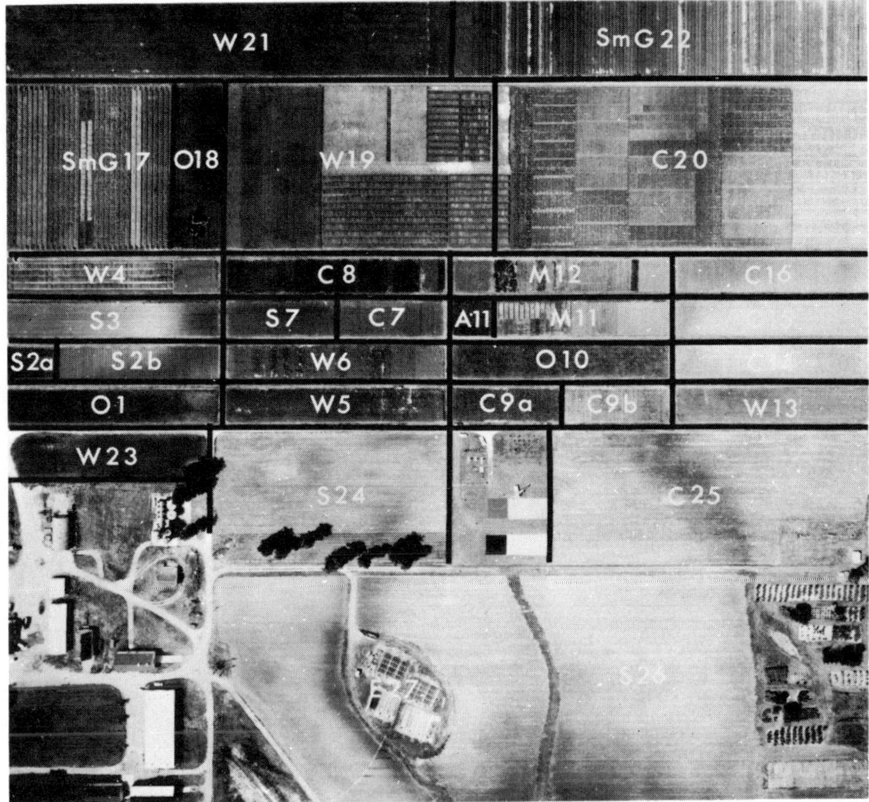

Key: W = Wheat
O = Oats
C = Corn
S = Soybeans
A = Alfalfa
SmG = Small Grains
M = Misc.: Legume, Soybeans, Sorghum, Oil Crops, or Sudan

Figure 6B Ground truth map for portion of Purdue agronomy farm, late June, 1966. (University of Michigan)

such distortion. The panchromatic photograph which serves as the background for the ground-truth map (Figure 6B exhibits very little difference between fields of wheat and oats, yet by using data in several bands simultaneously these fields are easily distinguished from each other. The only potential false detection, field *SmG17*, was identified only as small grains and probably also contained wheat. A maximum likelihood decision rule with 12 channels of data in the region 0.4 to 1.0 μ was used in the processing.

This multispectral approach may also aid in the mapping of underwater features and may enhance the study of aquatic ecosystems (Polcyn, 1968). Figure 7 shows selected imagery taken September 7, 1967, from 10,000 feet of an area near Elliot Key on the southeast coast of Florida. Light penetration into the water was good in the 0.52 to 0.58 μ region. As expected, the best water/land boundary was mapped in the 0.8 to 1.0 μ region where water strongly absorbs the incoming light. In the 0.62 to 0.66 μ band, only the shallowest parts of the coral reefs are seen because of relatively stronger absorption of water with respect to the 0.55 to 0.58 μ region. In the region 0.40 to 0.44 μ an image of lower contrast was produced because of the greater scattering by the water in this region. The capability for identifying underwater objects will be impared because of the wavelength filter effect of the water and light attenuation caused by suspended materials. Nevertheless, the roles that the multispectral sensors can play are being developed for pollution studies, for depth measurements in shallow water, and for measurements of chlorophyll concentrations in water.

The capability of new techniques for processing single-channel data can be illustrated by a thermal image (Figure 8A) of a power plant on the Clinch River in Tennessee (Kauth, 1968). Because of the calibration of the sensing system, apparent temperature levels or differences can be measured and displayed by special procedures. Figure 8B is the analog video map quantitized into 16 gray levels. Enhancement of each of the 16 levels by superimposing a spike at the decision point between levels resulted in a temperature contouring for which we know the temperature interval represented by each gray tone and the spatial extent of surface in that temperature interval (Figure 8C). Similar schemes can be done with a digital computer with typed symbols on a printout instead of a gray tone produced on a film.

Summary

The potential of multispectral remote sensing for ecologic research lies in its ability to provide information

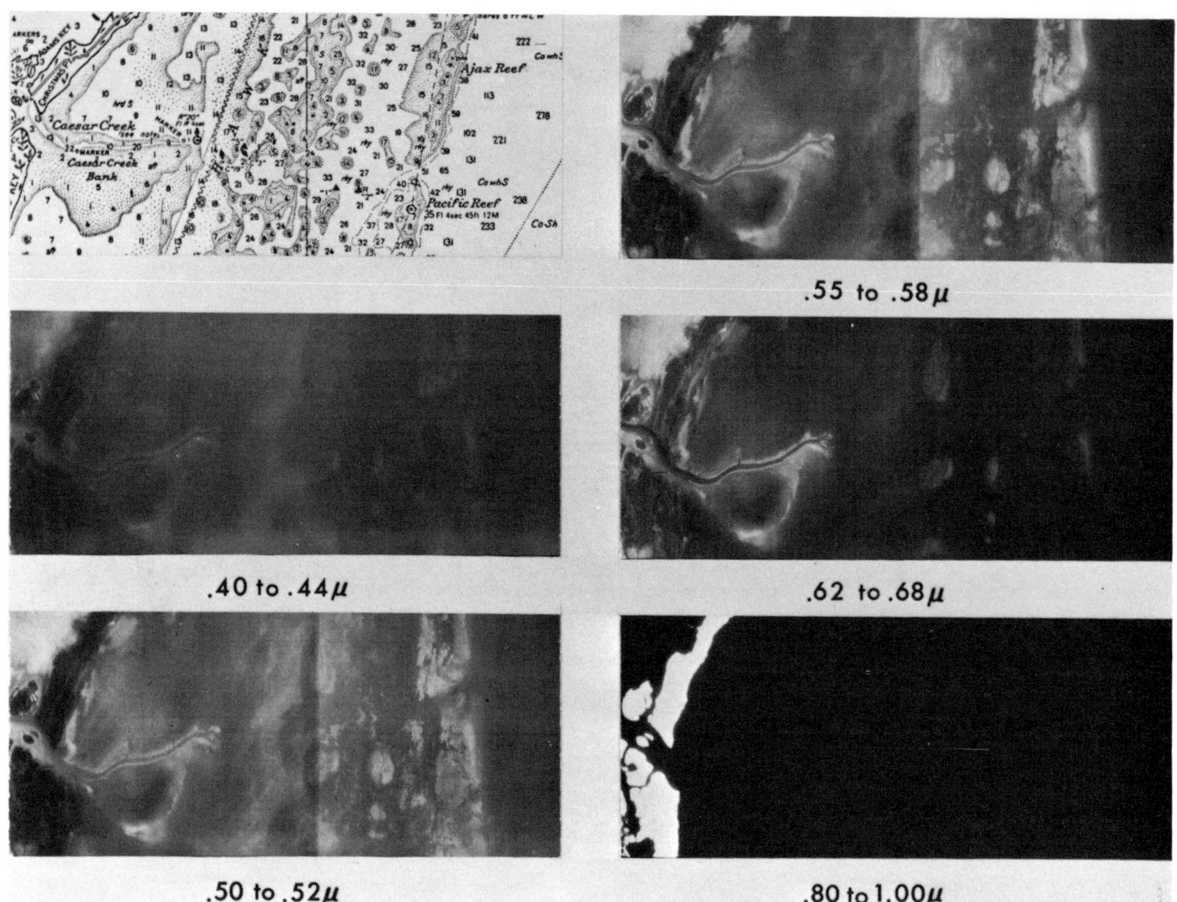

Figure 7 Multispectral comparison of light penetration of water.

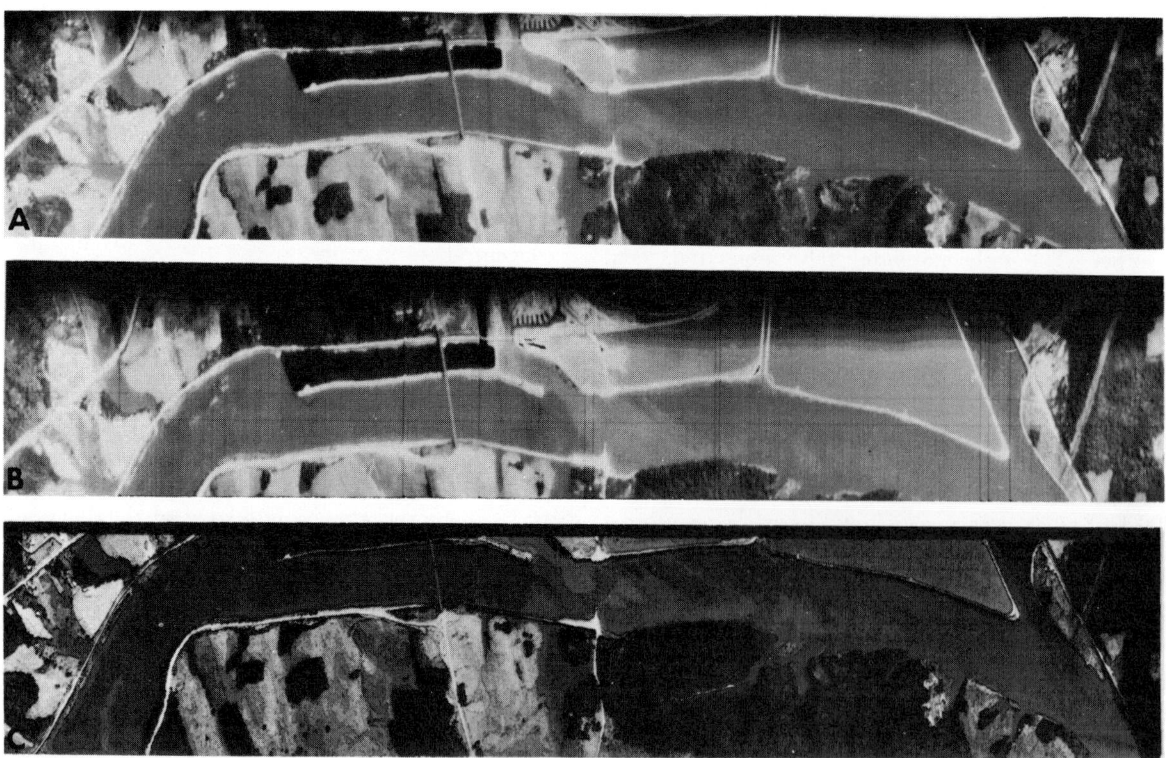

Figure 8 Map of temperature contours on the Clinch River, Tennessee. (University of Michigan)

about an object without necessarily resolving its spatial properties. This is particularly true for the recognition of a plant species by its multispectral signature, or the detection of physiologic changes of plants by observing differences in their multispectral responses.

In order to accomplish this form of data collection reliably, new calibrated, synchronized, multichannel sensors are being developed. Results thus far show that whenever spectral differences occur and can be related to a parameter of interest about an object, multispectral sensing can aid in its detection, the mapping of its distribution, or the analysis of its energy budget characteristics.

A wide variety of special signal processing techniques are available to aid in the recognition and mapping of objects of interest and the measurement of energy budget parameters. Much work remains to understand the influence of instrumental and environmental limiting factors upon the statistical reliability of the automatic signal processing techniques. What is needed now is a close cooperation between the instrument and processor developers and the ecologists to determine by experiment the degree to which multispectral sensing can contribute to a better understanding of the biologic world.

REFERENCES

Hasell, P. G. Investigations of spectrum matching techniques for remote sensing in agriculture. Report No. 8725-13-P, Willow Run Laboratories of the Institute of Science and Technology, Univ. Michigan, Ann Arbor, 1968.

Heller, R. C. Previsual detection of ponderosa pine trees dying from bark beetle attack. Proc. Fifth Symp., Remote Sensing Environment. Univ. Michigan, Ann Arbor, 1968.

Horvath, R. Multispectral survey of arctic regions. (Final Report), Report No. 1248-1-L, Institute of Science and Technology, Univ. Michigan, Ann Arbor, 1968.

Kauth, R. J. Preliminary analysis of TVA multispectral data. Report No. 1195-2-L, Institute of Science and Technology, Univ. Michigan, Ann Arbor, 1968.

Lowe, D. S., Braithwaite, J. G., and Larrowe, V. L. An investigative study of a spectrum-matching imaging system, (Final Report), Report No. 8201-1-F, Institute of Science and Technology, Univ. Michigan, Ann Arbor, 1966.

Malida, W. Multispectral techniques for contrast enhancement and discrimination. *Photogram. Engng.* 34:566, 1968.

Polcyn, F. C. Investigations of spectrum-matching sensing in agriculture Report No. 6590-7-P, Institute of Science and Technology, Univ. Michigan, Ann Arbor, 1967.

Polcyn, F. C. Remote sensing techniques for the location and measurement of shallow-water features. Report No. 8973-10-F, Institute of Science and Technology, Univ. Michigan, Ann Arbor, 1969.

Steiner, D., and Hafner, H. Tone distortion for automated interpretation, *Photogram. Engng.* 31:269, 1965.

Thomson, F. J. Multispectral discrimination of small targets. Report No. 6400-135-T, ECOM-00013-135, Institute of Science and Technology, Univ. Michigan, Ann Arbor, 1967.

SUGGESTED READINGS FOR PART NINE

Baker, L. R., and Slater, P. N. Study of Advanced Multispectral Sensor Systems. Interim Report I. University of Arizona at Tucson, Optical Sciences Center, 1968.

Barringer, A. R. The Use of Multi-parameter Remote Sensors: An Important New Tool for Mineral and Water Resource Evaluation. *Proceedings of the Fourth Symposium on Remote Sensing of Environment.* Ann Arbor: University of Michigan, Institute of Science and Technology, Willow Run Laboratories, 1966. Pp. 313–325.

Carneggie, D. M., and Lauer, D. T. Uses of multiband remote sensing in forest and range inventory. *Photogrammetria* 21:115–141, 1966.

Coker, A. E., Marchall, R., and Thomson, N. S. Application of Computer-Processed Multispectral Data to the Discrimination of Land Collapse (Sinkhole) Prone Areas in Florida. *Proceedings of the Sixth International Symposium on Remote Sensing of Environment.* Vol. II. Ann Arbor: University of Michigan, Institute of Science and Technology, Willow Run Laboratories, 1969. Pp. 65–78.

Colwell, R. N. Some practical applications of multiband spectral reconnaissance. *Am. Scientist* 29:3–36, 1961.

Colwell, R. N. Uses and Limitations of Multispectral Remote Sensing. *Proceedings of the Fourth Symposium on Remote Sensing of Environment.* Ann Arbor: University of Michigan, Institute of Science and Technology, Willow Run Laboratories, 1966. Pp. 71–100.

Frost, R. E. The program of multiband sensing research of the U.S. Army Snow, Ice, and Permafrost Research Establishment. *Photogrammetric Engineering* 26:786–792, 1960.

Hoffer, R. M. Interpretation of Remote Multispectral Imagery of Agricultural Crops. Research Bulletin 831, Purdue University Agr. Expt. Station, Lafayette, Indiana, 1967.

Holter, M. R. Infrared and multispectral sensing. *Bioscience* 17:376–383, 1967.

Holter, M., and Polcyn, F. Comparative Multispectral Sensing. In Johnson, P. (ed.), *Remote Sensing in Ecology.* Athens, Ga.: University of Georgia Press, 1969. P. 232.

Jerry, D., and Thorley, G. A. A Multispectral Photographic Experiment Based on Statistical Analysis of Spectrometric Data. NASA Report. University of California, Berkeley (3 parts), 1966.

Lancaster, C. W., and Feder, A. M. The multisensor mission. *Photogrammetric Engineering* 32:484–494, 1966.

Latham, J. P., and Witmer, R. E. Comparative waveform analysis of multisensor imagery. *Photogrammetric Engineering* 33:779–786, 1967.

Lauer, D. T. The Feasibility of Identifying Forest Species and Delineating Major Timber Types in California by Means of High-Altitude Multispectral Imagery. Annual Progress Report to NASA on Remote Sensing Applications in Forestry, 1967.

Miller, L. D., and Cooper, C. F. Analysis of Environmental and Vegetative Gradients in Yellowstone National Park from Remote Multispectral Sensing. In Johnson, P. (ed.), *Remote Sensing in Ecology.* Athens, Ga:. University of Georgia Press, 1969. Pp. 108–131.

Molineux, C. E. Aerial Reconnaissance of Surface Features with the Multiband Spectral System. *Proceedings of the Third Symposium on Remote Sensing of Environment.* Ann Arbor: University of Michigan, Institute of Science and Technology, Willow Run Laboratories, 1965. Pp. 399–421.

Richardson, A. J., Allen, W. A., and Thomas, J. R. Discrimination of Vegetation by Multispectral Reflectance Measurements. *Proceedings of the Sixth International Symposium on Remote Sensing of Environment,* Vol. II. Ann Arbor: University of Michigan, Institute of Science and Technology, Willow Run Laboratories, 1969. Pp 1143–1156.

Tanguay, M. G., Hoffer, R. M., and Miles, R. D. Multispectral Imagery and Automatic Classification of Spectral Response for Detailed Engineering Soils Mapping. *Proceedings of the Sixth International Symposium on Remote Sensing of Environment,* Vol. I. Ann Arbor: University of Michigan, Institute of Science and Technology, Willow Run Laboratories, 1969. Pp. 33–64.

PART TEN

SOCIAL IMPLICATIONS OF REMOTE SENSING

Will the "Spy in the Sky" become the Orwellian Big Brother?

REMOTE SENSING is considered vital to the military and of increasing importance to civilian agencies, but it presents many difficult social, legal, and policy problems. The startling speed of technical innovation has outstripped our social and legal processes. Advances are being made so rapidly that social and legal systems, which by their very nature react more slowly, are left behind, ill-equipped to deal with the technical breakthroughs. The recent and continuing furor over telephone wiretapping is a perfect example. How do we resolve the difference between scientific ability and implementation and the legal and social rights of the individual? One way would be to slow down scientific research and allow society time to absorb the advances already made. There is evidence that this process has begun—although not by conscious design. Another method of integrating technology into society is by changing the social and legal procedures that finance, manipulate, and control it. This process also has begun, albeit more slowly than one would hope.

Slowing scientific progress and speeding social change will close the gap—but never entirely; social change will always lag behind. One would hope that changing social values might make it easier to integrate scientific change, or perhaps lessen the demand for scientific innovation. This would mean a reordering of national and personal priorities and this could be accomplished only through major redirection of our educational and economic goals.

At this point in time, remote sensing provides us with information in a quantity and detail that we are not equipped to handle. Consider these examples. It is now possible to monitor a private conversation from a distance of several hundred yards or more, as long as there are no intervening objects. A farmer who plants more than his wheat or cotton allotment can have the extra acreage detected from the air. An industrial plant discharging waste into a river can be located from aloft. With such technology available, how do we protect our personal and collective privacy and freedom of action? How do we adjust our social and legal values to deal with offenders? There is nothing inherently immoral in gathering remote sensing information, but the potential exists for human abuse of this information. The problem is one of controlling information generation and dissemination without restricting its significant use and application. One thing is clear from modern technology. There is no privacy *from a well-financed, technically adept person or agency determined to gain personal information about an individual, group, or country. The scientific community which has been responsible for remote sensing technology has not directed its attention or concern to the social and legal ramifications of remote sensing. Who will determine how these new tools will be used?*

41-Legal and Social Policy Ramifications of Remote Sensing Techniques

SAMUEL D. ESTEP

THE PAPERS presented at these symposia are sufficient if only part of the proof of the dramatic success scientists and engineers have achieved in developing sophisticated remote sensing techniques. Even to a nonscientist, it is apparent that work and imagination have created fantastic tools. For the most part the papers presented at these symposia have dealt with these technical accomplishments. This paper, on the other hand, at least in emphasis if not wholly, is dealing with a different, although surely equally important, problem, that of the social policy and legal ramifications resulting from putting to practical uses remote sensing techniques. Most of our technologic advances since World War II have been fairly startling with respect to speed of development, complexity of technical requirements, and variety of potential uses. Remote sensing is just beginning to appear of importance to the general public, but in some ways presents the most difficult policy and legal problems.

This is not because remote sensing techniques are destructive in the direct sense of the word, but because of the kind of information that may be obtained in this way, usually without knowledge of those being observed. One cannot avoid completely some Orwellian concerns. There is nothing inherently wrong in any moral sense with the scientific discoveries or the remote sensing tools they have made possible. Rather it is potential human abuse that creates problems.

At the outset let me make it clear that those who have discovered and developed these remote sensing techniques are not guilty of moral turpitude any more than blame should be directed at the scientists for discovering the atom bomb and its progeny, Professor Goddard and his many successors in the development of rockets, or those who discover nerve gases. On the other hand, it is unrealistic if scientists, or others for that matter, refuse to recognize the difficult and sometimes frightening social consequences that could flow from these discoveries. The purpose of this paper is to attempt a selective, therefore spotty, and consequently partially unfair appraisal of the problems that can arise along with the benefits that surely will be derived from use of remote sensing capabilities.

From *Proceedings of the Fifth Symposium on Remote Sensing of Environment.* Ann Arbor: University of Michigan, Institute of Science and Technology, Willow Run Laboratories, 1968. Pp. 197–217. Reprinted with permission of the author and the publisher.

In presenting these ideas it is important that those who hear or read them know of the limitations inherent in this presentation, at least to the extent that I, myself, can identify them. It is for this reason that I am here undertaking to state what might be termed assumptions which in one way or another have shaped the ideas here presented. I am not hereby seeking to avoid criticism but only hoping to make it easier for those who note the thoughts here expressed to make their own evaluation of the worth of the suggestions here made. I wish first to set forth what one might term some negative assumptions and then state some of a more positive character.

In the negative or disclaimer vein would be the following:

1. I am not a trained scientist in the physics or engineering of remote sensing techniques. What knowledge I have of scientific matters in this area is secondhand, obtained from written reports or from talking with those who are experts. It is for this reason that we have other members on the panel who are fully conversant with the technical aspects of it.

2. I have read no secret reports, I have no access, legal or illegal, to classified information, and I have talked to no one about secret work going on here at the University of Michigan or elsewhere. In a real sense . . . this has its advantages. On the other hand, this necessarily creates real dangers that some of the conclusions I draw, particularly those based on technical evaluations, could well be wrong, and yet nobody would be in a position to agree or disagree because of the limitations that impose themselves upon those who have access to secret information. Therefore I should make it abundantly clear that these ideas are my own, arrived at without consultation, not only with the panel members participating with me in this discussion, but also any others within or outside the government or the University of Michigan. I am indebted to them in an immeasurable degree for making it easier for me to find unclassified information. They cannot be held responsible, however, directly or indirectly, for either inaccuracies of technical assumptions, or conclusions, or for the value judgments that underlie some of the suggestions here made.

3. Certainly more so than with respect to space technology generally and probably nuclear energy now, very important areas of remote sensing technique are shrouded in the great gray cloak of secrecy. Therefore, one must piece together and then use imagination, attempting to evaluate just what kinds of possibilities are presented by the use of remote sensing techniques. It may be wise

politics, international as well as domestic, to avoid mention of the U-2 flights, the Pueblo incident, and spy satellites. I doubt it. History teaches us, I think, that we get in more trouble by ducking these questions in the early parts of any technologic development than we could possibly get into by perhaps premature revelation.

4. It follows that because of this political delicacy it becomes almost impossible to involve in these evaluations people from certain government agencies and possibly even some scientific people whose contributions would be extremely helpful. On the other hand, it perhaps will be of some help to these people if an outsider, unrestricted by secrecy, attempts an evaluation no matter how incomplete it may be. At least be advised that I realize my limitations, and that in some cases they may seriously undercut the validity of conclusions drawn and suggestions made here. I do not think that this diminishes the responsibility that the serious lawyer and scholar surely must accept in just exactly these areas of great government sensitivity.

On the more positive side I would list the following assumptions:

1. I have no desire to go as far back as Aristotle's definition of the ultimate goal of organized society, i.e., happiness. This seems both too broad and, at least at the present time, a bit too unattainable, even if definable. I do think, however, that the United States, at the present time, is in a peculiarly appropriate position to make some fundamental value judgments and reflect them in such programs as those concerned with remote sensing techniques. This is so important that I dare burden you with at least some philosophizing about the role of a government and particularly one which is clearly one of two dominating powers in the world today.

I would like to define the goal of organized world society, towards which the U.S. itself should be working, as that of providing a decent condition of living for all the peoples of the earth. I used the word "decent" because it connotes more than minimal and yet is sufficiently modest to be obtainable in the foreseeable future with judicious use of the marvelous technologic advances you scientists are making. The word "condition" rather than "standard" was used because the term "standard of living," rightly or wrongly, has come to mean an economic standard of living which I believe is not sufficient to describe what our ultimate goal should be. Admittedly an economic standard of living is an awfully important part of it but there are crucial psychological needs also. They can be met perhaps only by the kind of dramatic search for the unknown that is exemplified in much of our space program. Such programs call upon the imagination and energies of creative people in ways that a direct welfare program to help the poor unfortunately never will call on, even among the great masses who themselves perhaps might be classified as poor. It is for this reason that I emphasize a decent condition of living as including these psychological challenges with which even the masses may identify, even though they themselves, perhaps, would never participate in a direct way in these programs. In such endeavors, however, the United States must keep in mind that as a world power it is crucial that we concern overselves with this ultimate goal of achieving a decent condition of living for the peoples of the world and conduct our foreign policy accordingly.

2. I believe that the U.S. has a very affirmative obligation to make the technology, or at least the fruits of the technology it is developing in so many areas, as fully available to the world as possible, even if this means that some people whom we rightly or wrongly feel are our competitors, may also benefit from it. Any other policy at best only retards those who realistically are our competitors, and in doing this we create suspicion about our own good intentions and deprive the rest of the world—whether on our side or only neutrals—from benefits we could well afford to share with them at relatively small cost to ourselves.

3. In addition, the closer we can come to a truly open society, even in our most sensitive technologic areas, the better off we will be psychologically as a nation. Past denials of this principle, because of international security concerns, perhaps have been realistic but have also been stultifying. I contend today that an unrealistic emphasis upon a quid pro quo for revelations in certain areas has tended to dominate our thinking, perhaps on the international scene even more than the domestic. Insufficient attention has been given to the long-run goals and the means to reach them. In most cases, in my opinion, what is good for the rest of the world is also good for the U.S.

There is a third set of assumptions which perhaps should be set forth, those involving technical capabilities of the remote sensing apparatus we now have. Actually I suspect all this would do would be to reveal in somewhat more detailed form what I have already admitted, a personal lack of detailed technical knowledge. It would certainly not be telling the scientific audience any new facts about remote sensing. It is important, however, that I indicate in general the assumptions I have made in the following analysis of social and legal problems.

The first assumption that I make is that today even those instruments which have been declassified are quite sensitive, even when very remote from the target. I also assume that those instruments which have not been declassified are considerably more sophisticated and sensitive and therefore would reveal considerably better and more information than is available from unclassified devices. Infrared techniques undoubtedly are some of the most important pieces of equipment that can be used for remote sensing, but we must not forget that there are other types of detectors such as gravity meters, seismographic instruments, ultrasonic detectors, ultraviolet devices, and, of course, radar, now supplemented by much more sophisticated equipment such as the side-looking type which has new capabilities and creates new problems. I also assume that, although some of these devices (such as radar and the sonic detectors) depend upon sending a beam of some kind to the target and recording the reflection, there are others (such as infrared) which depend only upon signals that radiate naturally from the target itself without even being triggered by the sensing device. One should also not forget that in the traditional field of detective work and industrial spying, even the popular literature is full of not only stories but advertising about various kinds of "bugging" techniques which in one sense or another are remote sensing devices. Emphasis here, however, is placed upon the more sophisticated ones such as those involved in infrared sensors.

One additional assumption probably is crucial to much of the analysis that follows. I am assuming that in most cases these remote sensing devices are physically harmless, at least if properly operated. Equally crucial is the fact that many of these devices can be used for remote sensing without detection by those in the target area. Obviously this is not true of those such as radar and ultrasonic detectors which do have to send out energy beams. With infrared techniques and many listening devices it is possible to do the remote sensing without knowledge on the part of the target itself. Even with some of those devices which depend upon emitted energy, only most sophisticated countries can really detect their use.

With these assumptions in mind we can turn to an analysis of the policy and legal problems.

Any two-category system of classification is likely to be a dangerous overgeneralization. This is particularly true when the subject is a complex one and perhaps even more so when it is a social policy and legal one. Nevertheless, for our purposes, it should be safe to divide the problems we are discussing into two categories: (1) international, and (2) domestic. We shall turn first to the international problem and then look at the domestic.

Although I do not believe this to be the most important international use of remote sensing devices, the one that is most likely to be headline material and to raise the greatest amount of public commotion here and abroad is their use to gather information about other countries, or, to put it less euphemistically, to spy. We will discuss this problem first and then move to some less dramatic but perhaps in long-range terms more important uses of remote sensing techniques.

The lore of spying in the international arena, both in fiction and in fact, is dramatic in some aspects, and probably everybody, while perhaps repulsed in some sense, also finds himself somewhat titillated by the whole idea. All of us have enough of the "curious cat" about us that we want to know the other person's secret, preferably without that person realizing that we know it, or at least not until we are ready to reveal the information. Yet, for practical purposes, there is no significant body of international law dealing with the problem of spying. There are certain practices which nations do not protest about, but to a large extent this has been dictated undoubtedly by the fact that each one knew that it was engaged in much the same kind of activity and in any event did not want to reveal complicity by making any kind of formal protest in most cases. Again there is a well-recognized custom of war about executing spies out-of-hand. Likewise, there is a general, unwritten rule that spies are not to reveal that they are spies and that they should prefer death at their own hand rather than suffer capture and possibly compromise the spy mission. If one can believe the newspaper reports (and this is all the information I have), this was true of our U-2 flights over Russia some years ago. Whether spies use the self-destructive devices or not is always another question.

Certainly there are no international cases of one nation suing another in international courts of justice, nor are there treaties about the way in which the "high-contracting parties" shall handle spies. The answer probably is that everybody is afraid to admit publicly that it goes on. So long as this is the case, drafting any kind of legal rule or developing the principles to govern nations in such cases is very difficult.

The problem today with our remote sensing techniques is quite different, however, and perhaps the U-2 incidents simply dramatize the gray area that exists between old-fashioned spying, which customarily all nations felt fully justified in resisting by the fullest possible force (including destruction of machinery and death for the operators), and the new techniques which make it very difficult for the spied-upon nation to realistically object or in any way impede the spying nation's activities.

With respect to flights of air-lifted craft an international convention not only regulates commercial flights but also undertakes to set forth the principle that no state shall fly its state craft over the territory of another nation without asking permission. Although there is nothing specifically providing that such aircraft may be shot down, it has generally been assumed that this is a legally acceptable action to enforce this stated principle of the inviolability of air space. Mention of the U-2 incident in discussing this question is not meant to suggest that the U.S. is the only one which has in effect violated this principle. Even from what little one can read in newspapers and piece together from incidental information, it is perfectly obvious that other countries, including Russia, have done similar things. I am not privy to government secrets and so do not know whether or not we have shot down such planes but I certainly would not be surprised if we have.

Any concept of territorial integrity actually involves two boundary questions. One is, how far out into the surrounding ocean does a nation's sovereignty extend? At the present time there is a significant international disagreement on where this international water begins, ranging from 3 to as much as 200 miles or more, depending upon the particular nation asserting the right.

There is a second and similar disagreement on where the air-space outer-space line should be drawn, even if one mistakenly assumes that all agree a line should be drawn at all. There is a growing recognition in international circles that a territorial sovereignty line should be drawn somewhere between the point at which there is insufficient aerodynamic lift to support flight of manmade vehicles and that altitude at which craft will orbit the earth, at least for a time. This probably means somewhere between 50 and 100 miles above the earth's surface.

As yet there is no international agreement with respect to this matter but serious discussions are underway under the auspices of the U.N. There at least is an agreement, now formalized in the outer-space treaty, which provides that celestial bodies and outer space shall not be used for military weapons purposes. Obviously, this does not deal directly with the question of spying.

In discussing remote sensing techniques and the possibility of international spying, one preliminary assumption should be stated about the significance of territorial jurisdiction. No nation under attack, or one which seriously thinks it is under attack, will worry much about international boundaries, whether they be 3, 12, or 200 miles at sea, or 20, 40, 70, or 100 miles out into space. If a nation thinks there is a serious danger, defensive actions probably will ignore boundary lines. I feel confident that the U.S. and Russia, as well as other nations, would undertake to defend against the attack by trying to destroy the vehicle carrying the offensive military threat even though this were well outside the traditional internationally defined territorial limits. One could say this is almost customary international law, I suppose. Therefore, this particular problem should be set aside. There remain significant problems concerning boundary lines and vehicles collecting information—spying, if you will.

It is possible to assert that within this last decade of orbiting satellites there has been an acceptance in international circles of the right of orbiting vehicles to pass over the territory of another country. Neither Russia nor the U.S. has seen fit to lodge any formal protests about such overflights and there is very little that any other country can do about it, even if they did desire to protest. If this is accepted as customary international law, then what is the significant difference between orbiting craft and overflights carried out by aircraft at 20 miles up, whether manned or unmanned? Assuming that cameras have been carried in aerodynamic vehicles to overfly the territory of other nations, it is clear that information could have been collected and undoubtedly was in many cases. Is this truly different from the kind of information that could be obtained by satellites with even optical cameras, let alone the sophisticated infrared sensors that have been developed in more recent years?

One possibility is vulnerability to destruction by the subadjacent nation. Although vulnerability probably is a closely guarded secret, satellites may be just as vulnerable and in some cases perhaps even more so than high-flying aircraft, or even low-flying aircraft. I suggest that vulnerability to destruction is not a valid test of legality of overflight or orbit.

Because of the inherent characteristics of infrared sensing techniques and their ability to sense movement of even relatively small heat sources, such as small groups of moving men, we can obtain information that would never be apparent to the optical camera, no matter how good the lenses. At lower levels the accuracy of infrared undoubtedly can be improved in many cases, but at least a great deal of information never heretofore available is now readily obtainable by overflights of infrared spying satellites. I have no hesitancy in speculating that both the U.S. and Russia are involved in such activities. Let me hasten to add to that, although only with the same accuracy with which Will Rogers used the statement, I only know what I read in the newspapers. Aside from any technical superiority we may have in infrared techniques, I suspect the only significant difference between ourselves and Russia is that at least a citizen in this country can safely (?) make such a public statement and in fact is willing to do so. I suspect this is not true in Russia. The problem then is what do we do about this as a matter of international policy?

One thing we can do is continue to pretend that officially we know nothing about such spying, by ourselves or the Russians. I doubt that we fool very many people, except possibly some of our own citizens, in following this policy. Being a constitutional lawyer, I therefore would quickly conclude that this is the last group of people that we should try to fool, or even allow to remain uninformed.

Another possibility is to make it known to the world that we are doing it, and that we have every reason to believe that the Russians are doing it. This certainly is the kind of talk that goes on in the corridors, if not in the public forums, not only in this country but in others as well. To make such an announcement might do something to enhance a now somewhat tarnished reputation for working aboveboard. It seems to me such an announcement about our activities, coupled with one that we know or strongly suspect the Russians are doing the same, is not giving any information of any value to the other side. If we do not harm ourselves significantly, it behooves an open society to let its citizens know that this is our official government policy, if that it is. I doubt that there is any militarily important information contained in this simple statement.

In any event I suggest that there are two alternatives, either of which is far better than the present posture of either ourselves or the Russians or any other country who enters the spy satellite business. The first, and probably most satisfactory, but I think unrealistic in terms of achievement in the near future, is to agree on an international treaty by which we would attempt to regulate this type of spying. I will have more to say about that shortly in connection with a conclusion about the spying problem.

The second alternative is one that may seem somewhat one-sided in favor of the opposition, but I seriously question whether this is the case. In any event, what loss we would suffer is not *nearly* as important as the psychological gain we might well derive from following such a policy. I would like to throw open for discussion the possibility of our making publicly available to the world in terms that can be understood by intelligent people in *any* country, just what kind of information we have obtained from our spy satellites about Russia, China, Cuba, or whatever country we happen to be spying on at the moment but with which we are not engaged in a hot war.

Those who place so much store on military secrecy and intelligence will be greatly disturbed by this suggestion. They feel the amount of information we have about other countries is itself very secret information. I assert the proposition, however, that it is more important that the people of this country know what we know about Russia or China, or whatever country you care to name, and that it is more important that the people of the world know as much as possible about what these other countries are doing to the best of our ability to find out, than it is that we keep from our opponents in the international conflict the information we know about them. At least in cold war as contrasted with hot war situations, it seems to me that the psychological posture that we take and the influence that we may have in the world would be much greater if we honestly made available this information for the whole world to see. If we can believe the claims of government leaders over the last 10 or 15 years, that we have the necessary force to overwhelm any country, even the size of Russia, if it should decide to attack us, then the details of how much information we have about them is not nearly as important as sometimes our censors would have us believe. With apologies to James Thurber, I would suggest that those of us who have such a neurotic fear of anything remotely communistic are suffering from a bad case of false insecurity.

I suggest we take a new approach to this ancient problem. In my opinion, this is so for two very basic and related reasons. It is almost impossible to keep either the good or the bad about our own country secret from the rest of the world at least for any significant period. I hope this continues to be the case because the very forces which threaten secrecy of this kind are those which make our kind of open society operate effectively. I think this not only because it is a matter of administrative efficiency; I also think our type of government needs sharp criticism, close surveillance, and crucial help from many outside the government. I say it also because I believe there is a psychological factor here that is a very important element in what success we have had as a nation and as a people. I think the attitude of openness must be manifested in as many ways as possible, including, in many respects, even our defense department operations. In my opinion, this is crucial for our own psychological health as a people—this is the essence of my argument here.

One then needs to ask the question, what's so different between an orbiting spy satellite and a U-2 flight? Part of the reason we have had no real objections to overflights or orbiting spy satellites undoubtedly is because nobody could do anything except object verbally, short of going to war with us. Admittedly this step seems fairly unrealistic, at least to any country other than Russia or China—and perhaps North Viet Nam! It may not seem silly for an Israel or an Egypt to be very concerned about the possible *offensive* military potential of an overflight of an airplane. I would maintain that today, with the kind of sophisticated missile weapons which Russia, the U.S., (and I suppose in the near future, China) have, the military threat of the overflight of an airplane is really not very great—or at least that of a single airplane certainly is not. Therefore, I would draw the conclusion that the real problem in such things as U-2 overflights is not the threat of *offensive* military weapons but rather dislike of information-gathering by a foreign nation. If this is the real basis, there is no realistic line to draw between an aircraft overflight and a spy satellite overorbit. One of these days the difference will not even be in terms of feasibility of knocking the spy vehicle out of the sky, though this may always be somewhat more difficult for the satellite than it would be for the airplane of the traditional sort.

Yet realism forces an admission that the psychological threat seems greater from a low-flying spy than there is from a high-orbiting one. In my view, however, this is but the vestige of the too-near immediate past when an enemy plane was something to be feared because of its offensive threat or possibly spying potential in a "hot war" situation.

In any event, we now have the capability of spying effectively at high altitudes and need a national policy decision. I suggest that we seriously entertain adopting the following policy: we announce publicly that we carry on surveillance and that we will make the information available not only to our own intelligence agencies but also to the world at large, for whatever use they wish to make of such information. I will return to this basic theme in connection with peaceful surveillance programs.

One of the obvious differences between the orbiting satellite and the low-flying airplane or the surface vessel is that there is a serious problem of traffic control on the sea and in the near adjacent air space, at least in the populated centers of major countries. This, however, is not the total concern that one nation has for the actions of another nation's vehicles, whether on the sea or in the air.

I do not wish to emphasize the Pueblo incident but one must not ignore the fact that it did happen, and it so well illustrates the kind of problems that could arise but which as yet have not been arising between the major powers of the world. They all have been carrying on such spy observations in one form or another, primarily from vehicles on the surface of the ocean, but also in many cases from low-flying radar or camera-carrying airplanes. It is only that the Pueblo incident does dramatically emphasize just what can happen when nations play the kind of game that all of us have been playing. Lest I be accused of pointing the finger only at the U.S., let me quickly remind you that it is perfectly obvious, as the newspapers have reported quite frequently, that Russian ships regularly carry on surveillance not only of our military operations such as those of the Mediterranean fleet and sometimes those in the Japanese Sea, but also off our own coastline. Without knowing any facts, I am confident (in both senses of the word) that we make similar observations off the coast of Russia, though we might be doing it by submarine rather than by surface vessel in many cases. I would assume that Russia took no more active role than it did in the Pueblo incident largely because it is itself vulnerable to similar actions. Just recently it was reported that a Russian vessel hovers off the mouth of the harbor at Kiev, Germany, and the superstructure of the vessel is such that it makes its spying mission clear. All people and countries play the game both ways, including that of never expressly admitting publicly that such things go on. In fact, nations carry the concept so far that in international conferences which allocate radio frequencies they even find a euphemistic synonym for military radar. It all seems so silly because countries know the game is being played. About the only people who really are at all fooled by this kind of thing are those at home who happen to think that surely the government must be doing the right thing, though they really don't know what it is the government does. The Pueblo type of incident poses for the major countries of the world the question of what they plan to do about this kind of activity.

I am not so naive as to think that we either will or should give up attempts to collect this kind of information. On the contrary, I think it is important that we continue to do so. It is equally important to other countries to do the same thing. A real danger exists, however, that some small country (acting independently or as a puppet) will change the rules on the game started by the big boys. I suspect this is exactly what happened in North Korea's case. Then there are extremely embarrassing, obviously delicate, and in fact life-and-death consequences, as I am sure is the case with the Pueblo incident. I mention it only as an awesome and dramatic example of exactly the kind of dangerous problem created by the big powers when they keep brushing the international situation under the rug. I submit that it is time we discussed it in the open. We should take some realistic steps toward a solution.

We should prepare a draft treaty and attempt to get agreement as to just what boundary lines are going to be established. We should agree on what kind of surveillance will be permitted off shore or above the surface of the earth without retaliation and at what distances. For these purposes, arbitrary, sharp lines are the only ones that will be satisfactory. Surely they can be determined so as to be satisfactory from a technical standpoint. Failing this, I do seriously suggest that we adopt the policy of making available to the world the information which we do discover from our own spying. If we do this to Russia and China, Russia in the immediate future, and I suppose China in the near future, will undertake to reveal the same kind of information about us. But is this intolerable to us? I trustfully assume that this would not violate any senatorial-type gentlemen's agreement about not revealing their dirty linen to the world if they keep mum about ours. If we believe a free and open society is best for us and the world, we should make this small contribution at very little risk to ourselves.

As I stated at the beginning, the use of remote sensing techniques for spying is merely the most dramatic policy and legal problem arising from this technologic development; it is not the most important for the future of mankind. Much more fascinating and in some ways much more complicated is that of remote sensing surveillance for completely or at least largely nonmilitary uses. Even if the policy problems are somewhat more complicated in some ways, this is an activity about which surely there can be a quicker consensus, even within our own country, as to what ought to be done.

Many of the papers being discussed at this symposium and the programs of concern to particular members of our panel fit into this category of peaceful surveillance. Within the limits of this presentation I can only generalize about the technical possibilities. Then we can consider some of the social policy and legal problems that must be solved. To anyone concerned with the health and economic welfare of the people of the world, there is no more fascinating possibility than that of using remote sensing surveillance for purposes of geology, geography, agriculture, and similar activities of importance to all mankind. Those of you who are deeply involved in this work must be extremely challenged by this kind of remote sensing activity. Even to the person who is not technically oriented in the engineering sciences or even in those involved in food production, mining, and similar activities, the potential good for all of mankind from much more complete information about the natural resources of the world is just plain exciting. The problem, of course, is, how do we make use of this information for the benefit of all mankind so that the greatest number can attain that decent condition of living I spoke about at the beginning?

Because you all are aware (or at least shortly will be as a result of the papers delivered at this symposium), of the benefits that can be derived from geologic studies of mineral resources throughout the world, of the existence of schools of fish and where the fishing fleets should search for them on the seven seas, of revealing much earlier and much more accurately what a country's crop yield for the year will be, let me turn almost immediately to the kinds of policy and legal problems we face in launching these programs. At the outset, however, and primarily for the nontechnically oriented, some of the technical aspects of this type of surveillance must be mentioned.

In many cases, particularly if infrared techniques are used, the country being surveyed does not need to be aware of this fact. Actually, unless they have very sophisticated devices, they really would have no way of knowing that the satellite was even present. Equally important is the fact that the raw data transmitted to the ground by radio signals (if this were the device used) would have no intelligible significance to most of the countries of the world who would be quite incapable of interpreting them to even the intelligent nontechnically trained person. Lastly, I assume that there really is practically no area of the world whose natural resources, at least of the kind mentioned above, could not be made known to the United States, or to Russia for that matter, if we want to obtain that information and put the necessary satellites in the required orbits.

One initial obstacle in carrying out programs of this kind is an important one, at least psychologically. Many countries of the world, particularly those that are short on capital and technical know-how needed to carry out their own survey of their natural resources, might also be the most sensitive about somebody else discovering this information without their permisson. Therefore, I suspect that most of the countries of the world would be pretty unhappy if the U.S. started collecting this information without their knowledge, even though they have no capability for physically interfering with this kind of surveillance. A recognition of this concern may well be the motivation for our bilateral treaties with countries such as Mexico and Brazil to develop programs of this character. This obstacle does not apply to these countries, therefore, but in most cases the same satellite that would do these things for these two countries obviously has the potential of doing it for many others. This will not be true for overflights of airplanes carrying such equipment, of course, but in the not-too-distant future I suppose it is quite realistic to expect that much of this kind of work could be done by quite remote sensing devices in orbiting satellites. With satellite surveillance possible, much of the world may not be satisfied with a simple assurance that we will not collect information about their natural resources without their permission.

This raises an initial policy question which we must

face, i.e., will we orbit such satellites and almost inevitably cross some countries which do not want the information collected? If we do orbit such satellites, shall we collect the information whether or not the other countries want it? These are not easy questions to answer. We cannot afford, however, to take the position that because the question is difficult we should let it rest until the full impact of this kind of very remote sensing is upon us. I think the potential good to be derived from this kind of information gathering is too great to deny it to the rest of the world. We must decide how do we go about regulating this type of remote sensing and disseminating the information that is to be gathered.

From a long-range point of view there surely is no answer but that we should collect this information whether or not it is desired. In the name of efficient utilization of such relatively expensive and potentially such beneficial equipment, we must collect the information, although it does raise a question of financing to be discussed in a moment. Before we reach that question, however, we first must face the basic policy issue of whether or not to even collect the information. I have stated my position but I recognize that there is room for argument.

In general I have believed that we should not force our kind of technology and our kind of social standards on other countries. I think I can still consistently say that we should collect this kind of information. Essentially it is similar to the kind of scientific research that we do all the time, primarily in the name of pure science and basically because of our search for knowledge, not for any kind of immediate political goals. I think the real dangers in this area are in the abuse of the information and it is to these that I will direct my attention here.

Before undertaking to discuss some of the potential legal and policy problems which could be created by these surveillance programs, I want to reiterate an earlier statement. The use of these techniques should not be discouraged! On the contrary, they have so much promise it is crucial that we develop them and use them to gather information of the kind suggested, not only for our own good but for that of the rest of the world, in particular the developing countries who need so much help. The problems I want to suggest and the potential abuses I would like to enumerate are there and we must learn to face them but I do not mean to suggest that we should not use this new technology.

Assuming that geologic surveys by remote sensing techniques could be very helpful in discovering mineral resources, what should we do with the information? One position could be justified along the following lines. We surveyed without invading the territory of another nation. Since we made the effort to develop the techniques, put the equipment in orbit, then recorded and analyzed the information received, we should be allowed to exploit the results. Leaving aside any military significance of natural resources data about other countries, we would have at least two alternatives based on this fairly hard-nosed approach. We could offer such information to American companies which might want to negotiate with these foreign countries for concessions leading to economical recovery of these mineral resources. If only our American company had this knowledge and the foreign country were not even aware that we had the knowledge, the company might obtain somewhat better concession rights from these countries by pleading grave risks of failure than if they also had the information. The history of our country is replete with examples of exactly this kind of early American let-the-horse-trader-beware ethical attitude. The American company certainly would not be lying, it would not necessarily even be giving misleading information, and it would be hiding nothing from the other country that they could not find out for themselves if they cared to commit enough of their economic resources to discover this information themselves.

Without being privy to knowledge of how American companies negotiating such mineral concessions with countries abroad actually operate, it seems fair to assume that the typical American business, at least in the past, has not been in the habit of giving out information it happens to have of a favorable character about just how good are the chances of finding some mineral resource. In the past this undoubtedly was the traditional approach of most countries in exploiting resources in other nations who were unable to develop their own natural resources. I submit that in today's world this smacks too much of the colonialism approach of the 19th and the first two-fifths of this century. I think *responsible* businessmen today, at least in America, do recognize that this kind of hard-boiled attitude in the long run breeds not just minor difficulty and suspicion but real trouble and perhaps seeds the attitudes and discontent which lead to revolutions of whole economic systems. It is my opinion, therefore, that we must not permit this scientific tool to be used by the U.S., at least, in this way. I would maintain that whatever our opponents in the world conflict do, we should not adopt this policy.

This leads to the second hard-nosed approach which at least is more palatable and, in my view, much more ethical. We could offer to sell such information to any country which desires to buy it. These programs do cost money. There is a maze of data collected and the raw data itself is of little value without the interpretation made possible only by the use of very sophisticated equipment and the time of very talented people who are in scarce supply. Therefore, we can make an excellent argument that we will not release the information to anybody except as they are willing to pay for it and we can give assurances that we would not abuse our custody of this information, although it is true many countries would not choose to believe us. If we announced this as our policy, it is crucial that we literally follow it. We as a nation must be able to live comfortably with our own conscience, even if other countries do not believe we are that ethical.

I would like to suggest a third alternative which seems to me more realistic under today's world conditions. We should collect such information, interpret it, and then disseminate our findings on the basis of the scientific value of the experiments, both for developing sensing techniques and for analyzing the natural resources of the world. I am not so naive as to think that political policy judgments, which are an inevitable part of the international diplomacy of this or any other major country, will ever be completely removed from such operations. I am asserting, however, that whatever information we do collect and whatever interpretations we do make should be made available for all the world to see. Scientifically, we can and will not neglect the development of these dramatic technological tools for surveying the world's resources. Our programs will need financial support in any event and with not too much more expense we can make this available to the rest of the world. It can be one of our contributions toward the optimum development of the world's natural resources. This could be a significant step toward the ultimate goal of providing the most people possible with a decent condition of living.

Even assuming we decide to make this contribution to other nations to help them develop their natural resources, there are some serious problems involved in such a policy. Inevitably, choices must be made as to which areas to survey and as to which data to process. The government officials who make the decisions of which programs to support cannot be immunized from the international as well as domestic policy concerns of this country. Therefore, their decisions will always be open to the usual charge that everything we do is done only to further our own political goals. Nevertheless, we must protect our scientists from such pressures as best we can and then accept philosophically the propaganda charges that such decisions are not made on purely scientific grounds. We have been subjected to a great deal of this since we became the major world power after World War II and we now show *some* signs of a maturing ability to accept such thrusts without so much anguish. If this program is to be successful, however, we must attempt to make the judgments of which programs to support on the basis of scientific value and the need for knowledge and not on the basis of which governments happen to oppose communism and be damned with the form of the opposition and the basic ethical characteristics of the government which is so willing to pursue the communists for us.

A much more serious objection to the proposal probably will come from countries who object to any surveillance of their natural resources without their permission. We cannot ignore the fact that with the present growth of nationalism within those areas of the world where there is such great need for development of natural resources we find extreme sensitivity about the overbearing omnipresence of an outsider, particularly a great one. It may be that our program of bilateral agreements is the best program to follow because at least in this way there will be agreement upon the surveillance program. This at least will obviate the kind of objection that comes from those who want to control the release of information about their own natural resources.

The U.S. may be subjected to even more criticism, however. Such bilaterals are very likely to be with those countries with quite advanced technologic development and not with some of those countries who need this help the most. Done on the bilateral basis, one cannot resist nearly so easily the temptation to support our "friends" only. If bilaterals are executed on a wholesale scale, on the other hand, there may be an unrealistic prestige race to participate which was so unfortunate in the early development of our Atoms for Peace program. Too many countries sought a reactor before they trained any people or even set up a program which would assure a continued supply of trained people to run these reactors. In a program of surveillance for natural resources we will do more good, and be subjected to less criticism, with a broad-ranging, scientifically based program which is not dependent upon getting agreement from another nation which commits money and manpower which is almost always in limited supply in many countries. At least such commitments must be carefully scrutinized and this can be diplomatically delicate.

One special problem can arise in this program of surveillance for natural resources. It rather dramatically demonstrates how difficult and delicate may be the question of dissemination of information collected by remote sensing techniques. Undoubtedly there will be other examples but this is the most obvious at this moment and will point up the legal and policy problems involved. With the kind of sensing devices we now have it often should be relatively easy to determine where schools of fish can best be found by the fishing fleets. Several treaties control fishing in international waters but none contemplates the kind of information as to fish concentration which ought to be available now through remote sensing techniques. This raises a very interesting problem which we will have to face if we undertake to make this type of surveillance: Should this kind of information be made available also to the world for all to use with no competitive advantage to the American fishing industry? We could get an advantage over the fishing fleets of other countries and surely our fishing industry often operates at a significant disadvantage. Remote sensing surveys of fish locations made available to American fishing fleets could be a valuable competitive advantage. Much fishing is done in international waters and, therefore, is not subject to the control of a particular nation under any of our traditional concepts of territoriality, although there are disputes with some countries as to just how far out this territorial jurisdiction goes for fishing purposes. These fishing operations usually are pursuant to treaty regulations aimed at preserving the supply of this vital source of food. Infrared sensing can make it much easier for fleets to get their catch the fastest and cheapest, quite aside from any quota system that might be imposed, so there would be no violation of these treaties. But this is a great competitive advantage.

This is a very difficult policy question to answer. The American fishing industry needs help, and if our government has made the information possible, arguably Americans should reap the primary benefit of it. To remain consistent with the basic policy of making our information collected by remote sensing techniques available to the world, this kind of remote sensing could be left to the fishing industry. Any other country in the world which cared to use the techniques we have already developed could do so but it might be difficult and expensive to duplicate the equipment. A greater difficulty arises from the fact that our natural resources surveillance might well make such automatically available. Then we must face the dilemma of what to do with this information. One easy out, of course, is simply to delay publication of the information until it is too late for fishing fleets of other countries to make use of them. This kind of subterfuge is not very palatable, and a great power trying to provide some kind of moral leadership for the world ought not resort to this stratagem.

Perhaps the only reasonable solution is to tell the American fishing industry that if it wants this information ahead of others they will have to carry out their own surveillance. Another alternative would be to have the government offer to sell such information to anybody who cares to pay for immediate processing, regardless of whether or not they are American. One could look to our weather satellite program as an analogy for this general position. There is not the same competitive factor present there, however, nor is there in the navigation satellite program. Those raise questions primarily of sharing expenses.

One other type of problem ought to be mentioned before we turn to domestic uses of remote sensing techniques and the policy problems inherent in their use. It seems quite possible that remote sensing techniques, even from orbiting satellites eventually, will enable us to gather a great deal of information about such things as the existence of forest fires otherwise undetectable by standard optical techniques, or the existence of diseased plants before they would even be discernible to the naked eye of an observer in the field, to mention only two dramatic examples. In the normal course of infrared surveillance such information could be collected. This raises the very serious policy question of whether or not we should undertake to make such information available to other countries where we find such conditions existing.

There are some common law tort concepts about the duties of good samaritans which might be applied by analogy. The force of these rules, however, would warn us to avoid involvement. I find these analogies not very apropos in the international area and their application here is not very likely. On the other hand, we must not ignore the implications of undertaking to transmit such information to other countries and then failing to do so sometimes, perhaps for reasons of scientific priority in processing data. Nevertheless, such information could be vitally important to another nation if we should discover that it is on the verge of having a devastating crop disease, famine condition, or the existence of some natural disaster such as a forest fire. Perhaps further development of some types of remote sensing equipment will help predict disastrous seismologic activity. This raises the serious problem of whether or not we should assume the responsibility of passing all such information on to the authorities in the possibly affected country as soon as possible.

We might be open to charges of sending misleading information, or failing to transmit it for political reasons. We could be subjected to accusations of favoring some nations as against others in the degree of diligence we use in collecting, processing, and disseminating this information. Because of the risks, therefore, clearly the government is the one to take the initiative and assume the responsibilities because it has the best resources and techniques for protecting itself in the event of some mistake. In any event, the information is so important that it ought to be made available whenever received and to whomever the information should go, regardless of their politics. If we can disseminate weather information which obviously can be wrongly interpreted and take the chances involved in this, then surely we ought not to be reluctant to pass on the other kinds of information suggested above. Such programs should be vigorously pursued and the widest possible dissemination should be our goal.

Unfortunately, here again, in some cases the affected country might well want to keep the information secret but it seems to me that we might take the traditional position of academic scientists; our purpose is not to censor information, but rather to make it available. It is important information, no ulterior motives are involved in making it available, and more harm will be done by suppression than by release.

I do not mean to be unmindful of the possible political ramifications within the halls of Congress and various government departments of some of the suggestions made. I would maintain, nevertheless, that this is a time for searching reevaluation of many of our post-World War II international policies. This area of remote sensing should give us an excellent opportunity to experiment with some different approaches based upon fundamentally different philosophic concepts than perhaps have guided us in some cases in the last quarter century.

If Americans have anything new to contribute to the conduct of world affairs that even aproaches uniqueness, it is not in the degree of power that we happen to hold in our hands at the present time. The degree of power is orders of magnitude greater than that of past world leaders but the genius of the American approach is in one of greater freedom, more directness, and less under-the-table diplomacy than has been true throughout recorded history. We must not ignore power politics but we must go beyond it. I would assert that such an approach is as important to our *own* psychological well-being within this country as to the rest of the world. We must take ethically justifiable positions and not conduct ourselves as do others with whom we happen to be in conflict. Survival is crucial. We should not knuckle under. But our position of power is such that we clearly can afford to worry about the ethics of our exercise of power.

The monetary cost to us of the suggestions made above are relatively minor. In the long run they might even be much more successful in influencing other nations than much of the aid programs we have been following the past 25 years. Billions of dollars have been spent and some have lined the pockets of a few local military leaders. The remote sensing area offers us a magnificent opportunity to take an entirely different posture and one that is relatively above politics, even if bureaucratic competition occasionally creeps into our own domestic management of such programs. It is hard to make much profit out of the kind of information we would be transmitting to others.

With this program we might well be able to regain some of the reputation we enjoyed with many of the developing countries prior to World War II. With some sad exceptions, in general we were not the exploiters, we were not the oppressors, we were not a colonial power. Mostly we were accepted as wanting to help with no particular expectation of any immediate return to ourselves. Remote sensing gives us another opportunity to help the rest of the world on the former basis. We can help the developing countries make use of their natural resources and provide a decent condition of living for all regardless of where they live and what their political, economic, or religious beliefs. Our problems in the international arena will not disappear, but remote sensing programs might be the catalyst toward the development of an entirely different psychological approach by the United States to handling world affairs.

This is not a complete catalog of all the international problems that can arise from the use of remote sensing techniques, nor is it a detailed presentation of just exactly how any one of these problems should be handled. It is enough, however, to raise for serious consideration some of the fundamental issues which this country faces in international relations. It is now time to turn to some of the problems that can be expected from the use of these same remote sensing techniques, within the boundaries of the U.S.

In discussing domestic uses of remote sensing some additional devices not used in surveillance programs discussed above should be mentioned. Even ultraviolet and ordinary optical cameras may be used at a significant distance from the target and quite often without knowledge of the target. Also, many of the now commonly known bugging devices are really in essence remote sensing equipment. These range from standard telephone bugging, through spike mikes, to much more sophisticated devices. Now we can add gamma ray probes, radar, ultrasonic equipment, neutron activation analysis, and even infrared.

The extent of misuse of some of this equipment in the United States is frightening, although we have no very good measure of just how extensive it is. It is used not only for private detective work in family conflicts cases, but also in the extensive industrial spying activity now rampant in this country. Some of the equipment would take a very competent electronics engineer to assemble but much of it can be done with fairly rudimentary knowledge of electronics. Clearly the equipment is readily available.

Brief stock should be taken of some of the things these remote sensing devices can do. For example, infrared sensors can detect an individual person from a considerable distance away such as from an airplane, so long as the person is not in a building. Such techniques might be useful in analyzing the extent of rioting and aid in its control. Most rioting goes on outside buildings. From a distance of 10 miles laterally and 20,000 feet altitude an airplane can track a single car. Think of law enforcement possibilities here. If one were concerned about illegal poachers, it may be possible to make use of devices which can measure temperature differences down to one one-thousandth of a degree centigrade under certain circumstances. This also opens whole new possibilities for law enforcement in this area.

Again, it might be possible to take meaningful infrared pictures for purposes of burglary detection, embezzlement situations, smuggling at night, etc., all without knowledge of the suspect. It is also possible to determine by infrared techniques whether a person at some distance and separated by a wall has even as little as a one degree temperature variation from normal. The possibilities for tracking the movement of people under these circumstances for certain legal purposes obviously are great. We can measure the radiation of pictures and thus determine whether we are looking at real masters or fakes. It is possible to measure radiation from coins and even be able to trace the mines from which the metal was derived. There seems to be no reason why we could not identify the dirt on gold bullion as coming from Fort Knox. By the same token it is possible to identify a piece of hair of a specific human being and even differentiate between the hair of twins. Even without access to information about capabilities of classified military equipment it is not hard to guess that temperature variations in water have been created by all kinds of disturbances and we can make these determinations some time after the disturbing force has left the vicinity. It might make it very easy to track smugglers this way or it might make it very simple to reconstruct paths of ships involved in marine accidents quite some time after the vessels have passed.

One other possible use of infrared sensing devices is to measure the pollution of air and water. This opens significant possibilities for detecting pollutants in both air and water that certainly could not, by older devices, be seen or otherwise sensed, and might not even be identifiable soon enough to be connected with a particular source. The technical literature indicates a great possibility for much more accurate surveillance of pollution sources.

One relatively minor legal problem which is unique to domestic uses should be mentioned at the beginning. With the degree of remoteness we were assuming in the international surveillance area there really was no danger involved in the use of remote sensing equipment. At the domestic level this may not be quite so true although in most cases there will be no problem of danger to the surveillance object whether it be a person or property. On the other hand, one must not ignore the fact that radar probably can fry an insect at several thousand feet and therefore harm could be caused to more valuable animals. If ultrasound can be used to cut through steel, it is not unrealistic to feel that it could cut through a man. If ultraviolet equipment is active, then it can blind a person. These are fairly unusual problems, however, and for the most part even these domestic, fairly close-range remote sensing devices will create no positive physical danger in and of themselves. But there are much more troublesome concerns.

Orwell's *1984* is not as far away as one might have liked to think. Domestic uses of remote sensing devices can shake one a bit. Even the above brief summary of potential uses of this equipment indicates how clearly we are moving into an era when we have the means to help law enforcement agencies investigate and prove their cases in a dramatically easier and more certain manner. As with the computer data center concept, however, one cannot consider even these admittedly desirable social goals without taking cognizance of the possible erosion of our concepts concerning the right of privacy we have come to cherish in this country. Obviously, even more serious problems arise when private uses are made of these devices to invade privacy. Let us first turn our attention then to the use of these devices by police, and the impact it might have on the right of privacy, whether a legal or only an ethical right.

First, it is clear that existing laws will not handle at all well the problems raised by these new techniques. Existing statutes, such as that making it illegal to bug or even record a telephone conversation except under certain limited circumstances, has no application to an infrared sensing device, or a long-range optical camera, or a long-range directional mike. Even by the most liberal interpretation it is unrealistic to expect courts to stretch the meaning of a statute that far. Obviously some common law tort concepts might be of some possible value for private abuses, but the analogic value is not very great, although in some cases it would be helpful.

When we consider police agency use, however, we are not left solely to common law principles or even statutory protections; we do have constitutional rights which we find in state constitutions as well as in the Constitution of the United States. The due process clauses, search and seizure concepts, and evidentiary rules give some protection against the invasion of what we like to call privacy. This is not the place to attempt a detailed analysis of the constitutional content of the right of privacy. Admittedly lawyers disagree as to just what and how much protection there is for this concept. But a brief summary will be helpful.

Under the federal constitution it is clear that limitations are imposed on both state and federal officials in conducting searches and seizures. Undoubtedly some of these limitations upon law enforcement officers will be applicable to new techniques for invading privacy. On the other hand, we must remember that these limitations on search and seizure are largely procedural in nature and do not give any absolute right of privacy. Suggestions by some justices in cases such as Griswold indicate the possible existence of sanctuary areas whose boundaries may not be violated. One of the problems most likely to arise from use of infrared sensors and similar devices is best indicated by cases which deal with the use of informers and the various bugging devices. The Supreme Court has drawn a distinction in some crude bugging device cases based upon the existence of a physical trespass upon the property or person being subjected to surveillance. With most of the devices we are talking about today there will be no necessity for any kind of trespass in the traditional legal sense. The camera, whether visual or infrared, the directional microphone, and similar devices simply record energy releases from the target. There is no physical invasion, not even as much as radar. One could argue radar beams physically invade the target but there is no physical invasion of the person or property with infrared or directional mikes. Yet equally intimate information may be obtained.

The question we must ask ourselves today is whether or not these new types of devices should be prohibited or at least closely regulated. Detailed surveillance which invades what a person might think is his private quarters without any indication of its presence is just as revealing even if there is no traditional trespass. We may need to develop new rules. It is possible that they will be developed from some of the "penumbra" of the first, third, fourth, fifth, and ninth amendments, as some of the justices stated in the Griswold case, but I do not believe this is sufficient to answer the problems raised. If we are adequately to balance the interests that are involved, Congress must accept responsibility for enacting legislation regulating the use of such equipment by various police agencies.

There will be some problem of Congress's power to regulate state police activities without dependence upon federal limitations found in the fourteenth amendment. I am of the opinion, however, that particularly under the power to regulate interstate commerce, Congress can largely control abuses. The most important question is just what policy conclusions Congress should enact into law. On the other hand, as a constitutional lawyer I cannot afford to minimize the difficulties that face Congress in attempting to regulate state as well as federal officials. Congress can regulate the use even by state trial authorities of evidence collected by federal officials. The difficulty comes in regulating what state officials do with such equipment when there has been no cooperation from the federal government. I would suggest here that through the use of licensing techniques for goods being shipped in interstate commerce Congress can go a long way toward achieving control if it desires to. In addition, one must not ignore the possibility of getting state legislators to enact laws limiting police agencies who are under their control. Let us turn then to the more basic question of just what policies legislatures should adopt.

The following suggestions for controlling use of such techniques by law enforcement agencies, both federal and state, are made tentatively and as a basis for discussion. Since most of this equipment is likely to be manufactured and distributed in interstate commerce, Congress may undertake to condition the interstate sale and transportation of this equipment upon compliance with Congressional standards, possibly along the lines here suggested. Secondly, Congress clearly should pass laws making it mandatory that federal officials live within these rules. Likewise, state legislatures should pass laws making these regulations mandatory for state police officials.

Again on a tentative basis, the following standards are suggested. Until we know more about what these devices can do and how they can be used, the legislature should impose a requirement upon all law enforcement agencies that they register their uses with a central bureau. In this way we can at least keep track of what uses are being made of these devices. The legislatures also should provide for a periodic review of these uses at intervals of no more than 5 years, and probably every 2 years. This review ought to be conducted by some non-law enforcement agency, such as a legislative committee or possibly an outside group of experts appointed especially for this purpose. I mean to suggest that no use of these devices should be made without first registering the proposed use with a central agency, but I am not suggesting that each individual use be registered. Only the type of use to be made should be registered and then only that kind of use should be made of the equipment. To provide for emergency situations, undoubtedly state legislatures should authorize governors, Congress, and the Attorney General to permit uses other than those already registered, but only on express written authority. Next, I would require enforcement agencies to report each actual use even though the type of use has already been registered. In this way we would have some idea of the prevalence of uses in various categories.

Additional standards undoubtedly should be developed to determine the types of cases in which such techniques should not be permissible. I think it would be unwise to attempt to devise an exclusive list of those uses which are permitted because this can be too limiting. It should be possible, however, to decide upon a minimum list of uses that simply should not be permitted.

In addition, limits should be placed upon the conditions under which even a permissible use is authorized. Whatever limitations are imposed on wiretapping and similar devices today surely should be applied. A heated argument is now going on about requiring a search warrant type of procedure to make use of such things as wiretaps. If this is a good policy for wiretaps, as I am inclined to think it is, then like limitations should be imposed upon use of infrared and similar devices. Possibly this procedural protection should be limited to uses of such devices in searching private homes. Obviously there is room for argument.

Policy decisions in this area call for a delicate balancing of needs in a modern, urban, complex society. We need sophisticated information-gathering devices if we are to make it at all tenable for members of our society to live with each other with some faith that wrongdoers will be discovered and restrained. On the other hand, the right of individuals to have some area where they are not the subject of public surveillance is a psychological problem existing in our kind of society in which so many people are jammed so close together. People must have some area where they can let off steam, discuss possibilities, think their thoughts out loud without having to account for them and to explain them to the world at large. Here again we have an area that needs considerable discussion. The new tools for remote sensing simply make the problem dramatically more pressing.

An almost equally important potential misuse of remote sensing devices can arise in connection with what has come to be described as industrial spying. The extent of this practice can never be accurately estimated but we do know that it has been going on almost as long as business has existed and that with the development of new devices in the modern competitive world the extent of it is great. What written information is available and interviews with private detectives indicate that there is a fantastic amount going on, partly in the form of some very sophisticated telephone bugging devices such as the one which permits a person to bug a telephone and then thousands of miles away dial that number and just before the first ring occurs play a certain harmonica note, thus preventing the phone from ringing but turning on the bugging device which picks up any conversation in the room and transmits it along the telephone line to the spy. (The one thing I did not find out was whether the telephone company has any way of at least charging the man for the call!) Some of the more sophisticated James Bond type of equipment permits one to listen in to all conversations in a room through hidden tiny radio transmitters.

Consider the consequences of such spying in even a fairly humorous situation. Today professional sports is big business and spying is prevalent, it is charged. Think what would happen if secret practices could be observed by infrared equipment in an airplane 20,000 feet in the air and some distance away. The movement of the players in executing plays should be obvious since football players, for example, surely produce measurable heat differences. Of far greater significance is the possibility that such devices might even be used to detect the movement of various materials, machinery, and personnel throughout a plant. It might be possible to discover process secrets extremely important to the company being surveyed.

This kind of detection potential raises some new problems for society in attempting to use legal rules to get at the problem of industrial spying. A dilemma is created. On the one hand we usually feel people should be free to look and listen so long as they do not trespass on the property of another. Trespassing on property is required to plant a bugging device or even to pay an informer to infiltrate the organization, both of which are at the present time generally actionable in the courts. The question that is raised with the new sensing devices is that the same information can be obtained by a spying device completely off the premises and without bribing anybody. The infrared techniques are getting us dangerously close to the x ray eyes of the science fiction comics. Very useful pictures can be taken in complete darkness so long as there are materials which are generating heat, and the people being spied upon would not be aware that they are under observation.

Conversations with private detectives also make it abundantly clear that various devices are very much in use for purposes of spying on spouses. Even advertisements in some of the popular magazines make it perfectly clear that these devices are sold for these purposes. The new possibilities for seeing through a thick wall and getting a definable image which would distinguish a man from a woman and would identify the various movements of the heat source as it moved about the room, should create even greater consternation among a fair share of the population.

All of these new techniques raise some serious questions as to whether or not there should be some attempt to regulate their use and hence their abuse. Obviously, there are great difficulties in drafting a fair set of regulations which would permit the exploitation of these devices for the many useful purposes that can be found for them and at the same time preventing their abuse. One approach to the whole problem, of course, is to deny altogether that there is a right of privacy. Perhaps people should not be afraid to have all things they say and do known to the world at large. In our present and imperfect world I doubt seriously that this is realistic. In fact, I doubt that very many of us would care to have many of the things we say and do in private made open to the public, not so much because what we say or do necessarily is in any way wrong or, if properly explained, would cause embarrassment, but because it does put a tremendous burden on us to be sure that things are not taken out of context. The human animal being as nastily interested in gossip and character assassination as he seems to be, the burden of explanation could be great.

In my opinion, we should recognize a right of privacy about a good many things. Such privacy should be protected, particularly from private spying, whether for commercial or noncommercial purposes. This is an important aspect of psychological health in our crowded society. If this is a sound conclusion, then the law must begin to take cognizance of the problem and attempt some kind of regulation. The regulation could take several forms but something should be attempted now. Obviously, regulation will put some obstacles in the freedom that manufacturers and users of such equipment might otherwise have but I doubt that this price is too great.

One simple standard can be agreed on, I believe. Any machine which through an emission of energy could cause serious physical harm should be subject to registration before it can be used. Included in this category might be (1) gamma ray probes which from an altitude of 1000 feet can analyze soil materials, (2) certain types of radar such as those which we think can fry insects at several thousand feet, (3) ultrasonic sound devices which are capable of cutting through steel, and also man, and (4) ultraviolet ray units which can blind a person. The development of x ray machines and radioisotope techniques has demonstrated the need for regulation. As a minimum requirement, all such devices should be registered. Those which are particularly dangerous perhaps should be licensed before use.

Much more difficult to decide, however, is the policy question of to what extent should we license inactive or passive devices such as infrared sensors which in no way cause physical harm but which are very useful for private spying purposes. At least registration of such devices should be required of any who choose to use them. Some of the same objections that have been made against a national firearms registration act by arms buffs might be made but if we move quickly there should be no large number of protestors against registration of remote sensors.

Enforcement will be very difficult because many of these devices can be assembled with not too sophisticated technical knowledge. Obviously, this is not true for some of the complicated devices but many would be capable of manufacture in small laboratories or even in basement workshops. Another approach, of course, is to make use of such equipment for private spying a crime and impose penalties such as imprisonment and fines. With the stakes so high in industrial spying, there is serious question that this will be much of a deterrent to those who care to use the equipment. National licensing of devices manufactured and distributed in interstate commerce with federal enforcement might help. Again, of course, this can be defeated by the manufacture of such devices in the private laboratory of the company wanting to use it without registration. The mere fact that it is not enforceable in every case, of course, does not necessarily militate against imposing such criminal penalties because obviously no criminal law is 100 percent effective in either deterring criminals or in punishing those who do break the rule.

These examples are sufficient to demonstrate again the problems as well as benefits that our technologic developments can create for society. The developments force an evaluation of some of our fundamental values by which our society operates. If privacy from some of the activities described here is an important value, as I think it is, we must then discover the means by which to prevent or at least reduce the amount of abuse when this equipment is in the wrong hands. It can mean red tape and some impediment to legitimate use. I do not believe myself that regulation is too high a price to pay.

In closing I would return to an assertion I made initially—I am not attempting to discourage the further development of remote sensing techniques. The advantages far outweigh the disadvantages, and some of the possibilities for benefitting mankind are truly exciting. Therefore, we need to push ahead vigorously. It is important, however, that we take stock of the kinds of policy and legal problems which are created by the advent of this remote sensing technology and reach some conclusions as to how best to minimize any possible harm to our basic values.

This paper is an attempt to stimulate discussion about these very problems and to make some suggestions which will possibly initiate this discussion, but I have one final suggestion to make. Because the policy and legal problems created by such technologies as remote sensing are so important in the preservation of the basic values by which we would live, I think it is high time that the federal government which has supported so much technologic research in recent years take cognizance of the need in much more concrete terms than has been done so far. It is essential that realistic financial support be furnished for legal research on these problems. Beginnings have been made in the form of recognition that law is a social science among categories that can be supported by the National Science Foundation. There are those in other groups such as the National Research Council, and even in some government departments, who recognize the need for analysis of these problems *while* the technical research is going on. You scientists and the general public can help insist on such financial support.

SINCE THE *days when Caesar Augustus sent out a decree that a census of the whole world should be taken, enumerating the population has been a traumatic experience, both for those being counted and for the census takers. It is increasingly difficult to count our larger and more mobile population. It has been estimated that the 1960 U.S. Census of Population may have failed to enumerate 5–7 million people, and there has been much criticism of the 1970 census. In our country, population numbers are particularly significant for they determine political representation and the amount of financial assistance from various state and federal programs. The population census is taken every ten years. Historically, with a smaller, less mobile population and a slower rate of growth, a ten-year census was adequate. But our modern population is increasing so rapidly and is so mobile that major population increases and shifts occur in much less than this interval. Are there better ways of counting population? Could enumeration be done more frequently? Can remote sensing be used?*

Research suggests that remote sensing can generate some types of census information comparatively easily and inexpensively. Remote sensing would be particularly valuable in measuring major population shifts during intercensal periods by monitoring urban change. Data could be gathered for communities located in areas in which data are very scant because of remoteness, political obfuscation, or insufficient resources to conduct census enumerations. Remote sensing will probably never be able to provide data on personal characteristics needed by the census, but it can provide accurate measurements of built-up urban areas and the number of people who live in them.

42-Urban Spatial Structure Based on Remote Sensing Imagery

ROBERT K. HOLZ
DAVID L. HUFF
ROBERT C. MAYFIELD

Purpose

The general purpose of this study was to determine whether observable phenomena determined solely from imagery secured by remote sensing could be incorporated into formal constructs designed to predict structural characteristics of urban areas. The particular structural characteristic that was selected for the present study was urban population.

The specific objective of the study was to attempt to estimate the existing population as well as to predict the future population of selected urban areas for which imagery was available. If the results of this initial investigation proved promising, then subsequent studies of other aspects of urban structure also would seem feasible.

From *Proceedings of the Sixth International Symposium on Remote Sensing of Environment.* Ann Arbor: University of Michigan, Institute of Science and Technology, Willow Run Laboratories, 1969. Pp. 819–830. Reprinted with permission of the authors and the publisher.

Ultimate benefits accruing from such studies would be enormous. Intercensal estimates could be generated comparatively easily and inexpensively. Furthermore, it would be possible to estimate population and other structural characteristics of communities located in areas for which data are very scant because of remoteness, political obfuscation, or insufficient resources to conduct census enumerations.

Method

The Tennessee River Valley region was selected as the test site for this study. The urban centers used here were chosen from the Tennessee Valley Authority region, as indicated by maps issued by the Authority of its primary watershed and service area (Fig. 1). There were a number of reasons for selecting this area. First, the Tennessee Valley Authority has on file a sequential record of air photographs. Permission to use those photographs was granted by TVA. Second, the Tennessee Valley is an

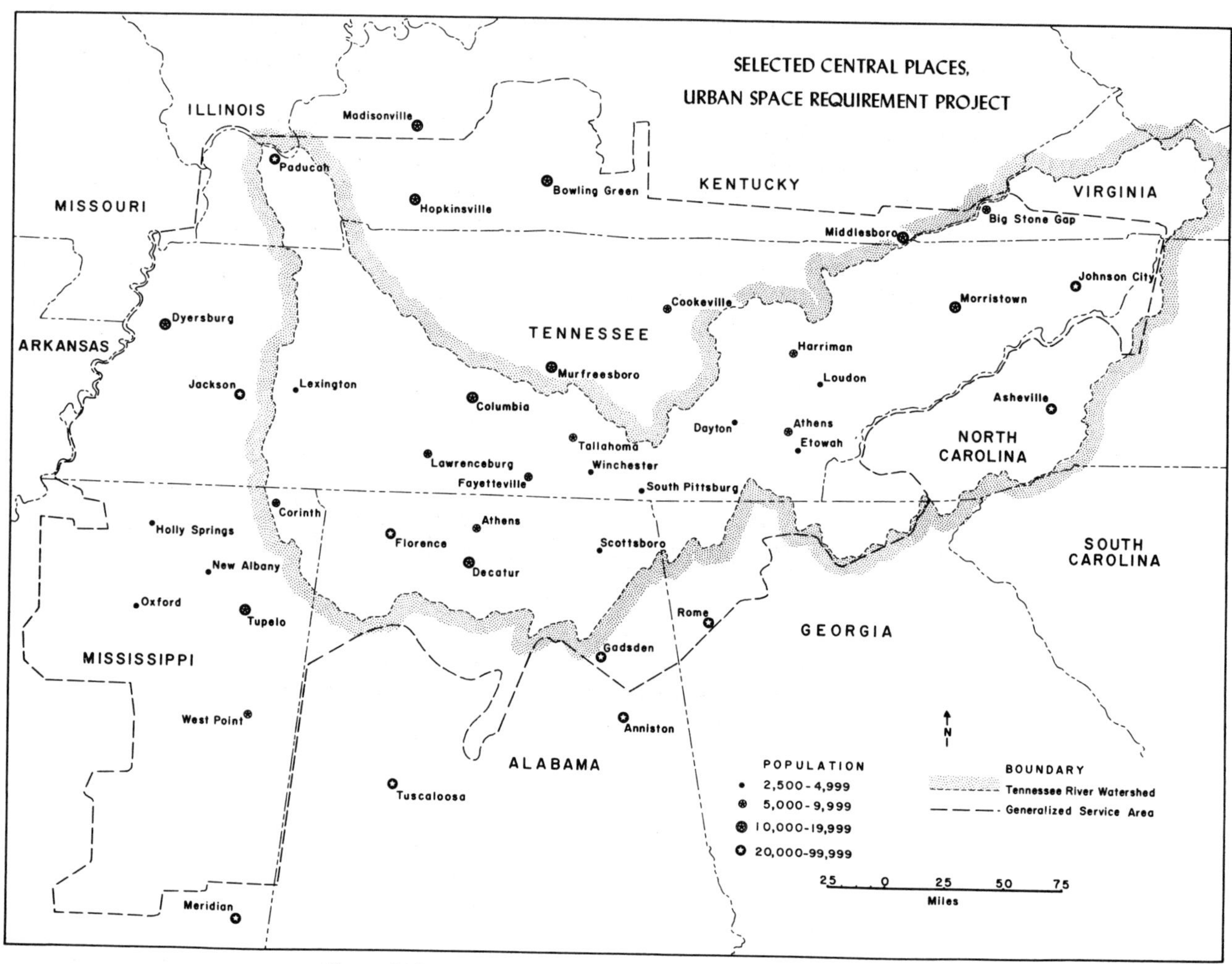

Figure 1 A map of selected central places in the TVA area.

important test site for the Earth Resources Orbiting Satellite (EROS). The information developed in this study will be of particular value and application to future studies of this area utilizing imagery generated from this satellite. Third, the Tennessee Valley contains a large number of urban places of varying size. These places are scattered throughout this farming/forest region and most exhibit a well-defined urban/rural interface.

Forty urban places were selected from this region, ten in each of the four strata listed below:

i.	2,500– 4,999
ii.	5,000– 9,999
iii.	10,000–19,999
iv.	20,000–99,999

The cities selected provided a broad regional distribution within the study area. They also represented the best set of places for which sequential aerial photo coverage was available. Two time periods were chosen: 1953 and 1963. These time periods were chosen so that data extracted from the imagery of the urban places could be coordinated with the U.S. Census of Business for these years. This facilitates possible subsequent studies dealing with other aspects of urban structure. Only urban places were used for which there was photo coverage within 1 year of these base dates, 1953 and 1963. The photo coverage for the first group was secured from 1952 to 1954, inclusive, and for the second group, 1962–1964, inclusive.

Data derived from the aerial photography were selected in order to test hypotheses based on central place theory and empirical evidence. These hypotheses were:

1. The population of an urban area is positively related to the number of links it has with other urban areas.
2. The population of an urban area is positively related to the population of the nearest larger urban area.
3. The population of an urban area is inversely related to the distance to the nearest larger urban area.
4. The population of an urban area is proportional to the observable area of occupied space of such a population.

Each photograph or group of photographs was covered with a piece of clear mylar. The outline of the built-up urban area was then traced from the photograph. The name of the city, scale of the photograph, and number of transportation links counted on the photographs were recorded on this sheet of plastic. The road and rail transport linkages were not required to reach another central place. Rather, all extensions beyond the built-up area were enumerated. This was done twice for each urban place—once for each time period. These mylar sheets were then brought back to the cartographic laboratory at The University of Texas where a specially designed template of 1-inch squares, subdivided into tenths, was used to calculate the area of each urban place for both time periods.

The calculated area of each place was recorded along with the number of transportation links counted on the imagery. Next, the population of the place and the population of the nearest larger town or city were recorded for each time period. Then the distance to the nearest larger town was computed from state highway maps. A sample data sheet for South Pittsburg, Tennessee, is shown in the next column.

City — South Pittsburg, Tennessee

	1953	1963
Area (in square feet)	20,680,000	23,200,000
Transportation links	4	4
Population of nearest larger neighbor (P)	131,041	130,009
Distance of nearest larger neighbor (D)	31	29
Population	2,573	4,130

The relevant data for each of the 40 urban areas in 1953 and 1963 are indicated in Tables 1 and 2.

Table 1 Basic Data for Selected Central Places within the Tennessee Valley Authority in 1953

I.D.	Size Class	Area Sq. Ft. (000)	Rail and Highway Links	Population NLN[a]	Distance NLN[a]	Population
1	1	20,680	4	131,041	31	2,573
2	1	28,578	6	11,527	26	3,680
3	1	16,359	6	8,618	15	3,191
4	1	32,850	6	8,618	12	3,261
5	1	15,034	5	4,199	14	3,567
6	1	26,791	5	11,527	50	3,956
7	1	22,303	7	3,680	35	3,276
8	1	29,720	7	5,253	28	4,731
9	1	22,624	7	30,207	28	3,566
10	1	32,060	5	7,562	15	3,974
11	2	53,583	6	12,605	27	8,618
12	2	25,740	7	9,456	27	5,447
13	2	38,699	7	17,172	23	6,432
14	2	82,756	8	11,527	46	9,785
15	2	65,800	10	19,974	15	6,309
16	2	56,070	7	10,911	36	5,442
17	2	63,680	12	7,913	52	6,924
18	2	33,408	5	19,571	42	5,173
19	2	116,190	9	9,456	16	7,562
20	2	61,650	5	8,618	51	6,389
21	3	53,583	7	17,172	61	11,527
22	3	63,563	7	30,207	57	10,885
23	3	47,393	9	174,307	32	13,052
24	3	72,968	10	124,769	42	13,019
25	3	104,265	9	174,307	43	10,911
26	3	146,790	6	23,879	51	19,974
27	3	79,779	4	124,769	64	14,482
28	3	112,629	6	16,246	26	12,526
29	3	114,906	4	174,307	65	18,347
30	3	56,036	9	12,526	37	11,132
31	4	314,437	7	124,769	111	53,000
32	4	143,760	8	326,037	113	23,879
33	4	126,270	11	53,000	47	27,864
34	4	264,343	8	174,307	145	32,828
35	4	130,981	10	396,000	84	30,207
36	4	249,790	9	326,037	65	55,725
37	4	192,361	8	331,314	67	29,615
38	4	303,038	8	98,271	93	41,893
39	4	204,070	9	326,037	30	31,006
40	4	212,712	9	326,037	69	46,396

[a] NLN = Nearest Large Neighbor.

Table 2 Basic Data for Selected Central Places within the Tennessee Valley Authority in 1963

I.D.	Size Class	Area Sq. Ft. (000)	Rail and Highway Links	Population NLN[a]	Distance NLN[a]	Population
1	1	23,200	4	130,009	29	4,130
2	1	38,321	7	17,221	26	5,151
3	1	34,920	6	12,103	15	3,500
4	1	47,120	7	12,103	12	3,223
5	1	23,160	6	4,145	14	3,812
6	1	62,636	6	17,221	50	5,283
7	1	49,353	9	497,524	45	5,621
8	1	43,740	7	6,592	28	6,449
9	1	49,409	9	34,376	28	3,943
10	1	44,351	6	12,242	15	4,760
11	2	77,170	8	16,196	27	12,103
12	2	43,360	7	10,466	27	6,804
13	2	64,970	9	24,771	23	8,550
14	2	75,475	10	17,221	46	11,453
15	2	80,960	11	29,217	15	9,330
16	2	67,720	8	17,624	36	8,042
17	2	91,320	13	10,512	52	7,805
18	2	24,947	5	26,314	41	4,688
19	2	133,120	9	18,991	42	12,242
20	2	63,280	5	27,169	48	5,931
21	3	134,859	9	24,771	61	17,221
22	3	96,427	9	34,376	57	12,499
23	3	161,481	9	170,874	32	18,991
24	3	133,470	10	111,827	42	21,267
25	3	137,200	12	170,874	43	17,624
26	3	240,720	8	72,365	25	29,217
27	3	96,289	4	111,827	61	12,607
28	3	232,704	8	22,021	26	19,465
29	3	213,030	6	170,874	57	28,338
30	3	112,323	11	19,465	37	13,110
31	4	350,251	9	111,827	96	60,192
32	4	578,340	12	72,365	65	31,649
33	4	155,880	11	60,192	47	31,187
34	4	313,848	10	170,874	145	34,479
35	4	153,240	10	497,524	84	34,376
36	4	345,027	12	340,887	56	58,088
37	4	398,548	9	487,455	67	32,226
38	4	303,510	10	144,422	93	49,374
39	4	358,310	11	340,887	30	33,657
40	4	452,292	11	340,887	59	63,370

[a] NLN = Nearest Large Neighbor.

Analysis

The hypotheses cited earlier were tested using a stepwise linear regression. The principal advantage of such a procedure is that it orders the independent variables in terms of the amount of variation in the dependent variable that is accounted for by each independent variable. The dependent variable in the regression equation is the population of an urban area. However, since we know from central place studies that population is highly correlated with retail and service functions, these latter variables could also have been used as the dependent variable. Disregarding the order of the independent variables, the regression equation is as follows:

$$P_i = a + b_1L_i + b_2P_j - b_3D_{ij} + b_4A_i$$

where P_i = the population P of urban area i

L_i = the number of direct links L between i and the other urban areas

P_j = the population P of the nearest larger urban area j

D_{ij} = the distance D between urban area i and the nearest larger urban area j

A_i = the observable occupied dwelling area of urban area i

For each of the two time periods 1953 and 1963, appropriate values were substituted for each of the independent variables as well as the actual population values of each urban area. The b values, i.e., the constants, in the regression equation were subjected to two different statistical tests. The first was to determine for a given year whether or not the variable with which each constant was associated accounted for enough of the variation in the dependent variable to be explanatory. The second test involved a comparison of these constants for each year to determine if they were the same or whether they were significantly different. The results of this latter test were of particular interest. If the constants were not significantly different from one time period to another, then it would be possible to use the regression equation to predict future populations of urban areas, given appropriate values for each of the independent variables.

The results of the statistical analysis revealed the following order of independent variables and the values of the coefficients:

1953 $P_i = -1711.896 + .142(A_i) + .027(P_j) + 174.997(L_i) - 8.650(D_{ij})$

1963 $P_i = -4628.394 + .079(A_i) + 122.846(D_{ij}) + .016(P_j) + 484.192(L_i)$

It can be seen that except for A_i (area), the order of the independent variables differed from one time period to the other. Furthermore, the values of the coefficients also differed significantly, and for one variable (D_{ij}) the direction (sign) of the coefficient was reversed.

Despite the lack of correspondence in the two regression equations, each was able to explain remarkably well for that particular year variations in the populations of the selected urban areas. For example, the multiple correlation coefficients for 1953 and 1963 were .95 and .88, respectively. This indicates that over 91 percent in 1953 and over 77 percent in 1963 of the variations in population of urban areas can be explained on the basis of the selected independent variables. The correspondence between the actual and expected population values for each of the 40 urban areas for each time period are shown in Tables 3 and 4.

Table 3 Relationship Between Actual and Estimated Values Using 1953 Data

Actual Population	Estimated Population	Difference
2,573	5,199	– 2,626
46,396	38,320	8,076
31,006	37,428	– 6,422
41,893	44,634	– 2,741
29,615	35,410	– 5,795
55,725	43,628	12,097
30,207	28,630	1,577
32,828	40,734	– 7,906
27,864	19,195	8,669
23,879	27,958	– 4,079
53,000	46,640	6,360
18,347	19,473	– 1,126
19,974	20,418	– 444
3,680	3,489	191
3,191	1,768	1,423
3,261	4,139	878
3,567	1,293	2,274
3,956	2,852	1,104
3,276	2,482	794
4,731	3,639	1,092
3,566	3,304	262
3,974	3,797	177
5,447	3,195	2,252
6,432	5,281	1,151
6,309	9,805	– 3,496
5,442	7,470	– 2,028
6,924	9,208	– 2,284
5,173	4,079	1,094
7,562	16,504	– 8,942
6,389	7,722	– 1,333
8,618	7,065	1,553
9,785	11,371	– 1,586
11,527	7,069	4,458
10,885	8,875	2,010
13,052	11,032	2,020
13,019	13,421	– 402
10,911	19,025	– 8,114
14,482	13,149	1,333
12,526	15,570	– 3,044
11,132	7,850	3,282

Table 4 Relationship Between Actual and Estimated Values Using 1963 Data

Actual Population	Estimated Population	Difference
4,130	4,715	– 585
5,151	5,240	– 89
3,500	3,057	443
3,223	4,134	– 911
3,812	1,885	1,927
5,283	9,619	– 4,336
5,621	16,864	– 11,243
6,449	5,748	701
3,943	7,593	– 3,650
4,760	3,802	958
6,804	5,655	1,149
7,805	15,409	– 7,604
4,688	5,202	– 514
5,931	9,094	– 3,163
8,550	8,056	494
8,042	9,274	– 1,232
12,103	8,891	3,212
11,453	12,075	– 622
9,330	9,370	– 40
12,242	15,667	– 3,425
17,221	18,228	– 1,007
12,499	14,859	– 2,360
18,991	19,029	– 38
21,267	17,619	3,648
17,624	19,921	– 2,297
29,217	22,396	6,821
12,607	14,120	– 1,513
19,465	21,107	– 1,642
28,338	24,707	3,631
13,110	14,391	– 1,281
60,192	40,841	19,351
31,649	55,835	– 24,186
31,187	19,681	11,506
34,479	45,394	– 10,915
34,376	30,321	4,055
58,088	40,523	17,565
32,226	46,910	– 14,684
49,374	37,781	11,593
33,657	37,890	– 4,233
63,370	48,854	14,516

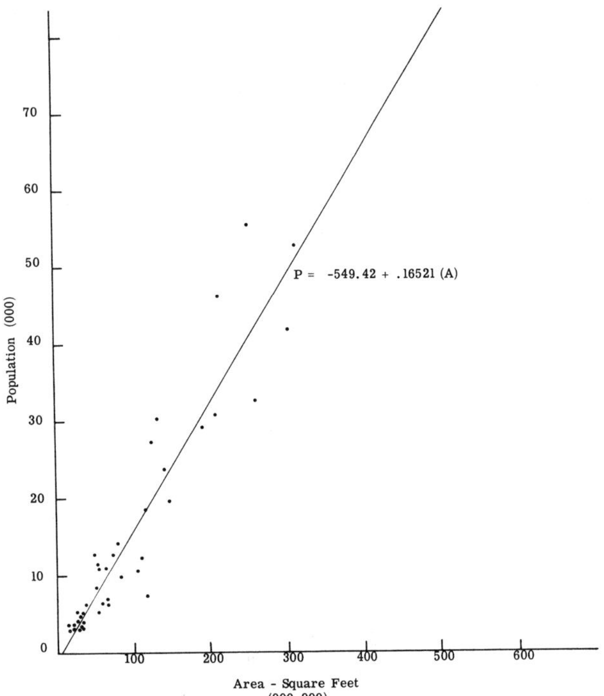

Figure 2 Relationship between population and area, 1953.

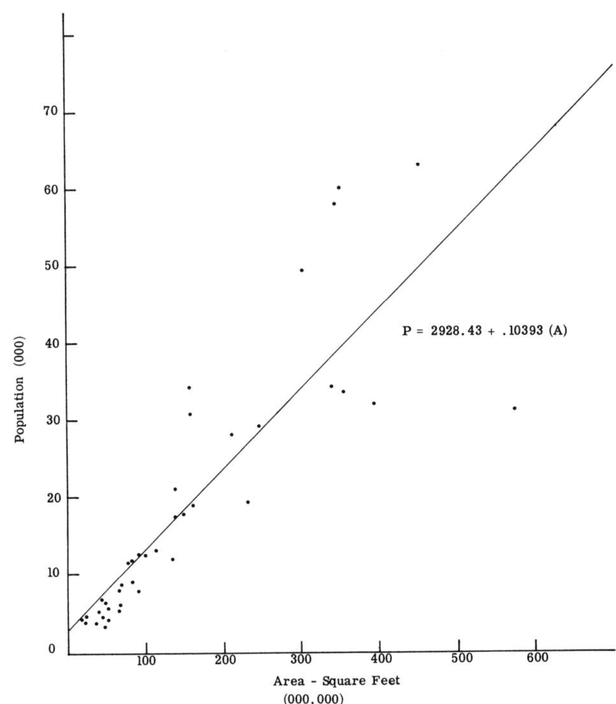

Figure 3 Relationship between population and area, 1963.

Even though the results of the analysis were encouraging, there existed a number of relationships that needed more detailed study before any general conclusions can be made. Time did not permit such an analysis in the present study. Nevertheless, an examination of one disturbing relationship was attempted: the marked variations between population and area of different-sized urban areas. Such variations can be seen quite vividly in Figures 2 and 3. Notice that there is little variation for the smaller urban areas and that just the opposite is true for the larger urban areas. Because of high correlation between area and population, the inclusion of the other three independent variables added little in explaining the remaining variation, even though each of the four variables was significant at the 1 percent level. (The coefficients for these variables are shown in Table 5.) Thus, these relationships suggest that the larger an urban area's population, the more important are variables other than area.

If resources of time and funding had permitted, structural changes in the research design would have been made. For example, the built-up urban area of the nearest larger neighbor would have been substituted for its population. This change will be incorporated in a follow-up study. Furthermore, the sample size will be increased to improve the stratification of city size, linkage data will be refined to reflect differences in type and quality, and time elements will be incorporated into distance variables.

Table 5 Values by Order of Variables: Correlation Coefficients and Coefficients of Determination

Order of Variables	Correlation Coefficient		Coefficient of Determination	
	1953	1963	1953	1963
Area	.7923	.8491	.6278	.7209
Links	.8424	.8687	.7096	.7547
Population NLN	.8599	.8769	.7394	.7689
Distance NLN	.8737	.8787	.7634	.7722

The age of Aquarius is also the age of environmental awareness. Environment and ecology have become commonplace words in our vocabulary. From the highest government officials to the farmer, factory worker, and elementary school child, we have become concerned with what is happening to the life support systems of spaceship Earth. We are turning slowly and painfully toward concern for environmental quality. Industrial technology has become increasingly complex, there are more people to support, and they demand a life style made comfortable by the products of technology. The great mass of industry and population creates pollution at a scale we are now not capable of handling. Even the monitoring of sources of pollution is beyond our present capabilities—but monitoring is the first step in pollution control and it can be accomplished by carefully designed remote sensing systems.

In our frantic search to find and produce more raw materials to be shoveled into the maw of industrial furnaces, we have probed into areas we know little about. The Antarctic littoral is the area most remote from major population centers on the surface of the earth and not even this isolated location is safe from man's influence. Whales and fish caught in these waters have measurable amounts of DDT in their tissues.

The quality of sea water along the land/water interface is declining rapidly. We are fast reaching the point where the ocean will be unable to absorb the varied kinds and amounts of pollution dumped into it. It is increasingly important that environmental chains of cause and effect be discovered and traced with ever-shorter reaction time. We must acquire new data rapidly and be able to process and interpret them almost immediately. Remote sensing offers the hope that this can be accomplished and that pollution can be monitored quickly and effectively and then brought under control.

43-Monitoring Environmental Pollution

JOHN E. ESTES
BERL GOLOMB

In the 1970s, we are finally entering the age of environmental awareness. It is now becoming imperative to be concerned with environmental quality. The need for concern is great, because our burgeoning technology has made it possible for us to modify major parameters of the planetary ecosystem in extremely short periods of time. No sane person or industry deliberately wants to pollute. Pollution is a byproduct. The increasing complexity of our technology and the expanding scale of industrial activities have only recently made these undesirable byproducts obvious to all. At the same time that the scale and scope of our activities have been increasing at an exponential rate, the time delay between cause and effect (between development of a new process or procedure and the discovery of undesired environmental side effects), has been drastically reduced. We must now trace environmental chains of cause and effect within increasingly short periods of time. Our need now is to acquire vast amounts of data rapidly, and to interpret and process them almost immediately. We must determine which environmental parameters are to be measured and the sensor systems with which they are to be monitored. And we must devise sensor effectiveness while constantly thinking of the cost effectiveness of long-term sequential monitoring of environmental pollution sources.

From *Journal of Remote Sensing* 1:8–13, March/April, 1970. Reprinted with permission of the authors and the International Remote Sensing Institute.

The Santa Barbara Oil Spill is one of the major events which stimulated the new political awareness of the quality of the environment. It is an example which clearly demonstrates the needs discussed above. Surprisingly, it is also an example in which the systematic application of remote sensing technology has been conspicuous by its absence.

The Santa Barbara Oil Spill, now entering its second year, began when well No. 21 on Union Oil Company's platform A blew out of control on 28 January 1969. The platform is located on Federal lease parcel 402, some 5½ miles offshore from Santa Barbara. The well was capped after 10 days, but shortly thereafter oil began to leak out of the fractured sea floor near the platform. This sea-floor leak has continued, but with decreased flow, to the present. Drilling, which had been halted by order of Secretary of the Interior Walter J. Hickel to provide time to study the problem, was ordered resumed on the advice of a presidential panel—to relieve pressure in the porous formations beneath the sea floor and thus reduce the rate of seepage. The resumption of drilling activities raised a storm of protest in Santa Barbara.

Despite the uproar generated by the continuing oil spill, there are aspects of the problem about which we still know very little. Even the amount of spillage is uncertain. While all parties agree that since October, 1969, the leakage has been on the order of 400 to 500 gallons per day, the estimated flow rates during the first 10 days range from General Research Corporation scientist Alan A. Allen's 210,000 gallons to the oil company's ten times smaller estimate of 21,000 gallons (500 barrels) daily.

Such lack of agreement only serves to point out the essential need for effective monitoring of areal extent and amount of pollutants produced by such incidents. The Santa Barbara spill continues to pose a major pollution problem close by the densely inhabited Southern California coastline. Lawsuits involving several billions of dollars are already in litigation. Yet, for all intents and purposes, no system for monitoring and measuring the spill has been developed. Even papers pertaining to the application of remote sensing technology, for the monitoring problem, are notably absent from the literature.

Many types of systems were employed to image the spill. No moneys, however, were made available by either private industry, nor local, State, or Federal agencies to develop an effective system to monitor the Santa Barbara spill—or subsequent spills for that matter. Various governmental agencies flew sporadic overflights, many for visual observations alone. Several private companies obtained thermal infrared scans on a nonsystematic basis. Mark Hurd Aerial Survey's Santa Barbara operation carried out consecutive photoreconnaissance flights of the channel area on its own initiative (though, to date, there seems to be few requests for the imagery). Even the Apollo 9 astronauts took pictures of the spill during their mission. A number of individuals (including the authors) have privately overflown the spill sites and obtained aerial imagery. No concerted effort, however, was made to develop, test, and put into operation an efficient, effective, economical monitoring system. Based on an analysis of imagery deposited in the University of California, Santa Barbara, Oil Archives, the authors feel that such an effective monitoring system can be devised, employing thermal infrared scanners.

Standard panchromatic minus blue photography does not lend itself well to resolving the extent of the oil spill. As shown in Figure 1, contrast between the actual area of active spill and the adjacent water is very low. High contrast occurs only where a favorable sun angle accentuates reflectance from the oil. Even where present, this oil reflectance is difficult to distinguish from sun glitter

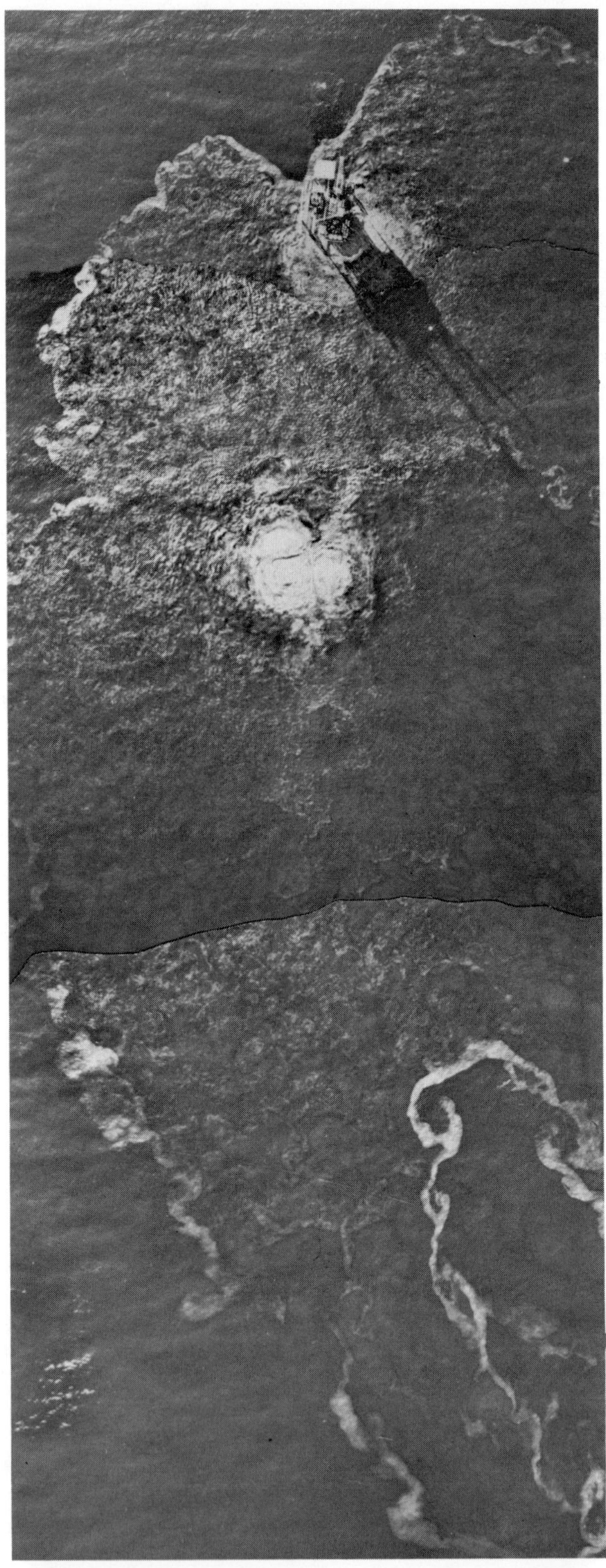

Figure 1 Conventional panchromatic minus-blue photography of Union Oil Company's Platform A, taken Jan. 29, 1969, by Mark Hurd Aerial Surveys. The contrast even between the actual site of the active spill and the surrounding water is quite low, making it difficult to determine the oil spill's real extent. Three airphotos were used to make this uncontrolled overlay; note the contrast falloff at the edge of each print used.

on oil-free waters. This is especially true in areas where upwellings, currents, and localized airflows create a streaked, alternately choppy and smooth sea surface.

Color imagery is similarly ineffective. Object-to-background contrast levels between the oil film and the adjacent water is again very low. Except for areas where wave action has rolled the oil into thick, ropy streamers that register as dark blue, the oil-polluted water shows a bluish black color, the margins of which are difficult to distinguish from the normal sea surface. Figure 2[1] demonstrates this effect. The graduations of the area of surface pollutants are difficult to determine. In Figure 2 the active area of the actual spill site is seen as the whitish bubble beyond platform A (the second tower in the foreground of the picture).

Another illustration of the difficulty in the perception of the oil slick on color photography is shown in Figure 3. Here almost the entire surface is covered with a layer of oil. The bow wave of the boat passing over the surface has skimmed the oil in its wake, leaving a swath of light blue to show the actual color of the underlying water. If it were not for the boat wake acting as a reference, we would accept this rendering by the oil slick as depicting the normal sea surface. (See also Fig. 4.)

[1] See color insert for Figures 2–4 and 6.

In contrast to the preceding photographs, the thermal infrared imagery (Fig. 5) shows the area of the spill clearly and distinctly. Because a water body normally has an essentially uniform emissivity, thermal mappers have been used to detect even small thermal gradients. Thermal infrared imagery should effectively indicate the areal extent of a pollutant, since any pollutant on the surface of a water body would alter that portion of the water body's emissivity. This may even be true if detergents are used to "sink" the oil. In cases where detergents are applied, the oil usually remains in suspension beneath the surface. Although thermal infrared will image only the top molecular layer of a water body, the emissivity of that top layer is affected (even if to a small extent), by the condition of the column of water beneath it. This potential to detect the effects of detergents in dispersing oil is an aspect which deserves further exploration.

The daytime thermal infrared image of the spill (Fig. 5) shows a mottled and streaked pattern, both lighter and darker than the background sea surface. The

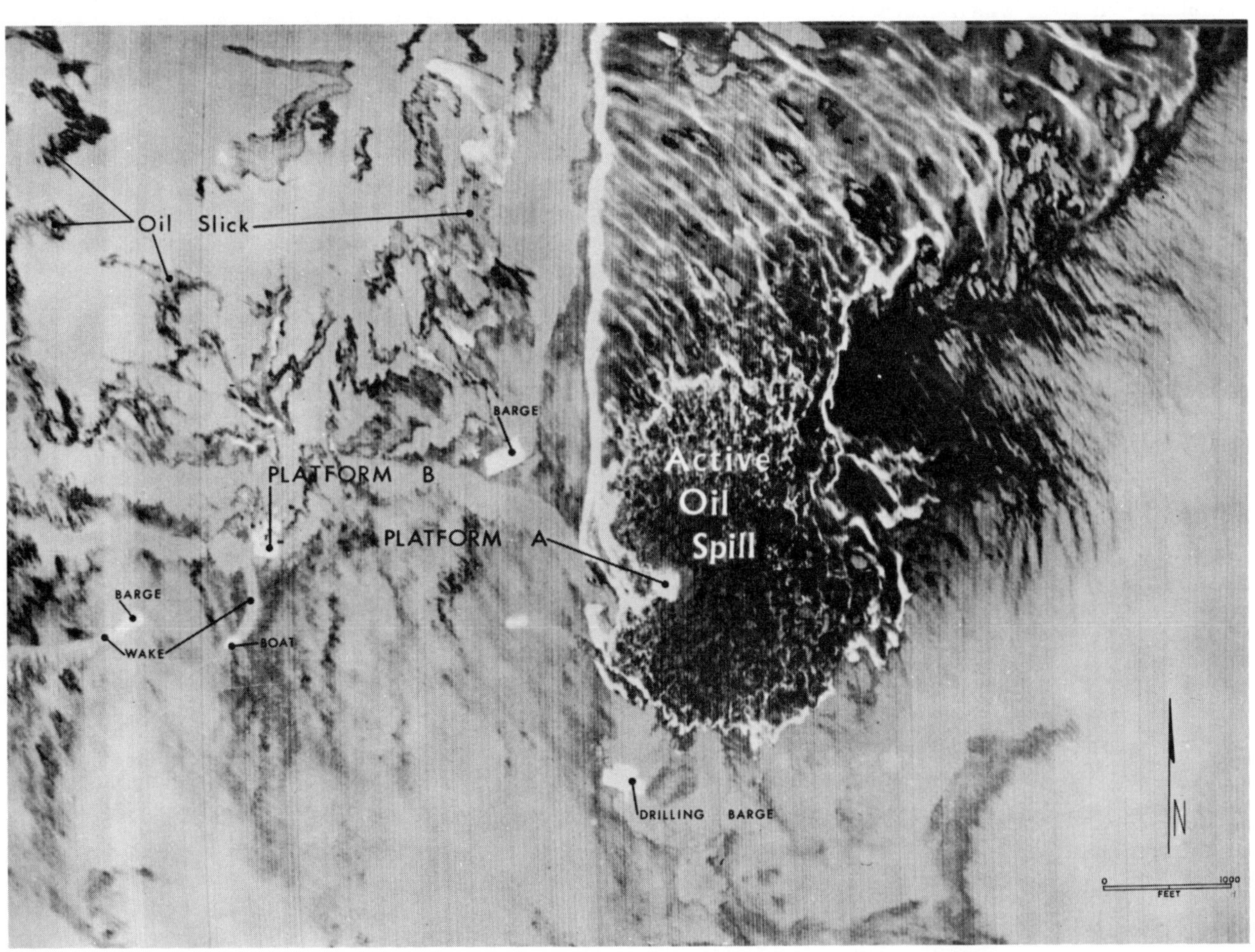

Figure 5 A daytime thermal infrared image of the Santa Barbara Oil Spill. This image, taken in early Feb., 1969, shows the area of the active spill and the extent of pollutants in the immediate area of Union Oil's Platforms A and B. The clear boat wakes seen on this image indicate that the oil outside the active blowout is only a thin surface layer. (North American Rockwell)

lighter areas, of higher infrared emissivity, seem to be areas where the oil has pooled in thick, ropy streamers. The oil streamers, being actually darker, absorb more heat and hence show a stronger infrared emission. Patterns where the oil-polluted surface shows darker than the background (in the active spill area) involve actual upwelling where cooler water mixed with oil from depth are displacing the warmer surface waters. Outside the main body of the blowout are diffuse patches which are occasionally lighter, but generally are darker, than the background. We would expect oil patches, being more efficient blackbodies, to register as uniformly lighter than the background water. The diffuse, dark patches show that this relationship is not universally true. Perhaps the dark scanned oil patches are areas covered by a thin film of oil whose iridescence makes it an efficient reflector and hence, these patches are poor absorbers of insulation. This interpretation of thin-film surfaces is supported by the imaging of the boat wakes, which make clean sweeps through the polluted surface layer. However, we lack data on the specific characteristics of this particular infrared imaging system to allow a more precise interpretation. Whatever the emissivity, the thermal infrared image clearly shows a distinctive pattern response for the oil-covered areas which allows these to be readily distinguished from the essentially unpolluted background surface. An operational thermal infrared monitoring system could be devised to accentuate even more clearly this object-to-background contrast ratio, either through refinement of the initial scanning system or through enhancement of the recorded image.

The enhanced image in Figure 6 was produced on a Datacolor film reader system manufactured by Spatial Data Systems, Inc., of Santa Barbara, California. This system, essentially a color densitometer, represents the density values of a given photograph as analog voltage levels of a video signal which can be displayed in color. The analog levels are converted to digital values using a very high speed (20 nano-second) analog-to-digital converter that continuously converts the analog voltage to discrete values. Each value is presented to a number of digital-to-analog converters (three for each color) that produce the red, green, and blue analog signals for the three electron guns of the color monitor. The three colors can be mixed on the screen of the color picture tube to produce desired colors.

Quantitative density readings may be obtained by placing a calibrated photographic step wedge in the picture for reference. The steps may then be made to appear in different colors by adjustment of the color controls. These same colors will appear in the picture on those areas having densities corresponding to those on the step wedge.

The Datacolor system, further, employs a digital planimeter. The planimeter provides a digital readout expressing the area of specific colored portions of the datacolor display as percentage of the total image area. A pushbutton selector enables the operator of the system to quickly measure any one, or any combination, of the ten basic colors. A digital voltmeter then indicates the percent area of the picture depicted in the selected color or colors.

The areas measured by the planimeter represent the proportion of the picture that has a density lying between the two iso-density contours framed by the edges of the selected color. When the image being analyzed represents emitted thermal patterns, the planimeter directly reads the percentage of the scene which is emitting at selected values.

Of the area imaged in Figure 5, approximately 62 percent is within the density range interpreted as water. Barges, boats, and drilling platforms represent a further 0.5 percent of the area. Approximately 37.5 percent of the sea surface imaged here was covered by oil spill pollutants.

Allowing for linear distortion in the infrared scan, the authors have calculated the total area depicted in this image to be 2.51 square miles. Thus, we have an area of approximately 1.6 square miles which falls within the density range interpreted as water, and 0.9 square miles within the density range interpreted as pollutants.

If we use a value of 0.01-inch as a representative average thickness of the oil layer in this part of the spill, some 155,000 gallons of oil are shown on this image.

The authors realize the limitations of assuming average thickness values without further research into this problem. The question of *amounts* of pollutants spilled, as distinct from areal extent of the pollution, may become one of great legal and financial significance in the future. Research here could be directed toward a twofold approach: appropriate sensor systems could be investigated to test their capability for automatic readout of oil thickness; there are a number of indications that this is feasible. More readily implemented, and initially more economical, may be the development of random sampling techniques to arrive at representative average thickness values for an entire oil slick or for specific image density values of an oil slick.

A monitoring system for sites of potential pollution may, in addition, serve as an important early-detection and warning system—especially in remote areas. Even in the Santa Barbara Channel there were significant delays between the initial event and public awareness of the blowout and the subsequent submarine pipeline rupture.

We envision a system of periodic overflights of active and potential pollution sources carrying a thermal infrared imaging system at the highest altitude commensurate with the acquisition of high-quality imagery. We have shown that imagery similar to that which could be acquired by such a system can yield a numerical measurement of the extent of the pollution, when processed through equipment described in this paper.

If we are to halt the growing degradation, or indeed are to improve the state of our environment, it is imperative that we develop methods to systematically monitor active and potential sources of environmental pollution. The Santa Barbara Oil Spill, as described here, is only one example of how remote sensing could effectively be applied to do the monitoring.

For many years geographers, economists, sociologists, and urban planners have relied heavily on census data or other governmental statistics to formulate an analytical approach toward solving urban problems. There is a danger in using this data—some are subject to statistical bias, they may be collected so infrequently that they do not truly reflect the current situation, and they are sometimes inaccurate due to inconsistencies in training and dedication of individual data collectors. Several recent studies have shown that remote sensing can be used in seeking answers to urban problems and that suborbital and orbital remote sensor imagery can contribute useful data for studying the urban environment.

Careful study of small-scale orbital imagery indicates that areas of new urban growth can be delineated. The direction and shape of that growth can be monitored. New subdivisions can be located in their earliest moment of development. Their stages and rate of growth can be measured from sequential imagery. This small-scale imagery also provides observable information on regional patterns such as transportation systems, land use, and surface geologic formations or structures.

Larger-scale suborbital imagery can be used to record and measure urban physical patterns directly. In the hands of a trained urban researcher these data are a valuable tool for assessing urban social structure. Using this larger-scale imagery, it is possible to infer population density and make fairly accurate statements about the general socioeconomic character of an urban population.

Policy planners working with urban systems are in need of constant inputs of current data. A city is a dynamic microcosm. It grows, and the shape and direction of growth changes. While cities serve as magnets for in-migration, they also serve as a source for out-migration. The city more than any other type of manmade land use system is subject to incredibly rapid change which can occur before policy planners can discover it and take effective social or economic action to control possible deleterious effects on the urban system.

The great advantage of remote sensing is that it can provide sequential coverage of an urban area easily and cheaply. In large measure an effective program for preventing urban blight depends upon better data collection. Remote sensing can be used as a method for delineating areas of substandard housing. Remote sensing imagery can not provide all the information that planners need about such areas but it can make available a good deal of useful information. Further, it can show where social or economic efforts should be concentrated by planners and it is also a valuable historic record of the urban scene.

44-Use of Remote Sensing To Determine Urban Poverty Neighborhoods

SHANE DAVIES
ALEX TUYAHOV
ROBERT K. HOLZ

An effective program of preventive action against urban blight depends upon new and improved means of data collection and analysis. The purpose of this discussion is to show how remote sensing methods can be used to facilitate the more precise delimiting of urban residential areas that exhibit characteristics of blight and how these methods can provide structural indicators of low-quality housing.

The definition of a substandard housing unit has been the subject of much discussion and the accuracy of the ground truth count by the census data collectors of much controversy. Present ground-based methods of housing data collection by the Census Bureau are too expensive, time consuming, infrequent, and inaccurate. Other deficiencies associated with this method of data collection are the failure to pay attention to the spatial and temporal properties of the data, the problem of data aggregation, the ignoring of environmental conditions, and the restricted nature of the census material. Utilization of photographic imagery provides a more efficient and less costly survey method for delineating areas of substandard housing units. It is a permanent record of an area's spatial pattern and human land use activities and is a useful tool in the quantitative and qualitative measurement of an area's characteristics.

Remote sensing is the recording of a phenomenon without having the sensing device in direct contact with the object. This implies information transfer through electromagnetic energy which creates a visible image of the world's surface. Remote sensing is not "new" since the eye, camera, radio, and television are remote sensors. The first air photographs of the earth's surface were probably generated by Gaspard Felix Tournachon in 1858 using a balloon as an aerial platform. By World War II remote sensing was used for accurate surveillance of enemy territory and its most recent advancements have been in space research.

Sensors respond to electromagnetic energy reflected or emitted from objects in the environment. This energy spectrum is a continuum, ranging from very short waves at high frequency to very long waves at low frequency. Our eyes and the conventional camera can perceive only a narrow band or "visible portion" of this spectrum. Just beyond the visible portion of this spectrum is a zone of reflected energy called the near infrared which is invisible to the eye. Special cameras and film emulsions have been designed to record these wavelengths.

From *Landscape*, January, 1973. Reprinted with permission of the authors and the publisher.

At still longer wavelengths very special sensors record the thermal or far infrared wavelengths which generate a heat image of the environment. Beyond the thermal infrared at still longer wavelengths, energy can be sensed *passively* by building a sensor which can record those microwaves which occur naturally in the environment or *actively*, by transmitting energy at a specific microwave wavelength. For example, radar is a sensor that generates energy at microwave wavelengths. These waves strike targets in the environment and are reflected or echoed back to a receiver which records the returned energy. Beyond the microwave region there is still a considerable amount of still longer wavelengths of electromagnetic energy for sensing purposes. At these longer wavelengths, television and short- and long-wave radios operate, although these are not *usually* used for remote sensing the environment.

The photograph, conventional and infrared, both color and black and white, is most useful in providing signatures for the determination of housing quality. In this study, conventional suborbital black and white photography at a scale of about 1:23,000 and suborbital color infrared photography at a scale of about 1:190,000 were the two types of imagery used. A middle- and a low-income residential area, both approximately eight city blocks in size, were selected in Austin, Texas, by sample ground survey using census tract and city block data (see Figs. 1 and 2)[1]. Identifiable signatures on housing quality were extracted from the photographs by stereoscopic viewing and computational techniques. The results show that not only can poverty areas be delineated from the imagery, but that the measurable variables are useful parameters in determining housing quality. The distinct difference in the signatures extracted from the imagery for the two housing areas which serves to exemplify the substandard housing area is summarized in Table 1 and categorized under the following three headings.

[1]Not reproduced here.

Table 1 Comparison of Housing Areas Based upon Selected Environmental Criteria

Criteria	Low-Income Areas	Middle-Income Areas
House size, (sq. ft.)	380 to 1220 (average, 731)	1110 to 1560 (average, 1305)
Placement of house and lot (distance from street in feet)	12 to 42 (average, 27)	34 to 45 (average, 40)
Potential landholding per housing unit (sq. ft.)	5670	7500
Building density (%)	14.2	17.7
Average lot size and frontage (sq. ft.)	4337 Narrow frontage	7376 Wide frontage
Image, pattern, and texture	Lack of uniformity Irregular	Uniform Regular
Houses with driveway (%)	8.3	97.0
Houses with garage (%)	3.0	97.0
Number of visible autos per house	0.20	0.76
Unpaved street (%)	65	0
Street width (ft.)	12 to 24 (average, 18)	24 to 31 (average, 28)
Quality of curbing	Generally lacking	Intact
Quality of vegetation	Not as vigorous	Cultivated, vigorous
Housekeeping	Presence of debris	Lack of debris
House orientation to street	Short side	Long side
City block pattern	Irregular, dead end, twisting streets	Regular
Vacant lots per city block	0.6	0
Proximity to manufacturing and retail activity	Close to manufacturing Remote from shopping outlets	Remote from manufacturing Close to retail outlets

Physical Structures

This includes all signatures which are produced by the appearance of the homes and other buildings in the areas, such as their size, diversity of building material, density, and geographic pattern and uniformity. The amount of space taken up by the physical structure of the house in the low-income neighborhood varied from 380 to 1220 square feet with a mean of 731 square feet. In the middle-income area, the variation was 1110 to 1560 square feet with a mean of 1305 square feet. Lot size, the shape of the lot, and its appearance can also be used as a supplemental indicator of poverty areas. The average lot size in the low-income neighborhood measures 4337 square feet while the mean in the middle-income area registers 7376 square feet. The lots in the latter area are uniform in size and shape and have wider frontage. The houses in the poverty area normally have their narrower side facing the street, while in the middle-income section, the longer side of the house parallels the street. This pattern reflects the trend for poverty areas to be subdivided into smaller tracts with less road frontage in order to avoid higher property taxes. The image signatures produced by the low-income area are irregular in tone, pattern, and texture, which is the result of the extremely varied shape and size of the residential dwellings.

Other factors producing the irregular patterns in the blighted areas are diversity in roof configurations, building material, and location of the house on the lot. The variation in building material is indicated by the difference in shades of gray and texture that the roof signatures manifest on the imagery. Photo evidence of the buildings in the poverty area suggests they are not part of an organized housing development whereas the similarity in configuration, material, and spacing in the middle-income area suggests a controlled development.

Another indicator, building density, is calculated as the percentage of the total ground area that is covered by roof area. The middle-income area has a building density of 17.7 percent as compared to 14.2 percent in the poverty areas. Although this indicates that more area in the middle-income development is occupied by buildings, or specifically roof cover, it does not indicate that there are more housing units per unit area. If all available space is distributed evenly among the buildings, each building has a potential land holding of 5670 square feet in the poverty area and 7500 square feet in the middle-income area. Consequently, although a greater percent of the middle-income area is occupied by roof cover, each areal unit of land holding is larger in middle-income neighborhoods.

To determine uniformity in the placement of a house on a lot, the distance from the street curb to the front of the residence is measured. In the low-income areas, this distance ranged between 12 and 42 feet and in the middle-income area between 34 to 45 feet. The smaller range in the more affluent neighborhood is indicative of a greater uniformity in layout, which is the result of a more organized housing development. Further, the greater distance between the curb and the house in the middle-income neighborhood also suggests a more extensive land holding.

Site

This category includes all signatures produced by the property upon which the house is situated, for example, the quantity and quality of the vegetation surrounding the house, the number of automobiles, garages, and driveways, and the degree of debris and litter. The upkeep or "housekeeping" of the immediate environs of the house offers significant signatures on the environmental quality of the neighborhood. Variations in the quality and cultivation of the vegetation in the two areas are easily detectable on panchromatic photography and color infrared photography. The latter is particularly helpful in measuring these variables because of the unusual response provided by the vegetation on the imagery. Healthy broad-leaf vegetation reflects a large amount of infrared radiation and records strongly as a bright red or magenta color. Less vigorous or unhealthy vegetation images in shades of greenish blue. The more carefully cultivated and more abundant shrubs, flowers, and trees of the better-watered vegetation of the middle-income areas record strongly in the bright red and magenta while the less-well-cultivated, sparser vegetation of the poverty areas, containing patches of exposed dirt as opposed to grass, render more shades of greenish blue.

The presence of the automobile and features associated with it also produce unique signatures which can be related to each study area. A distinctive signature of the poverty areas is the almost complete absence of garages and driveways and the considerable amount of on-street parking. Only 8.3 percent of the homes in the low-income neighborhood have driveways and only 3 percent have a garage. This formed a marked contrast with the middle-income neighborhood where 97 percent of the homes had both a private driveway and garage.

Where driveways did exist in the poverty areas, their appearance also served to further exemplify the condition of the environment. The driveways are unpaved, as evidenced by the lack of sharp definition in the driveway edges on the imagery. They lack uniformity in width and are generally not situated perpendicular to the house. Roadside curbing is missing or broken on most blocks. These features contrast with the middle-income area where all curbing is intact and nearly all the driveways paved, sharply delineated from the lawn and perpendicular to the house.

The individual lots in the poverty neighborhoods are not sharply delineated on the imagery but appear to be merged together. In the middle-income areas, the lot of an individual house is sharply delimited, their perimeters often emphasized by the existence of a fence, hedges, or bushes. This produces a well-defined checkerboard pattern with the individual blocks representing different ownership. A greater abundance of litter, more abandoned cars, and a variety of small sheds of various shapes and sizes are also very common to the poverty neighborhood.

Situation

This includes all signatures that are developed as a function of neighborhood characteristics, for example, maintenance of streets, configuration of street networks, and relative location to older areas in the city. The irregular city block pattern in the poverty area is formed by the narrow, twisting streets which are often unpaved and frequently terminate in a dead end. Sixty-five percent of the streets in the low-income area are unpaved, whereas all streets in the upper-income area are paved. The unpaved streets are recognized by the rendition of a light shade of gray on the black and white imagery and the lack of sharply defined street edges. Streets in the low-income area lack uniformity and are 12 to 24 feet in width as contrasted to 24 to 31 feet in the middle-income area. The poor area is adjacent to manufacturing plants and in close proximity to the central business district (CBD). There are fewer consumer retail outlets in this area and it experiences to a greater degree the external diseconomies of air and water pollution, noise, and traffic congestion. The existence of vacant lots also serves as a characteristic of the poverty neighborhood. Several vacant lots are scattered throughout the poverty area whereas no vacant lots are present in the middle-income area.

Ground photos illustrating some of the criteria are shown in Figures 3 to 5.

Figure 3 A middle-income and a low-income house and yard in the selected areas in Austin, Texas.

Summary and Policy Implications

In this study two separate housing areas were studied to assess variation in housing quality and to determine suitable indicators of urban blight. The environmental indicators of a substandard housing area can be further validated by studies in other urban areas and, if substantiated, used to delimit urban poverty pockets.

Some of these variables, however, suitable as indicators of blight in cities of this region may or may not be applicable to other urban areas. For example, unpaved streets and quality of vegetation may not be discernible in blighted sections of Chicago and Detroit. Certainly, quality of vegetation is not usable during a northern winter. However, deficiencies in thermal insulation due to the extreme age and inferior construction of buildings in low-income neighborhoods could be detected by thermal scanning devices and used as a blight indicator in northern cities. Further, in northern cities greater emphasis could be placed upon building density, traffic volume, street condition—width, grade, maintenance, lighting, street parking—degree of debris, vacant lots, city block patterns, inadequate basic community facilities, and so on.

This procedure may also encounter difficulties when attempting to delineate deteriorating conditions in the small, new, libertarian sectors of cities of the Southwest where rich and poor are residentially interspersed. Incipient blight in these cities of 10,000 to 50,000 is often masked and difficult to detect because the individual data bits (individual houses) in the sample are too small and scattered to be readily detected on remote sensing imagery.

The findings in this study demonstrate the information-gathering potential of aerial remote sensing in identifying quickly and economically the variables which designate a blighted urban environment. From the frequency of appearance and the level of association of the physical characteristics and site and situation of the housing units, specific areas of housing quality can be delineated. Identification of the spatial distribution of various grades of housing quality has practical implications in that it permits an assessment of those substandard housing areas where deprivation and its associated problems exist.

The location, spatial arrangement, density, and character of the housing units have ecologic meaning in that exterior characteristics are surrogate measures for interior conditions. Although the inferring of interior conditions from exterior characteristics is conceptually weak, it is a means of assessing the number of low-income families in a specific urban area. Since disadvantaged persons tend to concentrate in areas of substandard housing units, aerial photography in combination with ground truth data allows relevant socioeconomic and demographic data to be extracted on individuals residing in poverty neighborhoods. For example, this technique is presently being used by the authors to aid health workers in identifying significant environmental health problems in the Ozark Planning Region by helping trace and detect communicable diseases in rural areas.

Figure 4 Unpaved street in the poverty neighborhood.

Figure 5 Fenced yard in the middle-income neighborhood.

This method can also be used to monitor traffic flows, estimate manufacturing employment from the size of the industrial park, assess the number of shoppers using shopping malls from the density of cars in the parking lot, measure the degree of water pollution and urban sprawl, and monitor the movement of urban phenomena into rural hinterlands.

It provides a means of delineating urban-rural boundaries and monitoring the degree to which nonurban land is converted to urban uses. Likewise, it may be used to ascertain the loss of urban land and assess the decline of towns in depressed regions. It is an avenue through which the size and location of urban places can be discerned. When Texas, a progressive state, has poor and inadequate knowledge of the location, quality, and quantity of its housing stock, the monitoring utility of the technique becomes even more apparent. This point is particularly relevant to underdeveloped countries where urban developments mushroom overnight and where road and rail networks are poorly defined. Considerable use of this technique for just such purposes has been made by Latin American countries, especially Mexico, Brazil, Colombia, and Argentina. It further allows comparison of the urbanization processes taking place in these countries and the United States.

Further, remote sensing provides an overall view of a city's complex system of transportation links and provides a series of vignettes which captures the vital ongoing processes of a city. Day and nighttime surveillance can provide input for possible solutions to congestion problems. Since the volume and direction of transportation flows is a function of the spatial distribution of urban activities, an assessment of a city's internal land usage provides the basis for improved future transportation planning through controlled industrial and residential site selection. It provides the physical mosaic upon which a more efficient future allocation of land for urban purposes can be made.

Remote sensors expedite present data collections and have a promising role to play in future urban information-gathering and -storage systems at both the local and national levels. Other applications range from helping locate or predict impending storms to monitoring the origin and spread of air and water pollution and assessing the degree and intensity of blight among crops and orchards. Its applicability to ascertaining the extent of damage caused by natural disasters is particularly relevant with the advent of so many recent flooding and hurricane misfortunes (Rapid City, South Dakota, Pennsylvania, Florida, Mississippi). Its ability to provide a quick and reliable amount of the damage wrought is extremely important since the destruction often occurs in areas which are either made or are naturally inaccessible to ground workers.

The employment of sensors has obvious limitations and should only be used when their advantages exceed those of conventional data collection techniques. It is believed, however, as exemplified in this paper, that remote sensing can aid in finding solutions to certain basic urban problems.

SUGGESTED READINGS FOR PART TEN

Aukland, J. C., Conway, W. H., and Sanders, N. K. Detection of Oil Slick Pollution on Water Surfaces with Microwave Radiometer Systems. *Proceedings of the Sixth International Symposium on Remote Sensing of Environment,* Vol. II. Ann Arbor: University of Michigan, Institute of Science and Technology, Willow Run Laboratories, 1969. Pp. 789–795.

Cade, C. M. The industrial potential of the "heat camera." *New Scientist* 23 (No. 400):165–167, 1964.

Harrison, W. R. *Suspect Documents, Their Scientific Examination.* London: Sweet and Maxwell, 1958.

Kolipinski, M. C., and Higer, A. L. On the Detection of Pesticidal Fallout by Remote Sensing Techniques. *Proceedings of the Sixth International Symposium on Remote Sensing of Environment.* Ann Arbor: University of Michigan, Institute of Science and Technology, Willow Run Laboratories, 1969.

Mumbower, L., and Donoghue, J. Urban poverty study. *Photogrammetric Engineering* 33:610–618, 1967.

Newsweek. Sky spies: Nobody has a secret anymore. April 21, 1969.

O'Hara, C. E., and Osterburg, J. W. *An Introduction to Criminalists, the Application of the Physical Sciences to the Detection of Crime.* New York: Macmillan, 1949.

Wellar, B. S. Utilization of Multiband Aerial Photographs in Urban Housing Quality Studies. *Proceedings of the Fifth Symposium on Remote Sensing of Environment.* Ann Arbor: University of Michigan, Institute of Science and Technology, Willow Run Laboratories. 1968.

Whitehouse, W. M. Current and Future Needs for Remote Sensor Data in Medicine. *Proceedings of the Fourth Symposium on Remote Sensing of Environment.* Ann Arbor: University of Michigan, Institute of Science and Technology, Willow Run Laboratories, 1966. Pp. 55–61.

CDEFGHIJ–M–798765